Soft Soil Engineering International Conference 2015 (SEIC 2015)

IOP Conference Series: Materials Science and Engineering
Volume 136

Langkawi, Malaysia
27 - 29 October 2015

ISBN: 978-1-5108-2730-1
ISSN: 1757-8981

Printed from e-media with permission by:

Curran Associates, Inc.
57 Morehouse Lane
Red Hook, NY 12571

Some format issues inherent in the e-media version may also appear in this print version.

Printed by Curran Associates, Inc. (2016)

For permission requests, please contact the Institute of Physics
at the address below.

Institute of Physics
Dirac House, Temple Back
Bristol BS1 6BE UK

Phone: 44 1 17 929 7481
Fax: 44 1 17 920 0979

techtracking@iop.org

Additional copies of this publication are available from:

Curran Associates, Inc.
57 Morehouse Lane
Red Hook, NY 12571 USA
Phone: 845-758-0400
Fax: 845-758-2633
Email: curran@proceedings.com
Web: www.proceedings.com

Table of contents

Volume 136

Soft Soil Engineering International Conference 2015 (SEIC2015)
27–29 October 2015, Langkawi, Malaysia

Accepted papers received: 17 June 2016
Published online: 22 July 2016

Preface

Papers

IOP Conf. Series: Materials Science and Engineering 136 (2016) 011001 doi:10.1088/1757-899X/136/1/011001

Preface

The Soft Soil Engineering International Conference 2015 (SEIC'15) is jointly organized by the Research Centre for Soft Soil Malaysia (RECESS), Dept. of Infrastructure and Geomatic Eng., Faculty of Civil and Environmental Eng. (FKAAS), Office of Research, Innovation, Commercialization and Consultancy (ORICC), Universitas Katolik Parahyangan (UNPAR) Indonesia, Institute Technology Brunei (ITB), Langkawi Development Authority (LADA), Public Works Department Malaysia (JKR), Construction Industry Development Board Malaysia (CIDB) and Pertubuhan IGS Malaysia (MyIGS). The SEIC'15 invites submissions of full papers describing innovative work and state of the art sustainable developments in soft soil engineering and geotechnical engineering.

SEIC'15 will bring together leading geotechnical professionals, engineers, academics and researchers working in this domain of interest from around the world. The organising committee invites prospective authors to submit original work on innovations and new developments in the fields of soft soil engineering and geotechnical engineering. The topics of the conference are Soft Soil Properties, Geotechnical Modelling, Soft Soil Improvement, Construction on Soft Ground, Peat Technology, Geo-environmental Engineering and Geotechnical Instrumentation. SEIC'13 was a huge success with over 200 participants from various institutions and industries.

With Wisdom, We Explore.

Sincerely,
Prof. Dato' Dr. Ismail Hj. Bakar
Chairman

Dr. Aziman Madun
Co-Chairman

Soft Soil Engineering International Conference 2015 (SEIC'15)
Theme : Geotechnology of Problem TIC & Soft Soil Engineering
27, 28 & 29 October 2015, Resort World Langkawi, Malaysia.

IOP Conf. Series: Materials Science and Engineering **136** (2016) 011002 doi:10.1088/1757-899X/136/1/011002

Conference Committees

Organizing Committee

Patrons	:	Prof. Datuk Dr. Mohd Noh Dalimin
Advisors	:	Prof. Dr. Wahid Razzaly
	:	Prof. Dr. Hjh. Rosziati Abdullah

Working Committee

Chairman	:	Prof. Dato' Dr. Ismail Hj. Bakar
Co-Chairman	:	Dr. Aziman Madun
Secretary	:	Dr. Alvin John Lim Meng Siang
Asst. Secretary	:	Mrs. Salina Sani
Treasurer	:	Mr. Anuar Raze Abd. Rahim
Asst. Treasurer	:	Mrs. Latifah Nasir
Technical Publication & Printing	:	Dr. Saiful Azhar Ahmad Tajudin
		Mr. Mohd. Hazreek Zainal Abidin
		Assoc. Prof. Dr. Zaidi Embong

Publicity, Promotion & Certificate	:	Mr. Felix Ling Ngee Leh Mr. Lokman Hakim Jamil Mr. Mohd. Hazmi Kasmin
Invitation, Protocols & Gimmick Programme	:	Mrs. Hasrinas binti Husain Mr. Mohammad Zamri bin Ahmad Mr. Shafry bin Salim
Exhibition & Logistics	:	Assoc. Prof. Dr. Adnan bin Zainorabidin Mr. Mohd Fairus bin Yusof Mr. Mudzaffar Syah bin Kamarudin
Programme Book, Refreshment & Souvenirs	:	Ms. Maizatul Fahimah binti Miskam Dr. Ir. Agus Sulaeman Ms. Mardiha binti Mokhtar Mrs. Norsaliza Salleh
Participant Registrations	:	Secretariat
Sponsorships	:	Prof. Dr. Devapriya Chitral Wijeyesekera Ms. Ng Wei Fern Mr. Amir Zaki bin Salikin
Special Programmes & Tasks	:	Prof. Dr. Mohd. Idrus Hj. Mohd. Masirin Dr. Nor Azizi Yusoff

IOP Conf. Series: Materials Science and Engineering **136** (2016) 011002 doi:10.1088/1757-899X/136/1/011002

Broadcasting & Recording : Mr. Mohd Kjairi Zainal

External Advisors

Prof. Ian Jefferson, Univ. of Birmingham, United Kingdom.

Prof. A.F.L. Hyde, Univ. of Sheffield, United Kingdom.

Prof. B. Clarke, Leeds University, United Kingdom.

Prof. Buddhima Indraratna, University of Wollongong, Australia.

Prof. Nimal Seneviratne, Univ. Peradeniya, Sri Lanka

Prof. Yoshiaki Kikuchi, Tokyo Univ. of Science, Japan

Prof. C.F. Leung, National Univ. of Singapore, Singapore.

Prof. Masyhur Irsyam, Institut Teknologi Bandung, Indonesia.

Prof. Paulus P. Rahardjo, Univ. Katholik Parahyangan, Indonesia.

Prof. Sri Prabandiyani R Wardani, Diponegoro Univ., Indonesia.

Assoc. Prof. Noppadol Phien-wej, Asian Inst. Of Tech, Thailand.

Prof. Bujang Kim Huat, Univ. Putra Malaysia, Malaysia

Prof. Dato Roslan Hashim, Universiti Malaya, Malaysia.

Prof. Khairul Anuar Kassim, Universiti Teknologi Malaysia, Malaysia

Prof. Fauziah Ahmad, Universiti Sains Malaysia, Malaysia.

Prof. Mohd. Raihan Taha, Universiti Kebangsaan Malaysia, Malaysia.

Prof. Faizal Ali, Universiti Pertahanan Nasional Malaysia, Malaysia.

Prof. Aminaton Marto, Universiti Teknologi Malaysia, Malaysia

Prof. Dato Sabarudin Mohamed, Universiti College Technology Sarawak, Malaysia.

Assoc. Prof. Siti Noor Linda Taib, Universiti Malaysia Sarawak, Malaysia.

Prof. Ideris Zakaria, Infrastructure University Kuala Lumpur

Assoc. Prof. Low Kaw Sai, Universiti Tunku A Rahman, Malaysia

Dr. Mohamad Nor Omar, Jabatan Kerja Raya Malaysia, Malaysia

Dr. Mohd. Asbi Othman, Mohd. Asbi Assoc. SB, Malaysia.

Dr. Toh Cheng Teik, Dr. C.T.Toh Consultant, Malaysia

Dato' Dr. Gue See Sew, Gue and Assoc. SB, Malaysia.

Mr. Hidayat Mohd Yacob, Sapura Kencana Petroleum, Malaysia.

Ir. Asmaniza Asanan, MMD, PLUS, Malaysia.

IOP Conf. Series: Materials Science and Engineering **136** (2016) 011002 doi:10.1088/1757-899X/136/1/011002

Ir. Yee Thien Seng, Geotechnical Engr. Tech. Division, IEM, Malaysia.

Ir. Sofyan Effendi, Cakar Bumi Integrasi, Indonesia.

Peer review statement

All papers published in this volume of *IOP Conference Series: Materials Science and Engineering* have been peer reviewed through processes administered by the proceedings Editors. Reviews were conducted by expert referees to the professional and scientific standards expected of a proceedings journal published by IOP Publishing.

Embedded Empiricisms in Soft Soil Technology

D C Wijeyesekera[1,2], L M S Alvin John[2,1], Z Adnan[2,1]
[1] Research Centre for Soft Soil, Universiti Tun Hussein Onn Malaysia, 86400 Batu Pahat Johor, MALAYSIA
[2] Faculty of Civil and Environmental Engineering, Universiti Tun Hussein Onn Malaysia, 86400 Batu Pahat Johor, MALAYSIA

E-mail: dcwijey@gmail.com

Abstract. Civil engineers of today are continuously challenged by innovative projects that push further the knowledge boundaries with conceptual and/or ingenious solutions leading to the realization of that once was considered impossible in the realms of geotechnology. Some of the forward developments rely on empirical methods embedded within soft soil technology and the spectral realms of engineering in its entirety. Empiricisms unlike folklore are not always shrouded in mysticism but can find scientific reasoning to justify them being adopted in design and tangible construction projects. This lecture therefore is an outline exposition of how empiricism has been integrally embedded in total empirical beginnings in the evolution of soft soil technology from the Renaissance time, through the developments of soil mechanics in the 19th century which in turn has paved the way to the rise of computational soil mechanics. Developments in computational soil mechanics has always embraced and are founded on a wide backdrop of empirical geoenvironment simulations. However, it is imperative that a competent geotechnical engineer needs postgraduate training combined with empiricism that is based on years of well-winnowed practical experience to fathom the diverseness and complexity of nature. However, experience being regarded more highly than expertise can, perhaps inadvertently, inhibit development and innovation.

Keywords: Soft soil, geotechnology, geoenvironment.

1. Introduction

Collin's English Dictionary definition of empiricism is 'the doctrine that all knowledge of "matters of fact" derives from experience and that the mind is not furnished with a set of concepts in advance of experience'. The adjective empirical is derived from or relating to experiment and observation rather than theory, and particularly in the medical field, it relates to treatment based on experience rather than theory.

Over the last century, geotechnicians have been consistently made aware of the rigours of ever increasing complexity and uncertainty in the dynamisms within the lacustrine geo environment. The ever increasing technical impositions from the pace of development of infrastructure and construction are jettisoned by the increasingly volatile and often ambiguous societal, ecological, economic and political constraints. Such developments, progressive though may be, are ever attempting to embrace a wide backdrop of empirical geo environment simulations. Thesaurus claims empiricism to encompass pragmatism, experimentation, observation and practicality while a dictionary definition is the philosophical belief that all knowledge is derived from the experience of the senses. Thus empiricism may suggest that one can only have knowledge acquired through physically observable (empirical) evidence, but interestingly it is self-refuting that there is no empirical evidence that empiricism is true!! For instance, the Ptolemaic model of an earth centred solar system remained unchallenged for over 1300 years until 1543 when Nicholas Copernicus published the then controversial theory, without sufficient "scientific evidence" and "practice" on "the revolutions of the heavenly spheres" which stated that the sun is at rest and the earth is in rotation around the sun. Charles Morris, an author of the Encyclopaedia and Unified Science, specialised in scientific empiricism. He appreciated the continuous progress of scientific empiricism through the use of powerful tools of logical analysis which has been the underlying factor in much scientific advancement as that of Newton.

Geotechnology comprises of a combined application of soil and rock mechanics, engineering geology and other disciplines that bear on civil engineering construction and the geo environmental activities of extraction and use of geo materials with the conscious preservation and enhancement of the environment. Moreover, since all construction is necessarily built on or in the ground, geotechnical engineering plays a latent but key role in all civil engineering projects. The spectrum of geotechnical engineering challenges has grown immensely over the last century with the need to use ground that is naturally complex with the uncertainties of the current and future built environment. The complexity of the site geology often drives the geotechnical engineer to deal with variable materials, whose mechanical properties are usually dynamically degrading (such as that in peat) being affected critically by the local water chemistry and water pressures in the ground that change with rainfall and tidal / groundwater level effects.

Soft soil engineering problems find an appropriate solution through the adoption of the state of the art theoretical framework in the planning, design and construction phases. Present day design practice uses software models that are fast outdating spreadsheet calculations which in turn have superseded the practice of relying on design charts and tables. In the absence of established computational software models, designing turns to the adoption of rational empirical and observational methods such as full scale field or laboratory scale model load testing. Such practice not only provides circumstantial evidence to form the basis to review and endorse the design process, but its empiricism embeds itself to further develop theoretical frameworks through correlations established with data gathering. In the construction practice, engineers have recognised and used empirical phenomena such as Newtonian gravity, much the same way as the practice of alternative medicine.

Though geotechnical engineers do not possess a freehand in prescribing / specifying certain desirable properties or treatment of soils such as those of fills, they themselves are potentially complex by virtue of its nature and potential to change and degrade during its design life. All too often lessons are learnt from the extremely expensive consequences of inadequate and/or even lack of appropriate site investigation. Thorough ground investigation and site characterisation solicits the involvement of experienced geotechnical engineers that possess a good understanding of the geology, as well as the applied mechanics of the behaviour of soil and rock.

Additionally, in this era of technological innovation, with increasing and volatile ambiguity of the societal, ecological, economic and political constraints, there is an urgent need to harmonise with the typical life cycle of civil engineering project implementations. Most importantly this understanding must be combined with empiricism based on well-winnowed experience. In the early nineteenth century, civil engineering was not a highly specialised profession as it is today. The outstanding pioneering civil engineers such as William Smith, an active pioneering geologist; Robert Stephenson of railway construction eminence etc., relied on the observational geological science and empiricism to effectively complete major transportation projects during that time. It is inconceivable that civil engineers can be oblivious to the nature of the materials on which or in which they construct. This paper (and lecture) strives to trace the evolution of this integral discipline of geotechnology within constructional civil engineering from its ancient and empirically based beginnings, through the early

20th century developments of conceptual theories of mechanics of soils and rocks to the rise of modern computational geomechanics with multi physics simulation.

2. Folklores, Legends, Superstitions and Myths

Almost all aspects of our daily living are shrouded by superstitions with unknown origins. While some of them are logical, most of them are often silly. Human (and /or animal) minds extend beyond the anatomical brain to space and time consciousness that sensitively lead to telepathy where the mind acts like a magnetic field; images are not inside the brain but are a perception. Thus superstitions develop as empiricisms that are embedded in the mind and forebear on almost all aspects of daily living and most may have neither an origin nor a scientific justification. "Do not walk under a ladder" is such a superstition that is logical but often they sound just silly. Some sayings are unhealthily deemed superstitious and are most of the time not based on reason or a database of knowledge. Accordingly in a lighter vein, empiricism is also found within; superstitions, myths, legends, folklores, omens, old wives' tales, luck, sayings. There exists considerable intertwining and interchanging between these, that the Table 1 is an attempt to define and categorise them topically.

Table 1. Definitions of Superstitions, Myths, Legends, Folklore, Omens, Old Wives' tales, Sayings [1]

Term	Definition
Superstition	A belief or notion, not based on reason or knowledge, in or of the ominous significance of a particular thing, circumstance, occurrence, proceeding, or the like.
Myth	A popular belief or story that has become associated with a person, institution, or occurrence, especially one considered to illustrate a cultural ideal.
Legend	A non-historical or unverifiable story handed down by tradition from earlier times and popularly accepted as historical.
Folklore	The traditional beliefs, legends, customs etc., of a people; lore of a people
Omen	Anything perceived or happening that is believed to pretend a good or evil event or circumstance in the future.
Old wives' tale	A belief, usually superstitious or erroneous, passed on by word of mouth as a piece of traditional wisdom
Luck	The force that seems to operate for good or ill in a person's life, as in shaping circumstances, events , or opportunities
Saying	Something said, especially a proverb or apothegm

Folklore and legends are similarly non-historical, unverifiable stories and customs that are traditionally handed down through generations and sometimes even referred to as old wives' tales (notionally traditional wisdom). Myths and sayings (proverb or apothegm) illustrate cultural ideals matured from folklores and legends. All these form the hidden layer of neurons within the human mind (in the brain) to predict or predestine and perceive omens of good or ill luck as the outcome of the interaction between circumstances, events and opportunities.

Collin's English Dictionary definition of empiricism is 'the doctrine that all knowledge of matters of fact derives from experience and that the mind is not furnished with a set of concepts in advance of experience'. The adjective empirical is derived from or relating to experiment and observation rather than theory and particularly in the medical field, it relates to treatment based on experience rather than theory. Thesaurus claims empiricism to encompass pragmatism, experimentation, observation and practicality while the dictionary definition is the philosophical belief that all knowledge is derived from the experience of the senses. "Experimental data (empiricism) are believed by everyone but not necessarily by the one who did the experiment; while as a theory needs convincing, it is believed by nobody, excepting the one who developed it". Successful and pragmatic design needs to be based on scientifically sound conceptual models, supported by deduced empirical parameters that are substantiated by a sufficient data base of evidence, and can also be extrapolated outside the database.

3. Success (stability) and Failure (instability)

"Success" and "failure" are deemed through human perception of the eventless meeting of the circumstantial needs. In general terms, success and failure are at times considered to follow as a matter of fate (good luck/fortune or ill-luck). Superstition and culture shrouds this perception. As an example, there are cultural differences of opinion as to what numbers to avoid, or what are the acceptable and lucky colours. The particular superstition in the unlucky number 13 is well known. Notably, however, in the Chinese tradition the unlucky number is 4; Romans believed 13 to be symbol of death and destruction while Chinese traditionally consider number 4 as unlucky as its pronunciation can mean "to die". Hence, many apartment buildings do not have a 13th (or 4th) floor; therefore houses and apartments will not be so numbered; some people do not fly or buy things on the 13th; restaurants and homes will not favour 13 guests in a group to come or sit at a table. Moreover, Friday the 13th is considered as a day that attracts potential disasters. Contrastingly, seeing a new moon for the first time on a Monday will bring the observer good luck.

Lessons learned from both successes and failures necessarily enlarge the database and contribute to advances in all types of engineering. Emerging design standards and codes of practice have developed through the evolutionary process based on experience. Failures lead to new understandings of reviewing the empirically assumed / ignored soft soil properties. Construction industry lags behind other industries in the readiness and preparedness to systematically organise and disseminate failure information. Regardless of the size, all construction projects can benefit from the experience-based judgement of a competent geotechnical engineer.

4. Embedded empiricisms within "colour"

At the core of all good engineering science is the careful and diligent treatment of data; of particular significance is how the human visual system processes colour information in the data or its derivatives.

Eyes are the 'windows of the soul' and the colour leads one to differing beliefs;

- Dark blue eyes = delicate and refined souls.
- Light blue or grey eyes = strong and healthy ones;
- Green eyes = hardy souls;
- Hazel eyes = vigorous, deep thinking eyes.

Metaphorically speaking lawyers, like painters can convincingly change and manipulate the factual evidence of a case from "white to black" and vice-versa. If a black cat walks towards you, it means that good fortune is on its way to you. However, if it walks away, it takes the good luck away with it. From the perspective of weather predictions, it is empirically assumed that a red sky at night is a sailor's delight, while they brace themselves with cautious warning when they see a red sky in the morning.

Furthermore, colour has three perceptual dimensions of how bright it appears; luminance, saturation and hue. Colours can be scientifically coded (as exemplified in Table 2) using either HSB (Hue-Saturation-Brightness) followed by RGB (Red-Green-Blue) or a hexa alpha numeric code. Colour name is linked to one of the main colour hues; red, orange, yellow, green, blue, violet, brown, black, grey and white. Colour, lustre and streak are used to help identify rock forming minerals that give colour to the earthen materials (soils, rocks and their mixtures). Gold is the dust that blinds all eyes, but all that glitters is not gold.

Green depicts freshness in nature, and when it degenerates to brown and black it signify natural decay. Table 2 illustrates the colours associated with a range of typically met soils. HSB, RGB, and the hexa-alpha numeric code are also given beneath each colour as a more rigorous, scientific way of describing each colour. The soil characteristics described in the Table 2 gives a geo environmental dimension to its properties.

Table 2. Characteristics of soil based on colours

Image	Soil colour	Soil characteristics
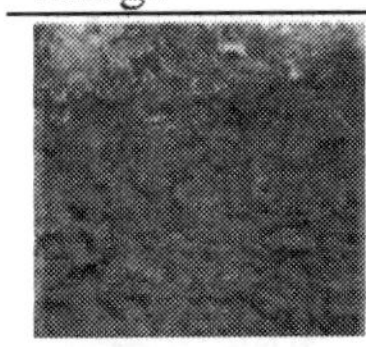	Black 7,28,0 0,50,0 000000	These soils are often associated with high levels of organic matter (peats).
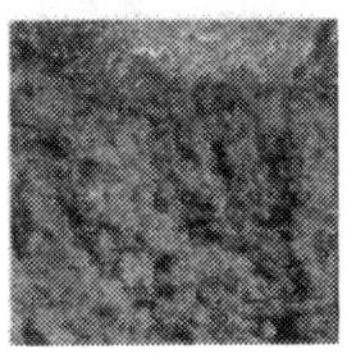	Brown 46,100,17 43,32,00 2B2100	Soils associated with moderate organic matter level and iron oxide
	Red 4,100,90 229,15,00 E50F00	This colour indicates good drainage. Iron found within the soil is oxidised more readily due to the higher oxygen content. This causes the soil to develop a 'rusty' colour. The colour can be darker due to organic matter. Blue London clay underlies the environmentally weathered brown London clay.
	Grey/Green 64,100,48 114,122,0 727A00	These soils are associated with very poor drainage or waterlogging. The lack of air in these soils provides conditions for iron and manganese to form compounds that give these soils their colour.
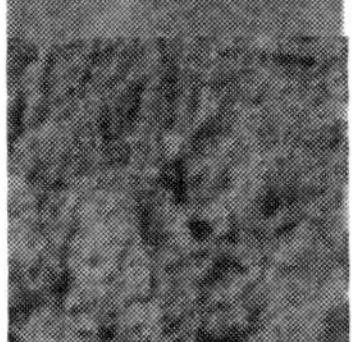	Yellow to brown 58,100,97 247,239,0 F7EF00	These soils often have poorer drainage than red soils. The iron compounds in these soils are in a hydrated form and therefore do not produce the 'rusty' colour.
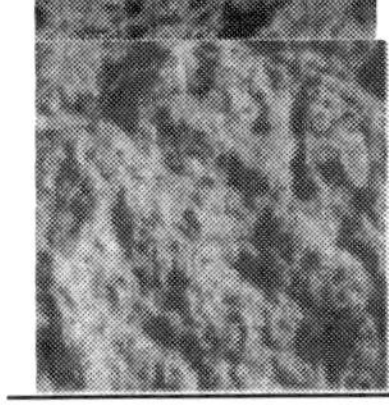	White/ bleached 60,4,100 255,255,244 FFFFFF	These soils are often referred to as being bleached or 'washed out'. The iron and manganese particles have been leached out due to excessive amounts of rainfall or drainage.

5. Manufacture of Kaolin (English China Clay)

Kaolinite is a common clay mineral with a general chemical composition of $2SiO_2Al_2O_32H_2O$ and is the consequence of weathering of the orthoclase feldspar in the parent igneous rock (granite). English China Clay, which is uniquely wash quarried (see Figure 1) extensively in the South West of England, is desirably pure and brilliant white in colour and is therefore a principal raw material used in the ceramic industry.

Figure 1. An English China Clay Quarry in St. Austell, Cornwall, UK

The uniquely elongated and white coloured orthoclase feldspar crystals (Figure. 1a) are readily identifiable with the unaided eye in the fresh Dartmoor Granite massif. Kaolinite quarrying using high pressure jet washing (Figure. 1c) of the hydrothermally weathered granite quarries have rivers of white kaolinite slurry directed to settling ponds (Figure. 1d) where the soft kaolinite sludge is separated and collected for further industrial treatment.

Intricate colour enhancement and image analysis can be effected using software PICASA 3 and analySIS@ package. The authors found this technique useful to assess or describe the fabric of peat samples to ascertain its composition. Figure 2 shows the typical images, normal and enhanced colour of undisturbed wet peat sample. The colour manipulation / image enhancement helps to show more clearly the fabric (a 2 dimensional fibre arrangement and orientation) in the fibrous peat composition.

Figure 2. An image of wet fibrous peat enhanced through Picassa 3 to facilitate image analysis

6. Empiricisms embedded in particle morphology

Particle morphology (size and shape) in geotechnics, can be related through topographic features to its genesis or the development of erosional forms. Morphology characterise the soil particles and help define the type of soil; from the rugged and massive boulders to platy clay particles. Soil mass classifications based on particle size distribution are adopted in various standards (BSCS, ASTM, USDS and AASHTO), particle shape, however has not been accounted for in these conventional soil classification systems [2]. Despite the many classification systems that are now in use, not one is totally definitive of any soil for all possible applications because of the wide diversity of soil properties [3]. The particle size distribution curve gives not only the range and distribution of particle sizes but from it, the uniformity coefficient and coefficient of gradation can be determined to class them as uniform (poorly graded), well graded or gap graded as illustrated in Table 3. These gradings influence the fabric of the soil. Table 3 further illustrates empirically typical ranges of angle of friction and permeability coefficient for the three different types of graded soils.

Table 3. The empirical deductions from the grading differences in particle size distributions

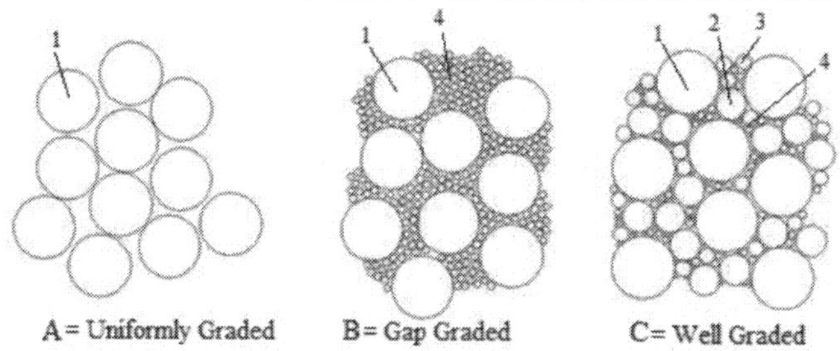

Note: The numbers indicates the scale of the particles

Increase in Shear Strength →

← Increase in Permeability

Type	Uniformly graded	Gap graded	Well graded
Typical Friction Angle values ϕ^0	40 – 45	35 - 40	30 – 35
Typical Permeability Values k m/s	$4.00x10^{-5} – 4.00x10^{-3}$	$2.55x10^{-5} – 5.35x10^{-4}$	$2.23x10^{-3}$

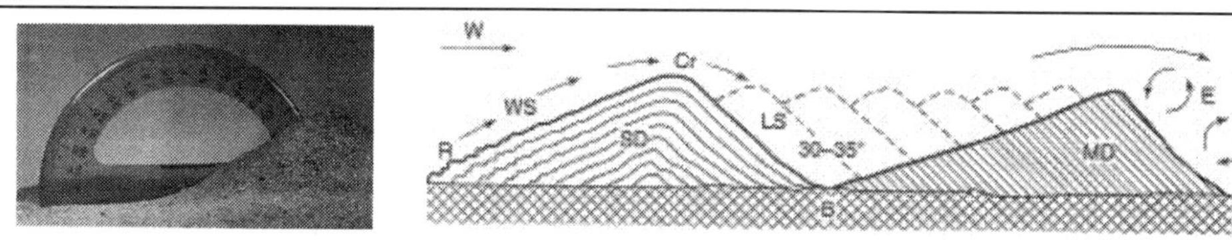

(a) (b)

Figure 3. a) An empirical method of illustrating angle of repose (angle of friction) [4], b) Formation of sand dunes and illustration of the angle of repose [4] Stationary Dune (SD), Migrating live dunes (MD), arrows indicate direction of air (W) and eddies(E) Windward slope (WS) Leeward slope (LS), Ripple marks (R), Crest of dune (Cr)

Sand dunes are so "soft" that they are often migratory landforms resulting from the wind as a geological process. Figures 3 a) and 3 b) show how sand dunes are nature's example of the angle of repose , which being the steepest slope for very loosely packed sands, it represents the angle of friction for grannular materials at its loosest state[4].

Table 4. Classification of particle shapes

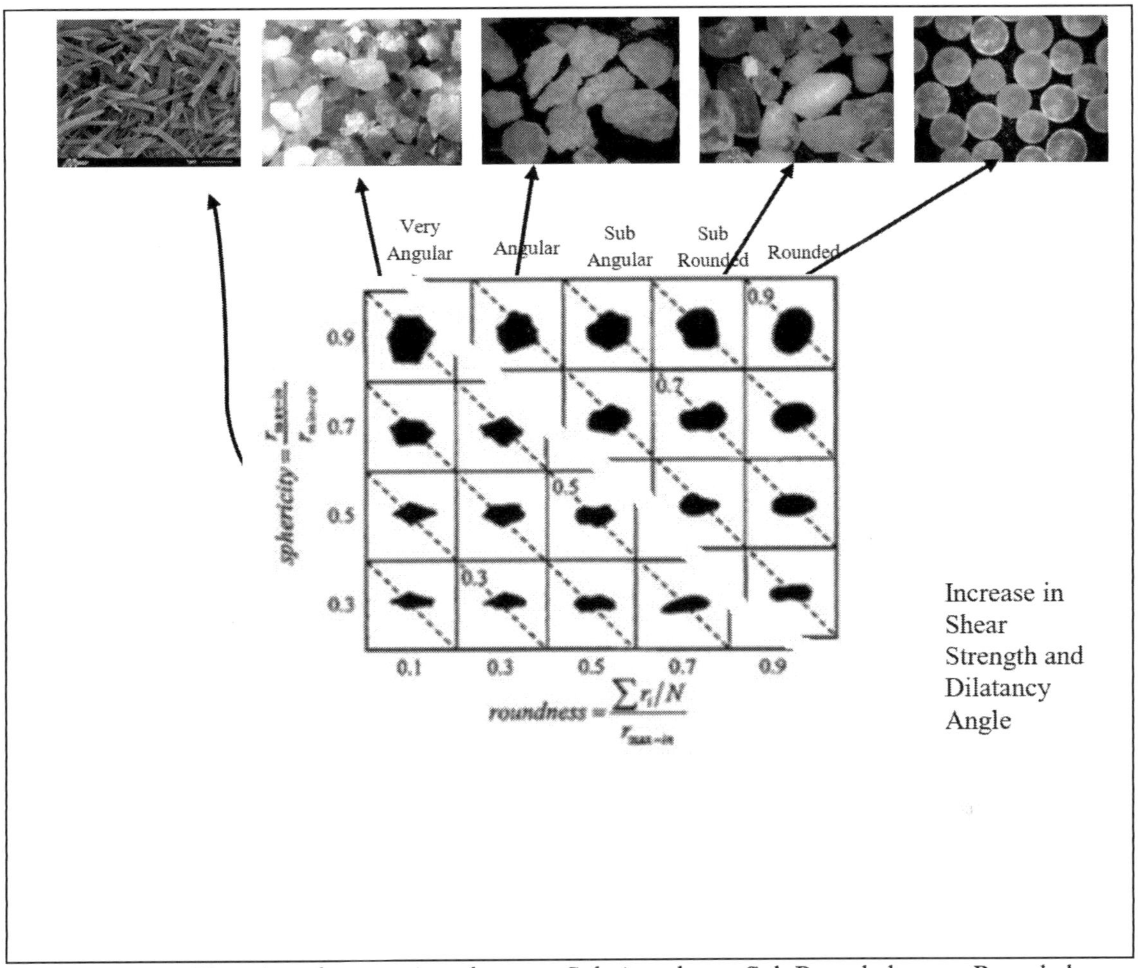

Type	Very Angular	Angular	Sub Angular	Sub Rounded	Rounded
Typical Friction Angle values ϕ	40 - 45	35 - 38		27 - 30	
	← Increase in friction angle value				
Typical Dilatancy angle	13.01 – 21.542	9.271 – 11.46		7.734 – 10.322	
	← Increase in dilatancy angle value				

As illustrated in Tables 3 and 4, the changes in the intrinsic angle of particulate friction can be made intuitively empirical. Particle aspect ratio further influences its development. The rare clay mineral, halloysite, has a tubular morphology akin to the decaying and contrastingly cellular morphology of roots in fibrous peat, which negates any empirical predictions. The latter microstructural feature in fibrous peat contradicts one of the founding assumptions in soil mechanics in that the "individual particles are incompressible". In a cohesionless soil, it was found that for a different types of sand based on its particle shape characteristics, not only the friction angle changes but the dilatancy angle also increases as the shape of the particle changes from rounded to very angular [5].

7. Empiricism in soil classification

In 1911, Atterberg, a Swedish soil physicist (and / agricultural engineer) recognised the four consistency states of liquid, plastic, semi solid, and solid that a fine grained soil can take as it interacts with water. The water content boundaries between these consistency states have been semi empirically assigned as liquid limit, plastic limit and shrinkage limit respectively. Soil classification systems provide, though empirical it may be, a universally accepted communication language to avert lengthy description of the soil. Many researchers have further established correlations between the consistency limits and strength characteristics of fine grained soils.

Soil investigations have then widely adopted these limits to classify soils using the charts as that of Casagrande. The chart displays empirically two boundary lines; the U line and the A line as shown in Figure.4. The corresponding equations for these lines are as follows;

U-Line; $PI = 0.9\,(LL-8)$ (1)

A -Line; $PI = 0.73\,(LL-20)$ (2)

Some of the AASHTO Soil classes [3] have also been embedded into the Casagrande Chart as included in the Figure. 4. The combination of these information enables further to present a "traffic light – red amber - green" zoning of the chart to obtain a perception of weak (challenging) to good (desirable) soils for construction purposes.

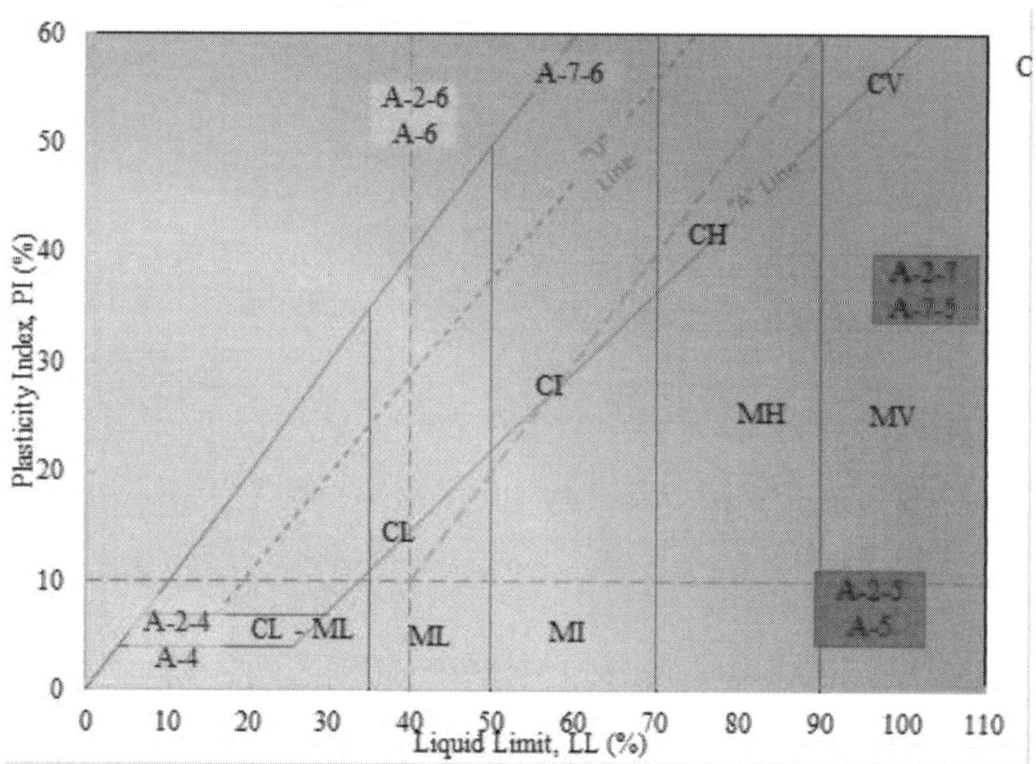

Figure 4. "Traffic light" zoned Casagrande chart with some embedded AASHTO soil classes

The effectiveness of a soil stabilisation technique can now be assessed through the locus of the "consistency state" of the unstabilised soil to that of the stabilised soil. As illustrated in Figure. 5, a locus towards the origin (downward and towards the left) conceptually signifies an improvement in the soil condition and vice versa. A vertical downward locus (an increase in PL with no change in LL) represents the effect of agglomeration due to the stabilisation process. A lateral locus towards the left of the Casagrande Chart (decrease in LL with no change in PI) suggests an increase in soil stiffness through cementation.

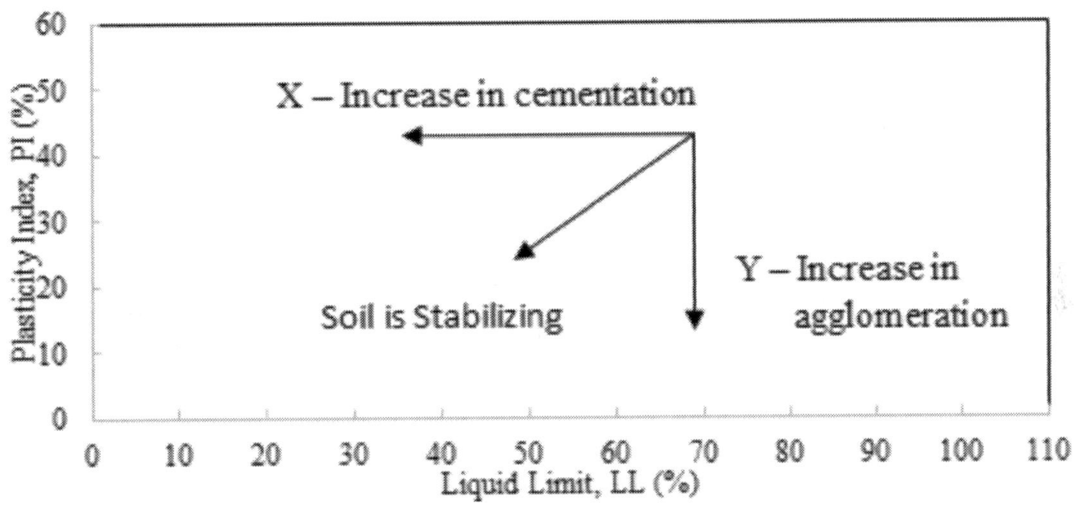

Figure 5. Soil stabilisation mechanisms explained through Casagrande Chart

8. Empiricism within Von Post Classification of Peat Soils

Table 5. Von Post Classification for Peat Soils and Squeeze Test (modified from [6])

Degree of Humification	Description	Squeeze test
H1	Completely undecomposed peat which releases almost clear water. Plant remains easily identifiable. No amorphous material present.	
H2	Almost completely undecomposed peat which releases clear or yellowish water. Plant remains still easily identifiable. No amorphous material present.	
H3	Very slightly decomposed peat which releases muddy brown water but for which no peat passes between the fingers. Plant remains still identifiable and no amorphous material present	
H4	Slightly decomposed peat which, when squeezed, releases very muddy dark water. No peat is passed between the fingers but the plant remains are slightly pasty and have lost some of their identifiable features.	
H5	Moderately decomposed peat which, when squeezed, releases very "muddy" water with a very small amount of amorphous granular peat escaping between the fingers. The structure of the plant remains is quite indistinct although it is still possible to recognize certain features. The residue is very pasty.	
H6	Moderately decomposed peat which a very indistinct plant structure. When squeezed, about one-third of the peat escapes between the fingers. The structure more distinctly than before squeezing.	
H7	Highly decomposed peat. Contains a lot of amorphous material with very faintly recognizable plant structure. When sequeezed, about one – half of the peat escapes between the fingers. The water, if any is released, is very dark and almost pasty.	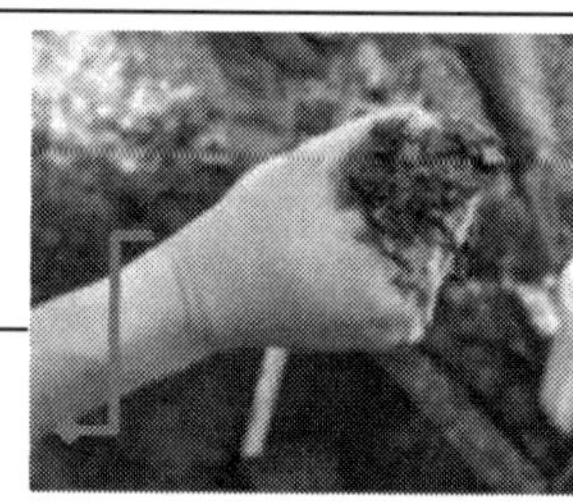
H8	Very highly decomposed peat with large quantity of amorphous material with very indistinct plant structure. When squeezed, about two thirds of the peat escapes between the fingers. A small quantity of pasty water may be released. The plant material remaining in the hand consists of residues such as roots and fibers that resist decomposition.	
H9	Practically fully decomposed peat in which there is hardly any recognizable plant structure. Fairly uniform paste when squeezed.	
H10	Completely decomposed peat with no discernible plant structure. When squeezed, all the wet peat escapes between the fingers.	

IOP Conf. Series: Materials Science and Engineering **136** (2016) 012001 doi:10.1088/1757-899X/136/1/012001

Table 6. Systematic outline of Von Post Classification

egree of umification VON POST)	Decomposition	Plant structure	Amorphous material	Colour of released water	Escape of material on squeezing	Nature of residue
H1	None	Easily identified	None	Clear	None	
H2	Insignificant	Easily identified	None	Clear or yellowish	~	
H3	Very slight	Still identifiable	~	Muddy brown	~	Not pasty
H4	Slight	Not easily identifiable	Very slight amount	~	None	Somewhat pasty
H5	Moderate	Recognizable but vague	~	~	Very small amount	~
H6	Moderately strong	Indistinct	~	~	About one third escapes	
H7	Strong	Very faintly recognizable	Lots	Very dark and almost pasty	About one half escapes	Fibres and roots more resistant to decomposition remain in hand
H8	Very strong	Very indistinct	Large quantity	~	About two thirds escape;	
H9	Nearly complete	Almost undiscernable	~	~	~	
H10	Complete	Not discernable	~	~	All the wet peat escapes	

IOP Conf. Series: Materials Science and Engineering **136** (2016) 012001 doi:10.1088/1757-899X/136/1/012001

Table 7. Summary of Published Geotechnical Properties for Peats

ee of Humification N POST)	Natural Moisture content (%)	Specific gravity	Organic content (%)	Fiber Content (%)	Liquid Limit (%)	Cc, Virgin compression	Cr, Re-compression ratio
H1							
H2				93[12]		0.021[12]	0.0025[12]
				40[12]			
H3	670(±20) [9]	1.30[9]	92[9]	75[9]			
	708(±10) [10]	1.49[10]	98[10]	74[11]	267[10]		
H4	480(±10) [10]	1.56[10]	92[10]	63[10]	162[10]		
	555[11]	1.24[11]	96.4[11]	90[11]			
H5	544(±10) [9]	1.40[9]	96[9]	63[9]	224[9]		
H6	472(±10) [9]	1.14[9]	92[9]	43[9]	137.2[9]		
H7	605[11]	1.82[11]	42.5[11]	31.98[11]			
H8				53[12]		0.012[12]	0.0030[12]
H9				48[12]		0.0030[12]	0.0020[12]
H10							

Table 5 shows an empirical visual assessment coupled with human subjectivity of a squeeze test that forms the basis of the Von Post classification of Peat into ten different "H" classes depending on the degree of humification [6]. The description of each Von Post class has been outlined systematically in Table 6 to delineate the differences between them, and thereby a logical transition between classes becomes apparent. Von Post's classification categorized peat into 10 "H" classes and this has now been further extended into 16 classes [7]. However these classes are empirical and do not provide an engineering perspective to its characteristics. Shear strength of peat is dependent on its moisture content, degree of humification and mineral content (higher the moisture content and humification, lower will be the shear strength; higher the mineral content, higher will be the shear strength [8] [9]. Table 7 presents geotechnical information from some research available in the public domain, on some peats described as H2 to H9. Some of the geotechnical parameters show appropriate changes with increasing Degree of Humification.

9. Empiricism observed through Animal Senses and Behaviour

Animals can sense disaster; they have a sixth sense. Many animals escaped the great 2004 Asian Tsunami. Elephants, Donkeys and dogs in Sri Lanka and Sumatra moved to higher ground before the giant waves struck. Similar peculiar behaviour of Elephants on that day was reported from Thailand; in that they even were trumpeting before they moved to higher ground. Reports from villagers in Bang Koey, Thailand gave the account of a herd of buffalo were grazing by the beach when "they suddenly lifted their heads and looked out to sea, ears standing upright"; They turned and stampeded up the hill, followed by bewildered villagers whose lives were thereby saved. At Ao Sane beach, near Phuket, dogs ran up to the hill tops, and at Galle in Sri Lanka, dog owners were puzzled by the fact that their pet animals refused to go for their usual morning walks on the beach. Similar reports from Cuddalore District in South India where buffalos, goats and dogs escaped and in addition a nesting colony of flamingos flew to a higher ground. In the "Andaman Islands "stone age" tribal groups moved away from the coast before the disaster, alerted by the behavior of the animals. Similar stories of animals "freaking out" and behaving in unusual manner long before (even up to 12 hours – early warning signal) are also reported from other past earthquakes (1995 Kobe, Japan; Assisi, Italy) Dogs have a remarkable sense of smell, birds can migrate using celestial cues and bats can locate food with echoes. Animals may sense unusual vibrations or changes in air pressure coming from one direction that suggest they should move in the opposite direction.

10.Empiricism in Ground Investigation

Ground investigation is a prerequisite to site characterization, which in itself needs to extract representative and reliable information from the earth which hosts a complex three dimensional distribution of (often non uniform) rocks/soils with its associated numerous structures and hydro-geomorphologies. A primary objective of site characterization is based on an empirical approach of acquiring a simplistic two dimensional (or 3D block) spatial arrangement description of topographical, hydro-geological, geotechnical, seismological and geo-environmental information that is relevant to the requirement of the particular project. This challenge needs to be approached by a coalition of hybrid disciplines [13] [14]. It has been said that, "one test result is worth a hundred expert opinions", but this is only true if such tests and their analysis are truly accurate, representative and relevant for the application [15].

Agricultural Engineers and Geotechnical Engineers can be found working on the same ground but with differing scopes and objectives. It is noteworthy, that one's gain may often be another's loss. While peat is considered a challenging problematic material to Geotechnical engineers, it is a material of prime importance to Agricultural engineers [16]. For instance, for the same amount of suction, tropical peat holds more water. The soil moisture at permanent wilting point (150 kPa matric suction) for Sapric / Hemic peat is reportedly still relatively high and prevalent at about 50 to 60% volumetric water content, which makes the ground to be still visually wet but lacks available water for plant use. This will necessitate maintaining the field water table optimally high to facilitate irrigation through capillary rise [17]. Such a requirement is met with mixed reception from a geotechnical perspective. The presence of a high water table reduces the permissible bearing capacity on one hand, while on the other hand, maintaining the water table constant will minimise shrinkage induced structural distress.

The most common problem that engineers face is whether the methods are appropriately suitable or not before carrying out the test. The selection of test method should be based on the information gathered from the site reconnaissance. Pioneering ground investigation methods used simple instruments such as pocket penetrometers etc. to solely identify and characterise the soil. Different equipment with varied appearances, shapes and purposes (such as the hoe, trowel and dibble etc. – Figure. 6 a) have been used in small scale farming. While some of these manual hand tools are intended for specific uses such as soil moving and soil conditioning, others can perform multiple functions. Figures 6b and 6c are an extension to the farming equipment that enable a ground investigator to help determine the depth of soft soil / depth to bedrock and obtain to study disturbed / in situ soil samples from a test pit. A rod penetrometer has dimensions which are approximately 1.2m long, 10 mm diameter and is a stainless steel rod with a 90 degree bend, 120 mm at the top to form a handle. Figures 5e and 5d are views of similar but empirical site investigation tools, Mackintosh probe and its modified JKR probe used to characterise the soil. Figure 5f illustrates the commonly used Standard Penetration Test (SPT). When the borehole is advanced to the desired depth, the drilling tools are removed and a split spoon sampler attached. A standard hammer weighing 623N is dropped through a distance of 0.762m to the top of the drilling rod. The number of blows required for driving the sampler through three 152.4 mm intervals is recorded. The sum of the number of blows required for driving the last two 152.4mm intervals is referred to as the standard penetration Number, N. This N value is very popularly used in design and has been empirically correlated by various researchers to give strength and stiffness parameters. The SPT has been further developed with electronic instrumentation to form the basis of Cone Penetration test and even further to the Piezo Cone test.

11.Empirical 45^0 Engineering Thumb Rule

A thumb rule is a principle with broad application, and by virtue of its empiricism, it is neither intended to be strictly accurate nor universally reliable. However its simplicity enables it to be easily remembered as well as learned and to be applied as a procedure for estimation or recalling some value for determination with first order accuracy. Table 8 outlines the empirical together with the conceptual failure surface inclinations for a variety of materials and situations. Ductile materials subjected to tension fail at 45^0 to the tensile stress direction. Brittle materials such as concrete, when subjected to unconfined compression fail at 45^0 and sometimes multiple parallel sets are visible. This as illustrated in the table is due to the failure along the plane of maximum shear stress. In the case of particulate soil samples, Mohr Coulomb failure criterion applies and the inclination of the failure surface will be greater than the 45^0 by an amount equal to one half of the angle of friction, φ. As the soil becomes softer, the failure mode changes from brittle to ductile, giving no well-defined failure surface but bulging. In both laboratory testing and field loading of foundations, a triangular wedge shaped zone of unfailed and passive soil mass moves monolithically. The angle of this wedge deviates from the 45^0 depending on factors such as the type of failure (ductile / brittle) and the roughness of the interface between the loading surface and the soil. Active and intermediate zones develop to various degrees depending on whether it is a general failure or local failure. These failure trajectories are associated with the heave of the soil at the ground surface. The failure surface in highly deformable soils such as soft peat occur with localised settlement followed by punching shear with no bearing on the 45^0 thumb rule.

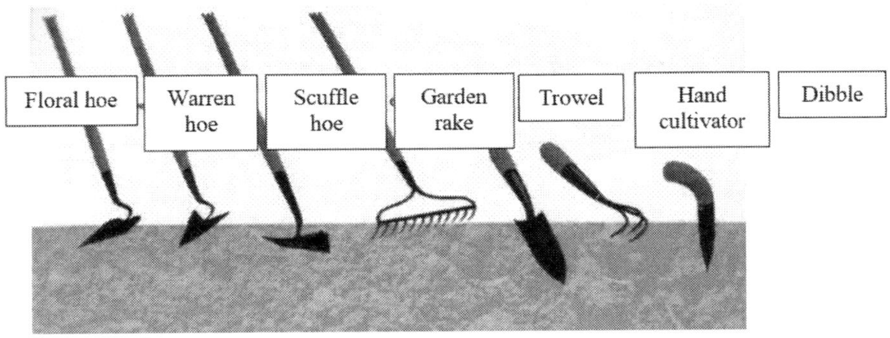

a) Types of Arable Agricultural Farming Equipment [18]

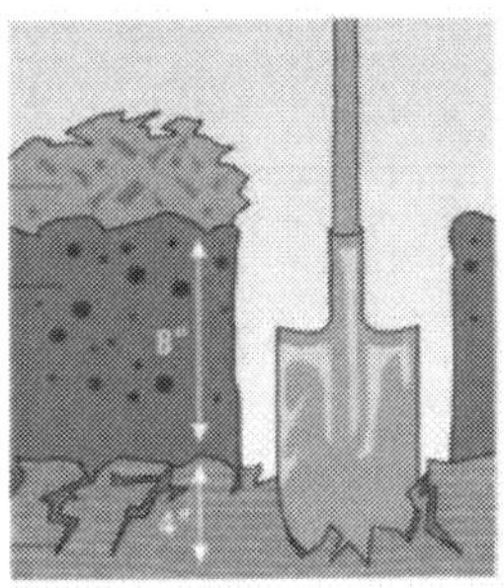

b) Use of shovel for topsoil

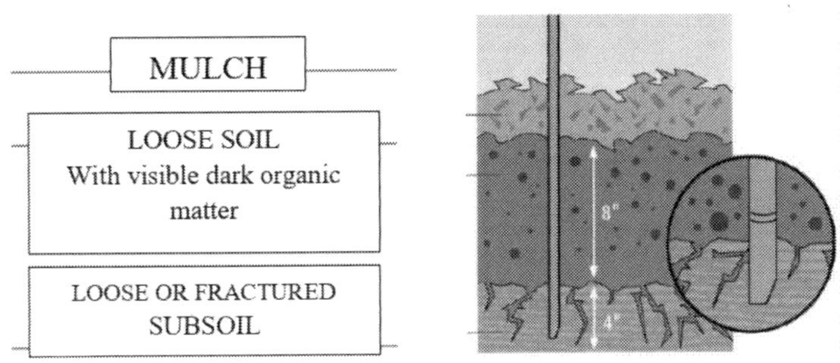

c) Rod penetrometer for soil investigation

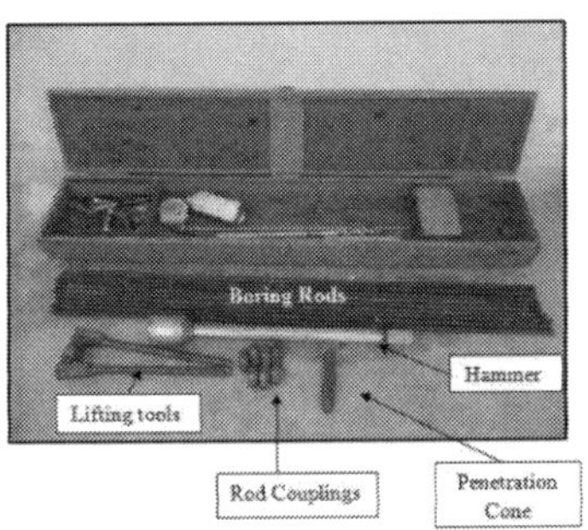

d) JKR probe

e) Mackintosh Probe

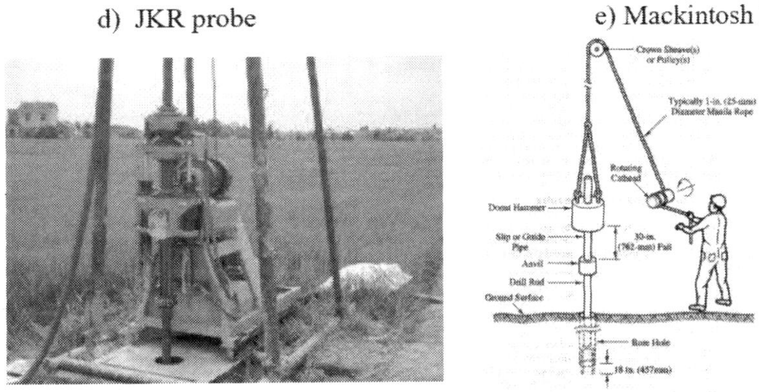

(f) Standard Penetration Test (SPT)

Figure 6. Some soil investigation tools [15]

IOP Conf. Series: Materials Science and Engineering **136** (2016) 012001 doi:10.1088/1757-899X/136/1/012001

Table 8. 45° Thumb Rule

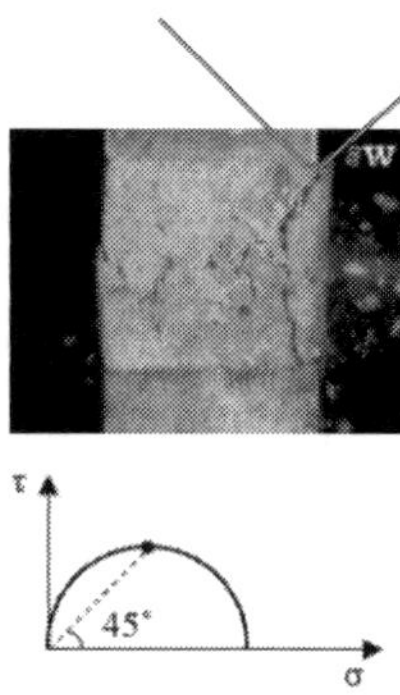	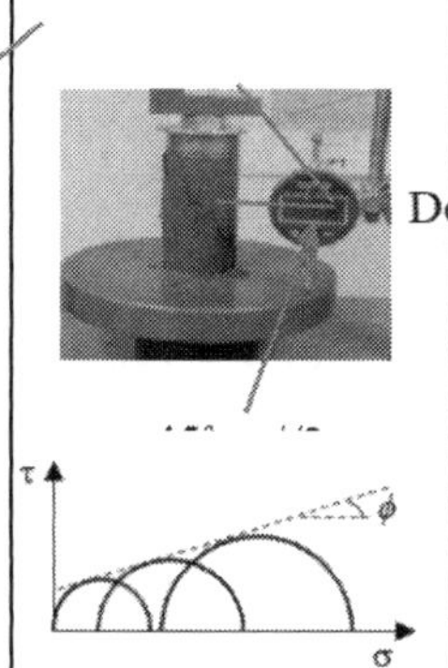	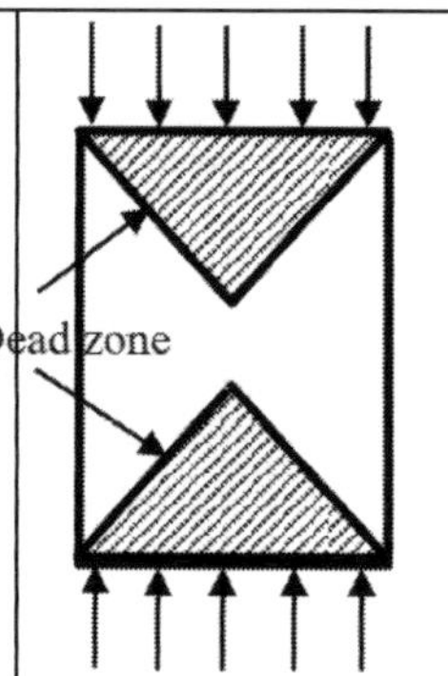	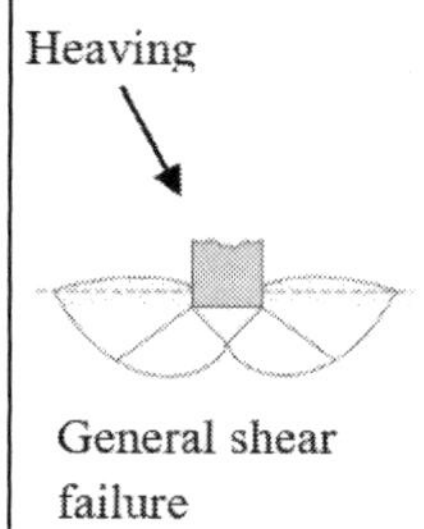	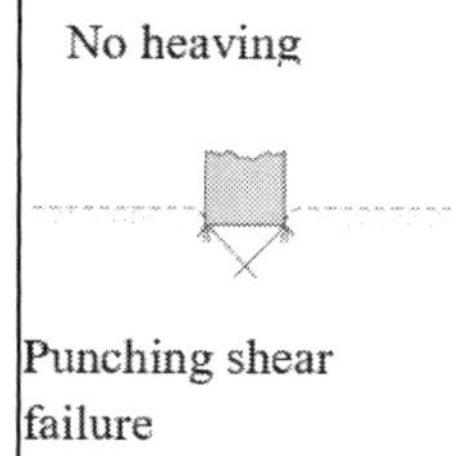
Concrete materials which are brittle	Particulate soil materials	Loading produces Triangular wedge shape zone (Dead zone)	Occurs on dense or medium dense soil	Highly deformable occurs on soft soil (Peat)

12. Empiricism through Experience

Experience based knowledge is not very common and "practical experience" does not always necessarily lead to sound intuitive judgement. A "practical person" is one who practices the theories developed at least thirty years ago or more. For every new project, civil engineer relies initially on some first-hand knowledge of the site that point through empiricism, experience of the senses to the need for site investigations which forms the basis for the complete contract drawings and specifications. Field investigations are based entirely on a systematic collation of prudent observations, ably coupled to deduce empirically may be to the presence and orientation of three-dimensional geologic structures from the limited facts explored. On the other hand the civil engineer through training, views the project three dimensionally to the all so important visible end product of the superstructure, but equally important, if not more so, is the knowledge leading to visualize the spatial character of the unseen subsurface geologic conditions. By virtue of the differences in their training, regretfully, geotechnical engineers and engineering geologists think differently; engineering science seeks a prediction for the future while geologists observe the present to interpret the past. These tabled predictions of the subsurface conditions affect, and often govern, the engineer's decisions in design, and also the construction technology to be adopted by the contractor. This empiricism needs to be continuously verified during the project construction stage and periodically inspected even thereafter to ensure safety of its function to be maintained in good order. Such documented records prove to be of inestimable value as a check on the assumptions made in the empirical predictions made at the onset and also as factual records for post construction contract litigations.

13. Empiricism and Public Safety

Historically, professional engineering has been established in response to threats to public safety. Empiricism is embedded into the public safety of modern structures such as bridges; however, in the 1860s American bridges were collapsing at the rate of 25 or more per year. Bridge failure and the associated loss of life precipitated the need for a stricter engineering approach to bridge design and construction. In Canada, engineering folklore holds that the collapse of the Quebec City Bridge in 1907 catalysed the establishment of higher standards in all branches of Canadian Engineering, which is symbolized today in the iron ring ceremony.

Engineering differs from other professions in that doctors, dentists, public accountants and lawyers generally provide their services to specific individuals / or defined corporations. Engineers design things with responsibility for specific societies rather than provide services to individuals. Engineering use of mathematics and science exposes it to the criticism that it is dry and it has sapped the artistic elements out of engineered structures. Art without engineering may be impossible, but engineering without art can be mundane and ugly. 13th century Reims Cathedral may lack art but 13th century builders were able to construct it. The Australian Sydney Opera House has its art dependent heavily upon engineering, as much that a 13th century builder would not be able to construct it. The first century Roman civil engineering of aquaducts and bridge construction was essentially a craft discipline. Arthur C. Clarke said that "any sufficiently advanced technology is indistinguishable from magic".

14.Experience & Empiricism, Expertise & Zeitgeist

Valuing experience (even though it be through empirical folklore or routine practice based on early theories and practice) more than expertise (innovative, sophisticated and the zeitgeist – the spirit of the age and adopting its novelties) is playing safe and being conservative though it can be considered to be unsustainable in wasting natural resources and overcharging the client. Professionals should not work beyond their own expertise, and must be considered competent and experienced only when they are chartered. Emeritus Professor John Atkinson rightly believes that geotechnical engineers should seek dual chartership from both a geological body and an engineering body. He also argues that the young tigers (new age students who can demonstrate and adopt the zeitgeist of geotechnical engineering, and also have been properly educated in engineering fundamentals) should be valued more than the experienced (the old heads) educated and trained in the 1960s and 70s have done the same thing slightly wrongly using old theories for a long time and have been lucky.

15.Conclusions

Geotechnical engineering that now encompass soil-water-environment phenomena and soil-structure interaction face challenges that necessitate the adoption of a multidisciplinary approach that are developed through embedded empiricisms. The occurrence of many progressive failures cannot be fully explained by classical soil mechanics concepts, but other factors such as chemical, physicochemical, micromechanical, and microbiological factors need to be considered, even though it may be empirically. The heterogeneous nature of both the soil and the prevailing environmental conditions make the engineering solutions at best to be an approximation.

Empiricism, as the name implies, lacks theoretical underpinnings; requiring as a dire necessity that the database of the relevant empirical relationships and assumptions be made explicit and be alerted to the instances where such proposed empirical relationships are at variance with the observed response. Professional engineers base their valued judgements in design on approximations, introduced and implied by these empiricism and/or experiential assumptions. From a solely practical viewpoint, in certain scenarios, applications based on such implications of empiricism and assumptions are inappropriate and inapplicable. On the other hand, in some other situations, one can ignore the distorted inaccuracies imposed by them to be negligible as they are of no practical significance. Caution needs to be therefore exercised in being one step away from reality, as the response evaluations are indirectly inferred.

Soft soils that embrace a wide variety of both mineral and organic soils, demand appropriately relevant empirical assumptions to be accommodated in order that the individual soil types can be appropriately assessed as an engineering material. Assumptions necessarily introduce approximations which in certain situations can distort the reality to an extent that the application of established theories may even produce unacceptable results.

References

[1] http://kentuckyfarmhouse.com/superstitions-myths-legends-folklore-omens-lucks-sayings/ accessed on 20/06/2015

[2] Wijeyesekera D C, Lim A and Zainorabidin A 2014 Particle Morphology in Geotechnics, *LAP Lambert Academic Publishing,* Germany.

[3] Das B M and Sobhan K 2014 Principles of Geotechnical Engineering, *SI Edition, 8th Edition, Cengage Learning, USA.*

[4] Holtz R D, Kovacs W D, Sheahan T C, 2011 An Introduction to Geotechnical Engineering, *2nd Edition, Pearson plc.*

[5] Lim A J M S 2014 Development Of A New Sand Particle Clustering Method With Respect To Its Static And Dynamic Morphological And Structural Characteristics, PhD Thesis, University Tun Hussein Onn Malaysia.

[6] Landva A O 1980 "Peat fabric and structure." *Canadian Geotechnical Journal 17*: 416-435.

[7] Macfarlane I C and Radforth N W 1968 Structure As A Basis Of Peat Classification. *Third International Peat Congress.* Quebec, Canada, National Research Council of Canada. 91-97, 1968

[8] Munro R 2005 "Dealing With Bearing Capacity Problems On Low Volume Roads Constructed On Peat", Highland Council, Transport, Environmental & Community Service, HQ, Glennurquhart Road, Inverness IV3 5NX Scotland.

[9] Zainorabidin A 2010 Static and Dynamic Characteristics of Peat with Macro and Micro Structure Perspectives, PhD thesis, University of East London, UK.

[10] Adon R (in Press), Laboratory Cone and Vane Shear Testing of Kaolin and Peats, PhD Thesis. Universiti Tun Hussein Onn Malaysia.

[11] Islam S and Hashim R 2008 Engineering properties of peat soils in Peninsular Malaysia, *J. Applied Science,* 8 (22).

[12] Price J S Cagampan J and Kellner E 2005 Assessment Of Peat Compressibility; Is There An Easy Way? *Hydrological Processes, 19(7)*; 3469-3475.

[13] Schnaid F, Lehane B M and Fahey M 2004 In-Situ Characterisation Of Unusual Geomaterials, *Proc. 2nd Int. Conf. on Site Charact., Milpress, Porto,* 1: 49-74.

[14] Schnaid F 2009 In Site Testing In Geomechanics. *Pub. Taylor & Francis, London and New York.*

[15] Adon R and Wijeyesekera D C 2015 Laboratory free Fall Cone and Shear Vane Testing in Soft Soil – Conceptual Research, *Proc. Soft Soil Engineering International Conference 2015,* Laangkawi, Malaysia.

[16] Katimon A and Melling L 2007 Moisture Retention Curve of Tropical Sapric and Hemic Peat, *Malaysian Journal of Civil Engineering 19(1)*: 84-90.

[17] Zainorabidin A and Wijeyesekera D C 2015 Why do Infrastructure Constructed on Peat Fail? , Keynote lecture, *International Civil and Infrastucture Conference,* Shah Alam, Selangor, Malaysia, September 2015.

[18] http://plantcaretoday.com/gardening-hand-tools-planting-cultivating.html

Challenges in Construction Over Soft Soil – Case Studies in Malaysia

N O Mohamad[1], C E Razali[1], A A A Hadi[1], P P Som[1], B C Eng[1], M B Rusli[1] and F R Mohamad[1]
[1]Geotechnical Engineering Branch, Public Work Department of Malaysia, Kuala Lumpur

E-mail: MNor_Omar@jkr.gov.my

Abstract. Construction on soft ground area is a great challenge in the field of geotechnical engineering. Many engineering problems in the form of slope instability, bearing capacity failure or excessive settlement could occur either during or after the construction phase due to low shear strength and high compressibility of this soil. As main technical agencies responsible for implementation of development projects for Government of Malaysia, Public Works Department has vast experience in dealing with this problematic soil over the years. This paper discussed and elaborate on the engineering problems encountered in construction projects that have been carried out by PWD, namely Core Facilities Building of Polytechnic Kota Kinabalu in Sabah and Hospital Tengku Ampuan Rahimah Integration Quarters in Klang, Selangor. Instability of the ground during construction works had caused delay and cost overrun in completion of the project in Selangor, whereas occurrence of continuous post construction settlement had affected the integrity and serviceability of the building in Sabah. The causes of failure and proposed rehabilitation work for both projects also will be discussed in brief.

Keywords: Slope instability, bearing capacity failure, excessive settlement.

1. Introduction

Soft soil typically characterized as a soil with low shear strength, highly compressible and low permeability [1,2]. The shear strength of the soil is reported to be less than 40 kPa and it can be physically moulded by light finger pressure. Generally, construction problems in this deposit are insufficient bearing capacity, excessive post construction settlement and instability on excavation and embankment forming. Theoretically, settlement problem can be defined as a deformation in the soil due to the applied stresses. As a result of settlement, the geometry of load carrying system will be changed, and if the ground water level is high, a part of the fill material will become buoyancy which will influence the total surcharge loading and the stability of the soil [3,4].

Generally, soft soil in Malaysia is considered as quaternary sediments consist of alluvial deposits and organic or peat soils [5]. According to Quarternary Geological Map of Malaysia, soft ground usually found in the coastal plains of the country covers large area of west coast and east coast of Peninsular and East Malaysia [6]. Due to significant proportion of soft ground to the country total land area and major economic activities and social developments are concentrating along the coastal area, construction projects on these problematic deposits are unavoidable. In the past, many failures related to high settlement and deformation to structure such as embankment which resulted from the high compressibility and low shear strength of soft soil have been reported either locally or internationally.

Due to the rapid development experienced by the nation in the last few decades, Public Works Department of Malaysia (PWD) had involved in many highway and building projects on soft soil. As a main technical agency for Government of Malaysia, involvement of PWD in any development project can be during construction or forensic investigation. Since 2010, statistics of geotechnical forensic investigation on problematic projects carried out by JKR is summarized in Table 1. There are 182 cases out of 252 forensic cases (approximately 72%) are related to the issue of ground settlement, the remaining 28% are caused by other factors such as vibration, erosion, foundation failures and so forth.

Therefore, it is clearly shown that the issue of ground settlement is the main engineering problem encountered in construction over soft ground area. A common settlement problem occurred in most of the project is:

- Excessive Settlement – Elastic or immediate settlement, Primary Consolidation Settlement, Secondary Settlement and;
- Differential Settlement – Occur where sudden change in the thickness of fill, e.g. at bridge abutment, or where the compaction of the fill is particularly difficult.

Table 1. Geotechnical forensic cases carried out by PWD

			Year				
Contributing Factors	2010	2011	2012	2013	2014	2015	Total
Vibration				2	1	3	6
Soil erosion				3	2	4	9
Foundation failures		2	1	6	3		12
Retaining structure failures			1	4	2		7
Slope stability			6	3	12	4	25
Ground settlement	1	8	10	63	70	30	182
Water infiltration				4	4	1	9
Collapsible soil			1		1		2
Total	1	10	19	85	95	42	252

2. Case Study 1: Hospital Tengku Ampuan Rahimah Integration Quarters in Klang, Selangor

2.1 Project Background

The construction of integration quarters for Hospital Tengku Ampuan Rahimah in Klang, Selangor was commenced in 2007. This project was implemented through design and builds concept, where the appointed contractor was responsible for the design and construction of the project. It is consist of the construction of three blocks of 13 and 14 storey building within the existing hospital compound. The site is underlain by quaternary alluvial deposit which consists of very soft to soft silty clay with the thickness ranges from 15 to 20m. This layer is followed by a layer of medium stiff to stiff silty clay/ sandy clay, whereas hard stratum is subsequently found at the depth of 18 to 28m.

Based on the site investigation report, 500mm diameter concrete spun pile with design capacity of 850kN has been proposed as foundation system for the buildings. Due to the constraint of the site, jacked – in method was adopted for the installation of the piles in order to minimize the noise and vibration which could affect the existing nearby structures. However, there was no ground treatment work proposed by the design consultant to improve the engineering behavior of soft soil at the site.

2.2 Construction Problems

In this project, engineering problems was encountered during the construction stage. Installation of piles was commenced over soft ground right after site clearing without preparation of proper working platform. Upon completion of piling work, most of the installed piles were found to be deviated from original position after some times. Piles were also found to be severely tilted and damaged at the location of the proposed lift pit and layout of the affected piles is shown in Figure 1.

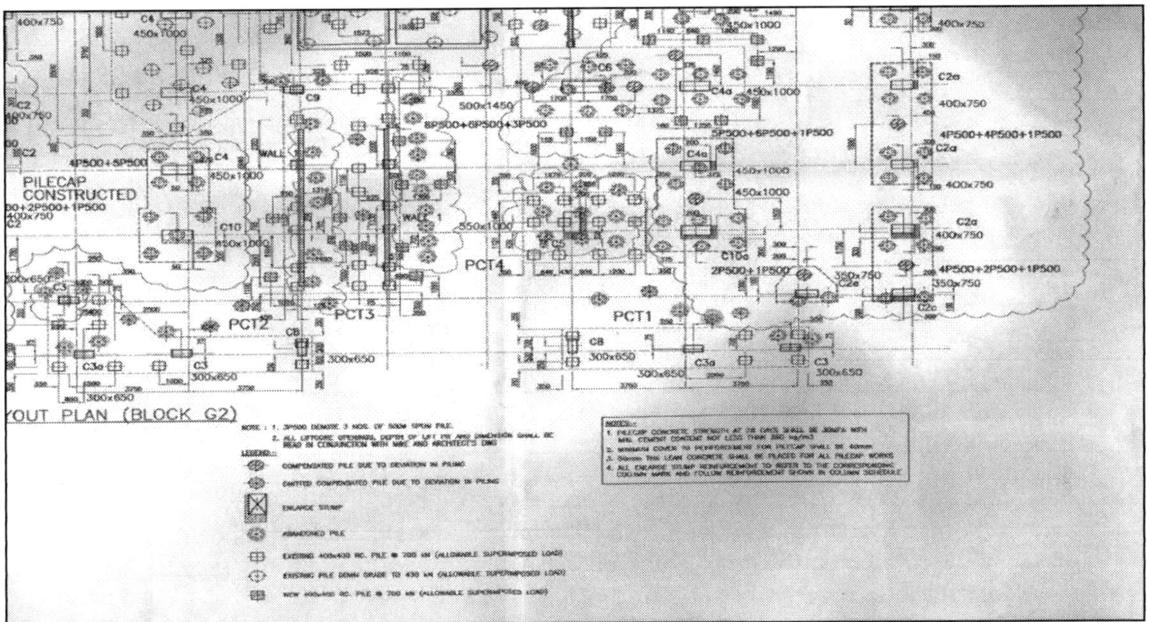

Figure 1. Layout plan of affected piles.

2.3 Causes of Problems

Based on the investigation carried out PWD, there are many possible causes contributed to the deviation of piles. The main cause was due to the displacement of soft ground as results of the moving jacked-in machine and excavation work for lift pits. In order to comply with standard requirement of PWD for pile installation using Jacked – in method, the total weight of the machine used must be approximately three times of the pile design capacity which is equivalent to about 260 metric ton. Such a heavy load that imposed on the surface of the ground had triggered vertical and horizontal displacement of the soil. Due to the existence of thick soft compressible layer at the site, the installed piles were expected to be pushed away by the displaced soil as it was lowly restrained in the horizontal direction.

The construction of lift pit involved an excavation work with approximate depth of 2.5m below ground level. For the purpose of excavation, a 6.0m sheet pile wall had been proposed as a temporary protection. During excavation work, the installed temporary sheet pile wall was found to be tilted as shown in Figure 2. An investigation carried out after the incidents revealed that the failure of temporary protection works was attributed to the shortfall in the design and improper construction practice adopted by the contractor in this particular project. Detailed analysis indicated that at least 9.0m depth of sheet pile wall is required to retain the 2.5m deep of excavation work, instead of 6.0m that was originally proposed.

Figure 2. Sheet piles and piles adjacent to the excavation pit tilted.

Probably due to the cost and time constraints, temporary sheet pile wall was installed at only one side of the excavation area. The excavated earth which was supposed to be transported out of the site was also temporary placed on the crest of the pit as shown in Figure 3. This improper construction practices adopted by the contractor had contributed to the failure of sheet pile wall as the placement of that unsuitable material had induced surcharge load to the ground that caused lateral movement of the unsupported side of the excavated pit.

Figure 3. A large quantity of excavated earth dumped at the crest of excavation pit.

The problem of pile deviation was also expected to be caused by negligent in carry out the setting out survey for the pile position. Movement of the ground as a result of the moving jacked – in machine may shift the pre – pegged pile position. Therefore, without confirmatory check on the pile position prior to installation had caused the piles to be installed at deviated position.

Subsequently, these problems had caused a great delay in completion of the project. It is also incurred huge amount of additional cost to the project because a lot of replacement piles were needed for the tilted or deviated piles.

3 Case Study 2: Core Facilities Building of Polytechnic Kota Kinabalu (PKK), Sabah

3.1 Project Background

Figure 4. Block A and B Polytechnic Kota Kinabalu

Polytechnic Kota Kinabalu is located at Kota Kinabalu Industrial Park (KKIP), Jalan Politeknik, Kota Kinabalu, Sabah. The construction of this project was commenced in 1999 and finally completed in 2003. Topographically, the project site is located on the low lying area which required extensive earthwork in order to achieve the proposed platform. Once completed, the first academic session at the new campus starts its operation on the 1 September 2003. Structural cracks were later detected at one of the main building in the campus, i.e. Core Facilities Building and the Teaching Restaurant. At the point of investigation, the drainage system such as perimeter drains had also experienced serious settlement.

In November 2004, Kumpulan IKRAM (Sabah) Sdn. Bhd. was appointed to conduct a structural assessment and geotechnical audit on the problematic building. The purpose and scope of work include the following:

(a) Physical investigation works include visits, visual inspection and necessary material and structural testing
(b) Monitoring works including defect and crack mapping, monitoring of settlement and movement of the Teaching Restaurant
(c) Proof checking of the structure of the building
(d) Carry out soil investigation works
(e) Carry out geotechnical analysis and ascertained the integrity of the pile foundation
(f) Recommendation of remedial works

IOP Conf. Series: Materials Science and Engineering **136** (2016) 012002 doi:10.1088/1757-899X/136/1/012002

The works that was carried out between September to November 2004 are as follows:-

(a) Condition survey and crack mapping
(b) Desk study of construction drawings (non-availability of as-built drawing)
(c) Material testing
(d) Preliminary structural analysis
(e) Field work for site investigation (2 nos. borehole, 4 nos. Macintosh Probe)

3.2 Geotechnical Investigation

The discussion in this section is based on soil investigation data, piling records and analyses conducted by Kumpulan Ikram Sdn. Bhd.

3.2.1 Subsoil Profile

From the boreholes result indicated the presence of fill material at the site with a thickness of about 7.2m. This layer is underlain by 10.6m of very soft to firm clayey sandy silt with some organic matters. A hard weathered siltstone and very poor weak sandstone was encountered at the depth of about 18.0m to 20.0m onward measured from the ground level (Figure 5).

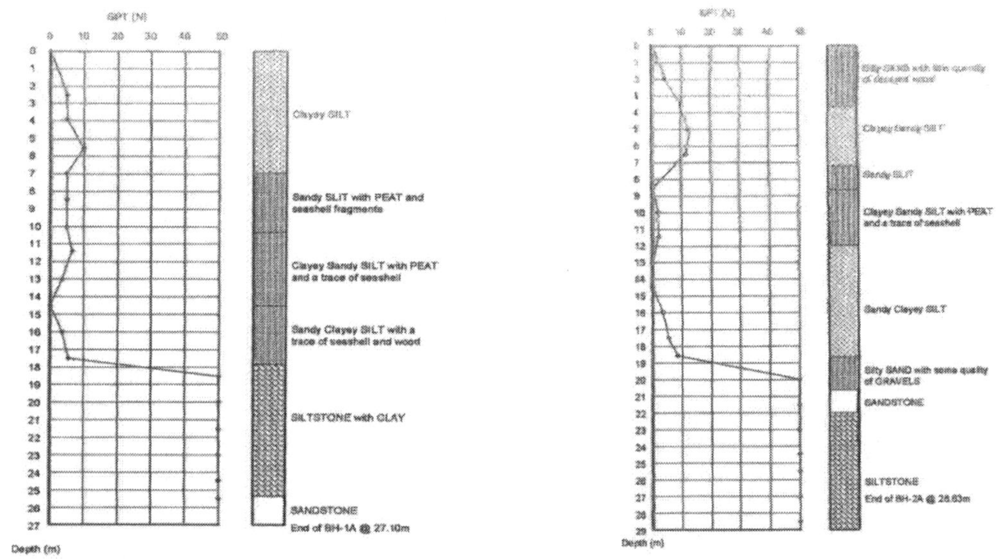

Figure 5. Cross – section of boreholes profile

3.2.2 Settlement Analysis and Monitoring

Back analysis was carried out by the investigation team in order to predict the magnitude and rate of consolidation process at the site. Compressibility parameters obtained from soil investigation work have been in the analysis with the assumption that 7.2m of fill acts as a surcharge load on 10.6m and 11.6m of soft compressible layer, respectively. The results of the analysis are summarized in Table 2 below.

Table 2. Magnitude and rate of anticipated settlement due to the 7.2m thick of fill material.

BH Location	Thickness of Compressible Layer (m)	Time Required For 90% Consolidation (year)	Predicted Post Settlement after 5 years (mm)
1A	10.6	9.06	243
2A	11.6	13.7	384

Based on the ground settlement monitoring records carried out from September – December 2005, a total ground settlement of about 130mm has been recorded, which equivalent to be about 103mm/ year. Therefore, it was concluded that the ground settlement within the building compound is still active at time of investigation. Depending on the actual thickness of soft compressible layer and height of fill at the site, a targeted 90% degree of consolidation expected to be achieved within 9 – 13 years after completion of the earthwork.

A separate settlement monitoring on columns also recorded a maximum reading of about 52mm settlement within a period of 15 months. The result of column settlement monitoring records had indicated that columns of Core Facilities Building had settled at different rates. Therefore, the major cause of the cracks is expected to be trigger by differential settlement that occurred on the foundation of the building.

3.2.3 Piling Records

Further inspection on the piling records also been conducted by the investigation team in order to ascertain the actual cause of failure. From the summary of the provided piling records, 250mm x 250mm reinforced concrete square piles have used as a foundation for the building. Average pile penetration depth for Block A and B of the Core Facilities Building is approximately 27.0m and 24.0m, respectively. Installation of the piles was carried out using hydraulic hammer with final set criteria was based on the blow count for the last 300mm of penetration. It was recorded that the installation of piles were terminated with blow count for the last 300mm is in the range of 57 - 84 to more than 100 blows. A simple calculation by dividing penetration of 300mm with the recorded total last blow counts indicated that piles may not properly set during installation. Evidently, some piles were proven to be settled trough settlement monitoring on column even though the pile driving records revealed that the piles had achieved set of 12mm/ 10 blows.

3.2.4 Pile Load Test

The proposed working load for 250mm x 250mm reinforced concrete pile used in this project is 600 kN/ pile. A static load test was conducted on the longest pile with penetration depth of about 27.0m in order to verify the design. However, no load test has been conducted on the shorter piles with average penetration depth of about 24.0m, which is mostly found at Block B of that particular building.

Based on the available data, pile capacity analysis was carried out by the investigation team. Taking into consideration of the negative skin friction, the geotechnical capacity for the proposed pile is found to be much lesser compared to the original design. The result of pile capacity analysis is summarized in the Table 3 below.

Table 3. Pile Allowable Bearing Capacity

Allowable Pile Bearing Capacity (kN)	Negative Skin Friction (kN)	Nett Allowable Bearing Capacity (kN)
629	217	412

IOP Conf. Series: Materials Science and Engineering **136** (2016) 012002 doi:10.1088/1757-899X/136/1/012002

The finding is in line with the design review carried out on the original design which found that the negative skin friction was not considered in the design. Theoretically, the effect of negative skin friction will induced an additional force on the piles thus reducing its effective capacity. Therefore, the main cause of the problems possibility due to the overloaded piles when subjected to the negative skin friction develops over time as a result of settling ground. This phenomenon has led the occurrence of differential settlement of the pile foundations that causing cracks to the building.

Conclusion

Soil contamination especially by hazardous heavy metals is pointed as a main contributor to environmental problems throughout the world. To ensure that these problems can be reduced and finally be solved, a systematic prevention and remediation methods should be taken. It is evident from this review that S/S remediation methods using various types of binders attempted of addressing these problems. S/S method using OPC are the most frequently studied for the remediation of heavy metals in soil. However, this review have revealed that S/S method using OPC incorporated with other potential additive such as fly ash, bottom ash and lime showed satisfactory results as well as help on reducing the remediation cost. This review also found that the use of recycle material from agricultural waste such as Rice Hush Ash (RHA) capable to improve the quality of soil treatment as well as contributed towards a sustainability remediation method.

References

[1] Huat B B K, Othman K and Jaafar A A, 1995 *Geotechnical Properties of Malaysia Marine Clays*, Journal – Institution of Engineers Malaysia.

[2] Huat B B K, Maail S and Mohamed T A, 2005 *Effect of Chemical on the Engineering Properties Of Tropical Peat Soil,* American Journal of Applied Sciences, Science Publication.

[3] Balasubramaniam A S and Brenner R P, 1981 Consolidation and Settlement of Soft Clay, Soft Clay Engineering (Development in Geotechnical Engineering), Elsevier Scientific Publishing Company, **20** 481 – 527.

[4] Tan Y C and Gue S S, 2000 *Design and Construction Control of Embankment over Soft Cohesive Soils*, Seminar on Ground Improvement – Soft Clay, Kuala Lumpur, Malaysia.

[5] Kaniraj S R and Josept R R, 2006 *Geotechnical Behavior of Organic Soils of Sarawak*, 4th International Conference Soft Soil Engineering, Vancouver, Canada.

[6] Jabatan Mineral and Geosains Malaysia, 2010 *Garispanduan Pemetaan Geologi Kejuruteraan Kawasan Tanah Gambut dan Tanih Lembut*, Kementerian Sumber Asli dan Alam Sekitar.

[7] Gue S S and Tan Y C, 2000 *Subsurface Investigation and Interpretation of Test Results for Foundation Design in Soft Clay*. Seminar on Ground Improvement – Soft Clay, Kuala Lumpur, Malaysia.

[8] Neoh C A, 1999 *Planning of Site Investigation and In – Situ Testing*, Short Course on Soil Investigation and Design for Slope, Kuala Lumpur, Malaysia.

[9] Omar R C, Jaafar R and Hassan H, 1998 *Engineering Geology and Earthwork Problem Associated with Highway Construction in Soft Soil at Sg. Rasau, Dengkil, Selangor*, Ninth Regional Congress on Geology, Mineral and Energy Resources of Southeast Asia, Kuala Lumpur, Malaysia.

[10] Pekeliling Unit Perancang Ekonomi Bil. 2 (2009), Garispanduan dan Peraturan Pemilihan Tapak dan Keperluan Peralatan Projek, Unit Perancang Ekonomi, Jabatan Perdana Menteri.

Development of Low Cost Soil Stabilization Using Recycled Material

F Ahmad[1], A S Yahaya[1] and A Safari[1]
[1]School of Civil Engineering, Universiti Sains Malaysia,Engineering Campus, 14300 Nibong Tebal, Penang, MALAYSIA

E-mail: cefahmad@usm.my

Abstract. Recycled tyres have been used in many geotechnical engineering projects such as soil improvement, soil erosion and slope stability. Recycled tyres mainly in chip and shredded form are highly compressible under low and normal pressures. This characteristic would cause challenging problems in some applications of soil stabilization such as retaining wall and river bank projects. For high tensile stress and low tensile strain the use of fiberglass would be a good alternative for recycled tyre in some cases. To evaluate fiberglass as an alternative for recycled tyre, this paper focused on tests of tensile tests which have been carried out between fiberglass and recycled tyre strips. Fibreglass samples were produced from chopped strand fibre mat, a very low-cost type of fibreglass, which is cured by resin and hardener. Fibreglass samples in the thickness of 1 mm, 2 mm, 3 mm and 4 mm were developed 100 mm x 300 mm pieces. It was found that 3 mm fibreglass exhibited the maximum tensile load (MTL) and maximum tensile stress (MTS) greater than other samples. Statistical analysis on 3 mm fibreglass indicated that in the approximately equal MTL fibreglass samples experienced 2% while tyre samples experienced 33.9% ultimate tensile strain (UTST) respectively. The results also showed an approximately linear relationship between stress and strain for fibreglass samples and Young's modulus (E), ranging from 3581 MPa to 4728 MPa.

Keywords: Recylcled tyre, fibreglass, tensile load, e modulus, stress-strain curve.

1. Introduction

Many countries are confronting the environmental problems posed by recycled tyres and are seeking to identify useful economic techniques for managing these tyres. Use of waste tyres as a construction material would be a desirable solution to avoid the technical, economic and environmental issues of discarding them. According to many studies which address the civil engineering applications of recycled tyres, these materials are normally grouped into three general categories: shredded, whole and bale tyres [1]. Many studies reported good results when using shredded tyre as construction material. A study to develop design procedures reported for using shredded recycled tyres as a light-weight fill material in highway construction [2]. Another study presented some examples of projects in which tyre shreds were used as light-weight fill for highway embankment construction, bridge abutment backfill, thermal insulation and drainage layers in landfills [3]. As for tyre shredded as light weight material, a full-scale project carried out using tyre shreds to reduce horizontal pressure in retaining walls [4]. Other researchers have focused on modifying clay soil with chip tyres and ascertaining the engineering properties of the clay-tyre composite [5-8].

Many others have performed studies on sand-tyre mixtures [9-11]. Some studies indicated that tyre chip-soil mixtures exhibit a significant initial plastic compression under loading and are highly compressible at normal low pressures [2,4]. Some other researches indicated that large strain ranged from 19.6% to 44 6% required to fully mobilizing the ultimate pull-out capacity [12].

To date there is no report on field application of fibreglass for soil improvement, soil erosion and slope stability. The unique properties of fibreglass including its affordability and cost effectiveness, strength, durability and light weight, make it suitable and desirable alternative for these purposes. Since reinforcement elements must provide additional stability for the soil mass, its tensile strength is a key parameter that needs to be measured. This paper describes an experimental investigation of the tensile properties of strip samples of fibreglass and recycled tyre. Then a comparison study carried out to show capability of fibreglass as an alternative to recycled tyre.

2. Experimental

To address the requirements of section 6 of ASTM D 4595 [13], the Instron 3690 series Actuator with 250 kN loading capacity was used for performing tensile tests on fibreglass samples. This machine is a double acting, equal area hydraulic piston which can exert a tensile and compressive force. This machine is able to apply load on the sample with the speed of 0.5 mm/min - 500 mm/min. Figure 1 is a photo of the Instron 3690. Tensile tests were performed on 1, 2, 3 and 4 layers fibreglass samples, with 1 mm, 2 mm, 3 mm and 4 mm thickness respectively. The samples were produced from chopped strand fibre mats (Figure 2) cured by resin and hardener. This type of fiberglass was chosen because of its low cost and properties which met all technical requirements (e.g. tensile stress and strain).

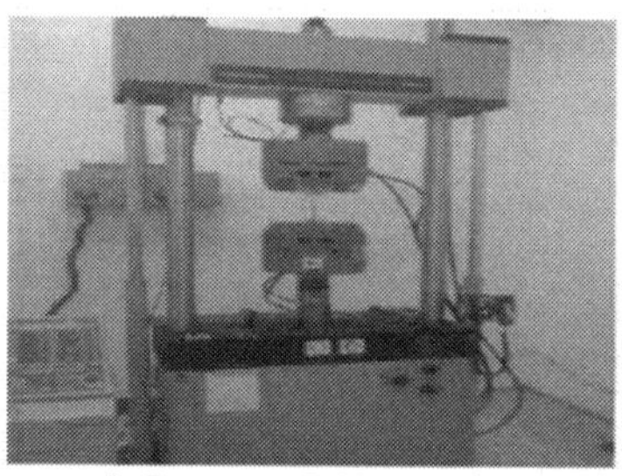

Figure 1. Instron 3690 series Actuator.

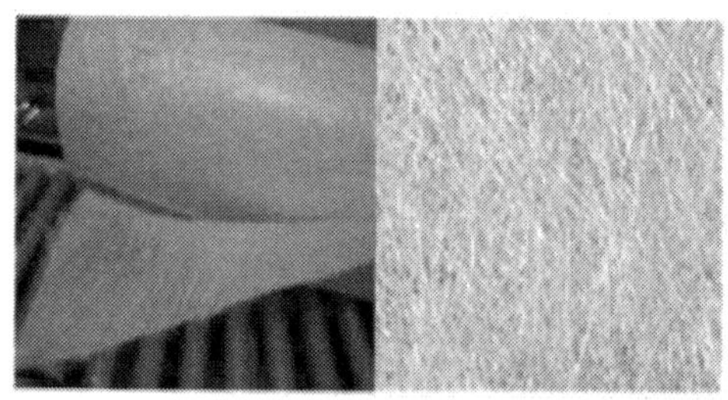

Figure 2. Chopped Strand Fibreglas.

Figure 3 shows fibreglass sample under tensile test,and The fibre strip were performed on developed fibreglas of 1, 2, 3 and 4 layers with 1 mm, 2 mm, 3 mm and 4 mm in thickness respectively. Tyre:For comparison purposes, strip samples of recycled tyre were also examined. Six samples of recycled tyre were cut into100 mm width by 300 mm length from whole recycled tyres No.175/70R13 (Malaysian "Good Year" brand)(Fig 4).

Figure 3. Fiber glass sample under tensile test.

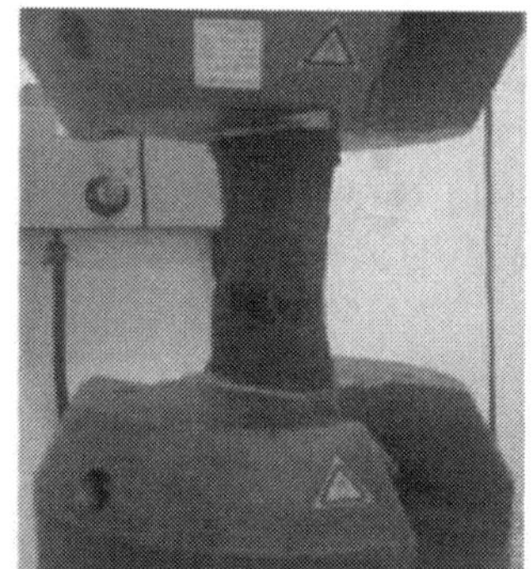

Figure 4. Tyre sample under tensile test.

3. Result And Discussion

The tensile tests were performed on strip samples where five parameters, maximum tensile load (MTL), elongation at break, maximum tensile (MTS) and tensile strain (TST) and Young's modulus (E) were measured. The results of tensile tests on fibreglass are shown in Table 1.According to Table 1, the, MTL increases as thickness increases, for all thicknesses except 4 mm. The maximum MTL value, 27.65 kN, was obtained with a thickness of 3 mm. Increasing the thickness from 1 mm to 2 mm and from 2 mm to 3 mm results in MTL increase of 276% and 146.7% respectively. Thus, increasing rate of MTL decreases with thickness, It should be noted that in the following of this study thickness of 3 mm of fibreglass samples as optimum thickness have been subjected to further study.

The mean MTL for 3 mm samples is 23.1 kN, with standard deviation 2.92 kN. Statistical probability analysis carried out on the results shows that the probability of MIL greater than 15.34 kN is 99.7% and greater than 19.53 is 88.88%. The same method carried out in a study showed the same results on recycled tyre strips [14]. A similar approach shows the mean value of TST at MIS is 0.02 mm/mm with 0 mm/mm standard deviation. This indicates that the probability of TST 0.02 mm/mm is 100%.

The stress-strain curves of samples numbers 13-18 from Table 1 and as shown in Fig. 5, indicate that the relationship between stress and strain is approximately linear and Hook's law is valid for this material. Thus, to find the best relationship between σ (stress) and E (strain), linear regression analysis was performed and the best fitting lines were obtained by zero value of intercept. The values of the coefficient of determination (R2) greater than 96 indicate good linear relationship between σ and E. Thus, the high tensile stress capacity in addition to low tensile strain characteristic of fibreglass could be a key parameter to control deformation in some geotechnical projects such as retaining-wall and riverbank projects where these can control the soft soil underneath.

Table 1. The result of tensile test on the strip sample of fiberglass.

Sample No	Thickness mm	MTL kN	Elongation at Break mm	MTS MPa	TST at MTS mm/mm	E modulus MPa
1	1	6.17	1.67	61.72	0.01	5249.67
2	1	3.3	2.6	33	0.01	5410.85
3	1	5.95	1.53	59.49	0.01	5834.75
4	1	5.66	1.45	56.57	0.01	5223.33
5	1	7.46	2.46	74.64	0.02	4783.91
6	1	5.7	1.47	57	0.01	5300.31
Mean	1	5.7	1.86	57.07	0.01	5300.5
Standard deviation	--	1.35	0.52	13.53	0.004	338.52
7	2	15.07	2.3	75.33	0.02	5069.27
8	2	18.46	2	92.3	0.02	6296.84
9	2	15.53	2.49	77.66	0.02	4759.39
10	2	17.14	2.63	85.68	0.02	5204.81
11	2	11.96	2.06	59.81	0.01	4687.17
12	2	16.3	2.52	81.5	0.02	5203.17
Mean	2	15.74	2.33	78.71	0.02	5203.4
Standard deviation		2.21	0.26	11.05	0.004	579.18
13	3	24.97	2.34	83.25	0.02	4768.65
14	3	23.31	1.52	77.7	0.02	-
15	3	20.86	2.28	69.52	0.02	4218.8
16	3	27.65	2.73	92.18	0.02	4438.63
17	3	19.53	2.44	65.09	0.02	3701.6
18	3	22.23	2.48	74.1	0.02	4324.12
Mean	3	23.1	2.30	77.0	0.02	4290.36
Standard deviation	0	2.92	0.41	2.75	0	388.47
19	4	18.68	1.56	46.71	0.01	3969.79
20	4	21.7	1.61	54.26	0.01	4139.66
21	4	18.57	3.7	46.43	0.01	4505.15
22	4	24.08	1.81	60.21	0.01	5241.41
23	4	25.88	2.01	64.7	0.01	5124.82
24	4	23.1	1.95	57.75	0.01	4673.50
Mean	4	22.00	2.11	55.01	0.01	4609.06
Standard deviation	0	2.95	0.80	7.36	0	511.83

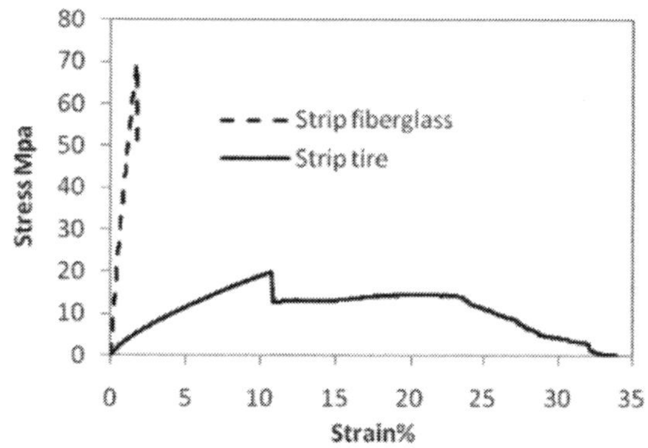

Figure 5. Stress- strain cure for fiberglass and tyre strip.

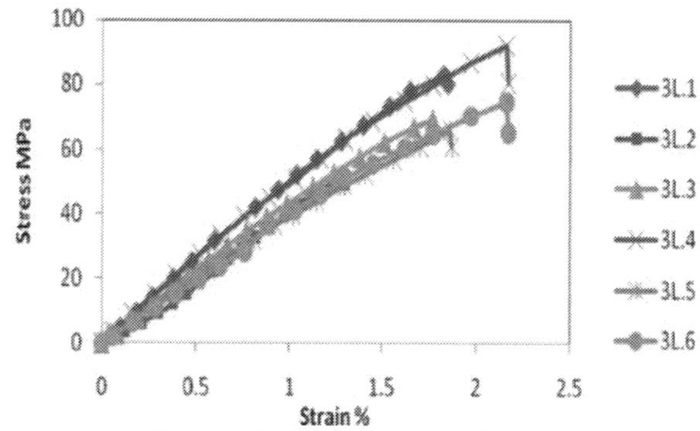

Figure 6. The stress-strain curve for fibreglass.

Conclusion

This study presents a new use development of fibreglass in geotechnical engineering. To determine the suitability of this material in this field, tensile tests were conducted on fibreglass strips. Since recycled tyres are a very useful material in geotechnical engineering field, a comparison analysis performed between tensile fibreglass and tyre samples. The results of this are as in Fig.5. By increasing the thickness from 1 mm to 2 mm and from 2 mm to 3 mm, MTL increases 276% and 146.7% respectively. Thus, the increasing rate of MTL decreases with thickness. The maximum value of MTL is 27.65 kN, obtained with a thickness of 3 mm. The MTL of fibreglass is 15.34 kN with 99.7 % probability. Its TST at MTS is 0.02 mm/mm. The stress-strain relationship is approximately linear as shown in Fig.6. The Young's modulus of fiberglass samples (3 mm thickness) ranges from 3581 MPa to 4728 MPa. The MTS is 68.4 MPa and 17.92 MPa for fiberglass and tyre respectively. In contrast, the tyre sample has experienced a very high value of strain, 33.9%.The high tensile stress capacity and low tensile strain characteristic of fibreglass will make it very suitable material to control deformation in geotechnical engineering projects such as retaining wall and riverbank projects in solving soft soil problem. This was then developed into 8 Shape Mat to solve problem in geotechnical applications.

References

[1] Zorenberg J G, Christopher R B and LaRocquc C J, 2004 Application of Tyre Bale in Transportation Projects. J. Recycled Materials in Geotechnics 42-60.

[2] Bosscher P J, Edil T B and Kuraoka S, 1997 Design of Highway Embankments Using Tyre Chips. Journal of Geotechnical and Geoenvironmental Engineering **123(4)** 295-304.

[3] Humphrey D N, 1999 Civil engineering application of tyre shreds, Proceedings of the Tyre Industry Conference, Clemson University, South Carolina, 3-5 .

[4] Humphrey D N, and Manion W P, 1992 Properties of Tyre Chips for Lightweight Fill. Proceedings of Grouting, Soil Improvement and Geosynthetics, 1344-1355.

[5] Akbulut S, Arasan S and Kalka E, 2007 Modification of clayey soils using recycled tyre rubber and synthetic fibers. Applied Clay Sci, **38** 23-32.

[6] Cetin H, Fener M and Gunaydin O, 2006 Geotechnical properties of tyre-cohesive clayey soil mixtures as a fill material. Engineering Geol, **88(1-2)** 110-120.

[7] Dai J C, and Huang J T, 1999 Surface of clays and clay-rubber composite Sci, **15(1-2)** 51-65

[8] Marefat V and Soltani-Jiagheh H, 2011 Laboratory Behavior of Caly-Tyre Mixture. Worl Applied Sciences J, **13(5)** 1035-1041.

[9] Christ M, and Park J B, 2010 Laboratory determination of strength properties of

frozenrubber-sand mixtures. Cold Regions Science and Technol, **60(2)** 169-175.
[10] Hataf N, and Rahimi M M, 2006 Experimental investigation of bearing capacity of sand reinforced with randomly distributed tyre shreds. Construction and Building Materials, **20(10)** 910-916.
[11] Rao G V and Dutta R K, 2000 Compressibilityand Strength Behaviour of Sand-tyre Chip Mixtures.J. Geotechnical and Geological Engineering, **24(7)** 11-724.
[12] O'Shaughnessy V and Garaga V K, 2000 Tyre-reinforced earthfill. Part 2: Pull- out behavior and reinforced slope design Canadian Geotechnic Journal, **37** 97-116.
[13] ASTM, D4595-05.Standard Test Method for Tensile Properties of Geotextile by the Wide-Width Strip Method (2007)
[14] Huat B B K, Aziz A A and Chuan L W, 2008 Application of Scrap Tyres as Earth Reinforcement for Repair of Tropical Residual Soil Slope, Electronic J. Geotechnical Engineering, **13**.

Road Maintenance Experience Using Polyurethane (PU) Foam Injection System and Geocrete Soil Stabilization as Ground Rehabilitation

A M M Fakhar[1] **and A Asmaniza**[1]
[1] Maintenance Monitoring Department, PLUS Berhad, PLUS, 47301 Petaling Jaya, MALAYSIA

E-mail: fakhar@plus.uemnet.com

Abstract. There are many types of ground rehabilation and improvement that can be consider and implement in engineering construction works for soil improvement in order to prevent road profile deformation in later stage. However, when comes to road maintenance especially on operated expressways, not all method can be apply directly as it must comply to opreation's working window and lane closure basis. Key factors that considering ideal proposal for ground rehabilitation are time, cost, quality and most importantly practicality. It should provide long lifespan structure in order to reduce continuous cycle of maintenance. Thus, this paper will present two approaches for ground rehabilitation, namely Polyurethane (PU) Foam Injection System and Geocrete Soil Stabilization. The first approach is an injection system which consists two-parts chemical grout of Isocynate and Polyol when mixed together within soil structure through injection will polymerized with volume expansion. The strong expansion of grouting causes significant compression and compacting of the surrounding soil and subsequently improve ground properties and uplift sunken structure. The later is a cold in-place recyclying whereby mixture process that combines in-situ soil materials, cement, white powder (alkaline) additive and water to produce hard yet flexible and durable ground layer that act as solid foundation with improved bearing capacity. The improvement of the mechanical behaviour of soil through these two systems is investigated by an extensive testing programme which includes in-situ and laboratory test in determining properties such as strength, stiffness, compressibility, bearing capacity, differential settlement and etc.

Keywords: Polyurethane foam (pu), geocrete, soil stabilization.

1. Introduction

Expressways or high speed road corridors have been one of the most vital infrastructures in the overall socio-economic development in Malaysia and the asset maintenance is basically the key toward preservation to restore and maintain at specified level of services and prolong the life span by reducing its deterioration rate. By neglecting or delaying maintenance activities may indirectly incurred the overall repair cost thus increase vehicle operating costs. With the existing road infrastructure that been served towards public road users, there will be time where deterioration may occur especially on soft ground and weak road base area (e.g settlement, depression, faulting, undulation, sinkhole and etc.) due to loss materials (wash away effect), poor compaction during construction, sunken structure, high traffic loadings, current global climate and many more which requires continuous round of maintenance thus increased the unnecessary budget spending.

In practice, conventional remedial method applied such as asphalt overlay and regulating, cementitious grout, sealing cracks only provide short-term solutions by addressing the symptoms, but not the underlying problem. These actually exacerbate the problem by placing additional weight on an already overstressed soil structure, thus will continuously require similar round of maintenance. Treatment reconstruction by full depth removal can be done however its time consuming, bigger space for closure and extremely costly. A practical of immediate measurement which able to sustain for some period of time, reduces frequency of remedial maintenance, shortening time of repairing is greatly needed [1]. The reparative pavement structure and soil improvement should have sufficient bearing strength and good flatness to ensure the vehicle's safety on the expressways [2]. A green, sustainable and cost-effective technology is hereby tabled to address the issues of insufficient bearing capacities and differential settlement for roads in soft grounds and weak base area. Soil stabilization can be defined as an approach or technique in improving the existing ground soil through modification properties of engineering characteristics behaviour such as bearing capacity, stiffness, strength, density, compressibility as well as hydraulic conductivity [3]. This is to address long term consolidation and settlement, stability and bearing capacities for the foundation. The following will describe further fundamental product for both soil stabilization systems.

1.1 Polyurethane (PU) Foam Injection System

A patented process that uses chemical grout (isocynate and polyol) combination product that able to uplift, realign, underseal, fill voids and strengthening ground soil structure which has been implemented with successful outcome at several States [4,5]. The injected resin within the soil will gain approximately 90% of its maximum compressive srrength (minimum compressive strength of 0.276 N/mm^2) within 15 minutes [6] and can be open to traffic the next day. Both chemicals are environmental friendly, non-toxic product and indefinite life span (refer Table 1 for chemical properties). The PU foam is based on three dimensional networked solid polymer with high strength that is produced with generation of heat and increase of volume when both liquid isocynate and polyol (contains little amount of water) compound is mixed [7,8,9]. Rigid PU (referred as "hard bubble") has characteristics of light weight and expansion foam ability with high specific strength, strong in bonding stress with other substances, anti-aging properties and curing in a short time.

Method of PU grouting is basically identical to the conventional cement grouting whereby the chemical resin is injected into the ground by drilling the hole, install the packer with tube and filling up the voids between the ground soil (see Figure1). In addition, with the use of PU the bearing capacity of the soil is improved due to the expansion of voids between the soil particles. [3]. Furthermore, the lightweight characteristics of PU allow a reduction of settlement rate and expansion characteristics of PU able to lift up the ground soil.

Table 1. Properties of Isocynate and Polyol Component.

Property	Polyol	Isocyanate	
Density, 25°C	1.10 g/cm^3	1.23 g/cm^3	
Viscosity, 25°C	1500 mPa.s	220 mPa.s	
	Units	**Test Method**	**Results**
Density	kg/m^3	ASTM D-1622	160
Ultimate Compressive Strength at 20°C	kPa	ASTM D-1621	
- Force parallel			2,290
- Force perpendicular			2,040
Thermal Conductivity at 20°C Fresh Foam	W/m.K	ASTM C-518	0.0379
Percentage of closed-cells	%	ISO 4590	91
Water Absorption at 20°C	% by volume	ASTM D-2842	0.33

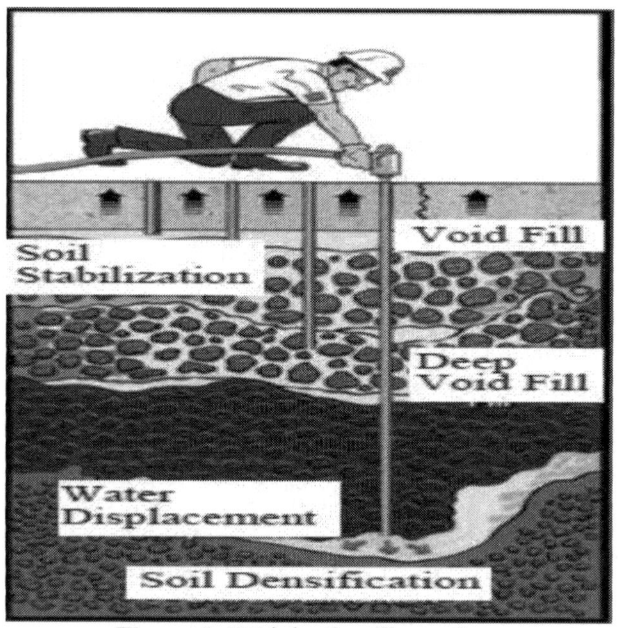

Figure 1. PU injection illustration.

1.2 Geocrete Soil Stabilization

Stabilization on existing ground soils using cement has been widely used through recycling approach in improving the properties and enhanced the span life usage. However, applications of cement with cohesive soils are difficult as serious cracks render, less durability and requires minimum of 7 days of curing prior to road opening. Stabilizing the soils with cement quite an issue as humus acids especially fulvic and carbon acids within the organic mass blocks the cement hydration process almost completely. Lime also has been introduced in stabilizing poor soils, however it results low strength and durability of soil. Moreover, the usage of foamed bitumen as stabilizing agent to waterproof soils and aggregates, keeping moisture contents low and high bearing capacity, however extremely costly with potential environment hazard [10]. Thus, Geocrete additive agent is introduced to countermeasures the issues highlighted especially curing process.

Geocrete (whitish powder consisting of alkaline earth element compound) is an additive to cement that able to neutralize the humus acids. The hydration of cement in the soil stabilization or immobilization when mix with water (see Figure 2) [11]. The powder and cement mix allow an environmental friendly transition of the existing soil into a newly build concrete like structure or in other word semi-rigid formation with impermeability characteristics. Unlike concrete where short crystalline structures are form and easily cracks. Geocrete allows long crystalline structures to be build and produce higher bearing capacities of soil, strength, elastic modulus and water impermeable. This can be illustrated using electronic microscopic as shown in Figure 3 & Figure 4. Besides improving the mentioned, Geocrete promotes the immobilization of pollutants such as heavy metals and organic parameters which get permanently embedded to the newly crystallized soil structure. Determination of the mixture will depend on the properties for in-situ soil. The next section will describe further the material preparation and scope of works for both systems that have been successfully implemented.

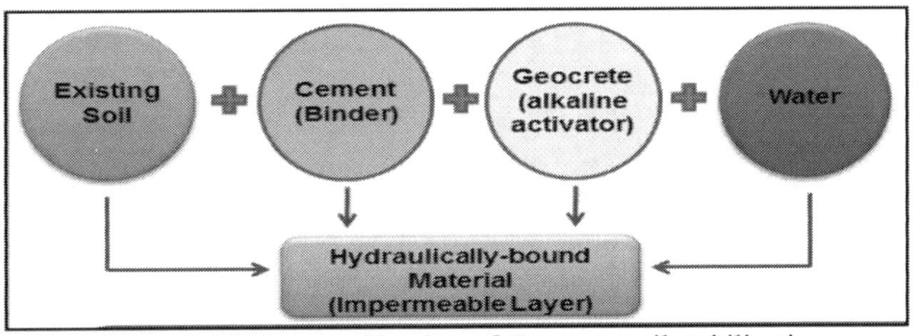

Figure 2. Mixing combination of Geocrete soil stabilization.

Figure 3. Cement soil stabilization – low crystal interlocking connectivity.

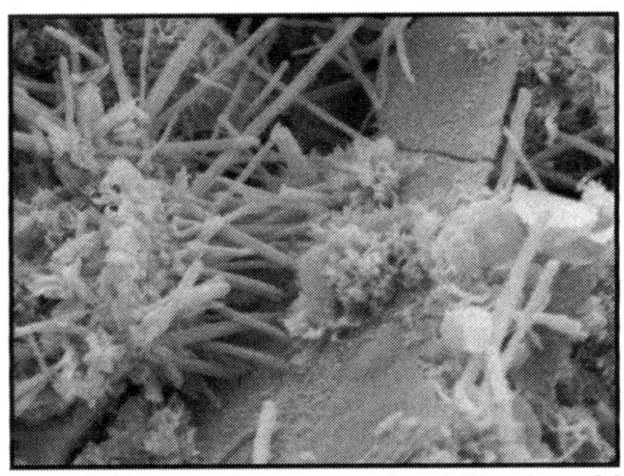

Figure 4. Cement-Geocrete soil stabilization - high crystal interlocking connectivity.

2. Materials and Methods

2.1 Material Preparation and Scope of Work for PU Foam Injection System

The case study presented herewith is at KLIA Toll Plaza, North-South Central Link of which experienced undulation road profile along concrete pavement. Thus, proposal of using 2 types of chemcial polyurethane, namely *Rapid Polyurethane* (soil stabilization and ground compacting) and *Dense Polyurethane* (densification and uplifting of concrete structure). Also, materials used the combination of Isocyanate and Polyol mixture with mix ratio of 1.1 : 1.0 respectively. Blowing agent has been incorporated into the resin, however additional agent (nitrogen based catalyst) will be added to improve foam's expansion.

The working area is measured 51 m length x 45 m wide which covers 4 Smart Tag, 3 Touch n Go and 2 Cash lanes as shown in Fig. 5. About 621 points were identified for grouting injection with drilled hole size of 16 mm diameter for *Rapid Polyurethane* injection and 32 mm diameter drilled hole for *Dense Polyurethane* injection at each point respectively. Each point was injected in two (2) phases; Phase 1 injection of *Rapid Polyurethane* up to the depth of 2 m and Phase 2 injection of *Dense Polyurethane* up to the depth of 1 m) below concrete structure. Works are done on lane per lane basis to compensate the operational highway requirement. Pre- and post evaluation and perfromance test were also carried out through using in-situ JKR Probe test for strength parameter. Coring soil samples were before and after injection to determine improved strength, stiffness and compressibility by performing Unconfined Compression Test and One-Dimensional Consolidation Test.

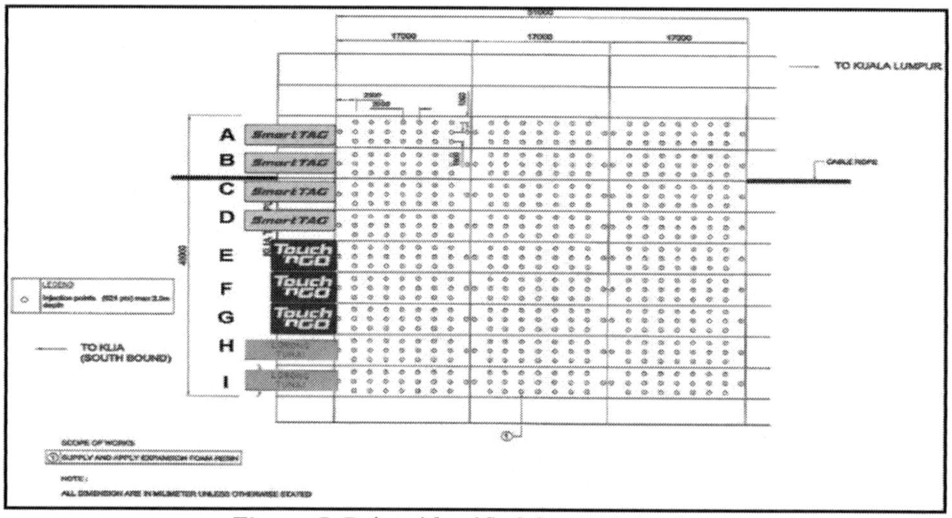

Figure 5. Points identified for PU injection.

2.2 Material Preparation and Scope of Work for Geocrete Soil Stabilization

Soil stabilization using Geocrete at KM 150.70 Southbound, Along Tangkak–Pagoh North South Expressway presented herewith. Initially, existing road experienced undulation with uneven surface, severe cracks and pumping effect at all lanes. Total area proposed for soil stabilization was 6215 m^2 (refer Figure 6). Prior to implementation, sample of existing road base was taken and conducted series of laboratory test through classification and pre-qualification test (e.g sieve analysis, proctor test, moisture content, density and unconfined compressive strength) in conforming Geocrete design mix. For this project, the proposed mix design was 8% of cement (Type 1), 2% of Geocrete powder (by weight of cement) and 10% of quarry dust (to improve the existing material nearer to an ideal particle size distribution curve). Proposed depth of soil stabilization is 300 mm which based on design traffic (e.g 10 to 30 x 10^6 ESAL) and specification used for recycling work is *JKR Standard Specification for Road Works, Section 4: Flexible Pavement—Cold in Place Recycling* and *Specification for Soil: Cement Agent Mixture* respectively. Evaluation test were carried out by means of Light Falling Weight Deflection (LFWD) and Unconfined Compressive Strength (UCS) test during and post construction. On site, the stabilization works are done through lane per lane basis.

Works commenced by initially scrapping off the existing asplhalt layer of 150 mm thickness. The Light Falling Weight Deflection (LFWD) are then carried out to determine the existing strength before stabilization. For Geocrete stabilization works, the existing road base material will be mixed with the designated cement (Type 1) + Geocrete powder mixture based on the approved mix design. The required quarry dust are spread uniformly over the existing base area prior spreading cement (Type 1) + Geocrete which are laid equally per unit area. Once complete, the CAT RM500 machine is then commenced and stabilized the soil with the required depth and ensure uniform blend of stabilization material. Water are spread over the soil-cement-quarry dust-Geocrete mix to start chemical reaction for stabilization process with an extra of 2% from Optimum Moisture Content (OMC). After stabilization, the area is compacted with 10 tones drum roller vibration for 2 passes and subsequently grading uniformly to the required profile depth. Final compactions are performed for another 3 passes with speed of not more than 3 km/hr. Once completed, curing process are carried out by using water tank sprinkler on the stabilized soil periodically for every 3 to 4 hours within 24 hours. This is to ensure to avoid any formation of premature cracks by controlling hydration of the stabilization process and risk of quick evaporation. After 24 hours curing, paving work commenced by laying 60 mm Asphaltic Concrete Base Coarse (ACBC 28), 60 mm Asphaltic Concrete Wearing Coasrse (ACWC – 50 mm) and regulating layer as final surface.

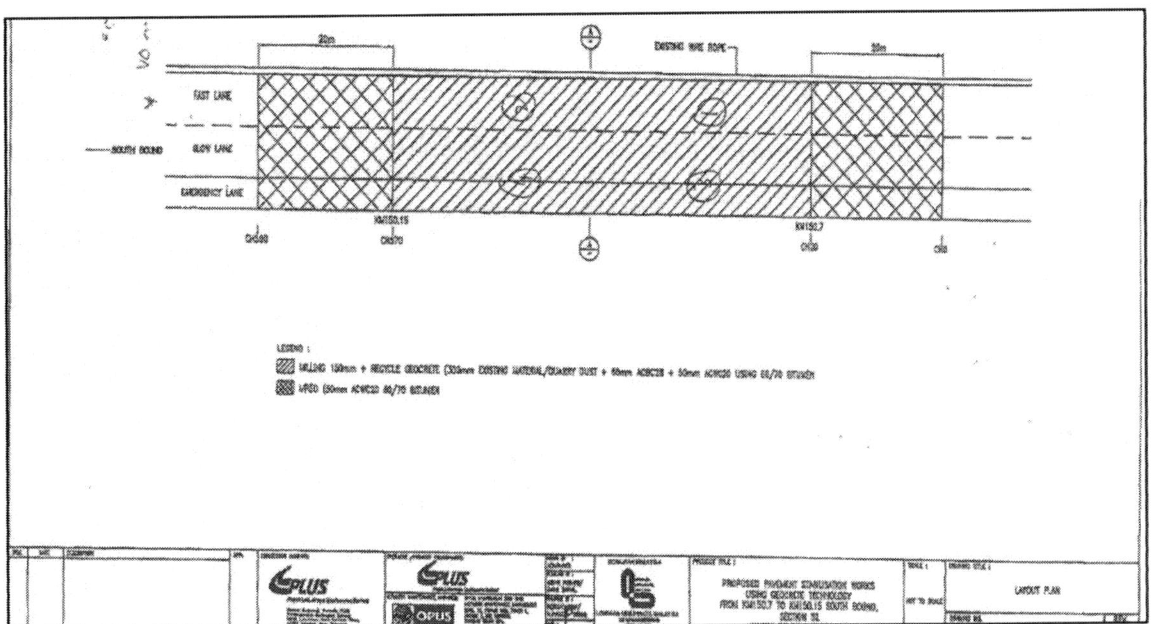

Figure 6. Layout plan for Geocrete stabilization work.

3. Results and Discussions

3.1 (PU Foam Injection System)

JKR Probe test measures soil resistance to penetration / stiffness by number of blows obtained per 300 mm penetration (150 blows per 300 mm for shallow foundation). Soil strength parameters correlated to JKR Probe is shown in Table 2.

Results blow count number per penetration depth for both natural (pre-injection) and modified (post-injection) soil displayed only for 2 lanes as shown in Figure 7 and Figure 8 as other area presents similar output. Observed that natural soil resulted low numbers of blows (less than 150 blows – blue line) and consistent with soil consistency ranging Very Soft to Stiff which suggests that the soil having loose soil formation with the presence of voids and cavities. As for modified soil (red line), it recorded 150 number of blows with maximum penetration depth up to only 350 mm and classified as Hard with strength value of 600 kPa. The injection foam successfully filled all existing voids, cavities and throughout the capillaries within the soil by expansion force thus improving the bearing capacity and strength of soil.

Table 2. Correlation for number of blows against strength of soil.

JKR Probe (Blows/300mm)	Consistency of soil	Unconfined Compressive Strength	
		kPa	ton/ft^2
0 - 10	Very soft	0.0 - 25	0.00 - 0.25
10 - 20	Soft	25 - 50	0.25 - 0.50
20 - 40	Medium (firm)	50 - 100	0.50 - 1.00
40 - 70	Stiff	100 - 200	1.00 - 2.00
70 - 100	Very stiff	200 - 400	2.00 - 4.00
100	Hard	400	4.00

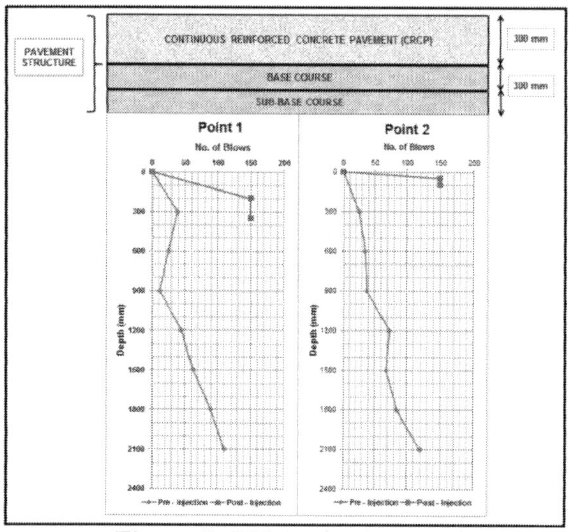

Figure 7. JKR Probe Test results.

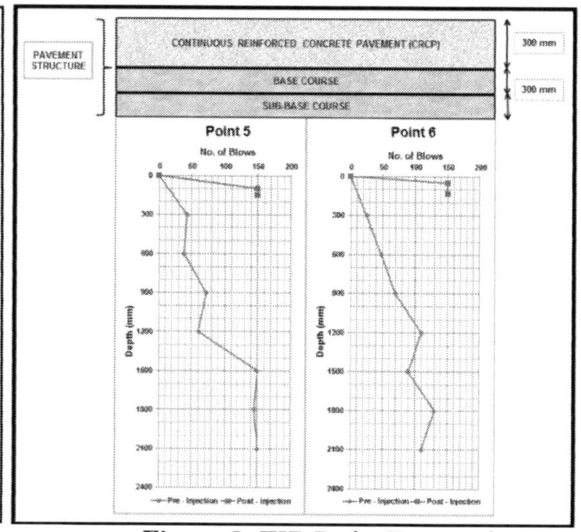

Figure 8. JKR Probe Test results.

As for strength and stiffness, results data are shown in Table 3. For reference purposes, natural soil specimens are labelled as KLIAud (undisturbed soil) and modified soils as KLIAms. The maximum compressive stress values for modified soil recorded between 190.30 to 199.30 kPa respectively, 1 to 3 times higher than the natural soil (62.10 to 145.00 kPa) and stiffness increased up around 5 to 6 times greater. This can suggest that the modified soil able to withstand bearing foundation of more than 150 kPa as specified for shallow foundation.

Table 3. Shear strength and stiffness for natural and modified soil specimens.

Soil sample	Maximum deviator stress (kPa)	Young's modulus (kPa)	Undrained shear strength (kPa)
KLIAud01	68.70	1,400	34.30
KLIAud02	145.00	6,000	72.50
KLIAud03	62.10	1,400	31.10
KLIAms01	198.30	5,000	99.15
KLIAms02	199.30	8,000	99.65
KLIAms03	190.30	7,000	95.15

For compressibility, both soil specimens (natural and modified) were analyzed to demonstrate the compressibility characteristics in terms of void ratio, pre-consolidate pressure, compression index and swelling index. Based on the results obtained (see Table 4), modified soil recorded 0.435 of initial void ratio (e_o), reduced about 31.5% than natural soil (0.635) and suggest that the PU foam has filled most of the void spaces and causing water or moisture entrapped within the soil to push out since it has water repellent ability. Furthermore, the swelling index also reduced by 60.9% (0.009) for modified soil than that of natural soil (0.023) which indicates that volume change, settlement rate and also plastic of permanent deformation are minimal.

Table 4. Compressibility characteristics between natural and modified soils.

Soil	Initial void ratio, e_o	Compression index, c_c	Swelling index, c_s	Preconsolidated pressure, p_c (kPa)
KLIAud04	0.635	0.076	0.023	52
KLIAms04	0.435	0.075	0.009	56

3.2 Geocrete Soil Stabilization

For unconfined compressive strength (UCS) test, specimens of 3 numbers were taken on site by means of coring for day 1, day 7 and day 28 respectively to determine the strength through average value. Based on specification, the minimum USC value is 2.5 N/mm^2 and results of the stabilized soil are displayed in Table 5. Observed that from day 1 to day 28, soil stabilization with Geocrete significantly improve the strength properties and increased its bearing capacity thus, surpass the minimum requirement.

This can be suggest due to its high crystal interlocking connectivity, crystalline during hydration process of cement, water and Geocrete powder (alkaline) whereby it increased the rate of strength exponentially and faster curing process. However it maintains avoidance of premature cracks compared to normal cement stabilization. The final product provides hydraulically-bound mixtures of hard yet flexible, durable and impermeable (waterproofing).

For light falling weight deflection (LFWD) test, it is used to monitor and measure the bearing capacity and behavior of strength and stiffness of stabilized soil with time. It is a quick method to evaluate the dynamic deflection modulus, E_{vd} in MN/m^2 which calculated from measured settlements. For comparison purpose, readings are taken before and after stabilize. Readings are taken at day 1, day 3 and day 7 respectively. Based on specification, the minimum value to surpass is 25 N/mm^2. Results for both untreated and treated soil are shown in Figure 9. From the results obtained, the dynamic deflection modulus E_{vd} for existing road base recorded an average value of 72.46 MN/m^2 up to 7 days. As for the stabilized soil, at 1 day age after stabilization, an average reading of 146.37 MN/m^2, 202% increase when compared to untreated road base. Similar trend at 3 days and 7 days age, the value continues improved significantly to 173.17 MN/m^2 (239%) and 221.95 MN/m^2 (306%) respectively. These suggest that the chemical reaction of geocrete (alkaline) and cement with correct amount of water and curing process improves the hydration through high interlocking connectivity, thus enhance the strength and stiffness characteristics prominently.

Table 5. Unconfined Compressive Strength results for stabilized soil.

Age of Soil Stabilized	Average Value of 3 samples	
	UCS, N/mm^2	Density, Mg/m^3
Day 1	4.27	2.17
Day 7	7.60	2.16
Day 28	11.00	2.16

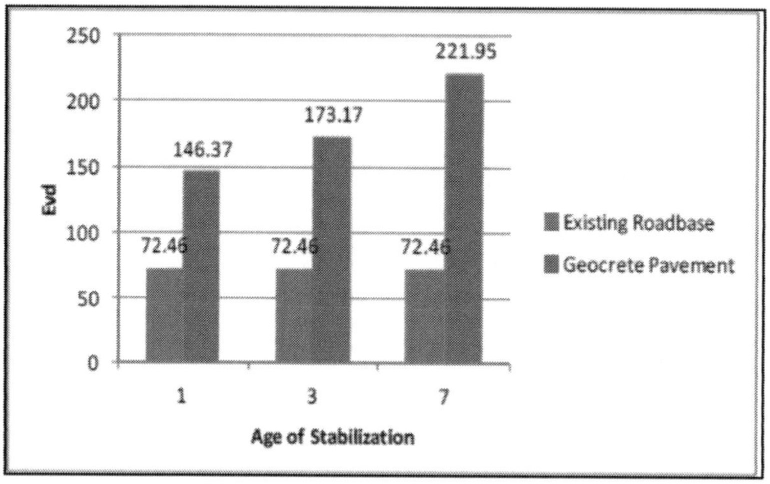

Figure 9. Dynamic deflection modulus (treated vs untreated soil).

4. Conclusion

This paper has presented 2 types of ground rehabilitation for maintenance that successfully comply the operational requirement (working window and duration) and yet maintained its quality to the upmost that met the requirements.

The use of both Rapid and Dense PU foam material with 1.1: 1.0 ratio (isocynate : polyol) for grouting injection can be concluded has significantly improved the characteristics soil such as stiffness and shear strength which makes the soil almost rigid yet flexible. In terms of compressibility, injection with PU foam impressively reduced the void and swelling index by filling up the voids throughout capillaries within the soil by expansion and increased strength through densification and potential uplift of the structure as well. A good technique and experience are essential in determining volume of voids within the soil thus able to estimate the right composition of PU and amount needed for stabilization and uplifting which will last longer period.

As for Geocrete soil stabilization, by assessing the test results obtained, it can be concluded that the mixture combination of cement (Type 1) of 8% and Geocrete additive of 2% (by weight of cement) with quarry dust in stabilization of the existing road base has resulted high values of strength, bearing capacity and resilient modulus at early age, also more effective approach than that of conventional method. It also offers low risk of premature cracks due to the control process of hydration. The use of Geocrete in soil stabilization also surpasses the specification requirement for roads and considered to be practical engineering solution for maintenance due fast construction and curing process which able to open for traffic within short period of time. In tandem with Geocrete design approach to provide tailor made pavement solution, the amount of material required are flexible which will be based on soil condition basis and suit the requirement. Other benefit is that savings from removal of existing weak soil and import new material with transportation. This, undeniably creates eco-green construction and towards to sustainability. By stabilizing the soil with correct mix design, the soil can withstand the direct load for longer life span and durable, thus less cycle of maintenance required in the near future.

In conclusion, both approach considered suitable and practical for ground rehabilitation when comes to maintenance on the operated infrastructure as it contribute less effect to operational. Work can be done fast, less curing time needed and significantly reduces operational hazards and costs of lane closure when compared to conventional method. Thus, it provides beneficial to the owner as it acquires lesser cost of future maintenance.

References

[1] Jong-Pil W, Joon-Mo K and Su-Jin L 2010 Mix proportion of high-strength, roller-compacted, latex-modified rapid-set concrete for rapid road repair. Construction and Building Materials. **25** 1796-180.

[2] Mounanga P and Gbongdon W 2008 Proportioning and Characterization of Lightweight Concrete Mixtures made with Rigid Polyurethane Foam wastes. Cement & Concrete Composites. **30** 806 – 814.

[3] Sidek N, Mohamed K, Jais I B M and Bakar I A A 2015 Strength Characteristics of Polyurethane (PU) With Modified Sand. Applied Mechanics and Materials. **773-774** 1508-1512

[4] Soltesz S 2002 Injected Polyurethane Slab Jacking. Final Report: SPR 306-261 Oregon Department of Transportation, USA.

[5] Gaspard K and Morvant M 2004 Assessment of the Uretek Process on Continuously Reinforced Concrete Pavement, Jointed Concrete Pavement and Bridge Approach Slabs. Technical Assistance Report Number 03-2TA, Louisiana Transportation Research Center.

[6] Puppala A J, Saride S, Acheewa E, Hoyos L R and Nazariam S 2009 Recommendations for Design, Construction and Maintenance of Bridge Approach Slabs: Synthesis Report. Sponsored by the Texas Department of Transportation (TxDOT) in cooperation with the U.S. Department of Transportation Federal Highway Administration (FHWA).

[7] Ulrich H 1982 Introduction to industrial Polymers. 56 – 79.

[8] Wolf H W 1956 Catalyst Activity in One-shot Urethane Foam. Technical Bulletin, Dupont, Wilmington. 30 - 65.

[9] Wood G 1982 Flexible Polyurethane Foams, Chemistry and Technology. Applied Science Publishers, London. 120 – 160.

[10] Ahmad S K and Wu D Q Over-Coming Differential Settlement in Soft Grounds Using 'Floating' Semi-Rigid Pavement.

[11] Gurp C V and Kroesen B Unsealed Geocrete – Road with High Bearing Capacity.

Use of Lightweight Cellular Mats to Reduce the Settlement of Structure on Soft Soil

R Ganasan[1,2], A J M S Lim[1,2] and D C Wijeyesekera[1,2]
[1] Faculty of Civil and Environmental Engineering, Universiti Tun Hussein Onn Malaysia, 86400 Batu Pahat Johor, Malaysia
[2] Research Center for Soft Soil, Universiti Tun Hussein Onn Malaysia, 86400 Batu Pahat Johor, Malaysia

E-mail: reven0207@yahoo.com

Abstract. Construction of structures on soft soils gives rise to some difficulties in Malaysia and other country especially in settlement both in short and long term. The focus of this research is to minimize the differential and non-uniform settlement on peat soil with the use of an innovative cellular mat. The behaviour and performance of the lightweight geo-material (in block form) is critically investigated and in particular the use as a fill in embankment on soft ground. Hemic peat soil, sponge and innovative cellular mat will be used as the main material in this study. The monitoring in settlement behavior from this part of research will be done as laboratory testing only. The uneven settlement in this problem was uniquely monitored photographically using spot markers. In the end of the research, it is seen that the innovative cellular mat has reduce the excessive and differential settlement up to 50% compare to flexible and rigid foundations. This had improve the stiffness of soils as well as the porous contain in cellular structure which help in allowing water/moisture to flow through in or out thus resulting in prevent the condition of floating.

Keywords: Ground modification, soft soil, geo-infrastructure.

1. Introduction

The construction and structure played an important role in Malaysia thus allowing many companies and firms of consultancy and contractors have begun to competitively and also cooperatively to design, build and make a building to be unique and ideal for local and foreign consumers. However, some of the constraints faced by every company that involved the construction of structures on soft soils which undergo settlement in the short or long term. Soft soils experience with low strength and rapid settlement for some foundations in unavoidable circumstances which lead to the ground failure. An example of soft soils is soft organic clays which have the characteristics of very low shear strength and lacks compressibility [1,2]. An extreme failure example of soft soil is organic peats [3,4]. Figure 1 shows the area covered by peat soils in Malaysia [5]. These challenges arise towards engineer facing in all sorts of problem to design and construct foundation of building, road and highway embankment. It is because structures construct on peat soil are often affected by stability due to high compressibility, low shear strength and high permeability [6,7]. The most critical geo-environment challenges are excessive settlement and differential settlement leading to hazardous and discomfort in road usage. Many conventional methods (pile, vertical drain, soil replacement, soil stabilization etc.) have been used to reduce these problems. However due to self-weight from the conventional methods used could not maintain the soil structure allowing them become as secondary subject towards excessive in soil settlement.

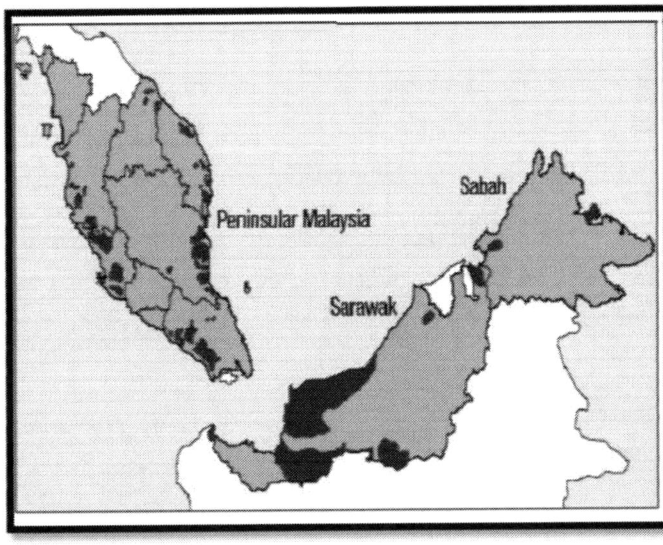

Figure 1. Location of peat swamps in Malaysia (Zainorabidin et al., 2007)

Settlements can be occurring in three different ways such as:

(i) Uniform settlement – The structure settled constant and uniformly
(ii) Tilt settlement – The structure experience in differential of settlement
(iii) Non-uniform settlement – The structure settled due to elongation

These three types of settlement can be seen according to the related places which sketched in Figure 2 [8]. The purpose of this research study is to investigate the settlement behaviour of peat soil and sponge which combine with and without the application of innovative cellular mat through physical model testing.

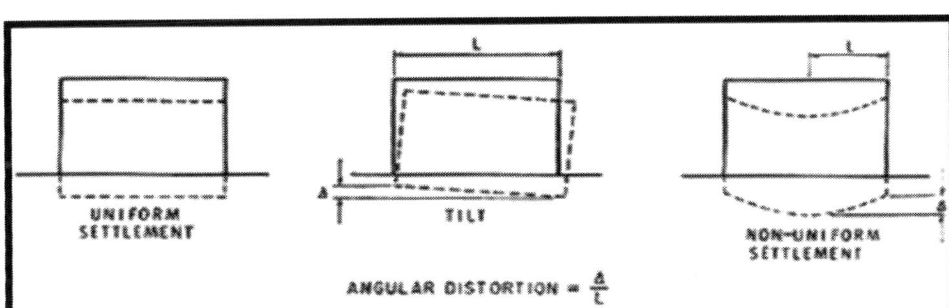

Figure 2. The diagram in types of settlement (WUS Austria, 2007)

2. Materials and methods

The primary step carried out in this research includes in computing the stiffness of sponge and the moisture content of the hemic peat soil where else the parameters of the hemic peat soil was taken through past literature review as shown in Table 1. A geo-composite material will be used as a main product in this study. This material act' as a lightweight fills material to exert little pressure to soft soil and also to reduce the self-weight of embankment. The physical modelling test was carried out in RECESS by monitoring the soil settlement behaviour in the view of 2-Dimensional (2-D). The laboratory model tests will be conducted in a steel tank of size 100cm (length) x 50cm (width) and 60cm (height). This model testing will be used to:

i. Investigate the settlement behaviour on peat soil and sponge only (without innovative cellular mat)
ii. Investigate the settlement behaviour of peat soil and sponge by application of innovative cellular mat.

Table 1. Physical properties of typical hemic peat soil in Johore, Malaysia (Adapted from Zainorabidin et.al, 2007)

Parameters	Value
Bulk Density	7.5 – 10.2 kN/m3
Water Contents	230 – 500 %
Organic contents	80 – 96%
pH	3 – 4.5
Specific Gravity	1.48 – 1.8
Shrinkage Limit	-
Plastic Limit	-
Liquid Limit	220 – 250%
Plasticity Index (%)	-
Undrained Shear Strength (kPa)	7 – 11
Compression Index, Cc	0.9 – 1.5

In this testing two different material used (sponge and the peat soil) was modelled under the three condition of foundations. They were in the form of flexible, rigid and finally in mats. The surface of the both materials was applied directly with increment loads up to 500N. The settlement data will be taken from the clear image of DSLR camera (Figure 3) then transfer them to Microsoft Excel for further analysis.

Figure 3. The constant position of DSLR camera during conducting the experiment.

3. Results and discussion

3.1 The Stiffness of Sponge

Stiffness is defined as the ratio of the stress along an axis over the strain along that axis in the range of elastic sponge behaviour. Based on this testing, the stiffness found in sponge was 35kPa. The stiffness test was not applied to the peat soil since the physical characteristics (oversaturated) could not be sustains the exerted forces. Thus the sponge was used as simulation having approachable similarity design of stiffness as in the peat soil.

3.2 Moisture Content of Hemic Peat Soil

The average moisture content of the disturbed hemic peat soil sample as shown in Table 2 from the experiment was 72.1% which taken from the site located at Parit Nipah Darat, Batu Pahat, Johor. The average value of this moisture content was less compare to the past researchers (more than 100%) as the temperature was set particularly to control the physical testing done (lab scale) for each foundation (flexible, rigid and mat) within the time period.

Table 2. Moisture content of disturbed hemic peat soil

Test No.	**1**	**2**	**3**
Mass of can (g)	9.687	9.563	9.773
Mass of can + moist soil (g)	44.982	59.454	55.486
Mass of can + dry soil (g)	30.326	38.505	36.233
Mass of water (g)	14.656	20.949	19.253
Water content (%)	71.0	72.4	72.8
Average (%)		**72.1**	

3.3 Critical Analysis in Settlement of Fill Loading Subjected on Sponge Foundation

The analysis subjected to sponge was done in flexible, rigid and mats foundation using the physical modelling box. Generally in this research, the test had been conducted in two different ways of sponge which were:

i. The settlement design on full sponge
ii. The settlement design on half sponge and half solid (Figure 4 and Figure 5)

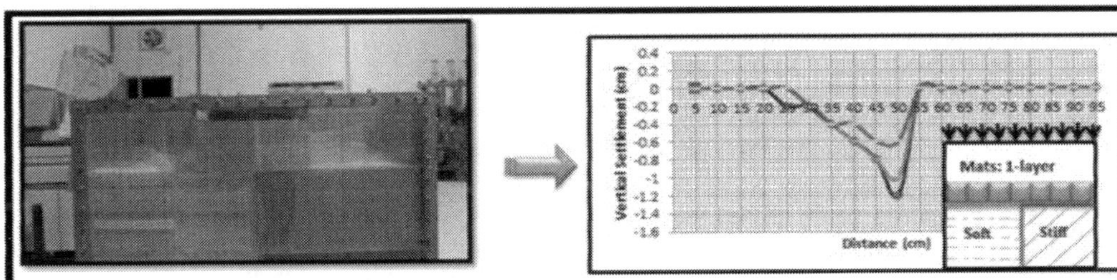

Figu **Figure 4.** The graphical result obtained from mat's 1-layer.

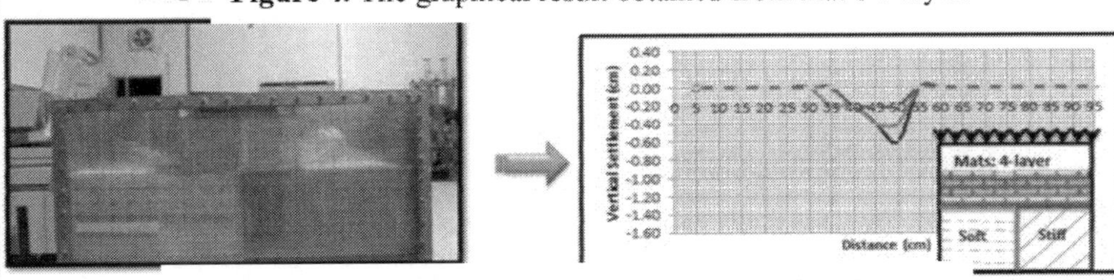

Figure 5. The graphical result obtained from mat's 4-layers.

- The mat design with 4-layers clearly has reduced the uneven settlement compare to other foundations (flexible, rigid and 1-layer mats).
- The Mat's effect demonstrated in Figure 5 which shows 50% reduction in settlement that occurs with the normal flexible full design shown in Figure 4.

3.4 Critical Analysis in Settlement of Fill Loading Subjected on Hemic Peat Soil

The analysis had been taken over in flexible, an innovative cellular and innovative cellular mats foundation using the physical modelling box. However during the analysis, it was hard to determine the value in settlement due to the colour of peat soil (dark brown) distinguish the transparency graph. Thus in order to evaluate the settlement occur in peat soil, the observation of analysis was took over at surface area of the physical modelling box shown in Table 3 whereby to predict the total required of mats (layers) to reduce the settlement in peat soil. Particularly the test was conducted on for half hemic peat soil with half solid.

Table 3. The differences in settlement design in foundations (half peat soil).

Foundations	Control Load	Applied Load 500N
Flexible	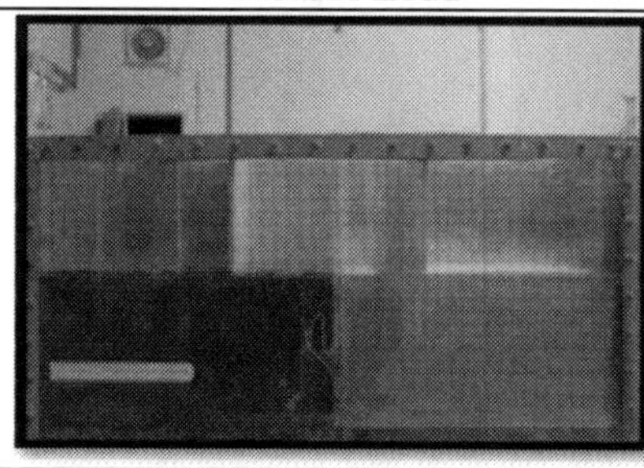	
Remarks: In the flexible form there is large amount of differential settlement occur as can be seen in the angle in position of loading moving towards inclined		
Rigid	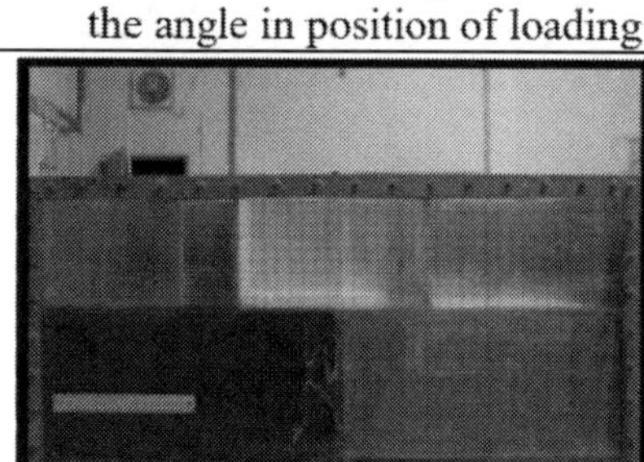	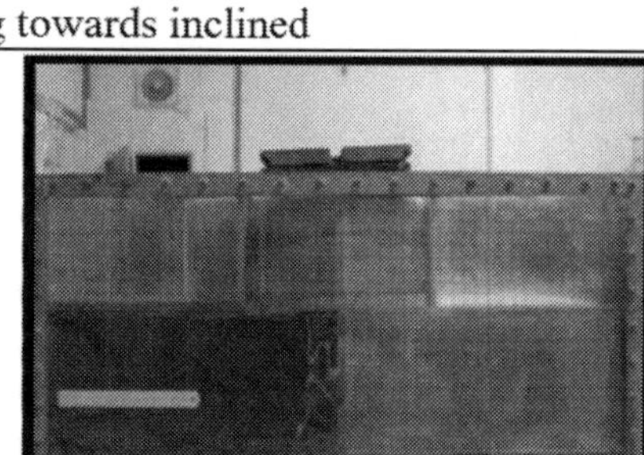
Remarks: There is differential settlement occur when using uncut piece of innovative cellular mats.		
Mats 3-layers		
Remarks: The settlement is being reduced compared to the others (flexible, rigid and 2-layers) thus according to the prediction if more layers are increase then the level between innovative cellular mats and the solids will become constant or at one line.		

4. Conclusion and recommendation

As for the conclusion, the comparison in settlement behaviour for both peat soil and sponge was done to achieve the objective of this research study. The settlement has been reduced to 50% by the use of mats in this research.

As for the recommendation, the physical modelling as shown in Figure 6 below can be used in the software modelling in the future. Apparently we can critically analyse this software modelling applied in real condition of field/site (full scale) in reducing and redistributing pressure exerted from loading to the soil.

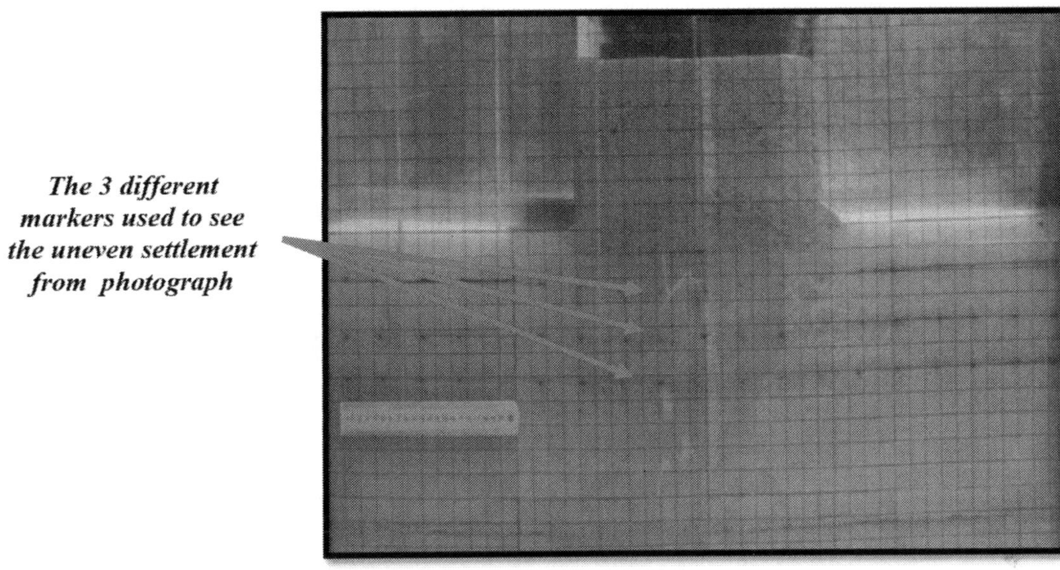

Figure 6. The settlement design/mode shown from the pointed markers in 3 different layers.

Acknowledgements
The author would like to greatly appreciate and kindly proceed with gratitude to the supervisory team for their guidance, the staff members from Faculty of Civil and Environmental Engineering (UTHM), Research and Soft Soil Centre (RECESS) and last but not least to the Research Innovation Commercialization and Consultancy Office (ORICC) in achieving the goal of this research study.

References

[1] Chan C M, & Abdullah S H 2008 Settlement Behaviour of A Cement-Stabilised Malaysian Clay, 6th International Conference on Case Histories in Geotechnical Engineering, Arlington, VA, August 11 – 16, 2008.
[2] Ho M H and Chan C M 2011 Some Mechanical Properties of Cement Stabilized Malaysian Soft Clay, World Academy of Science, Engineering and Technology 2011 **50**, pp. 24 – 25.
[3] Zainorabidin A & Bakar I 2003 Engineering Properties of Insitu and Modified Hemic Peat Soil in Western Johor. Proceedings of the 2nd Conference on Advances Soft Soil Engineering and Technology. Eds. Huat, B.B.K., Omar, H., Maail, S. & Mahsun, E. Putrajaya, Malaysia. 2-4 July 2003 173-181.
[4] Gabryś K & Szymański A 2010 The analysis of consolidation in organic soils, **42(2)**, pp. 261-270
[5] Zainorabidin A & Wijeyesekera D C 2007 Geotechnical Challenges with Malaysian Peat, Proc. of the AC & T, pp. 252 – 261. University of East London.

[6] Huat B B K, Maail S & Mohamed T A 2005 Effect of Chemical Admixtures on the Engineering Properties of Tropical Peat Soils. American Journal of Applied Sciences **2 (7)**: 1113-1120, ISSN 1546-9239.

[7] Kazemian S, Huat B K K, Prasad A & Barghchi M 2010 Effect of Peat Media on Stabilization of Peat by Traditional Binders. International Journal of the Physical Sciences Vol. **6(3)**, pp. 476-487.

[8] Information of the diagram in types of settlement, WUS Austria 2007 Retrieved on September 28, 2013.

The Development of Mathematical Prediction Model to Predict Resilient Modulus for Natural Soil Stabilized by Pofa-Opc Additive for the Use in Unpaved Road Design

Y M R Gamil[1], **and I H Bakar**[1]
[1]Faculty of Civil and Environmental Engineering, Universiti Tun Hussein Onn Malaysia, 86400 Batu Pahat Johor, MALAYSIA
[1] Research Center for Soft Soil, Universiti Tun Hussein Onn Malaysia, 86400 Batu Pahat Johor, MALAYSIA

E-mail: Yaseruthm@yahoo.com

Abstract: Resilient Modulus (Mr) is considered one of the most important parameters in the design of road structure. This paper describes the development of the mathematical model to predict resilient modulus of organic soil stabilized by the mix of Palm Oil Fuel Ash – Ordinary Portland Cement (POFA-OPC) soil stabilization additives. It aims to optimize the use of the use of POFA in soil stabilization. The optimization models enable to eliminate the arbitrary selection and its associated disadvantages in determination of the optimum additive proportion. The model was developed based on Scheffe regression theory. The mix proportions of the samples in the experiment were adopted from similar studies reported in the literature Twenty five samples were designed, prepared and then characterized for each mix proportion based on the MR in 28 days curing. The results are used to develop the mathematical prediction model. The model was statistically analyzed and verified for its adequacy and validity using F-test.

Keywords: Resilient Modulus, Soil Stabilization.

1. Introduction

A number of studies focused on the optimization use of POFA in concrete based on the UCS have been reported in the literature. This particular study aims to optimize the use POFA in soil stabilization of palm oil plantation soil road. The optimization seeks to find the optimum content of POFA in mixture to achieve the properties concerned which in this study is the Resilient Modulus. There are many mathematical methods of optimization which have been developed to simplify the analytical process. Scheffe [15] has developed a method of optimization for experiment with mixtures based on the regression theory. The pursued model in this study shall be used to predict the Mr and the amount of additives required to achieve certain strength of stabilized soil in which so that the stabilized soil can be used in road construction.

Mbadike and Osadeb [11] used Scheffe's simplex theory for the optimization of the compressive strength of lateritic concrete. [11] used modified regression theory to predict of concrete mix ratios with most economical and durable concrete that meet with certain properties as consistency. Okere et al. [6] have also used Scheffe method to optimize the concrete mix cost.

Several effluents are resulted from the production of palm oil. One of the wastes is POFA which is a by-product produced in palm oil mill [16] After palm oil is extracted from the palm oil fruit, both palm oil husk and palm oil shell are burned as fuel in the boiler of palm oil mill. Generally, after combustion about 5% palm oil fuel ashes by weight of solid wastes is produced [14]. Large quantity of ash is produced and creates problems of disposal [17]. POFA produced in Malaysian palm oil mill is dumped as waste without any profitable return

[9].Either in the 20th or 21st century, POFA is still considered as a nuisance to the environment and disposed without being put for any other use as compared to other types of palm oil by-products [10]. Since Malaysia maintains to increase yield of palm oil, more ashes will be brought out and a failure to find any solution in making use of POFA will create various environmental problems. This prompted a number of researchers to initiate studies on engineering characteristics and usability of POFA as cement replacement material. Palm oil waste is an agro-waste material [18].It was reported that, for every 100 tons of fresh fruit bunches processed, approximately 7 tons of fibers, 20 tons of nut shells, and 25 tons of empty bunches discharge from the mill. In 2009 the Malaysian palm oil industry produced around 60 million tons of total solid waste annually finally husk and fiber are used extensively as fuel for the production of steam in the palm oil mills [1].

The physical and chemical studies of POFA have indicated that the POFA is a pozzolanic material and can be used as a binder and has recently been accepted as a pozzolanic material [2, 3].In another review by Tangchirapat, Jaturapitakkul & Kiattikomol [19] it was shown that the ash has great potential in concrete technology advancement, when used in premixing with Ordinary Portland Cement. In terms of soil stabilization the addition of fly ash reduces the plasticity of expansive soils. The liquid limit decreases and the plastic limit increases with an increase in fly ash content. The free swell index of expansive soils can be effectively reduced by the addition of fly ash. Free swell index of expansive soil was reduced by about 50% by adding 20% fly ash [13]. Eldagal, A., & Elmukhtar, [7] have studied the behavior of high strength palm oil fuel ash (POFA) concrete. It was reported that POFA contains silica oxide which can react with calcium hydroxide (Ca(OH)2) generated from the hydration process; and the pozzolanic reactions produce more calcium silicate hydrate (C-S-H) gel compound as well as reducing the amount of calcium hydroxide .

2. Methodology

Fig. 1 shows the flow research work and the phases of the research.

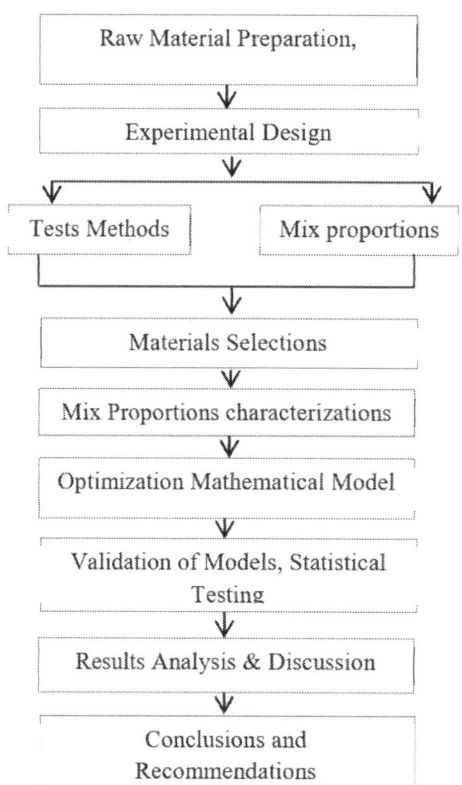

Figure 1. Flow chart of research work

3. Materials

The raw POFA was collected from palm oil mill located in Lot 835, Batu 3, Jalan Batu Pahat, Kluang, Johor Bahru, Malaysia. The laterite gravel was collected from Bukit Naning quarry in Muar, Malaysia. The cement used was ASTM Type I cement Portland cement is commonly used for general purpose.

Natural soil was collected from Ulu Tiram,Johor Bahru, Malaysia. The soil was basically collected from plantation site nearby in which a road is constructed by Probase Company. Disturbed sampling was collected at depth more than 0.4 m. The soil sample obtained was in slightly wet condition, and then the natural moisture content was taken besides the pH was also measured to test the acidity of the natural plantation soil. All of Mr Laboratory experiments were carried out in accordance to the specified standard of AASHTO T 307-99 (2003) [5].

4. Experimental Design

The proposed method of selecting the optimum mix design based on trial mixes having different combinations of five materials i.e.: POFA, water, OPC, natural soil and laterite gravel. The design content for each material is described in order to get a control design proportion that can be used to develop the mix proportions.

To adopt the use of Scheffe's method in optimization model. A simplex lattice is developed to be as structural representation of lines joining the atoms of mixture which is constrained with the theoretical findings. It means the values have to be within the factor space for a trial. Mixtures with proportion of outside the factor space were used for verifications of the developed model. The atoms indicate the constituent components of the mixture (POFA, Soil, Laterite gravel and OPC). It produced a simplex of a mixture with four components. Though, the simplex lattice of these four components mixture is three-dimensional solid equilateral tetrahedron. In a condition of Scheffe's method (1985), the components are subjected to the constraint that the sum of all the components must be equal to "one" which means:

$$x_1 + x_2 + x_3 + x_4 + \cdots \dots \dots . . x_q + \quad (1)$$

$$\sum_i^q x_i = 1 \quad (2)$$

Where "q" is the number of components of a mixture and Xi is the proportion of the ith component in the mixture, in the mix proportion it is impossible to use the, it is necessary to carry out a transformation from actual to pseudo components actual mix to develop the model since the sum of the mix component must be one. Hence, the actual components represent the proportion of the ingredients based in the theoretical constraints while the pseudo components represent the proportion of the components of the ith components in the mixture i.e. X1, X2, X3, X4. However, it has to be taking into account the four- component mixture tetrahedron simplex lattice, let the vertices of this tetrahedron (principal coordinates) be described by A1, A2, A3, A4.

The following arbitrary mixes are developed based on past practicing manual and literatures which are prescribed for the vertices of the tetrahedron in Fig. 2.The developed mixes is a 4*4 matrix which can be developed to experimentally design the mixes.

Ai :(POFA: Soil: Laterite Gravel: Cement)

A1 :(0.04: 1: 0.1: 0.04)

A2 :(0.024: 1: 0.15: 0.036)

A3 :(0.012: 1: 0.2: 0.028)

A4 :(0.004: 1: 0.3: 0.016)

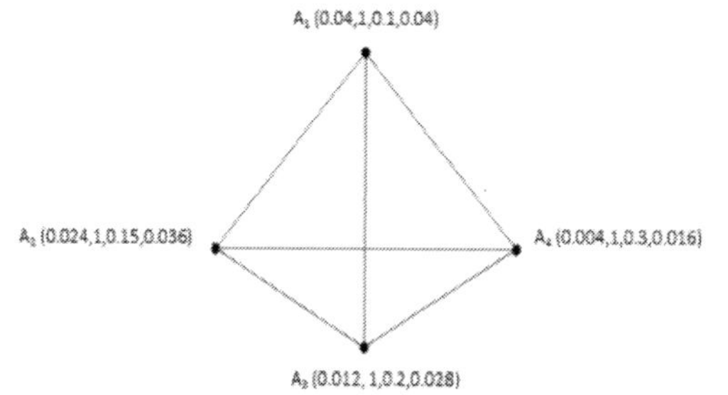

Figure 2. Vertices of Lattice (Represent Actual)

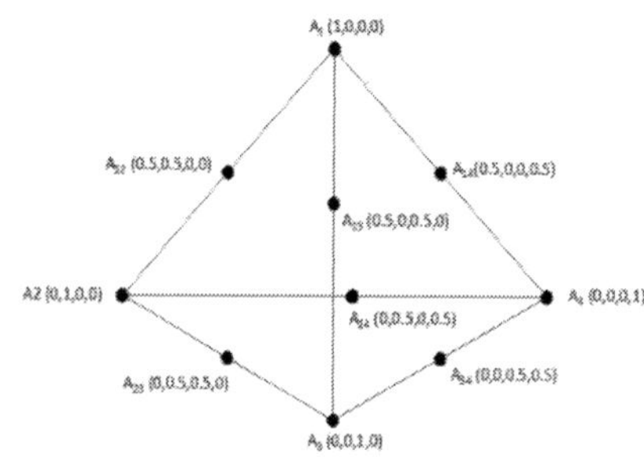

Figure 3. Vertices of Simplex Lattice (Represent Pseudo)

To continue analytical process let X represents the pseudo components and Z represents actual components then, for the processes of transforming pseudo to actual use the following equations:

$$Z = AX \tag{3}$$

In which A is a matrix whose elements are from the arbitrary mix proportions which will be selected when equation (3) is opened and solved mathematically however (A) is the inverse of matrix (X) times (Z) and X is chosen from the structure of pseudo components in Fig. 2 then, by expanding equations (3) the actual components of Z will be determined respectively. Table 1 shows the corresponding values of pseudo and actual.

The MR samples were 70 mm in diameter and 140 mm in height. Twenty five samples were prepared to develop the mathematical model. The compaction energy per unit volume was used to determine the number of blows to produce the samples with identical dry density the same as those achieved on the standard compaction test. After 28 days curing time the samples are tested for the Resilient Modulus.

5. Results and Analysis

Development of Mathematical Model

The use of the following polynomial equation developed by Scheffe [15]:

$$Y = \alpha_1 z_1 + \alpha_2 z_2 + \alpha_3 z_3 + \alpha_4 z_4 + \alpha_{12} z_1 z_2 + \alpha_{13} z_1 z_3 + \alpha_{14} z_1 z_4 + \alpha_{23} z_2 z_3 + \alpha_{24} z_2 z_4 + \alpha_{34} z_3 z_4 \tag{4}$$

Where:

αi and αij are coefficients and Zi are the pseudo components of mixture. The Y function is the response function at any point of observation, zi is the predictor and αi is the coefficient of the optimization model equations.

IOP Conf. Series: Materials Science and Engineering **136** (2016) 012007 doi:10.1088/1757-899X/136/1/012007

5.1. Determination of Optimization Model Coefficients

In the optimization model equation (4) the substitution of different values observations then there will be different response and different estimators at constant coefficient then, at nth observation point, Y (n) will correspond with Zi (n). That is,

According to equation (4):

$$Y^{(n)} = \sum \alpha_i z_i^{(n)} + \sum \alpha_{ij} z_i^{(n)} \ z_j^{(n)} \tag{5}$$

Where $1 \leq I \leq j \leq 4$ and n = 1, 2, 3, ...,10.

Equation (5) can be transformed in the form of matrix

$$[Y^{(n)}] = [Z^{(n)}]\{\alpha\} \tag{6}$$

rearranging equation (6) then results the following equation:

$$\{\alpha\} = [Z^{(n)}]^{-1}[Y^{(n)}] \tag{7}$$

By Letting the actual mixes in table 3 be represented by Zi and the corresponding fractional portions, Z(i)f are presented in Table 3. The table shows the actual mixes which were developed based on Scheffe simplex lattice and the corresponding fractional values which will be used to continue the mathematical calculations to determine the coefficients in the mathematical optimization model these values of the fractional portions Z(n) were used to develop Z(n) matrix and the inverse of Z(n) matrix. The values of Y(n) matrix are determined from laboratory tests from resilient modulus and unconfined compressive strength. With the values of the matrices Y(n) and Z(n) known, it is easy to determine the values of the constant coefficients of equation (5).

The resilient modulus response defers for each mix ratio hence to develop optimization model the first 10 mixes are chosen to fulfill the condition in Equation (4) The MR responses are:

$$[Y^{(n)}] = \begin{bmatrix} 15645000 \\ 15505420 \\ 15093100 \\ 15261300 \\ 15122960 \\ 15005460 \\ 14941830 \\ 15228080 \\ 15565240 \\ 15210120 \end{bmatrix}$$

The coefficient values of the model are:

$$\{\alpha\} = \begin{bmatrix} 33934348 \\ 1703582 \\ 16545683 \\ -1.1E+08 \\ -738592 \\ 376814.7 \\ 7784832 \\ -286502 \\ 843624.9 \\ 2177480 \end{bmatrix}$$

By substituting the values of α into Equation 4 then the optimization model at the response of MR is:

$$Y = 33934348z_1 + 1703582z_2 + 16545683z_3 - (1.1E + 08)z_4 - 738592z_1z_2 + 376814.7z_1z_3 + 7784832z_1z_4 - 286502z_2z_3 + 843624.9z_4 + 2177480z_3z_4 \quad (8)$$

5.2. Statistical Tests For the Adequacy of the Optimization Models

The optimization models are tested to evaluate the validity and adequacy based on the experimental results and to prove the agreement of the optimization models with the actual experimental results. In addition to that, the statistical tests most often used when comparing statistical models that have been fitted to a data set, in order to identify the model that best fits the population from which the data were sampled.

Letting H0 denotes the statistical Null Hypothesis and H1 represents the alternative Hypothesis.

The hypothetical conditions to validate the models are as follow: Null Hypothesis (H0): there are no statistical significant differences among the analytically observed results and the predicted results However, the Alternative Hypothesis (H1): There is a statistically significant difference between the analytical results and the expected results.

The use of control samples is adopted by taking the Y observed from laboratory results and Y predicted from the models by substituting the values of Zi .By the use of Microsoft excel it runs the F-test analysis. The mean is calculated for observed and predicted values and symbolized by (yobs) and (ypred) respectively. Mean is given by the summation of the samples divided by the number of samples.

The variance is given by:

$$v^2 = \left[\frac{1}{n-1}\right] [\Sigma(Y - y)^2] \quad (9)$$

$$y = \frac{\Sigma Y}{n} \quad (10)$$

Where n is the number of responses of the analysis.

According to table 5 the variances are calculated as follows:

$$v_{pred}^2 = \frac{0.53}{14} = 0.04$$

$$v_{ob}^2 = \frac{2.15}{14} = 0.15$$

By substituting the values into equation

$$F = \frac{0.04}{0.15} = 0.27$$

Referring to the F-distribution tables F0.95(14,14) = 2.48 which is higher than the calculated value 0.27 therefore the null hypothesis is accepted and the mathematical model at resilient modulus is considered adequate.

6. Conclusion

The use of Scheffe method to develop Resilient Modulus predication model for natural plantation soil stabilized by POFA has been prescribed and introduced. The mathematical prediction model can basically be used also to determine the optimum content of POFA required resulting in better stiffness at the same time eliminating the repetitions of testing and the arbitrary selection of additive percentage. For further analysis the mathematical model in Equation (8) can also be used for determining the optimum OPC required in stabilization of plantation soil.

Acknowledgements

This research was carried out at the research center for soft soil university Tun Hussein Onn Malaysia. The financial support from the university is highly appreciated. The researcher expresses his sincere gratitude and appreciation to the University Tun Hussein Onn Malaysia in providing all the necessary facilities to achieve the current outcomes.

References

[1] Altwair, N.M., Johari, Megat A Megat & Hashim, Syed F Saiyid, (2011). "Influence of Calcination Temperature on Characteristics and Pozzolanic Activity of School of Materials and Mineral Resources Engineering" , Engineering Campus. , 5(11), pp.1010–1018.
[2] Awal, A. S. M., & Hussin, M. W. (1997). "The effectiveness of palm oil fuel ash in preventing expansion due to alkali-silica reaction". Cement and Concrete Composites, 19(4), 367-372.
[3] Radin Sumadi, S., Chan, P. C., & Jasmin, L. S. (2006). "Blended cement for waterproofing applications".
[4] ASTM D2487 – 10 (1985). Classification of Soils for Engineering Purposes: Annual Book of ASTM Standards, D 2487-83, 04.08, American Society for Testing and Materials, 1985, pp. 395–408.
[5] AASHTO T307-99 (2000). Standard Method of Test for Resilient Modulus of Subgrade Soils and Untreated Base/Subbase Materials. American Association of State Highway and Transportation Officials.Washington, D. C.
[6] C.E. Okere, D.O. Onwuka, S.U. Onwuka, J.I. Arimanwa, (2013) "Optimisation of concrete mix cost using Scheffe's simplex lattice theory" Journal of Innovative Research in Engineering and Sciences 4(1), February, 2013. ISSN: 2141-8225 (Print); ISSN: 2251-0524 (Online).
[7] Dagal, A., & Elmukhtar, O. (2008). Study on the behaviour of high strength palm oil fuel ash (POFA) concrete (Doctoral dissertation, Universiti Teknologi Malaysia, Faculty of Civil Engineering).
[8] D.O. Onwukaa, Okerea, Arimanwaa and S.U. Onwukab (2011) "Prediction of concrete mix ratios using modified regression theory "Comp. Meth. Civil Eng., Vol. 2 No. 1, pp. 95-107.
[9] Hussin, Mohd Warid, Muthusamy, K. & Zakaria, F., (2010). "Effect of Mixing Constituent toward Engineering Properties of POFA Cement-Based Aerated Concrete"., (April), pp.287–295.
[10] K.Abdullah, M.W.H., (2006). "POFA : A Potential Partial Cement Replacement Material in Aerated Concrete.", (September), pp.5–6.
[11] Mbadike & Osadebe (2013). "Application of Scheffe's model in optimization of compressive strength of lateritic concrete". Journal of Civil Engineering and Construction Technology, 4(9).
Malaysia, European Union Report. (2012). "The Malaysian Palm Oil Sector –Overview." report (June): 1–53.
[12] Phani Kumar, B. R., & Sharma, R. S. (2004). "Effect of fly ash on engineering properties of expansive soils".Journal of Geotechnical and Geoenvironmental Engineering, 130(7), 764-767.
[13] V., Jaturapitakkul, C. & Kiattikomol, K., (2004). "Utilization of Palm Oil Fuel Ash in High-Strength Concrete"., (December), pp.623–628.
[14] Scheffe, H. (1958). Experiments with Mixtures. Journal of the Royal Statistical Society, Series. B., 20, 344 – 360.
[15] Tangchirapat, W., Saeting, T., Jaturapitakkul, C., Kiattikomol, K., & Siripanichgorn, A. (2007). "Use of waste ash from palm oil industry in concrete". Waste Management, 27(1), 81-88. Available at: http://www.ncbi.nlm.nih.gov/pubmed/16497498 [Accessed March 20, 2014].
[16] Tay, J., (1990). "Ash from Oil Palm Waste as a Concrete Material". Journal of Materials in Civil Engineering, 2(2), pp.94 105.
[17] Tay, J. H., & Show, K. Y. (1995). "Use of ash derived from oil-palm waste incineration as a cement replacement material, Resources, conservation and recycling", 13(1), 27-36.
[18] Tangchirapat, W., Jaturapitakkul, C. & Chindaprasirt, P., (2009). "Use of palm oil fuel ash as a supplementary cementitious material for producing high-strength concrete". Construction and Building Materials, 23(7), pp.2641–2646.

IOP Conf. Series: Materials Science and Engineering **136** (2016) 012007 doi:10.1088/1757-899X/136/1/012007

Appendices

Table 2. Pseudo Component with Their Corresponding Actual Component Values Where: Actual (Zi) and Pseudo (Xi) components

No of mix	X1	X2	X3	X4	Response	Z1	Z2	Z3	Z4
1	1	0	0	0	Y1	0.04	1	0.1	0.04
2	0	1	0	0	Y2	0.024	1	0.15	0.036
3	0	0	1	0	Y3	0.012	1	0.2	0.028
4	0	0	0	1	Y4	0.004	1	0.3	0.016
5	0.5	0.5	0	0	Y12	0.032	1	0.125	0.038
6	0.5	0	0.5	0	Y13	0.026	1	0.15	0.034
7	0.5	0	0	0.5	Y14	0.022	1	0.2	0.028
8	0	0.5	0.5	0	Y23	0.018	1	0.175	0.032
9	0	0.5	0	0.5	Y24	0.014	1	0.225	0.026
10	0	0	0.5	0.5	Y34	0.008	1	0.25	0.022
Control Points Calculated within the factor space									
11	0.5	0.25	0.25	0	C1	0.029	1	0.1375	0.036
12	0.25	0.25	0.25	0.25	C2	0.02	1	0.1875	0.03
13	0	0.25	0.25	0.5	C3	0.011	1	0.2375	0.024
14	0	0.25	0	0.75	C4	0.009	1	0.2625	0.021
15	0.75	0	0.25	0	C5	0.033	1	0.125	0.037
16	0	0.5	0.25	0.25	C6	0.016	1	0.2	0.029
17	0.25	0	0.5	0.25	C7	0.017	1	0.2	0.028
18	0.75	0.25	0	0	C8	0.036	1	0.1125	0.039
19	0	0.75	0.25	0	C9	0.021	1	0.1625	0.034
20	0	0.4	0.4	0.2	C10	0.0152	1	0.2	0.03
Control points Calculated outside the factor space									
21	0.5	0.5	0.5	0.5	C11	0.016	1	0.2	0.027
22	0.25	0	0.25	0.5	C12	0.015	1	0.225	0.025
23	0.5	0	0.5	0	C13	0.026	1	0.15	0.034
24	0.25	0.25	0.25	0	C14	0.019	0.75	0.1125	0.026
25	0	0.5	0.5	0.25	C15	0.019	1.25	0.25	0.036

Table 3. Actual Mix Proportions and Their Corresponding Fractional values

Mix No	Z1	Z2	Z3	Z4	Total of Si	Response	Z1f	Z2f	Z3f	Z4f
1	0.04	1	0.1	0.04	1.18	Y1	3.390	84.746	8.475	3.390
2	0.024	1	0.15	0.036	1.21	Y2	1.983	82.645	12.397	2.975
3	0.012	1	0.2	0.028	1.24	Y3	0.968	80.645	16.129	2.258
4	0.004	1	0.3	0.016	1.32	Y4	0.303	75.758	22.727	1.212
5	0.032	1	0.125	0.038	1.195	Y12	2.678	83.682	10.460	3.180
6	0.026	1	0.15	0.034	1.21	Y13	2.149	82.645	12.397	2.810
7	0.022	1	0.2	0.028	1.250	Y14	1.760	80.00	16.00	2.240
8	0.018	1	0.175	0.032	1.225	Y23	1.469	81.633	14.286	2.612
9	0.014	1	0.225	0.026	1.265	Y24	1.107	79.051	17.787	2.055
10	0.008	1	0.25	0.022	1.280	Y34	0.625	78.125	19.531	1.719
11	0.029	1	0.1373	0.036	1.202	C1	2.413	83.195	11.423	2.995
12	0.02	1	0.1875	0.03	1.238	C2	1.616	80.775	15.145	2.423
13	0.011	1	0.2375	0.024	1.273	C3	0.864	78.555	18.657	1.885
14	0.009	1	0.2625	0.021	1.293	C4	0.696	77.340	20.302	1.624
15	0.033	1	0.125	0.037	1.195	C5	2.762	83.682	10.460	3.096
16	0.016	1	0.2	0.029	1.245	C6	1.285	80.321	16.064	2.329
17	0.017	1	0.2	0.028	1.245	C7	1.364	80.257	16.051	2.247
18	0.036	1	0.1125	0.039	1.188	C8	3.030	84.175	9.470	3.283
19	0.021	1	0.1625	0.034	1.218	C9	1.724	82.102	13.342	2.791
20	0.0152	1	0.2	0.0288	1.244	C10	1.222	80.386	16.077	2.315
21	0.016	1	0.2	0.027	1.24	C11	1.20	80.41	16.10	2.28
22	0.015	1	0.225	0.025	1.265	C12	1.186	79.051	17.787	1.976
23	0.026	1	0.15	0.034	1.210	C13	2.149	82.645	12.397	2.810
24	0.019	0.75	0.1125	0.026	0.908	C14	2.093	82.599	12.390	2.863
25	0.019	1.25	0.25	0.036	1.555	C15	1.222	80.386	16.077	2.315

Table 4. The determination of Z n matrix values based on table 2

No	Z1	Z2	Z3	Z4	Z1Z2	Z1Z3	Z1Z4	Z2Z3	Z2Z4	Z3Z4	MR response *10^5 (kPa)
1	3.40	84.70	8.50	3.40	288.00	28.90	11.60	720.00	288.00	28.90	156.45
2	2.00	82.60	12.40	3.00	165.20	24.80	6.00	1024.20	247.80	37.20	155.054
3	1.00	80.60	16.10	2.30	80.60	16.10	2.30	1297.70	185.40	37.00	150.931
4	0.30	75.80	22.70	1.20	22.70	6.80	0.40	1720.70	91.00	27.20	152.613
5	2.70	83.70	10.50	3.20	226.00	28.40	8.60	878.90	267.80	33.60	151.2296
6	2.10	82.60	12.40	2.80	173.50	26.00	5.90	1024.20	231.30	34.70	150.0546
7	1.80	80.00	16.00	2.20	144.00	28.80	4.00	1280.00	176.00	35.20	149.4183
8	1.50	81.60	14.30	2.60	122.40	21.50	3.90	1166.90	212.20	37.20	152.2808
9	1.10	79.10	17.80	2.10	87.00	19.60	2.30	1408.00	166.10	37.40	155.6524
10	0.60	78.10	19.50	1.70	46.90	11.70	1.00	1523.00	132.80	33.20	152.1012
11	2.40	83.20	11.40	3.00	199.68	27.36	7.20	948.48	249.60	34.20	150.0034
12	1.70	80.80	15.10	2.40	137.36	25.67	4.08	1220.08	193.92	36.24	156.7584
13	0.86	78.56	18.66	1.89	67.56	16.05	1.63	1465.93	148.48	35.27	152.42
14	0.80	77.30	20.30	1.60	61.84	16.24	1.28	1569.19	123.68	32.48	148.7
15	2.80	83.70	10.50	3.10	234.36	29.40	8.68	878.85	259.47	32.55	155.91
16	1.30	80.30	16.10	2.30	104.39	20.93	2.99	1292.83	184.69	37.03	148.52
17	1.40	80.30	16.10	2.20	112.42	22.54	3.08	1292.83	176.66	35.42	150.261
18	3.00	84.20	9.50	3.30	252.60	28.50	9.90	799.90	277.86	31.35	153.52
19	1.70	82.10	13.30	2.80	139.57	22.61	4.76	1091.93	229.88	37.24	148.69
20	1.20	80.40	16.10	2.30	96.48	19.32	2.76	1294.44	184.92	37.03	152.04
21	1.20	80.41	16.10	2.28	96.49	19.32	2.74	1294.60	183.33	36.71	151.22
22	1.20	79.10	17.80	2.00	94.92	21.36	2.40	1407.98	158.20	35.60	146.36
23	2.10	82.60	12.40	2.80	173.46	26.04	5.88	1024.24	231.28	34.72	146.76
24	2.10	82.60	12.40	2.90	173.46	26.04	6.09	1024.24	239.54	35.96	143.222
25	1.20	80.40	16.10	2.30	96.48	19.32	2.76	1294.44	184.92	37.03	144.03

Table 5. The Determination of First 10 Mixes [Z n]-1 Matrix Inverse Values Based On Table 3

Z1	Z2	Z3	Z4	Z1Z2	Z1Z3	Z1Z4	Z2Z3	Z2Z4	Z3Z4
95.90	100.80	283.60	79.90	-196.20	-39.30	87.70	-136.70	-18.40	-257.30
1.20	0.30	-5.80	-0.60	-2.30	-2.30	0.10	8.10	-1.50	2.90
11.80	0.90	-46.10	-0.90	-24.90	-12.30	-2.60	66.30	-3.80	11.70
-141.40	96.10	266.50	-40.70	172.90	339.70	-62.10	-786.00	103.90	51.30
-1.30	-1.30	-1.60	-0.70	2.60	1.40	-1.10	-0.70	0.80	2.00
-0.20	-0.30	-4.80	-0.90	0.60	-1.50	-0.20	5.10	-1.10	3.30
5.80	1.70	-33.20	-4.60	-7.80	-21.00	3.00	48.50	-11.80	19.50
-0.20	0.00	0.90	0.00	0.40	0.30	0.00	-1.20	0.10	-0.30
1.30	-1.00	-1.50	0.60	-1.40	-3.10	0.60	6.60	-0.80	-1.30
2.00	-0.90	-6.30	-0.30	-2.90	-4.20	0.30	11.90	-1.20	1.60

Table 6. F-Test Analysis Optimization Model

Sample No	Y observed (kPa*106)	Y predicted (kPa*106)	Y(ob) - y(ob) (kPa*106)	Y(pred) - y(pred) (kPa*106)	(Y(obs) -y (obs))2 (kPa*106)	(Y(pred) - y(pre))2 (kPa*106)
C1	15.00	15.08	0.01	-0.05	0.00	0.00
C2	15.68	14.99	0.69	-0.14	0.47	0.02
C3	15.24	15.41	0.25	0.28	0.06	0.08
C4	14.87	15.19	-0.12	0.06	0.01	0.00
C5	15.59	15.01	0.60	-0.12	0.36	0.02
C6	14.85	15.24	-0.14	0.11	0.02	0.01
C7	15.03	14.70	0.04	-0.43	0.00	0.19
C8	15.35	15.38	0.36	0.25	0.13	0.06
C9	14.87	15.31	-0.12	0.18	0.01	0.03
C10	15.20	15.20	0.21	0.07	0.05	0.00
C11	15.12	15.13	0.13	0.00	0.02	0.00
C12	14.64	14.94	-0.35	-0.19	0.13	0.04
C13	14.68	14.91	-0.31	-0.22	0.10	0.05
C14	14.32	15.25	-0.67	0.12	0.45	0.01
C15	14.40	15.20	-0.59	0.07	0.34	0.00
Sum	224.84	226.93			2.15	0.53
Mean	14.99	15.13				

Physical Modelling on Detecting Buried Object Using Electrical Resistivity Imaging (ERI)

Z A M Hazreek[1,2], Z M Nizam[1,2], A T S Azhar[2,1], M Aziman[1,2] and M Z N Shaylinda[1]
[1] Faculty of Civil and Environmental Engineering, Universiti Tun Hussein Onn Malaysia, 86400 Batu Pahat Johor, Malaysia
[2] Research Center for Soft Soil, Universiti Tun Hussein Onn Malaysia, 86400 Batu Pahat Johor, Malaysia

E-mail: hazreek@uthm.edu.my

Abstract. This study focused on the evaluation of electrical resistivity method (ERM) for buried object detection and its relationship due to the different stiffness of material. In the past, the conventional method to detect the buried structure was face some limitation due to the time and cost. For example, previous approach related to the trial and error excavation has always expose to some risky outcome due to the uncertainties of the buried object location. Hence, this study introduced an alternative technique with particular reference to resistivity method to detect and evaluate the buried object with different strength of stiffness. The experiment was performed based on field miniature model (small scale study) using soil trial embankment made by lateritic soil and various concrete cube strengths (grade 20, 25 and 30) representing buried object with different conditions. 2D electrical resistivity test (electrical resistivity imaging) was perform using ABEM Terrameter SAS4000 during the data acquisition while the raw data was process using RES2DINV software. It was found that the electrical resistivity method was able to detect the buried concrete structures targeted based on the contrast of the electrical resistivity image produced. Moreover, three different strength of concrete cube were able to be differentiated based on the electrical resistivity values (ERV) obtained. This study found that the ERV of concrete cube for grade 20, 25 and 30 were 170 Ωm, 227 Ωm and 503 Ωm, respectively. Hence, this study shows that the ERV has a strong relationship with different stiffness of material thus applicable to be a useful alternative tool in underground structure detection.

Keywords: Geophysics, electrical resistivity, buried object detection.

1. Introduction

In the past, previous technique of bury object detection related to drilling and excavation suffer from several limitation due to the time and cost efficiency. Excavation machine which commonly heavy and bulky may not able to be operated in all site condition. For example, conventional method may suffer from difficulty relative to site workability especially when working at challenging and problematic site such as soft ground, swampy area, hilly terrain, limited working space, etc. Moreover, target location of the buried object was hard to be predicted due to the large unknown target area. Furthermore, unsuccessful excavation work based on trial and error may highly risk due to the unnecessary ground destruction thus give a bad impact to our environment. Hence, the alternative method needs to be discovered in order to minimize those previous limitations.

In multidisciplinary era, most researchers diversify their research area with an alternative method to improve the existing conventional method [1]. Nowadays, geophysical method has been increasingly used as an alternative technique in buried object detection. Geophysical method related to the electrical resistivity, seismic, ground penetration radar, gravity, magnetic etc. was highly adopted in engineering, environment, mining and archeological studies. Basically, geophysical method used to study an earth using physical properties such as electrical resistance, velocity, density, magnetic susceptibility, etc. Previously, geophysical method was commonly used to detect leachate migration [2,3,4], bedrock and overburden materials [5], boulder and cavity [6], groundwater [7,8,9], utilities, piling, archeology [10,11,12] etc. Generally, geophysical techniques contributes several advantages for example, it can be implemented more quickly and less expensively and has the ability to cover greater areas more thoroughly [13,14,15,16]. Moreover, the applications of geophysical method will applicable to minimize environmental destruction due to its non-destructive measurement thus preventing unnecessarily excavation. Furthermore as reported by [15], it provides a large-scale characterization of the physical properties under undisturbed conditions. As a result, site damageability was able to be minimized thus creating sustainable environment in buried object detection project. According to [17], geophysical method offers the chance to overcome some of the problems inherent in more conventional ground investigation techniques. However, the standard performance of individual geophysical method were still depends on fundamental physical constraints, e.g. penetration, resolution, and signal to-noise ratio [18].

This study used electrical resistivity method which one of the geophysical tool that heavily adopted in underground mapping. Electrical resistivity property was determined by measuring the potential difference at points on the ground surface which caused the propagation of direct current through the subsurface [19]. Good variation of the electrical resistivity contrast between the buried object and surrounding material need to be obtained for successful results. By determine the subsurface resistivity distribution, the purpose of electrical resistivity surveys can be determined by making measurements on the ground surface in order to estimate the actual resistivity of the subsurface [20]. Hence, this study performed a small scale study related to the buried object detection using electrical resistivity method with its relationship due to the different condition of the buried object strength. Finally, this study was able to prove and convince those related parties regarding the applicable of electrical resistivity method in underground structure mapping.

2. Materials and Methods

This study was divided into three phases; via laboratory, fieldwork and data processing. Laboratory works involve the preparation of concrete cube with different strength (grade 20, 25 and 30 MPa). Concrete cube with different strength was used as buried object due to this small scale study (field miniature model). Concrete strength of 20, 25 and 30 MPa was choose since it has been widely used in a construction industry. Hence, those stated assumption was used to simulate the underground buried structure with different strength/stiffness of material based on small scale study. A total of five concrete cube for each grade was designed in the laboratory. Three of five concrete cube was used to measure the compressive strength machine due to the strength verification purposes. In order to obtain good buried object prototype, the compressive strength of concrete need to achieve its matured strength before it was used as a buried object. Difference age of concrete strength contains difference strength of cement bond [21]. Then, the remaining two concrete cube for each grade was used as a buried object in the soil trial embankment model.

Fieldwork of this study has been divided into two phases via construction of soil trial embankment in miniature scale and electrical resistivity data acquisition (field measurement). A miniature model of soil trial embankments as shown in Figure 1, were built using lateritic soil. Dimensions of the model was 3.5 (length, m) x 1.0 (wide, m) x 0.5 (height, m) with all sides of the model edge shaped into a

IOP Conf. Series: Materials Science and Engineering **136** (2016) 012008 doi:10.1088/1757-899X/136/1/012008

gentle slope < 45°. During the construction of soil trial embankment miniature model, three different locations of concrete cubes were buried at center of the soil model as shown in Figure 1. After the completion of soil model construction, electrical resistivity test was performed using a single leveled line of 2D electrical resistivity survey (electrical resistivity imaging, ERI) on the top and center of the soil model. The ERI was performed twice since the first measurement was performed on soil model without buried object (control purpose) while the second measurement was performed under soil model with buried concrete cube. The data acquisition of ERI test was performed using ABEM SAS 4000 set of equipment. During the data acquisition (ERI measurement), two land resistivity cables were connected to 41 steel electrodes via jumper cables. Then, both resistivity land cables were connected to the electrode selector and Terramater SAS 4000 data logger for field setup. Finally, 12 volt battery was connected to the data logger to supply direct current (DC) for field measurement. This study used Schlumberger array due to its ability to produce dense surface data and deeper image penetration. Several considerations involving device and equipment setting, position of electrical resistivity line, ground condition, raw data processing etc. needed to be carefully considered and performed in order to determine the best electrical resistivity outcome. For example in order to reduce boundary effect that may reduce the electrical resistivity result accuracy caused by refracted and reflected current, the electrical resistivity line was placed at the center of the soil model.

Finally, raw data from field measurement (Terrameter SAS 4000 data logger) was transferred to the computer for processing. During the processing stage, smooth constraint least square method was used in order to produce a smooth boundary representing soil and concrete cube underneath. According to [22], smooth constraint least square method will produce a smooth boundary change which was considered more suitable for representing the soil material in contrast with rock and fractured material which preferred to analyzed using a robust method due to its sharp geomaterial boundary. RES2DINV commercialize software [23] was used to process and analyze the raw data thus producing an inverse model that estimate the actual subsurface structure. The inversion algorithm of RES2DINV was used to process the data, as proposed by [24] in order to obtain the 2-D resistivity section which replicate the subsurface profile of the soil model.

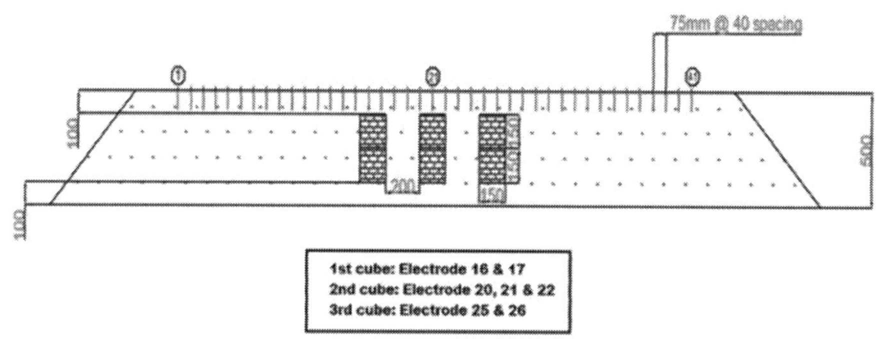

Figure 1. Arrangement of soil trial embankment with ERI (elevation view).

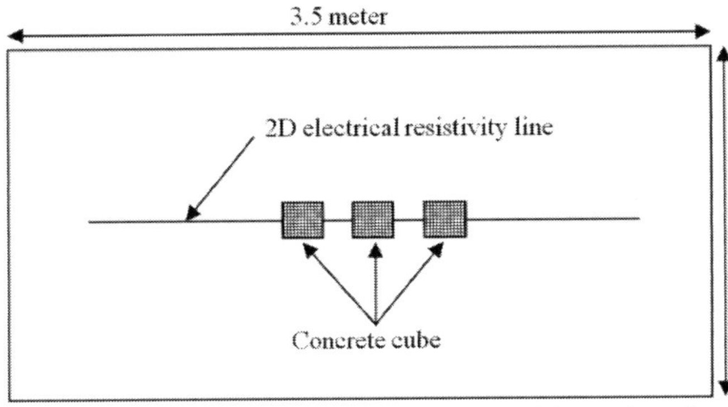

Figure 2. Position of the ERI and buried object (plan view).

3. Results and Discussions

All results presented and discussed are based on electrical resistivity value (ERV) and its relationship due to the existing of the concrete cube with different strengths. All results are presented in Figure 3, 4 and 10 representing electrical resistivity tomography section and its relationship due to the existing of the different strength of concrete cube.

According to Figure 3 and 4, it was found that the schlumberger array with 75 mm of electrode spacing was able to determine the resistivity section of the soil model with up to 0.6 m depth. Hence, the location of the targeted buried concrete cube was able to obtain as shown in Figure 4. According to Figure 3, it was found that the resistivity section has been dominated by low to moderate resistivity value (< 200 Ωm) representing soil model without buried object. As reported by [25] and [26], lateritic soil consist of silty and sandy soil has an electrical resistivity value from 15 – 152 Ωm and 100 – 250 Ωm respectively. Hence, it was prove that the resistivity section (electrical resistivity result or ERI result) of the soil model without buried concrete cube consist of homogeneous lateritic soil based on low to moderate of the electrical resistivity value (ERV) as shown in Figure 3. Furthermore, it was found that the homogeneous lateritic soil has produce some variation in term of ERV. This variation may due to the influence of soil gain and moisture content within the soil model. According to [27], the soil electrical resistivity data was observed to be very sensitive to the quantitative proportion of water and geomaterial particle fractions. As reported by [6], air filled void may influence the electrical resistivity variations since which commonly increased the electrical resistivity value compared with the water filled void. Furthermore, condition such as porosity, degree of saturation, salt concentration in pore fluid, grain size, size gradation, temperature and activity may influence to the electrical resistivity value variations [28].

Electrical resistivity section for soil model with buried object (concrete cube with different strength) was presented in Figure 4. It was found that the center of the resistivity section consist of high value of the ERV (1000 – 2000 Ωm). This anomaly has demonstrated that the existing of buried object with particular reference to concrete cube was able to influence the increment of the ERV. It was clearly observed that the surrounding buried concrete cube possess low to moderate ERV (50 – 200 Ωm) representing lateritic soil. Hence, this finding has proved that the electrical resistivity method was able to detect the buried object based on the variation of the ERV. However, it was found that the resistivity section was unable to produced real shape of the buried concrete cube due to several reasons. For example, the size (150 mm3) and spacing (200 mm) of the buried concrete cubes was too small thus unable to produce much resistivity plot based on the 75 mm of electrode spacing configuration. Hence, the contouring process in data processing may involve certain degree of interpolation thus producing rounded shape of the buried concrete cube. Furthermore, influence of resistivity array such as Wenner, Schlumberger, Wenner-schlumberger, Dipole-dipole, Pole-pole, Pole-dipole, Gradient, etc. may produce different outcome. Each array have their own strength and weakness relative to the several reasons. The consideration related to the targeted object, depth of interest, surrounding working area, etc. was crucial since it able to influence the array selection in order to produce the best subsurface profile result. According to [2,30], the ERV was largely influenced by types of array used due to the different geometry factor (K) derived from each different types of array used. Geometry factor, K describes the geometry of the electrode configuration used in data acquisition. As reported by [6], the best selection of array was based on signal strength, sensitivity of resistivity value due to the changing of vertical and horizontal structure, depth of investigation, type of structure which needs to be mapped and noise level. Apparent resistivity (ρa) is ERV estimated based on half-pace geometry assumption which refers to the field ERV. Apparent resistivity will be equal to the true resistivity provided the current and configuration was applied over the homogeneous isotropic ground [31]. However, the anomaly trend (electrical resistivity image) at certain same depth of the tomography may produce some similarity for most of the array used. This trend has proved that ERV was always subjective to some ranges of properties as given in the past reference charts and tables of ERV properties interpretation. The fundamental of the ERV based on Wenner, Schlumberger, Dipole-dipole and Pole-dipole array were presented with a geometry factor, K as given in Equations 1 – 4 which are derived from basic Equations 5 and 6. All the geometry factors, K was derived from Equation 6 based on basic four electrode system of measurement. The schematic diagram of field electrical resistivity configuration was given in Figure 5 – 8 while the schematic diagram for the basic four electrode system is given in Figure 9.

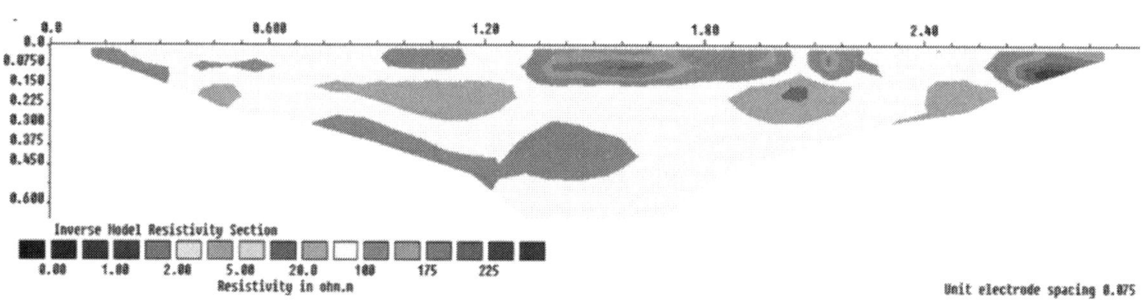

Figure 3. 2D electrical resistivity section of homogeneous lateritic soil.

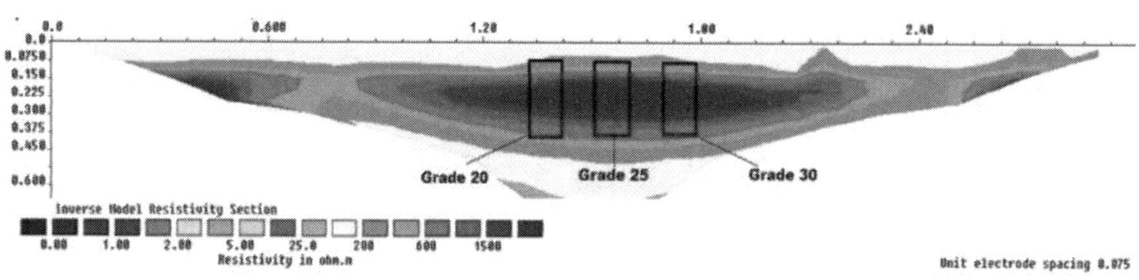

Figure 4. 2D electrical resistivity section with buried concrete cube.

$$\rho_a = 2\pi a * R \quad (1)$$

where R is a resistance term given by R=ΔV/I

$$\rho_a = ((2\pi\Delta V)/I)) * ((1/(1/r1 - 1/r2) - (1/r3 - 1/r4)) \quad (2)$$

where r1 = (L – x), r2 = (L + x), r3 = (L – x) and r4 = (L + x) - 1

$$\rho_a = (\pi an(n+1)(n+2)) * R \quad (3)$$

where R is a resistance term given by R=ΔV/I

$$\rho_a = ((2\pi ab)/(b-a))*R \tag{4}$$

where R is a resistance term given by R=ΔV/I

$$\rho_a = K*(R) \tag{5}$$

where R is a resistance term given by R=ΔV/I, K is geometry factor based on pole-dipole electrode configuration

$$\rho_a = ((2\pi\Delta V)/I))*((1/(1/r1-1/r2)-(1/r3-1/r4)) \tag{6}$$

where
$K = ((1/(1/r1-1/r2)-(1/r3-1/r4))$

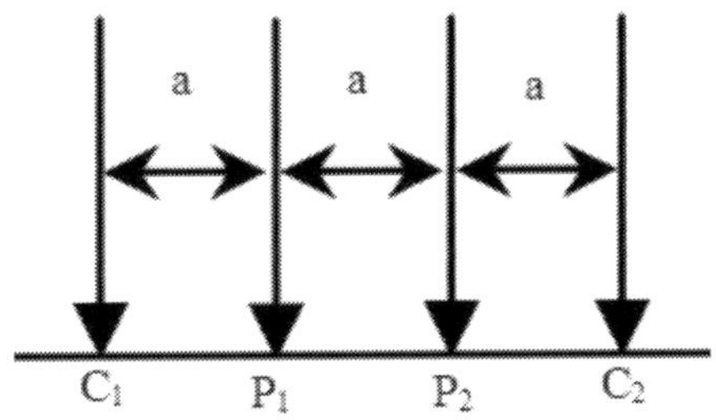

Figure 5. Wenner electrode array arrangement.

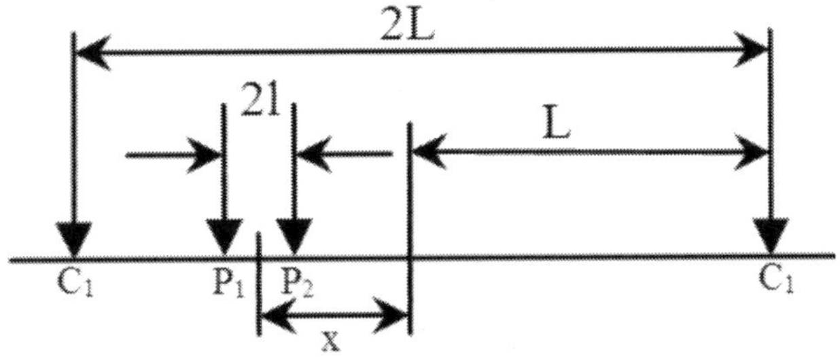

Figure 6. Schlumberger electrode array arrangement.

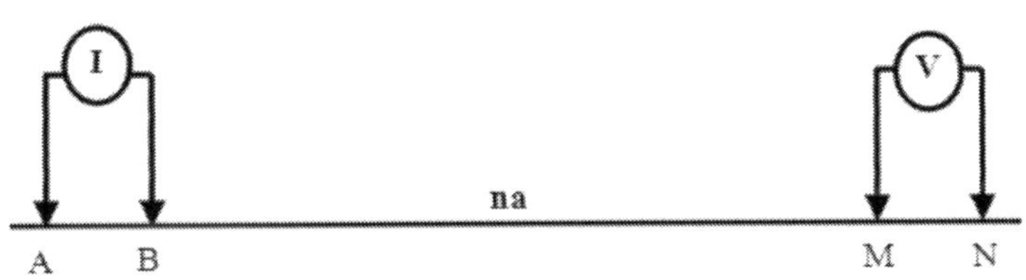

Figure 7. Dipole-dipole electrode array arrangement.

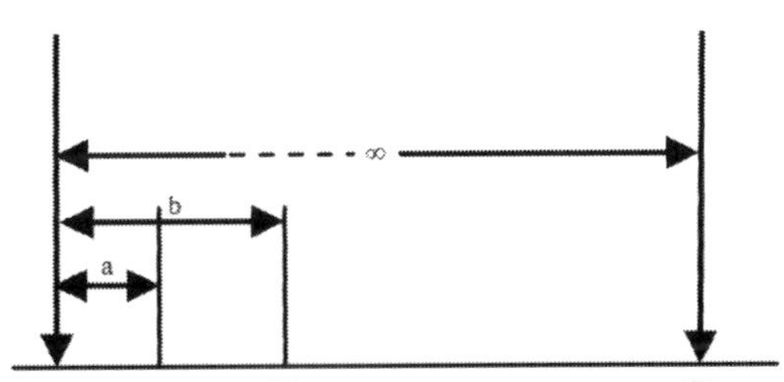

Figure 8. Pole-dipole electrode array arrangement.

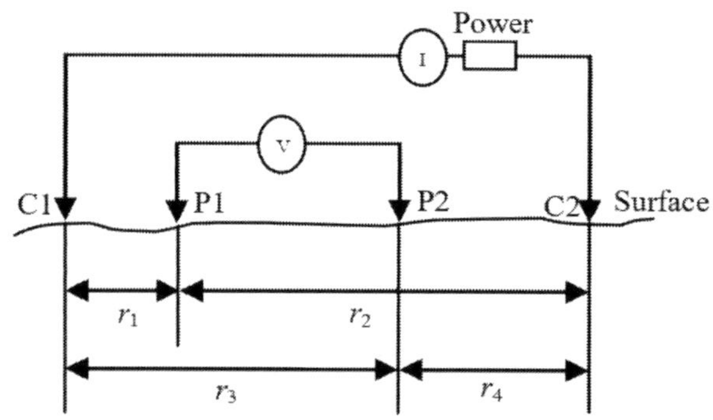

Figure 9. Four electrodes arrangement on the surface of homogeneous isotropic ground of resistivity.

Finally, this study has demonstrated the relationship of ERV due to the different strength of the concrete cube (grade 20, 30 and 35). The behavior of the ERV due to the different concrete cube strength were analyzed and presented using statistical bar chart as shown in Figure 10. This study found that the strength maturity of the concrete cube for grade 20, 25 and 30 were 22.3 MPa, 27.5 MPa and 32.5 MPa respectively. The localize ERV at the exact location of the buried concrete cube were extracted from the Figure 3 as 170.33 Ωm (grade 20), 226.82 Ωm (grade 25) and 502.88 Ωm (grade 30). Based on Figure 10, it was found that the concrete cube from grade 30 (32.5 MPa) has produce the highest of the ERV (502.88 Ωm) followed by the grade 25 (ERV = 226.82 Ωm) and grade 20 (ERV = 170.33 Ωm). Highest concrete cube strength (32.5 MPa) has rapid hydration process due to the sulphate reaction between the cement content and water [32] thus producing strong bond properties. Hence, those chemical reactions has produce low permeability of the highest concrete strength which promoting to the highest value of ERV. In other words, higher concrete strength produced a low conductivity which causes an increasing of the ERV. This phenomenon may due to the different composition of the concrete strength which relative to the concentration of cement, aggregate and sand. For example, concrete strength of 30 MPa was designed based on low coarse material compared to the others. This condition may cause the concrete to have lowest void thus decreasing its ability of conductivity. According to [31] electrical resistivity value was influenced by several factors such as the concentration and type of ions in pore fluid and grain matrix of geomaterials through electrolysis process where the current was carried by ions at a relatively slow rate. It was clearly observed that the ERV was greatly influenced by the stiffness variations of the material (S) which can be represented using general relationship of ERV ∞ S.

This study has successfully demonstrated that the different concrete strength will able to influence the electrical resistivity value as shown in Figure 10. Finally, the confidence level regarding the application of the electrical resistivity method in buried object detection can be enhanced due to the better understanding of electrical resistivity basic fundamental mapping performed in this study.

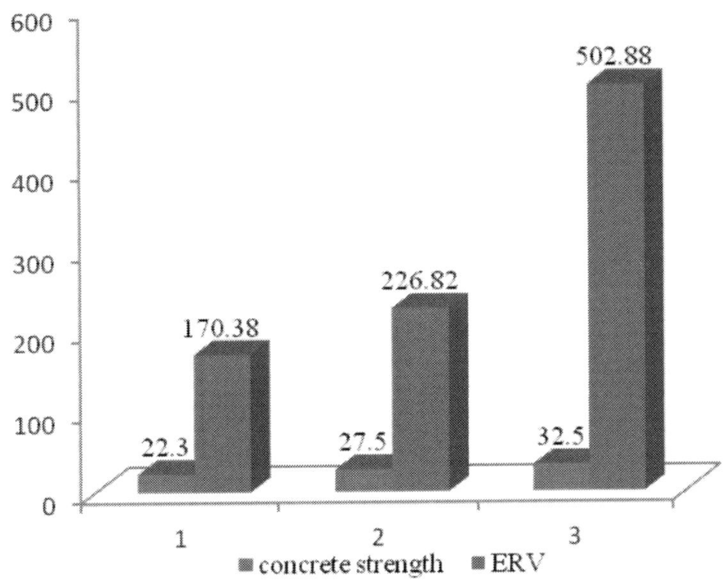

Figure 10. Relationship between ERV and concrete strength.

Conclusion

The electrical resistivity value from resistivity image was successfully performed on different strength of buried object. The influence on buried object detection using electrical resistivity method was successfully and methodically studied and presented. The electrical resistivity value was largely influenced by the variations strength of buried object under soil sample. The concrete cube was detected through the different value of electrical resistivity between surrounding material (soil model) and the buried object. The soil sample showed the resistivity value ranges from 50-200 Ωm while, location of concrete cube with different strength was presence by the resistivity value of 170-500 Ωm. The resistivity value ranges from 170-500 Ωm was found to be hard materials represented by concrete cube. Hence, this study shows that the ERV has a strong relationship with different stiffness of material thus applicable to be a useful alternative tool in underground structure detection. However due to the electrical resistivity configuration constraint, interpretation of the underground buried structure need to be supported by other comprehensive data in order to produce high accuracy and reliability of the result interpreted.

Acknowledgment

This work was funded by Universiti Tun Hussein Onn Malaysia (IGSP from Vote U258 and FRGS from Vote 1455). Many thank are due to all research members for their tremendous guidance and cooperation.

References

[1] Fragaszy R, Santamarina J, Amekudzi A, Assimaki D, Bachus R, Burns S, Cha M, Cho G, Cortes D, Dai S, Espinoza D, Garrow L, Huang H, Jang J, Jung J, Kim S, Kurtis K, Lee C, Pasten C, Phadnis H, Rix G, Shin H, Torres M and Tsouris C 2011 Sustainable development and energy geotechnology — Potential roles for geotechnical engineering *KSCE Journal of Civil Engineering*. **15** 611-621

[2] H.R. Wijesekara H R, Silvab S N D, Wijesundarab D T S S, Basnayakec B F A and Vithanagea M S 2015 Leachate plume delineation and lithologic profiling using surface resistivity in an open municipal solid waste dumpsite, Sri Lanka, *Environmental Technology*. **36** 2936-2943

[3] Popoola O I, and Fakunle M A 2014 Estimation Ofaverage Leachate Velocity Of Flow From Lapite Public Dump Site, Ibadan, Using Electrical Resistivity Method, *Nigerian Journal of Physics*. **25** 71-80

[4] Rosqvist H, Dahlin T T, Fourie A, Röhrs L, Bengtsson A and Larsson M 2003 Mapping of Leachate Plumes At Two Landfill Sites In South Africa Using Geoelectrical Imaging Techniques 2003 *Proceedings Sardinia, Ninth International Waste Management and Landfill Symposium S. Margherita di Pula, (Cagliari, Italy; 6 - 10 October 2003)*

[5] Saad R, Muztaza N M and Mohamad E T 2011 The 2D Electrical Resistivity Tomography (ERT) Study for Civil and Geotechnical Engineering Purposes *Electronic Journal of Geotechnical Engineering*. **16** 1537-1545

[6] Jusoh Z 2010 Application of 2-D resistivity imaging and seismic refraction technique in subsurface investigation for civil engineering, *M.S. thesis*, Sch. Physics, Science Univ., Penang, Malaysia

[7] Abidin M H Z, Baharuddin M F T, Zawawi M H, Ali N M, Madun A, Tajudin S A A 2015 Groundwater Seepage Mapping using Electrical Resistivity Imaging. *Applied Mechanics and Materials*, **773-774** 1524-1534

[8] Yeh H F, Lin H I, Wu C S, Hsu K C, Lee J W and Lee C H 2014 Electrical resistivity tomography applied to groundwater aquifer at downstream of Chih-Ben Creek basin, Taiwan, *Environmental Earth Sciences*, **73** 4681-4687

[9] Adiat K A N, Nawawi M N and Abdullah K 2010 Assessing the Accuracy of GIS-based Elementary Multi Criteria Decision Analysis as a Spatial Prediction Tool - A case of Predicting Potential Zones of Sustainable Groundwater Resources. *Journal of Hydrology*, **440-441** 75-89

[10] Cozzolino M, Mauriello P and Patella D 2014 Resistivity Tomography Imaging of the Substratum of the Bedestan Monumental Complex at Nicosia, Cyprus, *Archaeometry*, **56** 331–350

[11] Selim I E, Basheer A A, Elqady G and Hafez M A 2014 Shallow Seismic Refraction, Two-Dimensional Electrical Resistivity Imaging, and Ground Penetrating Radar for Imaging the Ancient Monuments at the Western Shore of Old Luxor City, Egypt, *Archaeological Discovery*, **2** 31-43

[12] Marchetti M, Sapia V, Garello A, Rita D D and Venuti A 2014 Geology and geophysics at the archeological park of Vulci (central Italy), *Annals Of Geophysics*, **57** 1-11

[13] Khatri R, Shrivastava V K and Chandak R 2011 Correlation between Vertical Electric Sounding and Conventional Methods of Geotechnical Site Investigation. *International Journal of Advanced Engineering Sciences and Technologies*, **4** 042-053

[14] Liu C and Evett J B 2008 *Soils and Foundation* (New Jersey: Pearson International)

[15] Godio A, Strobbia C and De Bacco G 2006 Geophysical Characterisation of a Rockslide in an Alpine Region, *Engineering Geology*. **83** 273-286

[16] Cosenza P, Marmet E, Rejiba F, Jun Cui Y, Tabbagh A and Charlery Y 2006 Correlations between geotechnical and electrical data: A case study at Garchy in France. *Journal of Applied Geophysics*, **60** 165-178

[17] Clayton C R I, Matthews M C and Simons N E 1995 *Site Investigation* (UK: Blackwell Science Ltd)

[18] Mauritsch H J, Seiberl W, Arndt R, Romer A, Schneiderbauer K and Sendlhofer G P 1999 Geophysical investigations of large landslides in the Carnic Region of Southern Austria, *Engineering Geology*, **56** 373-388

[19] Burger H R, Sheehan A F and Jones C H 2006 *Introduction to Applied Geophysics* (New York: W.W. Norton & Company)

[20] Loke M H 1995 Electrical imaging surveys for environmental and engineering studies

[21] Ali B A H 1991 Assessment of Concrete Compressive Strength by Ultrasonic Non-Destructive Test. *M.Sc. thesis*, The College of Engineering, The Univ. of Baghdad

[22] Abidin M H Z, Saad R, Ahmad F, Wijeyesekera D C and MBaharuddin M F T 2014 Correlation Analysis between Field Electrical Resistivity Value and Basic Geotechnical Properties, Soil Mechanics and Foundation Engineering, **51** 117-125
[23] Loke M H, Acworth I and Dahlin T 2003 A comparison of smooth and blocky inversion methods 2-D electrical imaging surveys, *Exploration Geophysics*, **34** 182–187
[24] Loke M H and Barker R D 1996 Rapid least squares inversion of apparent resistivity pseudosection using a quasi-Newton method, *Geophysical Prospecting*, **44** 131–152
[25] McCarthy D F 2007 *Essentials of Soil Mechanics and Foundations: Basic Geotechnics* (New Jersey: Pearson International Edition)
[26] Lee T S 2002 *Slope Stability and Stabilization Methods* (New York: John Wiley & Sons, Inc.)
[27] Abidin M H Z, Wijeyesekera D C, Saad R and Ahmad F 2013 The Influence of Soil Moisture Content and Grain Size Characteristics on its Field Electrical Resistivity, *Electronic Journal of Geotechnical Engineering*, **18/D** 699-705
[28] Rinaldi V A and Cuestas G 2002 Ohmic Conductivity of a Compacted Silty Clay, *Journal of Geotechnical and Geoenvironmental Engineering*, **128** 824-835
[29] Abidin M H Z, Saad R, Wijeyesekera D C, Ahmad F and Ismail N A 2014 The Influence of Electrical Resistivity Array on its Soil Electrical Resistivity Value, *Applied Mechanics and Materials*, **510** 185-192
[30] Abidin M H Z, Saad R, Wijeyesekera D C and Ahmad F 2013 Soil resistivity influence due to the different utilization of electrical resistivity array, *Electronic Journal of Geotechnical Engineering*, **18/X** 5643-5654
[31] W.M. Telford W M, L.P. Geldart L P and R.E. Sheriff R E 1990 *Applied Geophysics* (Cambridge: Cambridge University Press)
[32] Neville A M and Brooks J J 2001 *Concrete Technology* (Prentice Hall)

Convertion Shear Wave Velocity to Standard Penetration Resistance

A Madun[1,2], S A A Tajuddin[2,1], M E Abdullah[1,2], M H Z Abidin[1,2], S Sani[2,1] A J L M Siang[1,2] and M F Yusof[1,2]

[1]Department of Infrastructure and Geomatic, Universiti Tun Hussein Onn Malaysia.
[2]Research Centre for Soft Soil, Universiti Tun Hussein Onn Malaysia.

E-mail: aziman@uthm.edu.my

Abstract. Multichannel Analysis Surface Wave (MASW) measurement is one of the geophysics exploration techniques to determine the soil profile based on shear wave velocity. Meanwhile, borehole intrusive technique identifies the changes of soil layer based on soil penetration resistance, i.e. standard penetration test-number of blows (SPT-N). Researchers across the world introduced many empirical conversions of standard penetration test blow number of borehole data to shear wave velocity or vice versa. This is because geophysics test is a non-destructive and relatively fast assessment, and thus should be promoted to compliment the site investigation work. These empirical conversions of shear wave velocity to SPT-N blow can be utilised, and thus suitable geotechnical parameters for design purposes can be achieved. This study has demonstrated the conversion between MASW and SPT-N value. The study was conducted at the university campus and Sejagung Sri Medan. The MASW seismic profiles at the University campus test site and Sejagung were at a depth of 21 m and 13 m, respectively. The shear wave velocities were also calculated empirically using SPT-N value, and thus both calculated and measured shear wave velocities were compared. It is essential to note that the MASW test and empirical conversion always underestimate the actual shear wave velocity of hard layer or rock due to the effect of soil properties on the upper layer.

Keywords: Soft soil, Multichannel Analysis Surface Wave (MASW), SPT-N blow, stiffness profile.

1. Introduction

Seismic wave techniques, which depend on the modulus and density of the materials can be converted to very useful parameters for engineering purposes, such as elastic modulus, shear modulus and Poisson's ratio. The seismic method based results are empirically derived geotechnical properties such as maximum shear modulus, bulk modulus (B), Young's modulus (E), and Poisson's ratio [1,2]. In addition, seismic velocity can also correlate empirically to the standard penetration test number of blows (SPT-N) obtained from the penetration tests [3,4,5]. The seismic-based techniques have proved particularly useful in determining the shear modulus profile from site investigation [6,7]. There are two methods of obtaining seismic wave data that can potentially be used for site investigation (1) borehole methods and (2) surface methods [8]. The surface wave data collection uses the surface method, which is more versatile than other methods because it is not constrained by any ground models and considered more economical in terms of field operation [9].

Traditionally, the measurement of stiffness profile was carried out by using a combination of laboratory and in situ, invasive field tests. However, geophysical methods, such as seismic surface wave, offer a non-intrusive and non-destructive approach to carry out these measurements. Moreover, geophysical approaches such as this provide a cost effective way to assess site conditions while overcoming a key limitation of traditional investigative approaches. A comparison between geophysical seismic-based techniques and conventional geotechnical load-testing methods for the measurement of the ground stiffness profile were presented by Matthews et al. [10] and Clayton [11], drawing the conclusion that geophysical testing can deliver results of significant quality. However, care is needed not to overestimate what geophysics can achieve, by understanding geophysical techniques, in particular their limitations [12]. Thus, geophysical assessment of any site investigation must be carried out with physical soundings to ensure proper calibration and validation.

In situ field testing enables larger volumes of soil to be tested and therefore tends to be more representative of the soil mass compared to laboratory testing. In situ field tests have an advantage as samples do not need to be retrieved. Some field tests are considered as destructive tests as they involve preliminary work such as drilling or inserting instruments into the ground. The dynamic Standard Penetration Test (SPT) is used world-wide to measure the force that is needed for an open sampler to penetrate the soil. The SPT is, essentially, a simple test involving the dropping of a standard hammer of known weight from a specific height. The SPT is the most common in situ test method used and is usually accompanied by other complementary measurements, field or laboratory based, for the purpose of making comparisons and establishing the relevant correlations. There were many convincing empirical correlations between N-values and shear wave velocity which are accepted worldwide. However, these empirical correlations were developed based on a local scale in various countries, which are unique to geological conditions and yet to support firm theoretical formulation [13]. In this study, three empirical correlations were referred to are Uma Maheswari et al. [3], Tsiambaos and Sabatakakis [4] and Anbazhagan et al. [5].

Multichannel Analysis of Surface Wave (MASW) method adopted in this study has significant advantages over other surface wave techniques as all seismic wave energy, consisting of both body and surface waves, is recorded by multichannel receivers. Seismic waves propagate in the form of body waves and surface waves. The difference between the two is that body waves are usually non-dispersive. In a solid and homogeneous medium, the velocity of surface waves does not fluctuate significantly as a function of distance propagates. However, when the properties of the medium vary with depth, surface waves become dispersive such that the velocity of propagation varies with respect to wavelength or frequency. Multichannel Analysis of Surface Wave (MASW) method has an investigation depth shallower than 30 m whereas the passive method (source like traffic and tidal motion) can reach a few hundred meters. Sampling redundancy due to multi receivers provides flexibility in the signal processing approach to extract the dispersion curve. Many advantages have been stated above, and thus, the evaluation of this technique for site investigation is conducted to promote this technique. This study aims to investigate the conversion of shear wave velocity obtained from SPT-N value and shear wave velocity from MASW technique or vice versa. The location of this study is shown in Figure 1, on the campus of Universiti Tun Hussein Onn Malaysia and Sejagung Sri Medan. The campus of Universiti Tun Hussein Onn Malaysia possesses marine clay and silt deposits of Quaternary age. Meanwhile, Sejagung Sri Medan is located at the geological boundary between marine deposit and volcanic rock.

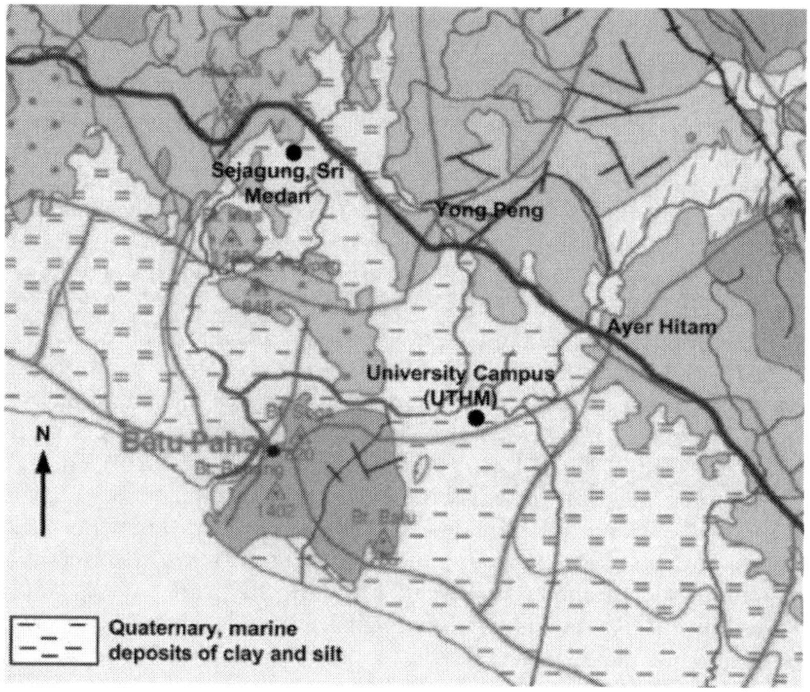

Figure 1. Test locations at university campus and Parit Jelutong.

2. Methodology

Multichannel Analysis of Surface Wave (MASW) method uses similar equipment in the seismic refraction method, but uses low frequency geophones. A 7 kg sledgehammer is used as source that impacts the metal plate. A 24-unit 4.5 Hz vertical geophone was used as a detector which was connected to a 24-channel cable and ABEM Terraloc MK-8 seismograph as a recorder. The seismograph setting for MASW test required longer record time, i.e. about 2 second to measure seismic data. The sampling interval is between 250 and 500 μs and the number of samples is between 4096 and 8192. Hammering the ground about 5 times produces active waves. The length of array and distance of seismic source to the first geophone are at 69 m and 15 m, respectively. MASW tests were conducted close to the borehole location.

3. Results and Discussions

Figure 2 shows the borehole data at two study areas, i.e. UTHM campus and Sejagung Sri Medan. At the UTHM campus, the soil layer consists of very soft layers (SPT-N <2) up to 10.5 m followed by soft layers (SPT-N 2-4) up to 19.5 m depth. Meanwhile, borehole data at Sejagung, Sri Medan indicated that the subsurface can be categorized as loose soil (SPT N 4 - 10) up to 6 m in depth, followed by very dense soil (SPT-N > 50) until 9 m in depth. The borehole continued to drill the rock core for 6 m and was terminated at 15 m. The conversion of the empirical shear wave velocity using SPT-N value via three equations suggested by Anbazhagan et al. [5] using $V_s = 68.96N^{0.51}$; Tsiambaos and Sabatakakis [4] $V_s = 105.7N^{0.327}$; and Uma Maheswari et al. [3] $V_s = 95.64N^{0.301}$ were presented in Figure 3 and Figure 4.

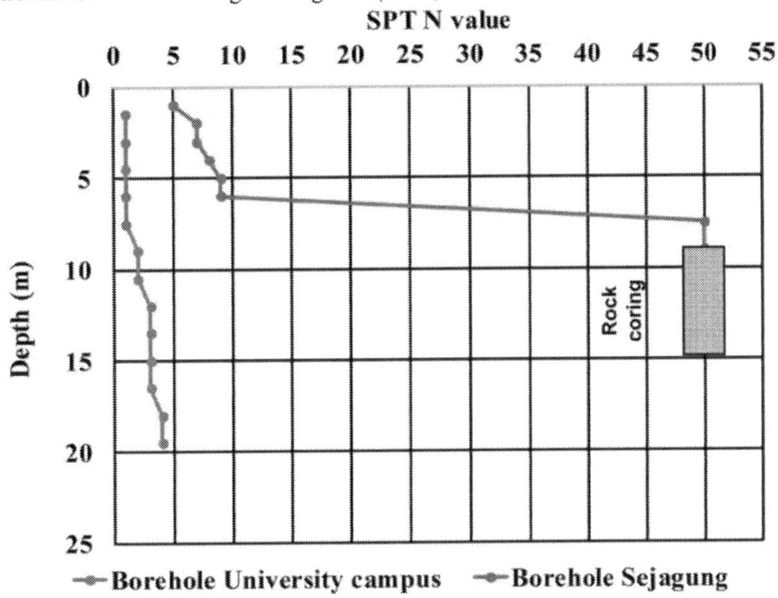

Figure 2. The borehole data at the university campus and Sejagung Sri Medan.

The MASW test to obtain shear wave profiles at the university campus and the Sejagung Sri Medan site was conducted and analysed using SeisImager software. The dispersive phase velocities then produced 1-dimensional shear wave velocity versus depth. The shear wave velocities for very soft soil was below 180 m/s and soft soil was between 180 m/s and 270 m/s at the university campus site as shown in Figure 3. Meanwhile, at Sejagung Sri Medan indicated the shear wave velocities for loose soil was between 220 m/s and 450 m/s, very dense soil was between 450 m/s and 630 m/s and rock was between 630 m/s and 680 m/s as presented in Figure 4. The shear wave velocities using MASW technique indicated good agreement with the converting shear wave velocities using empirical conversion for both cases, i.e. the university campus and Sejagung Sri Medan site. However, at a deeper layer, the differences between measured and empirical conversion shear velocities became significant. This is because the MASW technique measured across its array length of 69 m (geophone 1 to 24) and the frequencies were dispersed according to its depth. As a result, the shear wave velocities were the outcome of an averaged velocity at specified depth across the area. In contrast, the calculated shear wave velocities using an empirical conversion and borehole SPT-N data are specifically at a certain depth.

The surface wave technique utilized Rayleigh wave velocity which is not only dependent on the depth, but also on the horizontal spans of the soil properties. Hence, the depth and horizontal spans can be used to identify the sampling volume. The sampling volume is based on the elliptical particle movement suggested by Madun et al. [14,15]. The relationship between the shear wave velocity and the effective region for each depth and wavelength is important for analysing and interpreting the measured result. Figure 5 illustrates the MASW measurement at Sejagung Sri Medan, where almost 100 % of the elliptical effective regions sit within the first layer of loose soil at a depth of 6 m. This effective region grows and the effective region presents the largest change in averaged shear wave velocity at the stage where the elliptical begins to encroach into the second and third layer of very hard and bedrock, respectively. Thus, this can explain the upward trend in shear wave velocities with further increase in depth. This in turn is a function of its wavelength. All the elliptical-shaped sampling volumes shown in Figure 5 are likely to explain the measured shear wave velocity is lower than actual velocity. For example, at Sejagung, after 9 m to 13 m depth, the shear wave velocity is between 630 m/s and 680 m/s, indicating the rock layer via borehole data. It is worth noting that the shear wave velocity of rock is above 1500 m/s [16]. In practical applications for the understanding of the relationship between the wavelength and the effective region of measurement (the sampling volume) it is important to plot a graphical representation of the dispersion curve, which delineates the variation and defines the soil boundaries. Therefore, understanding the geophysical limitations is important to avoid overestimating what geophysics can achieve. It is essential to note that the MASW test and three empirical conversions suggested by Anbazhagan et al (2012), Tsiambaos and Sabatakakis (2011), and Maheswari et al (2010) always underestimate the actual shear wave velocity of hard layer or rocks due to the effect of soil properties on the upper layer.

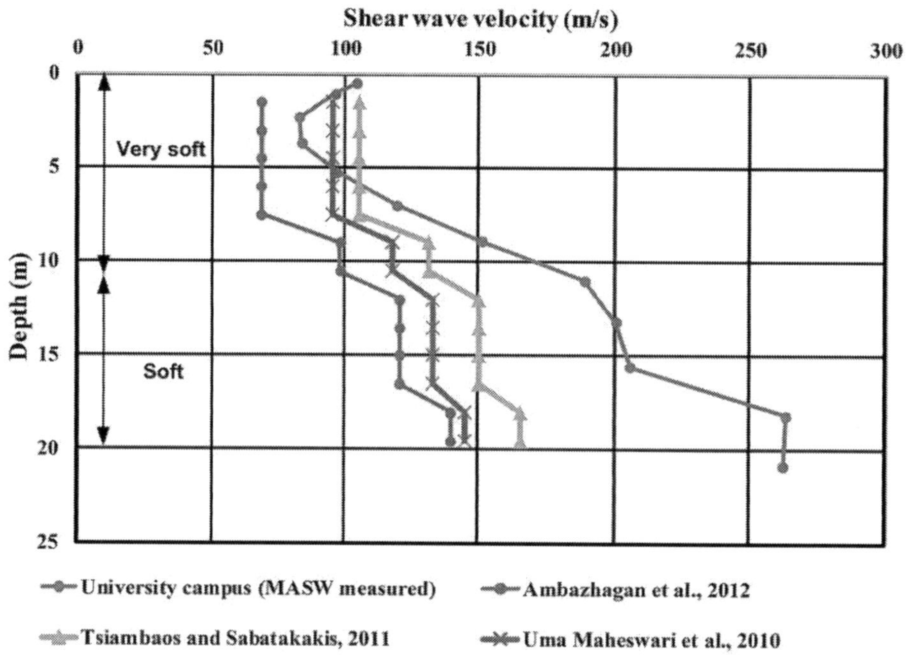

Figure 3. The shear wave velocities measured using MASW and empirical conversion for data at university campus

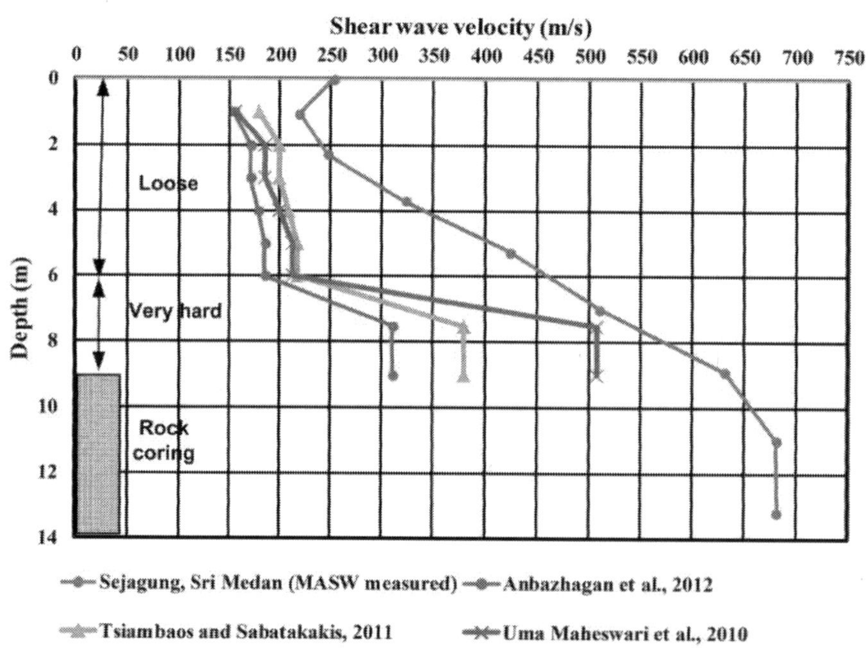

Figure 4. The shear wave velocities measured using MASW and empirical conversion for data at Sejagung, Sri Medan

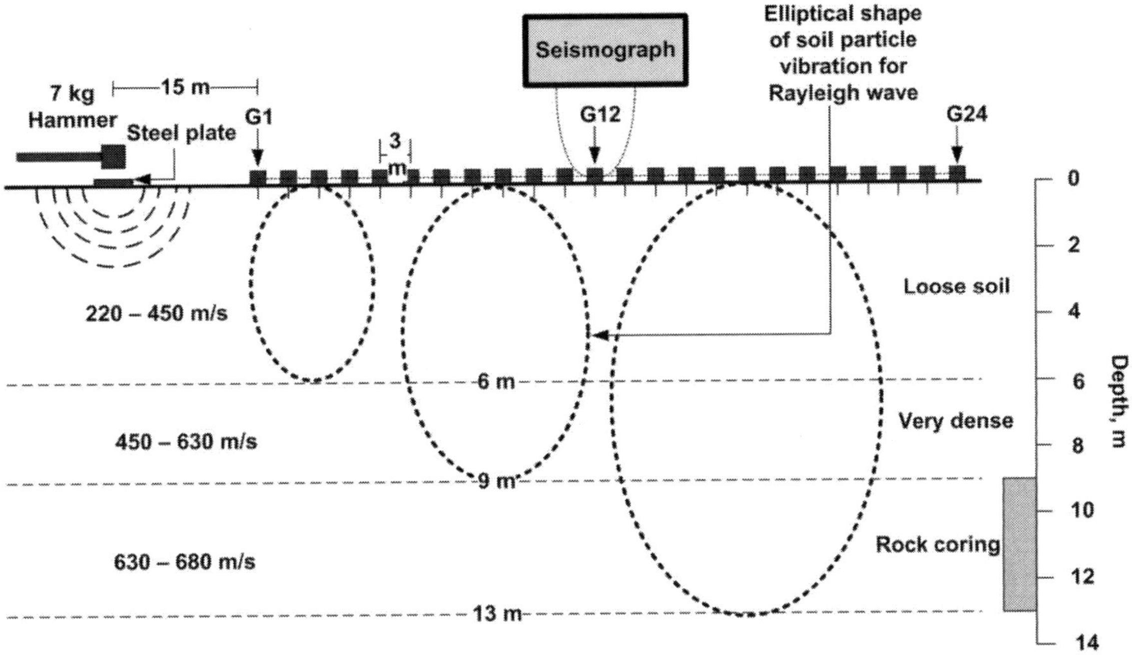

Figure 5. Illustration of volume measurement in elliptical shape to demonstrate that the shear wave velocity is lower than actual values.

4. Conclusion

The MASW seismic profiles at the university campus test site and Sejagung were at 21 m and 13 m in depth, respectively. The shear wave velocities were also calculated empirically using SPT-N value, and thus both calculated and measured shear wave velocities were compared. The shear wave velocity obtained using MASW technique represents the average velocity at specified depth across the lateral length of the array. Thus, it is expected that the velocity was empirically converted using the SPT-N value slightly deviated to the variability of soil horizontally and vertically. Therefore, understanding the geophysical limitations is important to avoid overestimating what geophysics can achieve. It is essential to note that the MASW test and empirical conversion usually underestimates the actual shear wave velocity of hard layers or rock due to the effect of soil properties on the upper layer.

Acknowledgement

The authors would like to thank University of Tun Hussein Onn, Malaysia, GPP vot. B048 for generous sponsorship of this research.

References

[1] Charles J A and Watts K S 2002 Treated ground engineering properties and performance, London: Construction Industry Research and Information Association, CIRIA C572.
[2] Crice D. 2005 MASW: the wave of the future. *Journal of Environmental and Engineering Geophysics*, 10(2) pp. 77–79.
[3] Maheswari R U, Boominathan A and Dodagoudar G R 2010 Use of surface waves in statistical correlations of shear wave velocity and penetration resistance of Chennai soils, *Geotechnical and Geology Engineering*, 28(2) pp.119-137.
[4] Tsiambaos G and N. Sabatakakis, Empirical estimation of shear wave velocity from in situ tests on soil formations in Greece, Bulletin of Engineering Geology and the Environment, 70(2011) 291–297.
[5] Anbazhagan P, Parihar A and Rashimi H.N 2012 Review of correlations between SPT N and shear modulus: A new correlation applicable to any region, *Soil Dynamics and Earthquake Engineering* 36 pp. 52-69.

[6] A.L. Moxhay, R.D. Tinsley and J.A. Sutton Monitoring Of Soil Stiffness During Ground Improvement Using Seismic Surface Wave. Ground Engineering Magazine, January (2001) 34-37.

[7] Redges J D, Moxhay A L, Ghataora G S and Jefferson I 2008 Case Histories Of Settlement Performance Comparisons On Ground Improvements Using Stiffness, Seismic Waves And Traditional Methods: *Proc. 6th International Conference on Case Histories in Geotechnical Engineering, Arlington, VA,* pp. 11-16, Expanded Abstracts.

[8] Menzies B K 2001 Near-surface site characterisation by ground stiffness profiling using surface wave geophysics, Instrumentation in Geotechnical Engineering, *H.C.Verma Commemorative Volume*, pp. 43-71.

[9] Matthews M C, Clayton C R I, and Own Y 2000 The Use Of Field Geophysical Techniques To Determine Geotechnical Stiffness Parameters. *Proceedings of the Institution of Civil Engineers Geotechnical Engineering*, 143 pp. 31-42.

[10] Matthews M C, Hope V S and Clayton C R I 1995 The geotechnical value of ground stiffness determined using seismic methods: Geological Society, London, *Engineering Geology Special Publication,* 12 pp. 113-123.

[11] Clayton C R I 2011 Stiffness At Small Strain: Research And Practice. *Géotechnique,* 61(1) pp. 5 – 37.

[12] Clayton C R I, Matthews M C and Simons N E 1995 Site Investigation. 2nd edition. *London Blackwell Science* Ltd, 1995.

[13] Marto A, Tan C S, Kasim F and Suhatril M 2013 A Correlation Of Shear Wave Velocity And Standard Penetration Resistance, *Electronic Journal of Geotechnical Engineering, 18c* pp. 463 – 471.

[14] Madun A, Jefferson I, Foo K Y, Chapman D N, Culshaw M G and Atkins P.R 2012 Characterization And Quality Control Of Stone Columns Using Surface Wave Testing, *Canadian Geotechnical Journal,* vol. 49 pp. 1357 – 1368.

[15] Madun A, Jefferson I, Chapman D N, Culshaw M G, Foo K Y and Atkins P R 2010 Evaluation Of The Multi-Channel Surface Wave Analysis Approach For The Monitoring Of Multiple Soil-Stiffening Columns, *Near Surface Geophysics,* 8(6) pp. 611 – 621.

[16] Bray J D and Rodriguez-Marek A 1997 Geotechnical Site Categories, *Proceedings of First PEERPG&E Workshop on Seismic Reliability of Utility Lifelines,* San Francisco, CA.

Characterization of Soil Organic Matter in Peat Soil with Different Humification Levels using FTIR

T I Tong[1], N L L Felix[2], S Mohd[3], A Sulaeman[4]

[1] Faculty of Civil and Env. Eng., Universiti Tun Hussein Onn Malaysia (UTHM), MALAYSIA.

[2] Research Center for Soft Soil, Universiti Tun Hussein Onn Malaysia, 86400 Batu Pahat Johor, MALAYSIA

[3] School of Engineering & Technology, Universiti College Technology Sarawak (UCTS), MALAYSIA.

[4] Faculty of Engineering Technology, Universiti Tun Hussein Onn Malaysia (UTHM), MALAYSIA

E-mail: teongingtong@gmail.com

Abstract: Peat soil is defined as an accumulation of the debris and vegetative under the water logging condition. Soil organic matter of peat soil was affected by the environmental, weather, types of vegetative. Peat soil was normally classified based on its level of humification. Humification can be defined as the transformation of numerous group of substances (proteins, carbohydrates, lipids, etc.) and individual molecules present in living organic matter into group of substances with similar properties (humic substances). During the peat transformation process, content of soil organic matter also will change. Hence, that is important to determine out the types of the organic compound. FTIR (Fourier Transform Infrared) is a machine which is used to differential soil organic matter by using infrared. Infrared is a types of low energy which can determine the organic minerals. Hence, FTIR can be suitable as an indicator on its level of humification. The main objective of this study is to identify an optimized method to characterization of the soil organic content in different level of humification. The case study areas which had been chosen for this study are Parit Sulong, Batu Pahat and UCTS, Sibu. Peat soil samples were taken by every 0.5 m depth until it reached the clay layer. However, the soil organic matter in different humification levels is not significant. FTIR is an indicator which is used to determine the types of soil, but it is unable to differentiate the soil organic matter in peat soil FTIR can determine different types of the soil based on different wave length. Generally, soil organic matter was found that it is not significant to the level of humification.

Keywords: Soft soil, peat soil, soil organic, ftir, humification.

1. Introduction

In geotechnical field, peat soil is defined as a problematic soil due to its physical properties. Peat soil which has very high moisture content, high fiber content, low permeability and low shear strength cause the soil cannot support high loading and not suitable for construction.

Research such as that conducted by [1,2,3] have shown that peat is an accumulation of the debris of plant, trunk and roots in water logging condition. Besides that, in research [4] stated that peat soil is an organic matter which is usually in light brown or dark colour. It has high organic matter and therefore it can be classified into different types of peat soil based on different level of humification of peat.

According to [5,6], humification can be defined as the transformation of numerous groups substances (proteins, carbohydrates, lipids,etc.) and individual molecules present in living organic matter into groups of substances with similar properties (humic substances). After that, the humic substances are changed into carbon deposits. In the carbon biogeochemical cycle, transformation from living organic to the carbon substances, in terms of humification is affected by the high variability of environmental conditions and affected by degradation and synthetic reaction.

Based on [7,8] in their research stated that the organic matter inside peat soil will have some differences in its physical properties during the living organic matter transformation. During peat transformation process, the degree of decomposition, peat's specific gravity and compaction are increased while the peat's moisture content is decreased. In addition, the colour of peat will also change. According to [9], the colour of peat will change into dark brown and black colour due to calorific value. Other than that, [10,11] also mentioned that the behavior and types of organic matter of peat soil will be affected by the environmental condition. There are several factors which will lead to the affection such as the types of forest, vegetative, temperature and weather as well as bog hydrological conditions. Hence, different level of humification will have different quantity and different types of soil organic matter during the transformation process.

To characterize the peat soil, level of humification is one of the indexes link that rate the transformation of organic matter. Von Post classification are widely used by the most of the researchers to determine the level of humification on site [12]. This is because this method can directly obtain result of the level of humification on site based on the squeezing of peat soil sample.

Nowadays, Fourier Transform Infrared (FTIR) is a popular and simple equipment that can be used to determine and characterize the soil organic matter. This is because FTIR spectroscopy can show the direct information of functional groups in the soil organic matter fraction analyzed [13,14]. Based on [15], used the FTIR spectroscopy to determine organic surface layers of forest while [16,17] use it to determine or evaluate the humic extracts. Based on [23], there are different types of mineral in different wave length. For range less than 1500 cm^{-1} is fingerprint area (mineral) while the soil organic matter is in the range of 1500 cm^{-1} to 3500 cm^{-1}. For the clay mineral, it is more than 3500 cm^{-1}. Based on the differences in wavelength, the type of the mineral content can be determined.

2. Method and Methodology

In this study, there are 2 investigated peat soil sites. The first investigated site is located at UCTS Sibu, Sarawak. Sibu is a place which has largest peat soil land in Malaysia. Some of the geo properties have been partly described by [7]. For another site, it is located at Peninsular Malaysia. Peat soil lands in Peninsular Malaysia are mostly located at west coastal and east coastal. The site that has been chosen for the research is located at UCTS, Sibu as shown in figure 1 (a) and Parit Sulong Batu Pahat Johor as shown in Figure 1(b).

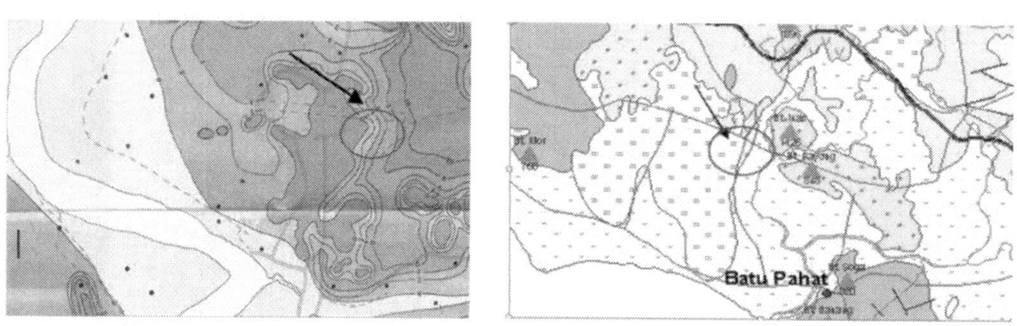

(a) (b)

Figure 1. (a) shows the location at Sibu and figure 1 (b) show the location at Parit Sulong.

There are 2 different types of land activity on site. In Parit Sulong, the peat soil land is mainly used for palm oil plantation while in Sibu, there is no any land activity. According to the map, it is shown that the depth of the peat soil in Sibu is around 4-6m while for Parit Sulong, it shows that the location is at the boundary between peat soil and other types of soil. This means that peat soil in Parit Sulong is shallow and most probably has been mixed with other types of soil such as clay.

The peat soil was taken from every site until reached the clay layers, this is because peat soil is types of soil which affected and classify by the humification level. For the clay soil which have own mineral to separate out the different types of clay and clay soil would not happen any humification process. Peat soil was taken by every 0.5m depth. The depth for the peat soil in Sibu was recorded at 5.5m. For Parit Sulong site, the peat soil depth was recorded at 3m. There are several tests used to determine the level of decomposition of the peat soil. In this study, Von Post classification is chosen to determine the level of humification on site [12]. This is because it can directly obtained the result on site. Degree of humification of peat is identified based on the color of water escape from fingers when squeezing the fresh peat soil by hand and quantity that left inside hand after squeezing. The range of scale for the humification of peat is from 1 to 10. H1 represents the least humification and the H10 is the most humification.

In this paper, peat soil samples were tested. This is because peat soil properties was affected by the environmental and the different level of humification. Although there are different depth of two different points, but in this paper, the main focus was soil organic matters in different level of humification.

Table 1. Classification types of peat according to the humification level.

Level of Humification	Types of Peat
H1 to H3	Fibric peat
H4 to H6	Hemic peat
H7 to H10	Sapric peat

Based on the [18,22], there are 3 types of peat was classified due to different level of humification and it is general and common standard to classify types of peat. During sample preparation for FTIR, the sample must be dried in advance at the temperature of 50°C. This is because higher temperature or temperature of 105°C will destroy the fiber. After that, the sample is grinded into powder form. Fourier Transform Infrared (FTIR) is carried out by using the low energy, infrared to identify the structures of molecule of soil.[19,20] In this study, FTIR spectroscopy that is used is Mid FTIR which has wave length of range from 4000 to 200 cm-^{1}. The spectroscopy must be ensured to be under the low temperatures to reduce the moistin order to prevent it from affecting the result of FTIR. Furthermore, the stage must be cleaned and there must not have any samples on it. This is because different types of samples will have different composition that may affect the result.

3. Result and analysis

The result of the laboratory Mid FTIR and in-situ testing (Von Post classification) are summarized in different types of format. The Mid FTIR result is shown in Figure 2 and Figure 3 while the result of level of humification is summarized in Table 2 and Table 3. Data that have been summarized is subdivided based on its location, namely Parit Sulong, Batu Pahat and UCTS, Sibu.

According to Table 2, the result showed the humification level of the site at the Parit Sulong. From the data obtained, 0-0.5m which has a record of H3 is the lowest humification on this site. For the second layer, it is recorded H6 while the rest are maintain at the same level of the humification. According to the reseacher [21], H1 to the H4 is the fibrous peat, H5-H7 is pseudo- fibrous peat and H8-H10 is amorphous peat. Based on table 1, peat on the depth 0-0.5m was fibrous peat, 0.5-1m was pseudo- fibrous peat and 1-2.5m was amorphous peat.

Data in Table 3 shows that humification of peat soil is almost the same. However, there is a slightly different result at the depth of 2m to 3 m. For the depth of 2-2.5m, it was H7 while for the depth of 2.5-3m, it has a record of H6. Other layers of peat soil are recorded H8.

Table 2. Von post classification of peat soil (Parit Sulong, Batu Pahat).

Depth (m)	Degree Of Humification
Top Soil	N/A
0 - 0.5	H3
0.5 - 1.0	H6
1.0 - 1.5	H8
1.5 - 2.0	H8
2.0 -2.5	H8
2.5 - 3.0	N/A

Table 3. Von post classification of peat soil (UCTS, Sibu).

Depth (m)	Degree Of Humification
0-0.5	H8
0.5-1.0	H8
1.0-1.5	H8
1.5-2.0	H8
2.0-2.5	H7
2.5-3.0	H6
3.0-3.5	H8
3.5-4.0	H8
4.0-4.5	H8

3.1 FTIR result

FTIR result of Parit Sulong is shown in Figure 2. According to FTIR machine software, it shows that the peat soil of Parit Sulong is Slvent Impurities, group methanol. According to [13], 500–1500 cm-^{1} is fingerprint region for FTIR. It usually contains a very complicated series of absorptions. This is mainly due to all manner of bending vibrations within the molecule.

The higher point is recorded at 1606cm-^{1}. The second higher peak which is dwi peak is at 2850.24cm-^{1} and 2919.66 cm-^{1} While, the last peak is recorderd at the 3351 cm-^{1}. From the result, there are some slightly differences between various layers. At the peak of 1600 cm-^{1}, layers 2-2.5m and layer 2.5-3m is lower than others layers, layer 2.5-3m is obviously has very low intensity. Furthermore, the layer of 2.5-3m also has the lowest intensity at the peak 2800-3000 cm-^{1}. It does not have any peak at the range of 3000-3500 cm-^{1}, but have a peak after 3500 cm-^{1}. Based on the table of charateristic of IR absorptions, functional group at different peak can be determined. For the peak which is range of 1600-1585 cm-^{1}, it is aromatics. For the range 1760–1665 cm-^{1} is carbonyls (general). For 3000–2850 cm-^{1} is alkanes. 3500–3200 cm-^{1} is alcohol and for more than 3500cm-1 is categorized as clay mineral.

Based on the Parit Sulong humification result, samples of the depth 0-0.5m and 0.5-1m were different with the others layers samples. But from the FTIR result, there are no different of the peak or any change with the other layers. This show that organic compound of the samples still in the same group of the mineral. There are slightly different for the layer 2.5-3m, because this layer is clay layer. From the FTIR, result, it have a peak after 3500cm-^{1} which is categorized as a clay mineral.

IOP Conf. Series: Materials Science and Engineering **136** (2016) 012010 doi:10.1088/1757-899X/136/1/012010

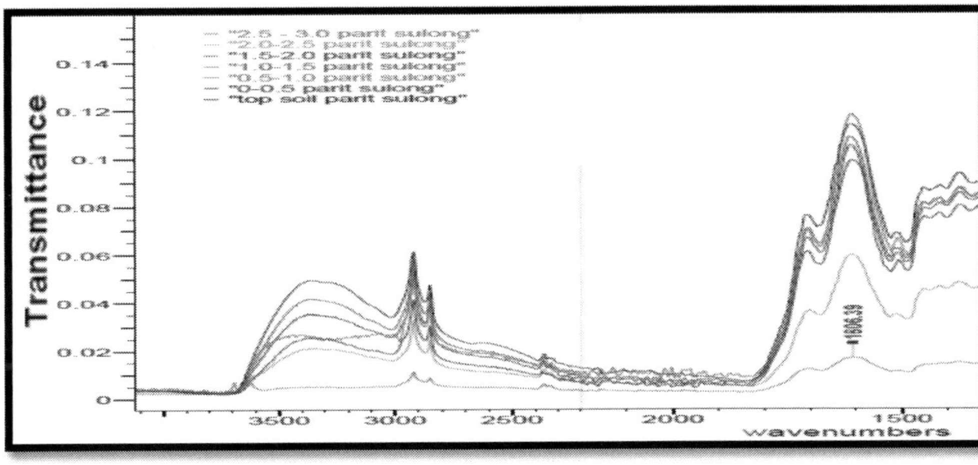

Figure 2. FTIR result of the Parit Sulong, Batu Pahat.

For the UCTS Sibu FTIR result, there is a total of 5 peak point which is located at 1508cm-[1], 1606 cm-[1], 1702 cm-[1], dwi peat at 2850 cm-[1], 2919 cm-[1] and last at 3351 cm-[1]. According to the humification result, peat soil samples in Sibu get the almost same result which is H8. While based on FTIR result, although the humification of all samples are same, it still have some differences. For the peak before 3500 cm-[1], all the samples are basically do not have any changes. However after 3500 cm-[1], there are 3 layers of soil are consisting of peaks which have the depth of 3-3.5m, 3.5-4m and 4-4.5m respectively.

According to this FTIR result, although Parit Sulong peat samples have different level of humification, the result of FTIR for organic matter is still shown in same pattern and do not have changes. For Sibu FTIR result, all the samples are almost in same humification level, H8. The range before 3500 cm-[1] is still the same pattern but there is slightly different after 3500 cm-[1]. From this, the level of humification of peak soil could not be directly observed from FTIR spectroscopy. The group of soil organic matter does not change much and it could be affected by the quantity of mineral.

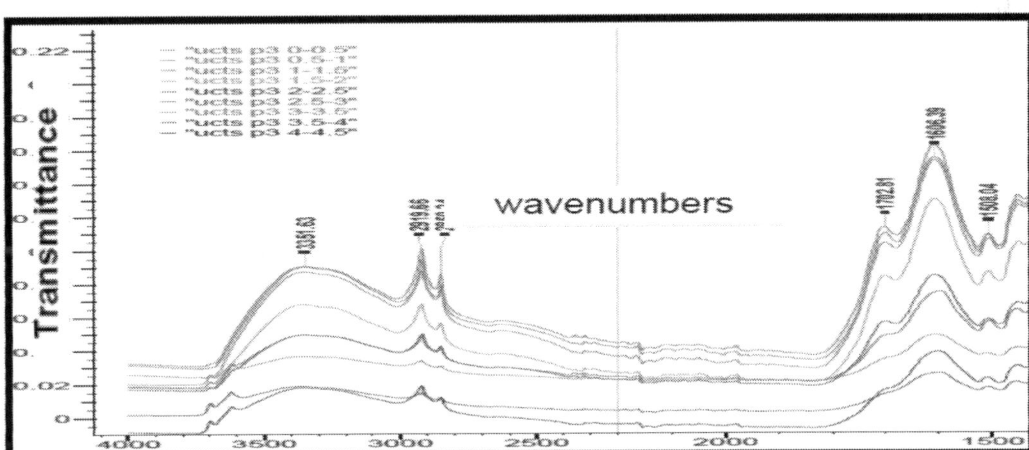

Figure 3. FTIR result of the UCTS, Sibu

4. Conclusion

In general, soil organic matter in different types of soil is different. This is because of the soil organic matter and the minerals in soil are different. Peat soil which is affected by environmental and the humification level. For the clay soil, there are different types of minerals which is giving different properties in physical and chemical. In different level of humification, the quantity and types of the organic compound will also be different. In this study, FTIR can only classify the types of organic matter and determine the mineral content inside peat soil. From the research result of Parit Sulong, FTIR can directly identify that the layer of 2.5-3m is clay layer. For the UCTS, Sibu result, FTIR has showed that some of the layers have mixture of clay. This is because they have peat more than 3500cm^{-1}. According to the result, FTIR can used for separated peat soil and clay soil. Besides that, FTIR has several limitations. It cannot directly classify the level of humification of peat soil. The range for humic acid is very wide and the level of humification could be affected by the quantity of organic matter. From this, another conclusion can made is FTIR cannot used for separated out the humification of peat soil. Furthermore, von post classification procedures also has some limitations. The samples that are used to carry out the test through squeezing them by hand are taken from a certain part of 0.5m undisturbed soil samples. Therefore, the result cannot represent the rest of the soil since there is a large range of soil and that may contain other composition which leads to different result. As a conclusion, FTIR is not suitable for classify level of humification for the peat soil.

Acknowledgements
The authors wish to gratefully acknowledge the financial support from Ministry of Higher Education (MOHE), Malaysia and Universiti Tun Hussein Onn Malaysia (UTHM) through FRGS Vote no. 1415.

Reference

[1] Mohamed R M S R, Chan C M, Senin H, Kassim A H M 2014 Feasibility of the direct filtration over peat filter media for bathroom greywater treatment, J. Mater. Environ. Sci. **5** (6) 2021-2029.

[2] Adon R, Bakar I, Wijeyesekera D C, Zainorabidin A 2012 Overview of the Sustainable Uses of Peat Soil in Malaysia with Some Relevant Geotechnical Assessments,Iternational Journal of Integrated Engineering – Special Issue on ICONCEES Vol. **4** No. 3 p. 38-46.

[3] Salimin M I, Gandaseca S, Ahmed O H 2010 Comparison of selected chemical properties of Peat swamp soil before and after timber Harvesting, American Journal of Environmental Sciences **6** (2) 164-167.

[4] Ling F N L, Kassim K A, Karim A T A, Tiong K & Tan C K 2013 Geochemistry Properties of Southern Malaysian Organic Soil., Applied Mechanics and Materials, 284-287 1340–1344.

[5] Domeizel M, Khalil A and Prudent P UV spectroscopy: a tool for monitoring humification and for proposing an index of the maturity of compost, Bioresource Technology **94(2)**, (994) 177–184.

[6] Lu X Q, Hanna J V and Johnson W D 2004 Evidence of chemical pathways of humification: a study of aquatic humic substances heated at various temperatures, Chemical Geology. **177(3-4),** 249–264.

[7] Mutalib A A, lim J S, Wang M H, and Koonvai L 1991 Characterization, distribution and utilization of peat in Malaysia., In Proceeding of International Symposium on Tropical Peatland 6 – 10 May, 7-8.

[8] Ellerbrock R H, Kaiser M 2005 Stability and composition of different soluble soil organic matter fractions – evidence from d13C and FTIR signatures, Geoderma 128 28–37.

[9] Cocozza C, D'Orazio V, Miano T M 2003 Characterization of solid and aqueous phases of a peat bog profile using molecular fluorescence spectroscopy, ESR and FT-IR, and comparison with physical properties. Organic Geochemistry **34** 49–60.

[10] Mohamed R M S R, Chan C M, Senin H, Kassim A H M 2014 Feasibility of the direct filtration over peat filter media for bathroom greywater treatment, J. Mater. Environ. Sci. **5** (6) 2021-2029.

[11] Borgmark, A 2005 The colour of climate: changes in peat decomposition as a proxy for climate change – a study of raised bogs in South-central Sweden. PhD thesis, Stockholm: Stockholm University

[12] Biester H, Knorr K.-H, Schellekens J, Basler A 2014 Comparison of different methods to determine the degree of peat decomposition in peat bogs., Biogeosciences **11**, 2691–2707.

[13] Bhaskar J Saikia1, Gopalakrishnarao Parthasarathy. 2010 Fourier Transform Infrared Spectroscopic Characterization of Kaolinite from Assam and Meghalaya, Northeastern India, J. Mod. Phys. 206-210.

[14] Tatzber M, Stemmer M, H Spiegel., 2007 FTIR-spectroscopic characterization of humic acids and humin fractions obtained by advanced NaOH, Na4P2O7, and Na2CO3 extraction procedures., J. Plant Nutr. Soil Sci. **170** 522–529.

[15] Preston C M, Newman R H. 1995 A long term effect of N fertilization on the 13C CPMAS NMR of the de-ashed soil humin in a second-growth Douglas-fir stand of coastal British Columbia, Geoderma **68,** 229–241.

[16] Derek P., George W, Donald L 2003 ATR-FTIR spectroscopic studies of boric acid adsorption on hydrous ferric oxide, Geochimica et Cosmochimica Acta, Vol. **67**, No. 14, pp. 2551–2560.

[17] Song J, Peng P, Huang W 2005 Characterization of humic acid-like material isolated from the humin fraction of a topsoil, Soil Sci. **170**, 599–611.

[18] Paramananthan S 2010 Keys to the identification of Malaysian soils according to parent materials (Mimeo)., Param Agricultural Soil Surveys, Selangor

[19] Oskars P 2013 Comparative study of peat humic acids by using UV spectroscopy., 1st Annual International Interdisciplinary Conference, AIIC

[20] Changwen D, Jianmin Z, Huoyan W 2009 Determination of soil properties using Fourier transform mid-infrared photoacoustic spectroscopy Vibrational Spectroscopy **49** 32–37.

[21] Huat B B K., Organic and peat soils engineering. 2004 Universiti Putra Malaysia Press, Serdang, pp 146–150.

[22] Jarrett P M 1995 6. Site investigations for organic soils and peat. J.K.R. Document 20709-0341-95. Institut Kerja Raya Malaysia, Sarawak

[23] Robertson A H J, Hill H R, Main A M. 2003 Analysis of Soil in the Field using portable FTIR, International Workshop Soil Spectroscopy: the present and future of Soil Monitoring

Strength Development of Lime Treated Artificial Organic Soil

S W Yeo[1], F N L Ling[1], A Sulaeman[2], V S Low[1], K L Toh[1]
[1] Research Center for Soft Soil, Universiti Tun Hussein Onn Malaysia, 86400 Batu Pahat Johor, MALAYSIA
[2] Faculty of Civil and Environmental Engineering, Universiti Tun Hussein Onn Malaysia, 86400 Batu Pahat Johor, MALAYSIA

E-mail: shiweiyeo8@gmail.com

Abstract. In over many years, considerable research has been carried out on organic soils which consists of various components of organic matter but the effect of particular organic matter is less reported. Thus, some of contributing factors for each organic matter are not fully understood yet. Hence, the aim of this study is to determine the effect of organic acid concentration on the strength of artificial organic soil. There are four types of artificial organic soil created by mixing kaolin (inorganic matter) and organic acid (a kind of humified organic matter) in different concentrations. Unconfined Compressive Strength test (UCT) was carried out for all soil samples after being cured for 7 and 28 days under room temperature and 50°C. Soil samples shows highest strength when cured for 28 days under 50°C compared to those cured under room temperature. However, when the organic acid concentration decrease, the strength increased for soil 2 after 7 and 28 days cured under room temperature and 50°C. Apart from this, soil 3 and soil 4 that were cured under room temperature shows decrease in strength when the organic acid concentration decreasing but different result shown for both samples when cured under 50°C.

Keywords: Soft Soil, kaolin, organic acid, fulvic acid, humic acid, lime.

1. Introduction

Organic soils have not received attention in civil engineering due to the presence of organic matter in soils associated with high compressibility, high water content, low permeability and low shear strength. These characteristics have caused organic soils become unsuitable for engineering construction purpose. A suitable improvement method is needed to prevent excessive settlement during construction work also to enhance the shear strength and bearing capacity of the soil.

Nowadays, soil stabilization is one of the very importance knowledge in the civil engineering field because all building structures and infrastructures will stand upon the soil as the base and platform. So, strength of the soil must be strong enough in order to ensure the properties can withstand safely without damages. With the high level efficiency of soil stabilization, it does not only protects against those structures mentioned above from harmful but also against many things like erosion in many different places, economic design for sub-grade that build on soft soil and environmental concerned.

In general, lime can be used to treat soils in order to improve their workability and load-bearing characteristics. Lime is an excellent choice for short-term modification of soil properties. Pe This reaction produces stable calcium silicate hydrates and calcium aluminate hydrates as the calcium from the lime reacts with the aluminates and silicates solubilized from the clay.

This paper will discuss about the result of tests carried out on artificial organic soil mixed of kaolin (inorganic matter) and organic acid (organic matter) in different concentration. At the same time, the strength of artificial organic soils cured under room temperature and 50°C were determined. Hence, the correlation of organic acid concentration on the strength of artificial organic soils is summarized.

2. Material and Methods

The soils used in this study were artificial organic soils. The soil samples were prepared by mixing kaolin grade S300 as an inorganic component and mixed of 15% of organic acid (containing 0.37% of humic acid and 46.60% of fulvic acid) as the organic components. These artificial organic soils were prepared in 4 different dilution factors. Table 1 below shows the mixed proportion for the soil samples used in this study. Nonetheless, lime were the addictive for remolding samples. The curing time 7 and 28 days are required for the lime-treated soil samples. Not only than that, the curing temperature used was different for the remolded sample which were under room temperature and 50°C. It is believed that samples that is cured under 50°C will accelerated the hardening process.

Table 1. Type of soil.

Material	Percentage required for each samples (%)			
	Soil 1	Soil 2	Soil 3	Soil 4
Kaolin S300	85	85	85	85
Organic acid	15	7.5	5	3.75
Distilled water	0	7.5	10	11.25

Initial consumption of lime was carried out according to [1] to indicate the sufficient amount of lime and to ensure the pH during the reaction is 12.4 at 25°C. Initial consumption of lime (ICL) test was conducted as a primary step to identify the minimum amount of modification of lime added to artificial organic soil had to be determined. This was essential to ensure the amount of lime added to organic soil is sufficient for the occurrence of modification stabilization process.

Standard Proctor Compaction test was conducted in accordance to BS 1377: Part 4: 1990 with the aim to identify the maximum dry density and optimum moisture content of the artificial organic soil. Then, test specimens were remolded based on [2] with dimension 50mm of diameter and 100mm in height. There were a total of 48 samples remolded in this study. 24 units of the samples were cured under room temperature and another 24 units were cured in an oven with constant temperature of 50°C. Lastly, unconfined compression test (UCT) was conducted for all the test specimens after 7 and 28 days for both curing temperature. This testing used for determine the shear strength of the soil.

3. Results and Discussion

Based on Figure 1, percentage of lime 1.5% was chosen for the stabilization process in this study as more than 1.5% of lime content has been achieved pH of 12.4 among every ICL test data and to ensure a long term strength gain achieved from the stabilization process. Besides that, constant lime content for all specimens was for the purposes of eliminating multiple variables occurred in test and effect of physical and geochemical properties to the strength might hard to found.

Results of Unconfined Compressive Test are shown in Table 2. According to the data collected, overall strength of soils showing that after curing for 28 days has higher than 7 days curing. It is clearly shown in Figures 2 and 3. Either the soil specimens is placed on room temperature or 50°C, the unconfined compressive strength for specimens cured after 28 days are higher than 7 days.

Nonetheless, based on the Figure 4, for soil samples cured under room temperature recorded that the strength of soil samples decrease when increase the dilution factor of organic acids, but only soil 2 had the highest strength among the four samples. However, for the samples cured at 50°C, soil 2 had the highest strength also. By referring to the Figure 5, the graph shows the strength for soils samples after curing 28 days under different temperature. It can be clearly seen that strength for soil samples cured under 50°C are higher than cured under room temperature. According to [3], strength of soils can be improved when increasing the temperature and extended the curing time. The author had found that the strength of the soil-lime mixtures increased substantially when the curing temperature is doubling from 20°C to 40°C. This is because the pozzolanic reactions are promoted. Another study from [4] also stated that, rate of formation cementitous compounds that binding soil aggregate together will be affected by the temperature.

Table 2. Result of Unconfined Compression Strength Test after curing.

Soil Sample	Curing Temperature	Unconfined Compressive Strength (kPa)		
		7 Days Curing	28 Days Curing	Increment (%)
Soil 1	Room temperature	168.83	168.10	-0.43
	50°C	161.51	239.60	78.09
Soil 2	Room temperature	171.07	226.07	32.15
	50°C	284.75	354.36	69.61
Soil 3	Room temperature	143.43	191.97	33.84
	50°C	136.09	204.42	68.33
Soil 4	Room temperature	120.06	157.67	31.32
	50°C	177.16	322.03	144.87

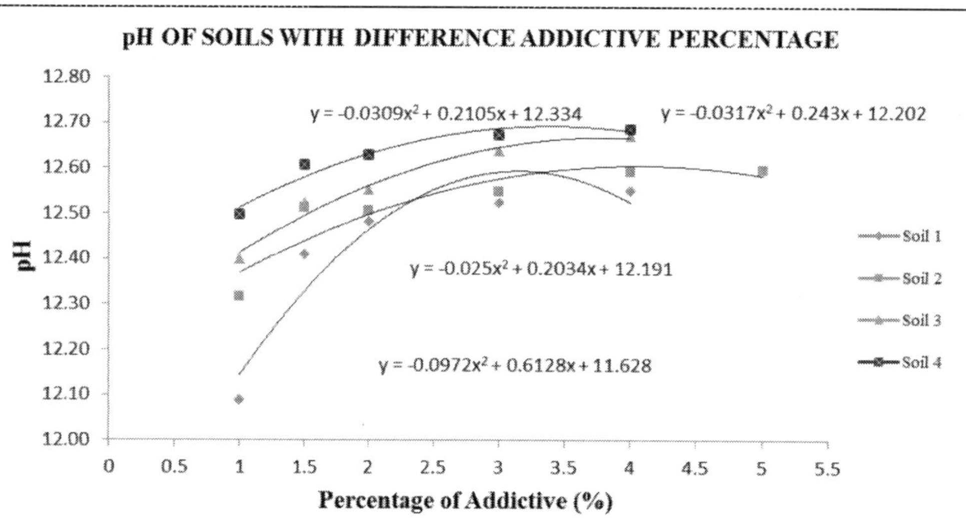

Figure 1. pH of soils with different addictive percentage

IOP Conf. Series: Materials Science and Engineering **136** (2016) 012011 doi:10.1088/1757-899X/136/1/012011

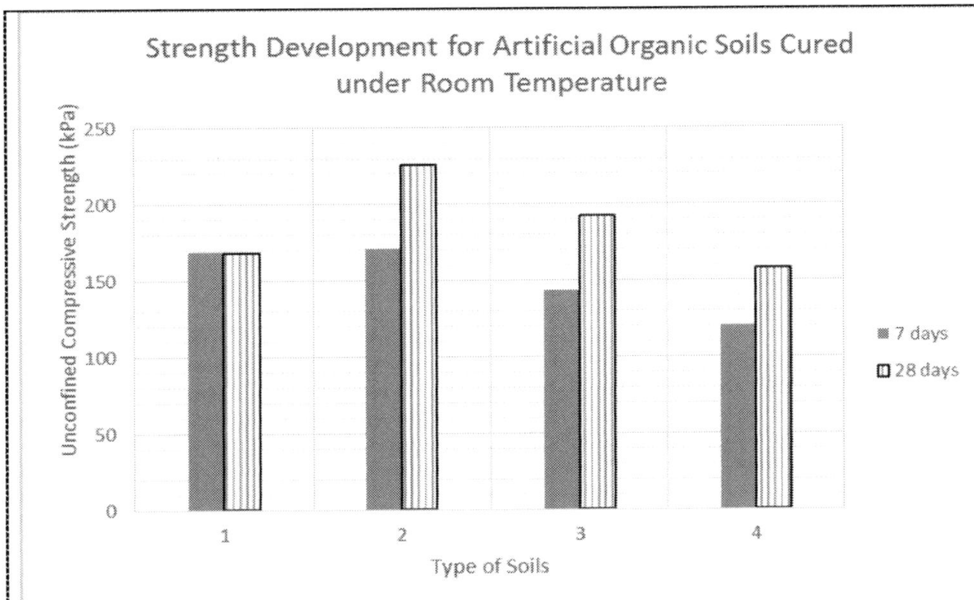

Figure 2. Uncomfined Compression Strength for artificial organic soils cured under room temperature after 7 and 28 days.

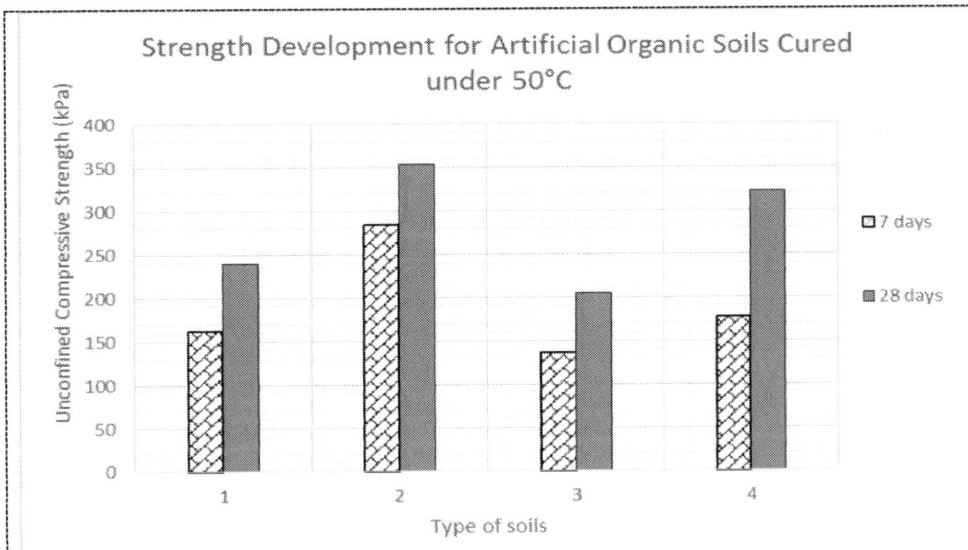

Figure 3. Uncomfined Compression Strength for artificial organic soils cured under 50°C after 7 and 28 days.

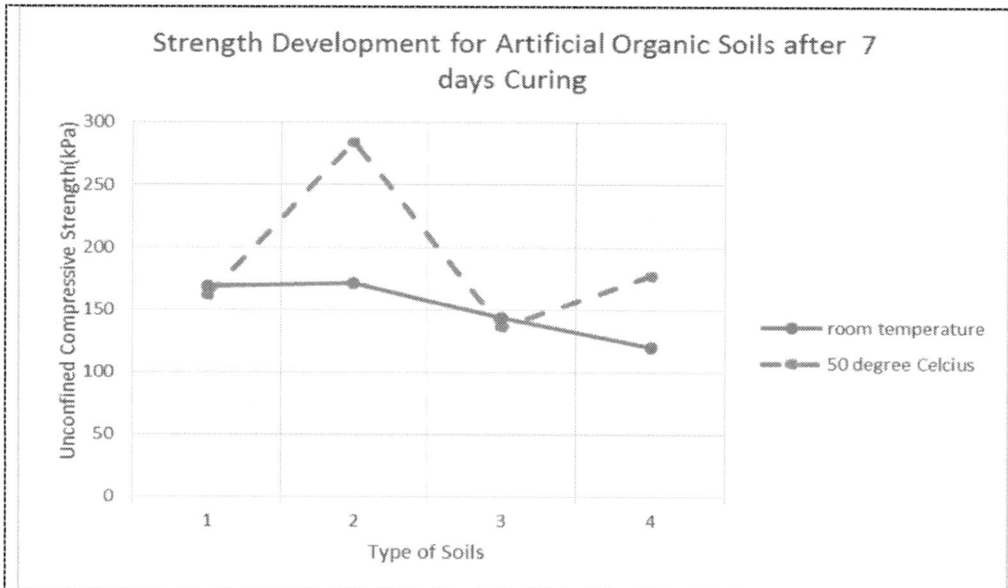

Figure 4. Uncomfined Compression Strength for artificial organic soils cured under room temperature and 50°C with additive after 7 days curing.

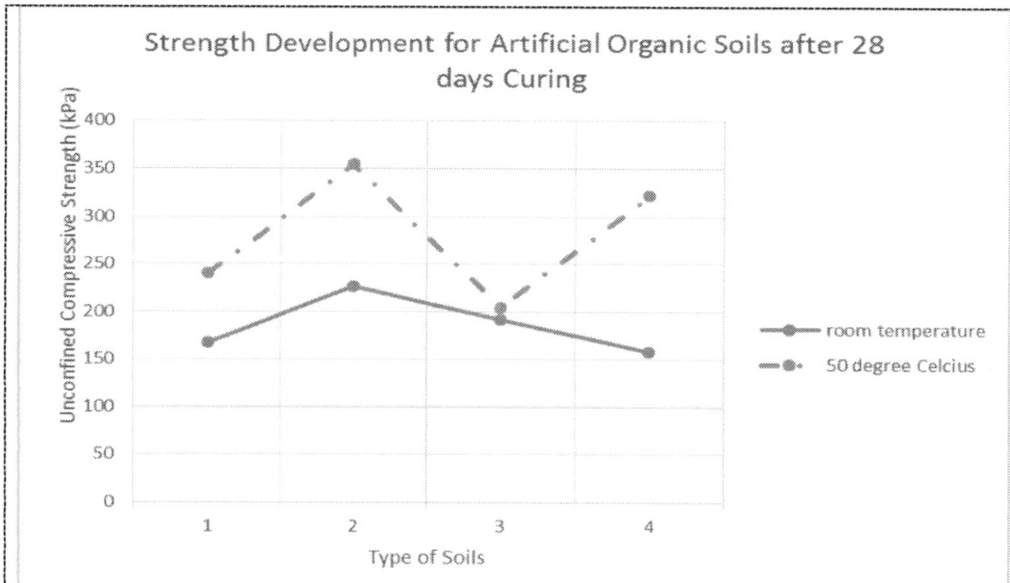

Figure 5. Uncomfined Compression Strength for artificial organic soils cured under room temperature and 50°C with additive after 28 days curing.

5. Conclusion

The curing periods had contribute in the soil strength enhancement. Soil samples cured under 28 days have given higher strength compare to 7 days curing. Soil 2 had contribute to higher strength among the 4 samples. However, soil 3 and 4 had lower strength compared to others when cured under 50°C and room temperature respectively. Besides that, among the two condition curing temperature of soil samples, room temperature and 50°C shown the significant differences in term of strength after curing for a given period where the strength of soil samples that cured under 50°C was higher than samples cured under room temperature. Hence, different dilution factor for organic acid used will influenced not only the physical and chemical properties of soils, it will also influenced the strength of the soil. Although the strength of soil samples do not shows a significant increasing or decreasing in term of dilution factor increasing, but it was believed that this may cause by the behavior of soil itself when reacts with the organic acid. The internal bonding of the soil particles have been affected through the concentration of organic acid added.

Acknowledgement

The authors wish to express grateful and thankful appreciate to the financial support from University Tun Hussein Onn Malaysia through the Centre for Graduate Studies. The authors also thank to all laboratory technicians in Research Centre of Soft Soil (RECESS), Department of Infrastructure and Geomatic, Department of Water and Environmental in University Tun Hussein Onn Malaysia. Their advise, knowledge and technical support had made this research possible.

References

[1] British Standard Methods of Stabilized Materials for Civil Engineering Purposes, BS 1924 1990 Methods of test for cement-stabilized and lime-stabilized materials

[2] British Standard Methods of Unbound and Hydraulically Bound Mixtures, BS EN 13826-53 2004 Methods for the Manufacture of Test Specimens of Hydraulically Bound Mixtures using Axial Compression

[3] Bell F G 1996 Lime stabilization of clay minerals and soils. Enginerring Geology 42 223-237

[4] Kolawole J, Charles M O 2006 Compaction Delay Effects on Properties of Lime-Treated Soil Journal of Materials in Civil Engineering © ASCE 250-258

[5] Mohd Yunus N Z, Darius W, Rod S, Aminaton M, Rini Asnida A and Nord M 2011 A Short Review of Factors Influencing Lime-Clay Reactions Electronic Joural of Geotechnical Engineering 19 8305-8318

[6] Sudhakar M R, Shivananda P 2003 Role of Curing Temperature in Progress of Lime Reaction. Geotechnical and Geological Engineering © Springer 23 79-85.

[7] British Standard Methods of Tests for Soils for Civil Engineering Purposes, BS 1377. 1990 London: British Standard Institution.

Impact of Oriented Clay Particles on X-Ray Spectroscopy Analysis

A J M S Lim[1,2], R N Syazwani[1], D C Wijeyesekera[1,2]
[1]F Faculty of Civil and Environmental Engineering, Universiti Tun Hussein Onn Malaysia, 86400 Batu Pahat Johor, MALAYSIA
[2]Research Center for Soft Soil, Universiti Tun Hussein Onn Malaysia, 86400 Batu Pahat Johor, MALAYSIA

E-mail: [a]alvinjlms@gmail.com

Abstract. Understanding the engineering properties of the mineralogy and microfabic of clayey soils is very complex and thus very difficult for soil characterization. Micromechanics of soils recognize that the micro structure and mineralogy of clay have a significant influence on its engineering behaviour. To achieve a more reliable quantitative evaluation of clay mineralogy, a proper sample preparation technique for quantitative clay mineral analysis is necessary. This paper presents the quantitative evaluation of elemental analysis and chemical characterization of oriented and random oriented clay particles using X-ray spectroscopy. Three different types of clays namely marine clay, bentonite and kaolin clay were studied. The oriented samples were prepared by placing the dispersed clay in water and left to settle on porous ceramic tiles by applying a relatively weak suction through a vacuum pump. Images form a Scanning Electron Microscope (SEM) was also used to show the comparison between the orientation patterns of both the sample preparation techniques. From the quantitative analysis of the X-ray spectroscopy, oriented sampling method showed more accuracy in identifying mineral deposits, because it produced better peak intensity on the spectrum and more mineral content can be identified compared to randomly oriented samples.

Keywords: Clay particle, mineralogy, x-ray spectroscopy.

1. Introduction

Clay minerals are the most important chemical and dominating weathering product in the soil. They are formed by the alteration of existing minerals or by synthesis from elements when minerals weather to their elemental form [1]. Soils differ from other engineering materials because of its morphology characteristics which require a distributed particulate mechanics approach [2]. Specifically, clay mineral particles have commonly platy characteristics, the use of flat sample preparations for X-Ray Stereoscopy or X-Ray diffraction analysis are achieved from the naturally occurring random orientation of the clay particles. It is known that X-ray diffracted intensities from powdered samples depend on the orientation of the particles [3]. In theory, random orientation of powder form clay particles provides the maximum number of basal reflections but also gives a maximum overlapping reflections and this may cause reliable quantitative analysis inconveniently arduous [4]. Although analysis can be made with overlapping reflections, it is very convenient to use non-overlapping reflections. Non-overlapping reflections provides better peak intensities for X-Ray diffraction. However, there is still a dearth of published information on its effect on the elemental composition of clay particles.

This study aims to contribute by producing a well-crystallized and well oriented particle samples using appropriate techniques. The understanding of the quantitative evaluation of clay mineralogy is important for fundamental studies as it can be used to identify suitable methods for further civil engineering experiments. In this study, the oriented samples were prepared by placing dispersed clay in water and leave it to settle on a porous ceramic tile while applying a relatively weak vacuum suction to enhance the free fall to stable state of the particles. With that, the comparison on the elemental and mineral compounds with micro-imaging analysis between the random oriented and well oriented samples using X-ray stereoscopy and Scanning Electron Microscopy (SEM) was carried out. SEM image analysis can provide a better idea on how the clay particles are arranged between the two techniques. Ultimately, a thorough study on the impact of particle orientation on the elemental composition of the elements and its mineral compound can be achieved.

2. Materials and Method

In this study, three types of clay samples were used which are kaolin, bentonite and marine clay. The samples were first sieved through 63μm sieve (No. 200-mesh) to separate irrelevant non clay sized materials and turning it to a powdery form. This therefore maximized the amount of clay particles as it is defined as having sizes finer than 0.002mm [5]. The analysis was done on the Scanning Electron Microscope equipped with the Energy Dispersive X-Ray apparatus (SEM-EDX). This apparatus was used due to its capability of combining micro imaging and elemental compound distribution characterization analysis. Its characterization capabilities are due to its fundamental principle that each element has a unique atomic structure allowing unique set of peaks on its X-ray emission spectrum [6].

Sample preparation was done carefully to achieve the required orientation of the particles. The random oriented samples were attained by applying the clay powder directly flat unto the specimen holder of the SEM-EDX apparatus. Well oriented samples were prepared by adopting the method as described by Shultz [7] and Kinter and Diamond [8]. Although the methods and apparatuses have considerably changed throughout the years, the concept of the test is still the same. In this study, the vacuum filtration technique (Figure 1) rather than centrifugal settling was used for separating approximately 1 gram of clay solid from a solvent and collecting the fraction on a porous disc. The clay fraction is seen to have a thin smooth layer on top of the porous disc. The thickness of the oriented clay was less than 1 mm. It was then dried at room temperature and left carefully undisturbed before it is taken for the SEM-EDX analysis.

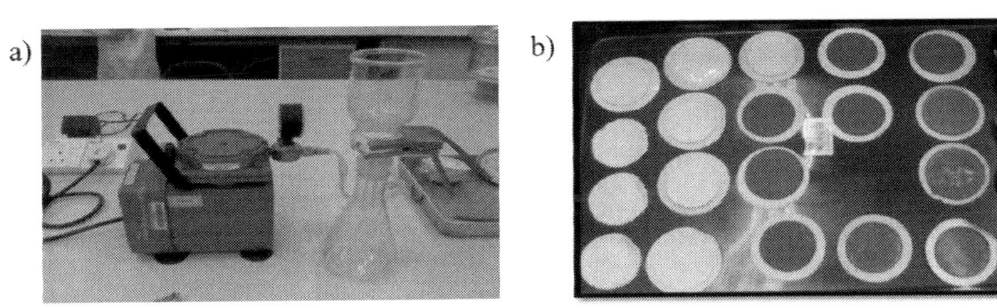

Figure 1. (a) Vacuum with porous disc for filtration (b) Smooth layer on porous disc.

3. Results and Discussions

The EDX analysis will produce results that shows the percentage by mass of the chemical elements and the mineral compounds respectively. Table 1 shows the SEM images of all the non-oriented and oriented clay samples. The magnification of the images was up to x850, where the non-oriented clay samples seem to show that the particle assemblage is unstructured. The oriented samples showed that the clay particles were more compacted and well structured. With that, the high-energy beam from the spectroscopy is assumed to be focused consistently on the smooth surface of the oriented particle of clay sample.

IOP Conf. Series: Materials Science and Engineering **136** (2016) 012012 doi:10.1088/1757-899X/136/1/012012

Table 1. SEM image analysis and the peaks of the X-ray Spectrum of the Element of Koalin, Bentonite and Marine Clay respectively.

Kaolin	**Bentonite**	**Marine Clay**
Non-oriented clay particles	Non oriented particles are unstructured and uneven	

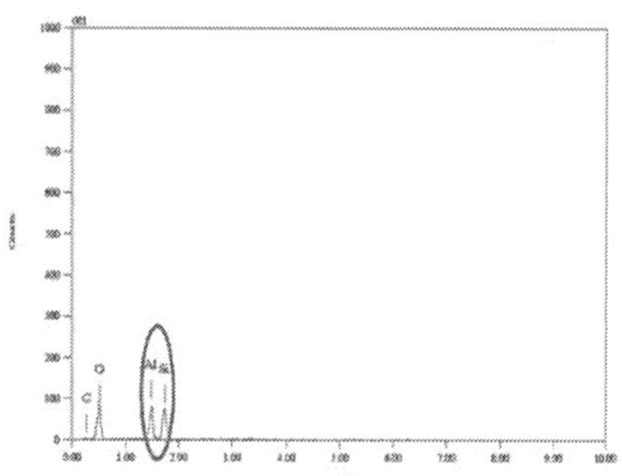	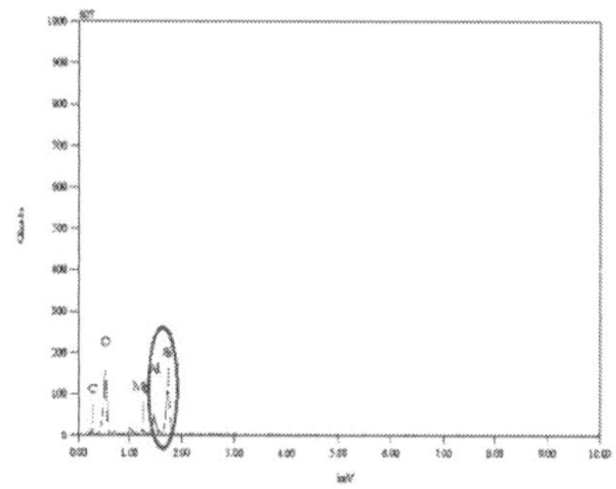	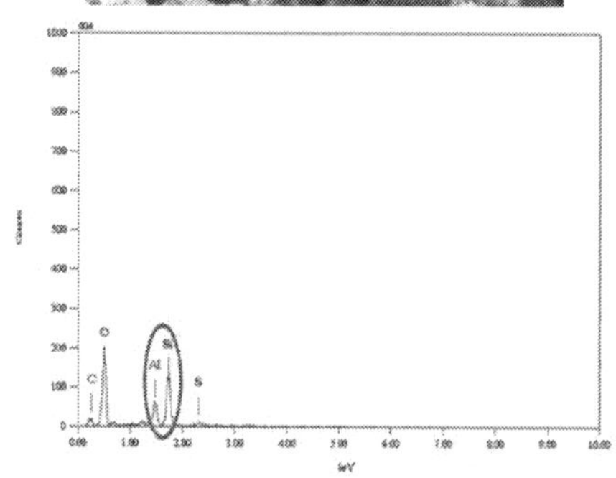
Oriented clay particles	Oriented particles are well structured and more compacted	
	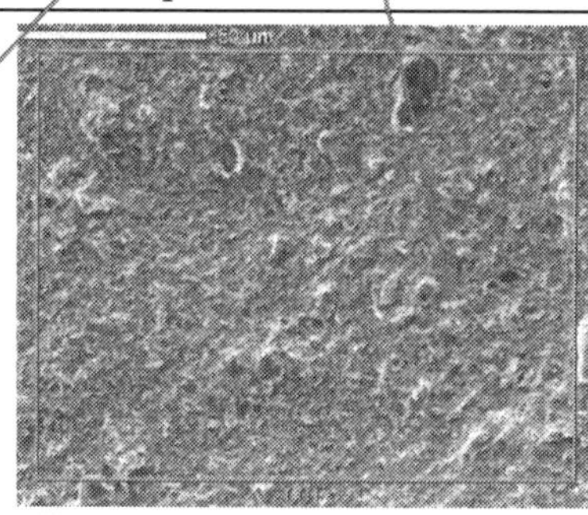	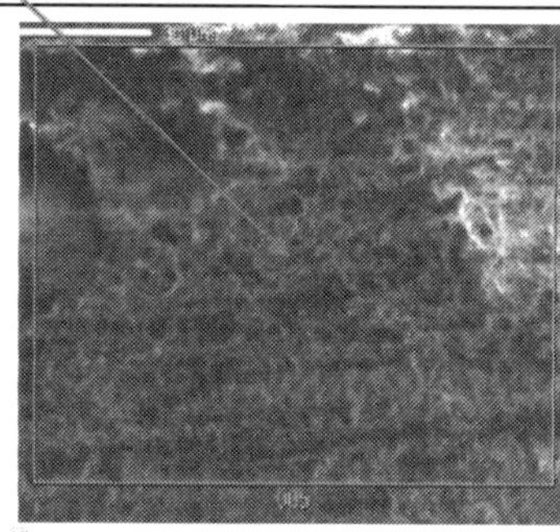
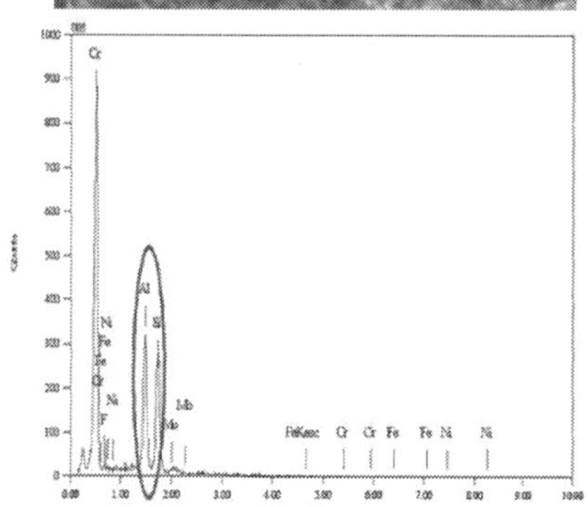	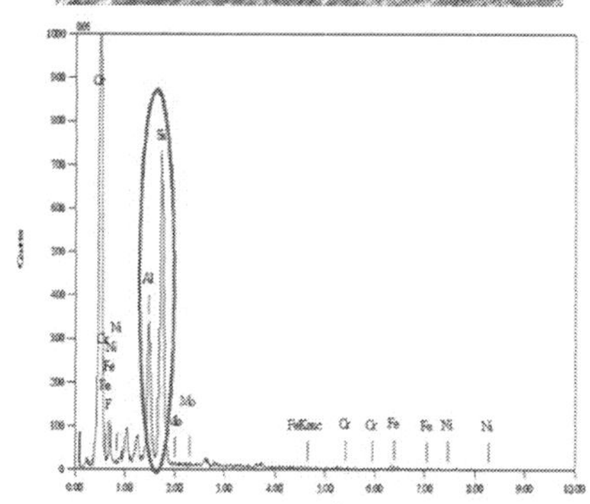	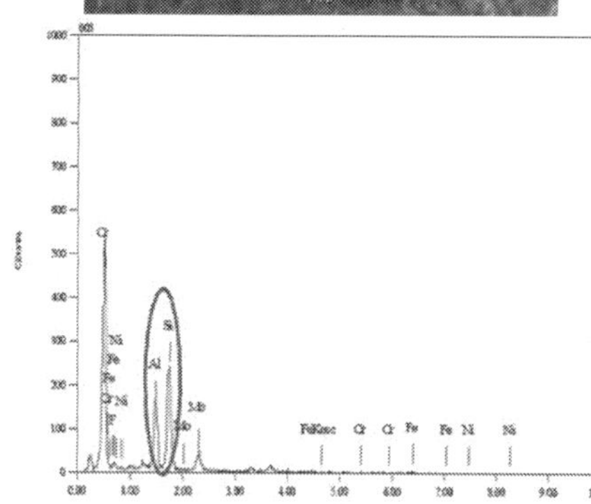

The table also confirms that the peaks of the X-ray emission spectra demonstrate convincingly that the oriented clay particles have an increased in the basal peak intensity of the Aluminum (Al) and Silicon (Si) elements as compared to the non-oriented clay particles. Figures 2 to 3 show the percentage by mass of the chemical elements and the mineral compounds found present within the Kaolin, bentonite and marine clay samples respectively.

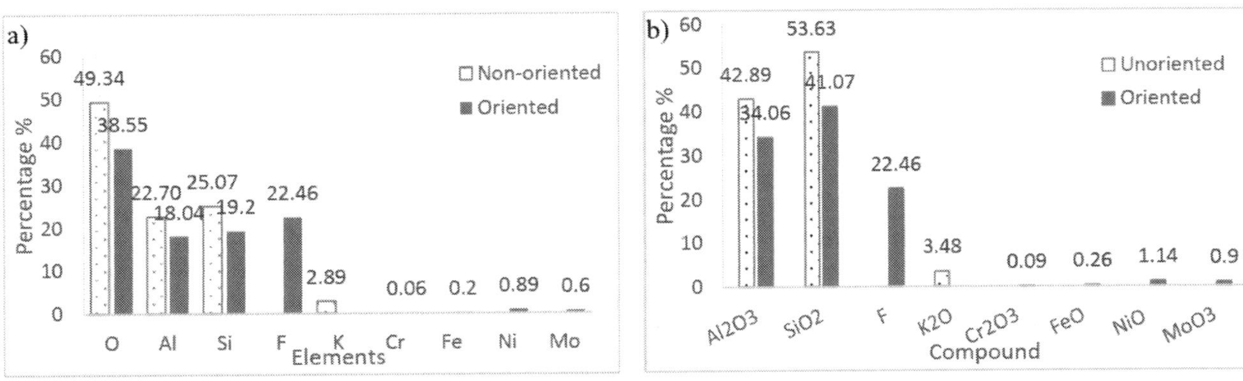

Figure 2. Percentage by mass of (a) chemical element and (b) mineral compounds of Kaolin.

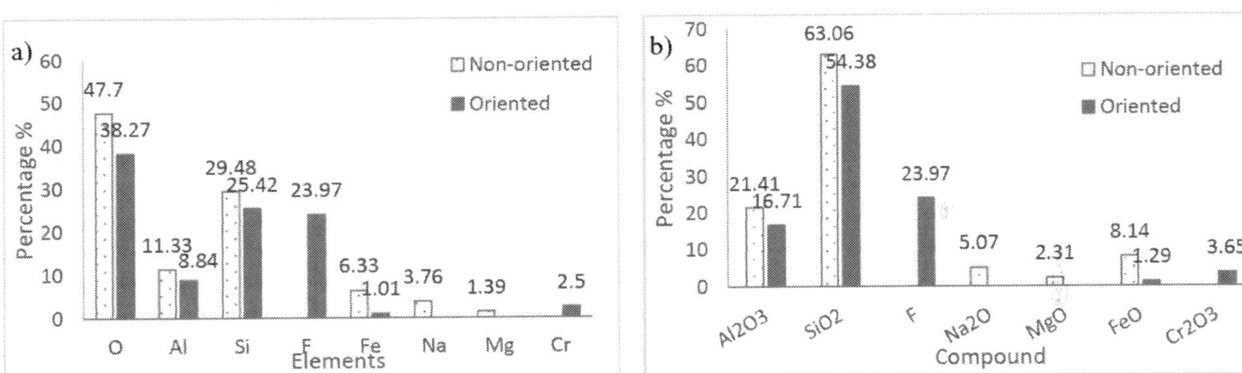

Figure 3. Percentage by mass of (a) chemical element and (b) mineral compounds of Bentonite.

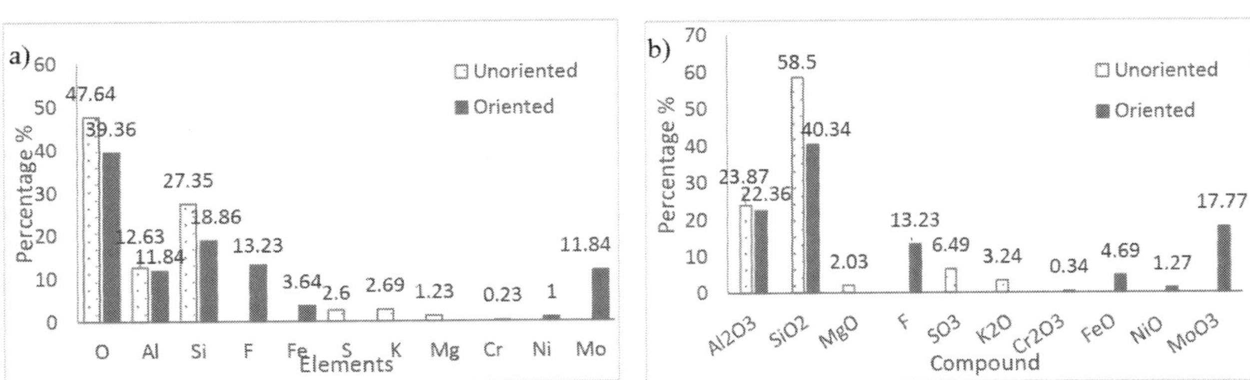

Figure 4. Percentage by mass of (a) chemical element and (b) mineral compounds of Marine Clay.

For all the samples, the most dominant elements by percentage of mass are Al, Si and Oxygen (O). These being the dominant elements in all clay minerals. However, oriented samples showed lower percentages of Al, Si and O to compensate for the small additional traces of other elements such as Chromium (Cr), Iron (Fe), Nickel (Ni) and Molybdenum (Mo). Furthermore, the amount of mineral compound by mass percentage can also be determined. The most dominant compound for all the samples was found to be Silicon Oxide (SiO2) and Aluminum Oxide (Al2O3). Again, additional mineral compounds were found for the oriented clays particles such as Chromium Oxide (Cr2O3), Iron Oxide (FeO), Nickel Oxide (NiO), and Molybdenum Oxide (MoO3). The analysis also showed that the dominant mineral compounds of SiO2 and Al2O3 for all the samples are almost similar where seemed to comprise of more than 15% of its mass. This is in accordance with the typical mineralogical content of clayey soils.

4. Conclusion

From this study, it was found that the vacuum filtration technique to create a solid form of clay solvent to settle on a porous disc with suction allows the clay particles to be in an oriented pattern. Free falling of sediments occur in the natural gravitational compaction of clay. This is often hindered by the presence of coarse grained particle and by a chemical environment that causes the electricity charged fine particles to flocculate rather than be dispersed and form a pack of cards type structure. Image analysis from SEM showed that the oriented particles have a sample surface of a more structured and compacted appearances, compared to the non-oriented clay particles which is seen to be more scattered and less dense. From the EDX analysis, it was found that both the oriented and non-oriented clay particles identified the dominant chemical element and mineral compound. However, the oriented clay particles showed additional elements and minerals in the samples. Therefore, in order to achieve a more accurate quantitative evaluation of clay mineralogy, a proper technique of obtaining oriented soil particles is essential in sample preparation for quantitative mineral analysis.

Acknowledgement

The author would appreciate the staffs of RECESS for their contribution and would also like to thank ORICC UTHM under the grant Vot U343 for their financial support.

References

[1] Mizota C and Reeuwijk L P v 1989. Clay Mineralogy and Chemistry of Soils Formed in Volcanic Material in Diverse Climatic Regions. Soil Monograph 2, ISRIC, Wageningen.

[2] Lim A J M S, Wijeyesekera D C and Shukri A Y 2013 Advanced Statistical Analysis for Relationships between Particle Morphology (size and shape) and Shear (static and dynamic) Characteristics of Sands. International Journal of Geosciences. **4(10A)** 27-36.

[3] Brindley G W 1980 Quantitative X-Ray Mineral Analysis of Clays. Crystal Structures of Clay Minerals and Their X-Ray Identification. Mineralogical Society, London, 411-438.

[4] Srodon J, Dritis V A, McCarty D K, Hsieh J C C, and Eberl D D 2001 Quantitative X-Ray Diffraction Analysis of Clay Bearing Rocks from Random Orientation. Clays and Clay Minerals, **49(6)** 514 – 528

[5] Das B M 2014 Principles of Geotechnical Engineering. Cengage Learning. USA.

[6] Goldstein J 2003 Scanning Electron Microscopy and X-Ray Microanalysis. Third Edition, Springer, USA.

[7] Schultz L G 1964 Quantitative Interpretation of Mineral Composition from X-Ray and Chemical Data for the Pierre Shale. Geological Survey Professional Paper 391-C.

[8] Kinter E B and Diamond S 1956 A New Method for Preparation and Treatment of Oriented-Aggregate Specimens of Soil Clays for X-Ray Diffraction Analysis. Soil Science, **81(2)** 111-120.

The Soil-Water Characteristic Curve of Unsaturated Tropical Residual Soil

M F Yusof[1], A S Setapa[2], S A A Tajudin[3], A Madun[2,1], M H Z Abidin[2,1] and A Marto[2]

[1] Research Center for Soft Soil, Universiti Tun Hussein Onn Malaysia, 86400 Batu Pahat Johor, MALAYSIA

[2] Faculty of Civil Engineering, Universiti Teknologi Malaysia, 81310 Skudai, Johor, MALAYSIA

[3] Faculty of Civil and Environmental Engineering, Universiti Tun Hussein Onn Malaysia, 86400 Batu Pahat Johor, MALAYSIA

E-mail: fairus@uthm.edu.my

Abstract. This study was conducted to determine the SWCC of unsaturated tropical residual soil in Kuala Lumpur, Malaysia. Undisturbed soil samples at five locations of high-risk slopes area were taken at a depth of 0.5 m using block sampler. In the determination of the SWCC, the pressure plate extractor with the capacity of 1500 kN/m2 has been used. The index properties of the soil such as natural moisture content, Atterberg limits, specific gravity, and soil classification are performed according to BS 1377: Part 2: 1990. The results of index properties show that the natural moisture content of the soil is between 36% to 46%, the plasticity index is between 10% - 26%, the specific gravity is between 2.51 - 2.61 and the soils is classified as silty organic clay of low plasticity. The SWCC data from the pressure plate extractor have been fitted with the Fredlund and Xing equation. The results show that the air entry value and residual matric suction for residual soils are in the range of 17 kN/m2 to 24 kN/m2 and 145 kN/m2 to 225 kN/m2 respectively. From the fitting curve, it is found that the average value of the Fredlund and Xing parameters such as a, n and m are in the range of 0.24-0.299, 1.7-4.8 and 0.142-0.440 respectively.

Keywords: Unsaturated soil, soil water characteristic curve.

1. Introduction

More than 70% of the land area in Malaysia covered by residual soil [1]. The properties of residual soil have been investigated by many researchers regarding saturated condition [2,3]. However, research work on unsaturated properties of residual soil in Malaysia is still lacking. Recently some investigations have been performed after very serious slope stability problems occurred [4]. The landslides often occur, especially during the rainy season, resulting in property damage, sometimes causing injury and fatality.

The high intensity of rainfalls infiltrates into the soil led to loss of the soil suction and hence reduce the additional shear strength that exists in unsaturated soils. Most of the designer, prefer to use saturated parameters in their design of the slope by assuming that the slope is in saturated condition due to their simplicity. But, this is not true in real situations where most of the slopes are in partially saturated. In designing the slope, they have to consider the effect of the soil suction which is the important properties of unsaturated soils. One of the basic parameters of unsaturated soil is the SWCC. From the SWCC, the unsaturated shear strength and unsaturated permeability can be estimated. This study has been conducted to determine the SWCC of unsaturated soil at high risk slope area. Then, the laboratory data were fitted according to Fredlund and Xing function [5].

1.1 Soil-Water Characteristics Curves (SWCC)

SWCC is the relationship between the moisture content and suction which is the key in the implementation of unsaturated soil mechanics in engineering practices [6]. Initially the SWCC was used in estimating the in situ soil suction by measuring its natural moisture content. Then, the SWCC has been used as an indirect method to predict the unsaturated soil properties such as unsaturated permeability, shear strength and volume change which are expensive and time consuming to measure in the laboratories [6]. The typical SWCC for drying curve is shown in Fig.1. The SWCC consist of air entry value –AEV (□a) and residual matric suction (□r) where the AEV is the suction that air start to enter the voids in the soil and the residual matric suction is the suction that water start to exist in discontinuous phase.

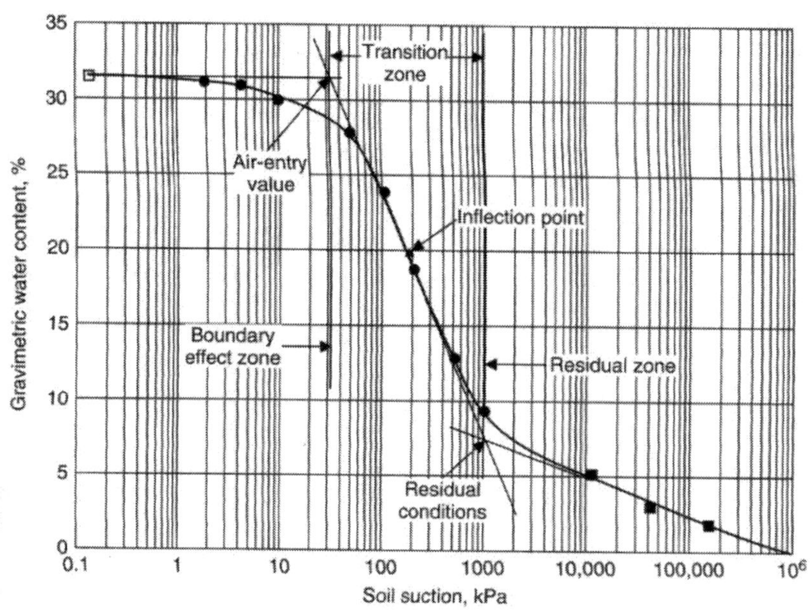

Figure 1. Typical SWCC – drying curve [7].

There are many mathematical equations have been proposed to describe the SWCC but the best SWCC model is Fredlund and Xing equation [5]. Therefore in this study, the SWCC data were fitted with the Fredlund and Xing equation.

2. Laboratory testing

Residual soil samples used in this study were derived from five locations of the high risk slope area in Kuala Lumpur, according to the map issued by the JKR National Slope Master Plan, (2009-2023). The exact locations of the soil samples were Kuala Lumpur tower, Bukit Damansara, Bukit Dinding, Desa Putra and Bukit Kiara. All the undisturbed soil samples were taken at a depth of 0.5 m using block sampler.

The index properties of the samples such as particle size distribution and Atterberg limits were performed according to BS 1377 at the Geotechnical Laboratory, Faculty of Civil and Environmental Engineering, UTHM. Three specimens for each location were prepared for the determination of SWCC using pressure plate extractor with the capacity of 1500 kN/m2 accordance to ASTM D6836-02 (2008).

3. Results and discussion

Table 1 show the results of the index properties of the soil samples from five locations in Kuala Lumpur. The samples were categorized as from low to high plasticity. The specific gravity are in the range of 2.51 – 2.62. The highest percentage of clay are from Kiara, which is in line with the highest value of the liquid limit.

Table 1. Index Properties of the soils

Parameters	Location				
	Pantai Dalam	Kiara	KL Tower	Bukit Dinding	Desa Putra
Liquid Limit	27	60	50.2	53.4	40
Plastic Limit	17	39	31	27	26
Plasticity Index	10	21	19	26	14
Specific Gravity	2.62	2.60	2.51	2.61	2.61
Gravel (%)	18	4	13	10	9
Sand (%)	46	47	45	47	36
Silt (%)	26	21	24	26	42
Clay (%)	10	28	18	17	13

The results of the SWCC from Pantai Dalam, Kiara, KL Tower, Bukit Dinding and Desa Putra are presented in Fig. 1, Fig.2, Fig.3, Fig. 4 and Fig.5 respectively. The average of measured SWCCs were fitted with the Fredlund and Xing function [5] as in Equation (1) by assuming that $C(\psi)$ is equal to 1 as recommended by Leong and Rahardjo [8].

$$\theta_w = \frac{\theta_s}{\left\{ \ln\left[e + \left(\frac{\psi}{a} \right)^n \right] \right\}^m} \quad (1)$$

where θ_w = volumetric water content; θ_s = saturated volumetric water content; ψ = soil suction (kN/m2); a,b and c are constants.

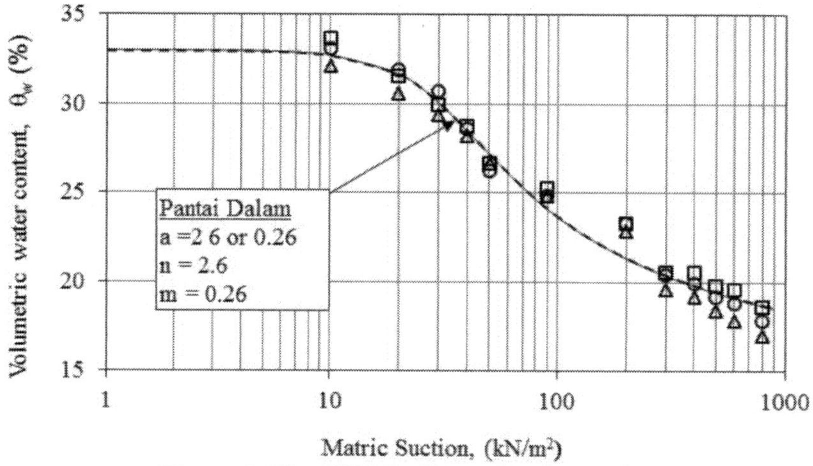

Figure 1. The SWCC of Pantai Dalam soils.

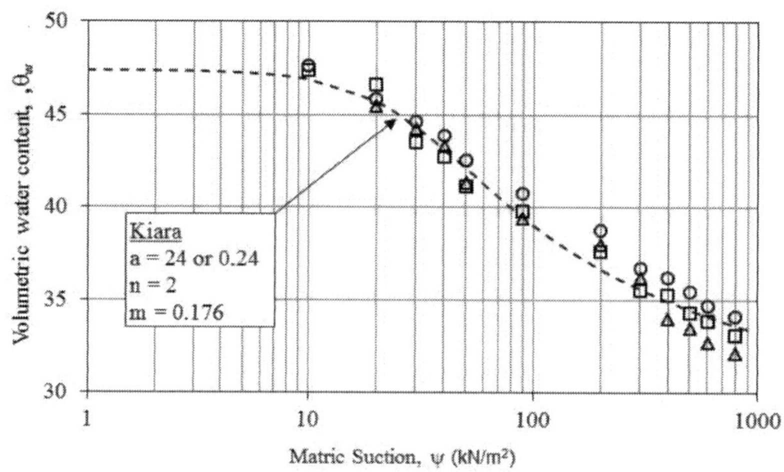

Figure 2. The SWCC of Kiara soils.

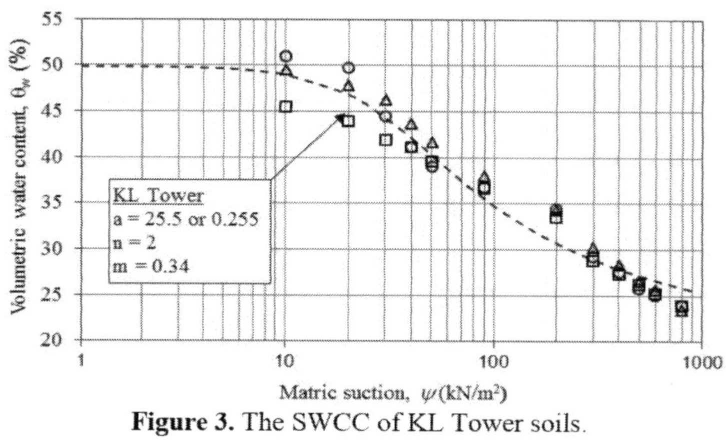

Figure 3. The SWCC of KL Tower soils.

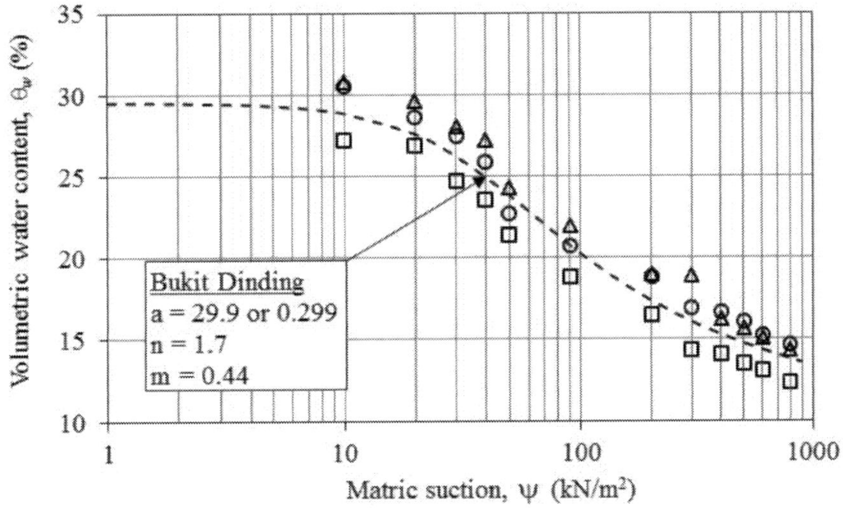

Figure 4. The SWCC of Bukit Dinding soils.

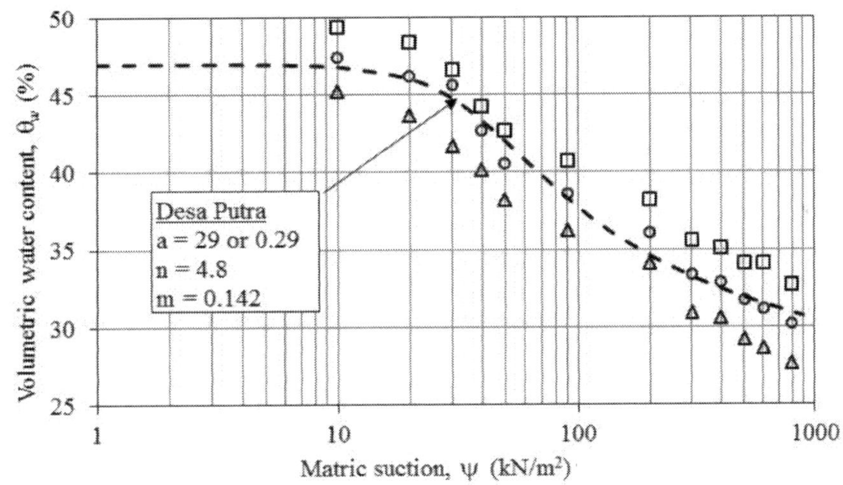

Figure 5. The SWCC of Desa Putra soils.

Table 2 shows the properties of SWCC based on the average value. The air start to enter the soil sample when the pressure are in the range of 17 kN/m2 to 24 kN/m2 and the residual matric suction are in the range of 145 kN/m2 to 225 kN/m2. In terms of volumetric water content, the saturated and residual values are in the range of 0.296-0.497 and 0.165-0.363 respectively. The results show good agreement in term of volumetric water content with Indrawan et al. [9] Where the samples were obtained from the Singapore residual soil.

Table 2. SWCC properties.

Location	Air entry value, ψ_a (kN/m^2)	Residual matric suction, ψ_r (kN/m^2)	Saturated volumetric water content, θ_s	Residual volumetric water content, θ_r
Pantai Dalam	19.5	145	0.329	0.213
Kiara	17.0	190	0.473	0.363
KL Tower	18.0	193	0.497	0.275
Bukit Dinding	17.0	225	0.296	0.165
Desa Putra	24.0	170	0.470	0.340
Indrawan et al [5]	4.687	14.88	0.432	0.232

Table 3 shows the value of fitting parameters (*a, n* and *m*). The a value is in the range of 0.24 – 0.299, lower than the Singapore residual soil [9]. The *n* value is quite similar with result from Indrawan et al [9] but has significant difference can be seen for *m* value.

Table 3. Summaries of fitting parameters.

Location	Fredlund and Xing fitting parameters		
	a	*n*	*m*
Pantai Dalam	0.260	2.6	0.260
Kiara	0.240	2.0	0.176
KL Tower	0.255	2.0	0.340
Bukit Dinding	0.299	1.7	0.440
Desa Putra	0.290	4.8	0.142
Indrawan et al [9]	0.321	2.468	0.937

4. Conclusion

In the present study, the soil water characteristic curve for five residual soils were investigated. The experimental data obtained from pressure plate extractor were best fitted with the Fredlund and Xing (1994) equation [3] and the fitting parameters were then collected. The main conclusions that can be drawn from this study are.

(1) The air entry value and residual matric suction for residual soils are in the range of 17 kN/m2 to 24 kN/m2 and 145 kN/m2 to 225 kN/m2 respectively.

(2) The fitting parameters such as a, m and n are in the range of 0.24 – 0.299, 1.7 – 4.8 and 0.142 – 0.440 respectively.

References

[1] B.B.K. Huat, G.S. Sew and F.H. Ali. Tropical Residual Soils Engineering, Taylor & Francis Group, London, 2004.

[2] A. Marto, F. Kassim and M.F. Yusof. Correlations of Engineering Properties of Residual Soils for Design Purposes. Proc. 5th Asia Pacific Structural Engineering and Construction Conference (APSEC 2003), Johor Bahru, 26-28 August 2003. pp 539-548.

[3] A. Marto, F.Kassim, M.F. Yusof and K.N. Yusof. Physical and Chemical Composition of Residual Soils at Eastern Region of Peninsular Malaysia. Proc. 2nd World Engineering Congress 2002, 22-25 July, Kuching Sarawak.

[4] F.H. Ali and H. Rahardjo. Unsaturated Residual Soil in B.B.K. Huat, G.S. Sew and F.H. Ali. Tropical Residual Soils Engineering, Taylor & Francis Group, London, 2004, pp. 57-71.

[5] D.G. Fredlund and A. Xing. Equation for the Soil-Water Characteristic Curve. Canadian Geotechnical Journal, 1994, 31(4). pp. 521-532.

[6] D.G. Fredlund. The Implementation of Unsaturated Soil Mechanics into Geotechnical

Engineering. The R.M. Hardy Address. Canadian Geotechnical Journal. 2002. 37(5). pp. 963-986.

[7] D.G. Fredlund, H. Rahardjo and M.D. Fredlund. Unsaturated Soil Mechanics in Engineering Practice. John Wiley & Sons. Canada. 2012.

[8] E.C. Leong and H. Rahardjo. Review of Soil-Water Characteristics Curve Equations. Journal of Geotechnical and Geoenvironmental Engineering. 1997. ASCE 23 (12). pp. 1106-1117.

[9] I.G.B. Indrawan, H. Rahardjo and E.C. Leong. Effects of Coarse-grained Materials on Properties of Residual Soil. Engineering Geology. 2006. pp. 154-164. compression from saturation to undrained shear, Soils and Foundations. 33, 2 (1993) 14-34.

Simulation of Sub-Drains Performance Using Visual MODFLOW for Slope Water Seepage Problem

M F T Baharuddin[1,2], S A A Tajudin[1,2], M H Z Abidin[2,1] and N A Yusoff [2,1]
[1] Research Center for Soft Soil, Universiti Tun Hussein Onn Malaysia, 86400 Batu Pahat Johor, MALAYSIA
[2] Faculty of Civil and Environmental Engineering, Tun Hussein Onn University, 86400 Batu Pahat, Johor, MALAYSIA

E-mail: mdfaizal@uthm.edu.my

Abstract. Numerical simulation technique was used for investigating water seepage problem at the Botanic Park Kuala Lumpur. A proposed sub-drains installation in problematic site location was simulated using Modular Three-Dimensional Finite Difference Groundwater Flow (MODFLOW) software. The results of simulation heads during transient condition showed that heads in between 43 m (water seepage occurred at level 2) until 45 m (water seepage occurred at level 4) which heads measurement are referred to mean sea level. However, elevations measurements for level 2 showed the values between 41 to 42 m from mean sea level and elevations for level 4 between 42 to 45 m from mean sea level. These results indicated an increase in heads for level 2 and level 4 between 1 to 2 m when compared to elevations slope at the level 2 and level 4. The head increases surpass the elevation level of the slope area that causing water seepage at level 2 and level 4. In order to overcome this problems, the heads level need to be decrease to 1 until 2 m by using two options of sub-drain dimension size. Sub-drain with the dimension of 0.0750 m (diameter), 0.10 m (length) and using 4.90 m spacing was the best method to use as it was able to decrease the heads to the required levels of 1 to 2 m.

Keywords: Numerical simulation, seepage, modflow.

1. Introduction

Occurrence of water seepage on slope face might be caused by presence of high pore water pressure in subsurface soil strata. Measurements of high pore water pressure always related with several times of heads observation on site that required a lots of number piezometers installation. This works involved time consuming, expensive cost and limited data coverage. Those limitations have created some gap of methods selection which be able to fulfill several considerations that related to cost, time and results quality [1, 2, 3].

Thus, in order to overcome those limitations mentioned, numerical simulation technique was used for investigating groundwater seepage occurs at Botanic Park of Kuala Lumpur. Furthermore, the sub-drain remediation technique approach for resolving a groundwater seepage problem by using simulation technique was also presented.

2. Method and Methodology

2.1 Site description and conceptual numerical model

The study conducted to understand groundwater seepage occurrence at Panggung Anniversary slope area as shown in Fig 1 and Fig 2. The conceptual model consists of a graphic representation of the groundwater flow systems, usually in the form of a block diagram or a cross section [4]. Conceptual model was developed with few assumptions on the aquitard (sandy silt) layers and hydraulic properties. The model boundary condition was extended to the north and south of the Panggung Anniversary location (Fig. 3). Boundary condition for numerical simulation in the north and south area were bounded with constant head boundary. The model was adjusted based on the measured water levels in the study area. Based on that data [5], the aquitard layer is generally formed by a medium to coarse grained SAND with dominant SILT material. The thickness of aquitard ranging from 13 to 25 m and assumed as unconfined layer, while at the bottom is the bedrock which consists of inter-bedded shale and sandstone that can be encountered at depths ranging from 13.0 to 25.0 m. Hydraulic conductivity in this study area ranging from 1.48 x 10^{-7} to 3.50 x10^{-8} cm/sec with an average of 9.25 x 10^{-8} cm/sec.

Figure 1. Study area and boreholes locations.

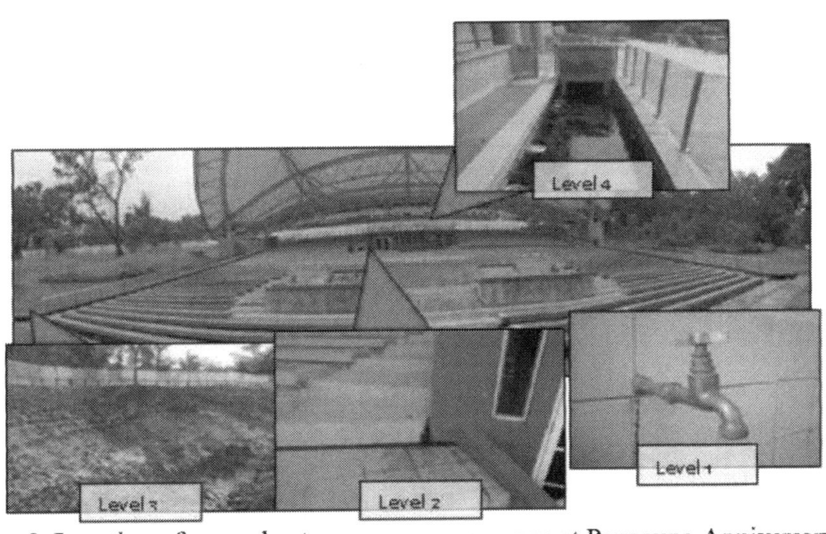

Figure 2. Location of groundwater seepage occurrence at Panggung Anniversary site.

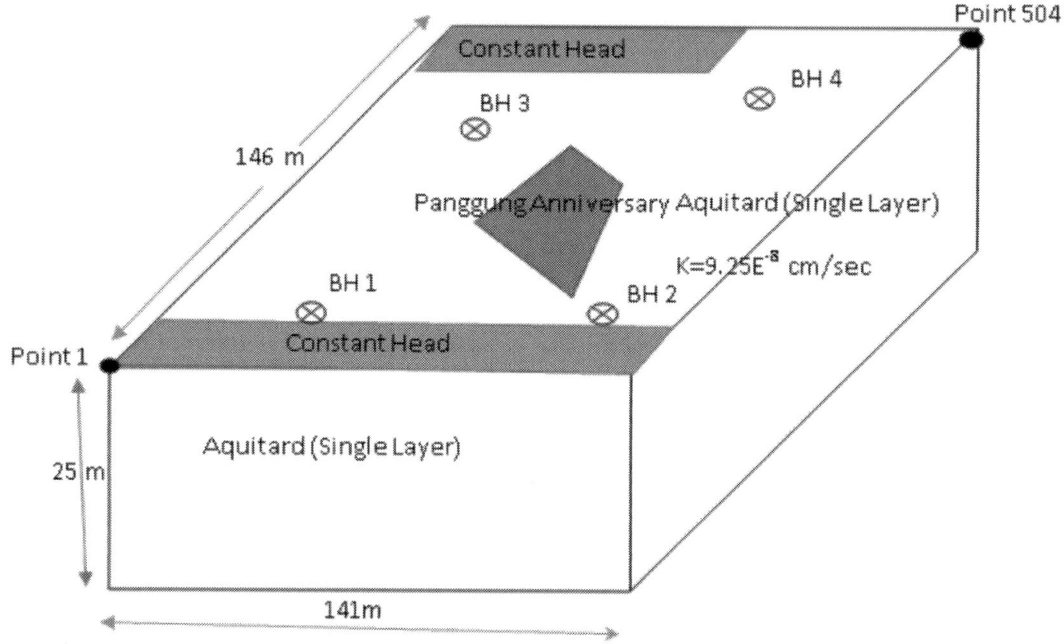

Figure 3. Conceptual model of numerical modelling at Panggung Anniversary site.

2.1 Numerical model setup

Three-dimensional representation of the site was created in Visual MODFLOW. The development of input files was compiled using Visual MODFLOW® (Version 2003.3.1.0.85), a commonly used data preprocessor to accelerate and facilitate the development of the MODFLOW model. This model was created as a 141 m x 147 m area in the x and y directions (east–west and north–south directions, respectively) (Fig. 4). Topography model (the first layer) was imported from elevation data for the wells (BH1-BH4) and surface elevations of the site generated by the digital elevation model and by field survey with the use of theodolite (Fig. 5). Relative elevations of water level monitoring points were determined using a total station theodolite. The water table in the model was referred to indicate sea level (rectified skew orthomorphic). This configuration allowed the placement of the piezometer screens in the model immediately at the terminated borehole depth.

2.3 Model parameters

Hydraulic conductivities of model layer was estimated from the analysed data of permeability tests from BH2 and BH3. Single layer model was assumed to have homogeneous hydraulic properties because the data were not enough to determine heterogeneity layer in this area. Model parameters data such as specific yields (Sy), specific storage (Ss), total porosity and effective porosity values were estimated from data interpreted by previous study (Table 1). A value between 5% and 20% of annual precipitation was recommended as an estimate of recharge when data were unavailable [6]. Hence, recharge was set equal to 10% of the average annual precipitation after the revision throughout model calibration and validation processes. Recharge parameter of surface water into groundwater system was estimated by 10% of total annual rainfall at open area [6]. Meanwhile, zero value of recharge parameter was imposed in closed area of Panggung Anniversary. Evapotranspiration values was estimated using 1.44 times by total annual transpiration at open area. Meanwhile, zero value of evapotranspiration was imposed in closed area of Panggung Anniversary [6].

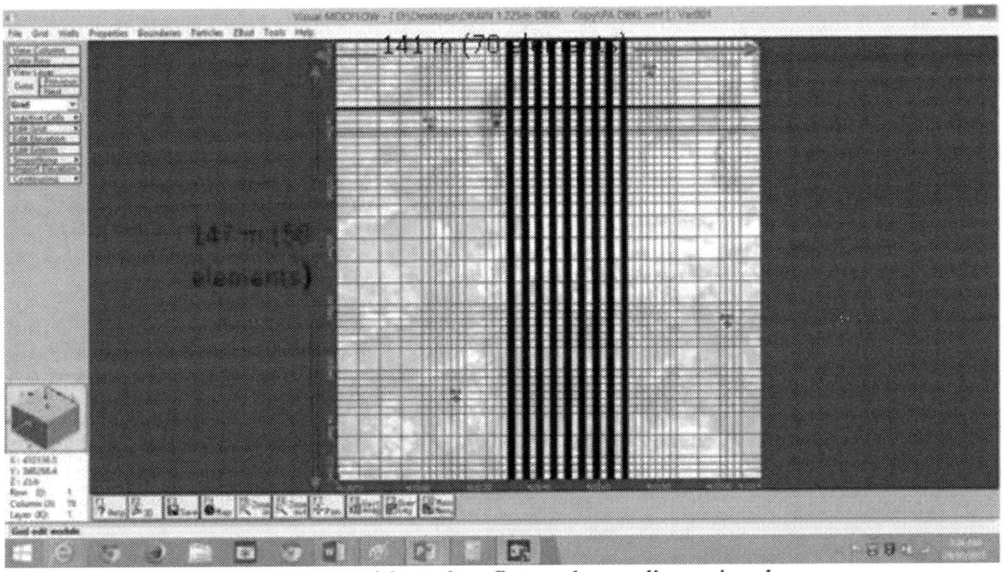

Figure 4. Grid mesh refine and area dimensional.

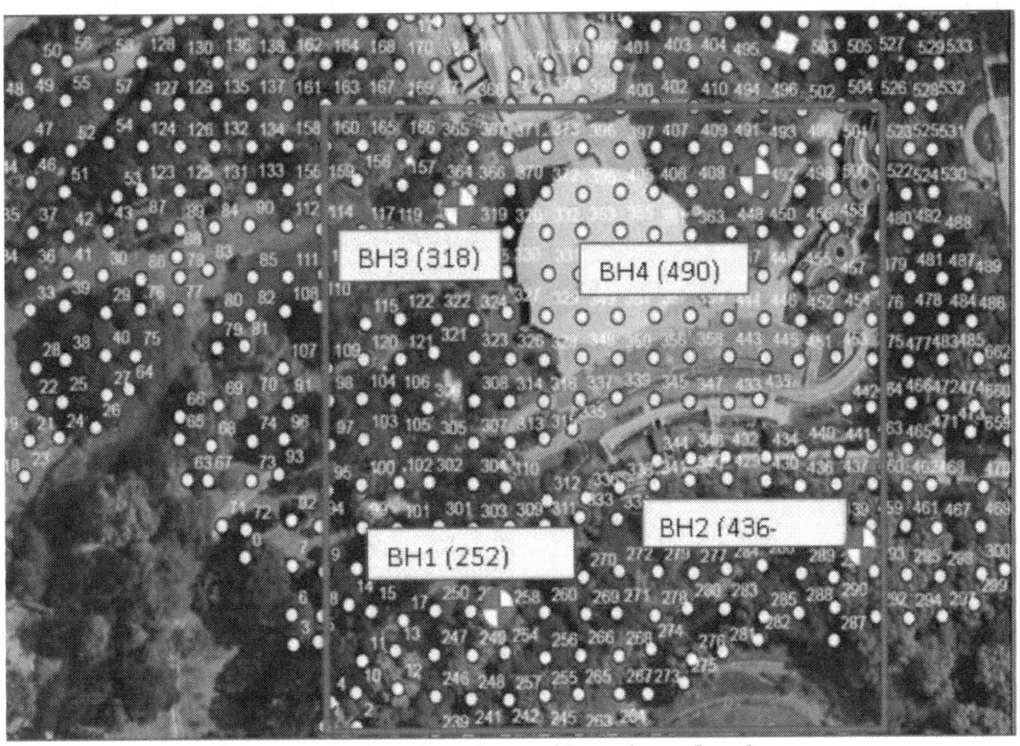

Figure 5. Surface elevation and boundary of study area.

2.4 Boundary Condition

Constant head boundaries were assigned for the northern and southern edges of the modelled area which are bounded by groundwater head of BH1-BH2 (south) and BH3-BH4 (north). The heads in the model was referred at mean sea level datum. This depth was on average 5 m below the ground surface. The highest heads located at the south area with an average head level of 48 m referred to mean sea level meanwhile the lowest heads located at the south area with an average 38 m referred to mean sea level. The model setup for the remediation technique were applied Drain, Wall and Pumping well Package where are contained in Visual MODFLOW software.

Table 1. Input parameters for the model and simulation strategies for this study.

Parameters	Value
K_x (cm/sec)	9.25e-8 (average value from permeability test average conducted at BH2 and BH3)
K_y (cm/sec)	9.25e-8 (average value from permeability test average conducted at BH2 and BH3)
K_z (cm/sec)	1e-9 [7]
S_s	1e-5 [8]
S_y	0.8 [8]
Total Porosity	0.3 [9]
Effective Porosity	0.15 [9]
Recharge (mm/year)	220 (10% of Annual Rainfall) and 0 at Panggung Anniversary [6]
Evapotranspiration (mm/year)	500 and 0 at Panggung Anniversary [6]
Conductance (m^2/day)	0.00094 – 0.00562

2.5 Transient groundwater flow simulation and calibration

The equation for groundwater flow through the porous media used in the model can be written as a partial differential equation:

$$\frac{\partial}{\partial x}\left(K_x\frac{\partial h}{\partial x}\right)+\frac{\partial}{\partial y}\left(K_y\frac{\partial h}{\partial y}\right)+\frac{\partial}{\partial z}\left(K_z\frac{\partial h}{\partial z}\right)=S_s\frac{\partial h}{\partial t}-W \qquad (1)$$

where Kx, Ky, and Kz are the hydraulic conductivities in the three orthogonal directions (m/s or ft/day), h is the head that drives the flow or the saturated thickness of the aquifer (m or ft),W is the volumetric flux per unit volume and represents the source/sink term for water or withdrawal (m3/s or ft3/day), Ss is the specific storage capacity of the porous medium (dimensionless), and t is the time (s or day). When the water withdraws, –W = R, where R is a general sink/source term, it is defined to be intrinsically positive to represent recharge (defines the volume of inflow to the system per unit volume of aquifer per unit of time (ft3/ft3/day or m3/m3/s)).The Visual MODFLOW model was run at transient state then calibrated to hydraulic heads recorded on a weekly basis from 16 April 2015 to 7 May 2015 at four standpipes (BH1-BH4) for calibration. For each transient model run, an analysis of the observed versus computed water levels was conducted to determine the accuracy of the simulation.

3. Result and discussion

3.1 Calibration results for groundwater flow simulation

The assessment of the model fit was quantified by the simulations, computed as the difference between the simulated and the observed hydraulic heads. For this model, the estimated average of mean absolute error (MAE) is 0.5 m, the root mean square (RMS) is 1.2 m, and the normalized root mean square error (NRMS) is 12%, which indicate a well calibrated model. Overall, the calibrated model is able to simulate water levels and water level fluctuations sufficiently to evaluate the options of remediation approach on groundwater seepage problem.

3.2 Sub-Drain

Sub-drain was proposed to overcome the water seepage problem at levels 2 and 4. Two options on the sub-drain dimensional design were suggested with taken consideration of materials is easy acquired in market.

Option 1:- 0.0750 m (diameter), 0.10 m (length), 4.90m (spacing)
Option 2:- 0.0375 m (diameter), 0.10 m (length), 4.90m (spacing)

Sub-drain using option 1 showed heads result between 41 and 44 m at level 2 and 4 (Fig. 6) respectively. Drawdown that is the difference in the head between the initial head (without sub-drain) and the head with sub-drain (option 1) showed difference of 1 – 2 m. Meanwhile for option 2, simulated head results showed heads between 42 m at level 2 until 45 m at level 4 (Fig. 7). Drawdown showed difference of 0 - 1 m for option 2 and also indicated that the drawdown was smaller compared to option 1. It was suggested that option 1 with the use of sub-drain of 0.0750 m (diameter), 0.10 m (length) and using 4.90 m spacing can be use in the remediation of the water seepage. This option can reduced the heads in optimal values to prevent settlement of the stage structure.

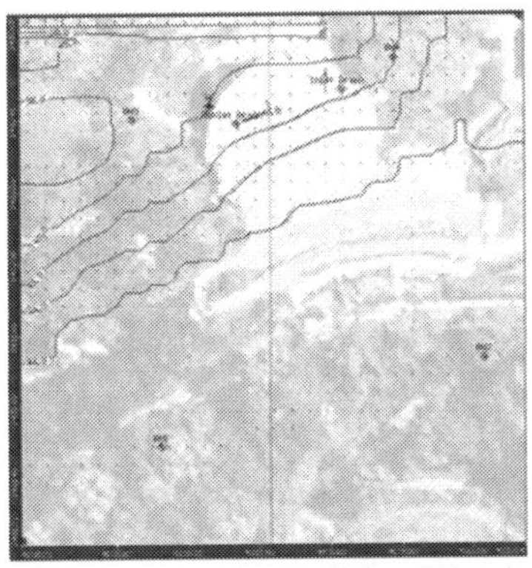

Figure 6. Option 1 (Plan View).

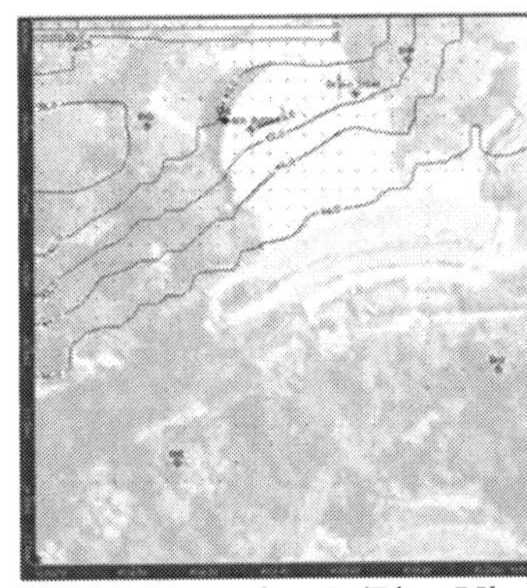

Figure 7. Option 2 (Plan View).

4. Conclusion

The results of the analysis using numerical simulation groundwater flow (Visual MODFLOW), has succeeded in determining the cause of water seepage problem in the study area. Water seepage occurrence at Panggung Anniversary was caused by the elevated groundwater heads compared to the elevation levels of the area. In order to overcome the water seepage occurrence caused by the increase of heads level, the reduction of the head levels between 1 – 2 m need to be done. The usage of the sub-drain with dimensional design of 0.0750 m (diameter), 0.10 m (length) and using 4.90 m spacing was able to reduce about 1 – 2 m of the heads level and is the best option of remediation to be considered. The usage of dimensional design remediation with sub-drain is the safest method to be use to prevent any structure settlement from occurring.

Reference

[1] Zainal Abidin M H, Tajul Baharuddin M F, Zawawi M H, Ali N M d, Madun A and Ahmad Tajuddin S.A. 2015 Groundwater Seepage Mapping using Electrical Resistivity Imaging, Applied Mechanics and Materials 1524-1534.
[2] Zainal Abidin M H, Saad R., Ahmad F, Wijeyesekera D C, Matarul J, Tajul Baharuddin M F, 2013 Mapping of Geotechnical Data Using Seismic Refraction Studies, AWAM International Conference on Civil Engineering & Geohazard Information Zonation 1-9.
[3] Zainal Abidin M H, Saad R, Ahmad F, Wijeyesekera D C, Tajul Baharuddin M F, 2015 Correlation Analysis between Field Electrical Resistivity Value (ERV) and Basic Geotechnical Properties (BGP), Soil Mechanics and Foundation Engineering 117-125.
[4] Anderson M P, Woessner W W, 1992 Applied groundwater modeling; simulation of flow and advective transport. Academic Press, San Deigo381.
[5] IKRAM, 2015Seepage Problem and Groundwater Studies at Panggung Anniversary, Taman Botani, Kuala Lumpur, IKRAM Preliminary Report,.
[6] Waterloo, Visual MODFLOWv 4/1 User's Manual. 2005 Waterloo Hydrogeologic Inc., Waterloo.
[7] Todd D K, 1980 Groundwater Hydrology (2nd ed) Wiley India Pvt Ltd, 556.
[8] Younger P L, 1993, Possible environmental impact of the closure of two collieries in County Durham. Journal of the Institution of Water and Environmental Management 7, 521-531.
[9] Freeze RA and Cheery JA, 1979 Groundwater Prentice Hall College Div. ISBN-13: 978-0130617309 604.

Influence of Relict Joints on Permeability of Residual Soil

Z A Talib [1,2], A Kassim,[2] and G H Yunusa[2]
[1] Faculty of Civil and Environmental Engineering, Universiti Tun Hussein Onn Malaysia, 86400 Batu Pahat Johor, MALAYSIA
[2] Faculty of Civil Engineering, Universiti Teknologi Malaysia, 81310 Skudai, Johor Bahru, Johor, MALAYSIA

E-mail: zaihasra@uthm.edu.my

Abstract. Weathering process of granitic material results in the formation of relict joint in lateritic layer of the weathering profile. The number and arrangements of the relict joints affects the permeability of the residual soil which invariably affects water flow and suction distribution in the residual soil. Although the permeability of residual soil without a relict joint can be determined using standard permeability test, it is difficult to be measured when a relict joint is incorporated due to limitation of size and area of the standard equipment. Hence, modified permeability test equipment is introduced in this study. Two arrangement of the relict joint in the equipment were considered. In the first arrangement one relict joint with various spacing were tested while the orientation and spacing of the relict joint were tested using two relict joints in the second arrangement. The results obtained shows that the permeability of the residual soil due to one and two relict joint varies by two orders of magnitude. Therefore, the number and spacing of relict joints modified the permeability of residual soil.

Keywords: Weathering, relict joint, residual soil.

1. Introduction

A landslide in residual soil normally occurs during or immediately after heavy rainfall. Previous studies have shown that rainfall infiltration increases the soil water content which decreases matric suction and hence decreases the additional shear strength provided by the matric suction which results in landslides [1– 4]. One of the factors that contribute to this phenomena is heterogeneity in the residual soil which greatly affects the permeability of the soil [3]. Residual soil is the final product of in-situ weathering process whether mechanical or chemical weathering of rock. This weathering process alters the characteristics and behaviour of the parent rock, and results in successive layers of weathering profile classified as Grade I to Grade VI according to six fold of weathering classification of International Society for Rock Mechanics (ISRM) [5]. The weathering process involves in the formation of residual soil introduces variation at material scale and also at field scale. At material scale, the weathering process causes the decomposition of the rock to Grade VI (lateritic layer) and Grade V (saprolitic layer). Therefore it introduces heterogeneity in the soil mass. These heterogeneity includes variations in grain size, porosity, mineralogy, lithologic texture, rock mechanical properties, structure and diagenetic processes [6, 7]. The heterogeneity of residual soil at field scales is due to discontinuities such as joint, fault and fold and it is more significant in saprolitic layer.

Furthermore, since permeability is a function of grain size distribution [8], the lateritic layer have less permeability compare to saprolitic layer, as a result. The difference in permeability between these two layers is more pronounced when relict joint appears in the saprolitic layer. Kassim et al. [3] have proven that increasing the number of relict joint, increases the permeability of the soil.

Previous studies such as [9-11] have shown that a relationship exist between soil discontinuity and suction distribution in residual soil. The existing of relict joint greatly influence the permeability of soil which results in changes in suction distribution during rainfall. The relict joint can prevent or allowed more water to infiltrate into deeper soil profile depending on its orientation and arrangement. Usually, the relict joints exist in two or three vertical to subvertical joint sets in residual soil. However, based on the previous studies, only few studies were performed on permeability of residual soil with more than one relict joint. Therefore, this paper compares the permeability of residual soil with one relict joint and the effect of relict joint orientation using two relict joints.

2. Methodology

The soil used in this study was collected from Balai Cerap, which is a sloping site located within the campus of Universiti Teknologi Malaysia, Johor Bahru. Modified permeability test was performed to study the effect of relict joint on permeability of saprolitic layer. Figure 1 shows the modified permeability equipment.

Figure 1. Modified permeability equipment.

The modified permeability tests conducted were performed according to the procedure outlined in [12], using an equipment as shown in Figure 1. This equipment was made from steel frame and acrylic sheet sidewalls. The artificial relict joints were formed using 1 mm aluminium sheet at difference spacing. As shown in Figure 2, only vertical orientation with various spacing was considered for one relict joint. A relict joint with the spacing of 50 mm, 100 mm, 150 mm, 200 mm, 250 mm, 350 mm, 400 mm and 450 mm were tested. However, for two relict joints, the orientations were vertical and slightly inclined at approximately 6°. Similarly, a spacing 100 mm, 150 mm, 200 mm, 250 mm, 300 mm, 350 mm, 400 mm and 450 mm were also tested. The soil was compacted according to soil density recorded at site. A 25 mm thick layer of gravel was placed at the bottom and upper part of sample.

Figure 2. The orientation of relict joint for one relict joint.

3.1 Calibration of Modified Permeability Equipment

Prior to the actual testing, the modified permeability apparatus was calibrated to determine the accuracy of the results obtained. The calibration was performed by conducting parallel permeability tests with standard and modified permeability equipment and comparing the results. The samples with same density were prepared and homogeneous soils without any relict joint were used for both tests.

3. Results and discussion

The soil collected from the site was divided into two major groups, sample obtained from 0 m to 0.5 m which is fully decomposed was considered as Grade VI soil, while other sample collected below 0.5 m were considered as Grade V. As mentioned earlier, the relict joint only appears in saprolitic layer which is Grade V layer. Thus, the soil sample was prepared according to Grade V properties.

The particles size distribution of the grade V shows that it possessed 48% gravel, 15% sand and the remaining 37% constitute the fine particle content. This soil was classified as silty gravel of high plasticity (GMH) according to British Soil Classification System (BSCS). The in-situ density and saturated permeability (ksat) tests shows that the soil possessed a density and permeability of 1805 kg/m3 and 3.68 x 10-6 m/s, respectively. The ksat of the soil obtained using modified permeability test is 4.52 x 10-6 m/s which indicate a difference of 23%.

Figure 3 shows the results of permeability test for one relict joint and two relicts joint. From this Figure, the permeability of soil with two relicts joint is higher than that of one relict joint. The variation in the permeability for the two relicts joint is almost in two orders of magnitudes which conformed to the previous findings by [5]. Similarly, the variation in the permeability due to the arrangements of the relict joint diminishes when the spacing increases. Furthermore, increasing the spacing between the relict joints indicates the samples almost replicate the homogeneous soil hence the permeability remained the same [3].

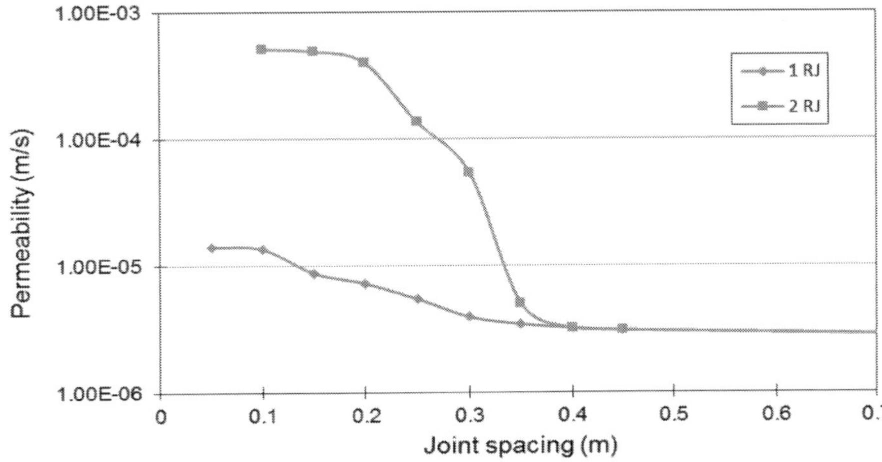

Figure 1. Permeability for 1 Relic Joint and 2 Relict Joints.

The permeability due to 1 and 2 relict joints are shown in Figure 4 and Figure 5, respectively. From both Figures the variation in the permeability can be divided into three parts. The curve starts with horizontal line then linear curve and finally ends with horizontal line.

For one relict joint, Part 1 is for spacing less than 100 mm, Part 2 for spacing greater than 100 mm but less than 300 mm and Part 3 for spacing greater than 300 mm. However, for two relict joint, Part 1 for spacing less than 200 mm, Part 2 for spacing greater than 200 mm but less than 350 mm and finally Part 3 for spacing greater than 350 mm. These results confirmed that, relict joint has significant effect on permeability of residual soil. The ksat of grade V (silty gravel) was clearly exhibited in Part 2 of the curve. Within this region an inverse relationship between permeability and spacing is observed which may largely contribute to the behaviours of the residual soil.

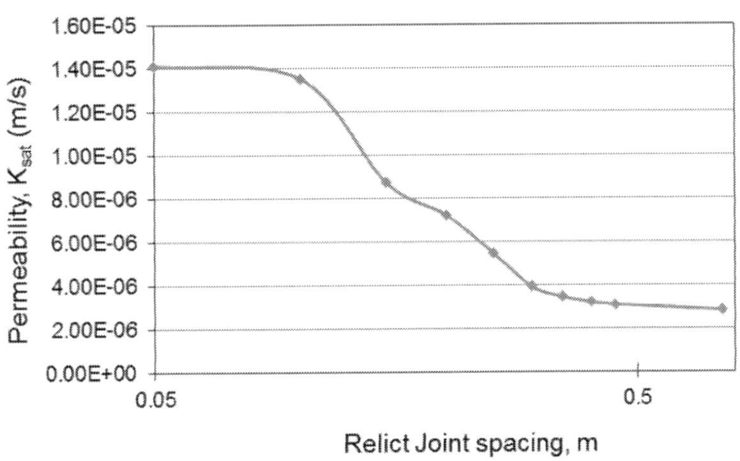

Figure 2. Permeability for different spacing for one relict joint.

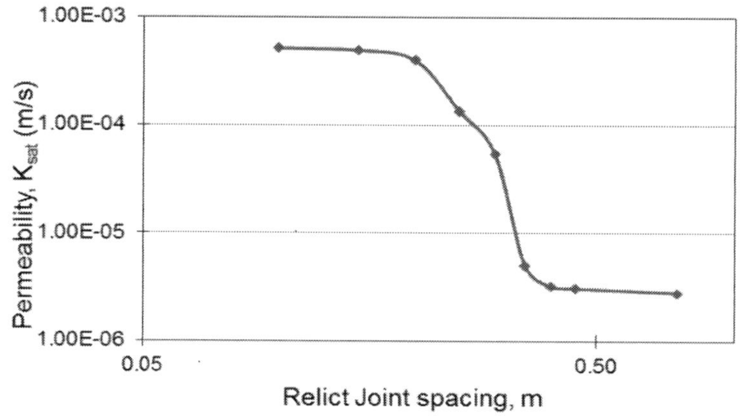

Figure 3. Permeability for different spacing for two relict joint.

4. Conclusion

The permeability of weathered granitic material varies significantly due to the effect of relict joint. This variation may occur in various order of magnitude depending on the orientation and spacing of the joints. These variations can be summaries as follows:

1. The weathering grade varies with depth. The trend of variation results in different soil properties which are difficult to measure using laboratory testing.
2. Modified permeability test is one method that can be used and determine the permeability of soil with discontinuity.
3. In Grade V layer, the presence of relict joint affects the permeability of the soil. By increasing the number of relict joint in vertical direction seems as increasing the permeability of the soil.
4. For 1 relict joint, spacing between 100 mm to 300 mm may control the permeability of the residual soil. However, for 2 relict joint the permeability is affected with the spacing between 200 mm to 350 mm.

References

[1] X. H. Hu, S. G. Cheng, and Z. W. Yu, "Influence of Matric Suction on the Stability of Unsaturated Slope," Adv. Mater. Res., vol. 594–597, pp. 18–22, Nov. 2012.

[2] L. Jinyu, Y. Qing, and L. Maotian, "Limit Equilibrium Analysis of the Influence of Matric Suction on Unsaturated Soil Slope Stability Engineering , School of Civil and Hydraulic Engineering , Dalian University of Technology , 2 . Sizes Objectives 3 . Relationship between water content and s," vol. 5, no. 10, pp. 167–175, 2010.

[3] A. Kassim, N. Gofar, L. M. Lee, and H. Rahardjo, "Modeling of suction distributions in an unsaturated heterogeneous residual soil slope," Eng. Geol., vol. 131–132, pp. 70–82, Mar. 2012.

[4] A. Rahimi, H. Rahardjo, and E.-C. Leong, "Effect of hydraulic properties of soil on rainfall-induced slope failure," Eng. Geol., vol. 114, no. 3–4, pp. 135–143, Aug. 2010.

[5] ISRM (1981). Basic Geotechnical Description for Rock Masses. International Journal of Rock Mechanics, Mining Science and Geomechanics. 18: 85-110.

[6] S. S. Agus, E.-C. Leong, and H. Rahardjo, "Estimating permeability functions of Singapore residual soils," Eng. Geol., vol. 78, no. 1–2, pp. 119–133, Apr. 2005.

[7] F. Arikan and N. Aydin, "Influence of Weathering on the Engineering Properties of Dacites in Northeastern Turkey," ISRN Soil Sci., vol. 2012, p. 1, 2012.

[8] H. Rahardjo, K. K. Aung, E. C. Leong, and R. B. Rezaur, "Effects of Pore-Size Distribution on Engineering Properties of Residual Soils," … World Eng. …, no. July, pp. 70–76, 2002.

[9] T. Y. Irfan, "Structurally controlled landslides in saprolitic soils in Hong Kong," Geotech. Geol. Eng., vol. 16, no. 3, pp. 215–238, 1998.
[10] J. Zhang, J. J. Jiao, and J. Yang, "In situ rainfall infiltration studies at a hillside in Hubei Province, China," Eng. Geol., vol. 57, no. 1–2, pp. 31–38, 2000.
[11] L. M. Lee, A. Kassim, and N. Gofar, "Performances of two instrumented laboratory models for the study of rainfall infiltration into unsaturated soils," Eng. Geol., vol. 117, no. 1–2, pp. 78–89, Jan. 2011.
[12] BSI. (1990). Methods of Test for Soils for Civil Engineering Purposes, (BS 1377: Part 5). British Standards Institution, London.

Investigation on Strength Development in RBI Grade 81 Stabilized Serian Soil with Microstructural Considerations

S N L Taib[1], S Striprabu[1], F Ahmad[2], H J Charmaine[1] and N E Patricia[1]
[1] Department of Civil Engineering, Faculty of Engineering, Universiti Malaysia Sarawak, 94300 Kota Samarahan, Sarawak, MALAYSIA
[2] School of Civil Engineering, Universiti Sains Malaysia, 14300, Nibong Tebal, Pulau Pinang, MALAYSIA

E-mail: tlinda@feng.unimas.my

Abstract. The aim of the research is to investigate the strength development of stabilised local Serian soil with RBI Grade 81 a chemical additive to enhance soil properties in term of strength. Serian is a town which is located about 60 km from Kuching city where the soil was chosen to be treated in this study. The soil sample was mixed with 2 %, 6 % and 8 % of RBI Grade 81 by weight of dry soil and added with water at optimum water content (OWC) to replicate field site conditions. The modified samples were cured in ambient air for 7, 14, and 28 days. Scanning Electron Microscope (SEM) was utilized to analyze surface morphology of the stabilized soil specifically on the formation of bonding between soils and RBI 81 particles. The experimental results show the highest average peak UCS strength achieved was 1071.6 kN/m^2 at 14 day curing period with 8 % of RBI Grade 81 which is higher than the untreated control sample which was 179.946 kN/m^2, showing increment by almost six folds. Hence the RBI 81 stabilization technique enhances the local soil structure by improving the inter-cluster bonding, reducing pore spaces in the soil and subsequently increasing the soil's strength.

Keywords: Soil stabilisation, microstructure, chemical additive.

1. Introduction

Soft Soils are characteristically known for their low strength, high compressibility and high groundwater table which cause large settlement. Typically due to sedimentary process in various environments, both physical and engineering properties such as void ratio, water content, grain size distribution, compressibility, permeability and strength exhibit substantial variation. In addition, soft soil also exhibit high compressibility (including the important secondary consolidation), reduced strength, low permeability which are unsuitable conditions for construction. Soft clay is also categorized as soft soil and is widely deposited in the state of Sarawak. Due to unavoidable reason such as population and economic growth, development and construction are forced to take place on poor ground such as soft clay. Soft clay has low load bearing capacity and undergo large settlement when loaded. Hence, construction of highways embankment on normally consolidated soft soil deposit has suffered from extravagant settlement and lateral displacement without proper ground improvement before construction implemented [1]. Therefore, ground improvement is needed to avoid these excessive settlements of structures constructed on soft ground.

Stabilisation is an effective technique in improving the engineering properties of soils especially in soft soils. Commonly, soil stabilisation is accomplished via blending and mixing additive with the soil that needs to be improved until the mixture achieves certain improvement in engineering properties [2]. Chemical stabilisation is one type of soil stabilisation that has been widely implemented in field where chemical additives such as cement, lime and fly ash have been used. Chemical stabilisation is very efficacious in strengthening soft and unstable ground in which fundamentally, the chemical additives will alter the mineralogical structure and ameliorate the inherent weak properties of soil for example in terms of its strength and stiffness [3].

Geophysical In this research, RBI Grade 81 (Road Building International Grade 81) was used as the chemical additives to stabilize the Serian soil. Serian soil is chosen as Serian is a newly formed division and is one of the main divisions to connect all the other towns in Sarawak via road. RBI Grade 81 is an odourless beige powder that is produced from a number of naturally occurring compound and has pH of 12.5 in saturated paste [4]. The chemical is insoluble in water, non UV degradable, inert, and chemically stable and work well by hydration reaction [5]. Moreover, it has capability to improve structural properties of wide range of soil especially with silty clay soil with low geo-mechanical qualities [4, 5, 6]. The hydration process takes place and filled the pore space with crystalline growth. In addition, usage of small dosages of RBI Grade 81 can effectively increase the volume stability of the soil [5, 6, 7]. The reaction between RBI Grade 81 and the soil will develop an inter particle matrix that binds the soil particle together into rigid mass. Via both chemical bonds and frictional forces, the binding soil particle serves to limit the pore volume of the created rigid stabilized soil system [5, 6, 7]. RBI Grade 81 form dust free surface, simple to use and harden fast. Benefits of RBI Grade 81 are durable, permanent, aesthetical and environment friendly [5]. Plus, soils treated with RBI Grade 81 exhibit increase of strength with time and even able to obtain strength after one year of application in soil [7]. RBI Grade 81 alters clay irreversibly into cementitious calcium silicate and aluminium hydrate where it also creates volume of stable layer with small capillary spaces [6] and reduces the permeability of soil mass [7].

Stabilisation Typically when the RBI Grade 81 added in the soil, there will be chemical reaction which takes place and is mainly of pozzolanic reaction. During the pozzolanic process, the Calcium Hydroxide released by additives will react with the clay mineral typically aluminous and siliceous which can induce the pozzolanic properties [8]. Hence high concentration of Calcium will produce free Ca^{+} for the chemical reaction with silica and alumina from the clay which leads to the formation of cementitious compounds; calcium silicate hydrate (C-S-H) and calcium aluminate hydrate (C-A-H) as shown in Eq. (1 & 2) [9].

$$Ca\,(OH)_2 + H_4SiO_4 \rightarrow Ca^{2+} + H_2SiO_4^{2-} + 2\,H_2O \rightarrow CaH_2SiO_4\,2\,H2O\ (\textit{C-S-H}) \quad (1)$$
$$Ca\,(OH)_2 + Al\,(OH)_4 \rightarrow C_4AH_{13},\ C_3AH_6\ (\textit{C-A-H}) \quad (2)$$

2. Materials and Methods

2.1 Soil Sample

The soil sample used for the present study was collected from Serian. The soil is classified as High Plasticity Silt (MH) according to the Unified Soil Classification Systems (USCS). Properties of this Serian soil are summarized in Table 1.

Table 1. Properties of Serian soil.

Property	Characteristic
Liquid limit, LL (%)	55
Plastic limit, PL (%)	32.3
Plasticity Index, PI (%)	22.7
Linear Shrinkage, LS (%)	10.4
Maximum Dry Density, MDD (Mg/m^3)	1.74
Optimum Water Content (%)	23.5
Specific gravity, G_s	2.6
Grain Size Distribution (%)	
Sand	24
Silt	54
Clay	22

2.2 Additives

RBI Grade 81 added at 2 %, 4 %, 6 % and 8 % of total dry mass.

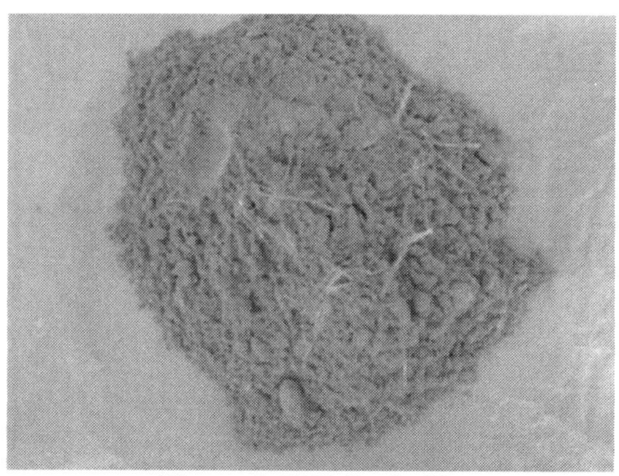

Figure 1. RBI Grade 81.

2.3 Sample Preparation

The collected sample was sun dried until totally desiccated. Next, the untreated samples were sieved passed through the 2 mm sieve. Then, the sieved untreated clay was mixed with RBI Grade 81 (at various percentages of total dry mass) and required amount of water. The water contents were at Optimum Water Content and MDD. The mixture was mixed thoroughly and compacted via a special miniature compacter until it reached required dry density and moisture content. All prepared samples have dimension of 35 mm internal diameter and height of 70 mm. The sample preparation process needs to be performed quickly to avoid any cementitious bonds in the soil to break due to delayed compaction. Hence in this study, sample preparation was done within 10 minutes to avoid low quality soil mixture. Next, the prepared samples were air cured for 7, 14 and 28 days in temperature and humidity of 25°C and 95% respectively.

3. Results and Discussions

3.1 Standard proctor test

The standard proctor test was done based on BS 1377: Part 4: 1990. The values in Table 2 show clearly that the maximum dry density increase with increase in RBI Grade 81 content.

Table 2. Summary of MDD and OMC of Soil with 2%, 4%, 6% and 8% of RBI 81.

Mix proportion	MDD (Mg/m^3)	OMC (%)
Untreated	1.74	23.46
Soil + 2% RBI Grade 81	1.81	24.45
Soil + 4% RBI Grade 81	1.82	24.60
Soil + 6% RBI Grade 81	1,84	25.73
Soil + 8% RBI Grade 81	1.90	26.46

3.2 Unconfined compressive strength test

The values of unconfined compressive strength of all the specimens with different amount of RBI Grade 81 added are presented in Figure 2 until Figure 4. As shown, increased amount of RBI Grade 81 induced improvement in term of strength for the specimen tested. Moreover, strength of specimen also improved proportionally to curing time and this tendency becomes more prominent with increase of RBI Grade 81 used. In Figure 2, 8 % of RBI Grade 81 shows the average maximum strength of 1071.6 kN/m^2 and is observed on the 14 day curing, which is more than six time the strength of untreated soil. Minimal reduction of strength occurs on 28 day curing perhaps caused by strength loss due to either shrinkage crack that was developed within stabilized soil or probably due improper curing. Excess of RBI content and insufficient water content to react with the excess RBI may also contribute to the condition.

Figure 3 shows the effect of curing time in increasing the strength of the samples. The graph shows that all samples exhibit strength improvement over time. As seen earlier, 8 % of RBI Grade 81 treated soil achieved maximum strength on the 14 day curing. From Fig 3, in general, it can be concluded that treated samples had greater strength after curing when compared with untreated sample.

As can be observed in Figure 4, increase of strength can be seen when the amount of RBI Grade 81 is increased in treating the soil. From the figure, 8 % of RBI Grade 81 achieved the greatest strength compared to 2 -6 %. This strength enhancement is shown as more amount of RBI Grade 81 will consist higher amount of free Ca^+ available for the reaction with silica and alumina from the clay which caused the formation of higher cementitious compounds - the calcium silicate hydrate (C-S-H) and calcium aluminate hydrate (C-A-H).

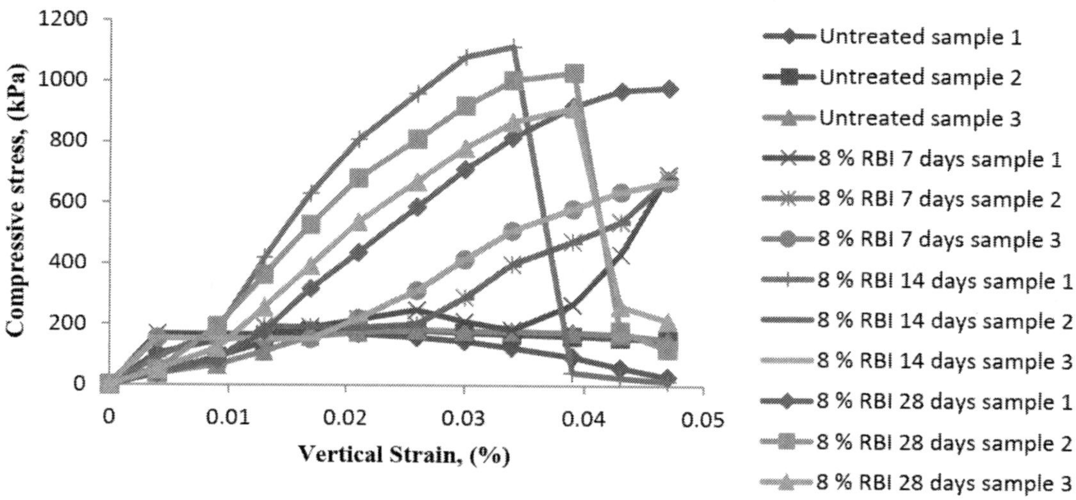

Figure 2. Graph of Untreated Soil and 8 % RBI Grade 81 treated soil.

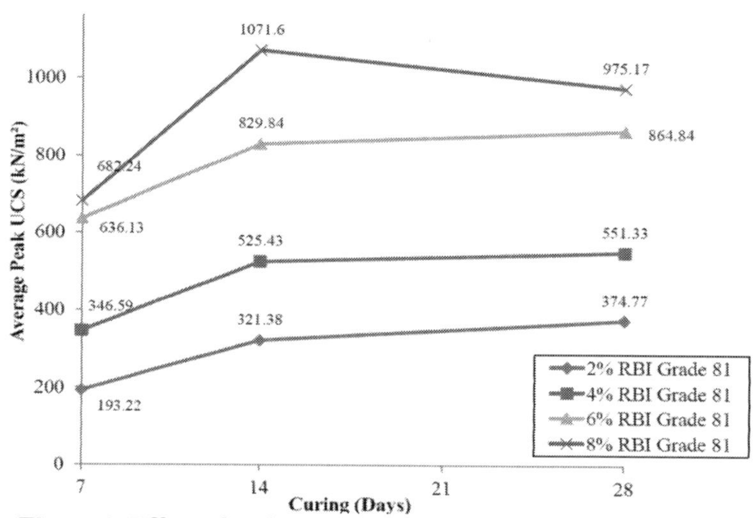

Figure 3. Effect of curing time on RBI Grade 81 treated soil.

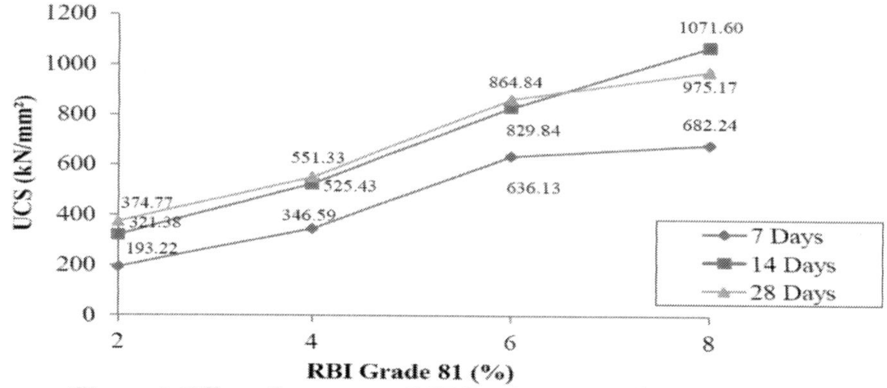

Figure 4. Effect of content on RBI Grade 81 treated soil .

3.3 Microstructural analysis

Model TM3030 Hitachi Tabletop Analytical Scanning Electron Microscope (SEM) had been used to investigate the microstructure of the natural soil and the RBI Grade 81 treated Serian soil. Figure 5 and 6 show the microstructure of natural soil under 5000× magnification and 2 % RBI Grade 81 7 day curing treated soil under 5000× magnification respectively. Figure 7 and 8 present the microstructure of 6 % RBI Grade 81 stabilised soil cured for 7 day and 28 day at 5000× magnification respectively. Fig 5 is the untreated soil which has many empty pores as indicated from the darker contrast area. When compared to Figure 6, most of the pores were filled with the cementitious products due to the chemical reaction within. However, Figure 7 shows more of cementitious products compared to Figure 6 due to the effect of higher amount of RBI Grade 81. The adsorption of higher amount of Ca^{2+} ions in the 6% sample onto the clay particle surface decreases the repulsion between successive diffused double layers and increases edge-to-face contacts between successive clay sheets [10] and the clay particle flocculate into larger particle [11,12]. The cementitious products improve the inter cluster bonding strength and filled the pore space significantly which directly reduce the total pore volume, also induced increment of compressive strength. The effect of curing was observed in Figure 7 and Figure 8. Figure 8 shows more densification of the morphological surface compared to Figure 7 due to growth of cementitous product over time.

Figure 5. Untreated Natural Soil.

Figure 6. 2 % RBI 81 treated soil (7 days Curing)

Figure 7. 6 % RBI 81 treated soil (7 days Curing).

Figure 8. 6 % RBI 81 treated soil (28 days curing)

4. Conclusion

This research was intended to study the effect of RBI Grade 81 as additive to stabilise the Serian soil. Samples were prepared at Optimum Water Content (OWC) and Maximum Dry Density (MDD) to replicate field site conditions. All samples were air cured for 7, 14 and 28 days. Results showed that the RBI 81 stabilised soils show increased in soil strength under various curing periods. The experimental results show the highest average peak UCS strength achieved was 1071.6 kN/m2 on 14 days curing period at 8 % of RBI Grade 81 which is higher than the untreated control sample which was 179.946 kN/m2, increment by almost six folds. Longer curing period should allow further enhancement in strength. Unconfined Compressive Strength (UCS) values increase with increment in RBI 81 addition suggests its suitability as good additive to improve performance of soft soils. SEM also provides reliable evidence on the effect of calcium hydroxide from RBI Grade 81 to the soil matrix via its morphological images. The presence of cementitious products in the microstructure of the RBI 81stabilised soil supported the result of UCS test. Formation of the cementitious products had increased the shear strength of the RBI 81 stabilized Serian soil.

Acknowledgment
The authors would like to thank the Universiti Malaysia Sarawak (UNIMAS) and Ministry of Higher Education for financially supporting the project. Highest gratitude is also directed to CMS Pavement Sdn. Bhd. for recommendations provided during the conduct of this research.

References

[1] Babu R D, Ramu K, Prasad S D, and Kumar K A 2013 Influence of geopolymer on the strength characteristics of sand mixed soft marine clay. Proceeding of Indian Geotechnical Conference. 22-24.
[2] Olufowobi J, Ogundoju A, Michael B, and Aderinlewo O 2014 Clay soil stabilization using powdered glass. Journal of Engineering Science and Technology. **9(5)** 541-558.
[3] Sing W, Hashim R, and Ali F H 2008 Influence of curing under stress on the engineering properties of stabilised peat soil. The international conference on construction and building technology. 159-180.
[4] Haricharan T S, Kumar K S V, Prashanth D L, Archana M R, and Ravishankar A U 2013 Laboratory investigation of expansive soil stabilized with natural inorganic stabilizer. International Journal of Research in Engineering and Technology. **2(13)** 201-204.
[5] Naseem A K A, Damgir R M, and Hake S L 2014 Effect of fly ash and RBI Grade 81 on black cotton soil as a sub grade for flexible pavements. International Journal of Innovations in Engineering and Technology. **4(1)** 124-130.
[6] Madurwar K V, Dahale P P and Burile A N 2013 Comparative study of black cotton soil stabilization with RBI grade 81 and Sodium silicate. International Journal of Innovative Research in Science, Engineering and Technology. **2(2)** 493-499.
[7] Singh E T and Riar E N 2013 Strengthening of subgrade by using RBI grade 81 a case study. IOSR Journal of Mechanical and Civil Engineering. **8(6)** 101-106.
[8] Xiao H and Lee F 2009 Curing time effect on behaviour of cement treated marine clay. International Journal of Engineering and Applied Sciences. **5(7)** 427-435.
[9] Gunturi M, Ravichandran P T, Annadurai R and Krishnan K D 2014 Study on strength characteristic of soil using soil stabiliser RBI-81. International Journal of Research in Engineering and Technology. **3(4)** 201-204.
[10] Chew S H, Kamruzzaman A H M, and Lee F H 2004 Physicochemical and engineering behavior of cement treated clays. Journal of geotechnical and geoenvironmental engineering. **130(7)** 696-706.
[11] Horpibulsuk S, Rachan R, Chinkulkijniwat A, Raksachon Y, and Suddeepong A 2010 Analysis of strength development in cement-stabilized clay from microstructural considerations. Construction and Building Material. **24** 2011-2021.
[12] Das S K and Parhi P S 2013 Stabilization of expansive soil using alkali activated fly ash. Proceeding of Indian Geotechnical Conference. 22-24.

Effect of Coarse Materials Percentage in the Shear Strength

B Alshameri[1,2], I Bakar[1], A Madun[3], L Abdeldjouad[3] and S Haimi Dahlan[4]
[1]Research Centre of Soft Soil-RECESS, Universiti Tun Hussein Onn Malaysia-UTHM, Batu Pahat, Johor, Malaysia
[2]Department of Exploration and Drilling, Yemen Company for Investment in Oil and Minerals-YICOM, Sana'a, Yemen.
[3]Faculty of Civil and Environmental Engineering, Universiti Tun Hussein Onn Malaysia-UTHM, Batu Pahat, Johor, Malaysia.
[4]Research Centre of Applied Electromagnetic, Universiti Tun Hussein Onn Malaysia, Batu Pahat, Johor, Malaysia.

E-mail: badee.alshameri@yahoo.com

Abstract. There are several factors that affecting the shear strength and shear strength parameters (i.e. cohesion and friction angle). In this study, the effect of coarse material percentage was tested. Six different mixtures of soils (clay and sand) with different coarse material percentages (i.e. from 80 % to 30% of coarse material percentage) were tested via using direct shear test under different moisture content percentage. The results indicated that the shear strength and friction angle were decreased by the increment of the percentage of coarse materials (sand). However, the cohesion results showed unique behavior. The cohesion (at every moisture content values) increased with the increment of the percentage of coarse materials until specific point then it started to decrease with the increment of the percentage of coarse materials.

Keywords: Soft soil, ground modification.

1. Introduction

Generally, several parameters gave effect to the shear strength of soils. These parameters such as: Particle size [1-4], particle shape [5-8], moisture content, density and percentage of coarse materials. However, in this study, the effect of the percentage of coarse materials was investigated. Table 1 shows the summary of findings from previous researchers regarding the relations between the shear strength and friction angle with particle size. Meanwhile, Table 2 shows the summary of some findings by the previous researchers regarding the relations between the shear strength, friction angle with percentages of coarse materials. In addition, there are some other parameters that can affect the shear strength such as: (1) applied normal stress, Liu et al. [9] and Li et al. [10], declared that the friction angle will decrease with the increment of the applied normal stress (or confining pressure). (2) Coefficient of uniformity, Liu et al. [9] indicated that, with the increment in the coefficient of uniformity, the friction angel would decrease. Kokusho et al. [11] mentioned the same thing (for coefficient of uniformity) if the soil did not contain crushable particles. (3) The size of specimen and the oversize particles have also affect on the shear strength [12-14]. (4) Shear rate has also influenced the shear strength [10, 15].

Table 1. Relations between the shear strength and friction angle with particle size.

parameter	Soil type	Relation to increase the particle size	References
Friction angle	Mixture of silt, sand and gravel	decrease	[3]
Undrained shear strength	Mixture of clay, silt, sand and gravel	Increase when the particle diameter > 20 mm	[9]
Friction angle	Mixture of clay, silt, sand and gravel	Increase	[16]

Table 2. Relations between the shear strength and friction angle with percentages of coarse materials.

Parameters	Soil type	Relation to increase the percentage of coarse materials	References
FA	Mixture of clay, silt, sand and gravel	Increase	[4]
FA	Mixture of clay and gravel	Increase	[6]
SH	Mixture of clay and gravel	Increase	[6]
FA	Mixture of sand and gravel	Increase	[7]
USH	Mixture of clay, silt, sand and gravel	Start to increase when the percentage of fine materials is less than 75%	[9]
FA	Mixture of clay, silt, sand and gravel	Increase	[16]
FA	Mixture of sand and gravel	Increase till specific point then decrease	[17]
USH	Mixture of clay and sand	High at 90% percentage of coarse material then decrease from 70% to 50% (the lower shear strength) and then decrease from 30% to 0%	[18]
USH	Mixture of sand and gravel (gravel <50%)	Decrease	[19]
FA	Mixture of sand and gravel	Increase	[20]
SH	Mixture of sand and gravel	Increase	[20]
SH	Mixture of sand and gravel	Increase	[21]
SH	Sand	Increase	[22]
FA	Simulation of mixture of soil	Increase	[23]
SH	Mixture of sand and gravel (gravel ≤60%)	Increase	[24]
FA	Mixture of sand and gravel (gravel ≤60%)	Increase	[24]
FA	Mixture of sand and gravel (gravel ≤50%)	Increase	[25]

SH is shear strength. FA is friction angle. USH is undrained shear strength.

2. Materials and test procedures

In this study, 141 compacted samples (compacted by using ASTM [26]) were tested by using direct shear test under ASTM [15] standard. The samples were classified to 6 mixtures soils (clay and sand) as the followings: (1) 80-20 (where 80 is 80% of coarse materials and 20 is 20% of fine materials < 75 μm), (2) 70-30, (3) 60-40, (4) 50-50, (5) 40-60 and (6) 30-70. Every soil mixture proportions was tested under three applied normal stresses, i.e. 10.5 kPa, 21 kPa and 31.5 kPa. However, due to the limitation of shear box size (100 × 100 mm), the maximum diameter size of coarse materials (sand) must be less than 3.35 mm. This was done to avoid soil particles to be oversized, thus to avoid overestimation in the shear strength and shear strength parameters values [22-24]. Meanwhile, the fine materials of kaolin were used due to the stable properties of kaolin compared with other clay minerals [27, 28]. Moreover, all the soil mixtures were compacted by using standard compaction test to achieve the maximum dry density. Therefore, the results on this paper included the optimum moisture content (OMC).

3. Effect of percentage of coarse materials in shear strength and shear strength parameters

The summary of shear strength results versus the percentage of coarse materials under different mixture and applied normal stress were tabulated in Table 3. The applied normal stresses in Fig. 1, 2, 3 were 10.5, 21.5 and 31.5 kPa, respectively. These plots indicate that the highest value of shear strength was at applied normal stress 31.5 kPa, while the lowest was at applied normal stress at 10.5 kPa. The mixtures 80-20, 70-30, 60-40, 50-50, 40-60 and 30-70 represent the percentages of coarse materials as 80, 70, 60, 50, 40 and 30 respectively.

Table 3. The range of shear strength for different mixtures under different applied normal stress.

	Range of shear strength in kPa					
The mixtures	80-20	70-30	60-40	50-50	40-60	30-70
Normal stress 10.5 kPa	19.3-44.2	26.9-53.4	32.6–63.2	25.6– 70.4	30.1–54.7	36.7–50.2
Normal stress 21 kPa	29.9–53.5	37.7-61.4	50.8-81.2	37.6-102	54.4-77.6	57.1-70.9
Normal stress 31.5 kPa	39.7-65.3	41.5-80.5	68.2-85.5	43.6-117	68.3-118	65.8-102

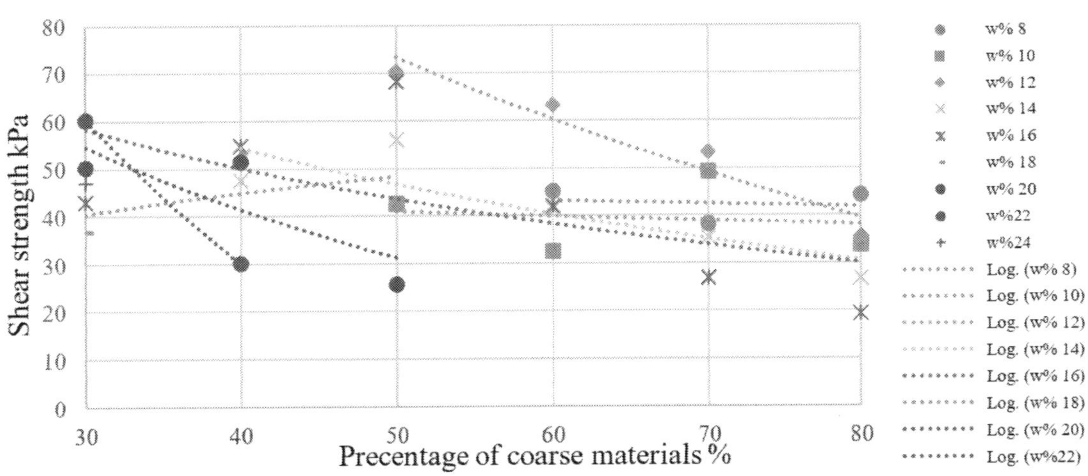

Figure 1. The plot of shear strength versus the precentage of coarse materials for applied normal stress equal to 10.5 kPa (where w% is the moisture content).

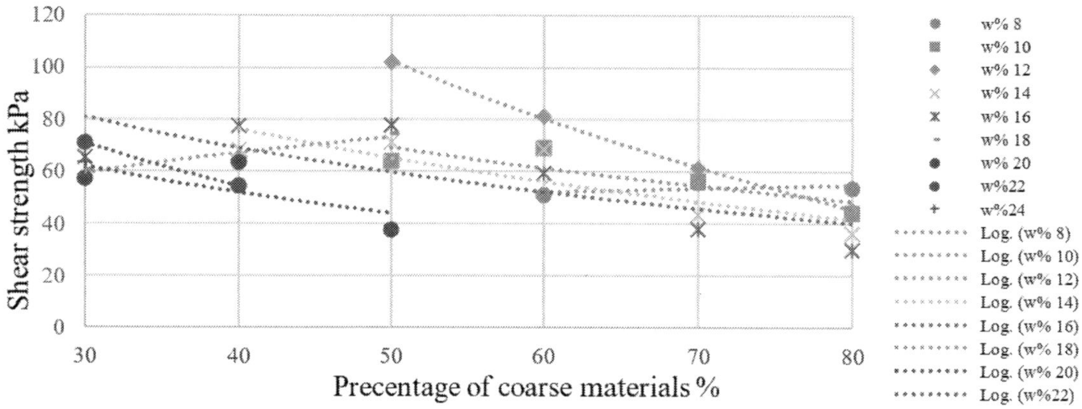

Figure 2. The plot of shear strength versus the precentage of coarse materials for applied normal stress equal to 21 kPa (where w% is the moisture content).

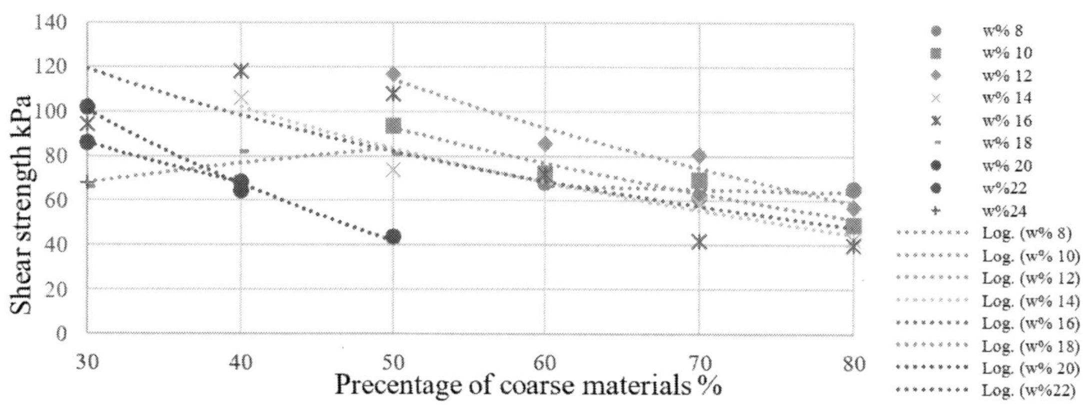

Figure 3. The plot of shear strength versus the precentage of coarse materials for applied normal stress equal to 31.5 kPa (where w% is the moisture content).

Moreover, Fig. 4 and 5 show the results of shear strength parameters (i.e. cohesion (c) and friction angle (Ø)) versus the percentage of coarse materials. Fig. 4 shows the results of cohesion versus the percentage of coarse materials. The results indicate that the cohesion tends to provide the concave down quadratic curve relation with percentage of coarse materials with most of moisture content points. On the other side, Fig. 5 shows the results of friction angle versus the percentage of coarse materials. The results indicate that the friction angle tends to provide linear relation with the percentage of coarse materials.

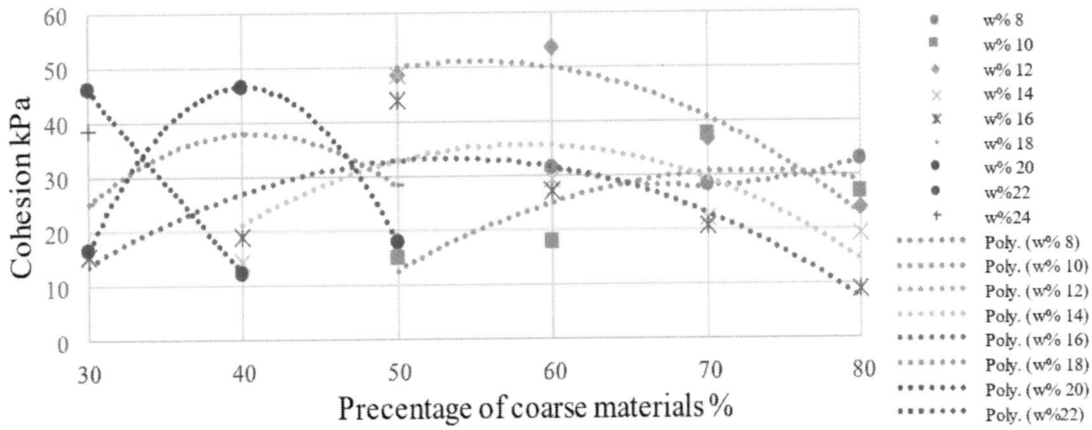

Figure 4. The plot of cohesion versus the precentage of coarse materials (where w% is the moisture content).

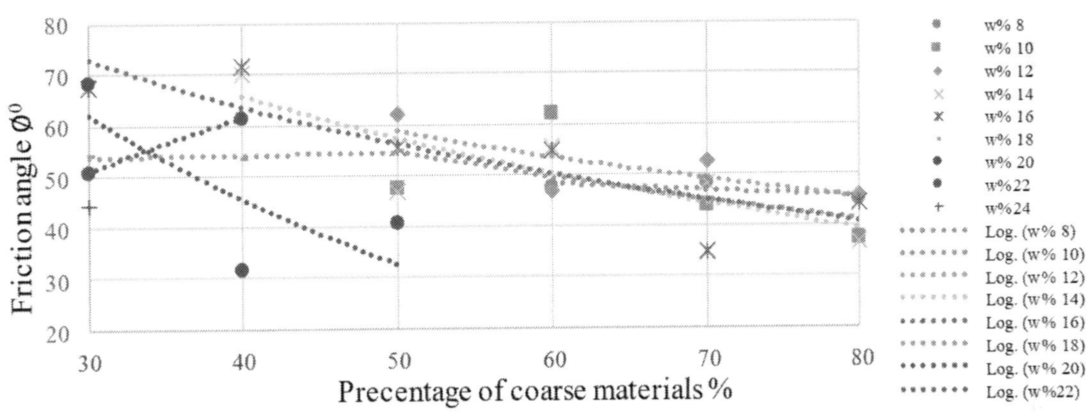

Figure 5. The plot of friction angle versus the precentage of coarse materials (where w% is the moisture content).

4. Discussion

The study indicates that the shear strength decreased with the increment of the percentage of coarse materials. This result has good agreement with previous researchers who studied on the mixture of clay and sand [12], and sand and gravel [13]. However, these outcomes disagree with other researchers who studied on the mixture of clay and gravel [6] and sand only [16]. The disagreement is due to the different shear rate, where a faster shear rate causes the pore water pressure to develop fast, and thus gives a low friction between the coarse particles. Thus, the decrement in the shear strength can be explained by the decrement of the friction angle with the increment of the percentage of coarse materials as shown in Fig. 5. Meanwhile, the cohesion shows different values at the same moisture content for all different mixtures proportions. Fig. 4 shows that the increment in the percentage of coarse materials led to the increment of the cohesion until it reaches maximum point. Then, with the increment in the percentage of coarse materials, the cohesion started to decrease, produced the concave down quadratic curve relation. On the other hand, with the increment of the applied normal stress, the shear strength increased. These results show an agreement with the findings by Liu et al. [9] and Yazdanjou et al. [20].

Conclusion

A series of tests were conducted by using direct shear box test to study the effect of the percentage of coarse materials in the shear strength and shear strength parameters. The results are as the followings:

- With the increment of the percentage of coarse materials, the shear strength and friction angle decreased.
- The cohesion trends increases with the increment of the percentage of coarse materials until maximum points, then it decreased with the increment of the percentage of coarse materials.
- With the increment of the applied normal stress, the shear strength increased.

Acknowledgments

Authors would like to express gratitude towards Ministry of Education of Malaysia-MOE, MTCP and Research Centre of Soft Soil-RECESS at University Tun Hussein Onn Malaysia-UTHM under grant Vot. U260 and MDR1320.

References

[1] Seminsky, L. 2013. The Shear Strength of Granular Materials with Dispersed and Non-Dispersed Oversized Particles (*Doctoral dissertation, University of Pittsburgh*).

[2] Hsieh, Chiwan Wayne, Gee Ham Chen, and Jeng-Han Wu. 2011. The shear behavior obtained from the direct shear and pullout tests for different poor graded soil-geosynthetic systems. *Journal of GeoEngineering* **6**, no. **1**: 15-26.

[3] Kim, Daehyeon, and Sungwoo Ha. 2014. Effects of Particle Size on the Shear Behavior of Coarse Grained Soils Reinforced with Geogrid. Materials **7**, no. **2** : 963-979.

[4] Li, Yanrong. 2013. Effects of particle shape and size distribution on the shear strength behavior of composite soils. *Bulletin of Engineering Geology and the Environment* **72**, no. **3-4**: 371-381.

[5] Cho, Gye-Chun, Jake Dodds, and J. Carlos Santamarina. 2006. Particle shape effects on packing density, stiffness, and strength: natural and crushed sands. *Journal of Geotechnical and Geoenvironmental Engineering* **132**, no. **5**: 591-602.

[6] Li, Y., Huang, R., Chan, L. S., and Chen, J. 2013. Effects of particle shape on shear strength of clay-gravel mixture. KSCE Journal of Civil Engineering. **17(4)**, 712-717.

[7] Yagiz, Saffet. 2001. Brief note on the influence of shape and percentage of gravel on the shear strength of sand and gravel mixtures. *Bulletin of Engineering Geology and the Environment* **60**, no. **4**: 321-323.

[8] Azéma, Emilien, Nicolas Estrada, and Farhang Radjai. 2012. Nonlinear effects of particle shape angularity in sheared granular media. *Physical Review* E **86**, no. 4: 041301.

[9] Liu, X. L., H. Loo, H. Min, J. H. Deng, L. G. Tham, and C. F. Lee. 2006. Shear strength of slip soils containing coarse particles of Xietan landslide. *Geotechnical Special Publication* **151**: 142-149.

[10] Li, Yanrong, Lung S. Chan, Albert T. Yeung, and Xiqiong Xiang. 2012. Effects of test conditions on shear behaviour of composite soil. *Proceedings of the ICE-Geotechnical Engineering* **166**, no. **3**: 310-320.

[11] Kokusho, Takeji, Tadashi Hara, and Ryousuke Hiraoka. 2004. Undrained shear strength of granular soils with different particle gradations. *Journal of Geotechnical and Geoenvironmental Engineering* **130**, no. **6**: 621-629.

[12] Wu, Po-Kai, Kenichi Matsushima, and Fumio Tatsuoka. 2008. Effects of specimen size and some other factors on the strength and deformation of granular soil in direct shear tests. *Geotechnical Testing Journal* **31**, no. **1**: 473.

[13] Cerato, Amy B., and Alan J. Lutenegger. 2006. Specimen size and scale effects of direct shear box tests of sands. *Geotechnical Testing Journal* **29**, no. **6**: 507.

[14] Fakhimi, Ali, and Hooman Hosseinpour. 2011. Experimental and numerical study of the effect of an oversize particle on the shear strength of mined-rock pile material. *ASTM geotechnical testing journal* 34, no. **2**: 131-138.

[15] ASTM D 3080. 2011. Standard Test Method for Direct Shear Test of Soils Under Consolidated Drained Conditions. West Conshohocken, PA: Am. Soc. Test. Mater. USA

[16] Wang, Jun-Jie, Hui-Ping Zhang, Sheng-Chuan Tang, and Yue Liang. 2013. Effects of particle size distribution on shear strength of accumulation soil. *Journal of Geotechnical and Geoenvironmental Engineering* **139**, no. **11**: 1994-1997.

[17] Simoni, Alessandro, and Guy T. Houlsby. 2006. The direct shear strength and dilatancy of sand–gravel mixtures. *Geotechnical and Geological Engineering* **24**, no. **3**: 523-549.

[18] Prakasha, K. S., and V. S. Chandrasekaran. 2005. Behavior of marine sand-clay mixtures under static and cyclic triaxial shear. *Journal of geotechnical and geoenvironmental engineering*. **131**: 213-222.

[19] Naeini, S. A. 2006. The ultimate shear behavior of loose gravelly sandy soils. *The Geological Society of London*. IAEG2006: 526

[20] Yazdanjou, V., N. Salimi, and A. Hamidi. 2008. Effect of gravel content on the shear behavior of sandy soils. *In Proc. of 4th National Congress on Civil Engrg*., Tehran University, Iran.

[21] Vallejo, Luis E., Sebastián Lobo-Guerrero, and Leanna F. Seminsky. 2014. Shear Strength of Sand-Gravel Mixtures: Laboratory and Theoretical Analysis. *In Geo-Congress 2014 Technical Papers@ Geo-characterization and Modeling for Sustainability*, pp. 74-83.ASCE.

[22] Shin, H., and J. C. Santamarina. 2012. Role of particle angularity on the mechanical behavior of granular mixtures. *Journal of Geotechnical and Geoenvironmental Engineering*. **139**: 353-355.

[23] Ueda, Takao, Takashi Matsushima, and Yasuo Yamada. 2011. Effect of particle size ratio and volume fraction on shear strength of binary granular mixture. *Granular Matter* **13**, no. **6**: 731-742.

[24] Hamidi, A., M. Alizadeh, and S. M. Soleimani. 2009. Effect of particle crushing on shear strength and dilation characteristics of sand-gravel mixtures. *International Journal of Civil Engineering* **7**, no. **1**: 61-72.

[25] Yagiz, Saffet. 2001. Brief note on the influence of shape and percentage of gravel on the shear strength of sand and gravel mixtures. *Bulletin of Engineering Geology and the Environment* **60**, no. **4**: 321-323.

[26] ASTM D 698. 2012. Standard Test Methods for Laboratory Compaction Characteristics of Soil Using Standard Effort (12400 ft-lbf/ft3 (600 kN-m/m3)). *ASTM International*, West Conshohocken, PA, USA.

[27] Das, B., 2007. *Fundamentals of Geotechnical Engineering*. Cengage Learning. pp 622

[28] Murthy, V. N. S. 2002. *Geotechnical Engineering: Principles and Practices of Soil Mechanics and Foundation Engineering*. CRC Press. pp 1029

Shear Strength of Remoulding Clay Samples Using Different Methods of Moulding

W Norhaliza[1,2], B Ismail[1,2], A T S Azhar [1,2] and N J Nurul[1,2]
[1] Research Center for Soft Soil, Universiti Tun Hussein Onn Malaysia, 86400 Batu Pahat Johor, MALAYSIA
[2] Faculty of Civil and Environmental Engineering, Universiti Tun Hussein Onn Malaysia, 86400 Batu Pahat Johor, MALAYSIA

E-mail: izawahab90@gmail.com

Abstract. Shear strength for clay soil was required to determine the soil stability. Clay was known as a soil with complex natural formations and very difficult to obtain undisturbed samples at the site. The aim of this paper was to determine the unconfined shear strength of remoulded clay on different methods in moulding samples which were proctor compaction, hand operated soil compacter and miniature mould methods. All the samples were remoulded with the same optimum moisture content (OMC) and density that were 18% and 1880 kg/m^3 respectively. The unconfined shear strength results of remoulding clay soils for proctor compaction method was 289.56kPa with the strain 4.8%, hand operated method was 261.66kPa with the strain 4.4% and miniature mould method was 247.52kPa with the strain 3.9%. Based on the proctor compaction method, the reduction percentage of unconfined shear strength of remoulded clay soil of hand operated method was 9.66%, and for miniature mould method was 14.52%. Thus, because there was no significant difference of reduction percentage of unconfined shear strength between three different methods, so it can be concluded that remoulding clay by hand operated method and miniature mould method were accepted and suggested to perform remoulding clay samples by other future researcher. However for comparison, the hand operated method was more suitable to form remoulded clay sample in term of easiness, saving time and less energy for unconfined shear strength determination purposes.

Keywords: Soft soil, ground modification, shear strength.

1. Introduction

The construction development on clay soil cannot be avoided because of the lack land space. Clay is a fined grained soil known as a very complex natural formations with high compressibility and low shear strength [1]. Soil in its remoulded state was commonly referred to as unstructured, and differences in mechanical behavior between a soil in its natural and remoulded states are considered to be the result of structure [2]. The fabric of soil is progressively distrupted when the soil was remolded and the behavior of the soil was altered. Depending on the size and strength of the soil particle the arrangement may alter the water retention and mechanical behavior of the soil and make it different from that of undisturbed soil of the same mineralogy [3]. There have been a lot of studies on developing constitutive models that consider the structure of the soil, such as those proposed by Nova and Gens, Liu et al. and Tanaka et al.[4,5,6]. They suggested new constitutive model for structured clays where its parameters can be readily determined in the laboratory.

The remoulded samples was very important to study because its very challenging and very hard to obtain undisturbed samples of soft soil at the site. Alshamib [7], state that to obtained the undisturbed samples was impossible due to the process of boring, driving the coring tool, raising and withdrawing the coring tool and extruding the soil specimens that can contributed some disturbance in the structure of the soft clay (Whitlow) [8]. Due to these problems, the remoulded samples was very important to study and understanding to overcome the difficulty process of taking undisturbed soil at the site.

2. Materials and Methods

2.1 Remoulding Sample

The disturbed clay soils from this study were collected at RECESS, UTHM, Batu Pahat, Johor. The disturbed clay was then remoulding through three different methods that were proctor compaction, hand operated and miniature mould. In this study, the proctor compaction method used as the based method for other methods that were hand operated method and miniature mould method. From the proctor compaction, the value of maximum dry density was 1525 kg/m^3 and the optimum moisture content (OMC) was 18 %. Then, the bulk density obtained was 1880 kg/m^3. Thus, to form the same density for all remoulded clay, 162 gram mass were kept constant. Then, the remoulded clay soils for all methods were formed with the specimens sizes 38 mm diameter and 76 mm of height. The procedure of samples preparation for proctor compaction method was according to (BS 1377: Part 8: 1990).

2.2 Remoulding Clay by Using Ordinary Proctor Compaction Method

Proctor test was designed to determine the dry density and moisture content relationship for compaction of soil based on BS 1377: Part 4: 1990. Figure 1 (a and b) shows the ordinary proctor compaction apparatus during and after soil was compacted. A cylindrical mould 1 liter or 1000 cm^3 volume filled with three layers of soil. This cylindrical mould has 105 mm internal diameter 115.5 mm effective height with removable extension collar. Each layer of soil were stamped with 25 blows per layer with a standard rammer 2.5 kg and a drop distance of 300 mm for each impact. The optimum moisture content was determined to be used in all other methods for mixing the soil.

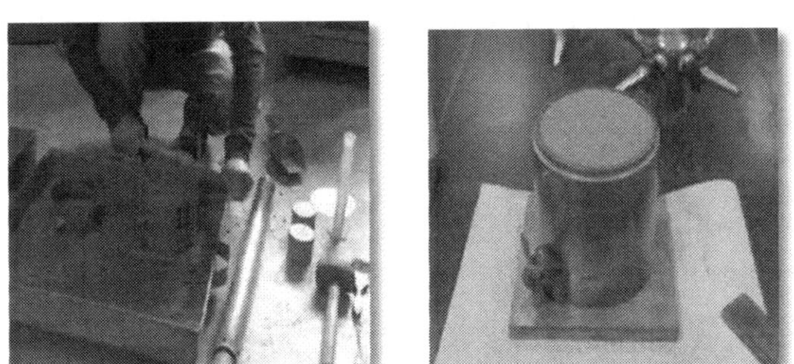

(a) During Compaction. (b) After Compaction.

Figure 1. (a and b) Ordinary Proctor Compaction During and After Compaction

2.3 Remoulding Clay by Using Hand Operated Soil Compactor Method

In this method, the soils were mixed with 18% of water content. Then, 162 gram of soils were transferred to a 38 mm diameter and 76 mm height a split mould and compacted in 4 layers of 40 g each layer used hydraulic hand operated compressor. The top of each layer was compacted properly to make sure the soil were bond together neatly. Then, the specimen was extracted out from the mould and marked to determine its orientation. Tang [9], state these method include tapping and static compaction where it's suitable to form the samples due to repeatability, workability and lower probability of soil sample to failure.

Figure 2. Hand Operated Soil Compactor.

2.4 Remoulding Clay by Customade Miniature Compaction (Miniature Mould)

Based on past researcher Mokhtar [10], the miniature mould apparatus was used in preparing remoulded samples for testing Unconfined Compression Strength test (UCS). In this method, the remolded clay soil was mixed with 18 % of water content and mixed well thoroughly with a spoon. Then, 162 gram mass of mixed soil was transferred to 38 mm diameter and 76 mm height of split mould and then compacted in 4 layers where 40 g for each layer. Each layer were tamped and compacted with customade miniature compaction tools: first tamped with Tool A for 40 times followed with Tool B for 40 times. Then, the specimen was extracted out from the mould and marked to determine its orientation. Mokhtar [10] stated that the determination of orientation was needed to standardize the orientation of top and bottom of the specimens during the tests and ensure exact duplicity in the specimens. Figure 2 shows the apparatus of hand operated soil compactor. The tools and compaction mould set was shown in Figure 3. The top of each layer of compacted properly before placing the following layers to improve the bonding between each layers. Then, the specimen was slid out from the 38 mm diameter and 76 mm height of the mould.

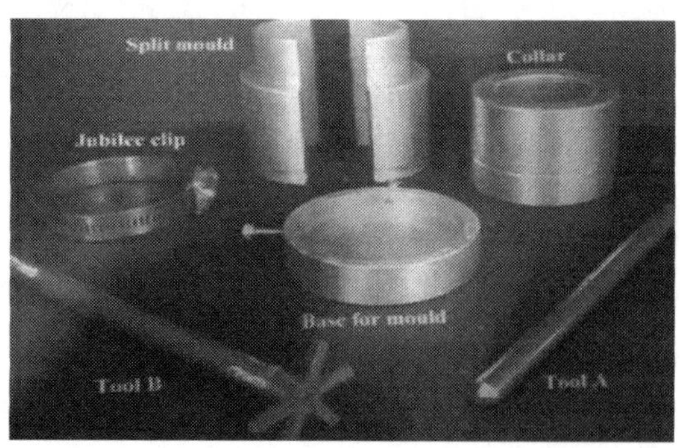

Figure 3. Customade Miniature Compaction (Miniature Mould) Equipment.

2.5 Unconfined Compression Strength Test

Unconfined compression strength test was accepted test to determine strength of more robust tropically weathered materials [11]. The purpose of unconfined compression strength test was to determine the unconfined shear strength of saturated clays quickly [12].The procedure of unconfined compressive strength test was defined as cylindrical specimen of cohesive soil where a steadily increasing axial load was subjected to the soil specimen until failure [13]. The main purpose of this test was to determine the unconfined compressive strength (q_u), which is then used to calculate the unconfined shear strength (c_u) value for all the methods.The reconstituted clay specimen for all methods were properly prepared and placed between the top and lower platens in the unconfined compression strength test machine that show in Figure 4. An axial load was then applied vertically at a standard strain rate of 1 mm per minute until the soil specimen was failed. Then the reading was taken and analysis.

Figure 4. Unconfined Compression Test Apparatus.

3. Results and Discussions

3.1 Clay Properties

The index properties of RECESS clay soil was 73.80 % of natural water content, liquid limit was 56 %, plastic limit was 37.19 % and specific gravity of clay soil was 2.6. The result of index properties was in the range with past previous study that shown in Table 1.

Table 1. Physical Properties of Typical RECESS Clay

	Author	Ibrahim and Chan [14]; Ho and Chan [15]	Alshamib [7]	Mohd Yusof [16]
Moisture Content (%)	73.8	80.5	-	53.95 – 74.65
Liquid Limit (%)	56.0	68.0	60.0	47.2 – 73.0
Plastic Limit (%)	37.19	31.7	32.27	19.83 – 41.1
Specific Gravity	2.6	2.62	2.6	2.52 – 2.66

3.2 Stress- Strain Behavior of Combination Method

Figure 5, shows the average of the unconfined compressive strength and the axial strain between all methods was 495.04 kPa – 579.12 kPa and 3.9 % - 4.8 % respectively. In this graph, the unconfined compressive strength of remoulded clay for proctor compaction method has record the highest value of 579.12 kPa with the strain 4.8 %. While, the unconfined compressive stress of miniature mould method has recorded the lowest value of 495.04 kPa with the strain 4.4% followed by hand operated method 523.32 kPa with the strain 3.9 %. The proctor compaction method recorded the higher strain failure because of the technique compaction makes the soil particle bind well to other soil particles. Das [17], state that the compaction effort can affect the soil particle.

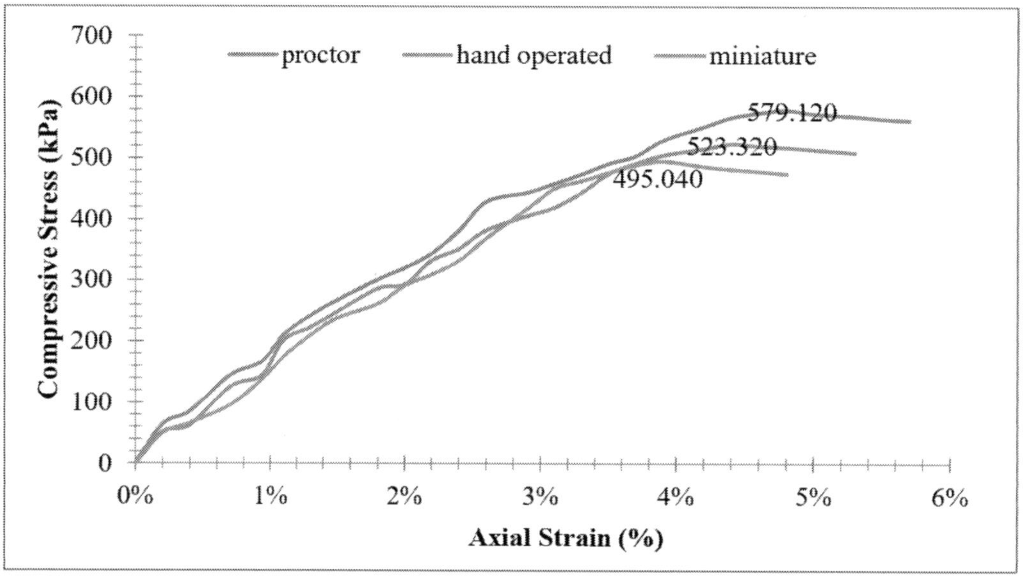

Figure 5. Axial Stress vs Axial Strain (All Three Methods).

Table 2 shows the average value of maximum unconfined compressive strength (q_u) and unconfined shear strength (c_u). The value of maximum unconfined compressive strength (q_u) for methods proctor compaction, hand operated and miniature mould were 579.120 kPa, 523.320 kPa and 495.040 kPa respectively. From this table, the lowest value of unconfined shear strength for miniature mould method was 247.52 kPa, followed by hand operated method, 261.66 kPa and the highest unconfined shear strength was by proctor method, 289.56 kPa. Based on the proctor compaction method, the reduction percentage of unconfined shear strength of remoulded clay soil for hand operated method was 9.66 %, and for miniature mould method was 14.52 %. This reduction percentage may cause from the type of methods conducted and the specimen handling manner. Yusoff et al [18], state the strength of soil specimen texture was affected since the samples was compacted. From these percentages, the hand operated method was more effective to form the remoulded clay samples because the percentage value of unconfined shear strength was lower than 10 %. However, the miniature mould method still can be accepted because it does not exceed than 15 %. Thus, it can be conclude that remoulding clay using hand operated method and miniature mould method were accepted and suitable to form the samples. However for comparison, the hand operated method was more suitable to form remoulded clay sample in term of easiness, saving time and less energy.

Table 2. The Average Value of Maximum Unconfined Compressive Strength and Unconfined Shear Strength.

Method	Unconfined Compressive Strength, q_u (kPa)				Unconfined Shear Strength, c_u (kPa)
	Sample 1	Sample 2	Sample 3	Average	
Proctor Compaction	589.187	574.079	574.079	579.120	289.56
Hand Operated	524.443	522.757	522.757	523.320	261.66
Miniature Mould	499.745	492.396	492.969	495.040	247.52

3.3 Types of Failure

Based on this study, failure mode of remoulding sample for proctor method is shearing as shown in Figure 6 (a). While failure mode on remoulding sample for hand operated method and miniature method are bulging as shown in Figure 6 (b) and (c) respectively. According to Das [17], the failure of soil specimens at the end of unconfined compression test are in two type that were shearing and bulging.

Figure 6a. Shearing.

Figure 6b. Bulging.

Figure 6c. Bulging.

Conclusion

In this paper, the remoulded clay sample for proctor compaction method used as comparison for other method that are hand operated method and miniature mould method. The unconfined shear strength results of remoulded clay soils for three different methods that were 289.56kPa (proctor compaction method), 261.66kPa (hand operated method) and 247.52kPa (miniature mould method). The unconfined shear strength readings for all methods were still in the range because the different of percentages of unconfined shear strength of remoulding clay between hand operated and proctor compaction method was 9.66% and remoulded clay for miniature mould and proctor compaction method was 14.52%. These three methods shows unconfined shear strength within to the range method that was proctor compaction method, and its proved that all three methods were suitable to used to performed the soil samples. Thus, it can conclude that remoulding clay of hand operated method and miniature mould method are accepted to perform remoulded clay samples. However based on the easiness, saving time and less energy used in preparing the remoulded samples, the hand operated method was more suitable for shear strength determination purposes.

Acknowledgment

The author greatly appreciate and express special gratitude to the supervisory team for the guidance, encouragement and valuable support and to all the staff members of Faculty Of Civil And Environmental Engineering Department (UTHM), Research and Soft Soils Centre (RECESS),

Research, Innovation, Commercialization and Consultancy Office (ORICC) and Centre for Graduate Studies (PS) who helped in many ways. Author also acknowledges the indebtedness and gratitude.

References

[1] Brand, E.W. and Brenner, R.P. 1981 Soft Clay Engineering. Elsevier Scientific Publishing Company. Amsterdam.

[2] Liu, M. C. 2011 Sydney Soil Model I: Theoretical Formulation. International Journal og Geomechanics, ASCE , 211-224.

[3] Liu, M. Z. 2013 Estimation of The Compression Behaviour of Reconstituted Clays. Engineering Geology , 84-94.

[4] Gens, A., and Nova, R. 1993 Conceptual bases for a constitutive model for bonded soils and weak rocks. Geotechnical engineering of hard soils- soft rocks. Edited by A. Anagnostopoulos, F. Schlosser, N. Kalteziotis, and R. Frank. A.A. Balkema, Rotterdam. Vol. 1, 485-494.

[5] Liu, M.D., Carter J. P. and Desai C. S. 2003 Modeling Com- Pression Behavior of Structured Geomaterials, International Journal of Geomechanics, ASCE, Vol. 3, No. 2, pp. 191-204.

[6] Tanaka, H., Hirabayashi, H., Matsuoka, T.& Kaneko, H. 2012 Use of Fall Cone Test as measurement of shear strength for soft clay minerals. Soils and Foundations , 590-599.

[7] Alshamib, A.M.M., 2010 Mechanical Properties of Remoulded Soft Clay. Master's Thesis. UTHM

[8] Whitlow, R. 2001 *Basic Soil Mechanics*. 4th Ed. England: Pearson Education Ltd.

[9] Tang, B. L., 2011 Reutilization of Organic and Peat Soils by Deep Cement Mixing. Master's Thesis. UTHM.

[10] Mokhtar, M. 2011 *Mechanical properties of Soft Clay Stabilized With Cement-Rice Husks (RH)*. Thesis Master's Universiti Tun Hussein Onn Malaysia.

[11] Bujang, H.K., Sew, G.S. and Ali, F.H.. 2004 Organic and peat soils engineering. University Putra Malaysia, Serdang, Malaysia,

[12] Budhu, M. 2007 Soil Mechanics and Foundations. (2nd ed.) Department of Civil Engineering and Engineering Mechanics, University of Arizona, USA. John Wiley and Sons, Inc.

[13] British Standard Institution (BSI) 1990 British Standard Methods of Test for Soils for Civil Engineering Purposes. Part 7:Shear strength Tests (total stress). BS1377.

[14] Ibrahim, K.A and Chan, C.M., 2007 Alternative road construction materials using clay admixed with natural wastes". *Proceedings of the Seminar on Geotechnical Engineering (SGE 2007)*, Batu Pahat, Johor, Malaysia.

[15] Ho, M.H and Chan, C.M., 2011 Some Mechanical Properties of Cement Stabilized Malaysia Soft Clay. International Journal of Civil, Environment, Structural, Construction and Architectual Engineering Vol:5, No:2

[16] Mohd Yusof, K. N., Lee, C.L., Phuai, P.G., Ahmad Tajuddin, S.A. 2006 The Correlations Between Chemical and Index Properties for Soft Clay of Peninsular Malaysia. *Technology and Innovation for Sustainable Development* (pp. 152-161).

[17] Das, B.M. 2007 Principles of Foundation Engineering. 7th Ed. United States of America: Christopher M. Shortt.

[18] Yusoff, S.A.N.M, Bakar, I., Wijeyesekera, D.C., Zainorabidin, A.,Madun, A., 2015 Comparison of Geotechnical Properties of Laterite, Kaolin and Peat. Applied mechanics and material 773, 1438-1442.

Development of Data Acquisition System for Consolidated Undrained Triaxial Test

L M Lee[1,2], T Yasuo[1], L C Wei[1] and L C Yuan[1]
[1]Universiti Tunku Abdul Rahman, Lee Kong Chian Faculty of Engineering & Science, 43000 Kajang, Selangor, MALAYSIA
[2]Faculty of Civil Engineering, Universiti Teknologi Malaysia, 81310 Skudai, Johor, MALAYSIA

E-mail: mllee@utar.edu.my

Abstract. Consolidated Undrained (CU) triaxial test is a common laboratory test used in practice for determining effective and total shear strength parameters of soil. This paper reported works carried out to develop a data acquisition system for a self-assembled triaxial machine. The developed system was capable of acquiring signals from the installed sensors (i.e. pressure transducer, load cell, LVDT), interpreting and presenting the data in real-time graphs. In addition, the study highlighted the advantages of performing double vacuuming method to saturate the soil specimen. The saturation can be obtained quicker and at a significantly lower cell pressure compared to the conventional stepwise increment of back pressure and cell pressure method.

Keywords: Triaxial test, residual soil, double vacuuming, effective stress, shear strength.

1. Introduction

Consolidated undrained (CU) triaxial test is a common laboratory test used for determining shear strength parameters of soil in practice. The CU triaxial test offers several advantages over direct shear test such as controllable drainage and stress conditions and measurable volume change [1]. The CU triaxial test with pore-water pressure measurement enables both the effective and total shear strength parameters (c', ϕ' and cu, ϕ_u) to be obtained, which makes it a popular test in practice. However, extra care should be given when selecting an appropriate triaxial test (e.g. Unconsolidated Undrained or Consolidated Undrained or Consolidated Drained) for a project as the selected test should replicate the actual soil conditions of the site.

For typical long term analyses, even though Consolidated Drained (CD) triaxial test may be more representative of the actual soil behaviors, the CU triaxial test with pore-water pressure measurement is normally preferred in practice due to its shorter shearing duration. Numerous researchers found that the effective shear strength parameters obtained from the CU test can be correlated well with those of CD test by applying a reduction factor of approximately 0.9 [2, 3]. The procedures of conducting the CU triaxial have been well documented in ASTM D4767-11 [4] and BS1377-Part 8 [5]. An operator is required to monitor and control the saturation, consolidation and shearing stages of the test which may span over several days. The testing time can be even longer when dealing with clayey soil with extremely low permeability, particularly during saturation stage. The main objective of the saturation stage is to ensure that all the voids of specimen to be fully filled with water without causing undesirable prestressing and swelling of the specimen. This process attempts to replicate a similar condition to that of actual field soil by saturating the soil followed by consolidation process in order to achieve equilibrium in a drained state under a known effective consolidation stress.

Recent advancement in computing and electronic technologies have seen introduction of fully automated triaxial test systems that can significantly shorten the testing duration of the CU triaxial test. The changing from one stage to another can be performed automatically without operation and monitoring of the operator. However, such system is considerably costly. Based on current practice in Malaysia, the soil saturation process of CU triaxial test is carried out by alternating the increase of back pressure and cell pressure. This normally results in an extremely high cell pressure. Eventually, the triaxial cell and other fittings of the triaxial machine need to be designed to sustain a considerably high pressure. Another drawback from the practice is that the high back pressure upon completion of saturation may destroy the soil structures and results in unrealistic testing results.

This study aims to improve the data acquisition and testing procedures of a self-assembled CU triaxial test machine. The detailed procedures of developing the data acquisition system are presented. In addition, the saturation duration of the soil specimen is shortened by adopting double-vacuuming method.

2. Materials and Methods

2.1 Soil Specimen

The soil specimen used in the present study was a typical tropical residual soil extracted from a site at Shah Alam, Selangor. The in-situ water content of the residual soil ranged from 19% to 22%, and the in-situ density was about 1.4 to 1.5 g/cm3. From the results of compaction test, the maximum dry density was 1.68 g/cm3 corresponding to the optimum moisture content of 19.6%. Based on Unified Soil Classification System (USCS), the soil was classified as Silty SAND.

2.2 Triaxial Test Setup

Fig. 1 shows the schematic diagram of the CU triaxial test setup while Fig. 2 shows the photograph of the actual setup. The instruments assembled in the present study have a similar setup with those typical CU triaxial tests, except that 2 units of convums were placed at the inlets of cell pressure and back pressure, respectively.

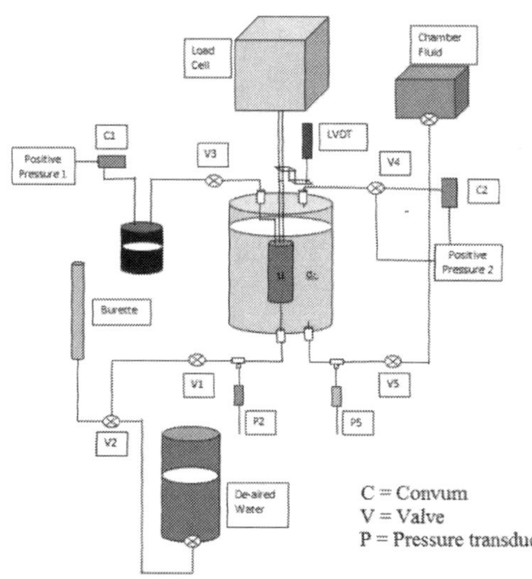

Figure 1. Schematic diagram of the Consolidated Undrained Triaxial setup.

Figure 2. Photograph of the actual Consolidated Undrained Triaxial setup.

Convum (Fig. 3) is a device used to convert positive pressure into negative pressure. In this study, double vacuuming method was used for saturating the specimen. The convum was required to generate negative pressure / suction in the cell and back pressure line. Calibration was required to obtain a correlation between the supplied positive pressure and the generated suction. Other sensors of the triaxial machine consisted of load cell for measuring the load increment during shearing, linear differential transformer (LVDT) for displacement measurement, and pressure transducer for measuring pressures in cell chamber and soil specimen.

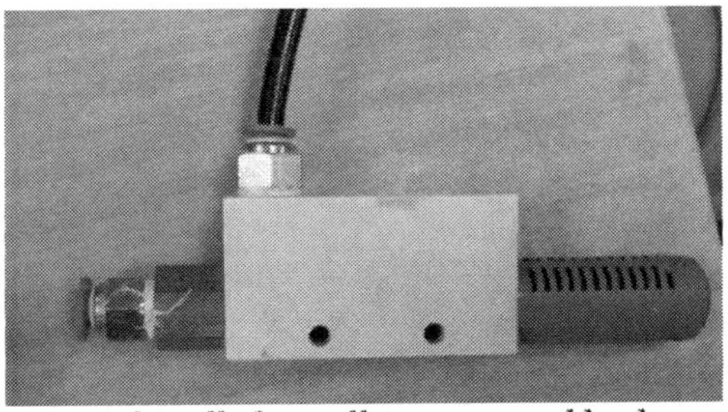

Figure 3. Convum installed at cell pressure and back pressure inlets.

3. Development of Data Acquisition System

All the sensors of the triaxial setup were connected to a connector box that consisted of 9 channels with 1.8V as a bridge excitation voltage. All the sensors used were of strain gauge type. This means that any physical changes during measurement will cause the strain gauges within the sensors to deform and generate an output voltage proportional to the strain. Each channel has two wires as output voltage terminals which have to be wired to the data logger to read the voltage.

The data logger used in the present study was Graphtec's midi LOGGER GL220. It consisted of 10 isolated channels which can measure and read the voltage from 0-50V, depending to the user's choice of amplitude range. Each sensor has a pair of output voltage terminals from the connector box, to be connected into the input terminals of a channel in the data logger. The data logger has instrument driver and software which were available online for connecting the data logger to a computer via a USB cable.

A set of programs or Virtual Instruments (VI) were written using Laboratory Virtual Instrument Engineering Workbench (LabVIEW) to bridge the communications between sensors, data logger and computer. LabVIEW is a system design software from National Instrument. In the present study, the developed program consisted of a main VI which configured the data logger with the sensors, and several sub-VIs for performing different data analysis tasks during different stages of triaxial test. In the main VI, data from the sensor channels were written into a global variable. Global variable can be imagined as a container that allows sharing of data of several items among several VIs, but under the same LabVIEW project. In the sub-VIs, the global variable will retrieve these data for specific processes. A sample of the developed program interface is shown in Fig. 4.

3.1 Saturation Stage

During saturation stage. Skempton's pore water pressure parameter (B value) was monitored. Several real-time graphs were developed in the sub-VI to monitor the changes of pore-water pressure and B value with time, as well as correlation between cell pressure and pore-water pressure.

3.2 Consolidation Stage

During consolidation stage, water was drained out from the sample. The volume of water drained from the sample is important to determine the degree of consolidation. A sub-VI was developed to present the volume change with time.

3.3 Shearing Stage

During shearing stage, a real time stress-strain curve was displayed in this VI to monitor and identify the failure point of the specimen.

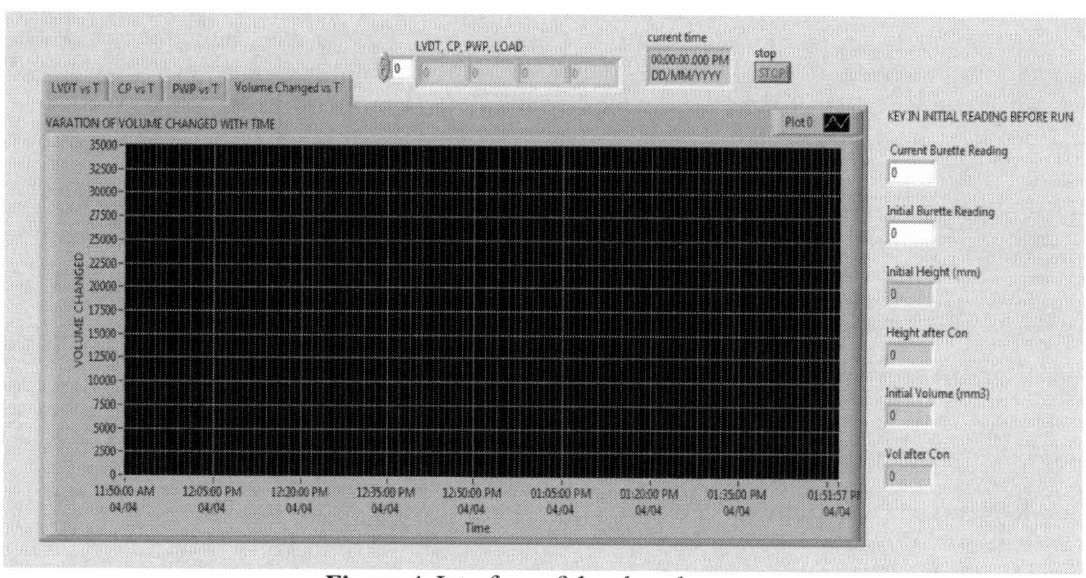

Figure 4. Interface of developed program.

4. Saturation by Double Vacuuming

Double vacuuming saturation method [6] was adopted in the present study. This method has been widely practiced in Japan. Two convums were used to perform the vacuuming process, i.e. one was used to vacuum the air in the specimen, while another was used to vacuum the air in the cell chamber. The difference in pressure between the specimen and the cell chamber was maintained at 20kPa, as detailed in Table 1.The pressures at each increment step were maintained for 5 minutes

Table 1. Cell and specimen pressures during double vacuuming saturation.

Step	Specimen pressure, u (kPa)	Cell pressure, σ_3 (kPa)
1	-10	0
2	-20	0
3	-30	-10
4	-40	-20
5	-50	-30
6	-60	-40
7	-70	-50
8	-80	-60
9	-90	-70

Upon applying the last step of negative pressures, the pressures were maintained for at least 30 minutes to achieve an equilibrium condition. Subsequently, de-aired water was introduced into the specimen. The water was allowed to flush through the specimen until the collected volume of water was about twice the specimen volume to ensure that all the voids had been filled with water. The entire process took about an hour.

The specimen was then subjected to Skempton's B test. The cell pressure was increased at an interval of 20 kPa until 120 kPa. The back pressure was applied accordingly to maintain the effective stress at 20 kPa. The cell pressure was then increased by an additional 100 kPa to monitor the increase of pore-water pressure. All the specimens successfully achieved Skempton's B coefficient of 0.95 or greater. By adopting the double vacuuming method, the specimen can be saturated quicker and easier compared to the conventional stepwise increment of back pressure and cell pressure. The entire saturation process was completed within 3 hours. More importantly, the final cell pressure upon completion of saturation can be maintained low (within 300 kPa). The subsequent procedures of conducting the consolidation and shearing stages were similar to that of typical CU triaxial test

5. Results

Fig. 5 shows the volume change against root time, deviator stress against strain, Mohr-Coulomb failure envelope at total stress, Mohr-Coulomb failure at effective stress obtained from the present CU triaxial test. The total shear strength parameters obtained were c = 16.5 kPa and ϕ = 22.5o, while the effective shear strength parameters were c' = 4 kPa and ϕ' = 31o. These parameters show good agreement with the testing results on similar residual soil reported by previous researchers [3].

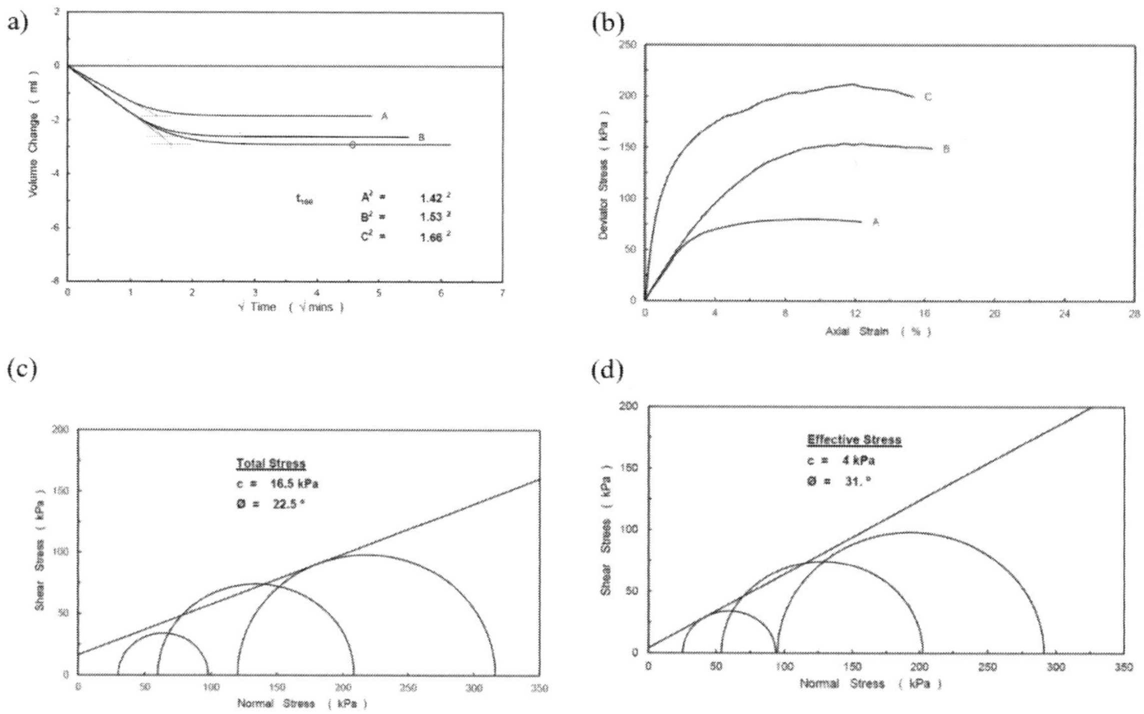

Figure 5. (a) Volume change Vs Root time, (b) Deviator stress Vs strain, (c) Mohr-Coulomb failure envelope at total stress, (d) Mohr-Coulomb failure at effective stress.

6. Conclusion & Discussion

This paper reported works carried out in the geotechnical laboratory of Universiti Tunku Abdul Rahman on developing a data acquisition system for a self-assembled Consolidated Undrained triaxial machine. The developed system was capable of acquiring signals from the installed sensors (i.e. pressure transducer, laod cell, LVDT), storage of the data, and presenting the data in real-time graphs. These features were useful for monitoring the measuring parameters at different stages of the CU triaxial test.

In addition, the study highlighted the advantages of performing double vacuuming method during saturation stage of the CU triaxial test. The saturation can be obtained quicker (within 3 hours) and at a significantly lower cell pressure compared to the conventional staging up back pressure and cell pressure method. The shear strength parameters of the residual soil obtained from the present study showed good agreement with the results reported by previous researchers.

Acknowledgement

The authors would like to thank Ministry of Education Malaysia for their financial support of the study under Fundamental Research Grant Scheme, grant no: 4450/L01.

References

[1] R.P. Brenner, V.K. Garga and G.E. Blight, Shear strength behaviour and the measurement of shear strength in residual soils, in: G.E. Blight (Eds.), Mechanics of Residual Soils, Balkema, Rotterdam, 1997, pp. 237.

[2] A.G. Salih and K.A. Kassim, Effective shear strength parameters of remoulded residual soil, Electronic Journal of Geotechnical Engineering. 17 (2012) 243- 253.

[3] R. Alias, A. Kasa and M.R. Taha, Effective shear strength parameters for remolded granite residual soil in direct shear and triaxial Tests, Electronic Journal of Geotechnical

Engineering. 19 (2014) 4559-4569.
[4] ASTM Standard D4767-11, Standard Test Method for Consolidated Undrained Triaxial Compression Test for Cohesive Soils. ASTM International, West Conshohocken, PA, 2011.
[5] BS 1377 Part 8, Effective Stress Triaxial Compression Shear Strength Tests. British Standards Institution, London, 1990.
[6] S.K. Ampadu, and F. Tatsuoka, Effect of setting method on the behavior of clays in trixial compression from saturation to undrained shear, Soils and Foundations. 33, 2 (1993) 14-34.

Soil-roots Strength Performance of Extensive Green Roof by Using Axonopus Compressus

N A Yusoff[1,3] M N Ramli[1], T N T Chik[1], H Ahmad[2], M F Abdullah[1], H Kasmin[1] and Z Embong[3]

[1] Faculty of Civil and Environmental Engineering, Universiti Tun Hussein Onn Malaysia, 86400 Parit Raja, Batu Pahat, Johor, Malaysia

[2] Horticulture Research Centre, MARDI Headquaters, P.O. Box 12301, 50774 Kuala Lumpur, Malaysia

[3] Research Center for Soft Soil, Universiti Tun Hussein Onn Malaysia, 86400 Batu Pahat Johor, Malaysia

E-mail: azizy@uthm.edu.my

Abstract. Green roof technology has been proven to provide potential environmental benefits including improved building thermal performance, removal of air pollution and reduced storm water runoff. Installation of green roof also involved soil element usage as a plant growth medium which creates several interactions between both strands. This study was carried out to investigate the soil-roots strength performance of green roof at different construction period up to 4 months. Axonopus compressus (pearl grass) was planted in a 1'x1' test plot with a designated suitable soil medium. Direct shear test was conducted for each plot to determine the soil shear strength according to different construction period. In addition, some basic geotechnical testing also been carried out. The results showed that the shear strength of soil sample increased over different construction period of 1st, 2nd, 3rd and 4th month with average result 3.81 kPa, 5.55 kPa, 6.05 kPa and 6.48 kPa respectively. Shear strength of rooted soil samples was higher than the soil samples without roots (control sample). In conclusion, increment of soil-roots shear strength was due to root growth over the time. The soil-roots shear strength development of Axonopus compressus can be expressed in a linear equation as: $y = 0.851x + 3.345$, where y = shear stress and x = time.

Keywords: Green roof, shear strength, soil-root interaction, axonopus compressus.

1. Introduction

Green roof is a roof of a building that is partially or completely covered with vegetation and growing medium. Green roof could serve several purposes in a building, such as absorbing or reducing rainwater, providing insulation layer, hence it could help to reduce urban air temperatures and in larger scale, it could combat the heat island effect. There are two types of modern green roofs which are intensive green roof and extensive green roof. Intensive green roof has a deeper soil medium and could support a wider variety of plants, but it has heavier load structure and requires more in maintenance.In contrast, extensive green roofs normally constructed in a shallower layer of soil, hence lighter vegetation is needed and are lighter than an intensive green roof [1].

Green roofs provided a number of important environmental benefits. It included improved building thermal performance, urban cooling, removal of air pollution and reduced storm water runoff [2,3]. In addition, green roofs can reduce the hydrological effect of vegetation function applied. It can withstand the hydrological effect by rainfall interception, surface water runoff, infiltration and evapotranspiration. Moreover, the green roof's vegetation could absorb carbon dioxide and through photosynthesis, oxygen is released, lead to reduction of greenhouse gases in the atmosphere [4].

The direct impact of raindrops on a thin layer of soil and the movement of water over the green roof's soil surface constitutes the major force initiating soil detachment. That case will cause structural bond between soil particles becomes weaker and lead to soil erosion [5]. As a result, the strength performance of green roof may become poorer.

Vegetation could protect soil from erosion of raindrops, runoff, water currents and wind. Plants could decrease an amount of rainfall that directly impacts the ground surface by intercepting and holding a portion of water on the leaves and stems. Vegetation could also influence the stability of soil slope when the roots act as reinforcement to the soil. Study done by [6] stated that the performance of slope stability is dependent on the types of plant, material used, the method of installation and its properties.

In this study, the geotechnical performance of soil-roots was assessed. This is a part of important element which significance in ensuring a longer life expectancy for the green roof. In addition, it is also significant to support its hydrological performance and insulation during the construction stage. This study investigated on the green roof strength performance at different construction period of less than a year. The relationship between shear strength and *Axonopus compressus* root growth of vegetated soil at different construction period was also presented in this study.

2. Method and Methodology

2.1 Plot Preparation
Six plots with a dimension of 1' x 1' were used (Fig. 1a) with five plots planted with the Axonopus compressus on a suitable soil medium. Soil medium used is a mixed of the following soil types: lateritic soil: organic soil: sand with a ratio of 5:3:1 respectively [7]. The sixth plot was not planted with any vegetation (bare soil as a control plot) as shown in Fig. 1b. The soil was self-compacted in order to resemble the common situation on-site.

2.2 Plot Maintenance
The vegetation was watered evenly with 2 kilograms of water, twice daily, to ensure that the planted grasses will grow well. However, the frequency of the plant watering is normally based on a few factors such as weather condition, the maturity of the plant, plant species and soil condition.

2.3 Direct Shear Test
Prior to direct shear test, it is important to ensure that the water content for each plot almost similar. A watering procedure had been set up by using 2 kilograms of water, which equals to half of the volume of the plot. After that, the plot is left under the roof area for 24 hours before the direct shear test can be performed.

2.4 Soil Sample Preparation
The mold with size 60 mm x 60 mm which penetrated into the soil in test plot was used to extract soil sample (Fig. 1c). Then, it was transferred into a direct shear apparatus that has the same dimension as previous mold. For the record, the sample thickness taken from the mold was 25 mm.

2.5 Shear Strength
Shear strength was measured at 100mm depth from the soil surface by using a Shear Trac II machine (Geocomp Inc. USA) at the Research Centre for Soft Soils (RECESS). This apparatus was fully operational supported by the software installed into a computer for data and results during the test. During consolidation phase, vertical stress used was 5 kPa. Direct shear was carried out with a shearing rate of 1.0 mm/minute and the maximum displacement of 20 mm (33.3% strain) was considered.

2.6 Root profiles

Once the direct shear test completed, the rooted sample was washed to remove soil from the clod leaving only the root specimens. This screening process enabled the root growth to be measured in term of mass at different construction period up to 4 months.

Figure 1. (a) Plot used for the study; (b) View of the six experimental plots; (c) A sampler has been penetrated into the soil.

3. Results and Discussion

3.1 Direct Shear Test Results

Figure 2 and Figure 3 show the shear strength test results for the control sample and rooted soil samples, respectively after a month of planting. For control specimen, three samples tested showed quite similar results of shear stress with an average of 3.72 kPa at 20mm displacement. In contrast, the average shear stress showed slight improvement of 3.81 kPa for the rooted soil sample in the first month.

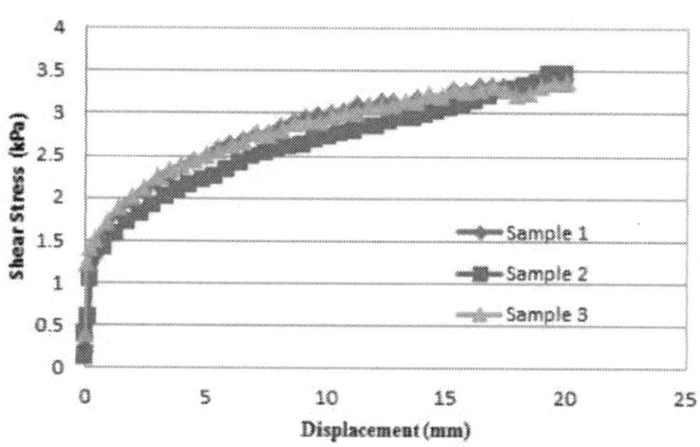

Figure 2. Shear stress versus displacement of the control sample.

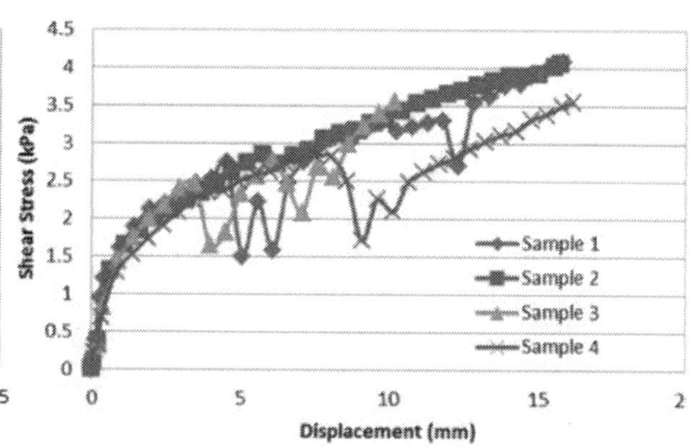

Figure 3. Shear stress versus displacement of month 1 rooted soil.

In addition, the results of the shear strength of rooted soil samples for the different construction period of 2nd, 3rd and 4th month after planting were presented in Fig. 4 to Fig. 6. From the results, the average shear stress for 2nd, 3rd and 4th months were 5.55 kPa, 6.05 kPa and 6.48 kPa respectively. Based on the results obtained for each experiment on different months, it can be noticed there was a gradual increment in shear strength reaching towards the end of the experiment at 20 mm displacement. The peak shear stress and residual shear stress cannot be clearly defined. It was similar to studies done by [8], where direct shear test results for several soil samples found to be continuously increased towards the end of the experiment at maximum displacement of 20 mm. In theory, it is understandable that the low compacted soil will behave as shown by this study.

Factor that leads to this trend was believed due to the increasing of vertical displacement during shearing phase. This is because the soil particles have been reoriented or dilatation occurred, resulting in increasing of shear stress. Another factor was due to the development of elongated and root growth beneath the soil. This situation caused the roots managed to hold soil particles in place. The mechanical reinforcement effect of soil-roots matrices prevent slippage of shear plane.

3.2 Comparison of Direct Shear Test Results

The comparisons performed on the results of direct shear tests for different construction time showed a linear increment in the average value of shear stress for different construction time, represented by a linear equation showed in graph Fig. 7.

Although there was a small improvement shown, but the results conformed with the predictions made by earlier studies. These results were parallel to a study conducted by [9], showing an increment of shear stress for different plant species at the age of 6 months and 12 months.

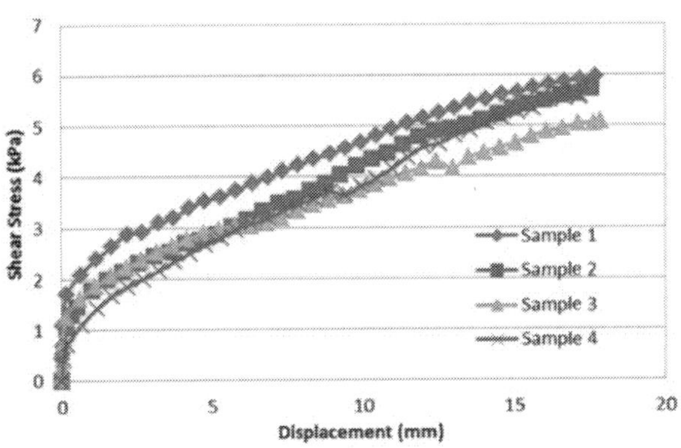

Figure 4. Shear stress versus displacement of month 2 rooted soil.

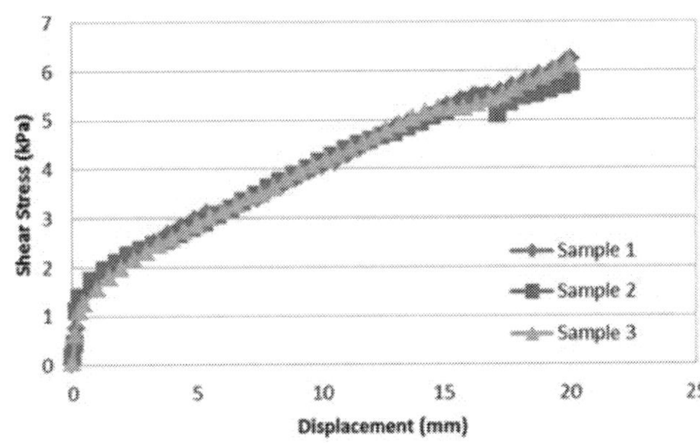

Figure 5. Shear stress versus displacement of month 3 rooted soil.

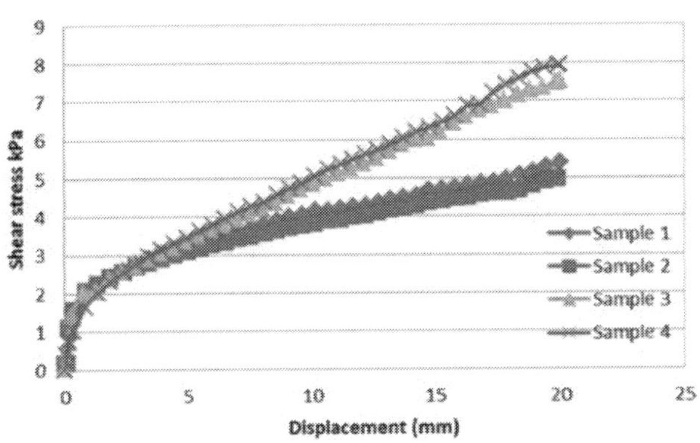

Figure 6. Shear stress versus displacement of month 4 rooted soil.

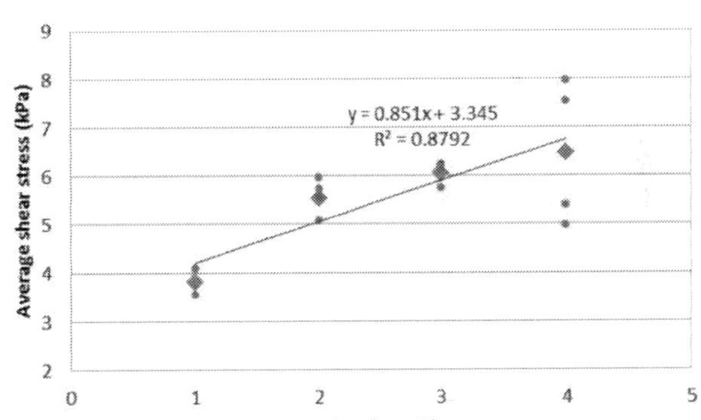

Figure 7. Average shear stress versus construction time.

3.3 Percentage of water content in soil samples

It is important to consider water content in soil samples during the direct shear test. Therefore, after completing the test, water content of the samples is recorded. It is observed, with the progression of time, the soil moisture content (%) was slightly difference between months (Fig. 8). This consistency would ensure the validity of shear strength of soil samples.

3.4 Effect of Root Development on Shear Stress

As expected at the beginning of this study, soil shear strength increased with the growth of roots. This was proven by studies [9,10] in which the mechanical effects of root development caused soil anchoring and shear strength increased. A graph of the root weight versus time is shown in Fig. 9. It shows the weight of the growing roots at the different construction time from month 1 to month 4.

Moreover, the increments of root weight also contribute to the increase of soil shear strength as shown in Fig. 10.

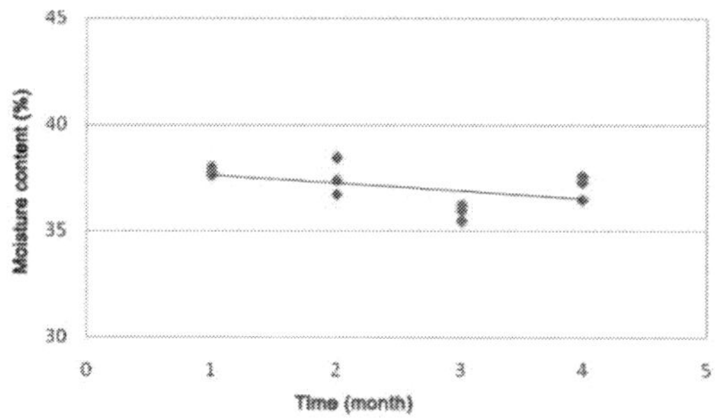

Figure 8. Percentages of soil moisture contain versus construction period.

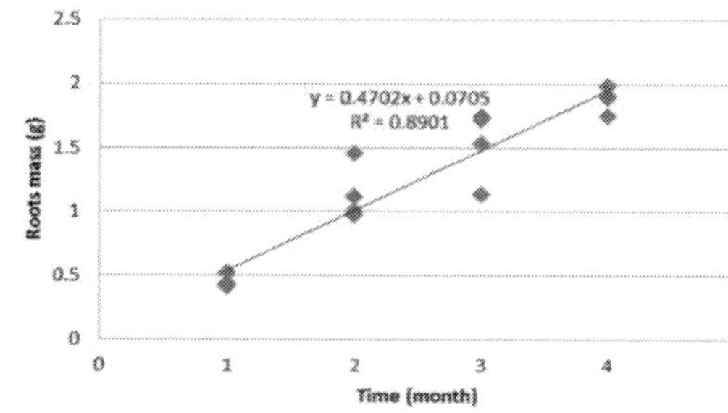

Figure 9. Root mass versus different construction period.

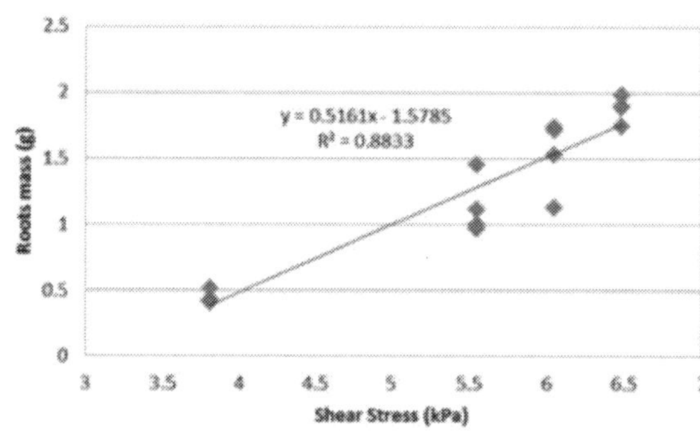

Figure 10. Root mass versus average shear stress.

4. Conclusion

The overall results have proven the shear strength of rooted soil samples is higher than the soil samples without roots (control sample). These results clearly demonstrated by the shear strength performance of soil-roots of Axonopus compressus (pearl grass) used in the study. Results show that shear stress for soil samples without root (3.72 kPa) was lower than shear stress for rooted samples (3.81 kPa). In terms of shear strength and root mass, shear strength of soil showed an increment due to plant growth in the root structure over the time. This is because the roots gripped into the ground resulted in strengthened of soil. This relationship can be expressed in a linear equation of shear stress and time; roots weight and time; and lastly root weight and shear stress are shown in Eq. (1), Eq. (2) and Eq. (3), respectively.

$$y = 0.851x + 3.345 \quad (1)$$

$$y = 0.470x + 0.070 \quad (2)$$

$$y = 0.516x + 1.578 \quad (3)$$

where y = roots weight, x = shear stress

References

[1] Getter K L, Rowe D B, 2006 The role of extensive green roofs in sustainable development, Journal of HortScience. 41(5) 1276 - 1285

[2] Bates A J, Mackay R, Greswell R B, Sadler J P, 2009SWITCH in Birmingham, UK: Experimental investigation of the ecological and hydrological performance of extensive green roofs, J. Reviews in Environmental Science and Bio/Technology. 8(4) 295-300

[3] Oberndorfer E, Lundholm J, Bass B, Coffman R R, Doshi H, Dunnett N, Gaffin S, Kohler M,

Karen K.Y Liu, Rowe B, 2007Green roofs as urban ecosystems: ecological structures, functions, and services. BioScience 57(10) 823-833

[4] Kasmin H, Yusoff NA, Samsudin M H, Atan N, Kassim R, Ahmad H, Sayuti Z, Abdullah F, 2015Preliminary overview on thermal performance of green roof, Applied Mechanics and Materials Vols. 773-774 1047-1052

[5] Agassi M, 1995Soil Erosion, Conservation, and Rehabilitation. Marcel Dekker, Inc., New York, United States of America

[6] Mafian S, Bujang B K H, Ghiasi V, 2009 Evaluation on root theories and root strength properties in slope stability. European Journal of Scientific Research. 30(4) 594-607

[7] Hung A Y, Kasmin H, 2011 Effectiveness of green roof in stormwater management: Rainfall runoff performance. UTHM Undergraduate Final Year Project

[8] Goldsmith W, 2006 Soil strength reinforcement by plant. The Bioengineering Group, Inc.

[9] Ali F H, Osman N, 2007 Soil-roots composite: correlation between shear strength and some plant properties, Electronic Journal of Geotechnical. 12D

[10] Ali F, 2010 Use of vegetation for slope protection: root mechanical properties of some tropical plants, International Journal of Physical Sciences. 5(5) 496-506

Numerical Modelling of Embankment on Soft Clay

M M Nujid[1] and M R Taha[1]
[1]Department of Civil Engineering and Structural Engineering, Universiti Kebangsaaan Malaysia 43650 Malaysia

E-mail: masyitahnujid@gmail.com

Abstract. This paper aims to predict deformation of embankment on soft clay of Muar. The prediction performance focusing on displacement at critical fill height of 5.5 m. The study was based on reported result in 1992. With the aid of computer intelligence, the advanced constitutive soil models could be adopted to analyze the soft clay behavior. The COMSOL Multiphysics (v4.4) has been used to simulate the problem with coupled physics available in the software. The vertical displacements are in good agreement close to published result.

Keywords: Embankment, soft clay, constitutive soil model, COMSOL multiphysics.

1. Introduction

Embankment has being used as a part of construction to build temporary structure on road, railway, dam and dike project in geotechnical engineering. Two components need to be taken seriously during initial stage of design are stability and deformation analysis. In present days, numerical tools provide solution in fast and accurate simulation of the geotechnical problems. The solution offers from closed form as simple to complex solution such as limit equilibrium method and finite element method. In theoretical solution four conditions should be satisfied in analysis and the basic input components are equilibrium, compatibility, material constitutive behaviour and boundary conditions [6].

Muar trial embankment on soft clay has become a subject since late 1980's to early 1990's to study the stability and deformation behaviour of the soft soil. The stability and deformation of the Muar soft clay of trial embankment were analysed and compared using Limit Equilibrium Method (total stress and effective stress analysis) and Finite Element Method (elastoplastic model) in computer programme [10].

The comparison also supported from previous data and presented by hand-calculation on similar problem and found some discrepancies had occurred. The discrepancies due to neglecting the fill strength, soil strength modulus and incomplete lateral movement data under the embankment. Due to neglecting fill strength during analysis, [2] took an effort to predict the fill strength and thickness of embankment at failure. It was found from back analysis at fill thickness of 5.4 m using uncorrected and corrected factor of 0.8 vane strength were 37 kPa and 51kPa respectively at internal friction angle, $\phi = 30°$. The study predicted well the failure strength of fill embankment from back analysis were around 42 kPa and 56 kPa and found that corrected factor of 0.8 gave high strength.

The vertical displacement, horizontal displacement and consolidation effects on soft clay of Muar embankment behaviour also been carried out by [7]. The loading was placed in stages construction with soil deformation behaviour was observed in each stage (early, during and end of construction). Undrained behaviour of soft clay developed in normal consolidated at early construction of embankment while the development of pore pressure in over consolidated clay which turn to normal consolidated clay during construction. The predicted embankment deformation failure also carried out in numerical simulation which effect of undrained, drained shear strength and effect of pore pressure incorporated in advanced constitutive soil models [8]. There are three ways to apply incremental embankment load in finite element analysis [3]:1):surface loading, 2) increasing the gravity of embankment element and 3) placing a new layer of embankment element.

In present study, this paper is attempted to predict soft clay behaviour using different constitutive soil models on trial embankment in Muar. The deformation is focus on displacement of fill embankment and presented results in pattern. The behaviour factors influence the deformation of embankment on soft clay are identified and investigated.

Theoretical and Mathematical Model

The soil behaviour can be described using stress and strain quantities in p-q space. The mean and the deviatoric stresses are given by:

$$p'=\frac{1}{3}(\sigma'_{11}+\sigma'_{22}+\sigma'_{33}) \quad (1)$$

$$q=\frac{1}{\sqrt{2}}\sqrt{[(\sigma'_{11}-\sigma'_{22})^2+(\sigma'_{22}-\sigma'_{33})^2+(\sigma'_{33}-\sigma'_{11})^2+6(\sigma'^2_{12}+\sigma'^2_{23}+\sigma'^2_{31})]} \quad (2)$$

In which σ'_{ij} are the Cartesian components of effective stress. The stress ratio η is given by:

$$\eta=\frac{q}{p} \quad (3)$$

The corresponding incremental volumetric and deviatoric strains are defined as:

$$d\varepsilon_v=d\varepsilon_{11}+d\varepsilon_{22}+d\varepsilon_{33} \quad (4)$$

And

$$d\varepsilon_d=\frac{\sqrt{2}}{3}\sqrt{[(d\varepsilon_{11}-d\varepsilon_{22})^2+(d\varepsilon_{22}-d\varepsilon_{33})^2+(d\varepsilon_{33}-d\varepsilon_{11})^2+6(d\varepsilon^2_{12}+d\varepsilon^2_{23}+d\varepsilon^2_{31})} \quad (5)$$

2. Methodology

2.1 Constitutive Soil Model

The embankment problem is assumed in plane strain condition and behave as an elastic-perfectly plastic material. Before yielding within the elastic range, the stress-strain relationship of the soil is described by Hooke's law. Beyond the elastic range, the soil is assumed to behave as perfectly plastic in accordance with Mohr-Coulomb, Drucker-Prager and Modified Cam Clay. All material are assumed to fail in von Mises yield criterion. In COMSOL Multiphysics (v4.4) programme [5] these three models are selected to be used in analysis.

IOP Conf. Series: Materials Science and Engineering **136** (2016) 012021 doi:10.1088/1757-899X/136/1/012021

2.2 *Modified Cam Clay Model*

Modified Cam Clay is proposed and applied in most geotechnical problems associated with soft clay to account for flow rule which originally from Cam Clay model based on Critical State theory. Characteristics of linear isotropic Modified Cam Clay (MCC) are [1]:

1) Logarithmic relationship between the mean effective stress, p' and the void ratio,e
2) Linear stress dependency of the stiffness
3) Distinction between isotropic primary loading, unloading and reloading

The model best in performing geotechnical application such as embankment or foundation on soft soil which involve loading condition. The yield criteria of MCC model is an ellipse and smooth in p-q plane, with a cross section independent of Lode angle and defined as a function of the pre consolidated pressure (p_c) is described by:

$$f_y = q^2 + M^2(p - p_c)p = 0 \quad (6)$$

It is assumed that the soil obey the normality condition and plastic potentials (g) are the same as yield function (*f*) in the p-q plane:

$$g = f = q^2 + M^2(p - p_c)p = 0 \quad (7)$$

In the Cam Clay model, hardening is controlled by the consolidation pressure, p_c which depends exponentially on the volumetric plastic strain ε_{pvol}.

$$p_c = p_{co}e^{-B_p \varepsilon_{pvol}} \quad (8)$$

$$B_p = \frac{1 + e_o}{\lambda - \kappa} \quad (9)$$

Here the parameter p_{co} is the initial consolidation pressure, and the exponent B_p is a parameter which depends on the initial void ratio e_o, the swelling index κ and the compression index λ. The initial void ratio, the compression index and the swelling index are all positive parameters and must fulfil.

$$0 < \kappa < \lambda, B_p > 0$$

2.3 Numerical Modelling

2.3.1 *Muar Embankment*

A case study of Muar trial embankment [8] has been chosen based on the level of available information on the properties of the foundation and embankment filling, and the amount, type, location, and available published results of instrumentation.

The local highway authority had decided to construct fifteen trial embankment on the Muar plain to build Malaysian North-South expressway crosses 10-20 m thick soft clay deposits having low undrained shear strength with high water content threats to surface structure failure, low bearing capacity and excessive settlement. The site of trial embankment is located 20 km inland from Muar and 50 km due east of Malacca on the southwest coast of Malaysia.

Malaysian Highway Authority (MHA) [7] had provided an in situ data such as soil condition, water contents, Atterberg limit, vane shear strength, cone penetration resistance tests of the soft Muar clay to a depth of about 17.5 m below ground level. Besides the in situ data, the laboratory testing were conducted such as direct shear test and triaxial test to obtain engineering properties of soft clay Muar. The water content is high as 100% (50-120%) and generally exceed the liquid limit (40-80%) and plasticity index in range 40-50% [2,8,10,7]. The vane shear strength has minimum value of 8kPa at 3 m depth and increase linearly with depth. The fill embankment properties were given as follow: E=5100kPa, ν=0.3,γ=20.5kN/m^3, c=19kPa, and ϕ=26°). Table 1 summarizes information data from soil condition, index and engineering properties of soft clay **from MHA.**

2.3.2 COMSOL Multiphysics

The COMSOL Multiphysics [5] finite element software offers real simulation problems with partial differential equation applied in the software with elements of multiphysics such as heat, particle tracing and water flow. In this study there are three physic branches have been selected: 1) Solid Mechanics; Geomechanic, 2) Fluid and Flow; Darcy Law and 3) Mathematics; ODE and DAE. Normally the processes of numerical simulation involve with assigning geometry, material, physic branch, mesh, analyze and viewing result [11]. The deformation behavior and failure embankment are obtained from displacement curve. The analysis is done in function of undrained strength of fill embankment and different constitutive soil models.

IOP Conf. Series: Materials Science and Engineering **136** (2016) 012021 doi:10.1088/1757-899X/136/1/012021

Table 1 Soil Properties of Muar Embankment [8].

Depth	Soil type	Unit weight ,γ	Grain size (%)				Coefficient of horizontal permeability, kh	Compression ratio,Cr	Pre-consolidation pressure,pc	Shear strength		Deformation
(m)	-	(kN/m³)	Clay	Silt	Sand	Gravel	(m/sec)	$Cc/1+e_o$	(kPa	c'(kPa)	ϕ'(°)	E_u(kPa)
0-2.0	Weathered crust	16.5	62	35	3	0	-	0.3	110	8	12.5	25500
2.0-8.5	Upper soft silty clay (very soft clay)	15.5	45	52	3	0	4×10^{-9}	0.5	40	14(2-5m) 22(5-8m)	14(2-5m) 7(5-8m)	6600(2-5m) 8933(5-8m)
8.5-18.5	Lower soft silty clay (soft clay)	15.5	50	47	3	0	10^{-9}	0.3	60	9(5-8m) 16(11-14m) 14(14-20m)	20(5-8m) 17(11-14m) 21.5(14-20m)	9120(8-11m) 6593(11-14m) 5884(14-20m)
18.5-19.0	Peat	-	-	-	-	-	-	-	-			
19.0-19.9	Sandy clay	16.5	20	36	44	0	2×10^{-7}	0.1	60			
19.0-22.5	Dense sand		4	20	71	5	-					

2.3.3 Geometry of Embankment, Layered Soils, Soil Properties, Boundary Condition and Mesh Discretization

The embankment and soil conditions were model in two dimensional plane strain with stationary effect (Fig.1). The foundation depth of 22.5 m was considered enough in geometry model because the dense sand layer below the soft clay while the lateral boundary of geometry model was three times of vertical height below the toe of embankment [8]. The boundary conditions was assigned as a fixity at bottom of the soil geometry and a roller on right of the model. Due to assumption symmetrical geometry of the fill embankment and soil condition, only half side of the geometry was modelled and the symmetrical boundary condition has been selected on the left side of the geometry problem. The free triangular mesh and extra fine element size with number of five were selected to the soil body (Fig.2). The fill loading was acting as pressure beneath the embankment. All soil properties were undrained parameters and the initial horizontal, vertical stresses and pore pressure were given in Table 2.

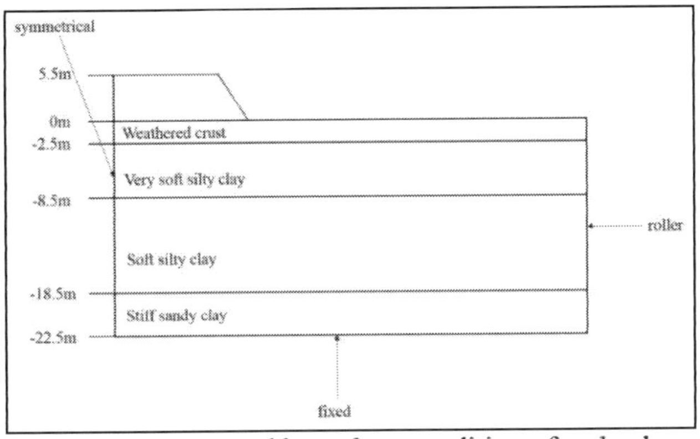

Figure 1. Geometry and boundary condition of embankment.

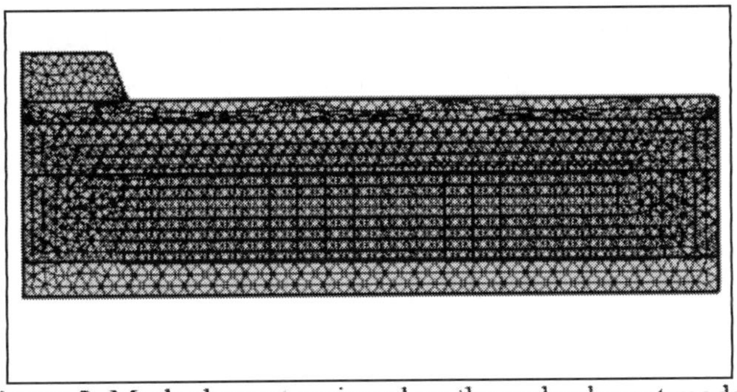

Figure 2. Mesh element assigned on the embankment model.

Table 2. In-situ stress condition and Modified Cam Clay (MCC) parameters at Muar Site.

Depth (m)	σ_{ho} (kPa)	σ_{vo} (kPa)	u (kPa)	κ	λ	e_{cs}	M	ν
0-2.5	13.2	22.0	16.7	0.05	0.13	3.07	1.19	0.3
2.5-8.5	33.7	56.1	75.5	0.05	0.13	3.07	1.19	0.3
8.5-18.5	67.9	113.1	173.6	0.08	0.11	1.61	1.07	0.3
18.5-22.5	81.5	135.9	212.9	0.10	0.10	1.55	1.04	0.3

3. Results and Discussions

For validity purpose, the deformation behaviour of embankment is conducted in three types of constitutive soil models from simple Mohr Coulomb to Drucker-Prager and finally Modified Cam Clay. The displacement results in Table 3 were obtained for fill height of 5.0 – 5.5 m. The Drucker-Prager model shows the highest total and vertical displacment of the embankment failure whereas the smallest horizontal displacement is Modified Cam Clay model. By referring to previous result the vertical displacement from Mohr-Coulomb and Modified Cam Clay models close to results presented in [8].

Table 3. Displacement result for embankment failure.

Displacement (m)	Mohr-Coulomb	Drucker-Prager	Modified Cam Clay	Modified Cam Clay (Indraratna et al. 1992)
Vertical	0.25	0.02	0.22	0.20

3.1 Deformation Pattern

Fig 3. shows displacement pattern for all constitutive model adopted in the study. The deformation pattern concentrate at 8.5-18.5 m depth below ground surface in soft clay layer. The Drucker-Prager model shows the highest difference compare to Mohr-Coulomb and Cam Clay which both the models have similarity with slightly difference. The deformation does not exceed beyond the soft clay layer where a state of K_o condition is approached toward the sandy soil deposits [8].

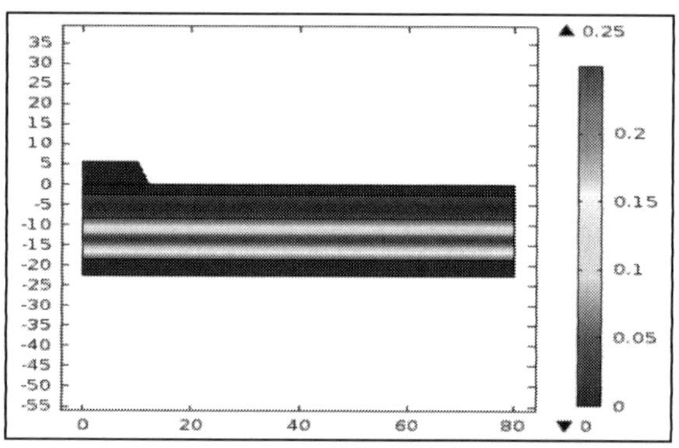

(a)

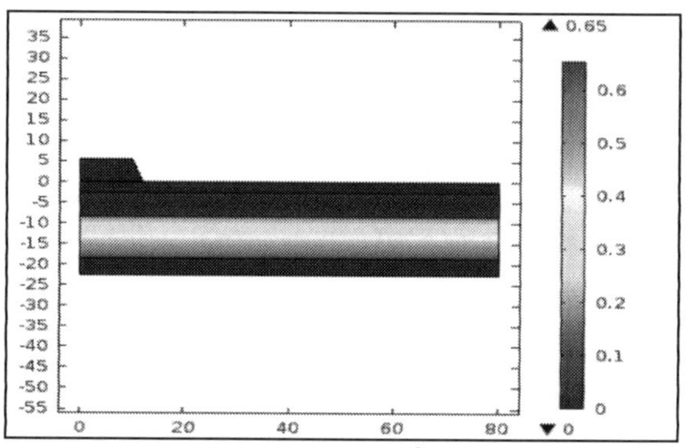

(b)

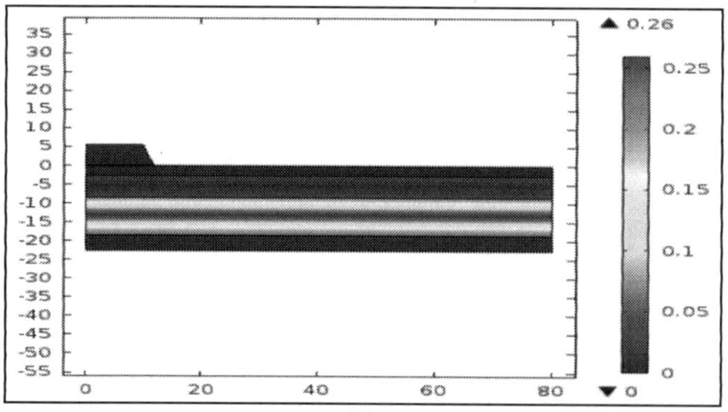

(c)

Figure 3. Displacement deformation pattern; (a) Mohr-Coulomb, (b) Drucker-Prager and (c) Modified Cam Clay.

3.2 Discussions

The deformation on displacement of fill embankment on soft clay has presented and several factors are identified due to discrepancy among the constitutive behaviour implemented in the analysis. Initially in numerical process input of material parameters play a significant role to the study. An uncorrected undrained vane shear strength of fill embankment has been assigned and basic properties were used accordingly to case study. A corrected factor of 0.8 on undrained vane shear strength increase the soil strength [4,2]. The significant difference is parameters for each soil models are not identical since the suitability in choosing the right soil model to be used in prediction the geotechnical problem behaviour accordingly to soil type. The Modified Cam Clay parameters on e – log p graph (virgin load/normal consolidated) give a parameter λ which change the displacement result influenced by initial stress condition and nature of stress path testing. Assumption in the associated flow rule overestimated the displacement and the MCC model most suitable for soft clay soil [8,1]. Assumption the absence of tension crack and shear band effect in the analysis could make some error in simulation and ignoring the importance of stages construction in embankment fieldwork and computer simulation [3] could lead to overestimated displacement to the deformation since the settlement occur in short term stability where undrained analysis take place in oppose to real in situ problem [9]. Besides this important assumptions on numerical analysis such as assigning boundary condition, loading and soil model could lead to differences in the result analysis. Even though Mohr-Coulomb and Drucker-Prager are elastic perfectly plastic, both soil models have different assumptions in their failure criterion in which Mohr-Coulomb defines an irregular hexagonal pyramid in the space of principal stresses, which generates singularities in the derivatives of the yield function and do not include the intermediate principal stress effect. The Drucker-Prager model neglects the influence of the invariant J3 (introduced by the Lode angle) on the cross-sectional shape but give smooth of the yield surface.

4. Conclusion

The paper predicted and summarized influence factors on displacement of embankment study on soft clay soil with different constitutive soil models. The parameter estimation for displacement prediction of embankment on soft clay are should carefully determine in consideration of in situ stress condition and stress path before using in the analysis. The undrained analysis in not suitable for embankment deformation behaviour study for all types constitutive soil model which gave unreliable displacement at specific depth (8-18.5 m). The associated flow rule in general overestimates the displacement. The three types of soil model have different assumptions adopted in the analysis eventually turn to dissimilar results. It can be observed that the finite element solution offer a better result for complex problem in soil engineering.

Acknowledgment

The first author would like to express gratefully acknowledgement to Ministry of Higher Education, Malaysia and Universiti Teknologi MARA for given a great opportunity in sponsoring the research study in the Geotechnical Engineering field.

References

[1] Brinkgreve R B J. 2005. Selection of soil models and parameters for geotechnical engineering application. *Soil Constitutive Models:Evaluation, Selection and Calibration*.ASCE

[2] Brand E W, 1991. Predicted and observed performance of an embankment built to failure on soft clay. *Geotechnical and Geological Engineering, 22.*

[3] Chai J C, & Bergado D T, 1993. Some techniques for finite element analysis of embankments on soft ground. *Canadian Geotechnical Journal, 30*(4), 710–719. doi:10.1139/t93-060

[4] Chai J-C, Miura N, & Shen S-L, 2002. Performance of embankments with and without reinforcement on soft subsoil. *Canadian Geotechnical Journal, 39*(4), 838–848.

[5] COMSOL Multiphysics v4.4.

[6] Potts M D, and Zdravkovic L. 1999.*Finite element analysis in geotechnical engineering:theory*.Thomas Telford
[7] Huat B B K. 1994. Behaviour of Soft Clay Foundation beneath an Embankment *2*(2), 215–235.
[8] Indraratna B, Balasubramaniam a S, & Balachandran S, 1992. Performance of Test Embankment Constructed to Failure on Soft Marine Clay. *Journal of Geotechnical Engineering, 118*(1), 12–1. doi:10.1061/(ASCE)0733-9410(1992)118:1(12)
[9] Loganathan N, Balasubramaniam A S, and Bergado D T 1994. Deformation analysis of embankment. *Journal of Geotechnical Engineering, 119*(8), 1185–1206.
[10] Pulos H G, Lee C Y. and Small J C 1991. Predicted and observed behaviour of a test embankment on Malaysia soft clays. *Austalian Geomechanics*,.
[11] Nujid M M & Taha M R. (2014). A Review of Bearing Capacity of Shallow Foundation on Clay Layered Soils Using Numerical Method. *Electronic Journal Geotechnical Engineering*.Vol (19).Bund.D.811-825.

Observation of pH Value in Electrokinetic Remediation using various electrolyte ($MgSO_4$, KH_2PO_4 and $Na(NO_3)$) for Barren Acidic Soil at Ayer Hitam, Johor, Malaysia

J Norashira[1], E Zaidi[2,3], M Aziman[1,3] and A T Saiful Azhar[3,1]
[1]Faculty of Civil and Environmental Engineering, Universiti Tun Hussien Onn Malaysia, 86400 Parit Raja, Batu Pahat, Johor
[2]Faculty of Science, Technology and Human Development, Universiti Tun Hussien Onn Malaysia, 86400 Parit Raja, Batu Pahat, Johor
[3]Research Centre for Soft Soil (RECESS), Universiti Tun Hussien Onn Malaysia, 86400 Parit Raja, Batu Pahat, Johor

E-mail: shafieera90@gmail.com

Abstract. Barren acidic soil collected at Ayer Hitam, Johor Malaysia was recorded at pH value of 2.36 with relative humidity of 86%. This pH value is not suitable for the growth of any plants especially for the soil stabilization purposes. Gradation weathering within the range of 4 to 6 indicates an incomplete/partial weathering process. The soil grade in this range is known as a black shale mudstone. Beside, this also influences to a factor of the high surface water run-off at this particular soil species. As the acidic pH become a major problem for soil fertilizing hence an appropriate technique was implemented *known as* using 'Electrokinetic Remediation', *EKR*. This technique has a great potential in changing the soil pH value from acidic to less acidic and also kept maintain the pH at the saturated rate of electrochemical process. This research study presents the monitoring data of pH value due to the effect of various electrolyte consist of 0.5M of $MgSO_4$, KH_2PO_4, and $Na(NO_3)$. Here, the distilled water (DW) was used as reference solution. The electric field was provided by dipping two pieces of identical rectangular aluminum foil as anode and cathode. The *EKR* was conducted under a constant voltage gradient of 50 V/m across the sample bulk at 0.14 m length measured between both electrodes. The data collection was conducted during the total period of 7 days surveillance. The variation of pH values at the remediation area between anode and cathode for various type of electrolyte indicates that there are a significant saturated value as it reaches 7 days of treatment. During the analysis, it is found that the highest pH value at the remediation area after 7 days treatment using $Na(NO_3)$, KH_2PO_4 and $MgSO_4$ was 3.93, 3.33 and 3.39 respectively. Hence from the last stage of pH value observation, it can be conclude that the best electrolyte for barren soil treatment is $Na(NO_3)$ whereby it contribute to highest pH value and turn the soil to be less acidic.

Keywords: Soft soil, geo-disasters, geo-remediation.

1. Introduction

Barren acidic soil is one of the major causing for surface erosion. This particular soil can be classified as sedimentary soil with weathering grade range of 4 to 6. The soil grade in this range is known as a black shale mudstone. This sedimentary soil become acidic after acidification process [1]. Acidification of soil is a complex process (changes) occur in soil properties such as decrease in pH value , release of base cations from the sorption complex involving rainfall and leaching, acidic parent rock material, organic decay and high yielding crops [2]–[4]. This process forming infertile and unfavorable soil [5]. The process of soil acidification took place when the subsoil or parent material containing acidic ion are exposed to the atmosphere.

Term "barrens" is used to describe regions that have poor nutrient to support vegetative species. Most vegetation species was adapted to their ecology. Barren soil can be express as poor farmland, or lacks in aesthetic landscape while the ecologist describes barren soil as edaphic drought and soil infertile [6]. Most barren soils contained pyrite mineral and the oxidation of pyrite resulting the formation of iron hydroxide minerals, sulfate and sulphuric acid. Chemical reaction of oxidation of pyrite minerals:

$$FeS_2 + 31/2\, O_2 + H_2O \rightarrow Fe_2SO_4 + 2\,H^+ + SO_4^{\;2-} \tag{1}$$

$$H_2SO_4 \leftrightarrow 2H^+ + SO_4^{\;2-} \tag{2}$$

From the Eq. 1, FeS_2 is the empirical formula for pyrite. The presence of excess oxygen and water produce a product of $FeSO_4$, which are ferrous sulphate and $2H^+ + SO_4^{2-}$ stand for dissociated sulphuric acid shown in Eq. 2. When this dissociated sulfuric acid rendering to the water catchment area, it will cause acidification process. These related reactions are accountable for producing an enormous amount of acidity thus reducing the presence of sulphur in the soil [7], [8]. Oxygen level increase by drainage or excavation of the ground surface. From Eq. 1 and Eq. 2, the formation of hydrogen ion has increased the availability to react with other acid functional group that then will be dissociated. Eq. 1 and Eq. 2 shows barren soil is an acidic soil where the chemical perction during weathering produce sulphuric acid after oxidation process and this acidic behavior eliminate the nutrients in soil for vegetation growth.

Electrokinetic remediation, *EKR* is a promising technology for the elimination of contaminated fine-grained soils has been studied broadly since the 1980s [9]–[11]. The low-level direct current, DC in the range of mA/cm^2 applies to soils within a particular treatment time. This effect will influence to a change of fundamental soil physic-chemical and hydrologic properties by chemical or non-chemical used. These chemical acts as an electrolyte that change the effect of the chemical speciation and transport of soil constituents, including metals [10], [12]. The electromigration takes places simultaneously with the electroosmosis. Here, the ions species at the anode are replaced by chemical electrolyte according to the cation exchange capacity, *CEC*. During this process, the bigger cations replacing smaller cations at the same valance and higher valance replacing smaller valance. This process occurs instantaneously with the migration of negatively charge ions to the anode and positively charge ions to the cathode [13], [14].

As the barren soil becomes a major problem causing soil infertility for stabilization purposes, therefore, the aim of this study is focus to the effect of pH change using various type of electrolyte for *EKR* treatment.

2. Material and Method

1.1. Soil Specimen Preparation

The soil specimens were prepared with a consistent method through all experiments. 1.5 time liquid limit, LL of deionized distilled water, DW was used to formed slurry soil specimen by mixing with the barren acidic clay soils with deionized water to achieved 90% moisture content to produce a homogeneous sample. Approximately, 4 kg of DW was used to mix with 20 kg of barren acidic soil by gradually added to ensure consistency of mixing. These slurry samples mixed using Kenwood mechanical mixer and blended thoroughly for 30 minutes with medium speed. Two mixtures of slurry samples needed to fill up the ¾ height of the tank with the total amount of 10 kg and 10 kg respectively achieving 40 % of moisture content for all test samples.

The initial height of the slurry is 236 mm for all tests samples before approximately 1 kPa of the surcharge was added for 24 hours to avoid soil particle migration or spill off via the drainage. Another 16 kPa of the surcharge was added to consolidate the slurry sample until reached final height of 180 mm. After pre-consolidation, the thin solid plastic walls were removed at both ends of the main compartment and replaced with aluminum electrodes. The soil samples were reconsolidated under the same loading as the last consolidation (17 kPa) for about 24 hours before released upon EKR testing. Subsequently, the electric current were turned and monitored throughout the test durations to make sure a good contact between soils and electrodes. Physical and chemical properties of the treated soil were carried out along the soil profiles that were divided into two sections from the anode to the cathode.

1.2. Basic properties of the Soil Sample

The physic-chemical properties of Ayer Hitam barren acidic soil was investigated using laboratory testing accordance with British Standard, BS 1377:1990 as shown in Table 2. The Ayer Hitam barren acidic soil was categories as clayey SILT with the percentage of silt higher than clay approximately 50.33 % and 49.67 %. The liquid limit, LL for Ayer Hitam barren acidic soil is 26.66 during the plastic limit, PL was 10.48 which give the plastic index, PI of 19.25. Average specific gravity identified was 2.53 with an initial pH value of 2.36 and electrical conductivity, EC of 2.86.

Table 1. Properties of Barren Acidic Soil.

Parameter	Value
Particle Size Distribution:	
Silt	50.33
Clay	49.67
Atterberg Limit:	
Liquid Limit	26.66
Plastic Limit	17.95
Average Specific Gravity (G_s)	2.53
Initial pH Value	2.36
Electrical Conductivity	2.86

1.3. Development of Experimental Apparatus

These remedial systems were set up in laboratory model using acrylic PVC-U to avoid current leaching. The EKR model shown in Figure 1 consists of the anode compartment, soil specimen compartment and cathode compartment with an overall dimension of 460 x 340 x 300 mm. The dimension of the soil specimen compartment (main compartment) is 180 x 320 x 300 mm and the other two small compartments is 100 x 320 x 300 mm. The thickness of the EKR model is 10 mm sealed with heavy duty silicon to prevent electrolyte leaking. Aluminium electrode used was with the dimension of 320 x 300 mm and polished both sites to minimize the expose of oxygen.

Table 2. The *EKR* Experimental Condition.

Exp. No.	Anolyte	Catholyte	Temperature, °C and Humidity, %
1	*DW*	*DW*	30, ≈80 %
2	0.5 M $MgSO_4$	*DW*	30, ≈78 %
3	0.5 M KH_2PO_4	*DW*	32, ≈ 68 %
4	0.5 M $Na(NO_3)$	*DW*	28, ≈ 86 %

Direct current, DC power supply used was from model GW INSTEK with a constant voltage gradient of 50 V/m for 7 days treatment with assessed of pH value and electric current every 24 hours. The pH value measured using Mettler Toledo Perfect Seven Pro and calibrate with the solution-temperature gradient graph.

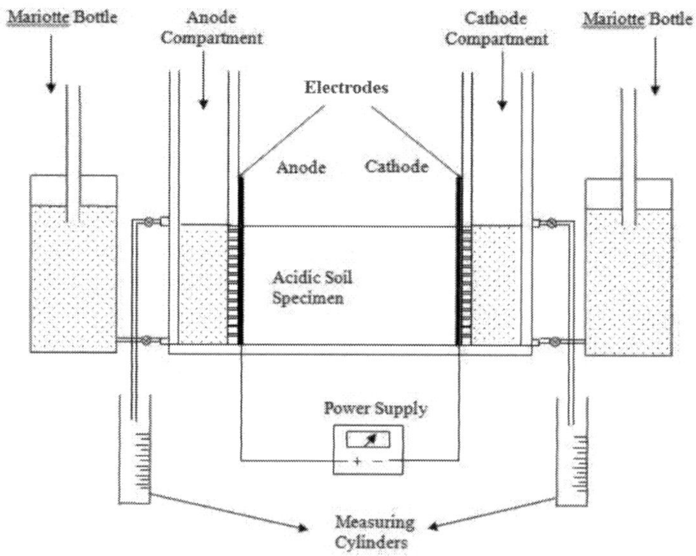

Figure 1. Schematic Drawing of *EKR* Testing Model [15].

3. Result and Discussions

Figure 2 shows monitoring data of pH value from different EKR system for 7 days. Overall observation shows at the beginning of the test indicate that the pH value recorded was fluctuated and slightly constant after days 3 onwards. Remediation of acidic soil using pure system shows an increasing of pH value from day 1 to days 4 with value of 2.36, 2.55, 2.67 and 2.99 and decreasing from days 5 to day 6 with value of 2.85 and 2.72 before increase to 2.88 at days 7. The graph pattern shows that the after the introduction of electric field by EK remediation using DW as an anolyte and catholyte increase the pH value of the soil sample from day 1 to days 4 and slightly constant pattern at days 5 to days 7 because of the adaption of electric field onto the soil sample.

The pH value of sample treated with 0.5 M of $MgSO_4$ as an anolyte and DW as catholyte shows decrement from day 1 to days 7 from 5.09, 4.57, 3.93, 3.67, 3.59, 3.42 and 3.39. The immediate change of pH value from 2.36 at the soil sample to 5.09 was due to influence of electric field by *EKR* and higher pH value of 0.5 M of $MgSO_4$ than soil sample. After day 1, the graph shows decreasing of pH value since the anolyte used altered the concentration of acidic soil sample. It is also can be conclude that the 0.5 M of $MgSO_4$ promote the addition of sulphate to the soil sample.

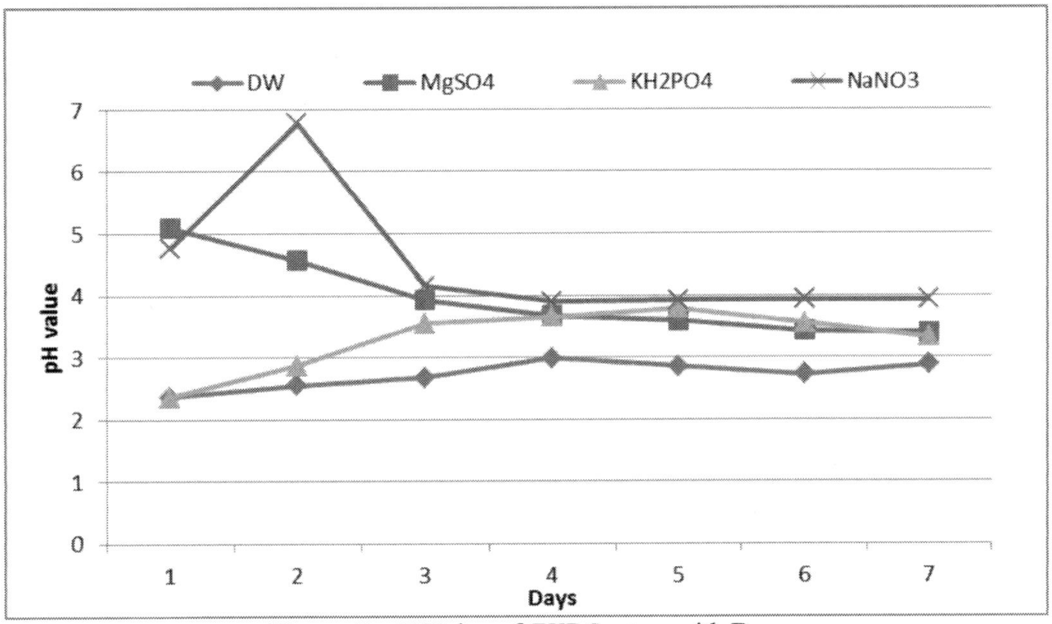

Figure 2. pH Value of *EKR* System with Days

The pH value at the 0.5 M of KH2PO4 shows increment from day 1 to days 5 with value of 2.36, 2.87, 3.55, 3.66, and 3.79 before decreased at days 6 and days 7 with value of 3.55 and 3.33. The increasing of pH value of soil sample was due to introduction of alkaline electrolyte with the presence of electric field that migrate the electrolyte throught the soil sample area. The similar pattern of pH value also can be observe by using 0.5 M of Na(NO3) which shows immediate increase on days 2 with value of 6.78 and drastically decrease at days 3 with value of 4.15 before maintaining at days 4 to days 7 with value of 3.90, 3.92, 3.93 and 3.93. based on the results, the optimum state of 0.5 M of KH2PO4 is at days 5 with value of 3.79 and the optimm state for 0.5 M of Na(NO3) is on days 2 with value of 6.78.

The inconsistency of this pH result might be due to the electrolysis process. This statement complies with Park et al., [6] which mention during electrolysis reaction, the H+ ions moves towards the cathode regions, and OH- ions move towards anode regions. Electromigration was affecting the pH value, electric current and voltage gradient. The ionic mobility of H+ ions is 1.78 times faster than the OH- ions. The OH- ions produced were neutralized by the catholyte. Giannis et al., [16] states that the acidic soil pH indicates that some of the H+ ions produced at the anode by electrolysis migrated into the soil column. The OH- ions produced was neutralize using catholyte.

Another factor that considered as mechanisms that affecting the variation of pH value is the soil-electrode contact. This factor varied upon insertion of the electrode during sample preparation. These electrode-soil contact interfaces were essential after pre-consolidation and post-consolidation after the exchange of thin-wall solid plastics wall with the electrode. These exchanging produces a gap between electrode and soil thus manipulated the volume of the direct current electric field across the soil sample. The post-consolidation should be carry out at least 24 hours or more with the same pressure of surcharge to minimize the gap between electrode-soil contacts.

4. Conclusion

From this observation, the summarize of the finding are as follows;

i. Electric field from the EKR influencing the changes of pH values by transport the ions across the soil sample. This electro-migration activities replacing the anions and cations according to its exchange capacity where influencing the deposition of H+ and OH- ions across the soil sample.

ii. The most suitable anolyte as a catalyst to increase the pH value of this barren acidic soil is 0.5 M of Na(NO3). This anolyte act as buffer providing baseline, supply nutrients and facilitates the pH value control thus it agree with the past research done by Li et al,. [9].

iii. The result presented for the other anolytes agree with the finding from past research done by Park et al., [6] and Giannis et al., [17] The results from observing data showed that there are several important mechanisms relating to the electrochemical effects. The pH value profiles were influenced by the ionic concentration of the fluid within the soil samples. *Style and spacing*

Acknowledgement

The authors would like to thank the Ministry of Higher Education and Universiti Tun Hussein Onn, Malaysia for supporting this research under Exploratory Research Grant Scheme (ERGS) Vot. E040.

References

[1] Brandy N C and Weil R R 1996 The Nature and Properties of Soils 14th Editi. Pearson Prentice Hall, 1996.
[2] Jamil N, Madun A, Ahmad Tajudin S A, and Embong Z 2014 An Overview of Electrokinetic Remediation Assisted Phytoremediation to Remediate Barren Acidic Soil
[3] Johnson G V Cause and Effects of Soil Acidity
[4] Hédl R, Petřík P, and Boublík K 2011 Long-term patterns in soil acidification due to pollution in forests of the Eastern Sudetes Mountains Environ. Pollut., vol. 159, no. 10, pp. 2586–2593.
[5] Salt D E, Blaylock M, Kumar N P, Dushenkov V, Ensley B D, Chet I, and Raskin I, Phytoremediation: a novel strategy for the removal of toxic metals from the environment using plants Biotechnology. (N. Y)., vol. 13, no. 5, pp. 468–74, May 1995.
[6] Edwards M A, Winslow M, and Blake R 2007 Assessing pine barrens soil moisture regimes using Synthetic Aperture Radar (SAR) techniques IEEE Int. Geosci. Remote Sens. Symp., pp. 1828–1831.
[7] Abubaker F, Lynsdale C, and Cripps J 2014 Laboratory study of the long-term durability of buried concrete exposed to Lower Lias Clay Constr. Build. Mater., vol. 64, pp. 130–140.
[8] Abubaker F, Lynsdale C, and Cripps J 2013 Investigation of concrete-clay interaction with regards to the thaumasite form of sulfate attack Constr. Build. Mater., vol. 67, pp. 88–94.
[9] Li D, Tan X Y, Wu X D, Pan C, and Xu P 2014 Effects of electrolyte characteristics on soil conductivity and current in electrokinetic remediation of lead-contaminated soil Sep. Purif. Technol., vol. 135, pp. 14–21.
[10] Yang J S, Kwon M J, Choi J, Baek K, and O'Loughlin E J 2014 The transport behavior of As, Cu, Pb, and Zn during electrokinetic remediation of a contaminated soil using electrolyte conditioning Chemosphere, vol. 117, pp. 79–86.
[11] Li D, Xiong Z, Nie Y, Niu Y Y, Wang L, and Liu Y Y 2012 Near-anode focusing phenomenon caused by the high anolyte concentration in the electrokinetic remediation of chromium(VI)-

contaminated soil J. Hazard. Mater., vol. 229–230, pp. 282–91.
[12] Acar Y B and Alshawabkeh A N 1993 Principles of electrokinetic remediation Environ. Sci. Technol, no. 27, pp. 2638–2647.
[13] Ahmad Tajudin S A 2012 Electrokinetic Stabilization of Soft Clay PhD Thesis University of Brimigham.
[14] Cameselle C and Gouveia S 2013 Advances in Electrokinetic Remediation for the Removal of Organic Contaminants in Soils.
[15] Ahmad Tajudin S A, Jefferson I, Madun A, Zainalabidin M H, Tajul Baharuddin M F, Adib M, and Razi M 2015 Monitoring of Electric Current During Electrokinetic Stabilisation Test for Soft Clay using EKG Electrode vol. 774, pp. 1560–1564.
[16] Park S W, Lee J Y, Yang J S, Kim K J, and Baek K 2009 Electrokinetic remediation of contaminated soil with waste-lubricant oils and zinc.," J. Hazard. Mater., vol. 169, no. 1–3, pp. 1168–72.
[17] Giannis A, Nikolaou A, Pentari D, and Gidarakos E 2009 Chelating agent-assisted electrokinetic removal of cadmium, lead and copper from contaminated soils Environ. Pollut., vol. 157, no. 12, pp. 3379–86.

Electromigration of Contaminated Soil by Electro-Bioremediation Technique

A T S Azhar[1,2], A T A Nabila[2,1*], M S Nurshuhaila[3], M Z N Shaylinda[2] and M A M Azim[2,1]
[1] Research Center for Soft Soil, Universiti Tun Hussein Onn Malaysia, 86400 Batu Pahat Johor, Malaysia
[2] Faculty of Civil and Environmental Engineering, Universiti Tun Hussein Onn Malaysia, 86400 Batu Pahat Johor, Malaysia
[3] Faculty of Engineering Technology, Universiti Tun Hussein Onn Malaysia, 86400 Batu Pahat Johor, Malaysia

E-mail: ainnabila782@gmail.com

Abstract. Soil contamination with heavy metals poses major environmental and human health problems. This problem needs an efficient method and affordable technological solution such as electro-bioremediation technique. The electro-bioremediation technique used in this study is the combination of bacteria and electrokinetic process. The aim of this study is to investigate the effectiveness of Pseudomonas putida bacteria as a biodegradation agent to remediate contaminated soil. 5 kg of kaolin soil was spiked with 5 g of zinc oxide. During this process, the anode reservoir was filled with Pseudomonas putida while the cathode was filled with distilled water for 5 days at 50 V of electrical gradient. The X-Ray Fluorescent (XRF) test indicated that there was a significant reduction of zinc concentration for the soil near the anode with 89 % percentage removal. The bacteria count is high near the anode which is 1.3x107 cfu/gww whereas the bacteria count at the middle and near the cathode was 5.0x106 cfu/gww and 8.0x106 cfu/gww respectively. The migration of ions to the opposite charge of electrodes during the electrokinetic process resulted from the reduction of zinc. The results obtained proved that the electro-bioremediation reduced the level of contaminants in the soil sample. Thus, the electro-bioremediation technique has the potential to be used in the treatment of contaminated soil.

Keywords: Soil contamination, electro-bioremediation, biodegradation.

1. Introduction

Environmental contamination caused by human activities has become a serious issue worldwide. It involves contamination in various mediums such as soil, water and air. The contamination is mainly due to the large number of industrial activities, disposal of municipal solid wastes, urbanization activities and agricultural waste [1]. Currently, governments are trying to minimize the adverse impact of contamination which generally affects physical and mental development in humans, even at the lowest level of exposure [2]. Nowadays, over 80% of hazardous wastes come from industrial activities [3]. Sludge, heavy metal oil and other hazardous wastes are found in abundance. Among these hazardous wastes, heavy metal (e.g. copper, zinc, cadmium and lead) contamination is considered to be the worst due to their harmful effects and long-term persistence in the environment [4]. Therefore, remediation of these contaminated soils becomes a great concern for both engineers and researchers.

Currently, there are several remediation methods that have been implemented. One of the promising methods is electro-bioremediation technique using microorganisms (e.g. yeast, fungi or bacteria) to degrade or remove the contaminants [5]. Electro-bioremediation is the combination of two technologies which are electrokinetic remediation technique (EK) and biological remediation. Generally, EK is a new remediation technique which mainly applies voltage at two sides of the soil and then forming electric field gradient. When electrolysis occurs in the sample, the oxidation and reduction process take place simultaneously and produce hydrogen (H+) and hydroxyl (OH-) ions. The equations representing the oxidation and reduction process are shown below [6]:

$$\text{Anode (oxidising)} \qquad H2O \qquad 2H+ + 0.5O2(g) + 2e- \tag{1}$$

$$\text{Cathode (reducing)} \qquad 2H2O + 2e- \qquad 2OH- + H2(g) \tag{2}$$

According to Syakeera et al. [7], this technique involves applying an electric current across the soil mass to boost the chemical migration from the injection point in order for it to react beneficially with the soil to bring improve its properties. On the other hand, biological remediation is a process of changing the physical and chemical characteristics through migration and the transformation process of contaminants by microorganisms. The selection of bacteria is due to the efficiency of bacteria in removing contamination found in soil [5]. Pseudomonas putida bacteria is naturally found in soil and can be present in dormant or slow-growing state [8]–[14].

Therefore, this study was conducted to investigate the ability of bacteria to remove heavy metals particularly zinc through the electrokinetic process. The aim of this study is to evaluate the effectiveness of bacteria as a medium to degrade and remove zinc in contaminated soil.minimize the time consuming, large data coverage and minimal ground destruction need to be discovered.

2. Materials and Methods

2.1 Soil sample preparation

Kaolin (Malaysia) Sdn. Bhd. supplied synthetic soil as a sample. To make synthetic soil, approximately 5kg of kaolin soil was mixed with 5g of zinc for 5 days. The soil sample took three days to dry. Then, the soil was grinded and sieved so that only soil particles measuring less than 2mm pass through. After that, to achieve 52% of moisture content similar to liquid limit, deionized water and soil were mixed. Lastly, the mixture was kept overnight.

2.2 Broth culture for Pseudomonas putida preparation

Pure Pseudomonas putida ATCC 49128 in dry culti-loop form was sourced from the United States of America (USA). Before the cultivation process, the Pseudomonas putida was stored in the chiller at 4°C. Approximately 16g of medium (nutrient broth) was filled into an Erlenmeyer flask containing distilled water and stirred. Then, the medium was heated at 100°C on a hot plate to dissolved it until complete dissolution for around 2 hours. Next, the solution was sterilized for around 15 minutes at 121°C in an autoclave machine. The solution was cooled for a few minutes around 35°C to 37°C. According to manufacturer instruction, one loop shaft was removed from the holder and put directly on the medium. Lastly, before the inoculation process, the mixture was stirred well and kept inside the chiller at 4°C for 2 days.

2.3 Bacteria counting (Spread plate method)

Firstly, nutrient media was prepared using Pseudomonas agar. The Pseudomonas agar was mixed with distilled water in the beaker and boiled. The agar was cooled up to 45 to 50°C and poured into half of the six petri plates. Then, bacteria dilution was prepared. 1mL from the sample was removed and blew into 9mL dilution in tube 1. Then, 1mL from tube 1 was placed into the 9mL dilution in tube 2 and the procedure is repeated until tube 6. After that, the solid medium in the plate was inoculated and spread over the surface evenly. All the petri plates were placed inside the incubator for 24 hours with a temperature of 37°C. The bacteria colonies grew on the surface of the medium. The petri plate was placed on the counting chamber to keep track of the bacteria count.

2.4 Electrokinetic bioremediation testing

Figure 1 and figure 2 shows the real model and cross section of electrokinetic bioremediation testing. At the anode, the reservoir was filled with Pseudomonas putida. The cathode of the reservoir was filled with distilled water. Direct electric current 50V was applied at the anode and cathode for 5 days. At the end of the duration, the samples of soil were taken from K1, K2 and K3 of the anode for XRF test and bacteria count test.

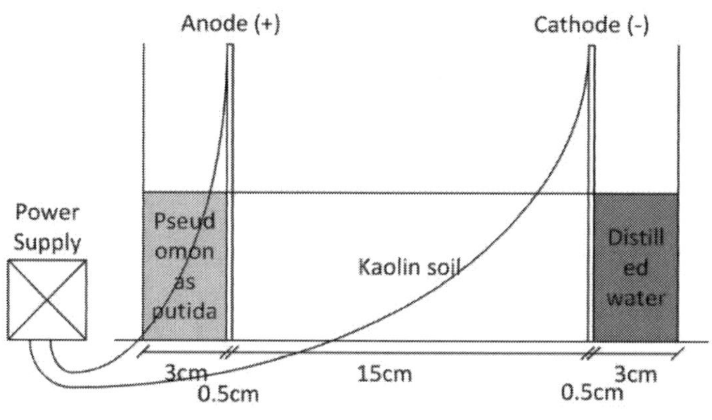

Figure 1. Electrokinetic bioremediation test.

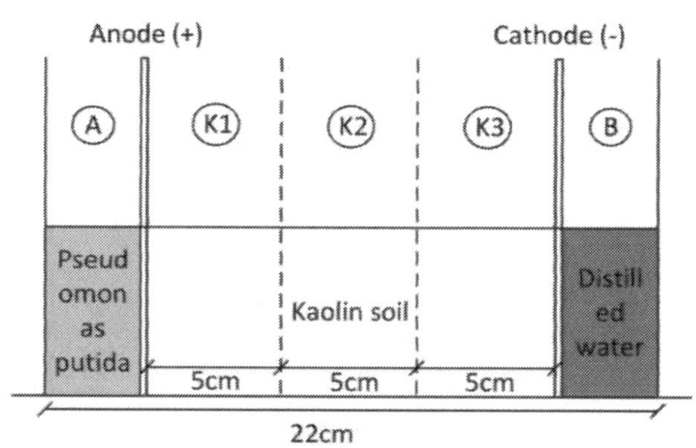

Figure 2. Cross-section of the physical modeling of the electrokinetic bioremediation tank.

Table 1. Description of electrokinetic bioremediation (code).

Section (code)	Description
A	Anode reservoir
K_1	5cm from anode reservoir
K_2	10cm from anode reservoir
K_3	15cm from anode reservoir
B	Cathode reservoir

3. Results and Discussions

3.1 pH development

Due to pH sensitivity during chemical interactions, it was necessary to evaluate the pH of soil before and after treatment. It was observed that the pH values are slightly similar as shown in Table 2. From the table, it was observed that soil pH before and after treatment using bacteria and distilled water was acidic which is 4.6 and 5.4 respectively. Meanwhile, pH of soil before and after treatment using distilled water stand-alone was also acidic at 4.6 and 5.8 respectively. This significant result indicates that the presence of bacteria does not affect the pH development in soil. This statement is supported by a research conducted by Zeng et. al [15] which stated that the mobility of heavy metals increase with the decrease in soil pH.

Table 2. pH values of kaolin soil before and after treatment using and without using bacteria.

pH Value	Electrokinetic bioremediation (bacteria and distilled water)	Electrokinetic remediation (both using distilled water)
Before Treatment	4.6	4.5
After Treatment	5.4	5.8

3.2 Concentration and bacteria

Table 3 shows the amount of concentration before and after treatment. The results indicated the different values of concentration from K1 to K3. The initial concentration of heavy metals before the treatment is different. The results indicated that there was a significant reduction of zinc concentration at K1 and K2, which was 80mg/L (89%) and 284mg/L (60%) respectively. However, the concentration of zinc at K3 increased to 1674mg/L (137%). The concentration increased to 1674mg/L because the zinc ions in the soil migrated and accumulated in this section.

The initial concentration for Pb, Cr and Cu was 255mg/L, 36mg/L and 7mg/L respectively. From the results, the heavy metals show a slight decrease at K1, which are 227mg/L (11%) for Pb, 28mg/L (22%) for Cr and 4mg/L (43%) for Cu. However, the concentration values of Pb, Cr and Cu at K2 and K3 were lower compared to zinc concentration. For this situation, the concentration of zinc was higher than the other three heavy metals because Pb, Cr and Cu have no additional chemicals added to the soil. This is because the soil sample was added with 5g of zinc oxide.

Figure 3 shows that the migration of bacteria from anolyte to K3. At the anolyte section, the recorded value is 1.0x108 cfu/gww. However, K1, K2 and K3 recorded values of 1.3x107 cfu/gww, 5.0x106 cfu/gww and 8.0x106 cfu/gww, respectively. It shows that the bacteria was still in good condition until the treatment was completed. The reduction of zinc concentration is caused by the bacteria. Duxbury [16] stated that bacteria used heavy metal as micronutrients. Overall, it was proven that the reduction of heavy metal occurs when the presence of bacteria in this treatment and the ions migrate toward the cathode [17].

Table 3. Concentration values for initial and after treatment.

	Distance from anode (cm)	Zn (mg/L)	Pb (mg/L)	Cr (mg/L)	Cu (mg/L)
Initial Concentration Before Treatment	0	706	255	36	7
Concentration After Treatment-5cm	5cm	80	227	28	4
Concentration After Treatment – 10cm	10cm	284	233	27	1
Concentration After Treatment – 15cm	15cm	1674	231	24	2

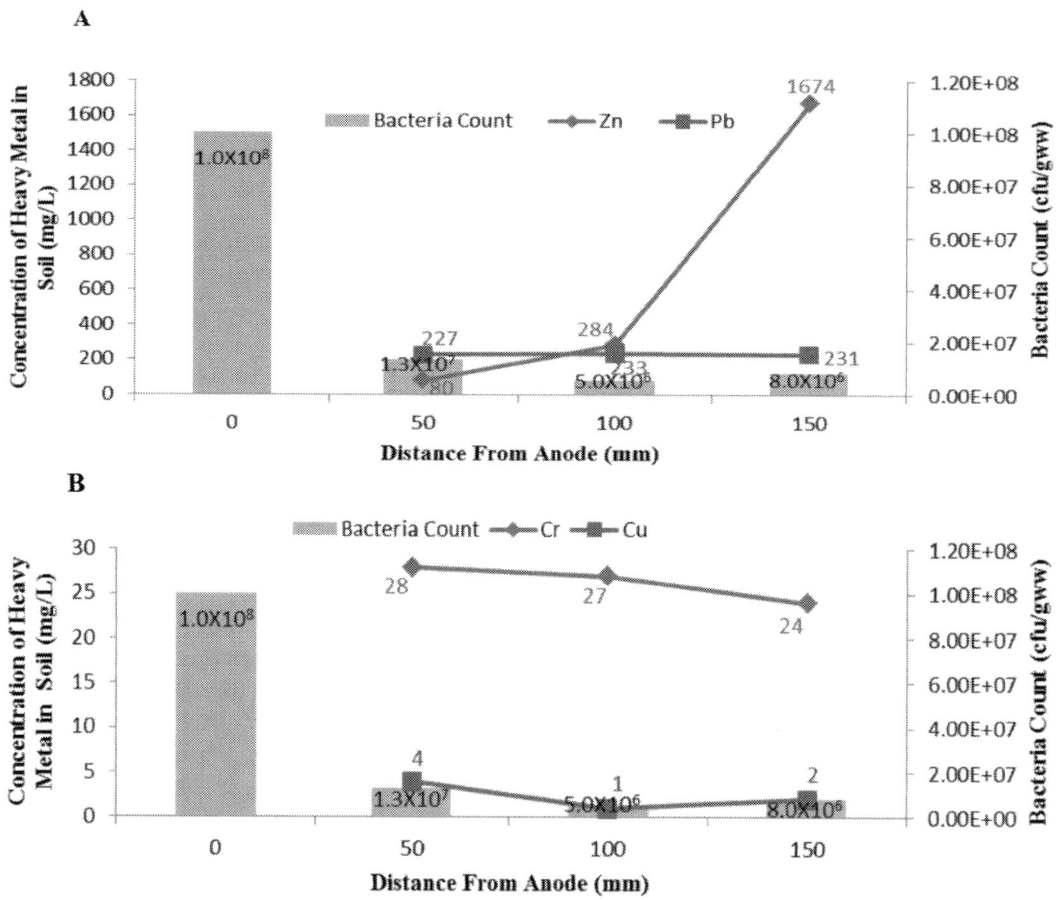

Figure 3. Bacteria migration and concentration of Zn and Pb (A) and concentration of Cr and Cu (B) after EK remediation.

4. Conclusion

Electrokinetic bioremediation is an alternative technology to remove contaminants in soil. This technology used a combination of bioremediation and electrokinetics. From this treatment and the results of zinc and bacteria count, it is proven that electro-bioremediation can remediate the contaminated soil up to 89%. This treatment only focuses on pseudomonas putida as an electro-bioremediation agent. Nevertheless, using various types of bacterial consortium is recommended for this treatment. To conclude, this technology is useful especially for industries to treat contaminated soil through green remediation technology.

Acknowledgment

The authors would like to thank the Ministry of Higher Education and Universiti Tun Hussein Onn, Malaysia for supporting this research under Fundamental Research Grant Scheme (FRGS) Vot. 1455.

References

[1] Azmi M A M, Tajudin S A A, Kadir A A, Hanif M N S M, Nordin N S, and Adnan M S 2015 Leaching Behavior of Lead Contaminated Soil Sample by Using Sugarcane Bagasse in Stabilization / Solidification Method. **774** 1481–1485.

[2] Hanif M N S M, Tajudin S A A, Kadir A A, Madun A, Azmi M A M and Nordin N S 2015 Leachate Characteristics of Contaminated Soil Containing Lead by Stabilisation /

Solidification Technique . **774** 1443–1447.

[3] Napia C, Sinsiri T, Jaturapitakkul C, and Chindaprasirt P 2012 Leaching of heavy metals from solidified waste using Portland cement and zeolite as a binder. Waste Manag. **32** 1459–1467.

[4] Taylor P, Reddy K R and Parupudi U S 1997 Removal of chromium, nickel and cadmium from clays by in-situ electrokinetic remediation. Soil Sediment Contam. **6** 391–407, 1997.

[5] Perpetuo E A, Souza C B, Augusto C, Nascimento O and Paulo C S 2011 Engineering bacteria for superior Chromate bioremediation. Environmental Microbiology. **28** 1–29.

[6] Gill R T, Harbottle M J, Smith J W N and Thornton S F 2014 Electrokinetic-enhanced bioremediation of organic contaminants: A review of processes and environmental applications. Chemosphere. **107** 31–42, 2014.

[7] Nordin N S, Tajudin S A A and Karim A T A 2013 Stabilisation of Soft Soil Using Electrokinetic Stabilisation Method. **1**.

[8] Sunar N M, Emparan Q, Karim A T A, Maslan M, Mustafa F and Khaled N 2014 Bioremediation of Biodiesel/Diesel Blend (B50) in Soil Contamination by Using Pseudomonas Putida. Appl. Mech. Mater. **699** 1012–1019

[9] Emparan Q, Sunar N M, Gani P, Ming E C and Parjo U K 2015 Survival of Pseudomonas Putida for Biodiesel Blend (B20) in Soil Bioremediation. Appl. Mech. Mater. **773–774n** 1324–1329, 2015.

[10] Sunar N M, Emparan Q, Karim A T A, Noor S F M, Maslan M, Mustafa F and Khaled N 2013 The Effectiveness of Bioremediation Treatment for Diesel-Soil Contamination. Adv. Mater. Res. **845** 146–152.

[11] Sunar N M, Emparan Q, Karim A T A, Noor S F M, Maslan M, Mustafa F and Khaled N 2009 Bioremediation of Biofuel-Soil Contamination by Using Pseudomonas putida. Adv. Mater. Res **845** 138–145.

[12] Sunar N M, Emparan Q, Karim A T A, Noor S F M, Maslan M, Mustafa F and Khaled N 2009 The Effectiveness of Pseudomonas putida ATCC 49128 as Biodegradable Agent in Biodiesel Soil Contamination.

[13] Sunar N M, Emparan Q, Karim A T A, Noor S F M, Maslan M, Mustafa F, Khaled N and Othman N 2014 Characteristics of polycyclic aromatic hydrocarbons emissions of palm-biodiesel blends. **18** 43–46.

[14] Sunar N M, Emparan Q, Karim A T A, Noor S F M and Maslan M 2014 Survival of Pseudomonas putida for Biodiesel Blend (B5) in Soil Bioremediation. Appl. Mech. Mater. **567** 139–143.

[15] Zeng F, Ali S, Zhang H, Ouyang Y, Qiu B, Wu F and Zhang G 2011 The influence of pH and organic matter content in paddy soil on heavy metal availability and their uptake by rice plants. Environ. Pollut.**159** 84–9.

[16] Duxbury P H 2000 Phytoremediation Systems for Treatment of Contaminant Mixtures in Soil. Macdonald Campus, McGill University.

[17] Kim B K, Baek K, Ko S H and Yang J W 2011 Research and field experiences on electrokinetic remediation in South Korea. Sep. Purif. Technol. **79** 116–123.

Dynamic Characteristics of Penor Peat Using MASW Method

A Zainorabidin[1,2] and M J M Said[1,2]
[1] Faculty of Civil and Environmental Engineering, Universiti Tun Hussein Onn Malaysia, 86400 Batu Pahat Johor, MALAYSIA
[2] Research Center for Soft Soil, Universiti Tun Hussein Onn Malaysia, 86400 Batu Pahat Johor, MALAYSIA

E-mail: adnanz@uthm.edu.my

Abstract. The dynamic behaviour of soil affected the mechanical properties of soil such as shear wave velocity, shear modulus, damping ratio and poisson's ratio [1] which is becoming important aspect need to be considered for structures influences by dynamic movement. This study is to determine the dynamic behaviour of Penor peat such as shear wave velocity using MASW and estimation its shear modulus. Peat soils are very problematic soils since it's have high compressibility, low shear strength, high moisture content and low bearing capacity which is very not suitable materials to construct any foundation structures. Shear wave velocity ranges between 32.94 – 95.89 m/s and shear modulus are ranging between 0.93 – 8.01 MPa. The differences of both dynamic properties are due to the changes of peat density and affected by the fibre content, organic content, degree of degradation and moisture content.

Keywords: Peat, shear wave velocity, shear modulus, MASW.

1. Introduction

The dynamic behaviour of soil affected the mechanical properties of soil such as shear wave velocity, shear modulus, damping ratio and poisson's ratio [1] which is becoming important aspect need to be considered for structures influences by dynamic movement. This study is to determine the dynamic behaviour of Penor peat such as shear wave velocity using MASW method and estimation its shear modulus.

The knowledge of the dynamic response of the soil-structure are required for dynamic structural analysis of the superstructures [1] and relies on dynamic soil properties. Soils tend to behave differently and very complex under dynamic loading such as earthquake, moving traffic, machineries and bomb blasting. The loadings will contribute to an accumulation of shear strains, and when the threshold shear strain has been exceeded, the shear modulus will be reduced [2]. Dynamic loading is distinguished by the magnitude of the loading fluctuating with time and the deformations of the soil consist of both recoverable and permanent displacement [3]. In many researches of soil properties, there are three typical dynamic properties of field test and laboratory test such as shear modulus, shear wave velocity and damping ratio. Researcher Shafiee et al., [4] also claimed that the dynamic properties of peat that typically investigated in prior work with cyclic shear strain are the differences of shear modulus and damping ratio [4].

Peat soils are very problematic soils since it's have high compressibility, low shear strength, high moisture content and low bearing capacity which is very not suitable materials to construct any foundation structures. This problem often faced by the engineers when attempted to construct roads for rural area or foundations which covered by very vast peat soil area and have difficulty to maintain the completed structures life span as peat continued to have settlement for long term period.

2. Properties of Peat

Peat is formed by the accumulation of organic material undergoing partially decomposition process with proper condition. Peat is a combination of fragmented organic and animal matter and formed in wetland under appropriate anaerobic conditions for long periods [5,6]. Peat soil is an organic with content more than 75 % [7] which caused a lot of problems for construction due to unpredictable behaviour of its properties. At the study location, the peat is known as hemic peat with fibre content 33.7 % [8] where, according to ASTM D 1997-91 [9] hemic has fibre content between 33% and 67 %. Hemic peat has moisture content 472 – 554 %, organic content 92 -96 %, and von post classification is H5 – H6 [6].

3. Multi-channel Analysis of Surface Wave (MASW)

Multi-channel Analysis of Surface Wave methods are used to determine shear wave velocity on the field for subsurface characterization by measuring the phase velocity of surface wave using an active source to create seismic waves at specified location along linear direction with spread length [10] then recorded the propagation of surface wave to generate shear wave velocities through Fourier transform and inversion dispersion curve of phase velocities in the frequency domain. The shear velocity, soil profile can determine when measure the phase velocity of Rayleigh waves of different frequencies [11].

This method is ideal to for effective identification and isolation of seismic noise such as body waves, scattered, and non – source generated surface waves and higher mode of surface waves according to distinctive trace to trace comparison of coherency in arrival time and amplitude. The MASW method can take out accurate phase velocities of Rayleigh wave ground roll [12].

4. Geological Setting

Study location actually near at Penor which located about 26 km north from Pekan town where the area is covered with natural tree and did not have any human activities. This research location has 1.5 meters depth of peat soil that observed through peat soil profiling using peat sampler. The geological feature is the research location where marine and continental unconsolidated deposits such as peat, humic clay and silt as shown in Figure 1.

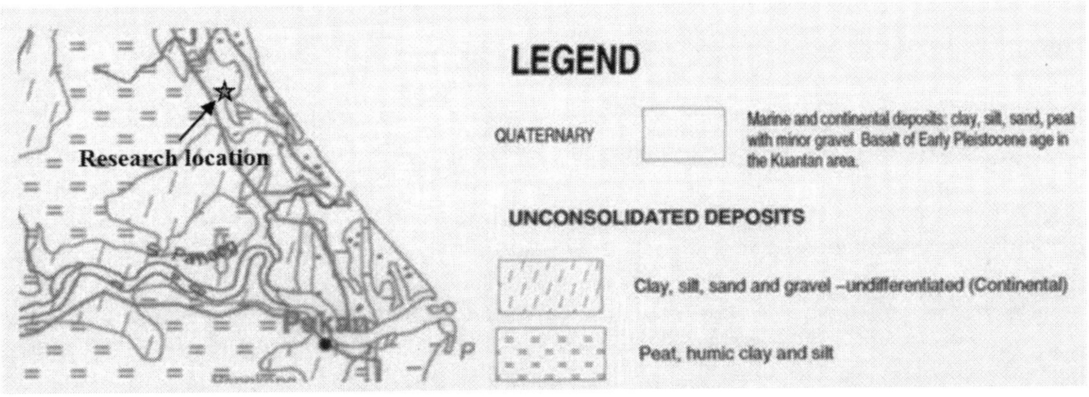

Figure 1. Geological map at Penor, Pahang (Jabatan Mineral dan Geosains Malaysia, 1985) [13].

5. Methodology

5.1. Determining peat soil density using peat sampler.

The density of peat was determined on site by using the peat sampler for every 0.5 meters depth. Peat sample has a semi-cylindrical shape after taken from specified depth. To ensure sample disturbance in minimum, the peat sample was measured it's diameter and length on site, then put into a sealed plastic bag to ensure minimum moisture reduction and measured its weight in laboratory. The soil sample was cut approximately 150 mm length and taken from the top part of soil sample for every 0.5 meters. The diameter and length of sample were measured to calculate its volume, then calculate the peat in-situ density after measuring its weight. Peat density determination for each study were repeated for three times to obtain average peat density.

5.2. Multi-channel Analysis of Surface Wave

This method just required a 7 kg of sledgehammer was used as a seismic source that impacted vertically to the aluminium plate. As a recorder, ABEM Terraloc MK-8 seismograph was used and connected with 24 units of 4.5 Hz vertical geophone is used as a sensor. Figure 2 shows the MASW field acquisition arrangement.

For record length, it requires time from 2 second until 4 second recording depend on the times requires to observe surface wave at the seismograph. Thus sampling interval need to be adjusted from 250 μs - 500 μs and number of samples is 4096 - 8192. Any filtering of noise turns off to prevent changes in the quality of any noise from the ground. Numbers of background noise record must not stack together. Array line cable is put across the research site in linear with a length of 23 meters long with 1 meter spacing between geophones and offset distance from 1st geophone is 5 meters. For the geophone should be placed in a clear area and approximately level with the ground. This is to ensure the accuracy of seismic data. For each study location, there are five lines and offset distance between lines is 0.5 meters.

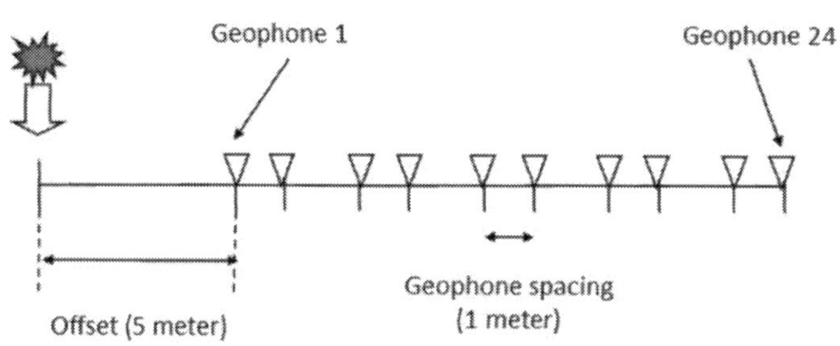

Figure 2. MASW field acquisition arrangement.

6. Result and Discussion

6.1. Shear wave velocity, soil profile

Surface wave seismic data were exported from seismograph then process it using SeisImager software. As shown in Figure 3 the shear wave velocity on peat from the surface to 1.0 m depth are ranging between 32.94 – 95.89 m/s. Peat mixed soft clay layer at 1.5 m depth where the shear wave velocity has decreased that are ranged from 41.57 – 69.73 m/s. When reached depth of 2.5 m of clay, the velocity continued to increase from 66.51 – 87.96 m/s.

The dispersions of surface wave are commonly affected by the variation formation of soil layer on the ground and shown variations in shear wave velocity. For peat, it contains a lot of organic matter and effected surface wave propagation on peat which makes the picking of the dispersion curve in phase velocity in the frequency domain are difficult to choose fundamental mode of the dispersion curve. The surface wave propagation also affected by void ratio of peats and void ratio are depends on the organic content and fibre content of peat. The shear wave velocity affected by void ratio through the fibre content [2]. In saturated soils, mass density will be directly related to the void ratio. This existence of peat void ratio caused the surface wave propagation have been absorbed before detected by the geophones. According to Beamish [14] seismic wave attenuation across the soil profile is controlled by moisture content in concurrence with the density and porosity of the materials especially for peat and organic soils.

6.2. Shear modulus estimation.

Shear modulus of peat has been estimated using the formula and obtained the in-situ density of Penor peat. In geophysical theory, shear modulus related to shear velocity as shown in Eq. 1. From the Eq. 1, it's show shear velocity is directly proportional with maximum shear modulus. Thus maximum shear modulus will increase if shear velocity also increases.

$$G_{max} = \rho V_s^2 \tag{1}$$

Where Gmax is small-strain shear modulus and ρ is density of soil

As shown in Figure 3, the shear modulus of peat were ranged between 0.93 – 8.01 MPa. Meanwhile Peat mixed soft clay layer at 1.5 m depth where modulus has decreased that are ranged from 2.01 – 8.38 Mpa. At a depth of 2.5 m of clay, the shear modulus maintains to increase from 7.16 – 12.52 MPa.

In geotechnical engineering properties such as the shear modulus for determination of peat stiffness, are mainly affected by the fibre content, moisture contents and degree of degradation. The percentage of fibre content can change the shear strength of peat where high fibre content will give high shear strength. The contents affected the density of peats where increasing of both contents will increase the peat density and vice versa. Meanwhile degree of decomposition is decrease of organic matter that causes the disappearance of the peat structure and changing the primary chemical composition of the peat [15]. As the degree of decomposition are high, the density of peat will decrease since the decomposition affects the weight of peat, organic matter which become lighter than low degree peat of decomposition. Peats with high moisture content makes it heavier compared with less moisture content and then can increase the density of peats. Peat soil contains more fibre or organic content and thus it has a high water absorption capacity since the soil fibre are very loose and hollow [7,15]. Even though most of organic matter in peats have been decomposed, there are few organic components such as small roots and partially decomposed organics are included during sampling the peat for measuring its weight and the components increase the weight and directly increase the density of peat.

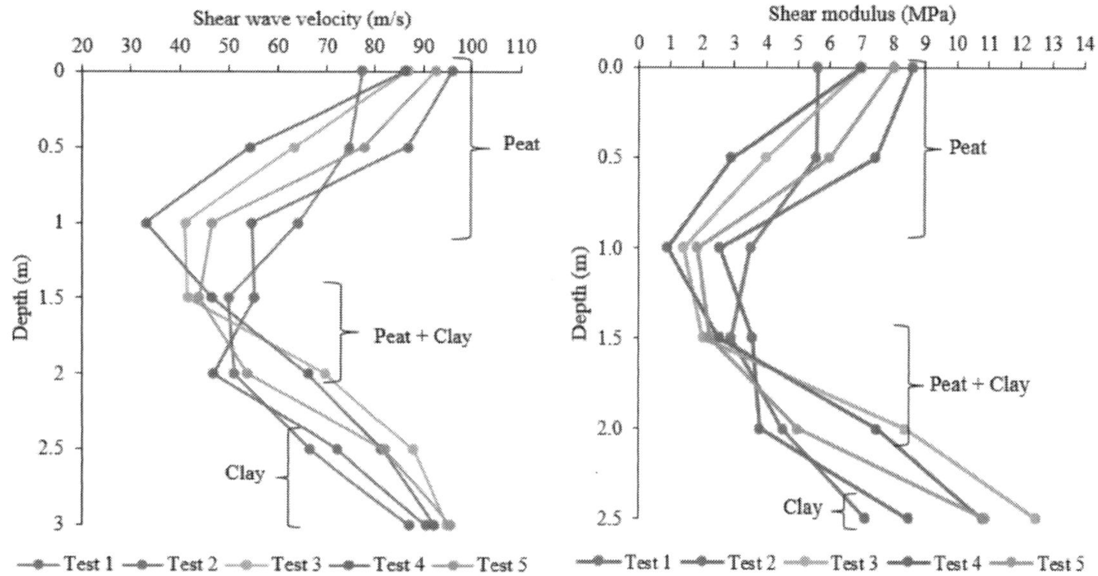

Figure 3. Shear wave velocity and shear modulus of Penor peat

7. Conclusion

The dynamic behaviour of Penor peat was determined using the MASW method. Shear wave velocity is ranged between 32.94 – 95.89 m/s and shear modulus are ranging between 0.93 – 8.01 MPa. The differences of both dynamic properties are due to the changes of peat density and affected by the fibre content, organic content, degree of degradation and moisture content.

Acknowledgement

The authors would like to thank the University of Tun Hussein Onn, Malaysia, and the Ministry of Education, Malaysia, for their generous granted of this research, Fundamental Research Grant Scheme Vot. 1224 "Undrained Dynamic Behaviour of Malaysian Peat Soil".

Reference

[1] Luna R & Jadi H 2000 Determination of Dynamic Soil Properties Using Geophysical Methods. Proceedings of the First International Conference on the Application of Geophysical and NDT Methodologies to Transportation Facilities and Infrastructure. St. Louis. Pp. 1 – 15.

[2] Schneider J A, Hoyos L, Jr. Mayne P W, Macari E J & Rix G J 1999 Field and Laboratory Measurements of Dynamic Shear Modulus of Piedmont Residual Soils. Behavioral Characteristics of Residual Soils. *Geotechnical Special Publication*. 92 pp. 12-25.

[3] Bødker L 1998 Behaviour of Soil Subjected to Dynamic Loads. *In Dynamics of Structures 1993 1997*. Denmark: Aalborg University. pp. 65 – 76.

[4] Shafiee A, Brandenberg S and Stewart J 2013 Laboratory Investigation of the Pre- and Post-Cyclic Volume Change Properties of Sherman Island Peat. *Geo-Congress*. 5 pp. 139 – 148.

[5] Hashim R & Islam S 2008 Engineering Properties of Peat Soils in Peninsular Malaysia. *Journal of Applied Science*. 8(22) pp. 4215 – 4219.

[6] Zainorabidin A, Wijeyesekera D C & Jayaratne R 2010 Fabric of Peat Soils Using Image Analysis. *Proceedings of Advances in Computing and Technology, (AC&T) the School of Computing and Technology 5th Annual Conference*. University of East London. pp 38 – 44.

[7] Kolay P K, Sii H Y & Taib S N L 2011 Tropical Peat Soil Stabilization using Class F Pond Ash from Coal Fired Power Plant. *International Journal of Civil and Environmental Engineering.* 3(2) pp. 79 – 83.

[8] Nordin S H 2013 The Shrinkage Measurement Peat Soil at Penor Site. Universiti Tun Hussein Onn Malaysia: Bachelor Degree Thesis.

[9] American Society for Testing and Materials, 2001 Standard Test Method for Laboratory Determination of the Fiber Content of Peat Samples by Dry Mass. United States. D 1997-91.

[10] Eker A M 2012 Local Site Characterization And Seismic Zonation Study By Utilizing Active And Passive Surface Wave Methods: A Case Study For The Northern Side Of Ankara, Turkey, *Engineering Geology* 151. Elsevier Journal. pp. 64 – 81.

[11] Neelima D S 2010 Multi-Channel Analysis Of Surface Wave (MASW) Testing For Dynamic Site Characterization Of Delhi Region, *Fifth International Conference on Recent Advances in Geotechnical Earthquake Engineering and Soil Dynamics.* Paper No. 1.01b

[12] Jianghai X, Richard D M, Choon B P, James A H and James B H 2000 Comparing Shear-wave Velocity Profiles from MASW with Borehole Measurements in Unconsolidated Sediments, Fraser River Delta, B.C., Canada, *Journal of Environmental and Engineering Geophysics.* 5(3) pp. 1 – 13.

[13] Jabatan Mineral dan Geosains Malaysia. 1985 Geological Map of Peninsular Malaysia. Kuala Lumpur: Ministry Of Natural Resources and Environment.

[14] Beamish D 2014 Peat Mapping Associations of Airborne Radiometric Survey Data. Remote Sensing Journal. 6 pp. 521 – 539.

[15] Huat B B K, Kazemian S, Prasad A & Barghchi M 2011 A Study of The Compressibility Behaviour of Peat Stabilized by DMM: Lab Model And FE Analysis. *Scientific Research and Essays.* 6(1) pp. 196 – 204.

Analysis on the Long Term Effect of Trial Test Road Constructed on Batu Pahat Soft Clay (BPSC) at Recess UTHM

M M M Idrus[1,2], A E Edayu[2], Z Adnan[2,1] and B Ismail[1,2]
[1] Research Center for Soft Soil, Universiti Tun Hussein Onn Malaysia, 86400 Batu Pahat Johor, MALAYSIA
[2] Faculty of Civil and Environmental Engineering, Universiti Tun Hussein Onn Malaysia, 86400 Batu Pahat Johor, MALAYSIA

E-mail: idrusmas@gmail.com

Abstract. The reinforcement used in soil for construction of roads on the soft clay is very important as it will determine the level of service of the road after it was built. Damage or defects on the road surface to be an indicator of the level of the road has dropped and shows the deformation of the road. For this research, an analysis has carried out on the long-term effect of trial test road constructed on Batu Pahat Soft Clay (BPSC) at RECESS UTHM. Through this research, the reinforcement using Rawell Geosynthetic Clay Liner (RGCL) was the best with the stability is 14964 N, a low flow is 2.69mm, stiffness modulus is 1766 MPa, the peak load is 739.4 N and a lower horizontal deformation which is 1.71 μm compared Woven Geotextile section and section without geotextile [1][9]. In terms of deformation can be seen clearly from physical observations that section without geotextile suffered significant damage than others. Settlement of road can also be analyzed by a longitudinal section that plotted based on the result of leveling work. After that, settlements are more visible way on the right side of the road trial. Through lab tests conducted, it indicate that the coring samples obtained from sites of each section meets the specifications set by the Jabatan Kerja Raya (JKR) in terms of stability, flow and stiffness [1]. Through this study, a trial road built on soft soil can be used as a test site because of the uniqueness of these roads which has three different types of reinforcements.

Keywords: Soft soil, ground modification, settlement.

1. Introduction

Research Center for Soft Soil (RECESS) UTHM was built to study the soft ground in Malaysia. The research on soft clay is mostly done on site RECESS UTHM and one of them is the construction of trial road that built on soft clay by using three different types reinforcement of soil. The main problem for the soft clay soil are has a high moisture content rate resulting settlement problems [7][6]. Soil type has characteristics such as moisture content between 23% to 69%, a specific gravity of between 2.18 and 2.65, plastic limit of 20% to 35%, liquid limit of 37% to 66%, the plastic from 17% to 31% and sensitivity of soil between 2.5 to 7.3 (m2/year). Soil particle size is less than 0.002mm. Therefore, before the road construction done on the ground, the soft clay has to be stabilized [1]. This research

IOP Conf. Series: Materials Science and Engineering **136** (2016) 012025 doi:10.1088/1757-899X/136/1/012025

was made to assess the settlement and deformation of road pavements constructed 11 years ago, when it was built in 2013. The type of pavement that was built in the trial road in RECESS UTHM is flexible pavement. Most of the roads in Malaysia were built of flexible pavement because of its low cost compared to rigid pavement [9]. The flexible pavement, each layer being built which is the highest load bearing capacity in the top and at the bottom has a low load bearing capacity [2].

2. Research Methodology

The information of research consists of the research location and plan layout of trial road. This information is provided to give understanding about trial road that constructed at RECESS UTHM. The research methodology carried out in order to ensure that a comparative study and analysis between previous and current conditions of the trial road is done to reflect the appropriate objectives.

2.1 Research Location

The study was conducted at the Research Center for Soft Soil (RECESS) UTHM. In 2003, this road was built at RECESS UTHM. The site is located on the campus of Universiti Tun Hussein Onn Malaysia (UTHM) and RECESS is the only research center on soft ground in Malaysia.The research site is located about 20 km from the center of Batu Pahat. The topography of the area is flat with the surface of the original height was about 1.35 until 1.80 m above the sea level. Ground water level for this area is from 0.5 m to 0.63 m from the ground [3].

2.2 Plan Layout of Trial Road

Figure 2.1 shows the plan layout of trial road that constructed on Batu Pahat Sofy Clay (BPSC) at RECESS UTHM. The road was built with two types of geotextiles which is Rawell Geosynthetic Clay Liner (RGCL) and Woven Geotextile (WG) while one part is being built without the use of geotextile [4]. The section that called as Buffer Zone (BZ) is a section that separates the two type reinforcement of soil. Besides that, the section without geotextile has three parts, 3a, 3b and 3c while WG section has two parts 2a and 2b. Section RGCL also has two parts which is 1a and 1b. The length of this road is 60 m and it has 12 sections where each section is 5m meanwhile the road width is 3.6m.

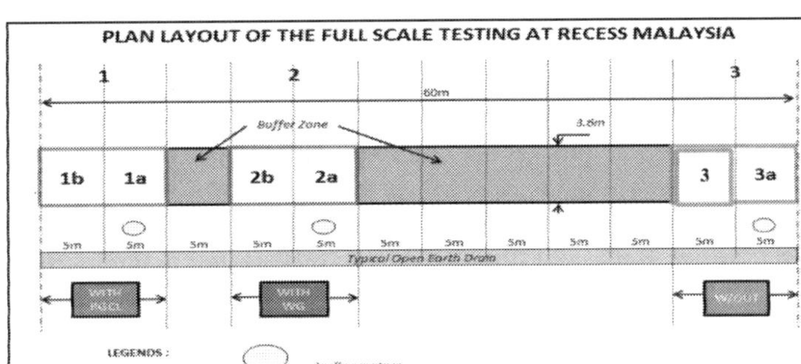

Figure 2.1. Plan Layout of Trial Road.

(a) Site Investigation. Site investigation are carried out to know the condition of the trial road by doing the site visit. During the site visit, found that each section of the road has longitudinal cracking. Furthermore, the location of inclinometer and bore holes are determined based on the plan layout of the trial road. The boundary between the section are cannot be seen during the site visit and need to measure it to ensure that every section position that have in this trial road.

(b) Levelling. Levelling is a process of measurement the height difference between points on the earth by referring to datum [5]. For carried out the leveling, the bench mark or temporary bench mark must be known in order to start the leveling work. So that, the bench mark at Parit

Jelutong are taken as the reference point with reference number J1512. After that the temporary bench mark are determined near to the research location by referring the bench mark. So that, the settlement of longitudinal section and cross section of the trial road can be known by carried out the leveling.

(c) **Sampling Process.** Coring method is used to get the sample for the laboratory testing. A total of 20 samples are taken where six sample of each section which are without geotextile, Woven Geotextile (WG) and Rawell Geosynthetic clay liner (RGCL). The two sample are taking in the buffer zone part. The figure 3.1 shows the samples that have been drilled on the road.

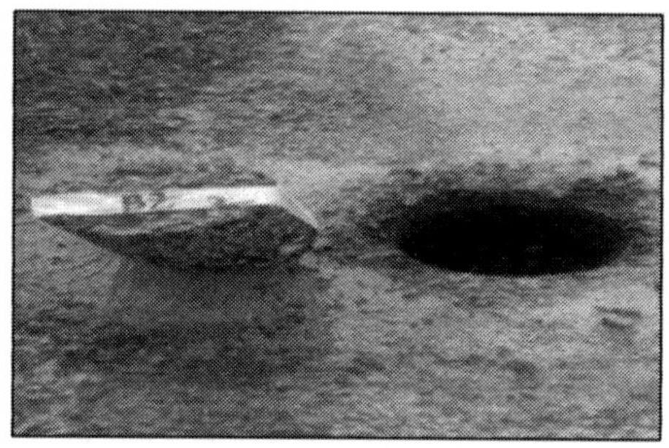

Figure 3.1. The coring sample.

(d) **Laboratory testing.** The experiments conducted in the laboratory are by using coring samples that taken on site. The experiments involved are Marshall test and Indirect tensile stiffness modulus test. **Marshall test.** The purpose of this test is performed to determine the stability and flow for each sample which has been in coring. Marshall test was based on ASTM D1559-59 (ATM, 19992A). In this test the sample is soaked at a temperature of 60 ° C for 30 to 40 minutes before the testing the sample. **Indirect tensile stiffness modulus test.** This experiment made is to identify stiffness modulus, peak load and horizontal deformation of coring samples that taken on the road trial in RECESS UTHM. The value obtained from this experiment will be compared for each section involved. The indirect tensile stiffness modulus is a non-destructive method and has been widely used for determination of stiffness modulus values. Two different temperature used for this test which are at a temperature of 20°C and 40°C for 2 hours. After that, the samples will be tested using a universal testing machine.

3. Results and Analysis

Result and analysis consists of physical observation, analysis between the different section of road, analysis on longitudinal section and analysis on Marshall test and Indirect tensile stiffness modulus test. Physical observation of the road that was built in RECESS trial involving the observation of the condition of the road. Overall, it appears at the beginning is the way it looks abandoned and overgrown with grass (Figure 4.1). This is because the road is no longer used for traffic and with any type of vehicle. The last time the road was resurfaced and the burden with traffic is in 2010. In addition, during rains the road will be stagnant with water (Figure 4.2). This is due to the existing drainage cannot cope with the flow of rain water.

Figure 4.1. Condition of the trial road.

Figure 4.2. Condition of road during rainy. day

3.1 Analysis between the different sections of road

Based on the observations, the Figure 4.3(a) are suffered severe damage between all of the sections. The red line shows the state of the road is no longer have pavement and overgrown with grass. Damage to roads is more to the left side of the road. In addition, transverse cracks can also be seen in this section. For the Figure 4.3(b) show on the left side of the road pavement layer is not seen as section without geotextile. After the road is cleaned, the grass begins to grow back visible on the pavement, which was covered with soil. Settlement can be seen on the right side of the road because there is stagnant water at the area. Figure 4.3(c) shows condition of trial road of WG section. In this section, the pavement still visible after cleaning the road. However, on the right of road shows the settlement occurred and stagnant water could be seen. In addition, the longitudinal cracks appear on the surface of the road for this section. Inconsistencies pavement layer can also be seen in this section because there is a layer of pavement that looks higher. Road conditions on RGCL section shown in Figure 4.3(d). In this section, the longitudinal cracks are seen on the surface of the road. Settlement clearly visible on the right side of the road that leads to stagnant water in that area. Pavement still visible in this section.

(a) W/out geotextile (b) Buffer Zone (c) WG (d) RGCL

Figure 4.3. Different condition of road between sections

3.2 Analysis on longitudinal section

Based on a longitudinal section of the graph shown in Figure 4.4 below, the highest position for the mid-point attempts in RECESS road is not uniform. There is a difference of settlement rate on these roads due to the different type of reinforcement of soil. Through a longitudinal section of the graph, the highest settlement rate experienced by road from the chainage 10 until chainage 35.

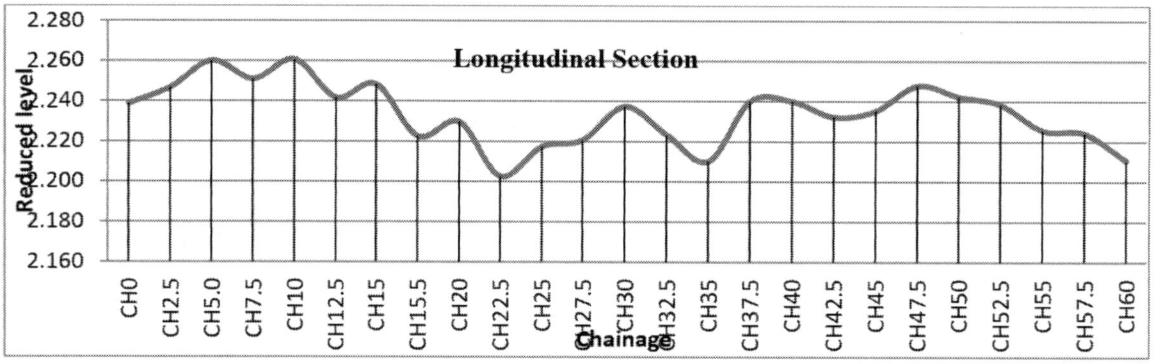

Figure 4.4. Longitudinal section trial road at RECESS UTHM.

3.3 Analysis on Marshall test and Indirect tensile stiffness modulus test

Based on figure 4.5, the results obtained from the Marshall test showed the stability and flow of the samples while the indirect tensile stiffness modulus test that showed the stiffness modulus, horizontal deformation and peak load behaviour. In terms of stability, the RGCL is highest compare than others which is 14,964 N meanwhile in terms of flow the RGCL is the lowest compared to other sections which is 2.69mm. The indirect tensile stiffness modulus test was carried out with two different temperature conditions i.e. 20°C and 40°C. The RGCL section showed the high stiffness modulus 1,766 Mpa and peak load of 739.4 N whereas the lowest value in terms of horizontal deformation is 1.71 μm. The results found from these two different tests showed that RGCL section has the ability to prevent water from seeping into the layers of road, thus giving higher stability, stiffness modulus and peak load, but with lower value in terms of flow and horizontal deformation.

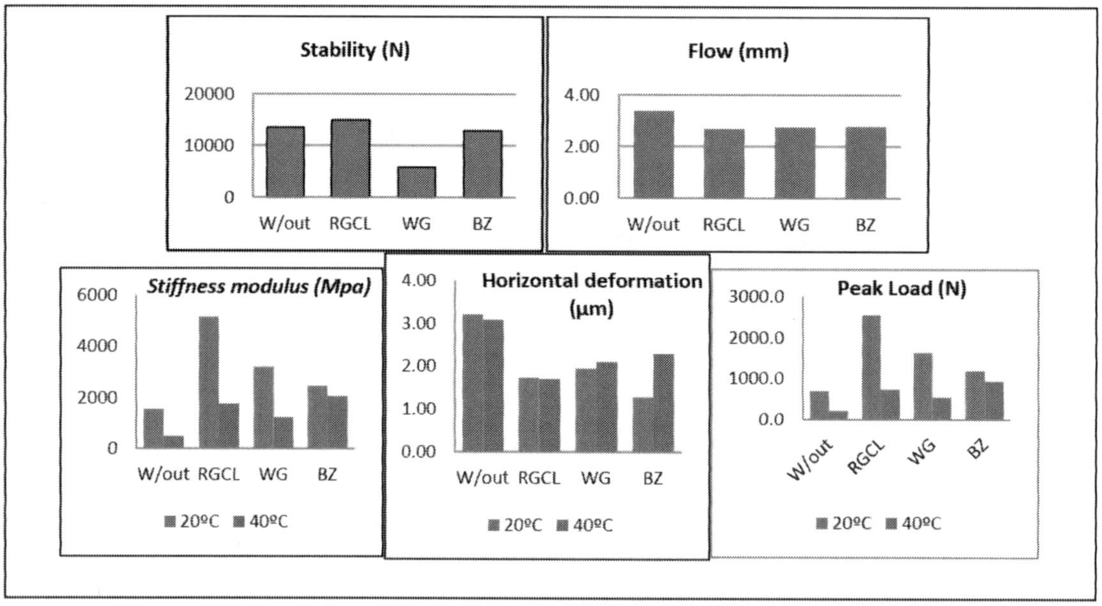

Figure 4.5. Result from Marshall test and Indirect tensile stiffness modulus.

4. Conclusion

After 11 years of construction, the defects can be seen on the pavement surface of the trial road constructed, During the rainy days, this trial road will be flooded with water because there were no proper drainage nearby the trial road area. The deformation of road surface can be seen at all sections and section without geotextile was adversely affected compared to the other sections. In terms of settlement, it was found that from the longitudinal section observations, the highest settlement rate experienced by the road was from chainage 10 until chainage 35, which is in the center of the road. In terms of laboratory testing, they showed that, the RGCL section have highest stability, stiffness modulus and peak load which are 14,964 N, 1,766 Mpa and 739.4 N respectively. The RGCL section was also found to have lower value of flow which is 2.69mm and lower value of horizontal deformation which is 1.71µm compared to section without geotextile and WG section. As a conclusion, the RGCL prevented water from seeping into the layer of road make the RGCL section the best section after 11 years of construction as a trial road on Batu Pahat Soft Clay (BPSC) at RECESS UTHM. This was proven after analysing the results obtained on the site and also through the laboratory testing results. It is hoped that this research will contribute towards better land transportation system which will be competitive and relevant with other systems especially the urban rail system. [8]. It was said that a better transportation system will depend on its infrastructural quality and its efficiency depends on the level of service it shows during its operation [1]. Thus, it is important to understand the long term effect of a structure constructed as it will ensure the sustainability, resilient and safety for the users and the community at large.

References

[1] Masirin M I M 2006 Performance Evaluation Of Encapsulated Road Pavement On Difficult Ground Condition. University of East London. Ph.D. British Library, UK.
[2] Azhari S 2012 Kitar Semula Campuran Asfalt Lama dan Asfalt Baru Dalam Turapan Lentur.Universiti Teknologi Malaysia (UTM). Tesis Ijazah Sarjana Muda.
[3] Baharin M S S 2010 Impak Laluan Kenderaan Terhadap Anjakan Mendatar Dan Menegak Untuk Jalan Raya Percubaan Dengan Pengukuhan Yang Berbeza.Fakulti KejuruteraanAwam dan Alam Sekitar (FKAAS), Universiti Tun Hussein Onn Malaysia (UTHM).Tesis Ijazah Sarjana Muda.
[4] Zakaria H 2008 Failure Mechanism Study of Different Road Reinforcement For Trial Road Embankment Constructed On Batu Pahat Soft Clay. Universiti Tun Hussein Onn Malaysia (UTHM). Tesis Ijazah Sarjana Muda.
[5] Zain R M 2013 Penilaian Integriti Struktur Robi Test Bed (Road Bridge Interfrace Test Bed) di Atas Tanah Lembut. Fakulti Kejuruteraan Awam dan Alam Sekitar (FKAAS), Universiti Tun Hussein Onn Malaysia (UTHM). Tesis Ijazah Sarjana Muda.
[6] Nor A H M N, Masirin M I M, Sanik M E 2014 Site Investigation Of Road Drains For Rural Road On Batu Pahat Soft Clay (BPSC). Journal of Mechanical and Civil Engineering, International Organization of Scientific Research (IOSR).
[7] Masirin M I M, Zain R M 2012 Differential settlement analysis of different approach slab design concept for road bridges. Elsevier Proceeding, 2012.
[8] Nordin N H, Masirin M I M, Ghazali I, Azis M I 2014 Appraisal on Rail Transit Development: Train services and safety. IIES 2014 Proceeding, UTHM.
[9] Hashim W, Aishah W, Masirin M I M 2008 Site investigation as a geotechnical forensic tool for road embankment performance evaluation. EPC Proceeding, UKM Malaysia.

Investigation of Stabilised Batu Pahat Soft Soil Pertaining on its CBR and Permeability Properties for Road Construction

M M Mohd Idrus[1,2], J S M Singh[1,2], A L A Musbah[1,2], D C Wijeyesekera[1,2]
[1]Research Center for Soft Soil, Universiti Tun Hussein Onn Malaysia
86400 Parit Raja, Batu Pahat, Johor, MALAYSIA
[2]Faculty of Civil and Environmental Engineering, Universiti Tun Hussein Onn Malaysia, 86400 Batu Pahat Johor, MALAYSIA

E-mail: idrusmas@gmail.com

Abstract. Soil stabilization by adding materials such as cement, lime and bitumen is one of the effective methods for improving the geotechnical properties of soils [11] Nano-particle is one of the newest additives and many studies about using nano-particle in soil improvement has been done but it was given less attention when soft clay soils stabilization is concerned. To evaluate the strength characteristics of stabilized Batu Pahat soft clay, laboratory investigation on early strength gained by the stabilized soil must be conducted to formulate a suitable and economical mix design [10]. To achieve such purpose, the study examined the effect of Nano-Clay on the California Bearing Ratio and the Permeability of soft clay. The results gained shows that the Nano-Clay is able to increase the strength of the soft clay [9]. The California Bearing Ratio of the soil is increase significantly where the results for the highest percentage of admixture is 14.4% while the permeability of the soil decreases significantly with increasing Nano-Clay whereby the results of the highest percentage of admixture is $2.0187x10^{-11}$ m/s. After doing this research, it is proven that Nano-clay can contribute towards better soil stabilization and enhance the quality of soil as subgrade and foundation at large.

Keywords: Soft Soil, soil stabilization, nano-clay.

1. Introduction

In Malaysia, the development of national road networks, residential and commercial properties have encroached into the areas underlain with very soft soils. The soft clay has created a challenge to the construction industry, particularly in road construction. The characteristic of soft soil are high compressibility, low shear strength and low permeability. General construction problems in this deposit are insufficient bearing capacity, excessive post construction settlement and instability on excavation and embankment forming. In this formation, usually the hard layer and bedrock are very deep and results in higher cost of foundation. Geotechnical works in deep deposits of highly compressible soft clay is often associated with problems such as excessive differential settlement, negative skin friction and bearing capacity failure

In order to counter these problems, one has to know the engineering properties of the soft clay. The conventional ground treatment methods such as soft soil replacement; expedite pore water dissipation and platform settlements through the insertions of prefabricated vertical drains (PVD) and surcharge fills; modify subsoil bearing capacity through the installation or stone column or combination of these techniques are widely used in Malaysia [10]. The applications of these methods are constrained by technical feasibility, space, time constraints and construction cost. Early selection and application of the most appropriate ground improvement techniques can improve considerably not only the design and performance of foundations and earth structures, including embankments, cut slopes, roads and railways but also result in their cost-effectiveness. Chemical stabilization methods are presented to provide soil strength improvement, mitigation of total and differential settlements, shorter construction period, reduced construction costs, and other characteristics which may impact on their utilization to specific projects on soft ground. This research addresses these deficiencies by creating Nano-clay by using Batu Pahat soft clay (BPSC) and performing testing laboratory stabilized soil using Batu Pahat soft clay (BPSC) Nano-clay as admixtures at Research Centre for Soft Soils (RECESS) [11].

Over the past 5 years, residential and commercial developments have increased in Batu Pahat. This development was constructed on soft clay. The civil engineering components of the project included construction of flood control, main drainage and access road. The construction on soft soil is increasing due to lack of suitable land for infrastructures and other developments. Imported soils from cutting of hills and highlands are used for various construction purposes. Many parts of Johor and other coastal areas consist of soft soils or peat soils. In this research, study is carried out in Batu Pahat district which is known to have abundance of soft clay. This type of clay called Batu Pahat soft clay (BPSC) is available up to a depth of 40 meters from ground level [2]. According to Hashim and Islam (2008)[3], roads in Batu Pahat district experienced many types of failures such as cracks, large surface deformation and structural deformation of pavement layers and the subgrade. They suggested that in order to reduce these failures, Batu Pahat soft clay needs to be utilized in order to reduce imported soil from other places and reduced the possibility of environmental damages. BPSC at Research Centre for Soft Soil (RECESS) has a plasticity index (PI) that range from about 36% to 46% in which the higher the PI, the greater the potential for problems [7]. Clays, especially highly plastic are subject to swell when their moisture content is increased. Moisture control is perhaps the most important single factor in the success of foundations on shrinking and swelling clays. The percentage of clay in a soil and the activity of clay minerals are reflected qualitatively by the value of the plasticity index. The larger content of clay minerals, and the more active the clay mineral, the greater is its potential for swelling, creep and changes in behavior. The Building research Establishment (BRE) suggests that the plasticity index over 35% provided an indication of volume change potential is very high. These volume changes can give rise to ground movements which can cause damage to buildings. Therefore, in order to prevent these problems, it is essential for engineers to stabilize the existing soil before commencing the construction activities. By stabilizing the soil, it is hoped that the soil will be more suitable as road subgrade and any road construction.

2. Soil Properties

Naturally occurring materials that are used for the construction of all except the surface layers of pavements (concrete and asphalt) and that are subject to classification test (ASTM D 2487) to provide a general concept of their engineering characteristics [1]. People describe soil types in all kind of ways such as heavy, light, clay and loam, poor or good. Soil scientist describes soil type by how much sand, silt and clay are present. This is called texture. It is possible to change the texture by adding different things. Changing texture can help in providing the right condition needed for plant growth. Sand is the largest particle in the soil. When you rub it, it feels rough. This is because it has sharp edges. Sand doesn't hold many nutrients. Silts are soil particle whose size is between sand and clay. Silt feels smooth and powdery. When wet, it feels smooth but not sticky. Clay is the smallest of particles. Clay is smooth when dry and sticky when wet. Soils high in clay content are called heavy soils. Clay also holds a lot of nutrients but doesn't let air and water through it well. Particle size has a lot to do with soils drainage and nutrients holding capacity. To better understand how this big three soil, we can imagine that if a particle of sand were the size of basketball, then silt would be the size of a baseball, and clay would be the size of a golf ball. Line them up, and we can see how these particle compare in size.

Construction of roadways over soft subgrade is one of the most common problems for highway construction in many parts of the world as well as in our country, Malaysia. The usual approach to soft subgrade stabilization is to remove the soft soil, and replace it with a stronger material of crushed rock. The high cost of replacement has caused highway agencies to evaluate alternative methods of highway construction on soft subgrade [10]

3. Nano-clay

The essential Nano-clay raw material is montmorillonite, a two to one layered smectite clay mineral with a platy structure. The thickness of each layer is about 1 nm, diameter from 10 nm to several microns, and the interlayer space around 1 nm depending on the modification methods. Due to its high aspect ratio and good physical and thermal properties, nanoclay has the potential for exceptional improvements in barrier, flammability resistance, thermal and mechanical properties. [5]

4. Stabilization

Stabilization is the process of blending and mixing with a soil to improve certain properties of the soil. The process may include the blending of soil to achieve a desired gradation or the mixing of commercially available additives that may alter the gradation, texture or plasticity, or act as a binder for cementation of the soil [1][3].

5. Soil Classification

Soil classification is a way of systematically categorizing soils according to their probable engineering characteristics. The classification of a soil is based on its particle distribution and, if the soil is fine-grained, on its plasticity (LL and PL). The most widely used classification systems used in road engineering are the unified soil classification system, AASHTO classification and British Standard Classification. Soil classification should only be regarded as a means of obtaining a general idea of soil behavior and it should never be used as a substitute for detailed investigation of soil properties [1].

6. Soil Engineering Properties

Engineering properties soil classifications have been evolved based on the suitability of a soil for use as a foundation material or as a construction material [8]. It was stated that engineering properties of soil are important as a preliminary guide to the engineering behaviour of the soil. Therefore, an engineering soil classification should be conducted in connection with the use of soil in any important project, since different properties govern the soil behaviour in different situations. Furthermore, the engineering properties are a function of the proposed end utilization [4]. The fundamental engineering properties of stabilized soil have been experimentally investigated by many researchers. The role of cement kiln dust and volcanic ash on the strength development in the blended cement admixed clay has been investigated for low-cost construction to build houses and road infrastructures [8].

7. Road Construction

Roads are built up in several layers, consisting of sub-grade, sub-base, base and surface layer. Subgrade is the uppermost part of the soil, consists of natural or imported soil to supporting the load transmitted from the overlying layers [6]. Therefore, sub-base course serves as an aid to disperse the load from the base course before transmitting it to the subgrade. The base course which is overlying the sub-base course plays a prominent role in the support and dispersion of the traffic loads. Surface course consists of binder course and wearing course. Binder course layer works as a supporting, dispersing traffic load and resists shear, while the topmost layer (wearing course) resists abrasion and prevent skidding. A soft subgrade in construction of roadways is one of the most frequent problems for highway construction in many parts of the world. These problematic soils do not possess enough strength to support the wheel loads upon them either in construction or during the service life of the pavement. The usual approach to soft subgrades stabilization is removes the soft soil, and replaces it with stronger materials likes crushed rock. The high cost of replacement causes highway contractors to explore alternative methods of highway construction on soft sub grades [8]. This soil must therefore, be treated to provide a stable subgrade or working platform for the construction of the pavement. One of the strategies to achieve this is soil stabilization. The soil stabilization includes both physical stabilization (such as dynamic compaction) and chemical stabilization (such as mixing with cement, fly ash, and lime). One of the most important layers of the road is the subgrade. Where the subgrade is founded in an inherently weak soil, this material is typically removed and replaced with a stronger granular material.

8. Laboratory Testing

8.1. Atterberg Limits
Atterberg limit is a laboratory test that is performed to determine the plastic limit and liquid limit of soil. The plastic and liquid limit of BPSC is determined by using cone penetrometer test. By obtaining the value of plastic and liquid limit of BPSC, the plasticity index (PI) of BPSC can obtain.

8.2. Permeability
Permeability test is a laboratory test that is performed to determine the coefficient of permeability, k or the rate of water flow through the soil sample. The type of permeability test used is falling head test as it is suitable for fine grained soil. In this case, BPSC is used. The test results provide the rate of water seepage through the soil. (Refer Figure 3.1).

8.3. California Bearing Ratio
It is a penetration test for evaluation of the mechanical strength of road subgrades and base courses. The test is performed by measuring the pressure required to penetrate a soil sample with a plunger of standard area with equipments as shown in figure 3.1. The CBR rating was developed for measuring the load-bearing capacity of soils used for building roads. The harder the surface, the higher the CBR rating.

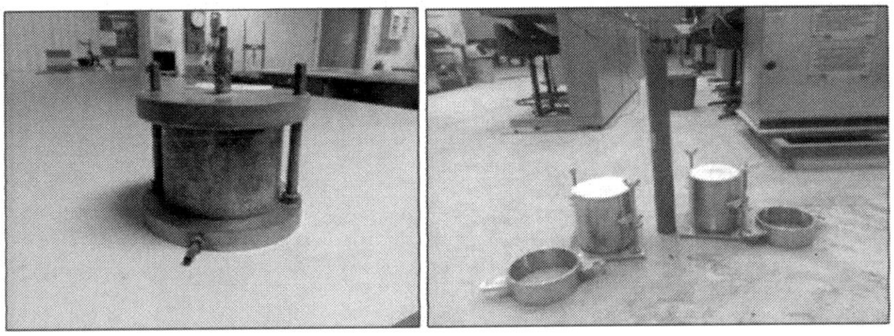

Figure 3.1. Core Cutter for Permeability & CBR Test Equipment

9. Results Analysis

The data gained from the completed experimental test were analyzed and discussed in this chapter. During the experimental process, the raw data, such as maximum dry density, the plasticity index, the coefficient of permeability, k and the CBR value for the soil were recorded. Their data were then processed and analyzed to determine the whether the soil gain more strength by using bentonite and nano clay as an admixture.

9.1. Optimum Moisture Content of BPSC
figure 4.1shows the optimum moisture content (OMC) of Batu Pahat soft clay (BPSC) is 16% and the maximum dry density is 1.774 Mg/m3. The compaction curve shown in the graph provides the relationship between the dry density and the water content for Batu Pahat soft clay subjected to a specific compaction effort. Formatting author affiliations.

9.2. Coefficient of Permeability, k of Different Percentage of Admixtures

figure 4.2 shows that the coefficient of permeability, k or the rate of seepage through BPSC with different percentage of admixture. The figure shows that the coefficient of permeability, k decreases from 0% of admixture to 15% of admixture. It shows that with increase of percentage of admixture in BPSC the rate of water seepage through the soil sample decreases almost linearly. Table 4.1 shows the coefficient of permeability, k for 0%, 5%, 10% and 15% of admixtures.

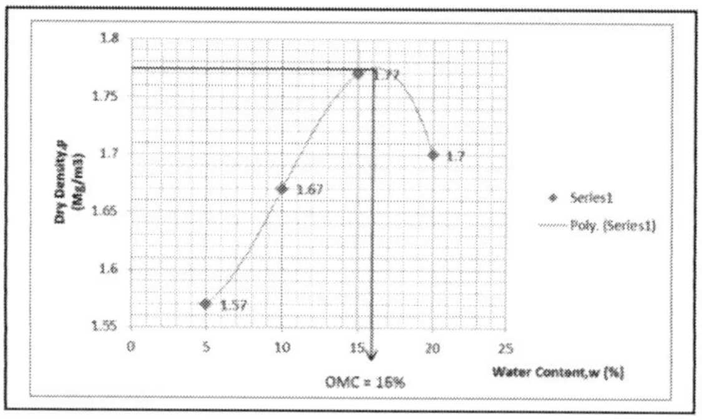

Figure 4.1. Optimum moisture content of BPSC.

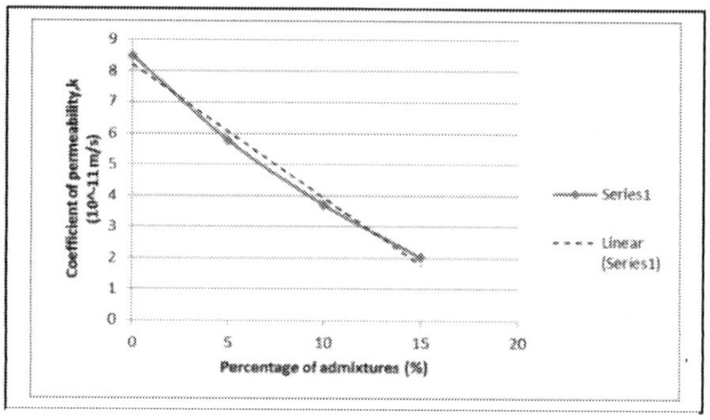

Figure 4.2. Percentage of admixtures against coefficient of permeability, k.

Table 4.1. Average Coefficient of Permeability, k.

Percentage Of Admixture (%)	k Value Average (x 10^{-11}m/s)
0	8.4750
5	5.7703
10	3.6950
15	2.0187

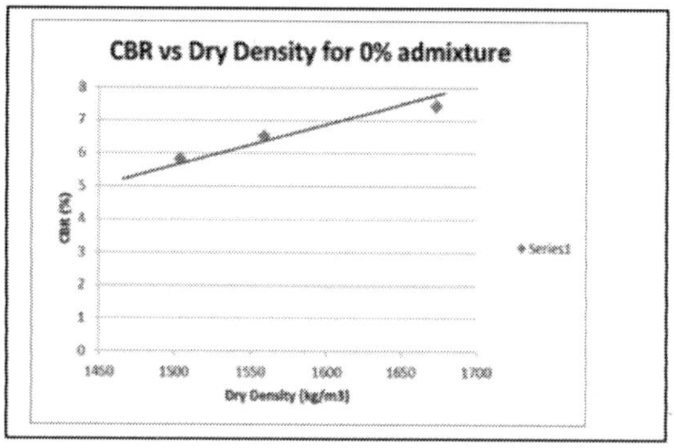

Figure 4.3. CBR against dry density for 0% admixture.

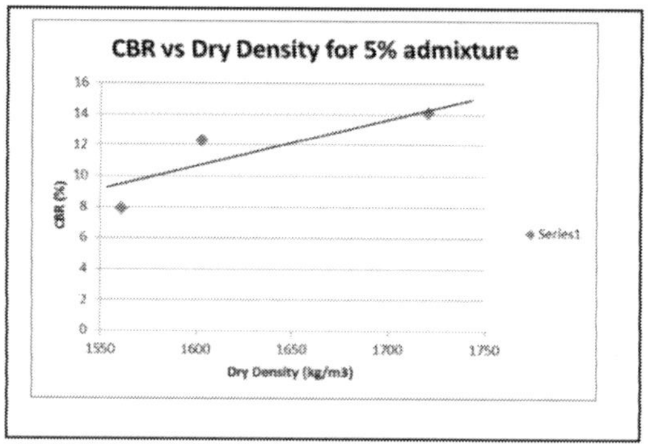

Figure 4.4. CBR against dry density for 5% admixture.

9.3. California Bearing Ratio Value for Different Percentage of Admixtures

California Bearing Ratio (CBR) test is conducted to evaluate the strength of BPSC nanoclay whether it is suitable for the construction of road subgrade. The test is performed by measuring the pressure required to penetrate a soil sample with a plunger of standard area. The CBR rating was developed for measuring the load-bearing capacity of soils used for building roads. Figure 4.3, 4.4, 4.5 and 4.6 shows the CBR against the maximum dry density of BPSC with different percentage of admixtures.

Table 4.2 shows the CBR value interpreted from figure 4.3, 4.4, 4.5 and 4.6. Based on the value, it shows that the CBR value increases as the percentage of admixture in BPSC increases. This result is then compared with Table 4.3 which is the general rating of CBR.

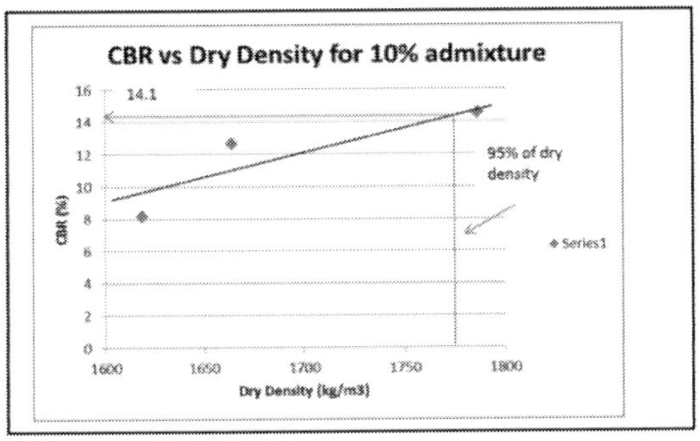

Figure 4.5. CBR against dry density for 10% admixture.

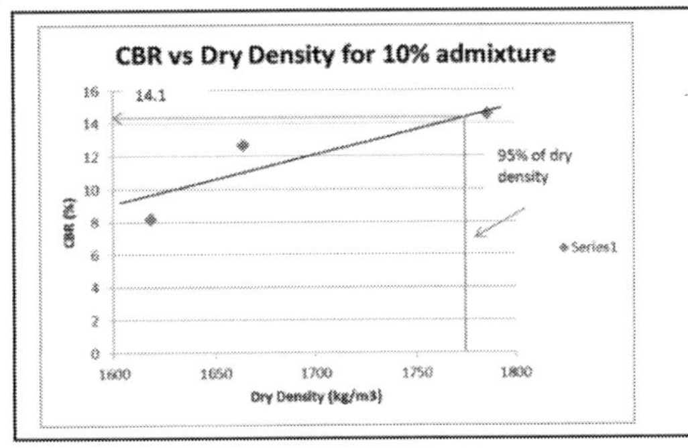

Figure 4.5. CBR against dry density for 10% admixture.

Table 4.2. Percentage of CBR for 95% of dry density.

Percentage Of Admixture (%)	Percentage Of CBR For 95% Dry Density
0	10.1
5	13.8
10	14.1
15	14.4

Table 4.3. General Rating of CBR.

CBR Value	Subgrade Strength	Comments
3% and less	Poor	" Capping is required
3% - 5%	Normal	Widely encountered CBR range capping considered according to road category
5% - 15%	Good	"Capping" normally unnecessary except on very heavily trafficked roads.

10. Conclusion

Based on the study, the following initial research conclusion of this project can be made.

1. The increasing of Bentonite and Nano clay decreases the void in BPSC. When the void in BPSC decreases, the rate of water flowing through the soil also decreases thus improving the soil and is suitable for road subgrade.
2. The CBR value for BPSC increases when the percentage of Bentonite and Nano clay increases thus increasing the strength of BPSC. High value of CBR shows that the strength of the soil is good.
3. The admixture used which is Bentonite and Nano clay are suitable additives for strengthening of soil and can be used to improve the quality of weak road subgrade.

Acknowledgment

The authors would like to express his acknowledgement to Universiti Tun Hussein Onn Malaysia, Department of Civil and Environmental Engineering and Research Center for Soft Soil (RECESS) for their financial and moral support.

References

[1] Ashraf, M. H. (2010). *Stabilization of Subgrade by using Fly Ash related to Road Pavement Thickness Design at Jalan Jaya Gading*. Universiti Malaysia Pahang: Degree of Bachelor Thesis.

[2] Chan, C-M. and Ibrahim, K.A. (2008). Using soft clay modified with cement agricultural waste as road construction materials. Proceedings of the 1st. International Conference on Transportation Geotechnics. Nottingham, UK, 25-27 August 2008 (University of Nottingham, UK).

[3] Hashim, R. & Islam, M. S. (2008). A Model Study to Determine Engineering Properties of Peat Soil and Effect on Strength after Stabilisation, European Journal of Scientific Research. 22(2), 205–215.

[4] Holt, C. (2010). Chemical Stabilization of Inherently Weak Subgrade Soils for Road Construction – Applicability in Canada. Annual Conference of the Transportation Association of Canada. 1-21. Retrieve from: http://www.tac-atc.ca/english/resourcecentre/readingroom/conference/conf2010/docs/d1/holt-2.pdf

[5] Neethu, S. V. (2013). Engineering Behaviour of Nano-clay Stabilised Soil. Proceedings of Indian Geotechnical Conference. Roorkee, India, 22-24 December 2013.
Public Works Department. Arahan Teknik (Jalan) 5/85.Standards Unit, Roads Branch, PWD Malaysia.

[6] Robani, R. and Chan, C-M. (2009). Reusing soft soils with cement-palm oil clinker (POC) stabilisation. Proceedings of the International Conference on Civil Engineering and Education in the 21st. Century (ICEE2009). Sarawak, Malaysia, 23-25 March2009 (Institution of Engineers, Malaysia).

[7] Rufaizal, C. H. (2013). Engineering Properties of Batu Pahat Soft Clay Stabilized with Lime, Cement and Bentonite for Subgrade in Road Construction. Universiti Tun Hussein Onn Malaysia: Degree of Master's Thesis.

[8] Ahmad Hakimi Mat Nor, Mohd Idrus Hj Mohd Masirin, Mohd Erwan Sanik. (2014). Site Investigation Of Road Drains For Rural Road On Batu Pahat Soft Clay (BPSC), Journal of Mechanical and Civil Engineering, International Organization of Scientific Research (IOSR).

[9] Mohd Idrus bin Mohd Masirin. Mohd Idrus Haji Mohd Masirin. (2007). Performance evaluation of encapsulated road pavement on difficult ground condition, PhD Thesis, British Library, UK.

[10] Mohd Masirin, Mohd Idrus, Rasimah Md Zain. (2012). Differential settlement analysis of different approach slab design concept for road bridges, Elsevier Proceeding, 2012.

The Potential of Soft Soil Improvement Through a Coupled Technique Between Electro Kinetic and Alkaline Activation of Soft Soil

G E Ahmed [1,2], H B Ismail[1,5], B K Huat[3], A Afshin[4] and A T S Azhar[1,5]
[1]Research Center of Soft Soil- RECESS, University of Tun Hussein Onn Malaysia, 86400 Parit Raja, Batu Pahat, Johor, MALAYSIA
[2]Department of Petroleum engineering, University of Garian, LIBYA
[3]School of graduate studies management, University Putra Malaysia 43400 UPM Serdang, MALAYSIA
[4]Department of Civil Engineering, Isfahan Science and Research Branch, Islamic Azad University, Isfahan, IRAN
[5]Faculty of Civil and Environmental Engineering, Universiti Tun Hussein Onn Malaysia, 86400 Batu Pahat Johor, MALAYSIA

E-mail: elkhebu2015@gmail.com

Abstract. Soil stabilization techniques have been in development for decades with different rates of success. Alkaline activation of soft soil is one of those techniques that has proved to deliver some of the best shear strength values with minor drawbacks in comparison with conventional soil stabilization methods. However, environmental considerations have not been taken into account, as major mineral glassy phase activators are poisoning alkaline solutions, such as sodium-, potassium-hydroxide, and sodium-, potassium-silicate, which poses serious hazards to man and environment. This paper addresses the ways of discarding the involvement of the aforementioned alkaline solutions in soft soil stabilization by investigating the potential of a coupled electro kinetic alkaline activation technique for soft soil strengthening, through which the provision of alkaline pH is governed by electro kinetic potential. Uncertainties in regard to the dissolution of aluminosilicate as well as the dominance of acidic front are challenges that need to be overcome.

Keywords: Soil stabilization, soft soil, electro kinetic.

1. Introduction

Sites that are considered abandoned due to their lack of bearing capacity have become a desired destination of contractors and governments all over the world. Their increased concern is related to the utilization of this given opportunity to launch huge projects on them [1]. However, clay soil poses problems to geotechnical engineers because of its complex nature. This derives from its low hydraulic conductivity, its plasticity, and therefore the time dependency of volume as well as pore water pressure change. Thus necessitating its treatment to modify its mentioned properties so as to stand firm constructions above it [2].

A common and significant approach applied is cement grouting to overcome the aforementioned problems by gluing the soil particles together mainly through chemical reactions. In terms of mechanical strength, cement-based binders usually deliver noteworthy results. However, concerns have been recently raised due to the high CO2 emission estimated at 7% of the total anthropogenic CO2. In addition to producing a soil cement basic binder, it is necessary to seek for a specific aggregate, which rarely occurs. Therefore it appears necessary to come up with a substitute for cement that can be easily sourced and also features the same mechanical characteristics. Binders such as carbon fly ash is a waste material containing aluminosilicate , which not only fulfill all above mentioned conditions but also its economic costs are considered very low compared with cement grouting techniques [3].

2. Alkaline activation of Soft soil

Alkaline activation of soft soil is a relatively new scientific field, utilized for concrete stabilization where alkaline solutions such as sodium hydroxide, potassium hydroxide, and sodium silicate are introduced to wastes such as carbon fly ashes, metallurgical slags, and to metakaolinte in order to activate the mineral glassy phases of the last mentioned precursors. Thereafter the mixture is added to concrete to improve its stability [4].

The potential of introducing synthetic product of alkaline solution and source binders such as fly ash, metakaolinite, etc. to soft soil to improve its mechanical properties has been investigated by plenty of researchers [3,5,6,7,8]. Alkali solutions such as sodium-, potassium- hydroxide and sodium-, potassium-silicate are very essential for the gel hardening reaction. They supply OH ions used to attack the sialate-siloxo bonds for the liberation of alumino and silicate to form a large network. This further turns into a gel, gets rearranged and finally condensate by releasing the pore water to produce a hardener [9]. To allow for such synthetic product to form, highly alkaline conditions are necessary pH>11 [10]. The aforesaid alkali solutions pose serious hazards to humans and environment. Such dangerous solutions are highly exothermic, which means large amount of heat is released upon reacting with water and can lead to severe consequences such as poisoning, when not properly handled. Poisoning can occur in several ways, such as through touching, swallowing, and inhaling, which can cause lung inflammation, throat swelling, vomiting, skin burning and possible bleeding [11].

Therefore it is needful to develop a new technique that does not make use of alkaline solutions but instead utilizes a proper pH provider to attain the required PH levels. One of the well-known soil stabilization methods is electro kinetic soil stabilization. Its advantages over conventional methods are those of low expense, rapid achievement, wide scope of application and a site accessibility to maintain or fix any soil failure arising either after or before project is terminated [12].

3. Electro kinetic soil stabilization

The principles of EK stabilization method involve applying a low direct current or a low potential gradient to electrodes embedded in a low permeable soil, from which water is not to be drained readily. The migration of charged species across the soil involves several complex mechanisms such as electrolysis, electro-osmosis, electro-migration and electrophoresis. The technique can also be enhanced by the employment of suitable nontoxic chemical lime or cement solutions to the soil by introducing them at the appropriate electrode. The combined effect of these mechanisms with other geochemical reactions alters the chemical composition of the soil porous medium and thus physicochemical properties of the soil are modified [13].

The movement of fluid and solute through soil depends upon precipitation and dissolution reactions, which in turn hinge on pore fluid alkalinity as well as on the concentration of ions. During EK treatment, the soil electrochemistry is substantially changed. This can cause oxidation and reduction of the species, which are pH dependent reactions. The resistance of the soil to pH changes can affect the geochemical reactions fundamentally. The movement of charged species across the soil comprises 4 complex mechanisms, namely electrolysis, electro-osmosis, electrophoresis and electro-migration (Fig.1). The electric DDL at the charged surface of clay particles gives rise to the EK phenomena, which includes electrophoresis, electro-migration and electro-osmosis. These electrochemical processes result in altering the hydrological physicochemical and engineering properties of the soils under electrical potential.

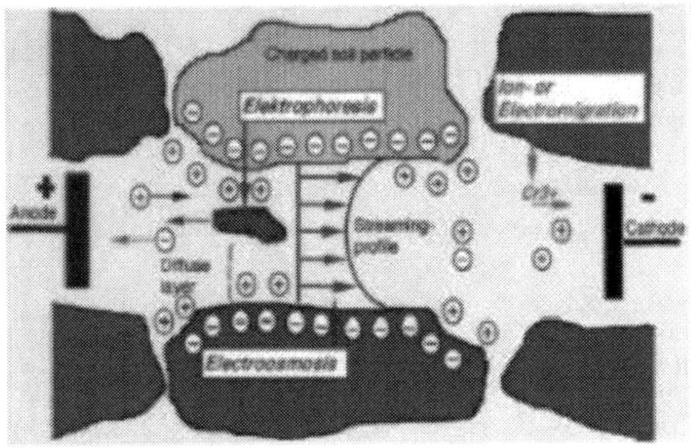

Figure1. Schematic diagram of various electrochemical processes under EK treatment technique after [14].

Electrolysis reactions occur at the anode and cathode zones, resulting in oxidation & reduction respectively. O_2 and H+ are generated at the anode (oxidation), while the electrical field employed induces H_2 and OH- to be released at the cathode (reduction) as follows:

$$\text{Anode: } 2H_20 - 4e- \longrightarrow O_2+4H \quad (1)$$

$$\text{Cathode: } 2H_20 + 2e- \longrightarrow H_2+2OH \quad (2)$$

Accordingly an acid front is formed in the vicinity of the anode and a base front dominates over the cathode zone. Both fronts move towards each other. The acid front moves faster than the base front due to the higher mobility of H+ compared with that of OH-. As a result, the acid front governs the chemistry that occurs all across the specimen except in the vicinity of the cathode. Electro-osmosis is the migration of the pore fluid from the anode towards the cathode by means of an electric field. Electrophoresis is the migration of charged particles under electrical field in pore fluid to either the anode or the cathode depending upon its charge, while electro-migration is the movement of the ions or charged electrical species towards the appropriate electrode under an electrical potential difference. Species, which have no surface charges, are dragged along with the electro-osmosis stimulated water flow [14].

4. Utilization mechanism of electro kinetic phenomena in the alkaline activation of soft soil

It is claimed that the absence of alkaline solution, which is the main trigger for the pozzolanic reaction of synthetic geopolymer product, is compensated by electrical gradient, induced by electro kinetic process. In the cathode vicinity a high PH is expected to develop as more OH ions are produced and hindered by the electroomotic flow when electrical potential is employed. These OH ions are anticipated to attack the sialate-siloxo bonds for the liberation of alumino-silicate species to further form the geoploymer hardener. As mentioned above, the electro kinetic process results in dissolution and precipitation of several ion species, which are absorbed on the soil surface, generating a high pH.

Released Al ions can be utilized in the formation of the geopolymer product. The introduction of a specific chemical solution, such as calcium chloride and potassium chloride, at appropriate electrodes can also enhance and stimulate the electro kinetic alkaline activation of soft soil

5. Challenges and limitations

Some challenges might be faced in regard with dissolution mechanism of aluminosilicate species as the former is dissolved between PH 11 and PH 14, depending upon the source material used. Even if these PH levels are obtained, it still remains uncertain, whether they can be maintained over long periods, given that availability of OH ions is dependent on several factors such as soil conductivity and electrical gradient. Therefor it is advisable to seek help in the chemistry to find out how aluminosilicate is otherwise dissolved. In addition it is known that H ions are as much as twice the OH ions and faster too. For this reason the dominance of acidic front across the specimen is likely to happen. Finally whether or not the hardener is obtainable by only involving alkaline water remain unknown before implementing trials. It is therefore strongly advised to direct the efforts towards overcoming these challenges before embarking on the electro kinetic activation of soft soil.

References

[1] Sherwood P 1993 Soil stabilization with cement and lime.
[2] Moayedi H, Huat B B, Moayedi F, Asadi A and Parsaie 2011 Effect of sodium silicate on unconfined compressive strength of soft clay Electronic Journal of Geotechnical Engineering. 16 289-295
[3] Cristelo N, Glendinning S, Fernandes L and Amandio T B 2012 Effects of alkaline-activated fly ash and Portland cement on soft soil stabilization
[4] Geopolymer concrete US department of transportation 2010
[5] Cristelo N and Glendinning S 2011Deep soft soil improvement by alkaline activation
[6] Cristelo N, Glendinning S, lisete F and T B. Amandio 2012 Effect of Calcium content on soft soil stabilization with alkaline activation
[7] Dayakar Babu R, Ramu K, Durga Prasad S and Ashok Kumar K 2013 Influence of geopolymer on the strength characterstics of sand mixed soft marine clay
[8] Cristelo N, Glendinning S, Miranda T, Oliveira D V and Silva R A 2012 Soil Stabilization using Alkaline Activation of Fly Ash for Self Compacting Rammed Earth Construction
[9] Duxson P, Ferna´ndez- ime´nez A, Provis J L, Lukey G C, Palomo A and van Deventer J S J 2007 Geopolymer technology: the current state of the art
[10] Catanescu I, Georgescu M and Melinescu A 2012 Synthesis and characterization of geopolymer binders from fly ash
[11] Agency for Toxic Substances and Disease Registry (ATSDR) 2013 Managing Hazardous Materials Incidents. Volume III – Medical Management Guidelines for Acute Chemical Exposures: Sodium Hydroxide. Emergency Medicine: A Comprehensive Study Guide, ed. Rosen's Emergency Medicine: Concepts and Clinical Practice
[12] Otsuki N , Yodsudjai W and Nishida T 2007 Feasibility study on soil improvement using electrochemical technique. Construction and building material 21 1046-1051
[13] Mosavat N, Oh E and Chai G 2012 A Review of Electrokinetic Treatment Technique for Improving the Engineering Characteristics of Low Permeable Problematic Soils. International journal of Geomat 2-2.
[14] Nasim M, Erwin Oh and Chai G 2012 A Review of Electrokinetic Treatment Technique for improving the Engineering Characteristics of Low Permeable Problematic Soils

Construction Shortfall and Forensic Investigation on Soft Ground

M Saravanan[1] and B N Ibrahim[1]
[1]Nexus EC Sdn Bhd, Kuala Lumpur, Malaysia

E-mail: vanan.nexus@gmail.com

Abstract. Recently, numerous construction activities performed on soft ground have undergone various kind of failures. Some of the failures were due to lack of understanding the nature of soft ground, inadequate site investigation data, adaptation of inappropriate type of ground improvement techniques or foundation system. Couple with inadequate desk study carried out to the historical condition of the site, chronological of activities carried out within the site prior to construction. Hence coordinated approach required to be executed for field investigation works with proper supervision would be the key to prevent any design or construction related failure in the future. This paper addresses some of shortfalls related to design and construction failures in soft ground.

Keywords: Site investigation, soft clay, stone column, ground improvement, forensic investigation.

1. Introduction

Carrying out project on soft ground requires careful measures to be undertaken systematic, namely:-

- Site selection, existing condition and historical even occurred at the site.
- Planning for adequate site investigation with suitable adaptation of cross verification of site investigation data.
- Selection of reliable, workable and cost effective solution to be implemented at the site.
- Stringent site supervision and quality assurance required to be executed.

With the above being implemented any potential failures related to construction activities of soft ground can be eliminated.

2. Shortfall Related To Site Investigate (SI) Work Carried Out

Selection of site for development in many situation could not be avoided. The wide corridor of soft clay deposit along Malaysian coastal line could not be avoided from development. In the event the site is located within soft clay deposit zone, careful planning measure for site investigation works required to be carried out. Some of the shortfalls or errors identified while handling site investigation (SI) works are:-

- Data mis-interpretation.
- Lack of supervision during site investigation carried out.
- Incompetent person to execute the works.
- Location identified for SI works.

With all the above mentioned shortfalls, engineers would not able to provide optimize design in which related to cost saving (economical) and environmental friendly.

3. Example Case Of Data Mis-Interpretation Due To Lack Of Supervision

Serious design error could be related mis-interpretation to SI works. Site supervisions are important as report in this example of 5m high embankment failures. The design of 5m high embankment were carried out on ground reported to be Medium Stiff CLAY / SILT without any indication of Soft CLAY/SILT deposit. The SI works performed in 2003 reported field data as shown in Figure 1.0.

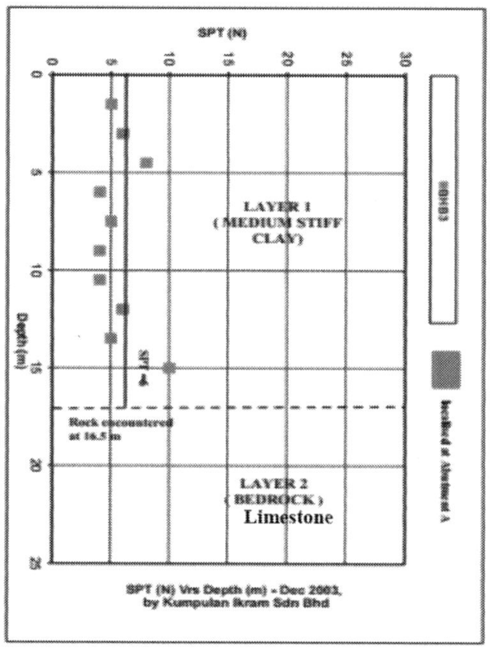

Figure 1.0. SPT-Nvalue (Blows/300mm) for Abutment A (SI in 2003).

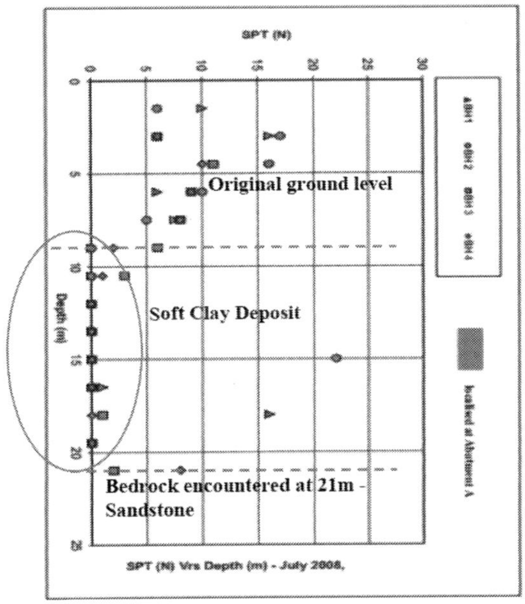

Figure 2.0. SPT-Nvalue (Blows/300mm) for Abutment A (SI in 2008).

The SI works performed were not cross check with the site geology. During the forensic investigation after failure, eleven (11) number of boreholes were carried out which are three (3) numbers (in year 2007) and eight (8) numbers (in year 2008) were performed. The new SI reported the presents of very soft ~ soft CLAY deposit as shown in Table 1.0 and Figure 2.0.

Table 1.0. SI comparison at Abutment A.

Date	12/10/2003-13/10/2003	8/5/2008 - 10/5/2008	11/5/2008 – 12/5/2008	22/4/2008-25/4/2008	26/4/2008-28/4/2008	
Year	2003	2008	2008	2008	2008	
Borehole	BHB3	BH1	BH2	BH3	BH4	
Location/ Depth (m)	On ground level	At 5.5m embankment fill (at non failed side of embankment)	At 5.5m embankment fill (at non failed side of embankment)	At 5.5m embankment fill (at non failed side of embankment)	At 5.5m embankment fill (at non failed side of embankment)	
0		10 Stiff	6 Firm	6 Firm		Fill Material
1.5		16 Stiff	17 Very Stiff	11 Stiff	6 Firm	
3.0		11 Stiff	16 Very Stiff	9 Stiff	10 Stiff	
OGL ∇ 4.5		10 Stiff	10 Stiff	8 Firm	9 Stiff	
6.0		6 Firm	5 Firm	6 Firm	8 Stiff	Original Ground
7.5	5 Firm	0 Very Soft	0 Very Soft	3 Soft	2 Very Soft	
9.0	6 Firm	0 Very Soft	0 Very Soft	0 Very Soft	1 Very Soft	
10.5	8 Stiff	0 Very Soft	0 Very Soft	0 Very Soft	0 Very Soft	
12.0	4 Soft	0 Very Soft	0 Very Soft	0 Very Soft	0 Very Soft	
13.5	5 Firm	0 Very Soft	22 Very Stiff	0 Very Soft	0 Very Soft	
15.0	4 Soft	1 Very Soft	50 Hard	1 Very Soft	0 Very Soft	
16.5	4 Soft	16 Very Stiff	Rock-Sandstone	0 Very Soft	0 Very Soft	
18.0	6 Firm	50 Hard		2 Very Soft	0 Very Soft	
19.5	5 Firm	Rock -Sandstone		Rock-Sand stone	8 Stiff	
21.0	10 Stiff				Rock-Sand stone	
22.5	Rock-Limestone					

* OGL – Assume original ground level.

The bedrock encountered and reported to be Sandstone and Limestone. Cross comparison with Geological Map of Malaysia the area reported to be Quaternary Marine and Continental deposit. The area were not likely to have Limestone as bedrock formation. The nearest Limestone found to be at

Gua Musang about 160 ~ 175km away (refer Figure 3.0 and 4.0). The bedrock reported in SI 2007 and 2008 were Sandstone. The present of competent geologist and supervisor are critical for the site data verification.

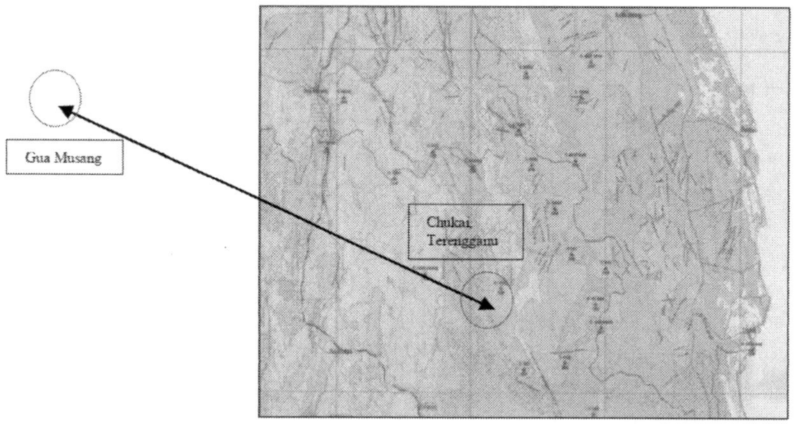

Figure 3.0. Geological Map showing site project (Chukai, Terengganu) and Gua Musang (Limestone area).

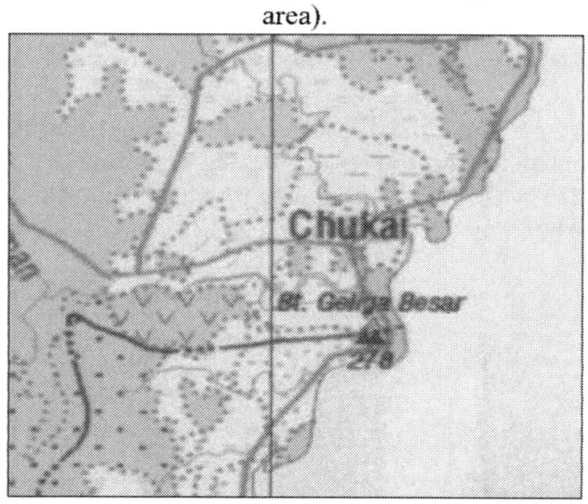

Figure 4.0. Geological Map showing Geological formation of project area.

4. Cross comparison with Standard Penetration Test (SPT), Cone Penetration Test (CPT) and Ground Penetration Radar (GPR).

Performing site investigation works using only SPT and insitu sampling will not be adequate enough to interpret the actual ground condition in area with soft CLAY deposit. Example on how data can be mis-interpretated if comparison were not made with other form of site investigation works. Three (3) types of SI data were cross compared, namely:-

- Boreholes with Standard Penetration Test (SPT).
- Cone Penetration Test (CPT).
- Ground Penetration Radar (GPR).

Boreholes (BH) only indicate the subsoil stiffness (SPT value) without indicate the other parameters which related to historical condition of the site. Whereas the CPT data were able to expose the presents material such as loosely deposited material which is not the original soil of the site. Example the filling over of old pond or river. This formation contains lot of voids that creates negative pore pressures (much lesser than natural water pressure with depth – refer Figure 5.0 and 6.0).

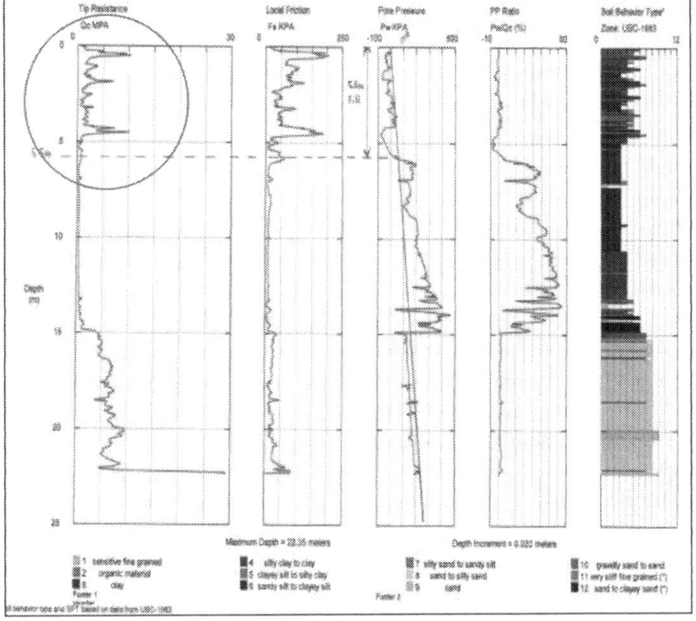

Figure 5.0. CPT data from site with pond filling.

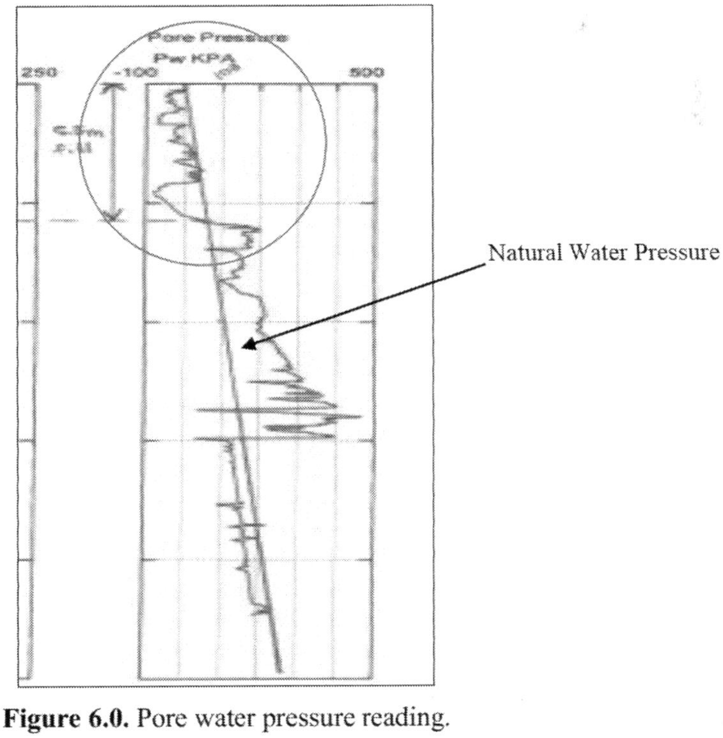

Figure 6.0. Pore water pressure reading.

Hence, as reported in the CPT data, this site consist of material most likely to be loosely deposited earth (transported and dumped) into natural pond. Presents of pond and historical closure of ponds could have direct consequences to the type of foundation to be adopted. Performing CPT test can be very useful to provide vital information to supplement boreholes data. Further cross comparison can also be made using Ground Penetration Radar (GPR). The performed GPR reported pockets of voids and also signs of ground subsidence as shown in Figure 7.0 and 8.0. The findings are relatively consistent with CPT findings. However, Boreholes data were not able to identify the loosely deposited earth. Subsoil stratum obtained from boreholes are shown in Figure 9.0.

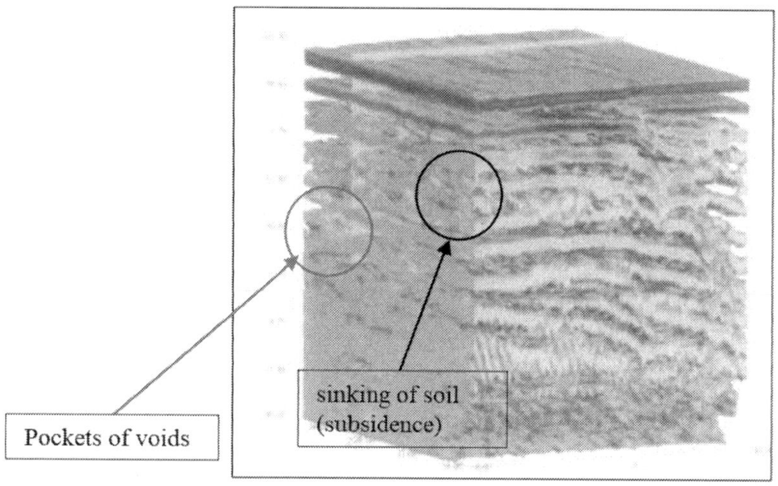

Figure 7.0. Localise pocket of voids and subsidence based on GPR scanning.

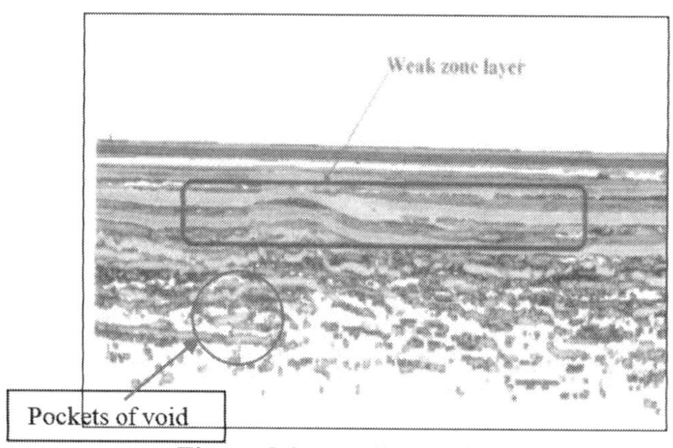

Figure 8.0. Localise pocket of voids.

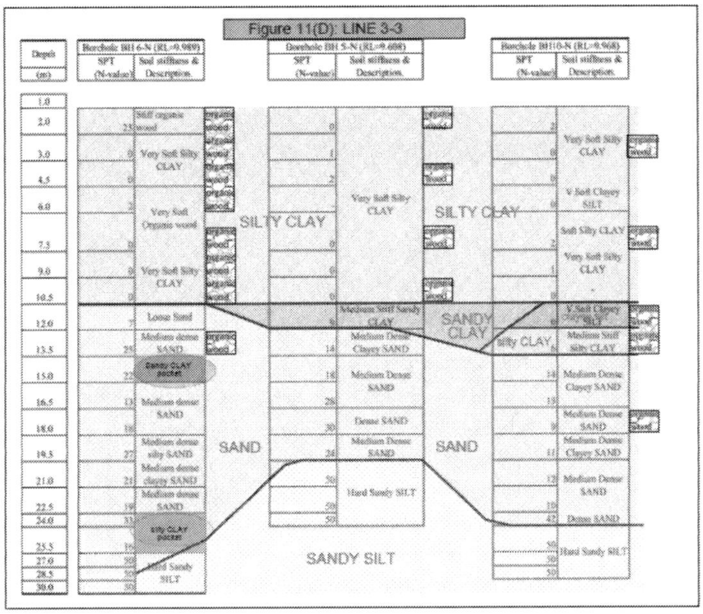

Figure 9.0. Subsoil stratum obtain by Borehole.

5. Conclusion

Performing adequate SI works are major way forward in preventing design failures especially on soft ground. Understanding why failures in soft ground occurs and not repeating the past mistake could also be useful on soft ground engineering. Example, some of key factors that many engineers overlook during stone column design are (refer **Plate A ~ D**):-

i. Due to shear strength.
   - The recommended lower limit to the undrained strength of Cu=15 kPa is suggested for treatment with stone column (Bryan, 2007) and (MCR Davies, 1997).
   - Although there have been few situations where softer soils have been successfully improved (Raju et al., 2004).
   - For very soft Clay with Cu between 4~15 kPa, soil with very low cohesion have relatively low radial support given to the stone columns by soil (Thornburn, 1975).

ii. Soil plasticity index.
   - The plasticity index (Ip) of soil reflects the potential for volume change, UK national Building Council (NHBC, 1998) suggests that stone columns not recommended to be used when Ip > 40% (Bryan, 2007).
   - However, only few successfully cases where high plasticity soils has been improved using stone column.

iii. Soil Permeability.
   - The major difficulties in forming uncontaminated stone columns in very soft soils are due to radial displacement seriously distorts and remoulds the surrounding soil inducing excess pore water pressure which cannot dissipate rapidly even in a laminated soil because the distortions destroy the natural horizontal drainage laminae (Thornburn S, 1975).

iv. Ground settlement.
   - With reference to the research work by Hughes and Withers (1941) indicates that the ultimate stress in the stone columns were obtained only when the vertical displacement of stone column was 58% of the column diameter. Field test results examined by S. Thorburn on full size stone columns indicates that the vertical displacement of the top of stone

columns at failure were only of the order of 10~15% of the diameter of stone columns (Thornburn, 1975).

Plate A

Plate B

Plate A and **B**: Photo showing failed stone column (non reinforced stone column) and column being clogged with soft Clay and preventing the pore water discharge during consolidation process and the infill material decentigrate.

The risk and project delivery responsibility for designing and constructing stone columns in ground weaker that the listed above, should not be shouldered by client and consultants. Consultant should not recommend the usage of stone column in such ground condition. Alternatively, instead of stone column, designer / consultant shall recommend more reliable ground improvement work (i.e. Pre-fabricated vertical drain or Encased stone column or pile foundation).

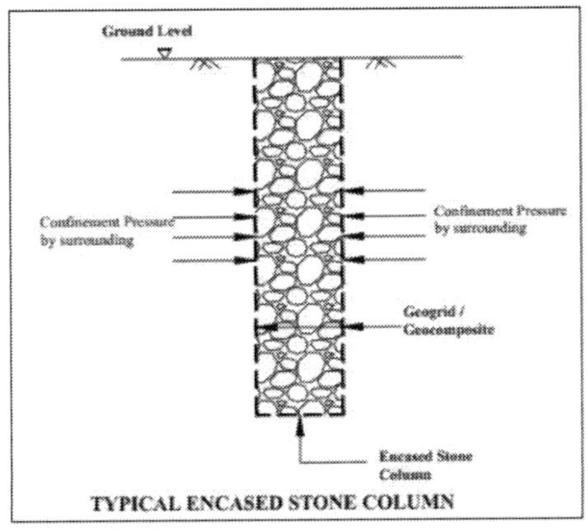

Plate C. Illustration of encased (reinforced) stone column.

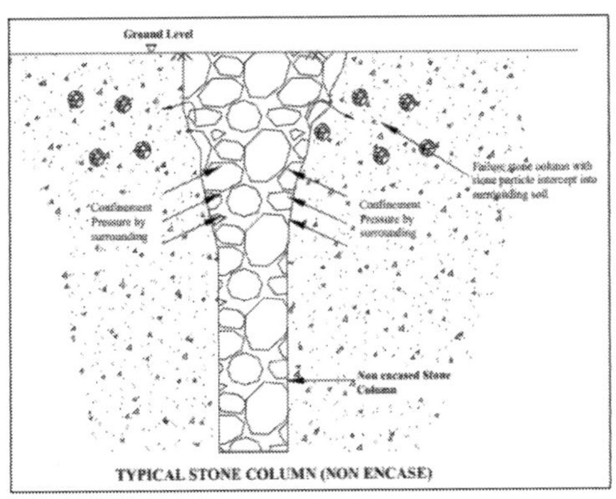

Plate D. Image showing decintigrade of stone column in soft ground due to lack confinement strength of surrounding soil.

References

[1] British Standards BS5930 1999 Code of Practice for Site Investigations. British Standard Institutions.

[2] Davies M C R, Francois S 1997 Ground Improvement Geosystem: Densification and reinforcement.

[3] Cabe B A 2007 Ground Improvement Using The Vibro-Stone Column Technique. Department of Civil Engineering, National University of Ireland, Galwa.

[4] Hughes J M O, Withers N J and Greenwood D A 1975 A field trial of the reinforcing effect of a stone column in soil. Geotechnique. **25(1)** 31-44.

[5] Krishna H, Raju V R and Wegner R 2004 Ground Improvement using Vibro Replacement in Asia 1994 to 2004 : A 10 Year Review. Proc., 5thInt. Conf. on Ground Improvement Techniques., Kuala Lampur, Malaysia.

[6] Raju V R 2002 Vibra replacement for high earth embankments and bridge abutment slopes in Putrajaya, Malaysia. International Conference on Ground Improvement Techniques, Malaysia. 607-614.

[7] Thorburn S 1975 Building structures supported by stabilized ground. Geotecnique **25(1)** 83–94.

A Tensile Strength of Bermuda Grass and Vetiver Grass in Terms of Root Reinforcement Ability Toward Soil Slope Stabilization

M N Noorasyikin[1] and M Zainab[2,1]
[1] Faculty of Civil Engineering, Universiti Teknologi Mara MALAYSIA
[2] Geotechnical Engineering and Georisk Management, Institute for Infrastructure Engineering and Sustainable Management, Universiti Teknologi Mara MALAYSIA

E-mail: shecan83us@gmail.com

Abstract. An examination on root characteristics and root properties has been implemented in this study. Two types of bioengineering were chose which are Vetiver grass and Bermuda grass as these grasses were widely applied for slope stabilization. The root samples were taken to the laboratory to investigate its classification, characteristics and strength. The root of both grasses was found grow with fibrous root matrix system. In terms of root anchorage, the root matrix system of Vetiver grass was exhibits more strengthen than the Bermuda grass. However, observation on root image from Scanning Electron Microscope test reveals that the root of Vetiver grass becomes non-porous as the moisture content reduced. Meanwhile, the root tensile strength of Bermuda grass was obtained acquired low value with higher percentage of moisture content, root morphology and bonding strength. The results indicated that the root tensile strength is mainly influence by percentage of moisture content and root morphology.

Keywords: Bioengineering, root morphology, slope stabilization.

1. Introduction

Nowadays, the slope failure is the most frequently phenomena occurred on Earth. The geotechnical engineer always faces a problem especially with shallow failure. Shallow failure which estimated about at depth about 1 to 1.5 meters is still under aware solving problem. There is none a good option to mitigate the slope failure until today.Apart from that, bioengineering technique which is vegetation cover was usually applied for mitigation the slope failure because its advantages. The technique is less cost, fast grows and easily planted. Generally,the vegetation mainly stabilizes theslope by mechanical effects through root matrix system [5,4, 13,9,8,7,12,11,14-1]. The soil shear strength is found increase through the mechanical effects of the vegetation root matrix system where the root acts as reinforcement toward the soil. [6,4,13,9,10,7,2,3,12,15-1]. In Malaysia environment, Bujang and Sina [3]stated that the vegetation can potentially provide immediate mechanical shear strength for slope remedial and long-term beneficial effects. There are three common types of root morphology; heart root system (fibrous root system), tap root system (horizontal and vertical root system) and plate root system (horizontal root system) [10]. The definition of root morphology is architecture, shape or structure of vegetations root system [2]. In this study, new concept of root matrix system was introduced shown in Figure 1.

The Vetiver grass is very strong with high mean tensile strength of 75 MPa at 0.7 to 0.8 mm root diameter which is approximately 1/6TH of strength of mild steel [5]. Bermuda grass also has being good vegetation for protection slope. However, to date the slope failure is still happen although the slope has been covered with these grasses. This study shall investigate the main factors influence the vegetation root toward slope failure.

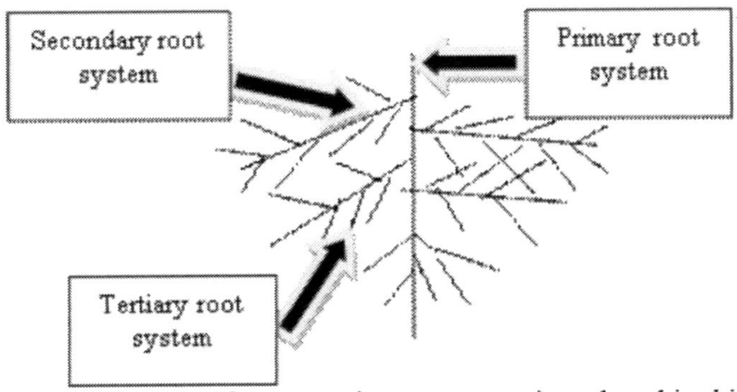

Figure 1. New concept of root matrix system was introduced in this study.

2. Materials and Methods

2.1 Site Characteristics and Grasses Types

The site investigation was carried out on well grown of Vetiver grass and Bermuda grass at cut slope gradient 450 which is at Maran area.The soil classification using Sieve analysis test was conducted based on British Standard 1377: Part 2: 1990: Clause 9.2.2. The soil and root samples were taken to the laboratory to study its characteristics and properties. Vernier Caliper was used to measure the root diameter and root length of both grasses as shown in Figure 2. The measurement unit was recorded in mm (millimeter). Besides that, the root morphology also was characterized by based on [10] findings.

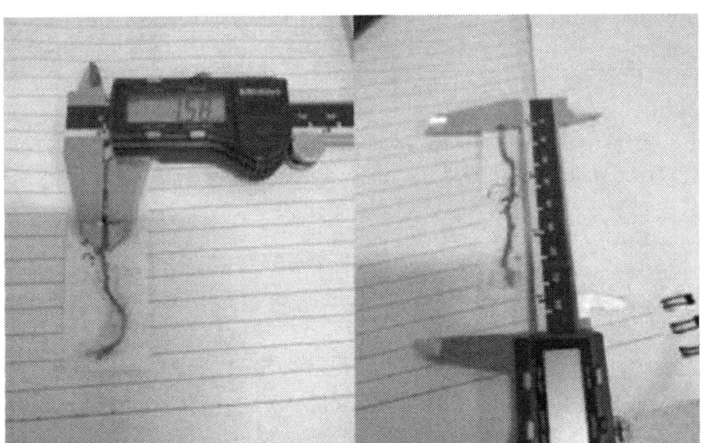

Figure 2. Photographs of root measurement using Vernier Caliper.

2.2 Root Tensile Strength Test

The Bermuda grass and Vetiver grass root tensile strength test was conducted using Shimadzu Universal Testing Machine as shown in Figure 3. The test was conducted to determine the bonding strength between root matrix system of vegetation and soil.There are two stages of conducting the tensile strength test. For first stage, the root tensile strength was carried out on individual primary root and secondary root matrix system which is a group of 5, 10 and 15 roots. For second stage, the root samples were kept in the plastic to retain the moisture content. Then, the root was taken out to carry out the tensile strength test with following days; 1 day, 5 day, 10 day, 15 day and 20 day.The test was conducted on single individual of primary root matrix system. Besides that, the percentage of moisture content alsowas determined based on British Standard 1377: Part 2: 1990: Clause 9.2.2.The unit tensile strength of root sample was recorded in MPa (MN/m^2).

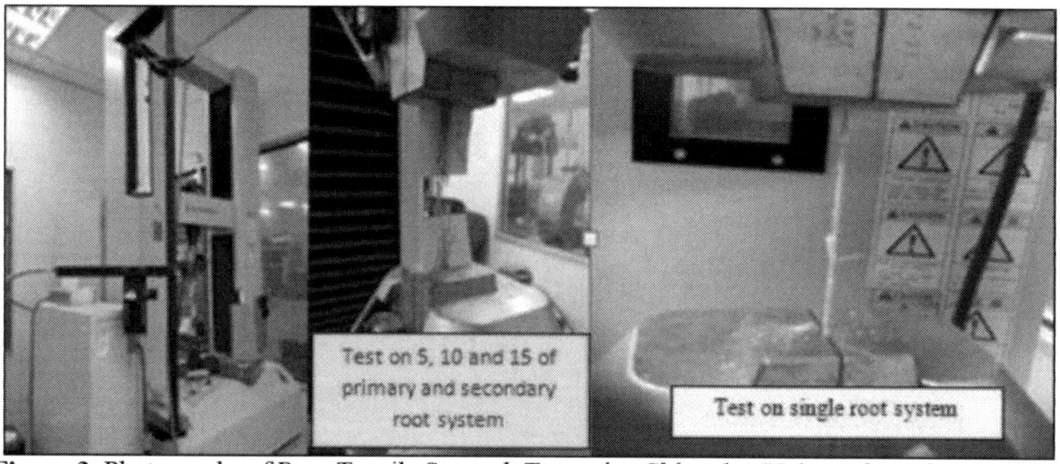

Figure 3. Photographs of Root Tensile Strength Test using Shimadzu Universal Testing Machine

2.3 Scanning Electron Microscope (SEM) Test

A scanning electron microscope (SEM) is an electron microscope that produces images of a sample by scanning it with a focused beam of electrons. The electrons interact with atoms in the sample, producing various signals that can be detected and display image of topography surface. SEM can achieve resolution better than 1 nanometer. The SEM test (Scanning Electron Microscope) was conducted to determine the texture of root after done second stage of root tensile strength test.

The root samples were examined with duration of 1 day, 5 day, 10 day, 15 day and 20 day. All samples must be of an appropriate size to fit in the specimen chamber and are generally mounted rigidly on a specimen holder called a specimen stub. Before begin the test, the root specimens were coated with gold (20nm) using a sputter coater (Emitech K550X) to prevent electron scattering as shown in Figure 4. Electron microscopic images of the root were recorded using a SEM (Hitachi) device with an accelerating voltage of 7 keV in high vacuum and secondary electron (SE) image mode.

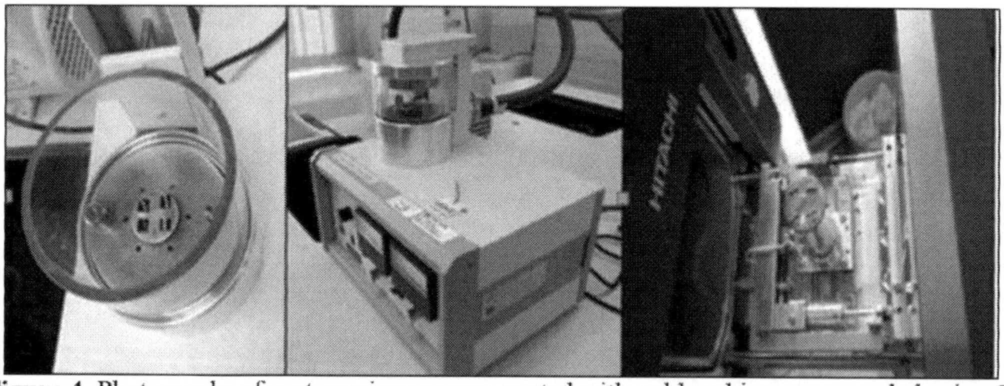

Figure 4. Photographs of root specimens were coated with gold and images recorded using SEM Hitachi.

3. Results and Discussions

3.1 Root Morphology

Based on [10], the root morphology was characterized as heart root system which is growing with fibrous root system as tabulated in Table 1. The pattern of heart root matrix system of Vetiver grass and Bermuda grass is similar with geometry which is in semi sphere form.As mentioned before, the new concept of root matix system has been introduced in this study shown in Figure 1.The root matrix system of Vetiver grass and Bermuda grass has three type of root system; primary root system, secondary root system and tertiary root system. It can be said that these grasses have a good anchoring toward slope.

3.2 Root Tensile Strength of Vetiver grass and Bermuda grass

The tensile strength test has been implemented on a group of 5, 10 and 15 for primary and secondary root system of Vetiver grass and Bermuda grass. The tensile strength toward root system was plotted in the bar graph as shown in Figure 6. It was observed that the Vetiver grass require maximum force rather than Bermuda grass to pull the root with results obtained; 5 primary root (138.56 MPa), 10 primary root (339.15 MPa), 15 primary root (431.07 MPa), 5 secondary root (181.91 MPa), 10 secondary root (238.52 MPa) and 15 secondary root (367.26 MPa) respectively represented by red bar from the graph. Meanwhile, for Bermuda grass the tensile strength for 5 primary root (181.12 MPa), 10 primary root (122.65 MPa), 15 primary root (460.74 MPa), 5 secondary root (52.98 MPa), 10 secondary root (71.74 MPa) and 15 secondary root (219.28 MPa).

Based on [5], it can be clarified that the values of root tensile strength for both grasses is more than 75 MPa which is can be considered very strong category of root system. To see whether the root matrix system for both grasses can be considered strong or not, the root system was further investigate through Scanning Electron Microscope test. The physical root system can be seen from the root texture image with respect to different percentage of moisture content. The root diameter and root length data were tabulated in Table 2.

Figure 7 and Figure 8 shows the relationship between root system and moisture content. For Bermuda grass, at day 1 the percentage of moisture content was recorded is 71.60% with tensile strength values ranged from 13.029 to 100.160 MPa. At day20, the tensile strength gradually decreases by 16% while the moisture content increases by 14.3%. For Vetiver grass, it shows contrarily. At day 1, the percentage of moisture content was recorded 79.29% with tensile strength values ranged from 12.61 to 88.52 MPa. The tensile strength was found decreases as the moisture content decreases starting from day 5 to day 20.

It was observed that the root diameter and root length is not much influence the tensile strength of root system. Based on data obtained, it can be summarized that the heart root matrix systems for both grasses have good bonding strength toward the sandy soil slope. The individual root matrix systems are characterized by vegetation type where the degree of rootdevelopment is varies within different type of vegetation. The root will grow through large poresif the soil has granular structure or sandy texture. Meanwhile, if the soil has clay or clay loam texture, the root will be confined to cracks. Drought condition and most nutrient deficiencies lead to limit root growth by slowing the growth of the shoot that supplies energy to the roots

Table 1. Root morphology of Vetiver grass and Bermuda grass.

Grass Types	Root Morphology	Description
Vetiver grass	 Heart Root System (Fibrous)	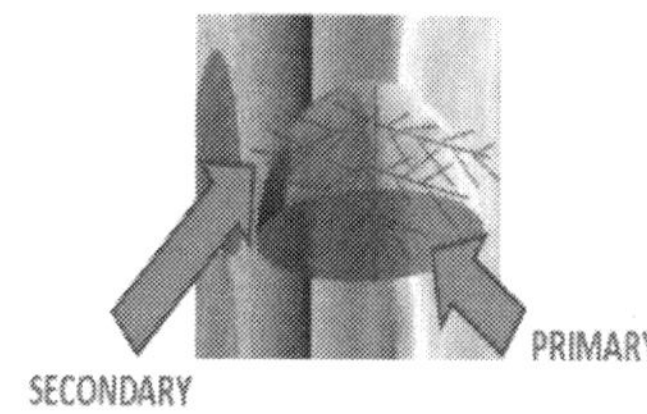 **Geometry pattern**: Semi-Sphere
Bermuda grass	 Heart Root System (Fibrous)	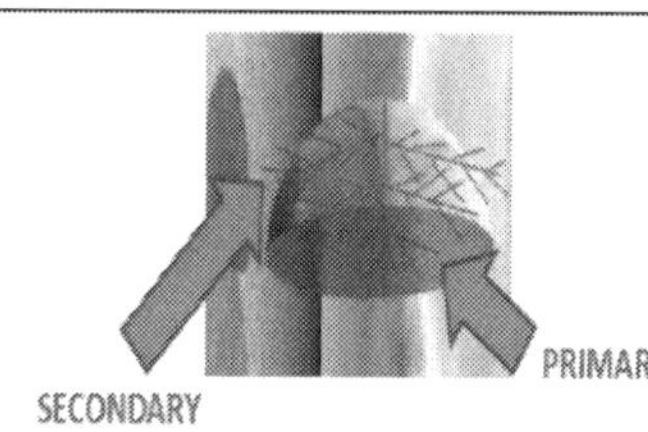 **Geometry pattern**: Semi-Sphere

Based on Scanning Electron Microscope images, the root of Bermuda grass was found become more porous as the moisture content increases. The moisture content is easily absorbed by the root system through the pore space. Meanwhile, for Vetiver grass the root system become less pore space as compared at day 1. The less moisture content made the tensile strength of root system acquired low value.

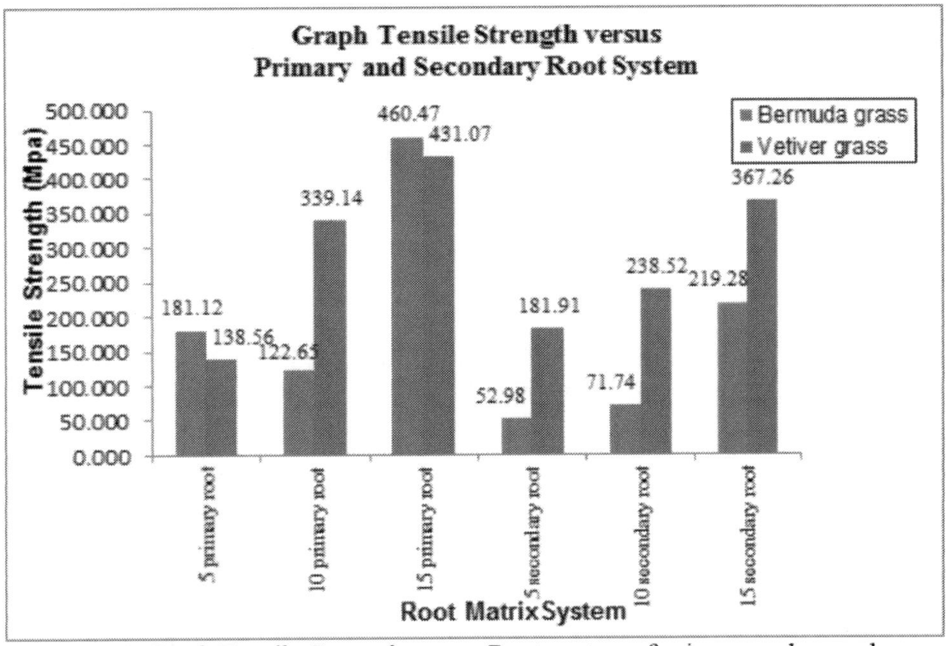

Figure 6. Graph Tensile Strength versus Root system of primary and secondary.

Table 2. Root Diameter and Root Length for each grass.

Grass type/ Growth Ranking	Root Matrix System	Root Diameter (mm)	Root Length (mm)	Soil Classification
Bermuda grass (Well grown)	5 primary root system	1.04	80	Sandy Soil Gravel 30.53 %, Sand 65.64% Silt 3.83% Slope gradient: 45^0 Soil pH = 5.55
	5 secondary root system	0.72	130	
	10 primary root system	1.01	140	
	10 secondary root system	0.79	140	
	15 primary root system	0.89	130	
	15 secondary root system	0.80	130	
	1			
Vetiver grass (Well grown)	5 primary root system	1.22	120	Sandy Soil Gravel 30.53 %, Sand 65.64% Silt 3.83% Slope gradient: 45^0 Soil pH= 5.55
	5 secondary root system	0.53	90	
	10 primary root system	1.24	120	
	10 secondary root system	0.52	110	
	15 primary root system	1.09	100	
	15 secondary root system	0.73	110	

Conclusion

The root morphology for Vetiver grass and Bermuda grass is fall into heart root system category which is growing with fibrous root system. It is a good root system where it has good bondingstrength between slope and root itself. The tensile strength of 5, 10 and 15 primary and secondary root system of Vetiver grass exhibits larger than Bermuda grass. However, from the Scanning Electron

Microscope image the root of Vetiver grass cannot able function well as the moisture content reduces. In drought condition, the root of Vetiver grass might have a problem to act as rootreinforcement toward the slope. While, in heavy raining season, the root of Bermuda grass cannot survive well since this grass planted on slope surface. The root system of Bermuda grass may break or fractures due to high absorbance of moisture content. It can be concluded that as the moisture content increases, the root matrix system of Vetiver grass has a good anchoring to the cut sloperather than Bermuda grass especially in heavy raining season in Malaysia tropical wet environment. The function of root matrix system of vegetation is mainly depending on moisture content and root morphology.

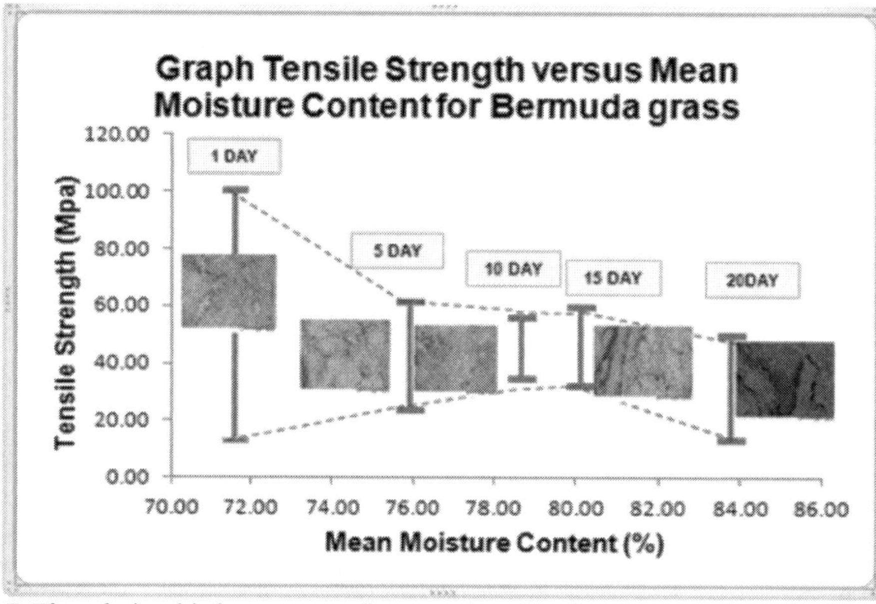

Figure 7. The relationship between tensile strength and moisture content included SEM image for Bermuda grass.

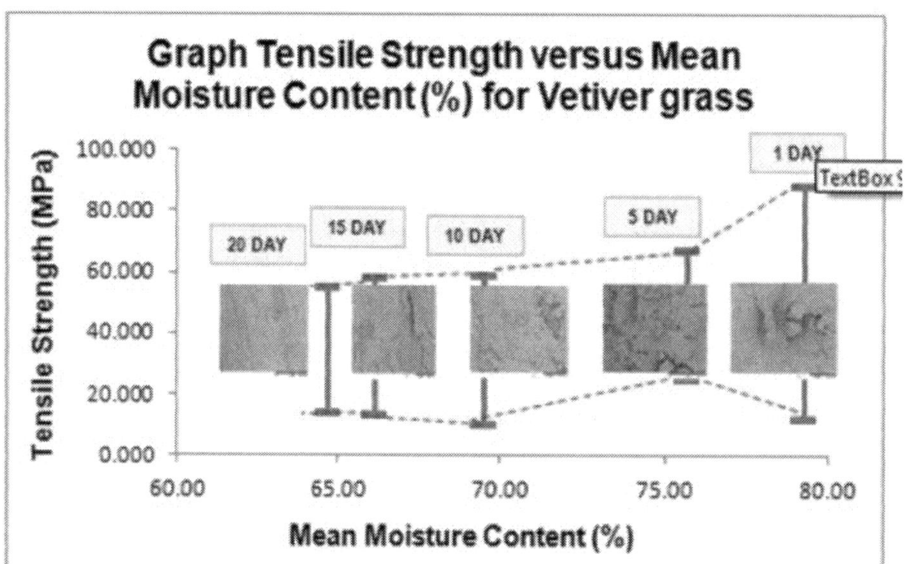

Figure 8. The relationship between tensile strength and moisture content included SEM image for Vetiver grass

Acknowledgement
Authors would like to thank Universiti Teknologi MARA for financial support under Research Intensive Faculty (RIF) grant [File No: 600-RMI/DANA 5/3/RIF (869/2012)].

References
[1] Ali,N.; Faeshchi,i.; Mu'azu,M.A.; and Rees, S.W. (2012). Soil-Root Interaction and Effects on Slope Stability Analysis. Electrical Journal of Geotechnical Engineering. Vol.17.[2012].
[2] Bibalani,G,H.;Golshani,A.A.; Zahedi,S.S.; and Bazhrang,Z. (2007). Soil Stabilizing Characteristics of RangelandsVegetation in Northwest Iran (Misho Rangelands Protected Location of Shabestar).Asian Journal of Plant Sciences. 6 (6).pp 1020-1023. ISSN: 1682-3974.
[3] Bujang, B.K.H.; and Sina, K. (2010). Study of Root Theories in Green Tropical Slope Stability. Electrical Journal of Geotechnical Engineering.Vol.15 [2012]. Bund Q
[4] Chok, Y.H.;Kaggwa,W.S.; Jaksa,M.B.; and Griffiths, D.V. (2004). Modeling the effects of vegetation on stability of slopes. Proceedings, 9th Australia New Zealand Conference on Geomechanics, Auckland.
[5] Diti, H. (1999). 15 Years of Bio-Engineering in the Wet Tropics. First Asia-Pacific Conference on Ground and Water Bio-engineering. Manila,April 1999.
[6] Elliott, M. (1998). Vegetation and Erosion – A Literature Survey. Conference at Oregon State University "Native Vegetations: Propagation and planting", December 9- 10,1998
[7] Faisal, H.A.; and Normaniza,O. (2007). Soil-Roots Composite: Correlation between Shear Strength and Some Plant Properties.Electrical Journal of Geotechnical Engineering.Volume 12/D,2007.
[8] Ghassem, H.B.;Abolfazl, A.G.; Shahriar,S.Z.;andZia,B. (2007). Soil Stabilizing Characteristics of Rangelands Vegetation In Northwest Iran (Misho Rangelands Protected Location of Shabestar).Asian Journal of Plant Sciences6 (6):1020-1023. ISSN 1682-3974. Sian Network for Scientific Information
[9] Joanne, E.N.; and John,R.G. (2006).Assessing the role of vegetation on soil slopes in urban areas.IAEG2006 Paper Number 744. The Geological Society of London
[10] Nomessi,K.; Thierr.F.; Kouami.K.; Kouma,N.; and Lac,P. (2006). Disaster Mitigation of Debris Flows, Slope Failures and Landslides. Universal Academy Press Inc Tokyo Japan,pp 561-569
[11] Rajesh,R.; and Shrivastva,B.K. (2011). Biological stabilization of mine dumps: shear strength and numerical simulation approach with special reference to Sisam tree. Environ.EarthSci J [2011] 63: 177-188. DOI10.1007/s12665-010-0682-4
[12] Sasan,M.; Bujang, B.K.H.; and Vahed, G. (2009). Evaluation on Root Theories and Root Strength Properties in Slope Stability.European Journal of Scientific Research.Vol.30 No.4.594-607.ISSN 1450-216X
[13] Wendi, S. (2006).Soil Strength Reinforcement by Plants. Most Distinguished Paper.
[14] Yoav,W.; Amran,E,; and Uzi, K. (2002). Plant Roots The Hidden Half. Third Edition. ISBN: 0-8247-0631-5
[15] Yong,W.; Qiang,X.; and Ying, Z. (2011).Mechanism Study and Numerical Simulation on Vegetation Affecting the Slope Stability.Electrical Journal of Geotechnical Engineering. [2011].Ppr11.054.Vol 16.

Effect of Reagents Concentration on Biocementation of Tropical Residual Soil

K T P Chiet[1], K A Kassim[1], K B Chen[1], U Martula[1], C S Yah[2] and A Arefnia[1]
[1]Faculty of Civil Engineering, Universiti Teknologi Malaysia, 81310, Johor Bahru, MALAYSIA
[2]Faculty of Engineering and Science, Universiti Tunku Abdul Rahman, Jalan Sungai Long, Bandar Sungai Long, Cheras, 41300, Kajang, SELANGOR

E-mail: [a]Tpckenny2@live.utm.my

Abstract. This study explores the feasibility of Bacillus subtilis and optimum reagents concentration used in Microbial-Induced Calcite Precipitation (MICP) treatment of tropical residual soil. Experiment was conducted to investigate the effect of cementation reagents concentration toward MICP treatment. The performance of MICP treatment was assessed by measurement of the soil shear strength and calcite content. Based on the experimental results, it is discovered that the cementation reagent concentration has significantly affected on the performance of MICP treatment. The results suggested that the most preferable MICP treatment reagents concentration is 0.25M with the presence of Bacillus subtilis; using these treatment parameters, both UCS value and calcite content of treated soil had increased about 38% and 65.6% respectively. However, the reduction in UCS value was manifested for those samples treated at higher reagents concentration (0.35M); this phenomenon is attributed to the salinity of reagents where high salinity is not favourable to the bacteria growth and microbial activity; subsequently, this resulted in a consequential decrease in shear strength of the treated soil.

Keywords: Microbial induced calcite precipitation (MICP), tropical residual soil, biocementation.

1. Introduction

Microbial-Induced Calcite Precipitation (MICP) method is relatively green, sustainable technique and environmental friendly tecnique in soil stabilization, which able to alter and improve the ground condition [1]. Soil stabilization using MICP technique can meet the green construction requirement as this treatment causes minimal disturbance to soil environment [2]. Bacteria and cementation reagents were percolated into the soil to form calcite precipitation which improve the engineering properties of soil [3]. In addition, many studies have been conducted to explore the potential and feasibility of the MICP technique in different kind of civil engineering applications such as: improved concrete strength and durability [4, 5] ; improve mortar and brick durability [6] ; increased soil strength [7,8] ; and sand impermeability [9].

In the past few years before introducing MICP technique, chemical grouting method has been applied for soil stabilization in civil engineering. The process of chemical grouting is achieved by adding different additives, some of the additives are very toxic and hazardous to the environment such as lime, asphalt, sodium silicate, acrylate, lignin, urethane, and resin [10]; these additives usually modify the soil pH, and contaminate the groundwater and soil [11-14].

In this case, MICP technique seems to be a more promising green and sustainable alternative as it utilizes biochemical process to improve the engineering properties of soil [15]. However, most of the MICP studies were conducted on the sandy soil rather than fine grained soils; the MICP studies on fine grain soils are limited [15, 16]. Due to the complexity in soil microbiology and chemical components, the MICP application on fine grain soils remains uncertain and challenging; more attempt and studies are need to provide more understanding and information on the MICP in fine grain soils. Hence, the objective of paper is to provide better understanding of this technique by investigating the effect of various concentrations of cementation reagents toward MICP treatment; as well as to determine the effectiveness and feasibility of *Bacillus subtilis* in this MICP treatment for tropical residual soil; Moreover, this research was designed to address the question arose in previous study [16].

The application of Microbial induced calcite precipitation (MICP) in this study focused on the bio-cementation. Bio-cementation is a process that produces binding materials in form of calcite within the soil particles through biochemical reaction between the introduced bacteria and cementation reagents in to the soil [15]. Bio-cementation results in the improvement of shear strength and stiffness of the soil. At the initial stages of the process, bacteria perform urea hydrolysis in which the supplied urea is decomposed into ammonium ion by its urease enzyme (Eq. 1); and provide an alkaline environment for the precipitation of calcium carbonate [17]. The formation of calcite precipitation causes the pH of the medium back to neutral and the final pH of the medium depending on the rates of reaction and substrate concentration [18]. The negative charged bacteria cells attract the carbonate ion (the by product from urea hydrolysis) and combine with the supplied calcium ions to form calcite as shown in Eq. 2 and Eq. 3.

2. Material and Methodology

2.1 Type of Bacteria

The urease producing bacteria used in current study was *Bacillus subtilis* (ATCC 55422) which is a type of facultative anaerobe bacteria. *Bacillus subtilis* is a gram positive and rod shaped bacterium with diameter about 2μm.

2.2 Cementation reagent

Cementation reagents are the essential ingredients to generate calcite precipitation in microbial processes. The cementation reagents comprise of urea, calcium chloride and 3g/L of nutrient broth as tabulated in Table 1. The nutrient broth was added provided additional nutrients for bacterial growth.

Table 1. Chemical composition of cementation reagents.

Chemical	Concentration of Cementation Reagent (M)		
	0.15	0.25	0.35
Urea, CH_4N_2O (g)	9.0	15.0	21.0
Nutrient broth (g)	3.0	3.0	3.0
Calcium chloride, $CaCl_2.2H_2O$ (g)	22.0	36.8	51.5

2.3 Laboratory Setup/Placement of microorganism

Fig. 1 illustrates the laboratory setup of the experiment. The setup comprises of the soil moulds, pressure tank, pneumatic compressor, effluent pipes, and effluent collector. The cementation reagents supplied continuously throughout the treatment. Corrosion resistance material, stainless steel was used to fabricate the sample mould of 50 mm in diameter and 200 mm in height (Fig. 2a). Before the soil sample was loaded into the stainless steel mould, a thin layer of lubricant applied at the inner surface of the mould to ease the sample extrusion process. All connections were checked to ensure no leakage of air in pressure tank and cementation reagents in the sample mould.

IOP Conf. Series: Materials Science and Engineering **136** (2016) 012030 doi:10.1088/1757-899X/136/1/012030

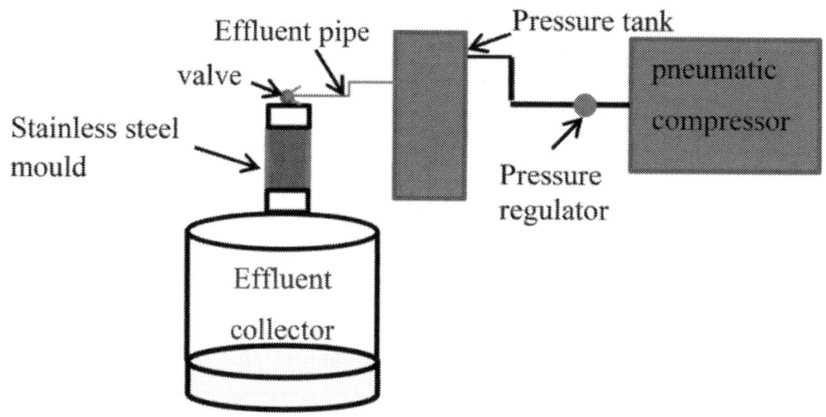

Figure 1. Schematic diagram of the laboratory set up

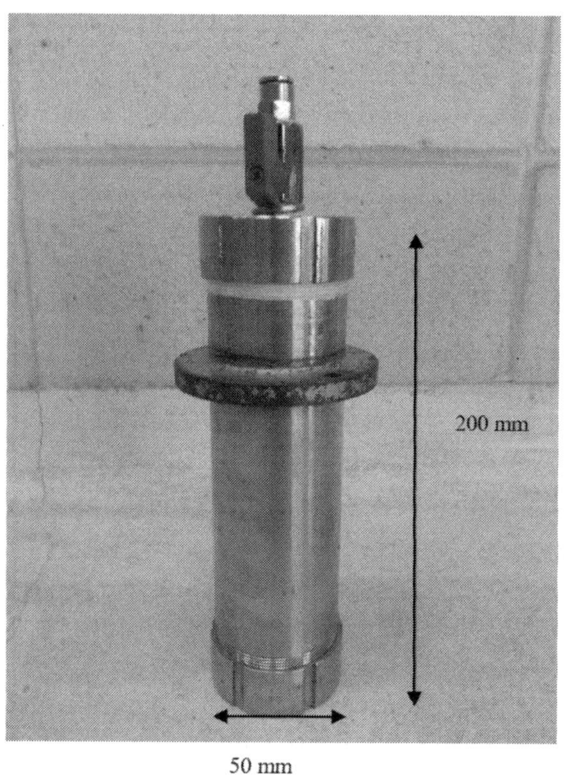

Figure 2a. Stainless steel mould

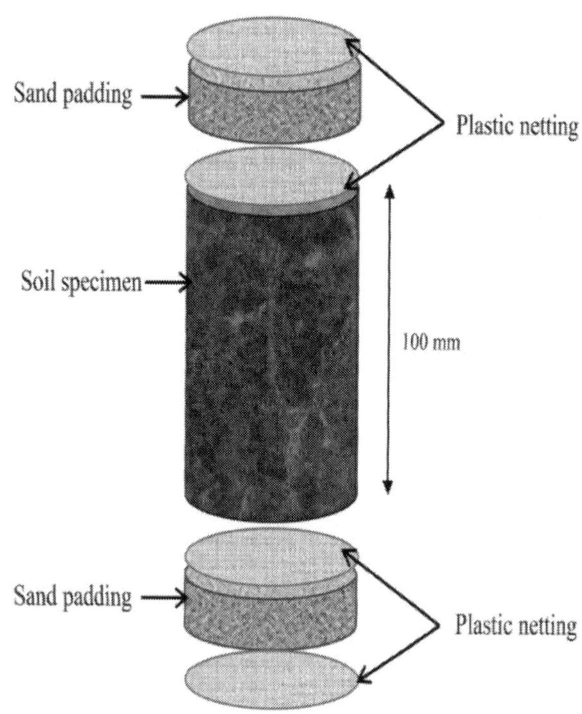

Figure 2b. Breakdown diagram of MICP sample and parts inside the mould

2.4 Soil Specimen

The soil material used in current study is tropical residual soil obtained from a site near Faculty of Electrical Engineering, Universiti Teknologi Malaysia. Table 1 tabulates the physical and index properties of the residual soil, which were determined in accordance to British standard soil properties test.

Table 2. Soil basic physical properties

Soil Index Properties	Description
Particle Size	15.3 % sand, 50.7 % silt, 33.6 % clay
Plastic Limit	48.95 %
Liquid Limit	80.80 %
Plasticity Index	31.85 %
Maximum Dry Density	1.322 /m^3

2.5 Soil Treatment Procedures

The bacteria were incubated at 30°C with 200 rpm for 24 hours. The bacteria were cultivated to a concentration of 1 × 10^8 cfu/ml. In prior MICP treatment, the desired density was achieved by mixing the soil specimen with the cultivated bacteria solution (about 87 ml) which is equavalent to the optimum moisture content of the soil. The soil sample was sprayed and mixed evenly with distilled water (control) or cultivated bacteria (MICP sample). After that, the mixed soil was poured and hand tamped into the sample mould. The soil specimen was compressed by compressor until it achieved the desired length of 100 mm. As shown in Fig 2b, the soil specimen was sandwiched between two layers of sand material to avoid turbulent flow and clogging at inlet. The plastic nettings were placed to separate each layer and as protection sheet for the top and bottom surface of the soil specimen. Then, the soil specimen was treated for 2 days with supply continuous flow of cementation reagents at a flow pressure of 0.2 bars. After the completion of MICP treatmet, the treated soil samples were extruded from the moulds for shear strength testing by using unconfined compression strength test (UCS).

2.6 Experimental Variables

There different reagents concentration were used in this study (0.15 M, 0.25 M and 0.35 M). A control soil specimen without bacteria was prepared at the same time and supply with continues flow of cementation reagent; tthe control sample was used to make comparison with result of treated soil specimen. All soil specimens treated for 2 days with cementation reagents flow pressure of 0.2 bars.

3.0 Result and Discussions

3.1 Shear Strength and Calcite Content Improvement

Fig. 3 shows the result of the effect of cementation reagents concentration on UCS and calcite content of the MICP treated sample. As seen in Figure 3, the optimum concentration was found to be 0.25M with the highest improvement of about 38% in UCS. While, the sample treated at 0.15M recorded improvements of about 23.7% and 19.4 % in UCS value and calcite content, respectively. The results shows that the UCS increased with an increased in the reagents concentration for both the samples treated at 0.15M and 0.25M; however, reduction in UCS value was observed for the sample treated at 0.35M. The UCS value for the sample treated at 0.35M was 94 kPa which is slightly lower than the control counterpart; a possible explanation for this observation is that, the improvement of UCS and calcite content could be affected as a result of the decrease in bacteria microbial activity due to high salinity of reagents; as high salinity impairs the nutrient and water availability for microorganism [19, 20]. As a result of that, the change of nutrient and water availability with the increase of the salinity might consequently affected on the growth and microbial activity of the bacteria [21]. Therefore, this could explain that the use of 0.35M reagents concentration in this study may have inhabited the growth of the bacteria. Since, bacteria act as the nucleon site for calcite precipitation [22]; the reduction in bacteria number might subsequently reduce the calcite precipitation and resulted in lower UCS value. excessive calcium chloride or further increase in the salt/ reagents concentration limits the enzyme activity, and results in slow production of urease enzyme from ureolytic bacteria [15].

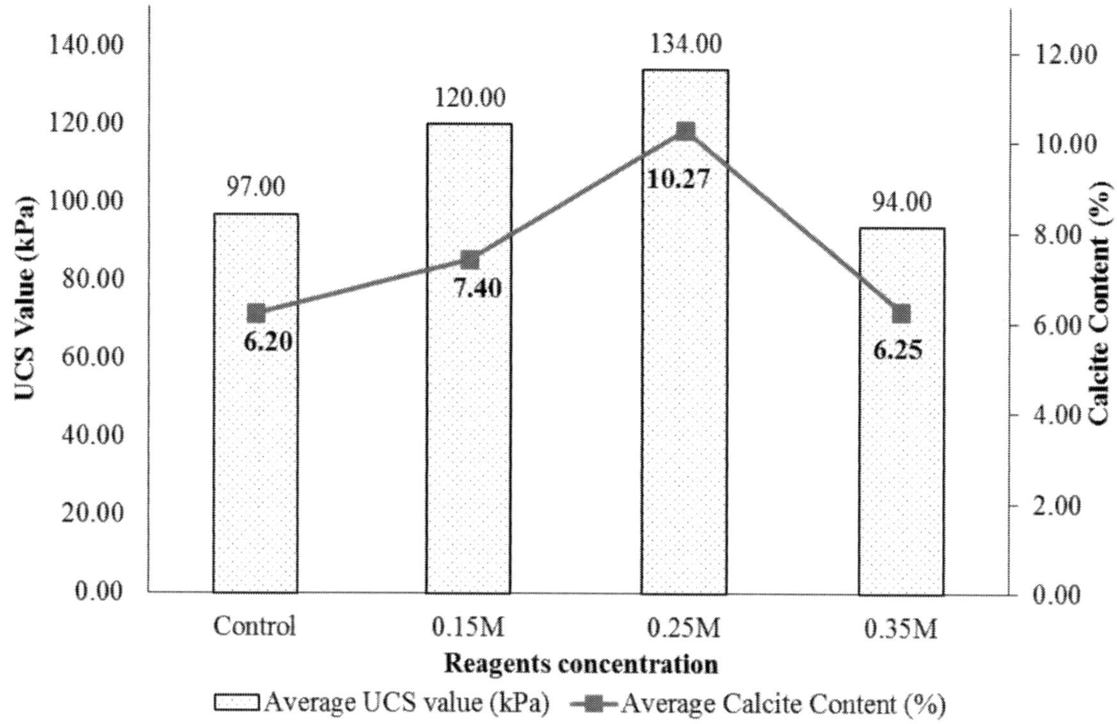

Figure 3. UCS and calcite content for various concentration of cementation reagent.

Furthermore, the use of calcium chloride in this MICP treatment might not be suitable, which impaired the enzyme produced by the bacteria and affected on the MICP treament. Esawy et al. [23] reported that the performance of enzyme produced by *Bacillus subtilis* in calcium chloride was very low. Other alternative to replace the calcium chloride might produce better outcome and performance. It is worth noting that calcite content of the soil sample treated at 0.35M (6.25%) was almost identical as compared with the control specimen (6.2%). This observation suggests no apparent calcite precipitation was formed as a result of the impaired microbial activity as mentioned earlier. The The result in this study also supports and in agreement with Qabany et al. 2013 [24] that the lower concentration (equal and lower than 0.25M) resulted in higher UCS value; this may associated with uniformity of the calcite precipitation in the treated sample. More investigation has to conduct to verify the uniformity of the calcite precipitation.

3.2 Relationship between Shear Strength and Calcite Content Improvement

As seen in Fig. 3, the shear strength of treated soil increased with the increase in calcite content. Similar observation was reported by Hammes et al. [25] in which the shear strength and calcite content tend to increase with the increase of reagents concentration during MICP treatment. This improvement in UCS is most probably due to the presence of calcite generated from MICP process, the formation of calcite tend to bind the soil particles together and subsequently improved the shear strength of soil. Calcite precipitation act as a bridges to bind the soil particles, which increase the shear strength of the treated soil. [26].

However, it is found The calcite precipitated become less effective in strength improvement when the calcite content was less than 8%; greater UCS improvement was manifested at those treated samples with a calcite content higher than 7.4%. This observation implies that more effective calcite bonding between the soils particles could be achieved when the total calcite content reached 7.4% in this study. Nonetheless, this may not be applicable to other MICP treatment which involve other type of soil and bacteria; as different bacteria and soil may require different quantity of calcite to achieve the same level of improvement. In addition, bacteria are living organism that subjected to uncertainties; other environmental factors such as temperature, and soil pH value might also affected on the MICP treatment. More investigations are needed to verify this point.

No correlation and relationship were found between the calcite precipitation and the reagents concentration; the calcite precipitation did not increase with increase of the reagents concentration. This observation implies that the calcite precipitation is governed and it is limited by the microbial activity and the urease enzyme production as discuss earlier.

4. Conclusion

From this study, the utilization of Bacillus subtilis in MICP treatment has shown improvement in UCS of tropical residual soil. This positive improvement has also proven the feasibility of Bacillus subtilis in MICP treatment for tropical residual soil. The results show that the UCS of the treated soil improved with increased concentration of cementation reagent and reached optimum at 0.25M of cementation reagent. However, no improvement was observable for those sample treated at 0.35M; This might associated with high salinity of the reagents concentration which might cause inhibitory effect to bacteria microbial activity, and subsequently, limit the enzyme activity and the production of urease enzyme. Furthermore, it is essential to determine the optimum reagents concentration as it may be vary with soil type and bacteria used. Nonetheless, more studies and experiment have to be conducted to confirm the points that arose in this paper.

Acknowledgement

This project is supported by Ministry of Higher Education (MOHE), Ministry of Science, Technology and Innovation (MOSTI), Malaysia and Universiti Teknologi Malaysia under E science Fund (R.J130000.7913.4S053). The authors would like to acknowledge Mr. Bong Mai Weng, and Miss Goh Yong Yuan for their participation and contribution in the experimental works of this study.

References

[1] Lee L M, Wei S N, Chew K T, and Siew L H. 2012 Bio- Mediated Soil Improvement under Various Concentrations of Cementation Reagent. *Applied Mechanics and Materials,* **204** 326-329.

[2] Wei S N, Lee L M, Chew K T, and Siew L H. 2014. Factors Affecting Improvement in Engineering Properties of Residual Soil through Microbial-Induced Calcite Precipitation. *Journal of Geotechnical and Geoenvironmental Engineering,* **140(5)**.

[3] DeJong J T, Michael B F, and Klaus N, 2006. Microbially induced cementation to control sand response to undrained shear. *Journal of Geotechnical and Geoenvironmental Engineering,* **132(11)** 1381-1392.

[4] Jonkers H M, Arjan T, Gerard M, Oguzhan C and Erik S, 2010 Application of bacteria as self-healing agent for the development of sustainable concrete. Ecological engineering **36(2)** 230-235.

[5] Reddy S, Rao M, Aparna P, and Sasikala C, 2010 Performance of standard grade bacterial

(Bacillus subtilis) concrete. *Asian J Civ Eng (Build Housing)* **11** 43-55.

[6] Vempada S R, Sunil S P R, Seshagiri R M V, and Sasikala Ch, 2011 Strength enhancement of cement mortar using microorganisms-an experimental study. *Int J Earth Sci Eng* **4** 933-936.

[7] Montoya, B M and DeJong J T, Healing of biologically induced cemented sands. *Geotechnique Letters* **3** 147-151.

[8] Chu J, Viktor S and Volodymyr I, 2012 Microbially induced calcium carbonate precipitation on surface or in the bulk of soil. *Geomicrobiology Journal* **29(6)** 544-549.

[9] Chu J V. Ivanov, Stabnikov V and Li B, 2013 Microbial method for construction of an aquaculture pond in sand." *Geotechnique* **63(10)** 871-875.

[10] Wei S N, Lee M L, Tan C K and Hii S L, 2013 Improvements in engineering properties of soils through microbial-induced calcite precipitation. *KSCE Journal of Civil Engineering,* **17(4)** 718-728.

[11] Anagnostopoulos C, and Hadjispyrou S, 2004 Laboratory study of an epoxy resin grouted sand. *Proceedings of the ICE-Ground Improvement,* **8(1)** 39-45.

[12] Basha E A, Hashim R, Mahmud H B and Muntohar A S, 2005 Stabilization of residual soil with rice husk ash and cement. *Construction and Building Materials,* **19(6)** 448-453.

[13] Karol R H, 2003 *Chemical grouting and soil stabilization, revised and expanded.* **12** CRC Press.

[14] DeJong J T, Brina M M, Brian C M and Douglas C N, 2010 Bio-mediated soil improvement. *Ecological Engineering,* **36(2)** 197-210.

[15] Ng W S, Min L L and Siew L H, 2012 An overview of the factors affecting microbial-induced calcite precipitation and its potential application in soil improvement. *World Academy of Science, Engineering and Technology,* **62** 723-729.

[16] Chiet P K T, Khairul A K and Siaw Y C, 2015 Biocementation Potential Of Tropical Residue Soil Infused With Facultative Anaerobe Bacteria. In *Applied Mechanics and Materials*, **773** 1412-1416.

[17] Ahmed A Q, Kenichi S and Carlos S, 2011 Factors affecting efficiency of microbially induced calcite precipitation. *Journal of Geotechnical and Geoenvironmental Engineering*, **138(8)** 992-1001.

[18] Leon A van P, Claudia M D, Marc S, Dimitri Y S, Willem V D Z and Mark C M V L, 2010 Potential soil reinforcement by biological denitrification. *Ecological Engineering,* **36(2)** 168-175.

[19] Canfora L, Giovanni B, Flavia P, Giuseppe L P, Carmelo D and Anna B, 2014 Salinity and bacterial diversity: to what extent does the concentration of salt affect the bacterial community in a saline soil? e106662.

[20] Tripathi, Sudipta, Sabitri Kumari, Ashis Chakraborty, Arindam Gupta, Kalyan Chakrabarti, and Bimal Kumar Bandyapadhyay, 2006 Microbial biomass and its activities in salt-affected coastal soils. *Biology and fertility of soils* **42 (3)** 273-277.

[21] Rietz D N and Haynes R J, 2003 Effects of irrigation-induced salinity and sodicity on soil microbial activity. *Soil Biology and Biochemistry* **35(6)** 845-854.

[22] Fujita Y, Grant F F, Daniel L R, Frederick S C and Robert W S, 2000 Subscribed content calcium carbonate precipitation by ureolytic subsurface bacteria. *Geomicrobiology*

Journal **17(4)** 305-318.

[23] Esawy M A, Sara H. M, Eman F A, Naziha M H and Hesham A E E, 2012 Characterization of extracellular dextranase from a novel halophilic Bacillus subtilis NRC-B233b a mutagenic honey isolate under solid state fermentation. *Journal of Chemistry* **9(3)** 1494-1510.

[24] Al Qabany A and Soga K 2013 Effect of chemical treatment used in MICP on engineering properties of cemented soils. *Géotechnique* **63(4)** 331-339.

[25] Hammes F, Nico B, Johan D V, Willy V and Steven D S, 2003 Strain-specific ureolytic microbial calcium carbonate precipitation. *Applied and environmental microbiology*, **69(8)** 4901-4909.

[26] Whiffin V S, Leon A V P and Marien P H, 2007 Microbial carbonate precipitation as a soil improvement technique. *Geomicrobiology Journal,* 24(5) 417-423.

Hydraulic Conductivity of Residual Soil-Cement Mix

P Govindasamy[1,] and M R Taha[1]
[1]Faculty Of Engineering and Built Environment, Universiti Kebangsaan Malaysia, 43600, Bangi, Selangor, Malaysia

E-mail: [a]gpam_21@yahoo.com

Abstract. In Malaysia, although there are several researches on engineering properties of residual soils, however study on the hydraulic conductivity properties of metasedimentary residual soils is still lacking. Construction of containment walls like slurry wall techniques can be achieved with hydraulic conductivity of approximately 5 x 10^{-7}cm/sec. The objectives of the study were to determine the physical properties of metasedimentary residual soils and to determine the influence of 1%, 3%, 5% and 10% of cement on hydraulic conductivity parameters. The coefficient of hydraulic conductivity of the soil naturally and soil-cement mixtures were determined by using the falling head test. According to the test, the hydraulic conductivity of the original soil was 4.16 x 10^{-8} m/s. The value decreases to 3.89 x 10^{-8} m/s, 2.78 x 10^{-8} m/s then 6.83 x 10^{-9} m/s with the addition of 1%, 3% and 5% of cement additives, respectively. During the hydration process, cement hydrates is formed followed by the increase in pH value and $Ca(OH)_2$ which will alter the modification of pores size and distribution. When the quantity of cement increases, the pores size decrease. But, the addition of 10% cement gives an increased hydraulic conductivity value to 2.78 x 10^{-8} m/s. With 10%, the pore size increase might due to flocculation and agglomeration reaction. The generated hydraulic conductivity values will indirectly become a guide in the preliminary soil cement stabilization to modify the properties of the soil to become more like the properties of a soft rock.1. Introduction

Keywords: residual soil, cement, hydraulic conductivity.

1.0 Introduction

Alternative areas for construction become more and more important since last decades, due to growing shortage of better quality soils for construction. These limitations could be overcome with the introduction of new mixing materials such as soil - cement mixture [1]. Soil mixing is also commonly used as a stabilization method for containing hazardous wastes and sludge. Containment walls can be constructed with hydraulic conductivity of approximately 5 x 10^{-7} cm/sec, similar to that achieved by most slurry wall techniques [2]. Thus, this study focused on cement, which was used as a modification material to be mixed with the disturbed sample of residual soil, which is classified as metasedimentary soil in various percentages. The modification was to decrease the hydraulic conductivity of the soil. For the aforementioned reason, laboratory testing was carried out in order to study the effect of inclusion of cement on the engineering behaviour of residual soil in reducing its hydraulic conductivity. Dense, compact, or cemented soil layers have very slow rates of hydraulic conductivity. Hydraulic conductivity of soil in its natural setting is highly variable and extremely difficult to measure. Soil hydraulic conductivity can be determined in a laboratory by measuring the rate of flow through a column of soil [3].

The aims of this study are to examine the original soil hydraulic conductivity actually encountered and comparing with those anticipated in the soil-cement modification, to carry out a laboratory testing in order to study the effect on hydraulic conductivity due to inclusion of cement in residual soil and to determine the influence of variability percentage of cement on hydraulic conductivity parameters of the residual soil.

In this study, disturbed sample were taken from hilly area at Bangi, Selangor as the sample is originally unpolluted. A study on the effect of hydraulic conductivity by mixing cement to residual soil was conducted. The hydraulic conductivity data was compared to the residual soil, which has no cement. A range of 1%, 3%, 5% and 10% of cement were used in this study. The laboratory tests that were done to determine physical analysis are particle size analysis, Atterberg limit, specific gravity, compaction and hydraulic conductivity parameter testing which was the falling - head permeameter. The entire laboratory tests were performed in Universiti Tun Hussein Onn (UTHM)'s geotechnical laboratory.

These methods of testing were to ensure that adequate mixing and treatment has been achieved. The study focused on finding the most efficient and the best percentage of cement mixture with soil, which can cause the properties of the soil to become more like the properties of a dense soil or soft rock [3]. This production is also for reduction of settlements by decreasing the hydraulic conductivity of residual soil.

Table 1. Typical strength and permeability characteristics of treated soils [3].

Soil Type	Cement Usage	UCS	Permeability
Sludge	240 to 400 kg/m3 (400 to 700 lbs/cy)	70-350 kPa (10-50 psi)	1x10-6 cm/sec
Organic silts and clays	150 to 260 kg/m3 (260 to 450 lbs/cy)	350-1400 kPa (50-200 psi)	5x10-7 cm/sec
Cohesive silts	120 to 240 kg/m3 (200 to 400 lbs/cy)	700-2100 kPa (100-300 psi)	5x10-7 cm/sec
Silty sands and sands	120 to 240 kg/m3 (200 to 400 lbs/cy)	1400-3500 kPa (200-500 psi)	5x10-6 cm/sec
Sands and gravels	120 to 240 kg/m3 (200 to 400 lbs/cy)	3000-7000 kPa (400-1000 psi)	1x10-5 cm/sec

Soil cement is a near impermeable material from the results of high-pressure hydraulic conductivity tests, and the hydraulic conductivity coefficient decreased as the mixed cement content of soil cement increased. Thus, through this study, it was concluded that the hydraulic conductivity of soil-cement decreases with the increasing cement content. Three types of soil cement have been categorized, which are, [4];

1. Normal soil-cement usually contains 5 to 14% cement by volume and is used generally for stabilizing low-plasticity soils and sandy soils.
2. Plastic soil-cement has enough water to produce a wet consistency similar to mortar. This material is suitable for use as waterproof canal linings and for erosion protection on steep slopes where road-building equipment may not be used.
3. Cement-modified soil is a mix that generally contains less than 5% cement by volume. This forms a less rigid system than either of the other types, but improves the engineering properties of the soil and reduces the ability of the soil to expand by drawing in water.

2. Methodology

2.1 Sample Preparation

The sampling of residual soil is essential in order to undertake laboratory testing. For this study, samples were excavated or dig using a hoe. These disturbed soil samples or remolded samples were obtained from a depth of 0.2m below ground surface. The disturbed natural soil were oven – dried in 105 °C before tested. Distilled water was used for all the laboratory tests. Distilled water should be relatively clean and free from harmful amounts of alkalies, acid or organic matter. With these preparations, the physical tests such as particle size analysis, Atterberg limit, specific gravity and compaction were done to classify the engineering properties of the natural residual soil. Then, for the hydraulic conductivity tests, a range of 1%, 3%, 5% and 10% of cement from the mass of residual soil taken were used in this study. The soil was mixed with these various cement percentages to make soil cement specimens. Table 1 shows the soil - cement sampling for the laboratory tests. The cement that was used is Ordinary Portland cement. Soil samples were tested for 0 % cement as a control sample and after mixing with the cement powder.

Before performing hydraulic conductivity test, soil – cement samples, which have been compacted using standard Proctor compaction method at optimum moisture content were immersed in the water for curing purposes. This process was carried on for 30 days for all the samples as stated in Table 1 except for the soil samples with 0% cement (control sample) that were submerged in the water only for 24 hours. Curing was carried out for saturation purpose as well as given enough time for cement to be fully hydrated. Fully hydrated cement takes up about 20 percent of its own weight of water. At high moisture contents in a cement-stabilized material the cement would have no difficulty in obtaining this water but as the moisture content is decreased the cement has to compete with the soil for moisture; if the soil has a high suction it may have a greater affinity for water than the cement and, the cement cannot fully hydrate. As the moisture content of a stabilized soil is decreased the degree of hydration of the cement is reduced [5]. Thus, proper curing process is a necessity before carrying out further testings.

Table 2. Samples of testing.

Soil + Cement Percentage	Total Mass (g)	Mass of soil (g)	Mass of cement (g)	Number of samples
Soil + 0% Cement	2000	2000	0	2
(control sample)	2000	1980	20	2
Soil + 1% Cement	2000	1940	60	2
Soil + 3% Cement	2000	1900	100	2
Soil + 5% Cement	2000	1800	200	2
Soil + 10% Cement				
	TOTAL			10

2.2 Basic Geotechnical Test

After preparing the soil samples, a series of physical test has been carried out as follows;

a. Moisture content -the determination of the moisture content of a specimen of soil as a percentage of its dry mass using British Standard (BS) 1377: Part 2:1990 Clause 3.

b. Particle size analysis (dry sieving and CILAS machine) -generally applied to the soil fraction larger than 75mm (dry sieving) and for soil fraction which passes the 2mm sieve size using CILAS particle size analyzer.

c. Atterberg Limits -to determine plastic limit, liquid limit and plasticity index of natural residual soil using BS1377 : Part 2 : 1990 : 4.3.

d. Specific gravity -for accurate measurement of the density of particles heavier than water using BS 1377:1990.

e. Proctor compaction test -to identify the maximum dry density and optimum moisture content for natural residual soil using BS 1377 : Part 4 : 1990 : 3.3.

2.3 Hydraulic conductivity Test

Hydraulic conductivity tests were done for natural residual soil and soil-cement samples from all the ranges of cement percentages as studied. After second stage of compaction involving soil – cement specimens were done; samples were cured for 30 days before hydraulic conductivity tests were conducted on them. For measuring the hydraulic conductivity of soil of intermediate and low hydraulic conductivity (less than 10^{-4} m/s), i.e. silts and clays, the falling head procedure was used. The procedure is generally an accepted practice as recommended [6].

3. Results

3.1 Physical Properties the Residual Soil

The distribution of soil is silt 49%, sand 39%, and clay 12%. Silt and sand consume almost half of the soil portion, whereas clay exists only in small quantity. From the grain distribution graph, the effective size, D_{10} is 1.577, uniformity coefficient, C_u is 21.88 and the coefficient of gradation, C_k is 0.842. The specific gravity for the studied soil is 2.63. The plastic limit was 24.75 ad liquid limit was 25.3. The soil can be classified as yellowish brown sandy SILT with slight clay and low-plasticity soil, where plasticity index is 0.55. In this study, the soil firstly categorized in a major group of fine-grained soils which is the L groups, which have LLs < 50. The symbol L has low and high compressibility. Fine-grained soils are further divided based on their position above or below the A-line of the plasticity chart. Typical soils of the (ML) and (MH) groups are inorganic silts. Those of low plasticity are in the (ML) group; others are in the (MH) group [7]. Since the studied soil is a silty soil with fine sands with slight plasticity, it is categorized in the **ML** group. Through the Proctor compaction test, maximum dry density obtained for the natural residual soil sample is 1760 kg/m^3 and the optimum moisture content for the sample is 14%.

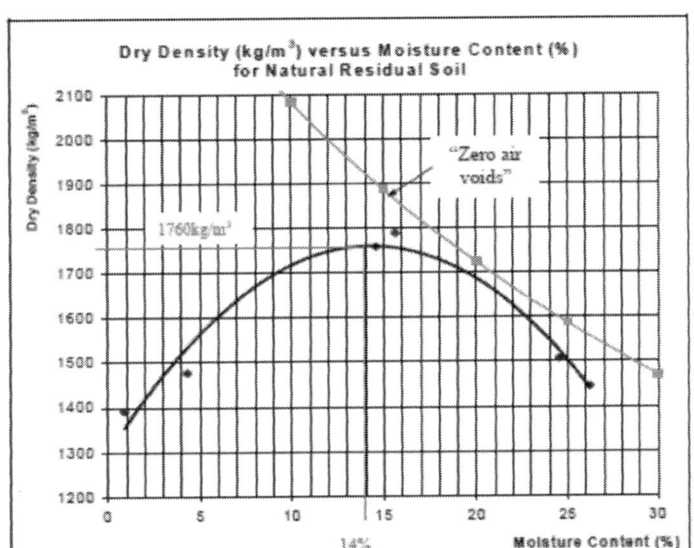

Figure 1. Determination of maximum dry density (kg/m^3) and optimum moisture content (%) for natural residual soil.

3.2 Hydraulic Conductivity Parameters

According to the falling head hydraulic conductivity test, the hydraulic conductivity of the original soil was 4.16 x 10^{-8} m/s. Figure 2 shows that the value of hydraulic conductivity decreases to 3.89 x 10^{-8} m/s, 2.78 x 10^{-8} m/s then 6.83 x 10^{-9} m/s with the addition of 1 %, 3 % and 5 % of cement additives, respectively. However, the addition of 10 % cement gives an increased value to 2.78 x 10^{-8} m/s.

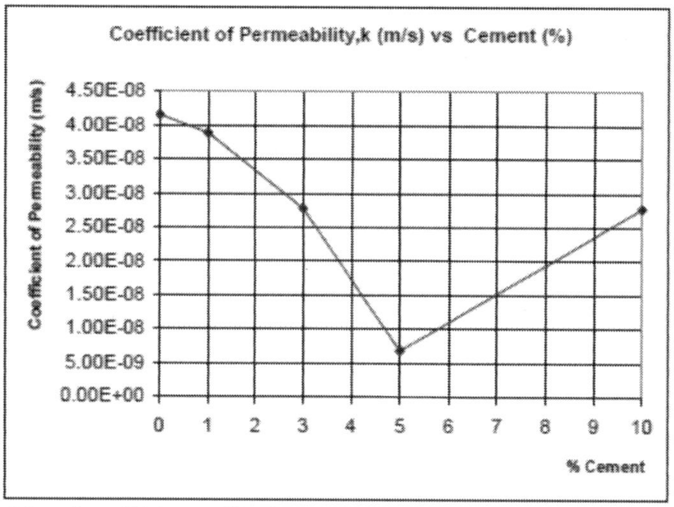

Figure 2. Relationship of coefficient of hydraulic conductivity, k (m/s) with cement additives (%).

4. Discussion

The distinctive feature of the residual soil is that it is a mixture of silt, sand and clay in varying proportions with silt 49%, sand 39%, and clay 12%. The particle size analysis indicated that residual soils have variable texture contain all fractions of particles. In residual soils consistent results are difficult to achieve. This was probably due to weathering process by which the structure of residual soil easily breaking down. The results were also influenced by time and condition of storage. The fresh samples were generally observed to be initially soft which further hardened with time. Hence particle size distribution for same sample in fresh state might have differed with that of in weathered state [8]. With Cu > 5, it indicates a well-graded soil and Ck between 0.5 and 2.0 indicates a well-graded soil too. The studied residual soil is a low – plasticity soil based on the Plasticity Index. It is categorized in the ML group as it is a silty soil with fine sands with slight plasticity. The specific gravity for the studied soil is 2.63. The specific gravity of the solid substance of most inorganic soils varies between 2.60 and 2.80 [9].

The dry density of soil produced by compaction depends mainly on the moisture content of the soil and the amount of compaction applied. Each soil has its own unique optimum moisture content, the value depending mainly on the amount and type of plastic fines that it contains. However, as the optimum moisture content is critically dependent on the compaction effort, the term always needs to be considered in relation to the type of compaction test used to define the property.

The hydraulic conductivity of the original soil was 4.16 x 10^{-8} m/s. Fine-textured soils have very tiny pores and very slow hydraulic conductivity rates (Figure 2). The hydraulic conductivity of the soil-cement is influenced by the increment of cement, where it decreases with the added content [10]. The value of hydraulic conductivity decreases to 3.89 x 10^{-8} m/s, 2.78 x 10^{-8} m/s then 6.83 x 10^{-9} m/s with the addition of 1%, 3% and 5% of cement additives, respectively. During the hydration process, cement hydrates is formed followed by the increase in pH value and $Ca(OH)_2$ concentration can break the clay particles and free the silica and aluminium to joint the calcium forming the second hydrate material [10]. This will alter the modification of pores size and distribution. When the quantity of cement increases, the pores size decrease. In this study, with 1%, 3% and 5% of cement additives, the pores size decreased. Moreover, in the curing process, pores size can be minimized. hydraulic conductivity will then decreases and soil – cement gains higher strength and stiffness.

But, Figure 2 also shows that the addition of 10% cement gives an increased value to 2.78 x 10^{-8} m/s. With cement percentages more than 9%, the pore size of soil-cement increases due to flocculation and agglomeration reaction. Hence, the hydraulic conductivity of the soil increases. Thus, flocculation and agglomeration reaction could have also influenced the increased hydraulic conductivity value of 10% cement addition to the soil in this study [11].

5. Conclusion

Properties of soil such as hydraulic conductivity can be altered by the addition of stabilizing agents as the main interest is usually in finding a means of increasing soil strength and resistance to softening by water. In this case, cement has been used as a bonding agent. It stabilizes soil by cementing the particles together so that the effect of water on the structure is lessened. The effectiveness of this type of stabilizer depends on the strength of the stabilized matrix; on a bond, which is formed between the soil and the matrix, and on individual particles or agglomerations of particles that are bonded together.

Based on the study, the best amount of cement to improve and stabilize the studied residual soil is 5% by weight compared to 1%,3% and 10% cement additives.The increase of 1%, 3% and 5% cement content reduces the hydraulic conductivity value with water curing process. However, 10% cement content has increased the hydraulic conductivity value. It is believed the flocculation and agglomeration reaction aggravates the pore size of soil-cement and thus increase the hydraulic conductivity.

References

[1] Pinto 2003 A guide to the formation, classification and geotechnical properties of residual soils, with advice for geotechnical design. Mechanics of Residual Soils. Balkema A A, Rotterdam.
[2] Nicholson P J, 1998 Cement Soil Mixing in Soft Ground, University of Houston.
[3] Huddleston J H 1996 How Soil Properties Affect Groundwater Vulnerability to Pesticide Contamination, Oregon State Extension Service.
[4] Bell F G 1993 Engineering Treatment of Soils, E & FN Spon.
[5] Sherwood P T 1993 Soil Stabilization with Cement and Lime, Transport Research Laboratory State of the Art Review, HMSO.
[6] Head K H, 1994 Manual of soil laboratory testing- Volume 1: Soil Classification and Compaction tests, London: Pentech Press.
[7] Information on http://www.eng-ips.com/viewthread.cfm *pcSITES User Manual Version 1.2.*, ACLEP Technical Report No. 5 (1997)
[8] Aminaton M, Fauziah K, Mohd F Y & Khairul N, 2002 Physical and Chemical Compositions of Residual Soils Of Eastern Region of Peninsular Malaysia. 2nd World Engineering Congress Sarawak, Malaysia, 22-25.
[9] Information on http://www.ems-i.com/wmshelp/Grids/GSSHA/Sediments "Sediments" GSSHA (2001)
[10] Ingles O G and Metcalf J B 1972 Soil Stabilization Principle and Practice, Butterworths, Melbourne, Australia.
[11] Basha E A, Hashim R & Muntohar A S, 2003 Effect of cement-rice husk ash on the plasticity and compaction of soil. EJCE Paper.

Site Characterization of Marine Clay Deposits in South Seberang Prai, Penang using Combined Active and Passive Multichannel Analysis of Surface Wave (MASW)

J Ariffin[1], M A M Ismail[2], C G Tan[3] and N M Murtadza[4]
[1, 2]School of Civil Engineering, Universiti Sains Malaysia, Penang, Malaysia
[3]Department of Civil Engineering, Faculty of Engineering, University of Malaya, Kuala Lumpur, Malaysia
[4]School of Physics, Universiti Sains Malaysia, Penang, Malaysia

E-mail: ajuliaariffin@hotmail.com

Abstract. The multichannel analysis of surface wave (MASW) has drawn considerable interest in subsoil investigation and characterization. This method has been studied and improved in the last few decades, thereby leading to several combinations of existing active and passive methods. This study aims to identify the presence of marine clay layer using combined active and passive MASW methods. The shear wave velocity profile obtained interpreting clay layer within the range of 70–150 m/s and sand layer within the range of 100–300 m/s. Results suggest that the combined MASW method significantly improves shear wave velocity profiling in shallow and deep soil layers and the results are validated using the cone resistance and shear wave velocity estimated from cone penetration test.Introduction

Keywords: Surface wave, passive and active MASW, marine clay.

1. Introduction

Construction over soft clay areas has increased in recent years because of the growth progress occurring in Malaysia. Many potential residential areas, such as Nibong Tebal, aim to be developed. The main concern in designing structures in such areas is identifying the presence of the clay layer and the extent of its depth by conducting costly and time consuming site investigation such as multiple points of close boreholes. Throughout the decades, bore logging has been a common method used to investigate and characterize subsoil conditions. However, obtaining samples from a borehole is an invasive method that will disturb sample properties.

Various non-invasive geophysical methods have been developed to investigate and characterize subsurface condition. Previous researchers have found that shear wave velocity could become a fundamental property in evaluating soil perimeters empirically [1,2,3]. In minimal disturbance, in situ geophysical methods have considerable advantages and accuracies over determining shear wave velocity empirically. Previous techniques, such as seismic reflection analysis, continuous surface wave analysis, and spectral analysis of surface wave, were applied before the multichannel analysis of surface wave (MASW) was developed in the mid-1990s. The limitations of individual active and passive MASW methods lead to the development of the combined active and passive MASW method. Some studies reviewed that combine active and passive method may be beneficial to improve resolution of penetration depth [7,8,9].

Reliability of MASW method has been assess with seismic cone penetration test (SCPT) on southeast Missouri in 2006 and Norwegian sites in 2007 [10,11]. The result reasonably agreed well with those obtained from CPT correlation. As cone penetration test (CPT) became popular, the economic value, continuous records and direct numerical values from various sensors incorporated within it credited CPT as the first rational soil profiling method after boring method [12]. This study aims to identify marine clay deposit using combined active and passive MASW method and the results will be validated with shear wave velocity and cone resistance estimated by CPT. This method shall

IOP Conf. Series: Materials Science and Engineering **136** (2016) 012032 doi:10.1088/1757-899X/136/1/012032

highlight the capability of the combined active and passive MASW method to be used as rapid and cheaper tool for subsurface characterization in a vast area.

2. Sites description

The surveyed sites are shown in Fig. 1. All the sites are located within a 1.2 km2 radius of the Engineering Campus of Universiti Sains Malaysia, which is located along the west coast of Peninsular Malaysia in the Nibong Tebal, South Seberang Prai, Penang. Active and passive MASW measurements together with the CPT were conducted on eight sites as shown in Fig 1(b).

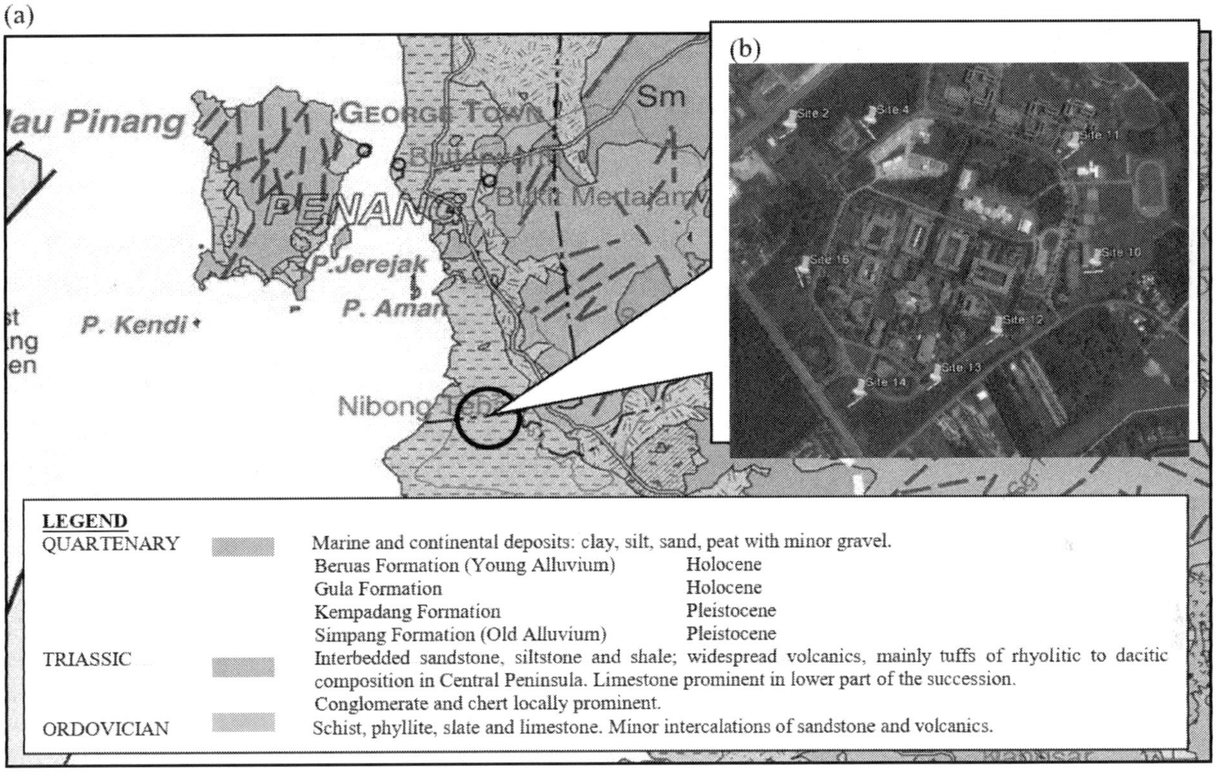

Figure 1. (a) Research location are shown in the Geological Map of Peninsular Malaysia (Geological Department of Malaysia, 1985) and (b) Location of the eight MASW sites.

According to the geological map of Peninsular Malaysia (Fig 1a), the soil along the west coast of Peninsular Malaysia consists of marine and continental deposits such as clay, silt, sand, and peat with minor gravel content. The research area which located within the boundary of Penang and Perak States was previously a mangrove area. Based on the quaternary geology report, the research area belongs to Gula Formation. The lithology of the Gula Formation is described as sand, silt, clay, gravel, and a small amount of peat. Sediments were fluvially deposited in estuaries and littoral zones, which shallower decades ago [13].

3. Methodology

MASW method were divided into passive and active method. As each method has limitation and advantage, combining both method will create an improved MASW version. In this study, MASW sites were measured and analyzed based on the steps in Fig. 2 by using SeisImager software. Surface Wave Analysis Wizard module of SeisImager automatically calls on functions such as Pickwin and WaveEq modules through the processing flows of individual active and passive method. Dispersion point are picked in the stable velocity range of phase velocity-frequency domain displayed in color contour. Dispersion curve will be displayed once WaveEq module lunches. This step applied for active and passive method only with exception on the survey geometry set up of each method. Dispersion curve of combined MASW method created by appending the dispersion curve of active and passive method. Next, the initial model of Vs with depth for one-third-wavelength approximation is set up. The software will iterate to converge on the best fit of the initial model with the observed data to create the final Vs curve.

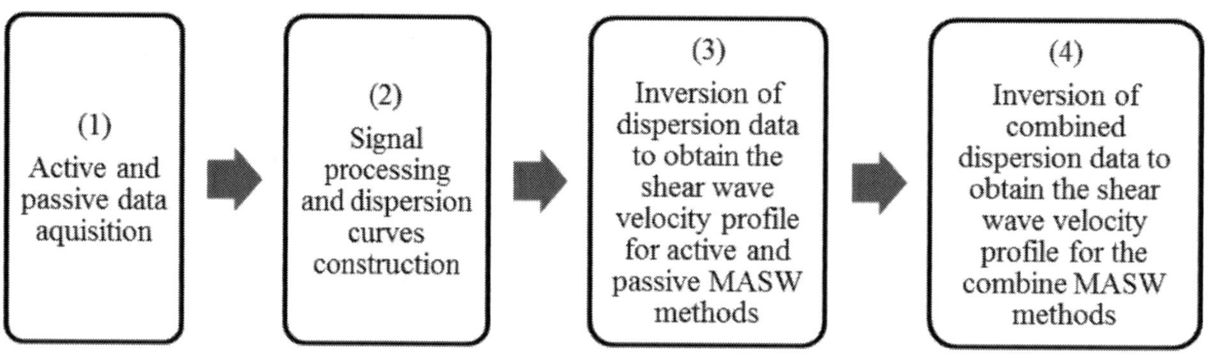

Figure 2. Steps in MASW measurement and analysis using SeisImager software.

Some practitioner are using shear wave velocities as parameter in evaluating soil strength and bearing capacity. This lead to seismic velocities estimation over soils and rock types as in Fig. 3. This study utilize the velocities estimation to estimate the types of soil and rocks from shear wave velocity (Vs) obtained.

Soil and Rock type	V_P m/s	V_S m/s
Gabro	6000	4000
Granite	5640	3760
Schist	5200	3500
Limestone	6190	3350
Mudstone	1900	700
Dilluvial gravel	2200	600
Gravel,dry sand	1000	300
Loose sand	1800	500
Alluvial gravel	1900	430
Diluvial clay	1800	350
Alluvial clay	600	150

Figure 3. The presumptive primary and shear wave velocity [14].

3.1 Active MASW

The type of seismic activity is categorized based on the type impulse source such as explosion, sledgehammer strike, and many others. In this study, sledgehammer was used. Offset was set at 5, 10, and 15 m from the first geophone with five time stacking to improve signal-to-noise ratio (SNR). This method was adopted from the conventional seismic refraction method. The produced impact would generate surface waves, which would be detected by the geophones. The wave forms shown in the seismogram and dispersion image in the Rayleigh fundamental mode. Shear waves velocity were derived from the inversion of the dispersion curve into a 1D shear wave profile.

3.2 Passive MASW

The passive method based on micro-tremor array measurement (MAM). Passive waves generated by industrial, vehicular, and wind noises were discovered to emit low-frequency wave with greater energy than those emitted by active sources. Researchers found that passive method may overcome the limitation of the active method via deeper wave intrusions. The use of passive waves was introduced when Louie [15] proposed the method in 2001. Common arrays used at present are L, triangles, circles, and semicircles. In this study, "L" arrays were used. A total of 20 records (32s record length each) were used in the study. The dispersion curves were calculated individually for all the 20 records and then stacked together.

3.3 Combined MASW method

Theoretically, high-frequency waves in the active method lack signal information when depth is deep, whereas low-frequency waves in the passive method lack information when depth is shallow. The investigation depth of the active method is less than 30 m, whereas that of the passive method can reach 100 m [4,7,16]. Combined MASW methods resolve the limitations in shallow depth of the active method and the insufficient near-surface signal of the passive method. Passive and active methods are used to measure data individually in sites, and then, the dispersion data from both methods are appended together. The deficiency of each method is improved by smoothening and averaging appended data. The simultaneous active–passive methods combine the dispersion curves generated before inversion to construct a combined MASW shear wave velocity profile. The combined profile covers a better range than both methods individually. Superimposing the dispersion curves obtained from both method create a better quality data.

The general schematic illustration of a multichannel configuration used in the study is shown in Fig. 4. The primary MASW equipment used in the study comprises of 24 geophones (4.5Hz) for active method and 9 geophones (4.5Hz) for passive method. The receiver spacing (2.5m for active and 5m for passive method), represent the shortest generated wavelength, which determines the minimum depth that will be used in the investigation. By contrast, the spread length of the receiver represents the longest wavelength, which determines the maximum depth that will be used in the investigation.

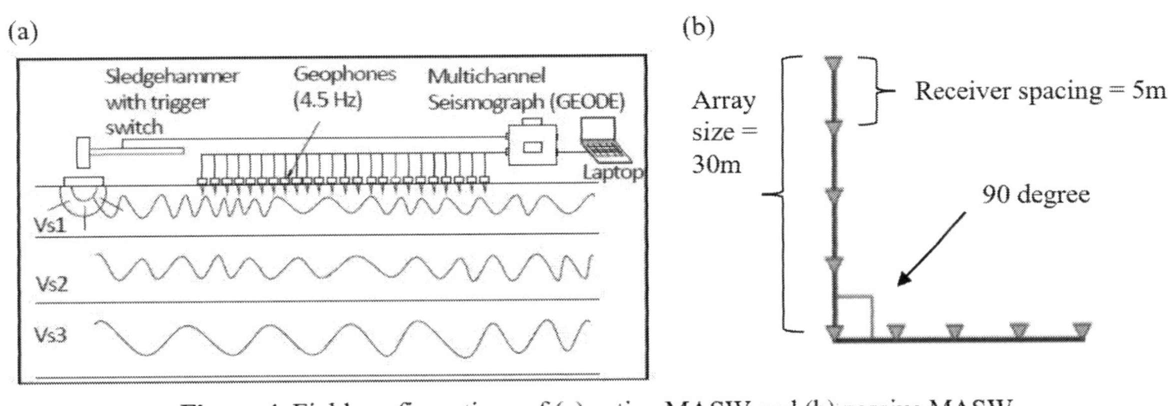

Figure 4. Field configurations of (a) active MASW and (b) passive MASW.

4. Result and discussion

Based on the data obtained from the phase velocity image in the frequency domain (Fig. 5), the frequency range for active method is within 5–25 Hz with phase velocity in 100–150 m/s range and the frequency range for passive MASW was observed to be at 3–14 Hz with phase velocity range in 50–300 m/s. The fundamental modes of the active and passive methods were clearly within the ranges of 3–25 Hz and 2–4 Hz, respectively. The dispersion curve was extracted by selecting minimum frequency of 4.5 Hz for the active method and 3 Hz for the passive method. The shapes of the both dispersion curves are similar, with an apparent bend around 4 Hz. However, the modal curve of passive MASW slightly shifts toward lower phase velocities and produce clear signals in low frequencies. This condition results in the deeper penetration of shear waves compared with the weak signals in the active method. The shear wave velocity profile for the passive method obtained using the inversion method matches well with that for the active method in comparison to its respective depth.

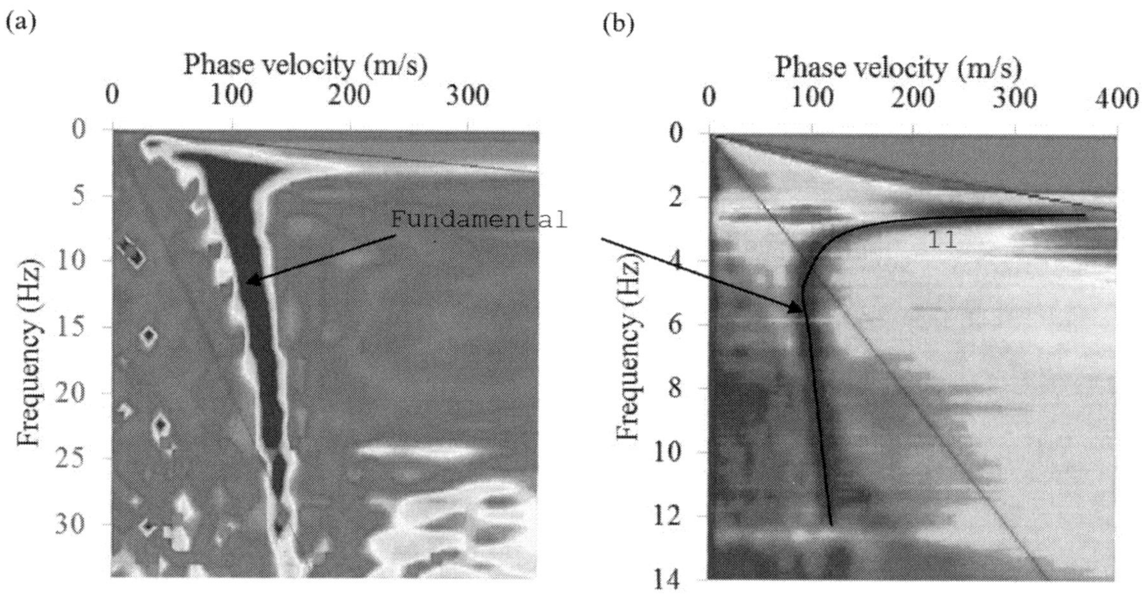

Figure 5. Phase velocity images in the frequency domain for (a) the active MASW method and (b) the passive MASW method.

Fig. 6 shows the data sets obtained using (a) active method, (b) passive method and (c) combination of active and passive methods. The figure shows the shear wave profile obtained after the inversion. The value of the shear wave velocity obtained using the passive method is slightly different from that obtained using the active method (Figs. 6a and 6b) in terms of penetration depth and velocity over depth pattern. Park et al. [7] suggesting combination of dispersion for a reliable modal identification. Vs profile of combine MASW method were improved where, the analyzable frequency range of dispersion broaden and better identification of dispersion trends modal identity. Fig. 6c supporting the facts that combine method is effective for determining shear wave velocity.

According to Tezcan et al. [14], Vs less than 750 m/s is the velocity range for soil. As observe in Fig. 6c, the shear wave velocity of combined active and passive MASW method was observed to be within the range of 50–380 m/s, which confirmed that the study area consist of soil. A clear increase in soil stiffness with increasing velocity is shown in Fig. 6c. Shear wave velocity was observed to decline

IOP Conf. Series: Materials Science and Engineering **136** (2016) 012032 doi:10.1088/1757-899X/136/1/012032

to 70–150 m/s at a depth of 13 m and increase to 350 m/s up to a depth of 30 m. In 2008, Watabe and Sassa [17] reported that the Vs of soft clay ranged from 80–175 m/s, and that of firm clay ranged from 100–300 m/s, while sand ranged from 130–300 m/s by Xia et al. [18] in 2002. From VS obtained in Fig. 6c, the soil type estimated in the first 13 m was the clay layer, whereas the sand layer located at a depth of 13 m to 40 m.

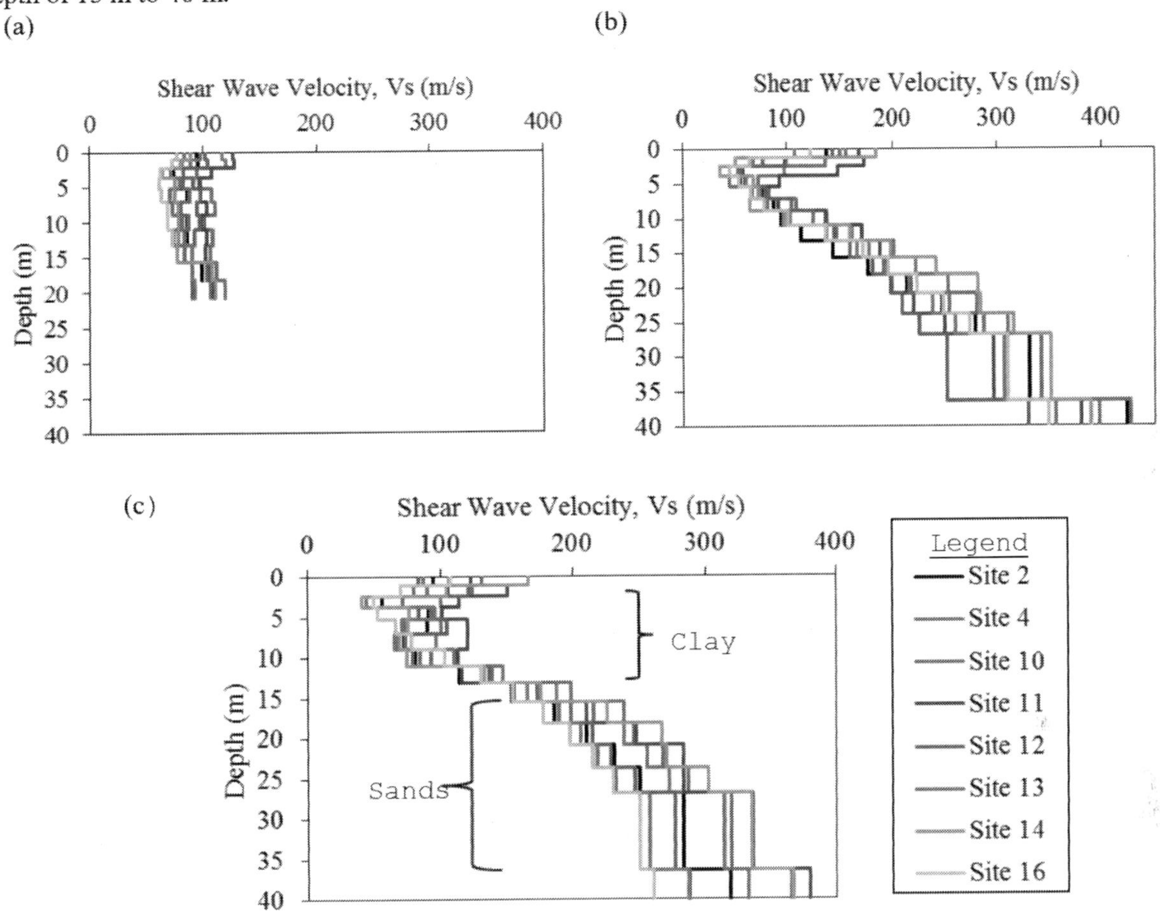

Figure 6. Shear wave velocity profiles of (a) active MASW, (b) passive MASW, and (c) combined MASW method.

A better explanation for combined method was illustrated in Fig. 7a. The three methods, active, passive and combined MASW was in good agreement in the first 13m, however start to diverge after 13 m depth. This may be because of the uncertainties with the low frequency dispersion curves [8]. Fig. 7b shows the shear wave profile overlay with one-third wavelength approximation obtained from the SeisImager software.

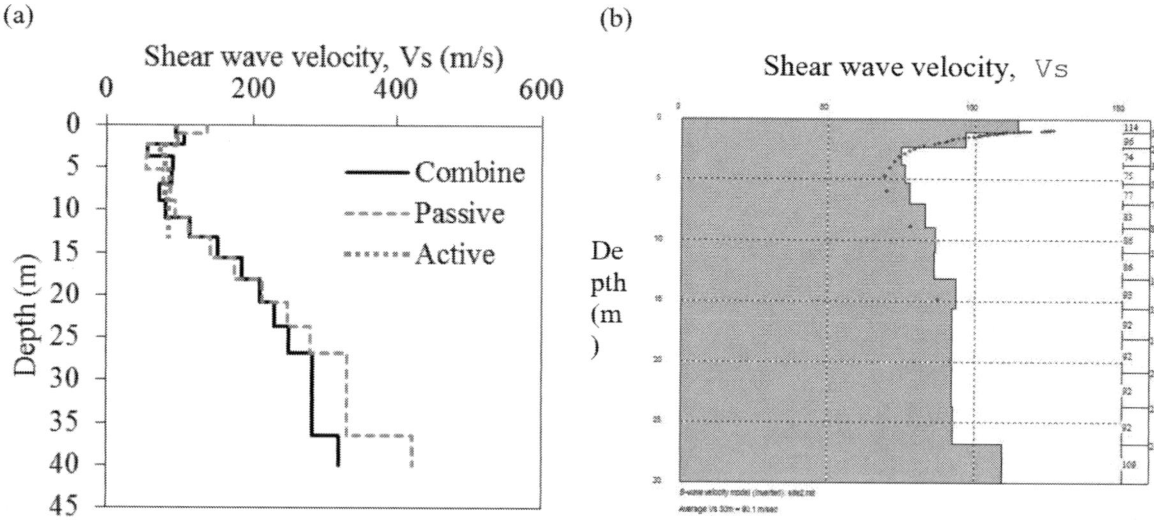

Figure 7. (a) Comparison of shear wave velocity profiles obtained using active, passive, and combined MASW and (b) 1D velocity model represented by actual depth range of penetration in Seismager.

Shear modulus reflect to material stiffness and is one of the most critical engineering parameters. Seismically, shear wave velocity is an important indicator to determine small strain shear modulus of soil as described in the equation below;

$$V_s = \left(\frac{G_o}{\rho} \right)^{0.50} \quad (1)$$

Where G_o is small strain shear modulus determined using Eq (2) and ρ is the density of the materials.

$$G_o = (q_t - \sigma_v)0.01888 \times 10^{0.55I_c + 1.68} \quad (2)$$

Where q_t is cone resistance; σ_v is vertical overburden stress and I_c is SBTn Index.

Shear wave velocity estimated from CPT using the Eq. (1) and cone resistance show a consistent trend with the shear wave velocity measured using combine MASW as shown in Fig. 8a. The soil type estimation as interpreted in Robertson's Soil Behavior Type (SBT), consist of clay layer in the first 15 meter followed by sand in the next layer (Fig. 8b). The CPT result validate the interpretation and result measured from combine MASW method as described in figure below.

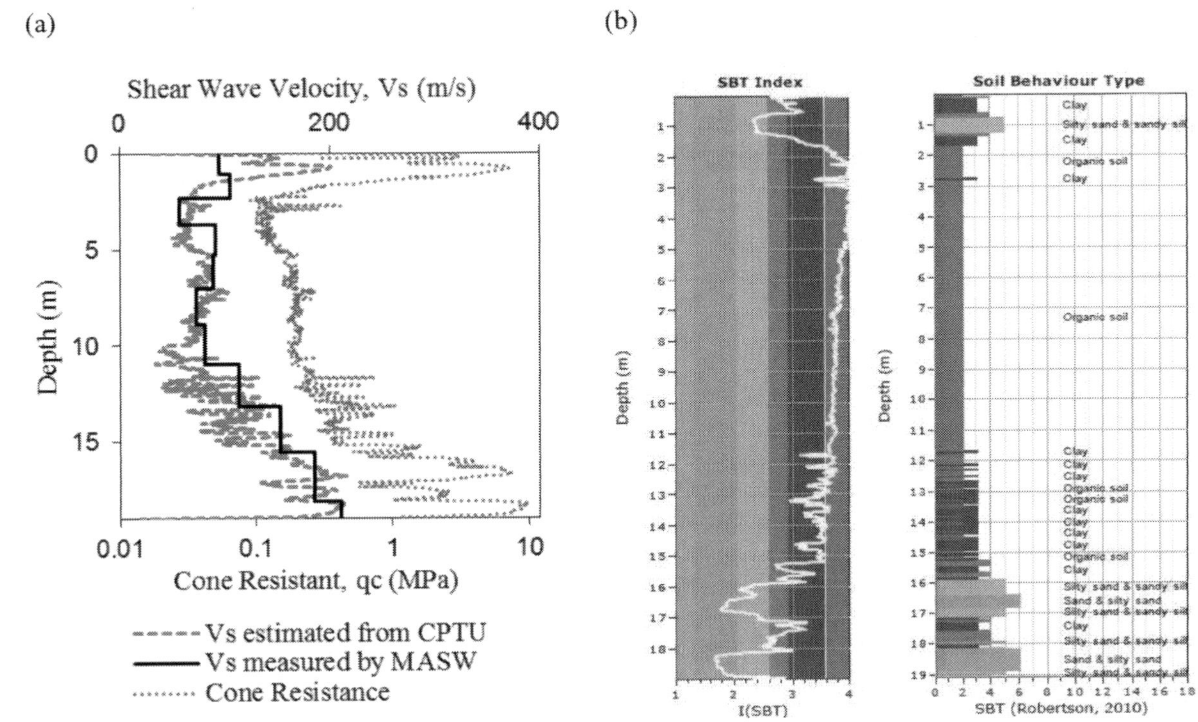

Figure 8. (a) Similar trend of shear wave velocity estimated from CPT and measured using the combined MASW method and (b) CPT estimation in the SBT index.

Conclusion

The measurements of the combined method exhibit a slightly different shear wave velocity from those obtained using the active and passive methods. The active method only covers the near–surface layer, the passive method covers the deep layer, and the combined method provides a better range than both the active and passive methods. The active, passive, and combined methods closely agree with one another in terms of the capability for soil investigation among surface-based methods. The results show that the combined method significantly enhances the capabilities of MASW for deep soil investigation. Moreover, the method is accurate for soil layer investigation, as validated via cone resistance and estimated shear wave velocity calculated empirically from cone penetration test in this study.

Acknowledgement

The authors would like to express their appreciation to Universiti Sains Malaysia Research University (RU) Grant (1001/PAWAM/ 814192) for the financial assistance to carry out this research.

References

[1] Donohue L, Shane M, 2010 Characterisation of Norwegian Marine Clays with Combined Shear Wave Velocity and CPTU Data, Canadian Geotechnical Journal 47(7) 709-718.
[2] Mayne P W and Rix G J, 1995 Correlation between Shear Wave Velocity and Cone Tip Resistant In Natural Clays, Soils and Foundations. 35(2) 107-110.
[3] Tezcan S, Ozdemir Z, Keceli A., 2009 Seismic Technique To Determine The Allowable Bearing Pressure For Shallow Foundation In Soils And Rocks, Acta Geophysica. 57(2).
[4] Park B, Miller R D, Ryden N, Xia J and Ivanov J, 2005 Combined Use of Active and Passive Surface Waves, Journal of Environmental and Engineering Geophysics. 10(3)

323-334.
[5] Bernard R, Wair J, Dejong T, Shantz D T, 2012 Guidelines for Estimation of Shear Wave Velocity Profiles Pacific Earthquake Engineering Research Center Peer Report.
[6] Taipodial J, Dey A, 2012 A Review of Active and Passive MASW Techniques, National Workshop Engineering Geophysics for Civil Engineering and Geo-Hazards.
[7] Park C B, Miller R D, Xia J, Ivanov J, 2007 Multichannel Analysis of surface wave (MASW)-Active and passive methods, in: The Leading Edge.
[8] Nazarian S, ASCE F, 2012 Shear Wave Velocity Profiling with Surface Wave Methods, Proceeding of GeoCongress.
[9] Luke B, Taylor W, Macias C, Jin X, Murvosh H, Wagoner J, 2008 Characterizing Anomalous Ground For Engineering Applications Using Surface-Based Seismic Methods, The Leading Edge.
[10] Long M, Donohue S, 2010 In Situ Shear Wave Velocity from Multichannel Analysis of Surface Waves (MASW) Test at Eight Norwegian Research Sites, Canadian Geotechnical Journal. 47 709-718.
[11] Thitimakorn T, Anderson N, Hoffman D, Ismail A, 2006 A Comparative Analysis of 2-D MASW Shear Wave Velocity Profiling Technique, Electronic Journal of Geotechnical Engineering.7598.
[12] Fellenius B H, Eslami A, 2000 Soil Profile Interpreted from CPTu Data. Geotechnical Engineering Conference.
[13] Hassan K, 1990 A Summary Of The Quaternary Geology Investigations In Seberang Prai, Pulau Pinang And Kuala Kurau, Geological Society of Malaysia. 26 47 – 53.
[14] Keceli A, 2012 Soil Parameters Which Can Be Determined With Seismic Velocities, Jeofizik. 16 17-29.
[15] Jin X, Luke B, Louie J, 2006 Comparison of Rayleigh Wave Dispersion Relations from Three Surface Wave Measurements in a Complex-Layered System. Geocongress.
[16] Louie J N, 2001 Shear-Wave Velocity to 100 Meters Depth from Refraction Microtremor Arrays, Bulletin of the Seismological Society of America. 91 (2) 347-364.
[17] Watabe Y, Sassa S, 2008 Application of MASW Technology to Identification of Tidal Flat Stratigraphy and Its Geoenvironmental Interpretation, Marine Geology. 252 79-88.
[18] Xia J, Miller R D, Park C B, Hunter J A, Harris J B, Ivanov J, 2002 Comparing Shear-Wave Velocity Profiles Inverted From Multichannel Surface Wave With Borehole Measurements, Soil Dynamics And Earthquake Engineering. 22 181-190.

Pile Spacing Optimization of Short Piled Raft Foundation System for Obtaining Minimum Settlement on Peat

S M Suro[1,2], I Bakar[2,1] and A Sulaeman[3]
[1] Faculty of Civil and Environmental Engineering, Universiti Tun Hussein Onn Malaysia, 86400 Batu Pahat, Johor, MALAYSIA
[2] Research Center for Soft Soil Malaysia, Universiti Tun Hussein Onn Malaysia, 86400 Batu Pahat, Johor, MALAYSIA
[3] Faculty of Engineering Technology, Universiti Tun Hussein Onn Malaysia, 86400 Batu Pahat, Johor, MALAYSIA

E-mail: [a]sajiharjo@gmail.com

Abstract. Short Piled Raft is a modified piled raft foundation system, which represents combination between raft foundation and pile foundation, but the length of pile is relatively shorter. The basic concept of the Short Piled Raft foundation system considers the passive soil pressure creating a stiff condition of slab-pile system. This means that the thin concrete slab floats on the supporting soil, while the piles serve as stiffeners concrete slab and also to reduce settlement of the foundation. Slab to pile ratio of such system has been mentioned by several researchers, however the optimum pile spacing of stability performance for obtaining minimum settlement on peat haven't been clearly discussed. In this study, finite element method to simulate the stability performance related to settlement of Short Piled Raft foundation system was used. Short Piled Raft foundation system with concrete slab of 7.0 m x 7.0 m square was assumed to be built on peat with the thickness of 3.5 m. The material properties of pile and raft were constant. The outer diameter of galvanized steel pipe as pile was 0.30 m; raft thickness was considered to be constant of 0.15 m and the length of pile was 3.00 m, while the pile spacing varied from 0.50 to 3.00 m. Point load varied from 0 to 100 kN with increment of 20 kN was also considered as a static load, acted on the centre of the concrete slab. Optimization was done by comparing each numerical result of simulations, thus conclusion can easily be drawn. The optimum pile spacing was 1.00 m which produced minimum settlement of 30.11 mm under the load of 100 kN.

Keywords: Soft soil, ground modification, pile raft.

1. Introduction

Currently, due to limited land available, many construction projects have penetrated into the problematic soil area, with some of the problems faced [1]. Completion of construction by using a conventional foundation system such as pile foundation system is still considered to be quite expensive [2]. To overcome these problems, several foundation systems have been developed, among others, is a piled raft foundation, which the concept of this system has received considerable attention in recent years [3] and even proves to be more effective on such conditions, increasingly recognized as a foundation more economical and effective on problematic soil [4]. Moreover, especially at the peat area, the construction method on peat is different for the different depth of peat [5]. For peat with depth less than 3 m, the removal and replacement method are usually used. For the depth 3 m to 10 m, engineers normally used sand drain, lightweight fills and stone column. While for the depth more than 10 m, the suitable method is deep stabilization techniques such as pile and dynamic compaction. This condition motivates to develop a foundation that can be directly applied on peat with the depth less than 3 m, neither using removal and replacement method nor soil stabilization.

In this study, a Short Piled Raft foundation system was introduced, built on peat which is known as problematic soil. Short Piled Raft foundation system is a modified piled raft foundation system, which is a combination between pile foundation and raft foundation, with the pile length relatively shorter, and considered as a reinforced concrete slab resting on a number of piles. The general plan and cross section of Short Piled Raft foundation system are shown in Fig.1. The basic concept of Short Piled Raft foundation system considers passive soil pressure creating a stiff condition of slab-pile system. This means that the thin concrete slab floats on the supporting soil while the piles serve as stiffeners slab concrete and also to improve stability performance by reducing settlement of the foundation. Slab to pile ratio of such system has been mentioned by several researchers [6], however the optimum pile spacing of stability performance for obtaining minimum settlement on peat haven't been clearly discussed.

In this parametric study, concrete slab of 7.0 m x 7.0 m square as a raft was assumed and the material properties of pile and raft were constant whereas peat was used. The raft thickness was considered to be constant of 0.15 m and the outer diameter of pile was 0.30 m; the length of pile was 3.00 m. To find out the optimum pile spacing of Short Piled Raft Foundation System, piles spacing varied from 0.50 m, 0.75 m, 1.00 m, 1.50 m, 2.00 m and 3.00 m were observed. Load varied from 0 kN to 100 kN with increment of 20 kN were considered as a static load, acted on the centre of the concrete slab. The settlements that caused by the varying point load have been investigated.

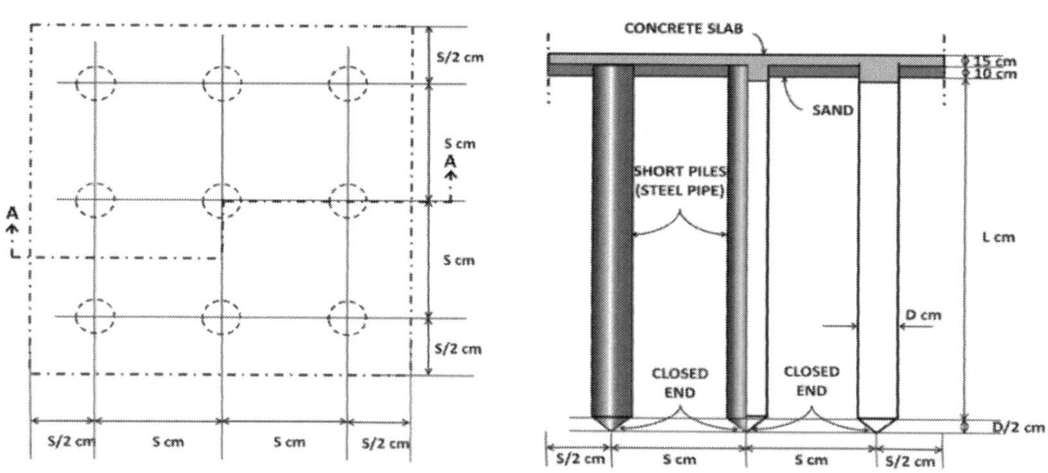

Figure 1. Short Piled Raft Foundation System; **(a)** general plan; **(b)** cross section A-A.

2. Scope and Limitation

The scope of this study is to determine the optimum pile spacing of Short Piled Raft foundation system by trial and error, using finite element method program for simulation, then comparing the results and finally to select the pile spacing that produced the minimum settlement.

The limitation of this study is only conducted numerically, with a certain condition e.g. on peat with the layer thickness of 3.5 m, constant ground water level, consolidated drained and considered static load. While the thickness of concrete slab, length and diameter of pile assumed to be constant.

3. Finite Element Method

In this study, a three-dimensional finite element method program, namely Plaxis 3D Foundation was used. Plaxis 3D Foundation is a special purpose three-dimensional finite element program used to perform deformation and stability analysis for various types of geotechnical applications. The program uses a convenient graphical user interface that enables users to quickly generate a geometry model and finite element mesh. With Plaxis 3D Foundation, complex geometry of soil and structures can be defined in two different modes. These modes are specifically defined for soil or structural modeling.

The finite element method provides a valuable analytical tool for the analysis and design of foundations. Since the piled raft is typical example of soil - structure interaction, a special type of element at pile – soil interface, simulating the displacement discontinuity between the pile and the soil mass is needed. Hence, PLAXIS 3D Foundation incorporates "Embedded pile" model, in which the pile is assumed as a slender beam element [7].

The pile-soil interaction is governed by relative movements between the pile nodes and the soil nodes. The connection between these nodes is established by means of special – purposed non – linear spring representing the pile - soil contact at the base. Based on the materials, linear elastic material model is used for concrete structure to simulate their stress – strain behavior, while the Mohr - Coulomb model is used for soft soil. [8]

4. Numerical Result

In this study, a Short Piled Raft foundation system with the size of 7.0 m x 7.0 m was considered as a model simulation. Concrete slab was set as 0.15 m thickness and point load varied from 0 kN to 100 kN with increment of 20 kN were considered as a static load, acted on the centre of the concrete slab. Parameters used for simulating the short piled raft foundation system as shown in Table 1.

The simulations were conducted by applying pile spacing of 0.50 m, 0.75 m, 1.00 m, 1.50 m, 2.00 m and 3.00 m in each simulation series respectively.

Table 1. Soil, pile and raft parameters used for Short Piled Raft foundation simulation.

Soil (Peat) Properties	
C	4.0 [kN/m^2]
ϕ	16 [^{0}]
E_{ref}	200 [kN/m^2]
γ_{unsat}	10 [kN/m^3]
γ_{sat}	11 [kN/m^3]
ν	0.12
Thickness of layer	3.50 [m]
Soil (Soft Clay) Properties	
C	5.0 [kN/m^2]
ϕ	25 [^{0}]
E_{50}	2000 [kN/m^2]
γ_{unsat}	16 [kN/m^3]
γ_{sat}	17 [kN/m^3]
ν	0.3
Thickness of layer	6.50 [m]

Short Piled Raft Foundation	
Thickness of raft	0.15 [m]
Point Load (varies)	0 to 100 [kN]
Outer pile diameter	0.30 [m]
Thickness of pipe	0.005 [m]
Pile length	3.0 [m]
Pile spacing (varies)	0.5 to 3.0 [m]

Illustration of Short Piled Raft foundation system in numerical model is shown in Figure-2, while generated 3-D mesh of Short Piled Raft foundation system is shown in Figure-3.

Based on the determined parameters, the results of simulation are shown in Table 2 and in order to obtain a more informative, the results are displayed graphically as shown in Figure-4. By comparing the numerical results of each simulation, it can be seen that the optimum pile spacing was identified to be 1.00 m. If the pile spacing increased (more than 1.0 m), in general the settlement will become greater, likewise, if the pile spacing decreased (less than 1.0 m), the settlement will also increase.

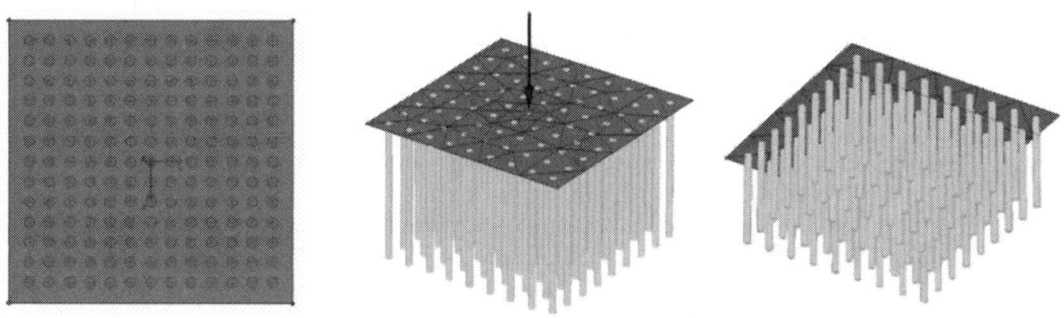

(a) (b) (c)

Figure 2. Model of Short Piled Raft foundation system with 7.0 m x 7.0 m concrete slab, **(a)** plan with pile spacing of 0.50 m; **(b)** & **(c)** upper & bottom view with pile spacing of 0.75 m.

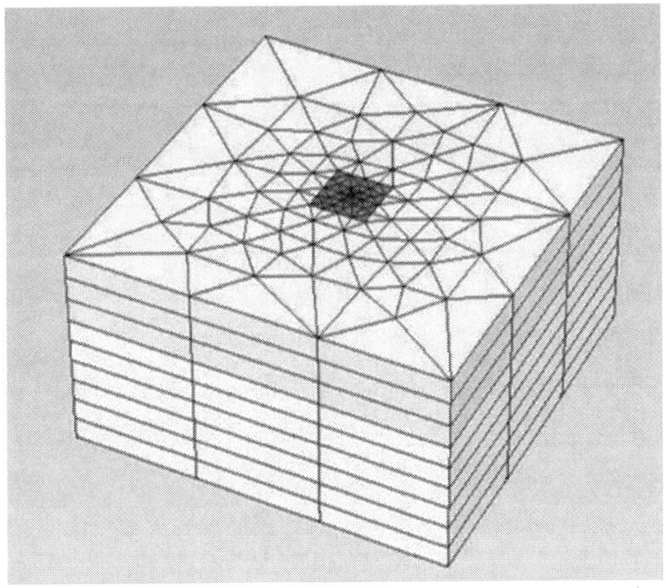

Figure 3. Generated 3-D mesh of Short Piled Raft Foundation System model.

Table 2. Results of Simulation for each Pile Spacing.

Pile Spacing [m]	**Settlement [mm]**					
	0 [kN]	20 [kN]	40 [kN]	60 [kN]	80 [kN]	100 [kN]
0.50	30.52	32.20	33.88	35.56	37.23	38.91
0.75	22.41	24.28	26.16	28.03	29.91	31.78
1.00	19.94	21.97	24.01	26.04	28.08	30.11
1.50	19.58	21.97	24.36	26.76	29.22	31.80
2.00	21.28	23.97	26.80	29.73	32.65	35.57
3.00	25.94	29.69	33.43	37.19	40.95	44.74

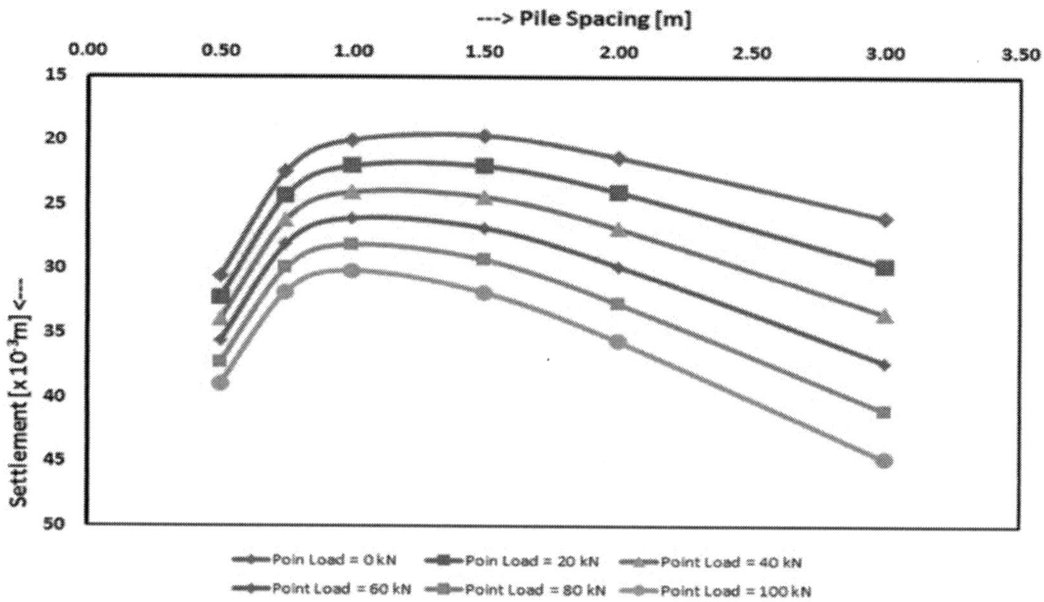

Figure 4. Numerical Results of Simulations for Different Pile Spacing.

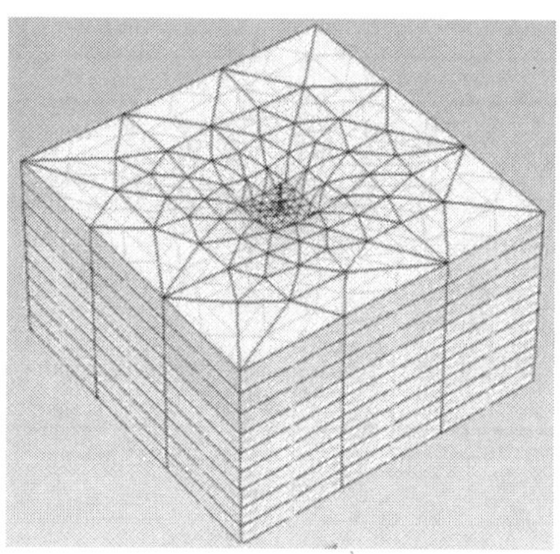

Figure 5. Deformation of 3-D mesh stage

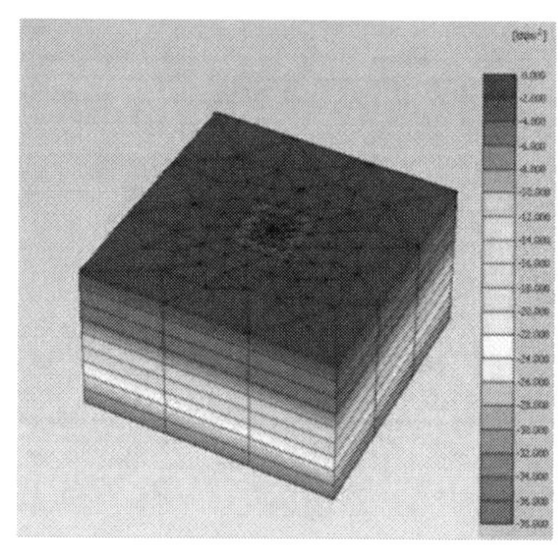

Figure 6. Normal stress of 3-D after final after final stage.

5. Discussion

Based on the numerical results shown in Fig. 4, the optimum pile spacing was identified to be 1.00 m. It can easily be seen when the pile spacing increased (more than 1.00 m), in general the settlement will become greater, likewise, when the pile spacing decreased (less than 1.00 m), the settlement will also increase.

This phenomena can be explained that if the pile spacing increased (more than 1.00 m), the effectiveness of total earth passive pressure decreased, followed by decreasing function of piles as stiffeners of concrete slab and will produce greater settlement. While if the pile spacing decreased (less than 1.00 m), although the total earth passive pressure increased but the total weight of foundation relatively heavier and lead to contribute a greater settlement.

6. Conclusion

Based on the result and discussion, it can be concluded that the optimum pile spacing of Short Piled Raft foundation system on peat with a certain conditioned as explained in this study is 1.00 m, which produced minimum settlement of 30.11 mm under the static load of 100 kN.
It is seemed that Short Piled Raft foundation system on peat is functioning effectively to reduce settlement produced.

Acknowledgements
The authors wish to express their gratitude to RECESS - UTHM for their supports.

References

[1] Patil J D 2013 et al. A Study on Piled Raft Foundation: State of Art. *International Journal of Engineering Research & Technology (IJERT)*, pp.1464-1470.
[2] Effendi S 2013 Cakar Ayam Soft Foundation System Revisited. *Soft Soil Engineering International Conference 2013, Kucing, Sarawak, Malaysia.*
[3] Prakoso W A and Kulway F H, 2001 Contribution To Piled Raft Foundation Design. *Journal of Geotechnical and Geoenvironmental Engineering*, pp.17-24..
[4] Srilakshmi G and Moudgalya D 2013 Analysis of Piled Raft Foundation Using Finite Element Method. *International Journal of Engineering Research and Science & Technology, Vol. 2, No. 3,* , pp.89-96.
[5] Bakar I 2014 Challenges in Peat Soil Research - Malaysian Experiences. *South East Asia Conference on Soft Soils Engineering and Ground Improvement*, Bandung.
[6] Tandjiria V 1999 Numerical Modelling of Chicken-Foot Foundation. *Dimensi Teknik Sipil Volume 1, no. 1 Maret.*
[7] Dao T P T 2011 Validation of PLAXIS Embedded Piles For Lateral Loading. *Master of Science Thesis, Delft University of Technology, The Netherlands.*
[8] Qaissy M A, Karim H H and Hameedi M K 2013 Behavior of Experimental Model of Piled Raft Foundations on Clayey Soils. *1st International Conference for Geotechnical Engineering and Transportation ICGTE in2415/4/2013*, pp.388-408.

A Review of Discrete Element Method Research on Particulate Systems

A A Mahmood[1] and M Elektorowicz[2]
[1]Dept. of Civil Engineering, University College of Technology Sarawak, 96000 Sibu, Sarawak, MALAYSIA
[2]Deptartment of Building Civil and Environmental Engineering, Concordia University, H3G 1M8, CANADA

Email: aali.mahmood@ucts.edu.my, bmariae@encs.concordia.ca

Abstract. This paper summarizes research done using the Discrete Element Method (DEM) and explores new trends in its use on Particulate systems. The rationale for using DEM versus the traditional continuum-based approach is explained first. Then, DEM application is explored in terms of geotechnical engineering and mining engineering materials, since particulate media are mostly associated with these two disciplines. It is concluded that no research to date had addressed the issue of using the DEM to model the strength and weathering characteristics of peaty soil-slag-Portland cement-fly ash combinations.

Keywords: Discrete element, numerical modelling.

1. Introduction

Although granular media are a multiphase particulate system, they have generally been modeled as a continuum. Problems occur with this assumption due to the known granular nature of soils and their multi-force multiphase nature. The problem is especially evident with non-linear stress-strain behavior or when significant local cracks and yielding appear in compression tests.

Numerical methods are desired over experimental setups in modeling particulate media since the latter are more expensive to build and operate. Numerical modeling has advantages over physical testing: parametric studies can be performed with ease and non-destructive and obtrusive sample testing can be performed.

Also, boundary conditions can be controlled explicitly, sample reproducibility can be guaranteed, and tests can be stopped and restarted to suit a particular need [1]. This paper discusses the implementation and use of the Discrete Element Method (DEM) for the analysis of particulate systems, with emphasis on soils.

2. Continuum Based Numerical Methods

Traditionally, continuum based numerical methods have been used qualitatively and quantitatively to analyze particulate and granular media. The most notable example in soil mechanics is the Finite Element Method and its various hybrids. Although the finite element method was used to study discontinuous problems, it has certain disadvantages. For example this method is continuous in nature, which means that it cannot efficiently mimic the discontinuity property. Also, as the accuracy of the model is as good as the assumptions used, reproducing complex behavior with continuous methods requires complex constitutive models, containing sometimes dozens of parameters and/or internal variables in order to capture discontinuous behavior. This can improve results, but on the other hand they are computationally intense and time consuming [2].

Due to its particulate nature, soil can experience significant local deformations and bifurcations. Continuum based methods cannot compensate for such anomalies without incremental analysis involving non-linear elasto-plasticity based constitutive relations. Even so this approach is lacking in adequately modeling the most rudimentary behavior of soil, including non-linear deformations and local yielding [1].

2.1 The Discrete Element Method

It was not until 1970s, that Cundall and Strack developed the Discrete Element Method, [3, 4, 5]. The method models particulate media as a discrete collection of particles. This random collection of particles interacts through contact forces. The method calculates the displaced positions and rotations of these particles at discrete time steps. The DEM-simulation is started by first generating a model, which results in the random orientation of particles with assigned initial velocities. The forces and moments acting on each particle are computed from the initial data and the relevant physical laws and contact models. Generally, the simulation consists of three parts: the initialization, the explicit time stepping and post processing.

In DEM the microstructure of the system is modeled rather than using constitutive laws or complicated elements. Using DEM the changes in microstructure, shape and deformations, dynamics and forces within the system can be captured in real time and in detail. Compared to conventional continuum methods, the DEM uses fundamental and much fewer parameters when modeling discontinuous behavior [2].

The method is capable of analyzing multiple interacting continuous, discontinuous or deformable interacting bodies undergoing large displacements and rotations. In the scheme developed by [5] the particles are not allowed to deform, instead they overlap and the method monitors each contact between particles and computes the new positions and orientations accordingly. Also since the algorithm can model dynamic stress propagation from particle to particle, it can be used to analyze dynamic as well as static soil behavior [6]. In addition, data evaluation at a macroscopic level can be achieved [1].

Matar [2] states that DEM is more suited for discontinuous materials for the following reasons:

1) Model is localized by nature, so there is currently no better tool to model discontinuous material than DEM,
2) Modeling discontinuous material is straightforward,
3) Any type of inter-particle forces can be incorporated,
4) Any particle shape can be considered,
5) Results obtained are more related to discontinuous materials than other methods,
6) It can be coupled with other methods to model continuous-discontinuous properties.

Cundall [7] made the case that continuum methods, such as finite element, for rock and soil might be completely replaced by particle models, such as DEM, in 10-20 years.

2.2 General Discrete Element Method Theory

The calculations performed in the discrete element method alternate between the application of Newton's second law to the discs that represent soil particles, and a force-displacement method at the contacts. Newton's second law gives the motion of a particle resulting from the forces acting on it. The force-displacement law is used to find contact forces from displacements [5].

The DEM simulation consists of an assembly of particles, which has a certain shape contained in a bounded area. Those particles interact with each other under certain conditions based on a specific relational equation. The bounded area can be two or three dimensional. The DEM method involves the cyclic calculation of the inter-particle force between particles in contact.

The simulation starts by assuming some initial configuration of particle positions and then contact detection module finds which set of particles are interacting. Based on type of interaction, forces are applied to these interacting particles. The simulation then proceeds by stepping in time. Velocities, orientations and positions are calculated based on Newton's second law of motion [2].

Velocities and incremental displacements are calculated by integrating the equation of motion. The equation of motion for the center of mass of body (i) is given by:

$$\text{mi}\ \frac{dvi}{dt} = \sum_{j=1}^{n} Fxij + mi\ g \tag{1}$$

Where:
mi = Mass of body i
vi = Velocity of body i
n= Total number of particles
Fxij = Inter-particle force between body i and j
g= Acceleration of gravity

The angular velocity is given by:

$$Ii \frac{d\omega i}{dt} = \sum_{j=1}^{n} \quad Lxij \tag{2}$$

Where:
Ii = Inertia of body i
ωi = Angular velocity of body i
Lxij = Angular momentum for body i and j

3. DEM Applications in Geotechnical Engineering

After the pioneering work of Cundall and Strack [3, 4, 5] in developing the first discrete element scheme for analyzing particulate and granular media, researchers started evaluating and improving the technique.
Dobry and Ng [8] performed a literature survey of publications which have used the DEM of compliant particles or blocks for simulations of granular media during the years from 1982-1992.

A major conference conducted by the American Society for Civil Engineers (ASCE) in 2002 in New Mexico, USA, presented several major trends in the DEM modeling and practice. Participants discussed and analyzed several topics in this field including the following:

1) Theory and algorithms: [9, 10, 11, 12, 13],
2) Model generation: [14, 15, 16],
3) Simulation environments: [17, 18, 19],
4) Solid continuum and discrete element methods [20, 21, 22, 23],
5) Fluid discrete element methods: [24, 25, 26, 27],
6) Experimental validation: [28, 29, 30],
7) Cohesive materials: [31, 32, 33],
8) Granular mechanics: [34, 35],
9) Powders and soils: [36, 37, 38],
10) Rock: [39, 40, 41],

Later on Potyondy and Cundall [42] presented a numerical model for rock using the DEM. They showed that the model reproduced many features of rock behavior including elasticity, fracturing, acoustic emission, damage accumulation producing material anisotropy, hysteresis, dilation, post peak softening and strength increase with confinement.

Matar et al. [43] modeled the evolution of particle subdivision in Montmorillonite clay using 3 dimensional DEM. They wrote a program using ANSI C++ that studied the swelling and swelling pressure response with various amounts of particle breakdown in a unidirectional swelling cell.

Chen et al. [44] performed a 3 dimensional modeling of sinkhole repair using the DEM. They showed that the DEM was a reasonable method to investigate sinkhole repair procedure.
More recently, researchers used the DEM to model geosynthetic reinforced soils [45], geotextile reinforced soils [46] and geogrid reinforced soils [47].

In the area of extraterrestrial geotechnical engineering, Schwartz et al [48] used the DEM in a simulation to model the contact forces between particles in granular materials. While, Lichtenheldt and Schafer [49] developed a simulation model for planetary exploration using the 3 dimensional DEM. They developed an inter-particle contact model and showed that the newly developed model is applicable to wheel soil interaction problems. Then, Kulchitsky et al [50] used the DEM in the numerical modeling of void ratio (packing density) to simulate lunar regolith properties. Later on, Kulchitsky et al. [51] used the DEM in the simulation of boulder extraction from an asteroid and developed an equation linking pulling strength to boulder diameter.

Manne and Satyam [52] reviewed in detail the discrete element modeling of dynamic laboratory tests for liquefaction assessment. They identified gap areas in these laboratory tests related to numerical modeling.

4. DEM Applications in Mining Engineering

Most DEM applications dealt with either rock or soil materials. However a few researchers investigated the use of mine tailings as granular media.

McBride et al. [28] performed an experimental and numerical study using a laboratory ball mill. In the experimental part they recorded the three dimensional trajectory of particles using an automated tracking technique and bi-planar X-ray filming. This was followed by 2 dimensional DEM modeling to determine the parameters needed for the 3 dimensional DEM simulation to avoid the extra computational overhead. Then 3 dimensional DEM simulation was performed. The methods presented provided rigorous benchmarking of DEM's predictive capabilities based on 3 dimensional trajectory data and the experimental measurement of material properties.

Sarracino et al. [53] performed another experimental and numerical study of mills in their granular form. In the experimental work, they measured incident and final velocities, final angular velocity and the coefficient of restitution of tumbling mills. The experiments were then modeled numerically in order to find contact and damping models that best reproduced the experimental results.

These two research studies represent a noticeable incursion into the area of DEM modeling of tailings in their granular form. However, beyond the granular form simulation of mills, no investigation of hardened tailings simulation and analysis had been performed.

Another study by Tannant and Wang [54] modeled using 2 dimensional Particle Flow Code (PFC), [55] spray-on, rapid setting polymeric liner materials for underground rock support in Canadian mines. Two dimensional tensile and block punch tests were modeled. The authors showed that the PFC models of liners were capable of exhibiting many features in the field including: progressive de-bonding, liner bending and elongation, isolated zones of high tensile load, progressive liner rupture in tension and eventual liner failure. The study did not investigate the mechanical properties of the actual mining residues or the mine rock.

5. Conclusions

This shows that application of the DEM method to various rock and soil materials was found to be most promising and adequate. The limited amount of work done on simulating mine tailings also appears to be going on the right track with initial good results.

The DEM method seems to be the most adequate for particulate systems. It should be noted that the DEM has never been tested for strength and weathering characteristics. It also has never been tested on Peat soils. Hence it is suggested to address these issues by doing the following:

1) Determining the unconfined compressive strength and weathering characteristics of several types of Portland cement-fly ash-slag stabilized/solidified Peaty soil combinations,
2) Determining the freezing/thawing and wetting/drying resistance of these combinations,
3) Simulating the strength and weathering characteristics of the Peaty soils using the Discrete Element Method.

References

[1] M. Khwaja (1996). Discrete Element Method: micro-mechanical and large scale modeling. M.Sc. thesis, University of Massachusetts Lowell, Massachusetts.

[2] M.I Matar (2005). Modeling of Montmorillonite clay-water interactions with particle subdivisions using three dimensional Discrete Element Method. Ph.D. thesis, North Dakota State University, Fargo, North Dakota.

[3] P.A Cundall (1971). A computer model for simulating progressive large scale movements in block rock systems. In Proceedings of the Symposium of the International Society of Rock Mechanics, Nancy, France. Article 8.

[4] P.A. Cundall (1974). A computer model for rock-mass behavior using interactive graphics for the input and output of geomechanical data. Report for Contract number DACW45-74-C-0066, for the Missouri River Division, US Army Corps of Engineers, University of Minnesota, Minneapolis, Minnesota.

[5] P. A Cundall, and O.D.L. Strack, A discrete numerical model for granular assemblies. Geotechnique, 29(1), (1979). 47-65.

[6] J.M. Ting, B.T. Corkum, C.R. Kauffman, and C.A. Greco, Discrete numerical model for soil mechanics. Journal of Geotechnical Engineering. ASCE, 115(3) (1989) 379-398.

[7] Peter A. Cundall, A discontinuous future for numerical modeling in soil and rock. In Proceedings of the 3rd International Conference on Discrete Element Methods, Discrete Element Methods: Numerical Modeling of Discontinua. 23-25 September (2002). Santa Fe, New Mexico. Edited by Benjamin K. Cook and Richard P. Jensen.

[8] R. Dobry and T.-T. Ng, Discrete modeling of stress-strain behavior of granular media at small and large strains. Engineering Computations, 9(2) (1992) 129-143.

[9] J.F. Favier, and M. Kremmer, Modeling a particle metering device using the Finite Wall Method. In Proceedings of the 3rd International Conference on Discrete Element Methods, Discrete Element Methods: Numerical Modeling of Discontinua. 23-25 September (2002). Santa Fe, New Mexico. Edited by Benjamin K. Cook and Richard P. Jensen.

[10] Y.T. Feng and D.R.J. Owen, An energy based corner to contact algorithm. In Proceedings of the 3rd International Conference on Discrete Element Methods, Discrete Element Methods: Numerical Modeling of Discontinua. 23-25 September (2002). Santa Fe, New Mexico. Edited by Benjamin K. Cook and Richard P. Jensen.

[11] Mark A. Hopkins, Discrete element modeling based on mathematical morphology. In Proceedings of the 3rd International Conference on Discrete Element Methods, Discrete Element Methods: Numerical Modeling of Discontinua. 23-25 September (2002). Santa Fe, New Mexico. Edited by Benjamin K. Cook and Richard P. Jensen.

[12] Matthew R. Kuhn, A torus primitive for particle shapes with the Discrete Element Method. In Proceedings of the 3rd International Conference on Discrete Element Methods, Discrete Element Methods: Numerical Modeling of Discontinua. 23-25 September (2002). Santa Fe, New Mexico. Edited by Benjamin K. Cook and Richard P. Jensen.

[13] T-T. Ng, Hydrostatic boundaries in Discrete Element Methods. In Proceedings of the 3rd International Conference on Discrete Element Methods, Discrete Element Methods: Numerical Modeling of Discontinua. 23-25 September (2002). Santa Fe, New Mexico. Edited by Benjamin K. Cook and Richard P. Jensen.

[14] D. Boutt, and B. McPherson, The role of particle packing in modeling rock mechanical behavior using discrete elements. In Proceedings of the 3rd International Conference on Discrete Element Methods, Discrete Element Methods: Numerical Modeling of Discontinua. 23-25 September (2002). Santa Fe, New Mexico. Edited by Benjamin K. Cook and Richard P. Jensen.

[15] Y.T. Feng, K. Han and D.R.J. Owen, An advancing front packing of polygons, ellipses and spheres. In Proceedings of the 3rd International Conference on Discrete Element Methods, Discrete Element Methods: Numerical Modeling of Discontinua. 23-25 September (2002). Santa Fe, New Mexico. Edited by Benjamin K. Cook and Richard P. Jensen.

[16] S. Johnson and J. Williams, Formation of packing structures in discrete element modeling with disks. In Proceedings of the 3rd International Conference on Discrete Element Methods, Discrete Element Methods: Numerical Modeling of Discontinua. 23-25 September (2002). Santa Fe, New Mexico. Edited by Benjamin K. Cook and Richard P. Jensen.

[17] Youssef M.A. Hashash, and J. Ghaboussi, Discrete element modeling for the development of a real-time soil model in a virtual reality environment. In Proceedings of the 3rd International Conference on Discrete Element Methods, Discrete Element Methods: Numerical Modeling of Discontinua. 23-25 September (2002). Santa Fe, New Mexico. Edited by Benjamin K. Cook and Richard P. Jensen.

[18] Petros I. Komodromos and John R. Williams, Utilization of Java and database technology in the development of a combined discrete and finite element multibody dynamics simulator. In Proceedings of the 3rd International Conference on Discrete Element Methods, Discrete Element Methods: Numerical Modeling of Discontinua. 23-25 September (2002). Santa Fe, New Mexico. Edited by Benjamin K. Cook and Richard P. Jensen.

[19] S. Masala, D. Chan, H. Lu and R. Chalaturnyk (2002). A Java-based graphical user interface for a 2-D Discrete Element program. In Proceedings of the 3rd International Conference on Discrete Element Methods, Discrete Element Methods: Numerical Modeling of Discontinua. 23-25 September 2002. Santa Fe, New Mexico. Edited by Benjamin K. Cook and Richard P. Jensen.

[20] T. Bangash, and A. Munjiza, A computationally efficient beam element for FEM/DEM simulations of structural failure and collapse. In Proceedings of the 3rd International Conference on Discrete Element Methods, Discrete Element Methods: Numerical Modeling of Discontinua. 23-25 September (2002). Santa Fe, New Mexico. Edited by Benjamin K. Cook and Richard P. Jensen.

[21] Jeen-Shang Lin, A unified framework for discrete and continuum analysis. In Proceedings of the 3rd International Conference on Discrete Element Methods, Discrete Element Methods: Numerical Modeling of Discontinua. 23-25 September (2002). Santa Fe, New Mexico. Edited by Benjamin K. Cook and Richard P. Jensen.

[22] M. Reza Salami and F. Amini, Numerical model for the implementation of discontinuous deformation analysis in Finite Element mesh. In Proceedings of the 3rd International Conference on Discrete

Element Methods, Discrete Element Methods: Numerical Modeling of Discontinua. 23-25 September (2002). Santa Fe, New Mexico. Edited by Benjamin K. Cook and Richard P. Jensen.

[23] Federico A., Plesha Tavarez, E. Michael and Lawrence C. Bank, Discrete Element Method (DEM) for modeling solid and particulate media. In Proceedings of the 3rd International Conference on Discrete Element Methods, Discrete Element Methods: Numerical Modeling of Discontinua. 23-25 September (2002). Santa Fe, New Mexico. Edited by Benjamin K. Cook and Richard P. Jensen.

[24] Benjamin K. Cook, David R. Noble, and John R Williams, A coupled DEM-LB model for the simulation of particle-fluid systems. In Proceedings of the 3rd International Conference on Discrete Element Methods, Discrete Element Methods: Numerical Modeling of Discontinua. 23-25 September (2002). Santa Fe, New Mexico. Edited by Benjamin K. Cook and Richard P. Jensen.

[25] Toshihiro Kawaguchi, Satoshi Kajiyama, Toshitsugu Tanaka, and Yutaka Tsuji, DEM simulation of 2-D fluidized bed using similarity model. In Proceedings of the 3rd International Conference on Discrete Element Methods, Discrete Element Methods: Numerical Modeling of Discontinua. 23-25 September (2002). Santa Fe, New Mexico. Edited by Benjamin K. Cook and Richard P. Jensen.

[26] Hiroshi Mori, Yoshimi Ogawa and Guoqiang Cao, Liquefaction analysis of river dike with Discrete Element Method. In Proceedings of the 3rd International Conference on Discrete Element Methods, Discrete Element Methods: Numerical Modeling of Discontinua. 23-25 September (2002). Santa Fe, New Mexico. Edited by Benjamin K. Cook and Richard P. Jensen.

[27] Colin Thornton and D. Kafui, 3D DEM simulations of gas-solid fluidised beds. In Proceedings of the 3rd International Conference on Discrete Element Methods, Discrete Element Methods: Numerical Modeling of Discontinua. 23-25 September (2002). Santa Fe, New Mexico. Edited by Benjamin K. Cook and Richard P. Jensen.

[28] A. McBride, I. Govender, M. Powell, and V. Balden, Three-dimensional validation of DEM using a laboratory ball mill. In Proceedings of the 3rd International Conference on Discrete Element Methods, Discrete Element Methods: Numerical Modeling of Discontinua. 23-25 September (2002). Santa Fe, New Mexico. Edited by Benjamin K. Cook and Richard P. Jensen. pp. 207-211.

[29] Takashi Murakami, Akira Murakami, Muneo Hori, and Hide Sakaguchi, Inverse analysis of stress developed in a granular assemblage under trap-door conditions and its validation using the Discrete Element Method. In Proceedings of the 3rd International Conference on Discrete Element Methods, Discrete Element Methods: Numerical Modeling of Discontinua. 23-25 September (2002). Santa Fe, New Mexico. Edited by Benjamin K. Cook and Richard P. Jensen.

[30] C. O'Sullivan, J.D. Bray and M.F. Riemer, 3-D DEM validation using steel balls with regular packing arrangements. In Proceedings of the 3rd International Conference on Discrete Element Methods, Discrete Element Methods: Numerical Modeling of Discontinua. 23-25 September (2002). Santa Fe, New Mexico. Edited by Benjamin K. Cook and Richard P. Jensen.

[31] A. Anandarajah and M. Yao, Three-dimensional Discrete Element Method of analysis of clays. In Proceedings of the 3rd International Conference on Discrete Element Methods, Discrete Element Methods: Numerical Modeling of Discontinua. 23-25 September (2002). Santa Fe, New Mexico. Edited by Benjamin K. Cook and Richard P. Jensen. pp. 237-241.

[32] J. Lechman, G.G.W. Mustoe, K.T. Miller, Ning Lu and K. Eccleston, Comparison of DEM simulations and physical experiments for direct measurement of strongly attractive particle-particle interactions. In Proceedings of the 3rd International Conference on Discrete Element Methods, Discrete Element Methods: Numerical Modeling of Discontinua. 23-25 September (2002). Santa Fe, New Mexico. Edited by Benjamin K. Cook and Richard P. Jensen.

[33] T.A. Newson, and A. Duliere, The compression behaviour of structured clayey soils. In Proceedings of the 3rd International Conference on Discrete Element Methods, Discrete Element Methods: Numerical Modeling of Discontinua. 23-25 September (2002). Santa Fe, New Mexico. Edited by Benjamin K. Cook and Richard P. Jensen.

[34] Hayley H. Shen, Regimes of granular shear flows. In Proceedings of the 3rd International Conference on Discrete Element Methods, Discrete Element Methods: Numerical Modeling of Discontinua. 23-25 September (2002). Santa Fe, New Mexico. Edited by Benjamin K. Cook and Richard P. Jensen.

[35] Duan Z. Zhang, Effects of particle interaction history in dense and slow granular flows. In Proceedings of the 3rd International Conference on Discrete Element Methods, Discrete Element Methods: Numerical Modeling of Discontinua. 23-25 September (2002). Santa Fe, New Mexico. Edited by Benjamin K. Cook and Richard P. Jensen.

[36] Y.P. Cheng, M.D. Bolton and Y. Nakata, The modelling of soil plasticity. In Proceedings of the 3rd International Conference on Discrete Element Methods, Discrete Element Methods: Numerical Modeling of Discontinua. 23-25 September (2002). Santa Fe, New Mexico. Edited by Benjamin K. Cook and Richard P. Jensen.

[37] Erfan G. Nezami and Youssef M. A. Hashash, The use of static Discrete Element Method to simulate biaxial compression test. In Proceedings of the 3rd International Conference on Discrete Element Methods, Discrete Element Methods: Numerical Modeling of Discontinua. 23-25 September (2002). Santa Fe, New Mexico. Edited by Benjamin K. Cook and Richard P. Jensen.

[38] Morched Zeghal, Tuncer B Edil and Michael E. Plesha, Discrete Element Method for sand-structure interaction. In Proceedings of the 3rd International Conference on Discrete Element Methods, Discrete Element Methods: Numerical Modeling of Discontinua. 23-25 September (2002). Santa Fe, New Mexico. Edited by Benjamin K. Cook and Richard P. Jensen.

[39] Mary M. MacLaughlin and Kathryn K. Clapp, Discrete Element Analysis of an underground opening in blocky rock: an investigation of the differences between UDEC and DDA results. In Proceedings of the 3rd International Conference on Discrete Element Methods, Discrete Element Methods: Numerical Modeling of Discontinua. 23-25 September (2002). Santa Fe, New Mexico. Edited by Benjamin K. Cook and Richard P. Jensen.

[40] Jon Narayanasamy Olson, Jon R. Holder, Alan Rauch and B. Comacho, DEM study of wave propagation in weak sandstone. In Proceedings of the 3rd International Conference on Discrete Element Methods, Discrete Element Methods: Numerical Modeling of Discontinua. 23-25 September (2002). Santa Fe, New Mexico. Edited by Benjamin K. Cook and Richard P. Jensen.

[41] David Potyondy, A bonded-disk model for rock: relating micro properties and macro properties. In Proceedings of the 3rd International Conference on Discrete Element Methods, Discrete Element Methods: Numerical Modeling of Discontinua. 23-25 September (2002). Santa Fe, New Mexico. Edited by Benjamin K. Cook and Richard P. Jensen.

[42] D.O. Potyondy and P.A. Cundall, A bonded-particle model for rock. International Journal of Rock Mechanics and Mining Sciences, 41(8) (2004) 1329–1364.

[43] M.I. Matar, D.R. Katti and K.S. Katti, Modeling the evolution of Montmorillonite clay particulate structure: A Discrete Element modeling study. In Proceedings of ASCE, GEO-Denver (2007): New Peaks in Geotechnics, GSP 173 Advances in Measurement and Modeling of Soil Behavior. 18-21 February 2007. Denver, Colorado.

[44] F. Chen, E.C. Drumm, and G. Guiochon, 3D DEM analysis of graded rock fill sinkhole repair: particle size effects on the probability of stability. In Proceedings of the 88th Annual Meeting of the Transportation Research Board. 11-15 January (2009). Transportation Research Board. Washington, DC.

[45] H.I. Park and S.R. Lee, Evaluation of bearing capacity for multi-layered clay deposits with geosynthetic reinforcement using Discrete Element Method. Marine Georesources and Geotechnology, 28(4) (2010) 363–374.

[46] A. Bhandari and J. Han, Investigation of geotextile–soil interaction under a cyclic vertical load using the Discrete Element Method. Geotextiles and Geomembranes, 28(1) (2010) 33–43.

[47] J. Han and A. Bhandari, Evaluation of geogrid-reinforced pile-supported embankments under cyclic loading using Discrete Element Method. Geotechnical Special Publication No. 188. Editors: J. Han, G. Zheng, V.R. Schaefer, M.S. Huang. In Proceedings of US-China Workshop on Ground Improvement Technologies, ASCE, Orlando, Florida. (2009) pp. 73-82.

[48] Stephen R. Schwartz, Derek C. Richardson and P. Michel. An implementation of the soft-sphere discrete element method in a high-performance parallel gravity tree-code. Granular Matter 14 (2012) 363–380.

[49] R. Lichtenheldt and B. Schafer, Locomotion on soft granular soils: A Discrete Element based approach for simulations in planetary exploration. III International Conference on Particle-based Methods-Fundamentals and Applications. PARTICLES (2013). M. Bischoff, E. Onate, D.R.J. Owen, E. Ramm & P. Wriggers (Eds).

[50] Anton V. Kulchitsky, J. B. Johnson and A. Wilkinson. Controlling bulk material packing density in COUPi DEM model to simulate lunar regolith: POSTER. The virtual 2013 Lunar Science Forum, July 16-18, (2013).

[51] A. Kulchitsky, J. Johnson, D. Reeves, and A. Wilkinson. Discrete Element Method simulation of a boulder extraction from an asteroid. Proceedings of the 14th ASCE Biennial International Conference

on Engineering, Science, Construction, and Operations in Challenging Environments, St. Louis, Missouri, USA (2014): pp. 485-494.

[52] A. Manne and N. Satyam. A review on the Discrete Element modelling of dynamic laboratory tests for liquefaction assessment. Electronic Journal of Geotechnical Engineering. 20.1 (2015) 21-46.

[53] R.S. Sarracino, H.-J. Dong and M.H. Moys, An experimental and theoretical study of ball-wall impact for DEM modeling of tumbling mills. In Proceedings of the 3rd International Conference on Discrete Element Methods, Discrete Element Methods: Numerical Modeling of Discontinua. 23-25 September (2002). Santa Fe, New Mexico. Edited by Benjamin K. Cook and Richard P. Jensen.

[54] Dwayne D. Tannant and Caigen Wang, Thin rock support liners modeled with Particle Flow Code. In Proceedings of the 3rd International Conference on Discrete Element Methods, Discrete Element Methods: Numerical Modeling of Discontinua. 23-25 September (2002). Santa Fe, New Mexico. Edited by Benjamin K. Cook and Richard P. Jensen.

[55] Itasca (1999), Itasca Consulting Group Inc., 111 Third Ave. South, Suite 450 Minneapolis, Minnesota 55401. www.itascacg.com.

Barren Acidic Soil Assessment using Seismic Refraction Survey

S A A Tajudin[1,2], M H Z Abidin[2,1], A Madun[2,1] and M H Zawawi[3]
[1] Research Center for Soft Soil, Universiti Tun Hussein Onn Malaysia, 86400 Batu Pahat Johor, MALAYSIA
[2] Faculty of Civil and Environmental Engineering, Universiti Tun Hussein Onn Malaysia, 86400 Batu Pahat Johor, MALAYSIA
[3]College of Engineering, University Tenaga Nasional, MALAYSIA

Email: saifulaz@uthm.edu.my

Abstract. Seismic refraction method is one of the geophysics subsurface exploration techniques used to determine subsurface profile characteristics. From past experience, seismic refraction method is commonly used to detect soil layers, overburden, bedrock, etc. However, the application of this method on barren geomaterials remains limited due to several reasons. Hence, this study was performed to evaluate the subsurface profile characteristics of barren acidic soil located in Ayer Hitam, Batu Pahat, Johor using seismic refraction survey. The seismic refraction survey was conducted using ABEM Terraloc MK 8 (seismograph), a sledge hammer weighing 7 kg (source) and 24 units of 10 Hz geophones (receiver). Seismic data processing was performed using OPTIM software which consists of SeisOpt@picker (picking the first arrival and seismic configureuration data input) and SeisOpt@2D (generating 2D image of barren acidic soil based on seismic velocity (primary velocity, Vp) distribution). It was found that the barren acidic soil profile consists of three layers representing residual soil (Vp= 200-400 m/s) at 0-2 m, highly to completely weathered soil (Vp= 500-1800 m/s) at 3-8 m and shale (Vp= 2100-6200 m/s) at 9-20 m depth. Furthermore, result verification was successfully done through the correlation of seismic refraction data based on physical mapping and the geological map of the study area. Finally, it was found that the seismic refraction survey was applicable for subsurface profiling of barren acidic soil as it was very efficient in terms of time, cost, large data coverage and sustainable.

Keywords: Seismic refraction method, primary velocity, barren acidic soil, soft soil.

1. Introduction

Subsurface profiling is an important data for soil investigation which is required for each proposed construction. The seismic refraction method is often presented to engineering and environmental professionals as an inexpensive and easy method for civil engineering projects [1].

Seismic refraction investigates the subsurface by generating arrival time and offset distance information to determine the path and velocity of the elastic disturbance in the ground. The disturbance is created by shot, hammer, weight drop, or some other comparable method for putting impulsive energy into the ground.

Detectors are placed at regular intervals to measure the first arrival of the energy and time needed. The data are plotted in time versus distance graphs from which the velocities of the different layers and their depths can be calculated [2]. On the other hand, a body wave is a seismic wave that moves through the interior of the earth, as opposed to surface waves that travel near the earth's surface. P and S waves are body waves. Each type of wave shakes the ground in different ways [3].In addition, the seismic refraction method provides the velocity of compressional P-waves in subsurface materials. Although the P-wave velocity is a good indicator of the type of soil or rock, it is not a unique indicator [4].

The velocity of sound travelling through the sub-surface varies according to material composition and compaction. Seismic energy from a source at the surface transmits waves through soils. It will then undergo refraction at boundaries between different media and eventually return to the surface. Seismic refraction surveying makes use of this phenomenon to determine ground structure by observing the time taken for energy to travel through the subsurface [5].Therefore, the objective of this study is to determine the subsurface profile of barren acidic soil.

2. Methodology

This study was performed at Ayer Hitam, Batu Pahat, Johor. A single line of seismic refraction survey was performed using ABEM Terraloc MK8 equipment set as shown in Figure. 1. According to Figure. 2, Ayer Hitam is located at the Triassic zone (blue pale) which consists of interbedded sandstone, siltstone, shale and volcanic rock. The raw data obtained from the field measurement was analysed using OPTIM software which consist of SeisOpt@picker and SeisOpt@2D software package.

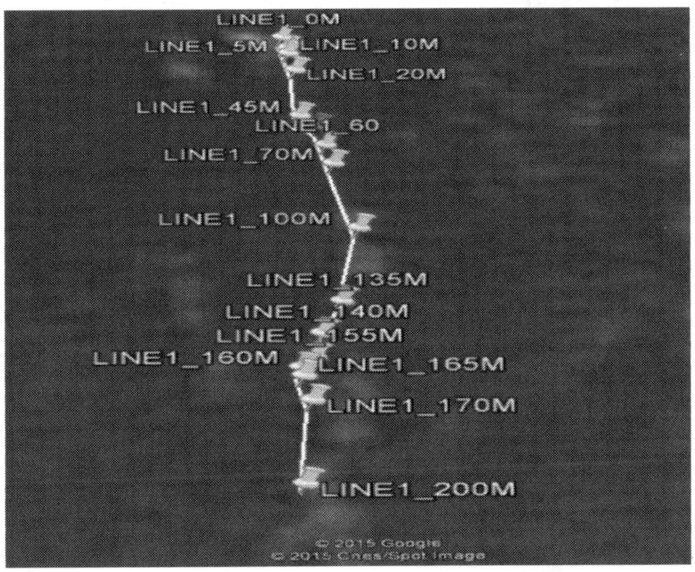

Figure 1. Spread line of seismic refraction survey.

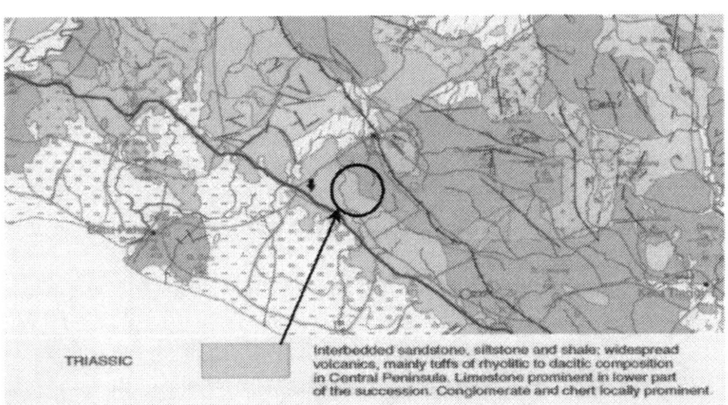

Figure 2. Geological map of the study area [6].

In the beginning, a site survey was conducted at the study area for data collection. The spread line location was determined prior to the data acquisition (field measurement). The seismic refraction survey was applied using a sledge hammer weighing 7 kg, hammering on a striker plate as a source. For receiver, a 24 channel vertical geophone was used based on 10 Hz of frequency. Meanwhile, ABEM Terraloc MK8 seismograph was used for field data recording. The spread line was arranged and the geophones were spaced at 5 m interval as given in Figure. 3. Seven shot points were performed along each profile representing five (in line) plus two (offset) of the shot point total number.

During the test setup, the geophone should be place at the best possible straight line alignment in order to obtain an optimum results during the data acquisition. Moreover, the geophones were placed on a clear area on top of ground surface due to the noise minimization purposes. The seismograph was placed at the centre of the spread line (G12 and G13) connected with both seismic land cables (cable 1 for G1-G12 & cable 2 for G13-G24). Seven shot point locations including two offsets was conducted at the interval of G1 and G2, G6 and G7, G12 and G13, G18 and G19, and G23 and G24 geophones as shown in Figure. 3.

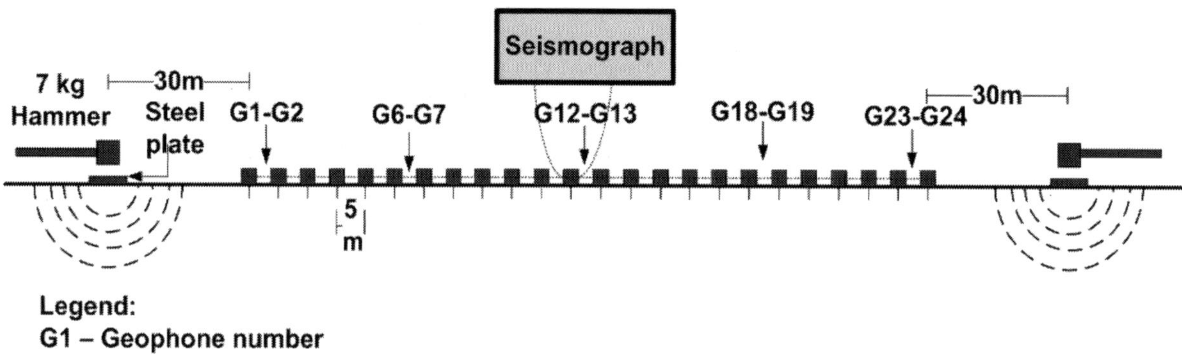

Figure 3. Seven seismic shot points along the spread line.

3. Results and Discussion

All results are presented in Figure. 4, Table 2, and 4 representing seismic refraction tomography, result summary of the seismic refraction survey and field density result. The seismic refraction tomography results that performed at the top of the barren soil profile was given in Figure. 4. Field observation regarding the seismic refraction survey performed at the barren soil outcrop was given in Figure. 5. It was found that this profile consist of three different range of primary velocity (Vp) representing three types of geomaterials with possible different characteristics as shown in Table 2. Those primary velocity (Vp) values in this study has been verified using Table 3 and was also supported by field observation, geological map (Figure. 1) and density results (Table 4 and 5). The core cutter method for the in-situ test was conducted specifically at the third layer and the result of field density test (core cutter) was given in Table 4. Consequently, the density value of common rock type was used for results verification as given in Table 5.

Based on the results obtained from Figure. 4, the first layer of the subsurface profile has the lowest primary velocity (Vp1: 200 – 400 m/s) which represents residual soil (Grade 6). The thickness of this layer was varied from 0 – 2 m from the ground surface. This weathering profile of Grade 6 was based on field observation which found that the original rock texture was completely destroyed together with the presence of vegetation. Furthermore, these velocities (Vp1: 200 – 400 m/s) were in good agreement according to ASTM [4] which stated that the residual soil has a lower velocity representing weathered rocks compared to the unweathered rocks. In addition, the lower primary velocity was also influenced by the stiffness of the geomaterials. The weathered fractured materials can disintegrate and decompose into fine grained materials such as mineral, sand, clay, silt, etc. The second layer of the subsurface profile has a primary velocity (Vp2: 500 – 1000 m/s) which represents highly to completely weathered soil (Grade 4-5). The thickness of this layer was approximately varied from 3 – 8 m. From the site observation, it was revealed that the soil is easily crushed by hand and the rock material plastic does not readily slake in water. It was found that the third layer from the spread line has the highest velocity (Vp3: 2100 – 6200 m/s) which represents slightly to moderately weathered shale (Grade 2-3). The thickness of this layer was approximately varied from 9 – 20 m. Based on site observation, it was found that the hammer blow gave a dull note and required more than one blow of the geological hammer to break the specimen.

Apart from the physical mapping, this study also performed a core cutter test in order to compare and correlate the geomaterials density results for verification purposes. As shown in Table 4, the average density for the third layer is 2.537 gcm^{-3} and is well within the range of density value for shale and sandstone as given in Table 5. Moreover, the interpretation result from seismic refraction regarding the third layer was also been supported by the existing geological map (see Figure. 1) which indicate that this area consist of sedimentary rock with particular reference to interbedded shale and sandstone. This result shows that seismic refraction is a good technique to be applied in geotechnical site investigation especially when the accessibility of site was complicated.

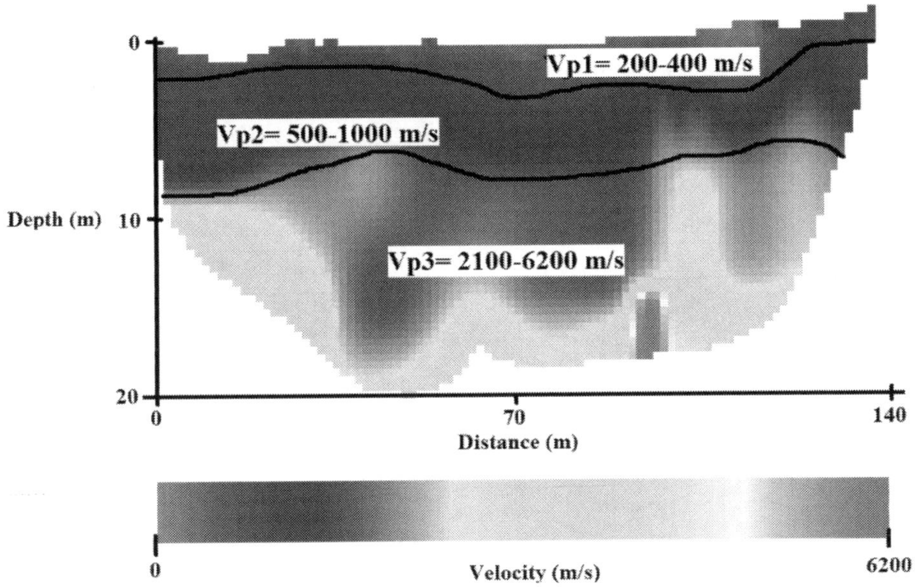

Figure 4. Seismic refraction tomography of the barren acidic soil.

Figure 5. Seismic refraction survey performed at the barren soil outcrop.

Table 2. Result summary of seismic refraction survey.

Layer 1 (Vp1)	Layer 2 (Vp2)	Layer 3 (Vp3)	Primary velocity, Vp (m/s)	Thickness layer (m)
Residual Soil (Grade 6)	Highly to completely weathered (Grade 4-5)	Slightly to moderate weathered shale (Grade 2-3)	Vp1: 200-400 Vp2: 500-1000 Vp3: 2100-6200	Vp1: 0-2 Vp2: 3-8 Vp3: 9-20

Table 3. Primary velocity (Vp) of geomaterials [4].

Materials Natural Soil and Rock	Primary velocity, Vp (m/s)
Weathered surface material	240-610
Gravel or dry sand	460-915
Soil	100-500
Sand (dry, loose)	200-2000
Sand (saturated)	1220-1830
Shale	2000-3500
Clay (saturated)	915-2750
Water1	1430-1665
Sea water1	1460-1525
Sandstone	1830-3960
Limestone	2134-6100
Granite	4575-5800
Metamorphic rock	3050-7000

Table 4. Field density test.

Average moisture content , wave	21.7	22.2	18.8	18.4
Density, gcm^{-3}	2.427	2.618	2.769	2.843
Average density, gcm^{-3}	2.66			

Table 5. Density of common rock type [10]

Type	Rhyolite	Rock salt	Sandstone	Shale	Slate
Density, gcm^{-3}	2.4-2.6	2.5-2.6	2.2-2.8	2.4-2.8	2.7-2.8

4. Conclusion

The seismic refraction survey is a good technique to determine an overview of the whole subsurface profile with particular reference to soil profiling and characterization. Geophysical methods are useful in determining a variety of physical properties of soil and rock. The subsurface profile of barren acidic soil was successfully investigated using 2D seismic refraction tomography. The geometry and primary velocity distribution at localize outcrop in Ayer Hitam, Batu Pahat, Johor was determined by analyzing seismic refraction data obtained along the physical mapping and geological map of the study area which showed some good agreement. Hence, the objective of this study was successfully achieved based on results obtained. Finally, seismic refraction survey was applicable to be a good alternative method in shallow subsurface profiling due to its effective in terms of cost, time and quality provided that the technique is properly done by experienced and trained personnel.

Furthermore, this method used a surface method during the data acquisition which enable the preservation of site destruction thus contributing to our sustainable environment in construction industry.

Acknowledgements
The authors would like to thank the Ministry of Higher Education, Universiti Tenaga Nasional, Malaysia (UNITEN), Universiti Tun Hussein Onn, Malaysia (UTHM) for supporting this research under the Fundamental Research Grant Scheme (FRGS) Vot. 1455 and Incentive Grant Scheme Publication (IGSP) Vot. U258.

Reference

[1] Riahi M A, Tabatabaei S H, Bevtollahi A, Ghaladarzzadeh A and Talebian M, 2012 Seismic Refraction and Downhole Survey for characterization of Shallow Depth Materials of Bam city, *southeast of Iran Journal of the Earth & Space Physic*. Vol. 37, pp. 41 – 58.
[2] Bery A and Saad R 2012 Tropical Clayey Sand Soil's Behaviour Analysis and its Empirical Correlations via Geo-physic Electrical Resistivity Method and Engineering Soil Characterization," *International Journal of Geo-sciences*, Vol. 3 pp.111 – 116.
[3] Society of Exploration Geophysics. (2002) Exploration Geophysics: Refraction Seismic. American Standard for Testing Materials (ASTM), Standard Guide for Using the Seismic Refraction Method for Subsurface Investigation, D 5777-00.
[4] Reynolds J M 2011 An Introduction to Applied and Environment Geophysics (John Wiley & Sons, Chichester 2011).
[5] Mineral And Geoscience Department Malaysia, Geological Map of Peninsular Malaysia, 8th Ed. Ministry of Natural Resources and Environment. (1985).
[6] Lutgens F K and Arbuck E J 2006 Essentials of Geology (Pearson International Edition, New Jersy.
[7] Abidin M H, Saad R, Ahmad F., D.C. Wijeyesekera, and M.F.T. Baharudin, in: Integral Analysis of geoelectrical (resistivity) and geotechnical (SPT) data in soil stability assesment. *International Journal of Arts and Sciences* 2012
[8] Saunders M K and Fookes P.G. , in: A review of the relationship of rock weathering and climate and its significance to foundation engineering. Engineering Geology.
[9] Andrew G (2010), Andrew Scott (1996). Field scale evaluation of the In Situ Permeable Flow Sensor and assessment of river-aquifer interaction at the Brazos River Hydrologic Field Site / by Andrew Scott Alden. Master's thesis, Texas A&M University
[10] Abidin M H, Saad R, Ahmad F, Wijeyesekera D C, and Baharudin M F T, 2012 Seismic Refraction Investigation on Near Surface Landslide at Kundasang area in Sabah, Malaysia. *International Conference on Advance Science and Contemporary Engineering*.

Fiber Effects on Compressibility of Peat

N N Johari[1,2], I Bakar[2,1], S N M Razali[1,2] and N Wahab[1,2]
[1] Faculty of Civil and Environmental Engineering, Universiti Tun Hussein Onn Malaysia, 86400 Batu Pahat Johor, MALAYSIA
[2] Research Center for Soft Soil, Universiti Tun Hussein Onn Malaysia, 86400 Batu Pahat Johor, MALAYSIA

E-mail: nnadiajohari90@gmail.com

Abstract. Fibers found in the soil, especially in peaty soil play an important role in the determination of soil compressibility. Peat soils are the results from the decomposition of organic matter and the type of peat can be classified based on the fibrous material in the soil. In the engineering field, peat soil was mostly known as soils that has a serious settlement with high compressibility index. From the previous research, fibers in the soil were influenced in compressibility in terms of size, shape, fibric, soil arrangement and etc. Hence, this study attempts the determination of fibers effects on the compressibility of peat using a 1-D oedometer consolidation test. The reconstituted peat samples of different particle sizes were used to determine the consolidation parameters and the results obtained from reconstituted samples were also compared with the undisturbed sample. 1-D oedometer consolidation tests were performed on the samples by using the load increment method. The results shows, the higher particle size (R3.35), give higher moisture content (w = 401.20%) and higher initial void ratio (e_o = 5.74). In settlement prediction, the higher the fiber content will results the higher the compression index, therefore, it will cause high of settlement.

Keywords: Soft Soil, ground modification, settlement.

1. Introduction

A soil can be classified as peat when the organic content of the soil exceed 75% as stated in ASTM D4427 [2]. Peat is mainly known as highly organic soils disintegrated primarily from plants remains, and undergo decomposition for a long time period. This type of soil has pale to dark brown or black in color, depends on its decomposition and time it was deposited. It can be classified into 3 types, which are, fibric (least decomposed), hemic (moderately decomposed) and sapric (highly decomposed) [9]. The classification of peat can be based on Von Post Classification system and the fiber content of the peat. Peat can be found all over the world, including Malaysia.

Malaysia Wetlands International [21], has reported that peat have covered 2,457,730 hectares from the total land surface in Malaysia. The largest area covered by peat was contributed by Sarawak (1,697,918 ha; 69.08% from the total), followed by Peninsular Malaysia (642,918 ha; 26.16%) and Sabah (116,965 ha; 4.76%). In Peninsular Malaysia, peat is located in most states and can be found in Johor, Selangor, Pahang, Perak, Terengganu, and Kelantan. Leete [11] has stated about 228,960 ha of peat were located in Johor and the selected location of this study were in West Johor.

In engineering purpose, peat soils are unsuitable for supporting or resist loads in its natural states [6]. Peat soils unable to perform well for most engineering structures and hence, pose serious settlement problems. However, development of peaty soil is inevitable since Malaysia has a deficiency on land surface to cater the population growth and rapid development moving towards a developed country. Whitlow [22]; Razali, Bakar and Zainorabidin [18], have highlighted that peaty type of soil has high compressibility with initial void ratio in the range of 5 – 15, low bearing capacity (5 – 20 kPa), high moisture content, which may up to 1000% depends on it compositions, high organic content (more than 75%) and high fiber content which leads to serious settlement issues. The soil experiences an excessive settlement for a short period and continues to settle for a long period of time. Based on a few research programs, compressibility of peat greatly affected from the decomposition of peat. The lower the decomposition, the higher the compressibility as it consists of fibrous organic matter like leaves and plant stems that have high porosity, which leads to high initial void ratio, high moisture and fiber content which was reported by Santaga, Bobet, Johnston and Hwang [19]. Besides those factors, the compressibility of peat also influenced by the size, shape, fibric and the arrangement of the constituent fibers, initial permeability and inter-particle of chemical bonding in the soil [13, 15, 23].

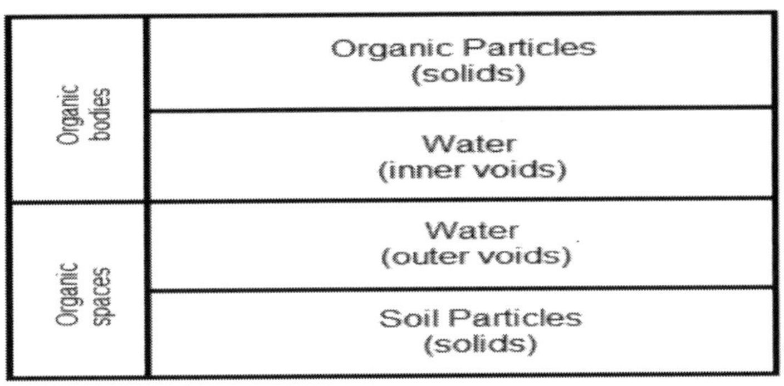

Figure 1. Physical peat model [23].

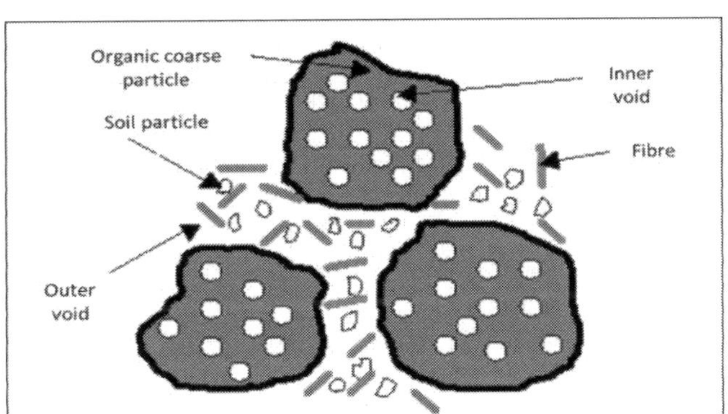

Figure 2. Schematic diagram of peat arrangement [23].

By the determination of fiber content in the peat, the type of peat could be known. Based on the ASTM D1997 [2], the determination of peat types is through the percentage of fiber, retained on 150μm sieve apertures over the total weight. Peat soil with a fiber content of more than 67% was classified as fibric peat, where less than 33% was sapric peat and 33 – 67% was hemic peat. Boelter [3] reported that fiber content gives a high impact on the difference of physical properties of peat soil especially in compressibility. The higher fiber content and low bulk density in decomposed peat contain many large pores which are yielding as much as 80% of their saturated water content for drainage and water movement. Hence, it increases the settlement of peat by itself. Fibrous peat typically made of the weakly decomposed plant remains which are not destroyed by rubbing, which is making it extremely acidic and has high fiber content [1]. It also has been reported by Kogure et al. [12] that the settlement on peat is higher compared to other soils such as clay and sand because of the inner and the outer void spaces in peat. Edil and Wang [14] added, by increasing the fiber content in the soil, it has influenced in increases the values of moisture content and void ratio.

By the presence of the fiber, it affected the structural arrangement of peat. Kogure et. al. [12] has developed and introduced to the physical model of peat as in figure 1, which peat has divided into two major components comprising organic bodies and organic spaces. The organic bodies consist of organic particles as solids and the inner voids filled with water, whereas, the organic spaces, including the soil particles as solids and the outer voids filled with water. Wong et. al [23] was drew a schematic diagram in figure 2, based from the findings in figure 1. The schematic diagram gives a clear explanation about the peat arrangement which consists of fibers and organic particles. In the organic particles, Hobbs [15,16] stated that, there had three main constituents, derived from plant stems and roots which are coarse fibers with particles sizes greater than 1 mm in diameter, finer fibers with sizes smaller than 1 mm in diameter and amorphous matter (completely decayed). At initial states, peat can hold a considerable amount of water due to the hollow and spongy of the particles and results high of moisture content and initial void ratio. Hence, depending by the fibers, size and particle arrangement, compressibility and the consolidation of soil could be predicted as well as the aim of this paper, to study the effects of peat fibers in compressibility by conducting 1- D oedometer test on the undisturbed sample and reconstituted samples with two different particle sizes.

2. Experimental Program

2.1. Sampling of Peat

The type of peat used in this study was a hemic peat or shall be known as pseudo-fibrous peat, collected from Parit Nipah Darat, Johor. In order to obtain the undisturbed and disturbed samples, the soil was excavated between depths of 0.3 to 1.2m below ground surface. The tube samplers of 100mm in diameter x 200mm in height with having a sharp edge at the lower end were used to collect the undisturbed peat samples. The tubes were immediately waxed after sampling on both ends, and sealed using layers of plastic film. This is to preserve the moisture content of samples while the disturbed sample was randomly collected at the same depth and placed in plastic bags. Then the peat samples were placed in a plastic container, covered with lid and kept in a laboratory at a constant temperature. Sampling of peat have been done quickly to avoid collecting more water due to the ground water level of peat was high, that is at the depth 0.3m from the ground surface. The Von Post classification test was also done at the site on fresh peat. About one-third of the soil was passed the fingers and the color of water released was very muddy dark brown. Hence, the peat sample taken was classified as H5 (moderately decomposed).

2.2. Sample Preparation

Types of samples used in this study were undisturbed sample (as original sample) and reconstituted samples. These samples were used to determine the consolidation parameter by conducting 1- D oedometer consolidation test. However, the disturbed samples were used to determine the index properties of the soil.

The reconstituted samples were prepared by the author in the Research Centre of Soft Soils, Malaysia (RECESS) with a specific procedure. The preparations of the sample were done by several steps: 1) Sieving; 2) Reconstituting; and 3) extrusion. At the sieving step, the disturbed sample was sieved using 0.300mm and 3.35mm sieve apertures with the aid of some water to remove the larger particles of organic matter and produce the slurry sample. The sizes selected were part of research conducted by the authors and the largest and smallest sizes have been chosen for this study. Then, the slurry sample of peat has been collected and placed in the reconstitution mould. For the preloading stage, the slurry peat of both sizes of peat particles was subjected to pressure of 50 kPa through deadweights as a preloading by using large strain consolidation equipment and maintained until the flow of excess pore water pressure stopped. After the reconstitution process, the reconstituted sample was extruded and trimmed into a consolidation ring with 75 mm in diameter x 20 mm height, for the 1-D oedometer test and the residue of the sample has been collected and tested for index properties as comparison purposes.

2.3. Index Properties of Original Peat and Reconstituted Peat

The properties of the natural peat sample and reconstituted peat of Parit Nipah Darat were as shown in the Table 1. The moisture content, *w* for peat samples was determined by using oven, drying over 24 hours at 105±5°C, which to remove the moisture in peat [17]. The moisture content for natural states of peat was 640.09%, while for the reconstituted samples for peat particles passing sieve 0.300mm (R0.300) and 3.35mm (R3.35) were 389.48% and 401.20% respectively. The values for the liquid limit (LL) determination were using 80g of 30° fall cone penetration apparatus [2]. As allowed and stated in the standards, the coarse fibers and organic particles in the natural sample were removed to determine the LL values. The specific gravity of natural samples (1.359) was higher than reconstituted samples with 1.281 (R0.300) and 1.334 (R3.35). The determination of specific gravity was conducted using small pycnometer method. The organic content was determined by the loss of ignition method, using muffle furnace (burned at 440±5°C). Fibers content determination was done by separation of fibrous materials from the soil specimen using 150μm sieve. The entire tests were conducted to all specimens to determine the differences between them and the results are as discussed in discussion part.

2.4. 1-D Oedometer Consolidation Tests

All the specimens of the peat samples were tested using a 1-D oedometer consolidation test with the purpose of the evaluation of the effect of fibers in different peat particle sizes and in natural peat in terms of compressibility. The sample specimens were prepared by specific procedure as mentioned in sample preparation and the method used in this testing program was incremental loading method with a stress range of 5 – 320 kPa. For each loading stage, the loading was maintained for 24hrs and the data were recorded as prescribed in ASTM D2435 [2] standards. The specimens were drained from the top and bottom. The tests were conducted in the same laboratory, Research Centre of Soft Soils, Malaysia (RECESS), UTHM.

Table 1. Properties of Peat

Peat Properties	Parit Nipah Darat Peat	Reconstituted Samples Passing 3.35mm (R3.35)	Reconstituted Samples Passing 0.300mm (R0.300)
Moisture Content, *w* (%)	640.09	401.20	389.48
Liquid Limit, LL (%)	322	325	305
Specific Gravity	1.36	1.33	1.28
Organic Content (%)	83.01	51.43	53.20
Fiber Content (%)	61.42	37.82	11.51
Initial Void Ratio, e_o	8.36	5.74	5.39
Peat density (kg/m^3)	1075	1010	1041

3. Experimental Results and Analysis

The experimental results obtained that have been highlighted were based on the determination of consolidation parameters of preconsolidation pressures, p_c, compression index, Cc, swelling index, Cs, coefficient of consolidation, c_v, coefficient of volume compressibility, m_v, and compression ratio, Cr, which collected from 1-D oedometer consolidation testing. In this study, the data obtained have been analyzed by using the Taylor's curve fitting method (square root time method) as the method was the most suitable and gives a better curve fitting for the determination of c_v values and have been reported by Wong et. al [24] and Hassan et. al [7].

3.1. Consolidation Parameters of undisturbed and Reconstituted Samples

The p_c value for undisturbed (UD) peat was 13 kPa and for reconstituted samples (R0.300 and R3.35) were 23 kPa and 26 kPa respectively. Based on figure 3, the Cc and Cs values for all the specimens were evaluated from the slopes of the graph. The Cc values were evaluated from the slopes of loading, curves where Cs values were evaluated by the slopes of unloading curves. Table 2 presents the results for Cc were 2.68 (UD), 1.88 (R0.300) and 2.40 (R3.35) while for Cs values were 0.17 (UD), 0.18 (R0.300) and 0.17 (R3.35). The Cc values were decreasing as the sizes of soil particles decreases. The undisturbed samples might have random of soil particles sizes since the peat sample was classified in moderately decomposed. The fibrous materials are contained in the sample has influenced the Cc values. As mentioned before, the fibrous materials have hollow structure and can hold a considerable amount of water. Hence, it will greatly affect in Cc values as shown between samples R3.35 and R0.300. However, for Cs values, the values show slightly different between the samples. The Cs values increase when the size of soil particles decreases. The fibrous materials, in sample R3.35 were unable to absorb the large amount of water, compared to the sample R0.300 which has less fibrous matter.

Compression ratio, Cr was calculated from $Cc/1+e_0$, that used to classify the compressibility of a soil [5]. As reported by O'Loughlin and Lehane [5], classification of peat soil were divided by four classes; very slightly compressible (Cr = 0 – 0.05), slightly compressible (Cr = 0.05 – 0.10), moderately compressible (Cr = 0.10 – 0.20) and very compressible (Cr = > 0.20). By referring to Table 2, the Cr values for three types of specimens were larger than 0.20 which indicates the peat specimens from Pt. Nipah Darat were very compressible.

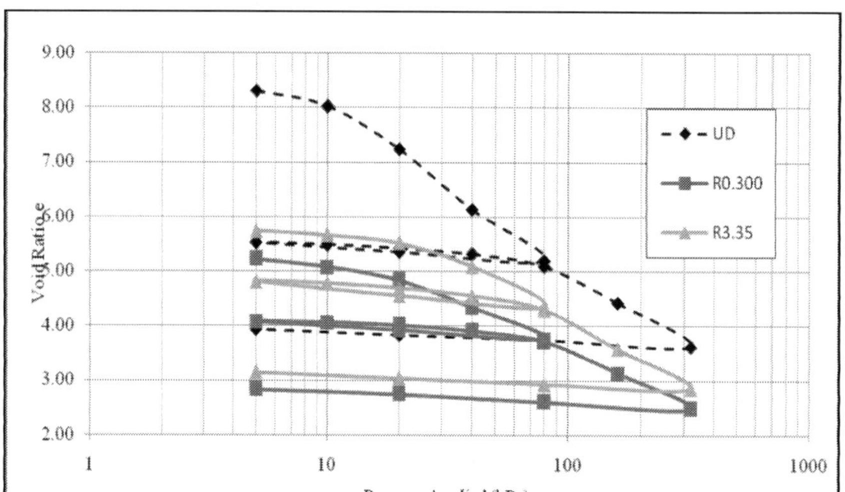

Figure 3. Graph of e against log σ for undisturbed peat and reconstituted peat samples.

Table 2. Consolidation parameters for peat specimens.

Parameters	Undisturbed Peat Sample (UD)	Reconstituted Samples Passing 3.35mm (R3.35)	Reconstituted Samples Passing 0.300mm (R0.300)
Initial Void Ratio, e_o	8.36	5.74	5.39
Preconsolidation Pressure, p_c, kPa	13	26	23
Compression Index, Cc	2.68	2.40	1.88
Swelling Index, Cs	0.17	0.17	0.18
Compression Ratio, Cr	0.29	0.36	0.29
Coefficient of consolidation, c_v (m^2/yr)	10.305 – 0.482	17.379 – 0.696	5.723 – 0.782
Coefficient of Volume Compressibility, m_v (m^2/MN)	16.594 – 0.887	3.512 – 0.361	5.589 – 0.799

The coefficient of consolidation, c_v decreases with increase of stress as shown in Table 2. For undisturbed sample, the c_v values were in the ranges of 10.305 to 0.482 m^2/yr while for R0.300 was 5.723 to 0.782 m^2/yr and for R3.35 was 17.379 to 0.696 m^2/yr. The c_v values for sample R3.35 were larger than sample R0.300. The sample of R3.35 has contain larger soil particle size than R0.300 sample, which means that the R3.35 sample has high initial void ratio and high in moisture content. Hence, the dissipation of water out from consolidation process which results the c_v value was high compared to R0.300 sample. On the other side, the coefficient of volume compressibility, m_v for all types of specimens were also decreases as the consolidation pressures of oedometer increase. The m_v values were depends on the void ratio and stress pressure and decreases when stress increases. The m_v values for undisturbed sample clearly larger than reconstituted sample, which were decreased from 16.594 to 0.887 where the specimen has a high void ratio compared to the other specimens meanwhile, the m_v values for reconstituted peat of R3.35 were 3.512 to 0.361 and R0.300 were 5.589 to 0.799.

3.2. Fiber Content, Water Content, and Consolidation Parameters (e_o, Cc, and Cs)

Table 3 presented the fiber content and consolidation parameters for peat specimens. The fiber content in UD sample was high with 61.42%, compared to reconstituted peat samples with 11.51% for R0.300 and 37.82% for R3.35. Based on the peat classification, the peat samples collected from Parit Nipah Darat was in a class of hemic with moderately decomposed peat since the fiber content in between 33% to 67%. The fiber content seems to decrease when the sizes of peat particles decrease as shown in reconstituted samples, R3.35 and R0.300 with 37.82% and 11.51%. The amount of fibers in specimens affected the moisture content and initial void ratio [14]. The fiber content of the sample will increase the moisture content and voids as shown in the Table 3. The undisturbed peat has 640.08% of water content and 8.36 of the initial void ratio, whereas for specimens of R0.300 was 389.48% and 5.39 and for R3.35 was 401.20% and 5.74.

As has been mentioned earlier, compressibility of peat has been affected by fibrous particles, mainly come from the fibers and organic particles. Figure 4 shows the illustration of hollow structures of fibers in peat by scanning electron micro-photographs. The photographs show, the peat sample have large pores where makes the sample spongy and can hold a considerable amount of moisture content in their structures which may up to 1000% and resulted the high of an initial void ratio. The higher the amount of fiber contents, the higher the moisture content and the void ratio. Hence, in compressibility, it will increase the compression index where it was one of the important indicators in prediction of soil settlement. These are proven by consolidation parameter of compression index, Cc. The Cc value of UD specimens was higher than other specimens with 2.682, followed by R3.35 and R0.300 with 2.400 and 1.882 respectively. The values were decreasing as the size of soil particles and the fiber content decrease. The fibers that reduced by sieving passing selected sizes of sieve apertures and clearly proved that samples with more fibrous materials have influenced in compression index, initial void ratio and water content. However, in Cs value determination, the values were not too affected by the fibers. The values between the samples were only show slightly different between the samples. The Cs values increase by decreasing the size of soil particles. The fibers in the fibrous specimen were unable to reabsorb the water that has been dissipated out by the consolidation process compared to the sample with less fibrous matter. This was shown that the peat sample with high fibrous material will experience less swelling after preloading phase compared to peat sample which has completely decomposed (high degree of humification) or known as amorphous peat.

Table 3. Fiber content and consolidation parameters for peat specimens.

Peat Type	Undisturbed Peat Sample (UD)	Reconstituted Samples Passing 3.35mm (R3.35)	Reconstituted Samples Passing 0.300mm (R0.300)
Fiber content, %	61.42	37.82	11.51
Moisture Content,%	640.08	401.20	389.48
Initial Void Ratio, e_o	8.36	5.74	5.39
Preconsolidation Pressure, p_c, kPa	13	26	23
Compression Index, Cc	2.68	2.40	1.88
Swelling Index, Cs	0.17	0.17	0.18

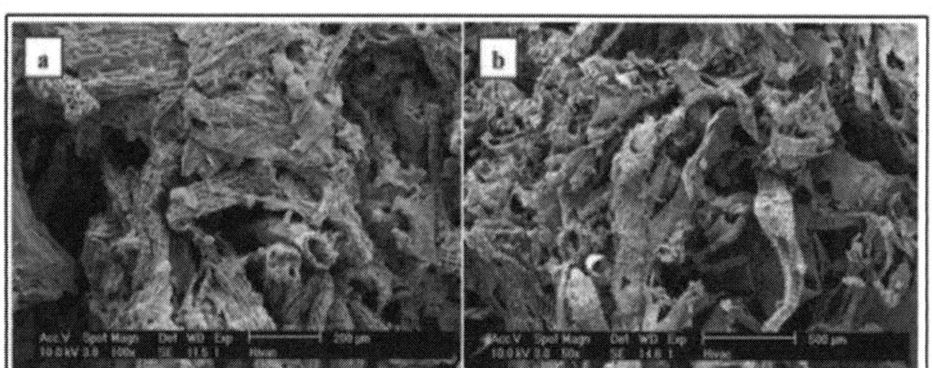

Figure 4. (a) and (b) Scanning Electron Microphotograph of peat specimen [13].

4. Conclusion

As a conclusion, fiber content in peat strongly affecting the compressibility of the soil especially peaty type of soil. The higher the fiber content in fibrous peat was increasing the water content, initial void ratio and its compression index. The high fiber content in undisturbed sample cause the high of an initial void ratio, moisture content and hence causing high compression index, Cc which indicates the high of settlement. The fiber content was distinguished by reconstituting samples with different sizes of soil particles passing two selected sieves (R3.35 and R0.300) and this study has proved that the different peat particle sizes have different values of fiber content. The higher the fiber content contains in the soil, the higher the soil compressibility and hence, the higher the settlement and these results were useful in gaining the information on predicting the settlement of peaty soil.

Acknowledgement
The authors wish to give special thanks and acknowledge to the research grant vote no. 1413 for research funding from ORICC UTHM. Many thanks to research grant vote no. U260 for giving funds in paper submission and lastly to the Research Centre for Soft Soils (RECESS) UTHM, Geotechnical Laboratory FKAAS, and staffs UTHM for providing the equipment and giving support during carrying out this research.

References
[1] Ali D, Kamarudin A and Nazri A 2013 Influence of Natural Fillers on Shear Strength of Cement Treated Peat, Gradevinar, 633-640.
[2] ASTM Standards, Construction, Soil and Rock (I): D420-D5611 2007 West Conshohocken, PA, ASTM Standards, **04.08**.
[3] Boelter D H. 1968 Important Physical Properties of Peat Materials, Proceeding of the Third International Peat Congress, Quebec, Canada.
[4] Duraisamy Y, Huat B B and Aziz A A 2007 Engineering Properties and Compressibility Behavior of Tropical Peat Soil, American Journal of Applied Sciences, **4(10)** 768-773.
[5] Duraisamy Y, Huat B B, Muniandy R and Aziz A A 2008 Compressibility behavior of fibrous peat reinforced with cements columns, International conference on construction and building technology (ICCBT), 93-110.
[6] Gofar N and Sutejo Y 2007 Long Term Compression Behavior of Fibrous Peat, Malaysian Journal of Civil Engineering, **19 (2)** 104-116.
[7] Hassan Z A, Daud M N M, Yahya M A and Shaharuddin M S, 2013 Effect of Granular Material on The Rate of Consolidation of Peat Soil, Electric Journal Geotechnical Engineering **18**, Bund. Z.
[8] Hashim R and Islam S 2008 Engineering Properties of Peat Soils in Peninsular, Malaysia, Journal Applied Science.
[9] Huat B B 2004 Organic and Peat Soils Engineering, Serdang, Selangor: Ampang Press Sdn. Bhd.
[10] Huat B B, Prasad A, Asadi A and Kazemian S 2014 Geotechnics of Organic Soils and Peat, London: CRC Press.
[11] Kazemian S, Huat B B, Prasad A and Barghchi M 2011 A State of Art Review of Peat: Geotechnical Engineering Perspective, International Journal of the Physical Sciences, **6(8)** 1974-1981.
[12] Kogure K, Yamaguchi H, Shogari T 2003 Physical and Pore Properties of Fibrous Peat Deposit, Proceeding of the 11th Southeast Asian, Geotechnical Conference, Singapore.
[13] Mesri G and Ajlouni M 2007 Engineering Properties of Fibrous Peat, Journal of Geotechnical and Geoenvironmental Engineering.
[14] O'Kelly C B and Zhang L 2013 Consolidated-Drained Triaxial Compression Testing of Peat, Geotechnical Testing Journal, **36(3)**.
[15] O'Kelly B C and Pichan S P 2013 Effects of Decomposition on The Compressibility of Fibrous Peat – A Review, Geomechanics and Geoengineering: An International Journal
[16] Pichan S P and O'Kelly B C 2012 Effect of Decomposition on The Compressibility of Fibrous Peat, ASCE GeoCongress 2012: State of the Art and Practice in Geotechnical Engineering, Oakland, California, USA, 4329-4338.
[17] Rahman J A and Chan C M 2013 Influence of Temperature on the Mass Losses of Tropical Peat at Different Decomposition Level, Soft Soil Engineering international Conference (SEIC 2013).

[18] Razali S N M, Bakar I and Zainorabidin A 2013 Behaviour of Peat Soil in Instrumented Physical Model Studies, Procedia Engineering, **56** 145 – 155.

[19] Santaga M, Bobet A, Johnston C T and Hwang J 2008 One-Dimensional Compression Behavior of a Soil with High Organic Matter Content, Journal of Geotechnical and Environmental Engineering, **134:1(1)**.

[20] Sridharan A and Nagaraj H B 2012 Coefficient of Consolidation and Its Correlation with Index Properties of Remolded Soils, Geotechnical Testing Journal, **27(5).**

[21] Wetlands International, A Quick Scan of Peat lands in Malaysia, Wetlands international-Malaysia: Petaling Jaya, Malaysia,

[22] Whitlow, R. Basic Soil Mechanics (Fourth Ed.), 2001 Harlow, England: Pearson Education Ltd.

[23] Wong L S, Hashim R and Ali F H 2009 A Review on Hydraulic Conductivity and Compressibility of Peat, Journal of Applied Sciences, **9(18)** 3207 – 3218.

[24] Wong L S, Hashim R and Ali F H 2008 Compression Rates of Untreated and Stabilized Peat Soils, Electric Journal Geotechnical Engineering, **13** Bund. F.

[25] Zainorabidin A and Wijeyesekera D C 2007 Geotechnical Challenges with Malaysian Peat, Proceedings of AC & T 252-261.

Vibration Response of Multi Storey Building Using Finite Element Modelling

T N T Chik[1], M F Zakaria[1], M A Remali[1] and N A Yusoff[1,2]
[1,] Faculty of Civil and Environmental Engineering, Universiti Tun Hussein Onn Malaysia, 86400 Batu Pahat Johor, MALAYSIA
[2] Research Center for Soft Soil, Universiti Tun Hussein Onn Malaysia, 86400 Batu Pahat Johor, MALAYSIA

E-mail: thayati@uthm.edu.my

Abstract. Interaction between building, type of foundation and the geotechnical parameter of ground may trigger a significant effect on the building. In general, stiffer foundations resulted in higher natural frequencies of the building-soil system and higher input frequencies are often associated with other ground. Usually, vibrations transmitted to the buildings by ground borne are often noticeable and can be felt. It might affect the building and become worse if the vibration level is not controlled. UTHM building is prone to the ground borne vibration due to closed distance from the main road, and the construction activities adjacent to the buildings. This paper investigates the natural frequency and vibration mode of multi storey office building with the presence of foundation system and comparison between both systems. Finite element modelling (FEM) package software of LUSAS is used to perform the vibration analysis of the building. The building is modelled based on the original plan with the foundation system on the structure model. The FEM results indicated that the structure which modelled with rigid base have high natural frequency compare to the structure with foundation system. These maybe due to soil structure interaction and also the damping of the system which related to the amount of energy dissipated through the foundation soil. Thus, this paper suggested that modelling with soil is necessary to demonstrate the soil influence towards vibration response to the structure.

Keywords: Natural frequency, finite element modelling, vibration.

1. Introduction

Excessive floor vibrations have become a greater problem for the occupants. These issues had become one of the most significant challenges produces by the traffic and occupants. Structure-borne sound and vibrations from traffic in dense urban environments can cause annoyance to the inhabitants of surrounding buildings besides being disruptive to the operation of manufacturing facilities, medical facilities, and research laboratories. Individuals can detect building vibration values that are well below those that can cause any risk of damage to the building or its contents. In addition, soil structure interaction is certainly a typical subject but nevertheless still new in it applicative aspects. The interaction may leads to an increment of the fundamental period of the structure compared to the fixed-base solution. In this study, the 3D finite element model in LUSAS was developed to perform the vibration analysis on the interaction of structure and foundation.

2. Reviews on Ground Vibration Theory

Some common sources of ground-borne vibration are trains, buses on rough roads, and construction activities such as blasting, pile-driving and operating heavy earth-moving equipment [1]. Ground borne vibration can be a serious concern for nearby neighbour of a transit system route or maintenance facility, causing buildings to shake and rumbling sounds to be heard. It is unusual for vibration from sources such as buses and trucks to be perceptible, even in locations close to major roads. The vibration being produced by the traffic propagated through the foundation of the floors and walls of a building will create resonance frequencies to the building. Hence, it will produce perceptible vibration, rattling of items such as windows or dishes on shelves, or a rumble noise. This will create the room surfaces act like a giant loudspeaker causing so called ground-borne noise. It will make the occupants in the building felt uncomfortable with the sounds in the room.

Ground borne vibration is a creation of man-made over decade ago due to the developments of road, highway and rail networks. Vehicles move on the road and the train on the railways will produce vibration to the ground. The vibration will propagate through the ground and into buildings, resulting in uncomfortable vibration and unacceptable levels of noise in the building. This is clearly investigated at the Wellington Hospital in London [2]. The building lies directly above two mainline railway tracks and within only a few metres of four underground railway tunnels and the main road. The problem related to this matters are the multiples sources of vibration produced by the train and cars propagates along the surfaces of the ground, and influenced the internal noise and vibration.

3. Soil Structure Interaction (SSI)

High vibration and unacceptable dynamic settlements could disturb sensitive devices and people and even be the cause of structural damage. The structural response to ground excitation depends on the soil response to waves propagated from the source, soil and structure [3]. Soil structure interaction is certainly a typical subject but nevertheless still new in it applicative aspects. Further studies are necessary for converting the research highlights in practical tools to be used in routine structural analysis [4]. In general, building-soil interaction consists of two parts; kinematic and dynamic (or inertial) interaction. In the inertial interaction, there are two types of dynamic analysis of soil structure interaction can be done [5]. The analyses are direct method which the soil and structure are modelled together in a single step accounting for both inertial as illustrated in Figure 1(a) and kinematic interaction. Inertial interaction develops in structure due to own vibrations give rise to base shear and base moment, which in turn cause displacements of the foundation relative to free field. Kinematic interaction develops due to presence of stiff foundation elements on or in soil cause foundation motion to deviate from free field motions as shown in Figure 1(b).

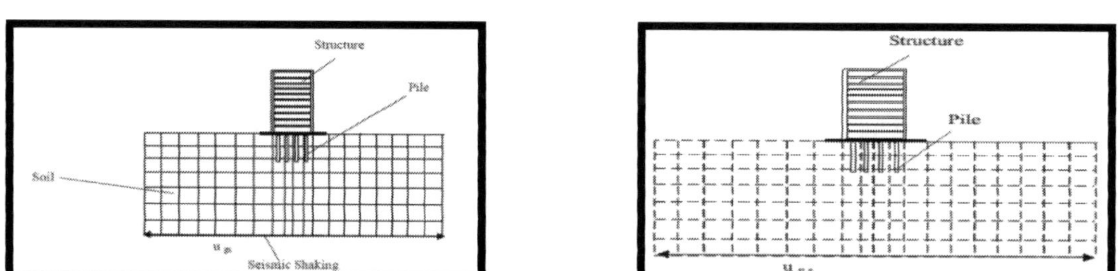

(a) (b)

Figure 1. (a) Direct method of soil structure interaction, (b) Substructure method for inertial interaction analysis [5]

According to Ceroni et.al. [4], soil structure interaction leads to an increment of the fundamental period of the structure compared to the fixed-base solution. It is also because of the damping of the system due to the amount of energy dissipated through the foundation soil (radiation damping). Cai et.al [6], Chore et.al [7] and Pulikanti and Ramancharla [8] found that three-dimensional (3D) non-linear finite element subsystem were developed to study the seismic soil-pile-structure interaction effects in which the plasticity and work hardening of soil have been considered as shown in Figure 2(a). Chore et.al [7] performed a methodology for the comprehensive analysis of building frame supported by pile groups embedded in soft marine clay using the 3D finite element method too as illustrated in Figure 2(b). In this study, the 3D finite element in LUSAS will be developed to perform the vibration analysis on the interaction of structure and foundation.

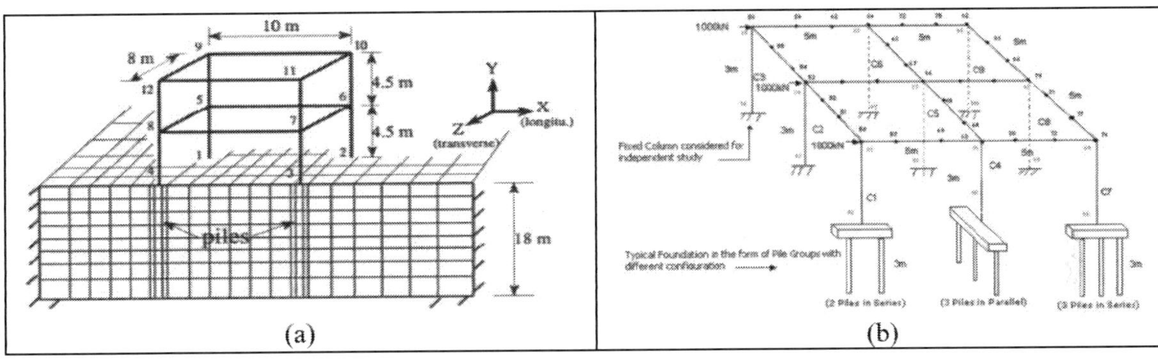

Figure 2. (a) Example of structure and pile foundation developed using FEM [6], (b) Typical building frame supported by group of piles [7].

4. Application of Finite Element Modelling by LUSAS software

Finite element method is a numerical method to find out an approximate solution variable in a problem which is difficult to obtain analytically [9][10]. The high-speed digital computer nowadays helps a lot in running and analyzing complex problems in continuum mechanics. A study on finite element method enable users to understand the synthesis of analysis, material behaviour, and constructability, environmental, social and economic realities that influence solutions in the practice of engineering.

LUSAS is finite element analysis software that can analyse all types of complicated models using linear or nonlinear analysis. It is an associative feature based Modeller. This means the model geometry is entered in terms of features which are then sub-divided into finite elements in order to perform the analysis. Increasing the number of elements usually increases the accuracy of the analysis but the time for the analysis to be done will also increased. In LUSAS, every element of concrete and soil must be assigned with material constant to define structure properties in term of mass density, poison ratio and modulus of elasticity. Details step by step in LUSAS are applied in this study to obtain the required natural frequency and mode shape and listed as follows [11]:

- Creating a new model
- Defining the geometry and groups
- Defining the mesh-reinforcement bars and mesh-concrete
- Defining the geometric properties and material properties
- Assigning attributes to the bars and concrete
- Assigning supports and loading
- Set the eigenvalue control and running the analysis
- Viewing the results of reformed shape and creating a load versus displacement

- Principal stress contour plots
- Viewing crack patterns and animating the results

5. Case Study: Multi Storey an Office Building

Registrar Office is a five storey building located in Universiti Tun Hussein Onn Malaysia (UTHM) campus. This office managed in the human resources and facilities for the staff [12]. UTHM campus area is surrounded by soft soil type which is low in shear strength, bearing capacity and suffers large settlements when subjected to loading. Referring to the index properties of the soil, it can be categorized as CH (Inorganic Clays of High Plasticity) according to Unified Soil Classification System [13]. The soft clay layer in the UTHM campus extends to a depth of 40m. The soft clay has high moisture content, low permeability, high compressibility, shrinks when dried and expands when wetted. Problem arises when structures are built on these soils, where under large superimposed load, significant settlements can occur if the soil is not being improved. A details structure and soil property for this model is listed in Table 1.

Table 1. Structure and soil properties used in the finite element modelling.

Type	Structure with rigid base	Structure with soil considered
	Size	Size
Beam size	200x600 mm	200x600 mm
Column size	400x400 mm	400x400 mm
Slab thickness	150 mm	150 mm
Young's Modulus	28 N/mm^2	25 N/mm^2
Poison Ratio (concrete)	0.2	0.2
Mass density (concrete)	2500 kg/m^3	2500 kg/m^3
Poison Ratio (soft soil)	-	0.3
Mass density (soft soil)	-	1.75 x 10^3 kg/m^3

The building has an estimated size about 54m long x 32.5m wide x 20m high. Figure 3 shows the different view of this office, which is the side view of office building, the view of the building with rigid base and also the view of the building considering the soil underneath the building. LUSAS software is used to model the existing structure using the original plan and perform the modal analysis to obtain the natural frequency and the mode shape.

(a) (b) (c)

Figure 3. (a) Office building with rigid base in LUSAS, (b) Side view of multi storey Registrar office building, (c) Office building considering the soil underneath the building in LUSAS [14].

6. Natural Frequency Analysis

The end results of the modal analysis by LUSAS were produced tenth modes with the natural frequencies and maximum deflection of each mode. The first mode of the structure with rigid base obtained the natural frequency of 1.203 Hz and the maximum deflection is 0.047mm as shown in the Figure 4(a), while the first mode of structure considering the soil underneath the building has the natural frequency of 0.7850×10^{-5} Hz and the maximum deflection is 0.04235 mm as shown in Figure 4(b). The other values of natural frequency and maximum deflection from mode 1 until mode 10 are shown in the Table 2. It can be seen that the natural frequencies are decreased when the structure is modelled considering the soil underneath the building. It shows that the influence of interaction between the structure and the foundation system produce the stiffer structure due to the mass is also increased. This is also due the combined effect of translational and rotational simulating soil deformability as stated by Ceroni [4]. The effect cannot be neglected; however it depends on the characteristics of the structure itself including stiffness, height and also the foundation of soil properties.

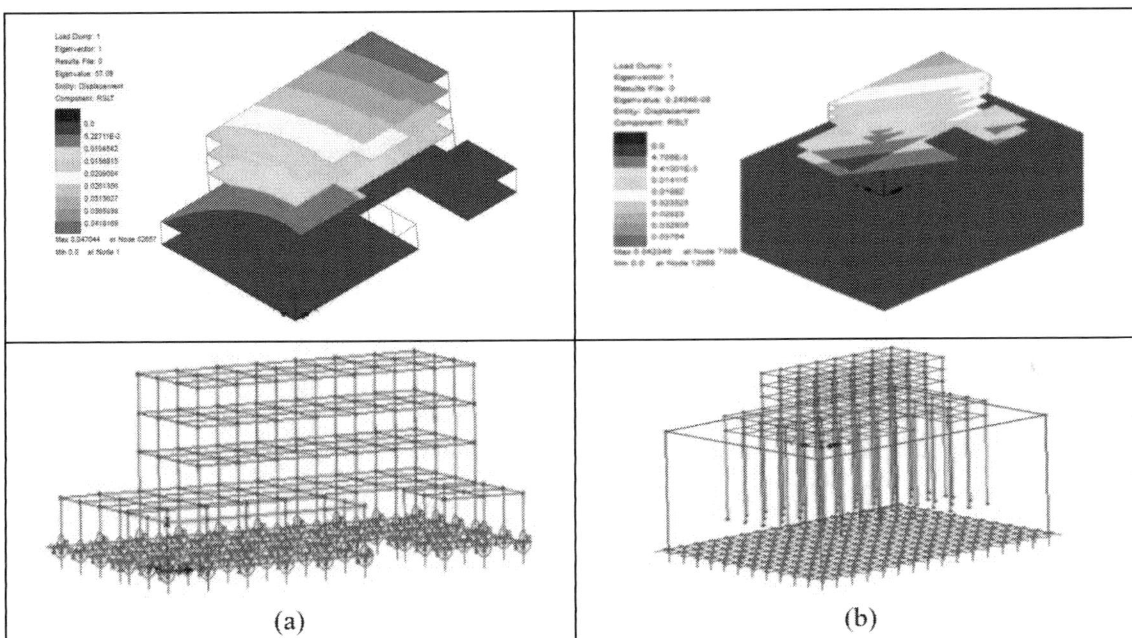

Figure 4. (a) Mode shape of the structure with rigid base, (b) Mode shape of the structure with considering the soil underneath the building.

IOP Conf. Series: Materials Science and Engineering **136** (2016) 012037 doi:10.1088/1757-899X/136/1/012037

Table 2. The first tenth mode for the system.

Mode	Structure with rigid base		Structure with soil considered	
	Frequency (Hz)	Max. Displacement (mm)	Frequency (Hz)	Max. Displacement (mm)
1	1.203	0.0470	$0.7850x10^{-5}$	0.04235
2	1.283	0.0324	$0.9188x10^{-5}$	0.06152
3	1.583	0.0652	$0.9627x10^{-5}$	0.04640
4	3.297	0.1235	$0.9783x10^{-5}$	0.04145
5	3.324	0.1242	$0.9971x10^{-5}$	0.04863
6	3.605	0.0303	$0.1118x10^{-4}$	0.04068
7	3.623	0.0306	$0.1128x10^{-1}$	0.01071
8	3.938	0.0596	$0.1208x10^{-1}$	0.0001065
9	4.004	0.2816	$0.1281x10^{-1}$	0.0001611
10	4.269	0.4517	$0.2232x10^{-1}$	0.0001751

7. Mode Shape Analysis

Ten Eigen value are analysed to get the different types of mode shapes, natural frequencies and maximum deflection. Mode shape analysis is used to determine the condition of the floor structure of the building when subjected to different frequency modes. The red area represented the highest peak displacement while the minimum displacement is represented by the blue area. Green areas were experiencing a lower displacement and not so obvious deformed. For areas with the maximum deflection, it is not suitable to accommodate the sensitive equipment and might be annoyed the occupants around this area. Figure 5 indicated tenth mode shape pattern of the building with rigid base and also with soil considered. It can be seen that obvious differences on structure deformation occurred for both systems. For lower modes and natural frequencies, the structure considering the soil underneath the building does not affected too much for the deformation of soil component. However, the soil starts to response and deform up to mode 7 and above, while the structure shows no response. The natural frequencies value also increased starts at mode 7 at 0.01128 Hz.

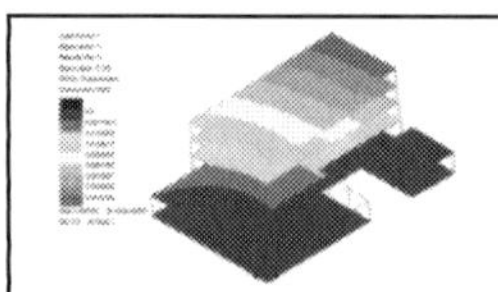	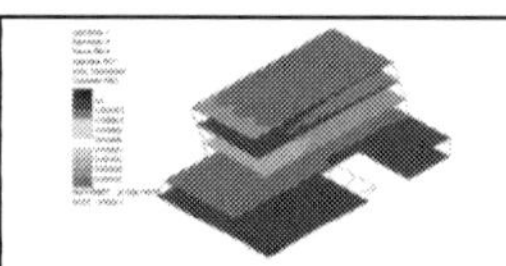	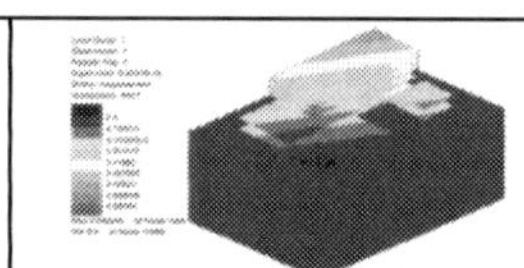	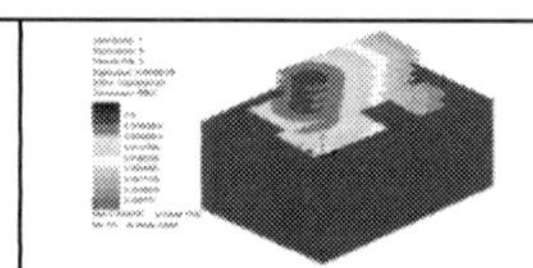
Mode 1: 1.203Hz	Mode 6: 3.605Hz	Mode 1: 0.0000079 Hz	Mode 6: 0.000011 Hz
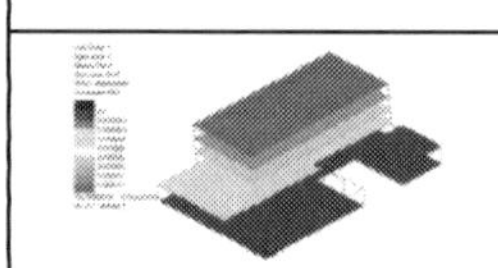	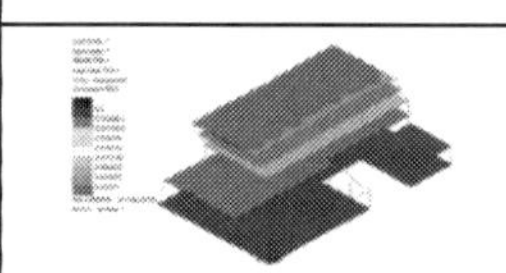	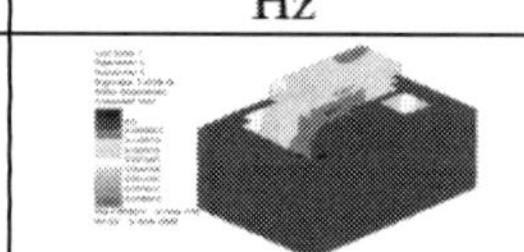	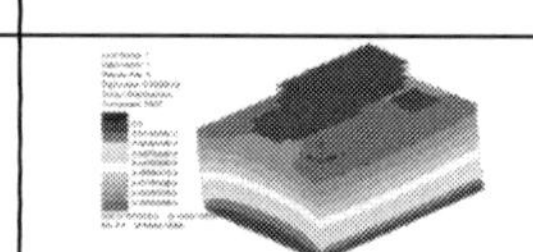
Mode 2: 1.283Hz	Mode 7: 3.623Hz	Mode 2: 0.0000092 Hz	Mode 7: 0.01128 Hz
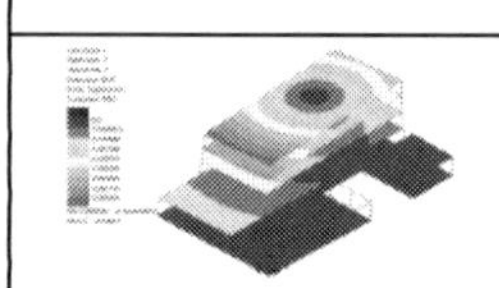	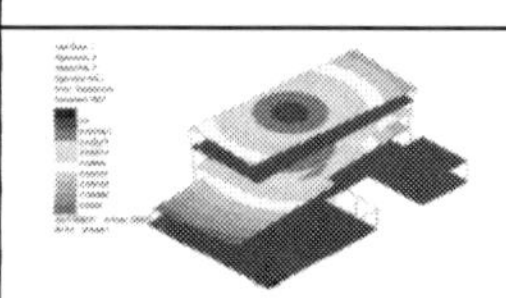	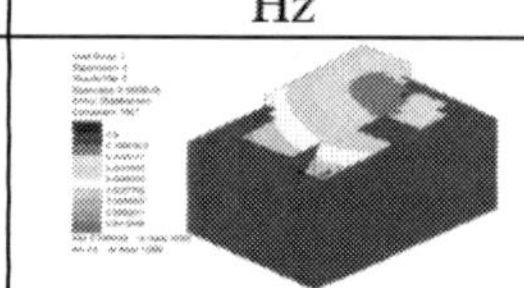	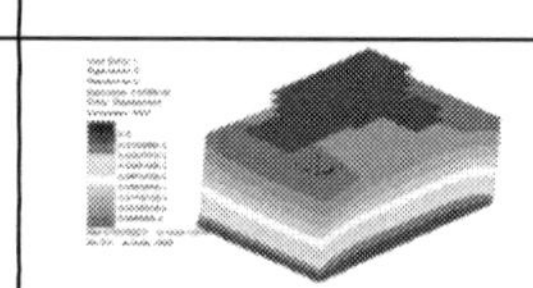
Mode 3: 1.583 Hz	Mode 8: 3.938 Hz	Mode 3: 0.0000096	Mode 8: 0.01208 Hz

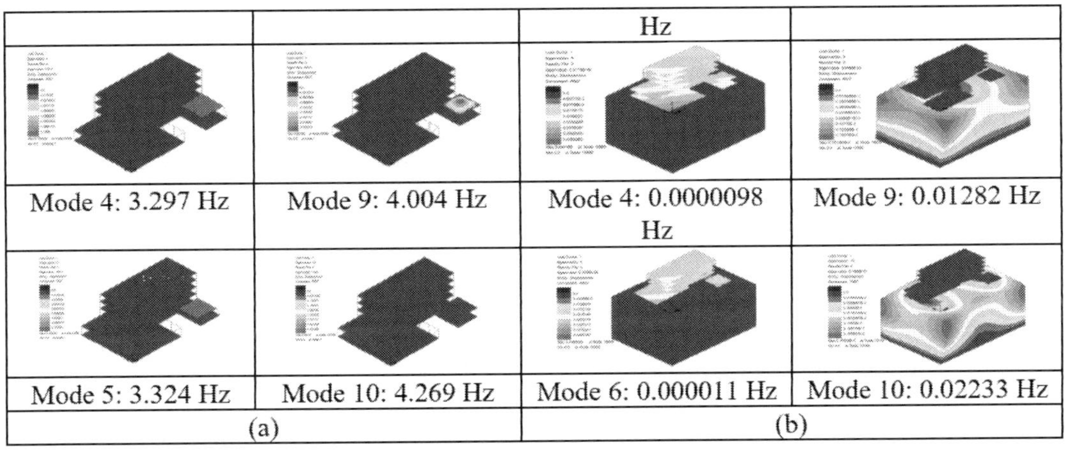

Figure 5. The mode shape pattern up to tenth mode of the structure (a) Structure with rigid base, (b) Structure with soil considered.

8. Conclusions

The main aim of this study is to investigate the vibration performance of multi storey office building considering the soil underneath the building. The comparison of natural frequencies and mode shape pattern also has been done for the structure with rigid base and with soil considered. From the results obtained, it can be finalised that the structure with soil considered are the main factors of the decreased value in natural frequencies upon the structure. The stiffness and mass of the structure is increased after modelling with the soil, hence it is reduced the value of natural frequencies compared to the structure with rigid base. As a structure was modelled and analysed as a one frame structure, it shows that the interaction between structure and foundation greatly influenced the stiffness, mass of the system and also the natural frequencies and mode shapes. Therefore, in conclusion, modelling with soil is necessary to demonstrate the soil influence towards vibration response to the structure.

Acknowledgment

The authors would like to thank to staff at the Development and Property Management Office (PPH) UTHM for supplying the structural drawing of Registrar office building and to all staff those who were involved directly or indirectly in this study. This research is funded by Multi-Disciplinary Research (MDR) research grant Vot 1310 at UniversitiTun Hussein Onn Malaysia (UTHM).

References

[1] Hanson C, Harris, Miller, Miller & Hanson Inc., United States, & Technology Sharing
[2] Program U.S. 1995 Transit noise and vibration impact assessment. Washington, D.C: U.S. Dept. of Transportation, Federal Transit Administration.J. Cohen, Statistical power analysis for the behavioral sciences 2nd ed., 1988.
[3] Grootenhuis P, Structural elastomeric bearings and resilient seatings in Polymers and Polymer Composites in Construction, L. Hollaway (ed.), Thomas Telford. 1990.
[4] Pal B K, Dey K, and Tandia P K 2010 Vibration on structure and soil due to industrial activities: unique case studies and mitigation measure. Journal of Environmental Research and Development **4(3)**: 834-853
[5] Ceroni F, Sica S, Pecce M, and Garofano A 2012 Effect of soil-structure interaction on the dynamic behaviour of masonry and RC building. 15 WCEE LISBOA 2012. Engineering Department, University of Sannio, Benevento, Italy.
[6] Pulikanti S 2013 Effect of soil – foundation – structure interaction on seismic behaviour of pile

supported frame building. Ph.D. Thesis. International Institute of Information Technology, Hyderabad.

[7] Cai Y X, Gould P L and Desai C S 2000 Nonlinear analysis of 3D seismic interaction of soil-pile-structure systems and application. Engineering Structures 22 191-199.

[8] Chore H S, Ingle R K, and Sawant V A 2010 Building frame-pile foundation-soil interaction analysis: a parametric study. Interaction and Multi Scale Mechanics, Vol. 3, No. 1, 55-79.

[9] Pulikanti S, Ramancharla P K 2013 SSI analysis of framed structures supported on pile foundations : a review. Frontier in Geotechnical Engineering (FGE) Volume 2 Issue 2.

[10] Hooi T E, Behaviour and strength study on steel semi rigid connection using LUSAS. Univeriti Teknologi Malaysia. Undergraduate Project.

[11] Tuan Chik T N, Brownjohn J M W and Petkovski M 2010 Finite element analysis of ground borne vibrations in sensitive buildings using inputs from free field measurements, Proceedings of International Conference on Noise and Vibration Engineering (ISMA), Leuven, Belgium, pp. 759-771.

[12] Vahid S Z, Osman S A and Badaruzzaman W H W Hands-on teching in finite element analysis to undergraduate and postgraduate students. Universiti Kebangsaan Malaysia, Selangor, Malaysia. Department of Civil & Structural Engineering.

[13] Pejabat Pembangunan dan Pengurusan Harta Bina UTHM. Online. Information on http:/pph.uthm.edu.com/

[14] Chan C M and Abdullah S H 2008 Settlement behaviour of a cement-stabilised Malaysian clay. Universiti Tun Hussein Onn Malaysia. Malaysia.

[15] Zakaria M F 2015 The vibration performance of registrar office building due to presence of foundation system, Bachelor Degree Project, Universiti Tun Hussein Onn Malaysia.

A Study of Geological Formation on Different Sites in Batu Pahat, Malaysia Based On HVSR Method Using Microtremor Measurement

M A M Noor[1], A Madun[1], A F Kamarudin[1] and M E Daud[1]
[1] Faculty of Civil and Environmental Engineering, Universiti Tun Hussein Onn Malaysia, 86400 Batu Pahat Johor, MALAYSIA

E-mail: alias88.uthm@gmail.com

Abstract. Geological formation is a one of information need to know during site reconnaissance. Conventional method like borehole has been known is very accurate to identify the formation of geology of a site. However, the problem of this technique is very expensive and not economical for large area. In the last decade, microtremor measurement has been introduced as an alternative technique and widely used in the geological formation study. Therefore, the aim in this study is to determine the geological formation underneath of surface in Batu Pahat district using microtremor measurement. There are two parameters have been carried out from microtremor measurement in term of natural frequency and HVSR curves images. Microtremor measurements are done conducted at 15 sites surrounding of Batu Pahat. Horizontal to vertical spectral ratio (HVSR) method was used for analyzing microtermor measurement data, to determine the natural frequency and also HVSR curves image. In this study, values of natural frequencies are used to classify the soil types with range in the between 0.93 to 5.35 Hz, meanwhile the pattern of HVSR curve images has been shown exists a few groups of soil types surrounding Batu Pahat district. Hence, microtremor measurement indirectly can be used as a one technique to add value in the site reconnaissance in the future.

Keywords: Geological formation, natural frequency, microtremor.

1. Introduction

During site reconnaissance, determination of geological formation is important for geologist and engineers [1]. It is very useful in the collecting various information of soil and very required in the foundation design. A direct technique such as borehole is a common technique and was used for a long time. However, this technique is very limited in providing a details information caused by factors of costing and time. As a fulfilling a high requirement to obtain a complete information of geological formation, microtremor measurement has been introduced as alternative technique [2].

Microtremor is a constant vibration of the earth surface and generated by natural sources (wind, rain and ocean waves) and man-made such as traffic and industry activities [3]. Microtremor measurement based on Horizontal to vertical spectral ratio (HVSR) method, which was employed by Nagoshi and Iragashi (1970, 1971) and widespread by Nakamura [4]. Nowadays, HVSR method is widely used in the various of application such as site effects evaluating, seismic microzonation and basin structure [5].

2. Geological Setting of Batu Pahat

Generally, the geological formation of Batu Pahat district is underlying on the unconsolidated soils deposits classified in Quaternary zone and intrusive rock. In addition, there are a few parts in Batu Pahat located on the Permian deposits (Northern Batu Pahat). Mostly the topography of Batu Pahat is relatively flat. Figure 1 shows the geological formation of Batu Pahat districts [6]. There are 15 sites where the microtremor has been done as a listed in the Table 1.

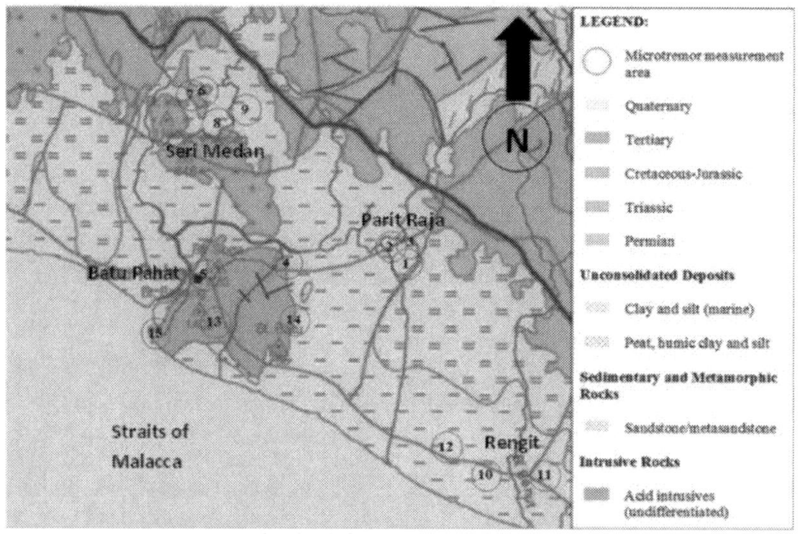

Figure 1. Geological formation at Batu Pahat district.

Table 1. Microtremor measurement locations at Batu Pahat.

No.	Name of Locations	Borehole
1	Rugby Field at Universiti Tun Hussein Onn Malaysia	Yes
2	Faculty of Civil & Environmental Engineering, Universiti Tun Hussein Onn Malaysia	Yes
3	Faculty of Mechanical & Manufacturing Engineering, Universiti Tun Hussein Onn Malaysia	Yes
4	Sekolah Kebangsaan Bukit Soga	Yes
5	Sekolah Menengah Kebangsaan Dato' Bentara Luar	Yes
6	Klinik Desa Sejagong	Yes
7	Sekolah Kebangsaan Sejagong	No
8	Sekolah Kebangsaan Seri Medan	No
9	Sekolah Menengah Kebangsaan Seri Medan	Yes
10	Sekolah Kebangsaan Sungai Tongkang	Yes
11	Masjid Pekan Rengit	Yes
12	Masjid Punggor Kechil, Rengit	Yes
13	Sekolah Menengah Kebangsaan Banang Jaya	Yes
14	Sekolah Kebangsaan Seri Molek	Yes
15	Pantai Minyak Beku	No

IOP Conf. Series: Materials Science and Engineering **136** (2016) 012038 doi:10.1088/1757-899X/136/1/012038

3. Methodology

Microtremor measurement was done conducted on 15 site surrounding Batu Pahat district. Only 13 from 15 sites have boreholes. All of location has been determine based on report as provided by IKRAM Selatan Sdn. Bhd. Meanwhile, microtremor measurement on the three locations without borehole reports are also conducted as a comparison data. Then, seismometer sensors were leveled on placed on the ground surface before recording. All of seismometer sensors are aligned to the True North as a benchmarking. Ten minutes of recording length for each measurement was taken with 100 Hz of sampling rate. The external noise disturbances such as extreme weather, transient noises, monochromatic sources and nearby structures were minimized to the lowest operational for the best result of natural frequency prediction as recommended by SESAME (2004) guideline [7]. Figure 2 shows the microtremor measurement equipment and positions of sensors on fieldwork in this study.

Geopsy software was used for HVSR analysis. A 10-sec of automatic window length selection with an anti-triggering algorithm and a cosine taper of 5% were used. Fourier spectra were computed for each window length and smoothed by a Konno Ohmachi smoothing constant of 40. The reliability and clarity for the significant peaks of mean HVSR curves were checked against the criterions recommended in the SESAME (2004) guideline.

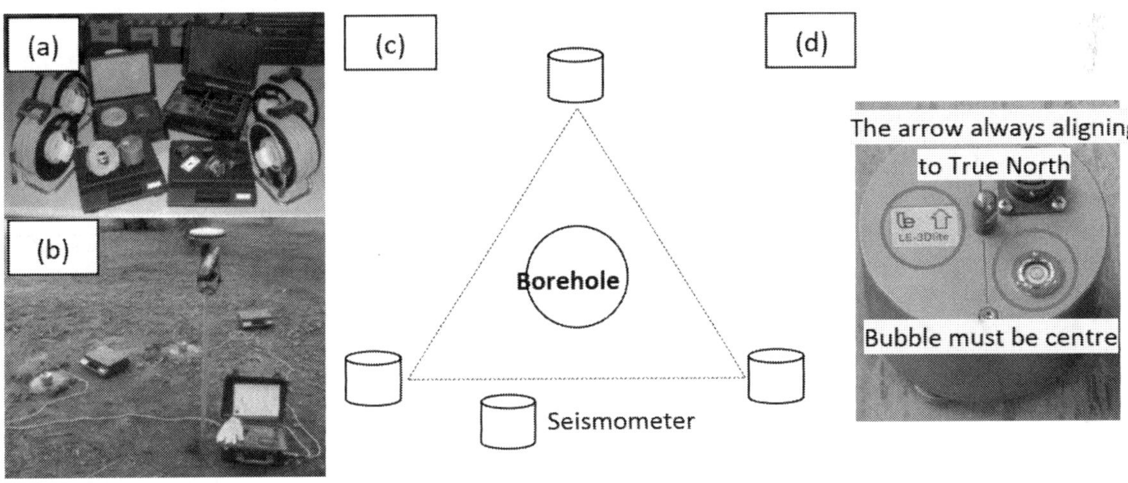

Figure 2. (a) Microtremor equipments, (b) Microtremor measurements on fieldwork (c) Arrangement of seismometer sensors (d) Seismometer sensors are leveled and always aligned to True North.

4. Result and Discussion

Mean HVSR curves of 15 locations of measurements in this studies area have been computed for natural frequencies in the average of directional energy. Natural frequencies are refer to ambient noise signal from surface and used to determine the types of subsurface conditions. Kanai and Tanaka (1954) were the founders to purpose a method of ground classification based on the characteristics of the ambient noise signals [8]. Table 2 shows the natural frequencies at Batu Pahat area.

Table 2. Result of natural frequency (F_o) at Batu Pahat.

Site No.	Natural Frequency, F_o (Hz)
1	1.26
2	1.61
3	1.61
4	2.51
5	2.27
6	4.51
7	4.51
8	2.19
9	1.91
10	0.93
11	0.96
12	0.99
13	4.55
14	2.69
15	14.49

From Table 2, mostly subsurface at Batu Pahat district can be classified in the soft soil group [9]. Based on study cases in the Almeria city, natural frequencies values are 16.67 Hz on hard rock, to 2.5 Hz on medium soil and 0.70 Hz on very soft [5]. It is consistence with the geological map (see Figure 1), where most of Batu Pahat area located in the Quaternary deposits. Comparison with boreholes reports are also shows a same thing happen, where most of soils at Batu Pahat are consists from marine and continental deposits likely clay, silt, sand, peat with minor gravel. At the meantime, most of sites at Batu Pahat are dominated by single peak of HVSR curves, with the natural frequencies in the range 0.93 to 4.55 Hz. Figure 3 shows the HVSR curves from microtremor measurement at Batu Pahat district.

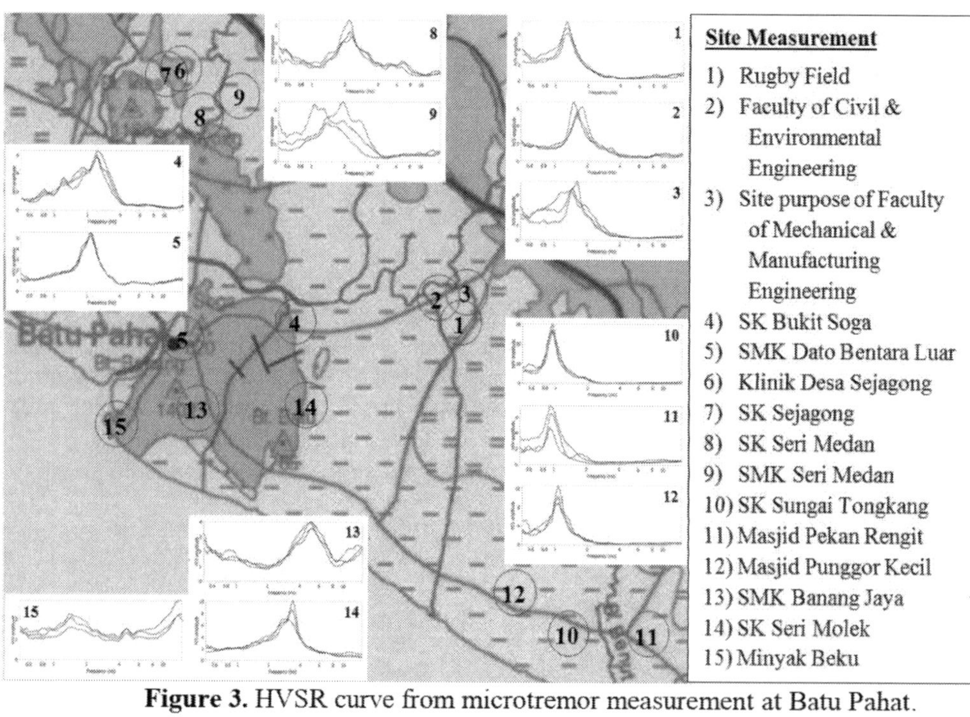

Figure 3. HVSR curve from microtremor measurement at Batu Pahat.

A single peak of HVSR curves are happen probably the existence of deep soil thickness with a large velocity contrast at the sites. In addition, a lower natural frequency (less than 2 Hz), at the most site is also contributes to exists a single peak of HVSR curve. However, there are 4 from 15 sites are shows the natural frequencies more than 3 Hz. Site No. 6, 7, 13 and 15 (see Table 2), the natural frequencies are high caused by geological factor. For site No. 6, existing of shallow bedrock (granite layer) has been founded at 7 to 10.5 m [6]. A same situations is happen for site No. 7 because this site located at the besides site No. 6. Slightly difference of soil composition between clay/silt, sand and gravel is a probably factor why the natural frequency at Sekolah Menengah Kebangsaan Banang Jaya (site No. 13) is high [7]. For site No. 15 (Pantai Minyak Beku), the natural frequency is 14.49 Hz and HVSR curve is multiple curves caused by the measurement is conducted on the outcrop of bedrock. Figure 4 shows the outcrop of bedrock at Pantai Minyak Beku.

Figure 4. Outcrop of bedrock at Pantai Minyak Beku.

5. Conclusion

Study of geological formation underneath of soil surface is very important to know the types and depth to hardlayer (bedrock). It is because these infomation are useful in the desinging works especially to design the types of foundation. Borehole is common technique is used to obtain these information, but this technique is not economical for the large scale like to develop a geological maps. Therefore, microtremor measurement of HVSR method has been introduced as an alternative method in the subsurface exploration. Although, this method is still debated until now, but it is very useful in the mapping works because this method is very simple.

Based on microtremor measurement at Batu Pahat, natural frequencies are distributed in the range 0.93 to 14.49 Hz. Location of Batu Pahat is located on the Quaternary deposits shows most of site at this site have a natural frequency less than 2 Hz. In addition, existing a single peak of HVSR curves is helpful to answer the geological formation at Batu Pahat. Supported facts from site investigation report and geological map as provided by IKRAM Selatan Sdn. Bhd and Department of Minerals and Geosciencs is evidence to prove the microtremor measurement of HVSR can be used is a one methods in the subsurface exploration.

Acknowledgment

This study is supported by Ministry of Education (Higher Education) under FRGS grant, Vot No. 1229 as the main sponsor of this study. A great appreciation to Faculty of Civil and Environmental Engineering, UTHM for the instrumentations. Special thanks to colleagues who participated in this study and for their priceless co-operation and commitment.

References

[1] Hellen M, Oubaince E H, Chatelain J, Machane D, Bensalem R, Guillier B and Cheikhlounis G 2012 Basement Mapping with Single –Station and Array Ambient Vibration Data: Delineating Faults under Bounmerdes City, *Algeria, Seismology Research Letters*. 83.

[2] Gok E & Polat O 2006 Microtremor HVSR Study of Site Effects in Bursa City (Northern Marmara Region, Turkey).

[3] Aswad S, Amiruddin S H, Suriamiharja D A, Maria, Suhendratman A P 2010 Mapping Seismic Vulnerability Index on Hasanuddin Area Using Spectral Ratio for Disaster Prevention.

[4] Motamed R, Ghalandarzadeh A, Tawhata I and Tabatabaei S H 2007 Seismic Microzonation and Damage Assessment of Bam City, Southeastern Iran, *Journal of Earthquake Engineering,* 11 pp. 110 – 132.

[5] Zhu C, Megawati K & Walling M Y, 2011 Site Characterization for Urban Seismic Hazard in Singapore from Microtremor Study.

[6] Department of Minerals and Geosciencs Malaysia, Ministry of Resources and Environment (1985). Geological Map of Peninsular Malaysia 8th Edition. Kuala Lumpur.

[7] SESAME 2004 Guidelines For The Implementation Of The H / V Spectral Ratio Technique On Ambient Vibrations Measurements, *Processing And Interpretation SESAME European research project*, 1–62.

[8] Beroya M A A, Aydin A, Tiglao R and Lasala M 2009 Use of microtremor in liquefaction hazard mapping, *Engineering Geology*. 107 pp. 140 – 153.

[9] Raj J K, Tan D N K and W. H., 2009 Geology of Peninsular Malaysia, *University of Malaya and the Geological Society of Malaysia* pp. 133-173.

[10] Kumpulan IKRAM Sdn. Bhd. (IKRAM Selatan) (2002). Laporan Akhir Penyiasatan Tapak untuk Cadangan Membina dan Menyiapkan Klinik Desa Baru Sejagong, Seri Medan, Batu Pahat, Johor Darul Takzim.

[11] Kumpulan IKRAM Sdn. Bhd. (IKRAM Selatan) (2002). Laporan Akhir Penyiasatan Tapak untuk Cadangan Membina dan Menyiapkan Klinik Desa Sejagong, Seri Medan Batu, Johor Darul Takzim.Pahat, Johor Darul Takzim.

[12] Kumpulan IKRAM Sdn. Bhd. (IKRAM Selatan) (2002). Laporan Akhir Penyiasatan Tapak untuk Cadangan Membina dan Menyiapkan Sekolah Menengah Kebangsaan Banang Jaya, Batu Pahat, Johor Darul Takzim.

Spatial Analysis in Determination Of Flood Prone Areas Using Geographic Information System and Analytical Hierarchy Process at Sungai Sembrong's Catchment

S M Bukari[1], M A Ahmad[1], T L Wai[1], M Kaamin[1] and N Alimin[1]
[1] Faculty of Civil and Environmental Engineering, Universiti Tun Hussein Onn Malaysia, 86400 Batu Pahat Johor, MALAYSIA

E-mail: saifulz@uthm.edu.my

Abstract. Floods that struck Johor state in 2006 and 2007 and the East Coastal in 2014 have triggered a greatly impact to the flood management here in Malaysia. Accordingly, this study conducted to determine potential areas of flooding, especially in Batu Pahat district since it faces terrifying experienced with heavy flood. This objective is archived by using the application of Geographic Information Systems (GIS) on study area of flood risk location at the watershed area of Sungai Sembrong. GIS functions as spatial analysis is capable to produce new information based on analysis of data stored in the system. Meanwhile the Analytical Hierarchy Process (AHP) was used as a method for setting up in decision making concerning the existing data. By using AHP method, preparation and position of the criteria and parameters required in GIS are neater and easier to analyze. Through this study, a flood prone area in the watershed of Sungai Sembrong was identified with the help of GIS and AHP. Analysis was conducted to test two different cell sizes, which are 30 and 5. The analysis of flood prone areas were tested on both cell sizes with two different water levels and the results of the analysis were displayed by GIS. Therefore, the use of AHP and GIS are effective and able to determine the potential flood plain areas in the watershed area of Sungai Sembrong.

Keywords: Flood management, geographic information systems, analytical hierarchy process.

1. Introduction

Flooding is an overflow of water covering the land which is usually dry. Flooding also defines as a groundwater cover. Floods usually caused by continuous rain and quantity of water in river rose than normal and overflow into the river [1]. Flooding also occur due to several factors such as high rainfall, low topographical area, riverbank erosion and rising water levels caused by global warming. Efforts to reduce the floods impact involve a lot of work especially in data collection. Heaps of data from multiple sources will complicate the process of selecting the appropriate data. In addition, the decision-making process will be much more complex if the situation still in below expectation. Compromise on this issue, Analytical Hierarchy Process (AHP) is used to managing data. AHP is a structured technique for controlling and analyzing complex decision, based on mathematics and psychology. AHP has been widely used in the process of producing results in various fields. AHP integrated with GIS applications is a systematic way to generate analysis of preliminary data related to the study area.

This study was conducted to identify areas of potential flooding in the catchment of the Sungai Sembrong in spatial methods. The study area is in Sungai Sembrong where Sembrong Dam which serves as a source of domestic water for the needs of the people around Batu Pahat. Topographically, Batu Pahat is flat and on average area is lower than 5 meter above mean sea level. Therefore, AHP is used to determine the weighting of criteria in causes of floods. GIS in the study used to analyze and demonstrate the potential flood area in Sungai Sembrong use all the criteria provided by the method of AHP.

Floods can bring a lot of problems for residents in the affected area. In Peninsular Malaysia, Johor was the second high-risk flood refers to flood Batu Pahat in 2006 [2]. Flood occured by the heavy rains in long period. High quantity of water that cannot be accommodated by the river and catchment areas will cause overflow into the lower regions. Higher sea levels tidal also contribute to flooding and coupled with the amount of rain water that causes the water to increase much above normal and then overflow to the river bank [3]. There are several other factors that cause floods. Among the factors are natural factors that cannot be controlled, such as rainfall, soil sedimentation in rivers and topography of the area. Due to the impact of human development is the process of urbanization, river erosion caused by construction and drainage systems that were not designed [2].

Choosing the right design concept stage of product development is not an easy decision. Inaccurate results will lead to a final result cannot be to believe. Then, Analytical Hierarchy Process (AHP) is one of the processes that can be used in the design stage. AHP has been widely used to solve various issues and criteria for decision making in research, particularly in engineering. AHP brings the concept of flexible, especially in determining the best results [4]. AHP provides a framework in which hierarchical multi-attribute decision problems can be structured. Using AHP scale ratio along with allowing relative comparison allows AHP to compare the features of insignificant. AHP method is now a widely used methodology in the field of research by developing a user-friendly software package to solve the AHP model [5].

Nowadays, most studies conducted were using GIS. The system's ability to store and process data necessary to refer to an area or space makes this system was used as a tool to solve problems in terms of management and planning. Mapping method used in this system can provide information visually to the user and can also be used as a reference to make predictions about something [6]. GIS is very effective in identifying the components of the flood in flood management. GIS shows the spatial distribution of the actual view of flood phenomenon. GIS is used to measure the location of flooded areas and provide a measurable estimation of the land and infrastructure affected by the floods [7]. The main advantages of using GIS for flood management are to generate the flood area figure and also to analyze the potential flood area. Analysis that carried out is to estimate the probability of damages caused by floods [8].

2. Analytical Hierarchy Process in GIS

In determining the areas of potential flooding in the catchment area of the Sungai Sembrong, several levels of analysis have been done. First is the process of determining the factors that caused flooding in the study area. Each factor will then be analyzed using Pair-wise Comparison Method (PCM) to get the weightage for each criterion. PCM is part of the AHP is used to simplify decision-making process using a matrix method. Each factor that causes floods are organized according to the weightage and will be included in GIS applications. The factors that have been prepared to facilitate the process of overlapping maps using GIS applications in order to determine areas of potential flooding in the Sungai Sembrong catchment area.

2.1 Selection and Evaluation of Criteria

The criteria considered in this study were selected according to factors that cause flooding in the study area. Selection Criteria causing floods may also be referred to the procedure for the determination of the flood area or process to build catchments using ArcGIS 10. The criteria used in this study were referred from previous studies. Factors to consider are digital terrain, rainfall, flow direction, flow accumulation and catchment areas. PCM is used to determine the relationship between the factors related to the flooding to determine areas of potential flooding in the catchment area of Sungai Sembrong. PCM is one of the methods used to determine priority criteria in decision making. PCM is pairwise comparisons were made on a scale of relative importance to the selection criteria set out in the ratio scale. Table 1 shows the scale of assessments for PCM.

Table 1: Continuous Rating Scale for Pairwise Comparison of Saaty's Method [9].

1/9	1/7	1/5	1/3	1	3	5	7	9
Extremely	Very Strongly	Strongly	Moderately	Equally	Moderately	Strongly	Very Strongly	Extremely
	Less Important					More Important		

IOP Conf. Series: Materials Science and Engineering **136** (2016) 012039 doi:10.1088/1757-899X/136/1/012039

Table 2: Continuous Rating Scale for Pairwise Comparison.

Criteria	DTM	Flow direction	Flow accumulation	Catchment area
DTM	1	3	3	3
Flow direction	0.3	1	5	3
Flow accumulation	0.3	0.2	1	3
Catchment area	0.3	0.3	0.3	1

Weightage for each criterion will be calculated using the matrix method. Refer with Table 2, all the criteria are organized by each weightage. Criteria which have a higher weightage values are very high priorities and followed by next criteria. The estimated weights are then calculated to produce the consistency ratio for scale banding pairs. If the consistency ratio >0.10, the scale of pairs should be calculated to be recomputed to get the consistency ratio <0.10. After the comparison process is completed, the priorities weighted that have been produced as refer to Table 3.

Table 3: The weightage for each criteria.

Criteria	Weightage
DTM	2.07
Flow direction	1.46
Flow accumulation	0.62
Catchment area	0.36

CR = 0.09 < 0.10 is accepted.

The consistent ratio data set in this study is 0.09, which is less than 0.10, and then set PCM data and methods used can be received and forwarded in the order of use in GIS. If the ratio is more consistently obtained from the 0.10, it indicates that matrix data set is unacceptable and needs to be reviewed.

2.2 GIS Database Design

ArcGIS 10 software is used to display the spatial data and attribute data that has been set up using the data needed to establish potential areas of flooding in the Sungai Sembrong catchment area. Catchment areas should be produced before the process of determining the potential flood areas. Design and development data in this study include the process as described below;

i. Develop the Digital Terrain Model (DTM) of the study area.
ii. Produce the flow direction.
iii. Produce the flow accumulation.
iv. Setting the position of pour points.
v. Produce the catchment area.
vi. Determination of potential flood areas.

Each criteria necessary data to establish potential areas of flooding should be changed to the shape file types that can be read and displayed by the software ArcGIS 10. Use of Raster Calculator function in the GIS application is to calculate and display the area at risk of flooding in the Sungai Sembrong catchment area. The height of the water level refers to the water level rises above sea level. The height of which is placed in Raster Calculator is in unit of metre. Then the potential flooded areas will be displayed by GIS. Refer to equation (1) to enter the height of the water level using the Raster Calculator to determine potential flood areas.

$$DTM <= height\ of\ water\ level \qquad (1)$$

3. Results and Analysis

The process of database development using GIS application is the process of overlap of the data according to criteria to form a spatial layer containing attribute data layer itself. Overlapping layers according to the criteria will be done until the flooded areas can be determined as specified in the objectives of study.

3.1 Cell Size Analysis

In this study, two cell sizes are used to view the end result of the difference between the sizes of the cell. The cell size of 5 and 30 were analyzed until the flood potential is formed. Table 4 is an explanation of the difference between the use of cell sizes of 5 and 30.

Table 4: Differences in the use of cell sizes of 5 and 30.

Characteristics	Cell size 5	Size cell 30
Figure of cell size	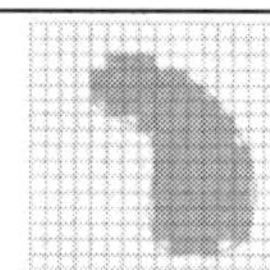	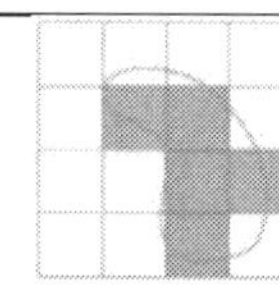
Resolutions	High	Low
Spatial accuracy	More accurate	Less accurate
Speed of display	Slow	Faster
Speed of data processing	Slow	Faster
File size	Larger	Smaller

3.2 Catchment Area Analysis

Determining size and the location of catchment area is dependent on the location of the pour points which is placed at all the point of intersection between the river and flow accumulation. Figure 1 is the location of Sungai Sembrong catchment in Batu Pahat. The analysis is related to the size of the cells used which are cell size of 5 and 30. The results reveal that the catchment area produced using cell size 5 is more detailed and closer to the actual terrain compared to the catchment areas resulting in cell size of 30. Figure 2 and Figure 3 shows the difference between the resulting catchment area using cell size of 5 and 30. Table 5 shows the difference of size between the catchments with cell size 5 and 30.

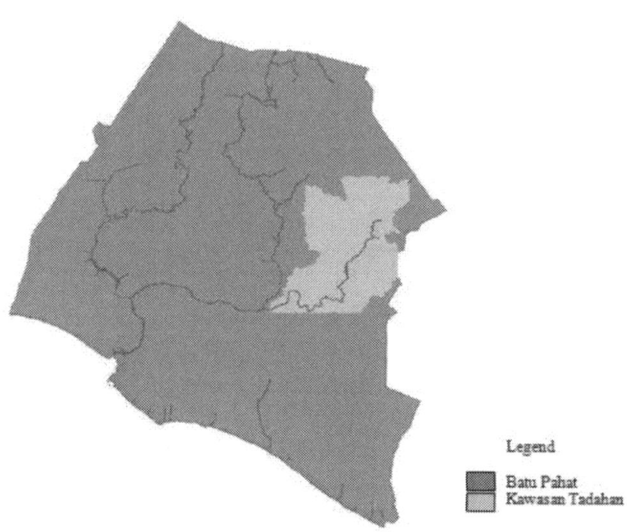

Figure 1: Sungai Sembrong catchment area.

Figure 2: Catchment area using cell size 30.

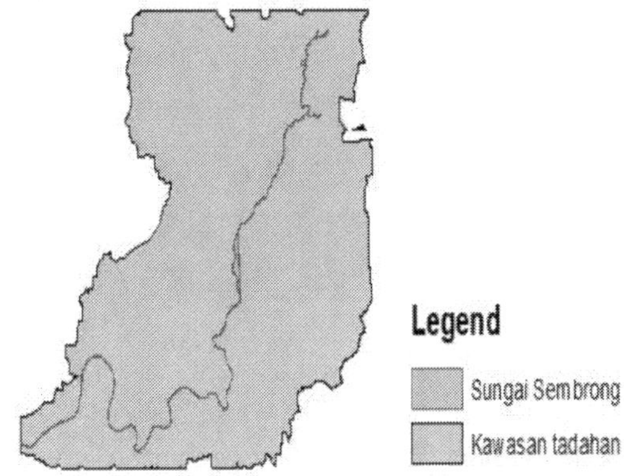

Figure 3: Catchment area using cell size 5.

Table 5: The size of the catchment area of the cell 30 and 5.

Cell size	Size of catchment area (m^2)	Percentage of catchment area (%)
30	11845808	6.5
5	12762114	7.0

3.3 Potential Flood Areas

Determination of potential flooding in the Sungai Sembrong catchment were calculated using the Map Algebra in ArcGIS. The analysis determines the extent to potential areas flooded when water from the river level rise caused by various factors. In the study area, high rainfall levels are key factors to lead to an increase in the water level. The ground water flow rate is also a factor because the existing drainage does not cope with large amounts of water in a short period and cause overflow. Water levels were set at 7 meters and 10 meters from mean sea level (MSL). The maximum water level is selected to determine the areas at risk of flooding in the catchment area of Sungai Sembrong. Potential areas of flooding were also analyzed using two different cell size which is the size of the cell 5 and 30. Figure 4 and Figure 5 shows the difference between a potential flood area produced by the cell size of 5 and 30. Table 6 shows the difference between the size of the flooded area with a cell size of 5 and 30.

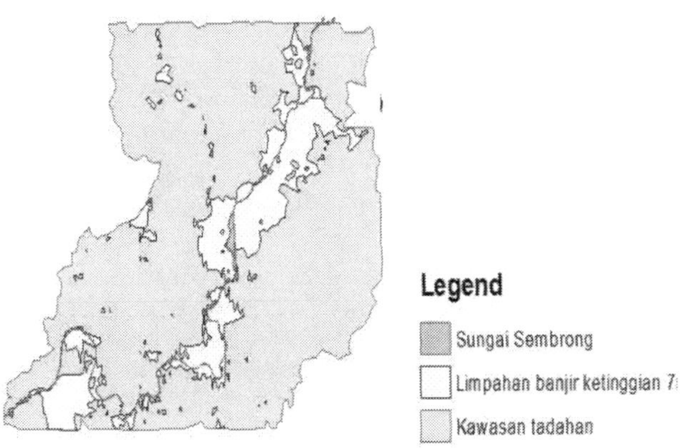

Figure 4: Flood area using cell size 30

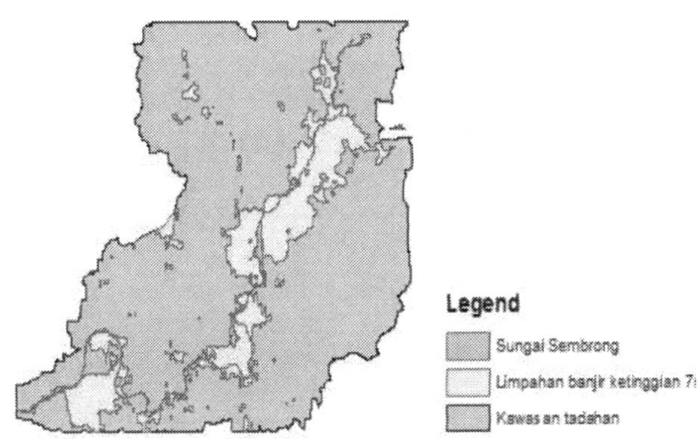

Figure 5: Flood area using cell size 5

Table 6: The size of the flood area of the cell 30 and 5

Cell size	Size of potential flood areas (m^2)	Percentage of potential flood areas (%)
30	1,962,493	16.6
5	1,987,963	15.6

3.4 Effectiveness of AHP with GIS in Determination of Potential Flood Areas

Previously, related study was conducted using GIS in the Tanjung Sembrong area [3]. The study using hydrological models and water balance model based on rainfall data to determine the location and size of the flood area. For this study, the location and size of flood area is determined using the maximum height of the water level. Tested water levels were referring to water levels during floods in 2006. Based on the results, the potential flood area in Sungai Sembrong catchment area can be determined. In conclusion, the use of AHP and GIS in determination of the potential for flooding in the catchment area of Sungai Sembrong is effective.

4. Summary

GIS is a tool for planners to analyze the planning region. Results of the determination of flood-prone areas are expected to be a consideration for decision makers so as not to focus settlement construction in areas prone to flooding. In addition , the need for flood mitigation efforts in this area such as the construction of dikes, dredging to increase the capacity of the river as well as perform other businesses from upstream to downstream to prevent flooding.

References

[1] Gasim, M. B., Surif, S., Mokhtar, M., Toriman, M. E., Rahim, S., & Bee, C. H.. Analisis Banjir Disember 2006 : Tumpuan di Kawasan Bandar Segamat , Johor. *Sains Malaysiana*, 39(3), (2010).

[2] Shafii, H., Musa, S. M. S., & Miskam, N). Kajian Kesanggupan Pemilik Rumah Untuk Mendapatkan Perlindungan Daripada Bencana Banjir. *Persidangan Kebangsaan Masyarakat, Ruang Dan Alam Sekitar*, 2011, 297–312. (2011)

[3] Hilaham, M. B. M. Aplikasi Sistem Maklumat Geografi (GIS) Dalam Menentukan Kawasan Limpahan Banjir di Sungai Sembrong. Universiti Tun Hussien Onn Malaysia: *Bachelor Thesis*. (2013)

[4] Nukman, Y., Ariff, H., Salit, M. S., & Ismail, Use of Analytical Hierarchy Process (AHP) for Selecting The Best Design Concept. *Jurnal Teknologi,* 49(A) Dis. 2008: 1–18 © Universiti Teknologi Malaysia, 49, 1–18. (2009).

[5] Leung, L. C. & Cao, D., On The Efficacy of Modeling Multi-Attribute Decision Problems Using AHP and Sinarchy. *European Journal of Operational Research* 132 (2001), 132, 39–49. (2001).

[6] Harun, N. E., Kajian Penentuan Kawasan Bukit Berisiko Menggunakan Sistem Maklumat Geografi (GIS). Universiti Teknologi Malaysia: *Master Thesis* (2006).

[7] Isma'il, M., & Saanyol, I. OApplication of Remote Sensing (RS) and Geographic Information Systems (GIS) in Flood Vulnerability Mapping: Case study of River Kaduna. *International Journal Of Geomatics And Geosciences*, 3(3), 618–627. (2013).

[8] Solaimani, K.. Flood Forecasting Based On Geographical Information System. *African Journal of Agricultural Research*, *4*(10), 950–956. (2009)

[9] Lawal, D.U., Maton, A.N., Hashim, A.M.,Yusof, K.W., & Chandio, I.A.. Detecting Flood Succeptible Areas Using GIS-based Analytic Hierarki Process. *International Conference on Future Environment and Energy*, 2012, vol 28. (2012).

Shoreline Delineation and Land Reclamation Change Detection Using Landsat Image

M I Rosli[1], M A Ahmad[1], M Kaamin[1] and M F N Izhar[1]
[1] Centre of Diploma Study, Universiti Tun Hussein Onn Malaysia, 86400 Batu Pahat Johor, MALAYSIA

E-mail: gf130012@uthm.siswa.edu.my

Abstract. This study is conducted on the usage of remote sensing images from several different years in order to analyze the changes of shoreline and land cover of the area. Remote sensing images used in this study are the data captured by the Landsat satellite. The images are projecting the land surface in 30 by 30 meter resolution and it is processed by the ENVI software. ENVI is able to change each digital number of the pixels on the images into specific value according to the applied model for classification in which could be used as an approach in calculating the area different classes based from the images itself. Therefore, using this method, the changes on the coastal area are possible to be determined. Analysis of the shoreline and land reclamation around the coastal area is integrated with the land use changes to determine its impact. The study shows that Batu Pahat area might have undergone land reclamation whereas in Pasir Gudang is experiencing substantial amount of erosion. Besides, the changes of land use in both areas were considered to be rapid and due to the results obtained from this study, the issues may be brought about for the local authority awareness action.

Keywords: Shoreline changes, land use, erosion, land reclamation.

1. Introduction

Coastal zone corresponds to the geomorphologic area on the part of the seashore in which the interaction between the marine and land parts begins in the form of complex ecological and resource systems composed of biotic and abiotic components coexisting and interacting with human communities and relevant socio-economic activities [1]. Its components may encompass river deltas, wetlands, coastal plains, beaches and dunes, mangrove forests, lagoons, reefs, and other coastal [2]. Coastal area around the world is a dynamic area with recurrently changing chemical, biological and geologic attributes which explain why they are frequently evolve over period.

Erosion is a naturally occurring process which is commonly made by rock and soil being released from the earth's surface over and over by water or wind before worned away [3]. The process is consistently slow and gradual that may take thousands of years but can be speeded up by human activities.

Thus, these makes the issue of environmental control on the development done in certain area to be vital in any academicals discussion. There are many different kinds of erosion. One of the common examples is coastal erosion which is a serious problem that has potential to breed catastrophic impacts. Land use denotes the physical and biological cover of the surface of land. Land use changes is the conversion of a type of land to another type of land. Typically it is the result from government policies but they also may occur naturally such as a consequence of natural hazards or local agriculture practices. Land reclamation is an act of acquiring new land which is an example of land use changes. After large-scale undertakings in the past decades, reclamation activities in few countries were already subjected to strict control. Thus, presenting the newest monitoring discoveries of the involved area are crucial for a better understanding by the parties involved for them to decide their next line of action in order to maintain the safest environment for the communities.

2. Methodology

2.1 Data Collection

Data from Landsat were obtained from USGS Global Visualization Viewer (GloVis) website. The maximum cloud allowable on the images was limited to 10% only as cloud is a distraction in data analysis. Images from the year 1996, 2011 and 2015 were seized for the study area in Batu Pahat while in Pasir Gudang, year 1990, 2007 and 2014 were attained. The selected temporal data are from Landsat 5 and Landsat 8.To process image from Landsat, all the spectral bands were combined in only one image file by layer stacking tool. Next, three bands that give the most natural looking image were set on display which areband 3, 5 and 7 for Landsat 5 data and band 4,5 and 6 for Landsat 7. All cloud that were present on some of the images were removed. Next off, classification was done on the image file to turn it into a land use map.

2.2 Land Use Map Creation

All land use maps generated from temporal data in both study areas undergone change detection analyses. Change detection statistics distinguished the contrasts between any pair of initial state and final state images. The change detection statistics produced a chart that displayed class-for-class image difference and also the land use changes. In order to analyze the coastal erosion and land reclamation, the changes in percentage of water and barren land are observed.

2.3 Land Use Maps Overlay

Land use map is a map that shows the types and intensities of different land uses in the particular area. Different colours were used to differentiate each classes in the land use maps. Blue was selected to represent the water body while green was for all kind of vegetation, includes plant with no chlorophyll. In addition, for barren land, orange colour was chosen to exhibited it whereas for urban and development, yellow color was picked.

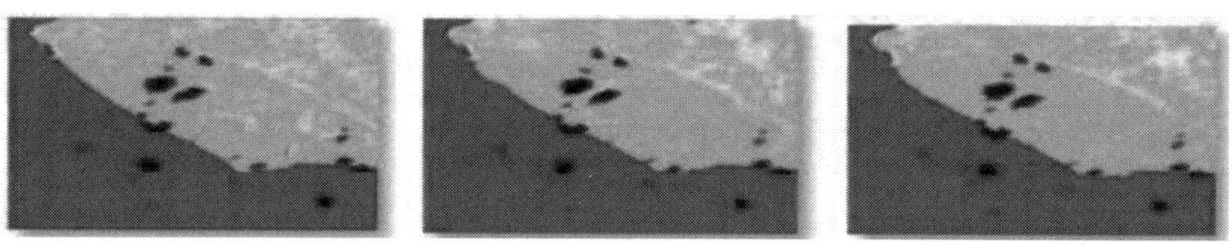

1996 2011 2015

Figure 1. Land use maps of Batu Pahat.

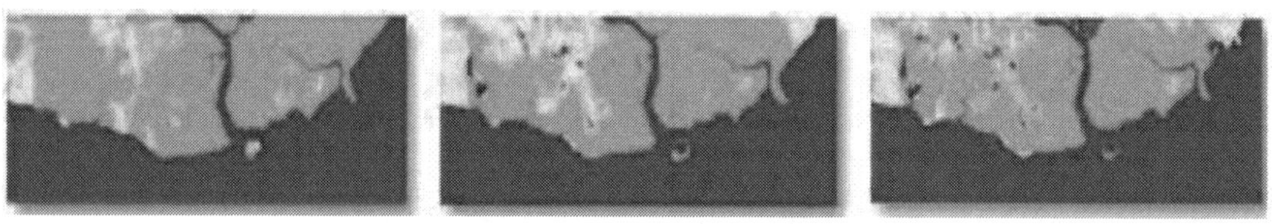

1990 2007 2014

Figure 2. Land use maps of Pasir Gudang.

2.4 Change Detection Analysis

The resolution in every picture taken by Landsat satellite is 30 m by 30 m. When ENVI process the image by change detection, it will output the total pixel count for each class in the area therefore the area in the study areas can be determined by using the given equation (refer Eq 1).

$$\text{Area} = \text{pixel count x } 30 \text{ m2 x } 30 \text{ m}^2 \quad (1)$$

3. Discussion and Analysis

Nowadays, growing number of worldwide sites have been affected by coastal erosion and land reclamation. Coastal ecosystems are one of the most impacted and altered regions worldwide because of the huge population that is concentrated along the coast [4]. On the other hand, land reclamation is the way to achieve more land. In recent decades, sea enclosing and land reclamation had become the best solution for the increasing demand of new land for living and development.

Urbanization, which mainly replaces permeable vegetated land surface with impervious surface areas, significantly alters the hydrologic fluxes of a drainage basin [5]. Physical alterations on shoreline or changes in coastal land use will affect coastal processes especially interfering adjacent marine communities [6]. The loss of biological diversity, erosion of beaches and loss of mangrove areas may influence ecosystem function, particularly shoreline stabilization. Humans receive valuable ecosystem services from coastal and estuarine ecosystems, including shoreline stabilization, protection of property, fisheries production and supporting biological diversity, but these services are lost with over-exploitation of coastal resources and loss of habitat [7].

Land Use Changes Analysis. On the year 1996 and 2011 in Batu Pahat study area, the water class grew by 0.4185 km2 (0.904%) while the vegetation class risen to 0.8478 km2 (2.398%) whereas the urban class increased by 1.4796 km2 (19.154%). Barren land class is the only class that are declining by 2.7459 km2 (77.437%). Whereas for the year 2011 and 2015, the water class decreased by 0.4761 km2 (1.019%) whereas barren land aroused by 6.6915 km2. Vegetation class is also declined by 0.1017 km2 (0.281%) and urban class falling off by 6.1137 km2 (66.422%).

Table 1. Pixel counts of Batu Pahat in 1996 to 2011.

	Water	Vegetation	Barren Land	Urban	Row Total	Class Total
Unclassified	0	0	0	0	0	**3827**
Water	51220	673	0	32	51925	**51925**
Vegetation	227	33164	1884	4942	40217	**40217**
Barren Land	0	293	414	182	889	**889**
Urban	13	5145	1642	3427	10227	**10227**
Class Total	51460	39275	3940	8583	0	**0**
Class Changes	240	6111	3526	5156	0	**0**
Image Difference	**465**	**942**	**-3051**	**1644**	**0**	**0**

Table 2. Pixel counts of Batu Pahat in 2011 to 2015.

	Water	Vegetation	Barren Land	Urban	Row Total	Class Total
Unclassified	0	0	0	0	0	**3827**
Water	51237	130	0	29	51396	**51396**
Vegetation	522	35636	149	3797	40104	**40104**
Barren Land	157	3823	107	4237	8324	**8324**
Urban	9	628	633	2164	3434	**3434**
Class Total	51925	40217	889	10227	0	**0**
Class Changes	688	4581	782	8063	0	**0**
Image Difference	**-529**	**-113**	**7435**	**-6793**	**0**	**0**

As for the year 1990 and 2007 in Pasir Gudang study area, the volume of the water in Pasir Gudang area between the year of 1990 to 2007 was boosted over 0.0774 km2 (0.572%) while for the vegetation class, 1.0755 km2 (10.082%) of area was lost.

In addition, the urban class was upsize by 2.2833 km2 (151.825%). For the year 2007 and 2014, the water volume was up surged by 0.5130 km2 (3.767%) and at the same time the vegetation class was withered by 0.5922 km2 (6.174%). Moreover, urban class was also decreased by 1.5336 km2 (40.494%).

Table 3. Pixel counts of Pasir Gudang in 1990 to 2007.

	Water	Vegetation	Urban	Row	Class Total
Unclassified	0	0	0	0	**0**
Water	14779	262	46	15087	**15131**
Vegetation	220	9511	402	10133	**10658**
Urban	46	2080	1223	3349	**4208**
Class Total	15045	11853	1671	0	**0**
Class Changes	266	2342	448	0	**0**
Image Difference	**86**	**-1195**	**2537**	**0**	**0**

Table 4. Pixel counts of Pasir Gudang in 2007 to 2014.

	Water	Vegetation	Urban	Row Total	Class Total
Unclassified	0	0	0	0	**0**
Water	15045	560	96	15701	**15701**
Vegetation	36	9475	489	10000	**10000**
Barren Land	1	157	1634	1792	**1792**
Urban	49	466	1989	2504	**2504**
Class Total	15131	10658	4208	0	**0**
Class Changes	86	1183	2219	0	**0**
Image Difference	**570**	**-658**	**-1704**	**0**	**0**

3.1 Shoreline Analysis

The size of the land use maps that were used in shoreline analysis were smaller as it were clipped so that only the water body from coastal area being processed in change detection. In the period from the year 1996 and 2011 in Batu Pahat, the region of water was deteriorated close to 0.1098 km2 (3.519%) whereas the vegetation zone marked up by 0.1566 km^2(5.303%). Besides, the amount of urban class was reduced too by 0.0414 km2 (38.333%). While in between the year 2011 and 2015, the class of water descended by 0.2403 km2 (7.982%) while the urban class also was ran low by 0.0567 km^2 (85.135%). In contrary, the vegetation class is the only class that was enlarged as it growth by 0.1143 km2 (3.676%).

Table 5. Pixel counts of Batu Pahat (clipped) in 1996 to 2011.

	Water	Vegetation	Urban	Row	Class Total
Unclassified	0	0	0	0	**98**
Water	3284	61	0	3345	**3345**
Vegetation	183	3157	109	3449	**3455**
Urban	0	63	11	74	**74**
Class Total	3467	3281	120	0	**0**
Class Changes	183	124	109	0	**0**
Image Difference	**-122**	**174**	**-46**	**0**	**0**

Table 6. Pixel counts of Batu Pahat (clipped) in 2011 to 2015.

	Water	Vegetation	Urban	Row	Class Total
Unclassified	0	0	0	0	**83**
Water	3072	1	0	3073	**3078**
Vegetation	182	3361	29	3572	**3582**
Barren Land	91	92	35	218	**218**
Urban	0	1	10	11	**11**
Class Total	3345	3455	74	0	**0**
Class Changes	273	94	64	0	**0**
Image Difference	**-267**	**127**	**-63**	**0**	**0**

On the other hand, from the year 1990 to 2007 in Pasir Gudang, all the classes were increased. The water class had risen by 0.0117 km2 (2.935%) and the vegetation class had increased by 0.0054 km2 (1.049%) the same with urban class which was 0.0054 km2 (12.5%). Furthermore, in between the year 2007 and 2014, the water class was roused by 0.0360 km2 (8.772%) while the vegetation class was lessen by 0.0468 km2 (8.997%). Similarly, the urban class was boosted over 0.0018 km2 (3.704%).

Table 7. Pixel counts of Pasir Gudang (clipped) in 1990 to 2007.

	Water	Vegetation	Urban	Row Total	Class Total
Unclassified	0	0	0	0	**0**
Water	432	24	0	456	**456**
Vegetation	1	513	46	560	**578**
Urban	10	35	2	47	**54**
Class Total	443	572	48	0	**0**
Class Changes	11	59	46	0	**0**
Image Difference	**13**	**6**	**6**	**0**	**0**

Table 8. Pixel counts of Pasir Gudang (clipped) in 2007 to 2014

	Water	Vegetation	Urban	Row Total	Class Total
Unclassified	0	0	0	0	**0**
Water	456	27	13	496	**496**
Vegetation	0	515	11	526	**526**
Barren Land	0	10	0	10	**10**
Urban	0	26	30	56	**56**
Class Total	456	578	54	0	**0**
Class Changes	0	63	24	0	**0**
Image Difference	**40**	**-52**	**2**	**0**	**0**

3.2 Analysis of Coastal Erosion and Land Reclamation

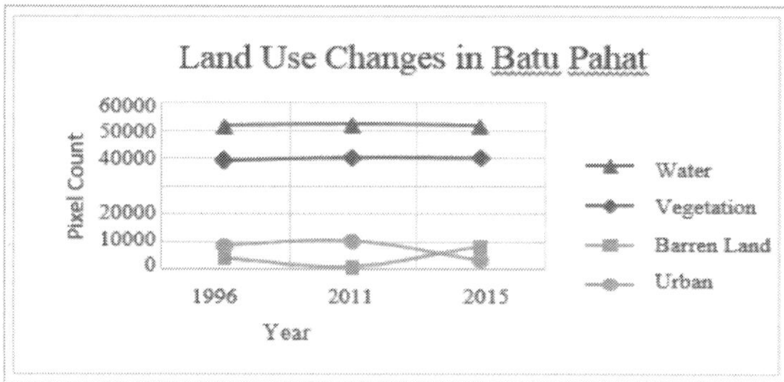

Figure 3. Graph of land use changes in Batu Pahat

In Batu Pahat study area, the water class was calmly increased from the year 1996 to 2011 but then it declined towards the year 2015 where the initial number of pixel from 51925 was lessen to 51396 which can be seen on Figure 3. Similarly, the urban class behaved the same way but towards the year 2015, it went down drastically. As for vegetation class, the pattern was also the same where the pixel count was accreted first before downturned after 2011 but it was indistinct. On the other hand, the barren land were the only that declined initially and started to went up following 2011. It can be conclude that the region in study area had undergone a rapid land use change since all classes were in uneven trend.

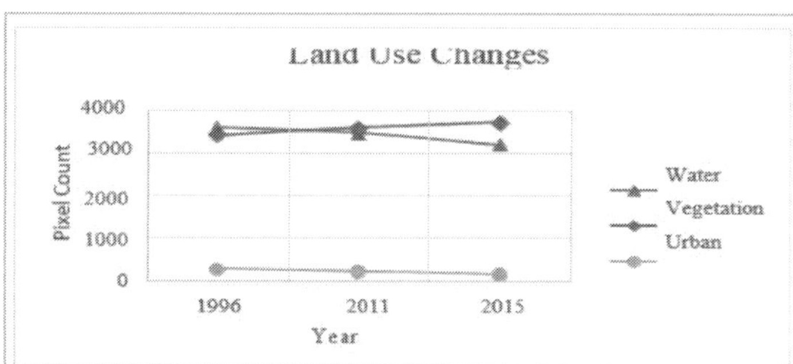

Figure 4. Graph of land use changes in Batu Pahat (clipped)

In shoreline analysis, the value of pixel count for vegetation was sequentially enlarged throughout the timeframe. The same patterned were followed by urban class. However, the vegetation class acted differently since the very beginning. The class progressively increased with the most prominent increment was from 2011 afterwards. The statistic were implying that this part of study area might undergone to land reclamation which can be analyzed from Figure 4.

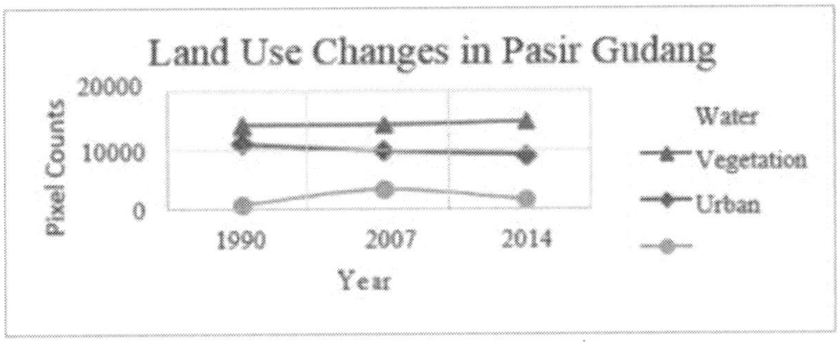

Figure 5. Graph of land use changes in Pasir Gudang

Based on Figure 5, in Pasir Gudang study area, the region of water continued to grew at a slow pace from the year 1990 until 2014. Whereas the vegetation class was descended gradually over the 24 year timeframe. Deviating from others, initially the urban class was risen progressively from 1990 to 2007 but it started to climb down thereafter. Based on these information, it seems like the region had undergone a rapid land use change too because the irregular pattern of urban growth and the continuous decrement of the vegetation class.

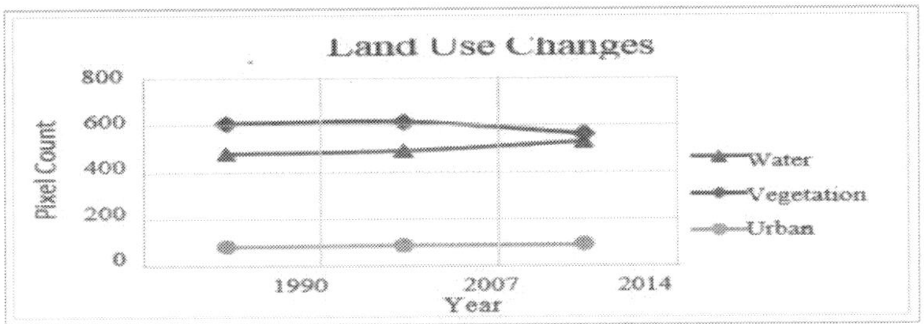

Figure 6. Graph of land use changes in Pasir Gudang (clipped)

As for its shoreline analysis, based from Figure 6, the urban class was risen continuously but at a very slow pace as if the changes was almost unnoticeable. These might implied that the development around the study area were carried out at a very slow speed. Furthermore, the pixel count for water went up crawlingly before 2007 but afterward it rouse dramatically. These might tell us that the region in study area was exposed to coastal erosion.

4. Discussion

In the study area, the change detection statistics proved that the area of water turned to vegetation and barren land are much more in comparison with the number of the other class that changed to water. At first only the coastal area near to the city was reclaimed but since there is erosion in adjacent area, land reclamation is extended throughout the area. The land reclamation rate is raised even more progressively after 2011. The purpose of it is maybe in order to prevent the flood as the last time it happened in Batu Pahat area was because the backwater from the sea into the Batu Pahat River. The reclamation is maybe delivered by the city council for future planning as reclamation to prevent erosion. Whereas in Pasir Gudang, the coastal erosion is much extensive. By referring to the change detection statistics, the area of water class kept growing throughout the interval as there is quite a vast portion of area from other class were exchanged with water class. The reason why erosion is vigorous in the area is most probably because of the hectic waterway. The region are bounded by two ports which are Pasir Gudang port and Tg. Langsat port. In fact, Pasir Gudang port is the biggest port in the southern of peninsular Malaysia.

In Batu Pahat study area, the rate of development is quite rapid. In the year 1996 to 2011 change detections statistics, there are 13.1% from total area of vegetation were swapped with urban land while 41.675% from area of barren land were exchanged with urban. Not to mention that there is 71.204% area of barren land were developed to urban between the year 2011 and 2015. The area of vegetation that is cleared to barren land between 2011 and 2015 was also vast which implies that there are more development in coming years. On the side, Pasir Gudang study area also shows a huge amount of development. In 1990 to 2007 change detection statistics, the percentage of the enlargement of urban area in the region are skyrocketed to 151.825%. Mainly the urbanization occurred in the study area are from the development of Tanjung Puteri Golf Resort and Tanjung Langsat port which mainly due to the development of three 18-hole of championship golf course built in the golf resort explaining the growing pixel count for vegetation and barren land classes in the area.

5. Accuracy Assessment.

With regard to accuracy assessment, a site verification was done in the study areas to confirm the accuracy of the land use map as the classification process might accumulate some errors. The site was verified by the aid of Global Positioning System (GPS) device. The coordinates from land use maps were extracted and it were compared with coordinates displayed by the GPS device on site in order to ensure each map produced are acceptable. For each study area, three random places were selected as checkpoints and for each checkpoints, all the classes available in vicinity area were inspected. For future study, few technique and methods should be taken into consideration such as by using seperability matrix or confusion matrix.

6. Conclusion

To conclude, the change detection done had successfully detected the changes between water body, vegetation, barren land and urban from medium spatial resolution remotely sense images provided by Landsat by performing a change detection analysis. From the results obtained, the rate of changes either impacted by the development processes or geomorphologic changes can be monitored based on the multi temporal images. This will help the organization and department involved in maintaining the sustainability of the study to reconstruct their plan of action in the upcoming future. In a nutshell, the land cover changes and also shoreline changes analyses were produced with a significant findings.

Acknowledgment

The authors would like to thank Ministry of Education Malaysia and Universiti Tun Hussein Onn Malaysia for their support with RAGS Vot U101 research grant.

References

[1] Pino G R, Marcello S, Raul M, Oscar G A. & Jurgi A. 2009. A contribution to the implementation of ICZM in the Mediterranean developing countries. Ocean & Coastal Management 52, 545–558.

[2] Mitra D. & Chanda P. 2000. Coastal Vulnerability Index for Parts of Orissa Coast, East Coast of India, Bay of Bengal. Journal of Remote Sensing & GIS, 5 (3), 2230-7990.

[3] Walker B, 1994. Global change extensive strategy agriculture options in the regions of the world. Clim. Change 27, 39-47.

[4] Rojas M L, Recalde M Y, London S, Perillo G M E, Zilio M I, & Piccolo M C, 2014. Behind the increasing erosion problem: The role of local institutions and social capital on coastal management in Argentina. *Ocean & Coastal Management*, *93*, 76–87.

[5] Zhou F, Xu Y, Chen Y, Xu C-Y, Gao Y, & Du J. 2013. Hydrological response to urbanization at different spatio-temporal scales simulated by coupling of CLUE-S and the SWAT model in the Yangtze River Delta region. *Journal of Hydrology*, *485*, 113–125.

[6] Sealey K, 2004. Large-scale ecological impacts of development on tropical islands systems: comparison of developed and undeveloped islands in the Central Bahamas. Bull. Marine Sci. 75 (2), 295-320.

[7] Jackson J B, Kirby M X, Berger W H, Bjorndal K A, Botsford L W, Bourque B J, Warner R R, 2001. Historical overfishing and the recent collapse of coastal eco- systems. Science 293 (5530), 629-638.

Strength and Stiffness Development in Soft Soils: A FESEM aided Soil Microstructure Viewpoint

D C Wijeyesekera[1,3], M H Ho[1], X Bai[2] and I Bakar[3,1]
[1] Faculty of Civil and Environmental Engineering, Universiti Tun Hussein Onn Malaysia, 86400 Batu Pahat Johor, MALAYSIA
[2] Department of Civil Engineering, Taiyuan University of Technology, Taiyuan 030024, China
[3] Research Center for Soft Soil, Universiti Tun Hussein Onn Malaysia, 86400 Batu Pahat Johor, MALAYSIA

E-mail: devapriya@uthm.edu.my

Abstract. This paper opens with an overview of the debatable definition of soft soil that goes beyond a (CH) organic / inorganic clay and OH peat to include weakly cemented periglacial deposits of loess and alike. It then outlines the findings obtained from stiffness test on cement-stabilised soft clay. The findings are complemented with a microstructure viewpoint obtained using field emission scanning electron microscope (FESEM). Research also comprised of making cylindrical stabilised clay samples, prepared in the laboratory with various rubber chips contents and cement, and then aged for 28 days. The samples were then subjected to unconfined compressive strength (UCS) test and observations were also made of its microstructure using the FESEM. The impact of the soil microstructure on the stiffness result was studied both with the stabilized soil and also of some of the natural undisturbed loess soils. Sustainability aspect and the potential of the use of rubber chips and sand as additives to cement stabilisation are also discussed. The overall test results indicated that rubber chips and sand contributed to the improvement in unconfined compressive strength (q_u). The derogatory influence of moisture on the stiffness of the stabilised clay was studied simultaneously. SEM micrographs are presented that show bonding of cement, rubber chips/ sand and soft clay, granular units and aggregated / agglomerated units in loess. The paper concludes with observations on the dependence of soil microstructure on the soil strength and deformability and even collapsibility of the loess. Current practices adopted as engineering solutions to these challenging soils are outlined.

Keywords: Soft soil, ground modification, microstructure.

1. Introduction

Advances in geotechnical engineering has not yet been able to totally eradicate the occurrence of failures during construction or after a long term, particularly in challenging and even genetically (geologically) different challenging soils such as highly sensitive clays, organic clays or stiff fissured soils. The softness of a soil may be due to it being weak in strength / highly deformable or both. Embedded artefacts such as shrinkage cracks, fissures, bedding surfaces and alike are visible macro structural features that influences more so the strength (bearing capacity) of the ground rather than its deformability.

On the other hand, the morphology (size, shape, roughness and compressibility of the particles), fabric (spatial arrangement of the particles) and the level of cementation (bonding between the particles) contribute to the microstructure that further characterizes a sensitive or soft ground condition. Water is a prime cause of softness in soils. Quick clays have a liquidity index greater than one and a metastable clay microstructure resulting from the exchange of the marine pore water with fresh water. Peat has an excessively compressible organic matter content, which contains intracellular free water, inter-particle and adsorbed water in the particle cell structure. The mechanism of loess collapsibility on flooding is attributed to the breakdown of the weak cementation bonds subjected to stress concentrations. Hence a "soft soil" has a low Unconfined

Compressive Strength and high compressibility, because of relatively high water content and a fragile macrostructure and microstructure.

Loess is a yellowish, open structured, quaternary deposit of wind-blown silty sands and ikt is widely distributed over the world, covering an estimated total area of 13,000,000 km2. China has the most integrated topographic occurrence (63,000 km2; and mainly in northwestern China) of loess and the thickest loess deposit in the world. Sixty percent of these deposits tend to settle uncontrollably due to collapse of its internal structure when wetted under the action of certain pressure. This phenomenon is defined as water collapsibility [1]. Water is one of the factors attributed to the settlement of the loess when wetted. Loess, which has high strength and low compression at low natural water content, will deform uncontrollably and the strength will decrease sharply when the water content is increased (or immersed in water) [2].

The water sensitivity of loess has been the subject of investigation for a long time. The moistening uniaxial compression test and dehumidifying uniaxial compression test were taken out to investigate the variation in tendency of collapse deformation with the variation of the saturation and axial pressure [3]. The mechanism of loess collapsibility is a problem that needs further detailed study. Amongst the many theories postulated, one that is most plausible is that when the loess is soaked in water, the chemical reactions and physico- chemical reactions occur within the cementing materials. These reactions break down the particle bonding to reduce the structural strength of soil, and hence the reason for collapse of the loess. The pore diameter is often greater than the diameter of the surrounding soil particle, this overhead structure in loess is the inner condition for the collapse. Water infiltration further catalyzes the collapse. Therefore, the micro-structure feature of loess is of importance for understanding loess collapsibility.

This paper presents some examples of microstructure of loess samples collected from Yucheng district of Shanxi province in China. The research study contradicts the popular belief that coefficient of collapsibility is proportional to the clay particle content.

Universiti Tun Hussein Onn Malaysia (UTHM) in general is located in an area consisting of soft soil deposits whose depths exceed 40 m. The low yield stresses, high compressibility, low shear strength and low permeability character of this soft soil deposit fall short of the desired construction specification criteria. Such properties can be improved by introducing a sustainable element within the existing various stabilization techniques. Rubber chips derived from waste rubber tyres and sand were used as additives together with cement to stabilize the clay.

Many researchers have studied the inclusion of tyre waste into soil or directly into concrete. Emiroglu et al[4] investigated mixture of tyre fibres and concrete. They studied the compressive strength and split tensile strength and both of them gave a lower strength compared to that of the plain concrete. They also concluded that increased rubber content will cause the rubberised concrete to be weaker in strength.

Segre and Joekes [5] used the Scanning Electron microscope (SEM) to characterize the rubber particles and to observe the interface between the rubber and the cement matrix. The micrographs were obtained using two different detectors: one with backscattered electrons (BEI), which are capable of distinguishing the cement (inorganic material) from the rubber (organic material) through the contrasting differences in the backscatter of the electrons, and the other using secondary electrons (SEI), which showed better surface detail.

2. Research Materials

Materials used in this study were coded, viz. loess (L), soft clay (SC), rubber chips (R), sand (S), and cement (C). Relevant information about the individual materials are given below.

2.1 Loess (L)

Microstructure images of loess samples from different regions in China can be classified into three representative types: grain structure, aggregation structure and grain-aggregation structure. The loess in Yuncheng of Shanxi province has the typical grain structure. The microstructure images of Yuncheng specimens obtained from different depths are given in Fig.1. From these images, it can be seen that the most of the solid composites in the specimen are single particles or fragmented grains. There is only a small quantity of cementing materials visible in the specimens from 14m depth. The micrograph Fig. 1a for specimen at a depth of 3m shows a coarser particle distribution as well as a slightly larger average aspect ratio of the grains in that they appear more angular than the grains in Fig. 1b (14m deep specimen).

a) Specimen from 3m depth

b) Specimen from 14m depth

Figure 1. SEM images of Yucheng Loess Specimens (vertical plane)

In Fig. 1a the particle or grain contact are varied in the way of contact; point to point, point to side or point to face. These cause some open voids to exist in the microstructure. Open void is formed by loosely arranged particles or grains in loess soil. The diameter of the open void is greater than the diameters of the particles around the void. The particles are likely to fall into the void when the fabric is destroyed caused by some destabilizing action such as loading. So the existence of the open void points towards microstructural instability. Thus the microstructural features of Yuncheng specimens in 3m depth can be concluded as: grain, contact connection, and open void. The collapsibility of the loess soil at 3 m depth is high.

Specimen from a depth of 14m (Fig. 1b) show microstructural features that imply grains with the coexistence of contact connection and cement connection (there are a light cement materials between the solid composites), and embedded voids. Hence the loess soil at this depth are different demonstrating them to be non-collapsible.

2.2 Soft clay (SC)

The soft clay was retrieved in a disturbed but bulk form from a depth of 1.5m in the ground at the Research Centre for Soft Soils (RECESS) site, located beside the university. The particle size distribution of this soft clay used in the study is shown in Fig. 2. The clay contained some organic materials, roots and small fragments of decaying wood. The size of all particles was less than 0.02mm with an effective particle size D10 of 0.001 mm.

The collected disturbed bulk samples were then remoulded in the laboratory as soon as it was collected from the site and then wrapped with a few layers of cling film. Remoulding process preceded the laboratory sample preparation to ensure that the entire soft clay will be uniform and constant in moisture content. It was then placed in a tightly sealed plastic container to maintain the original moisture contents, and was stored in a temperature controlled room, set at 25°C.

Some physical and chemical properties of typical RECESS soft clay are shown in Tables 1 and 2 respectively. This soft clay was classified as CH or OH in accordance with the Casagrande plasticity chart. The electron micrograph of the soft clay shown in Fig. 3 shows thin hexagonal particles and is composed of agglomerates of clay particles with thin platy flakes.

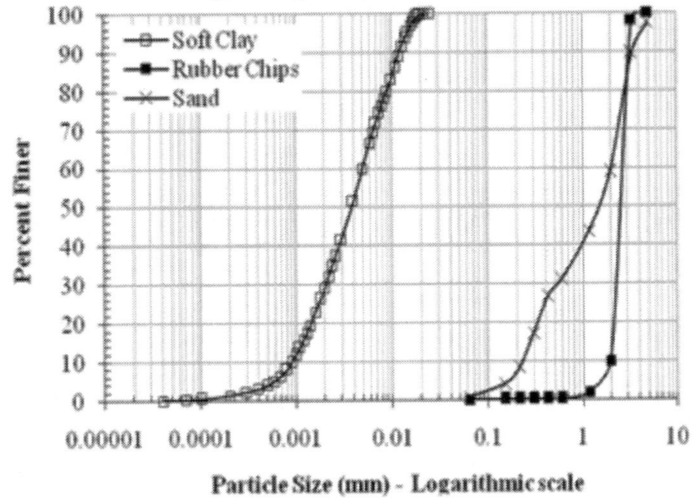

Figure 2. Grain size distribution of RECESS soft clay, rubber chips and sand.

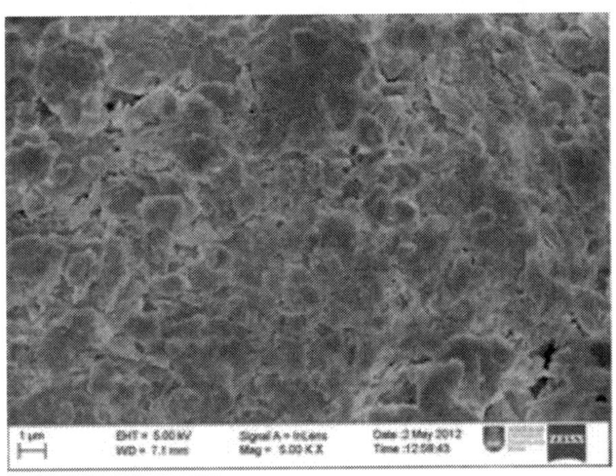

Figure 3. SEM Micrograph of soft clay (x5000).

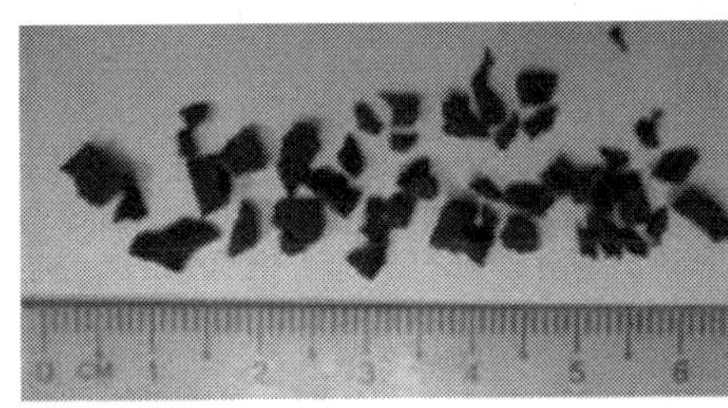

Figure 4. Rubber chips used in this study.

Table 1. Physical properties of RECESS soft clay

Properties	RECESS soft clay
Average water content before remoulding	± 80 %
Specific gravity (Gs)	2.60
Soil pH	3.5
Colour	Light grey
Liquid limit (LL)	68 %
Plastic limit (PL)	32 %
Plasticity index (PI)	36
Clay fraction (percentage by weight passing 2 μm sieve)	29 %
Activity (A) = PI/Clay faction	1.24

2.3 Rubber chips (R)

Rubber chips (Fig. 4) used in this study was obtained from discarded / used truck tyres by crushing and the subsequent removal of the textiles and metal fibers. The rubber chip sizes used were between 2 to 5 mm (refer to Fig. 2). It was obtained from Yong Fong Rubber Industries Sdn. Bhd., Klang, Malaysia which produces reclaimed rubber such as rubber powder, rubber chips and rubber shreds. Rubber chips are elastic material, with Poisson's ratio of 0.5 and elastic modulus is about 4 to 6 MPa (averaged over 0 % to 15 % strains) (Mitarai et al. 2006) [6]. Fig. 5 shows the FESEM of rubber chip that shows a characteristic rough and irregular surface. At lower magnification, viz. 1000 times, some longitudinal shapes can be seen. Thus though the particle size distribution depicts a uniform particle size, with a uniformity coefficient <3, the particles had different shapes with a wide range of aspect ratios.

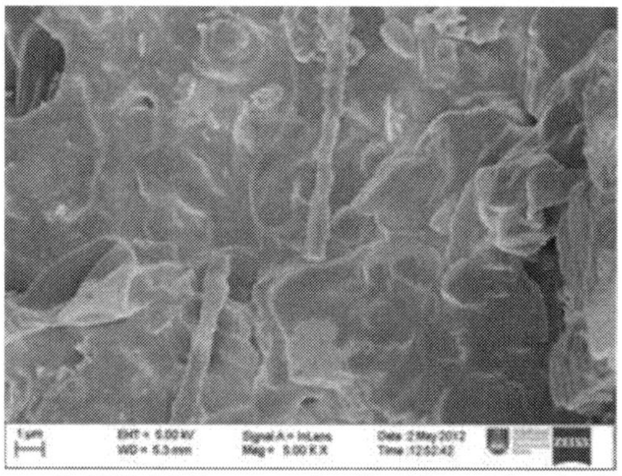

Figure 5. SEM Micrograph of rubber chips (x5000).

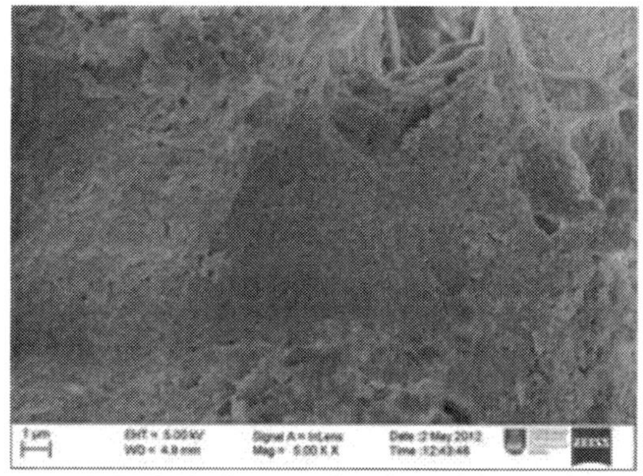

Figure 6. SEM Micrograph of sand (x5000).

2.4 Sand (S)

Clean river sand was used in this study and its gradation curve is also given in Fig. 2. This sand was obtained from a local building material distributor in Batu Pahat, Johor, Malaysia. The sand was oven dried before testing. The type of sand chosen for the study was easily available locally. It is noteworthy that the size distribution of the sand was chosen so as to be similar to the size distribution of rubber chips to ensure that both materials have a similar particle size range. Fig.6 shows the sand also to consist of a mixture of particle shapes; spherical and angular grains.

2.5 Ordinary Portland cement (C)

Ordinary Portland cement is a widely used stabiliser, either on its own or admixed with other additives. It has a specific gravity of 2.86 and was purchased from Holcim Malaysia Sdn. Bhd., Johor Bahru. Fig. 7 shows a micrograph observed for the cement. In cement systems, CH forms are comparatively much more finely dispersed particles throughout the microstructure. The thin platy CH crystals are frequently seen in Portland cement systems.

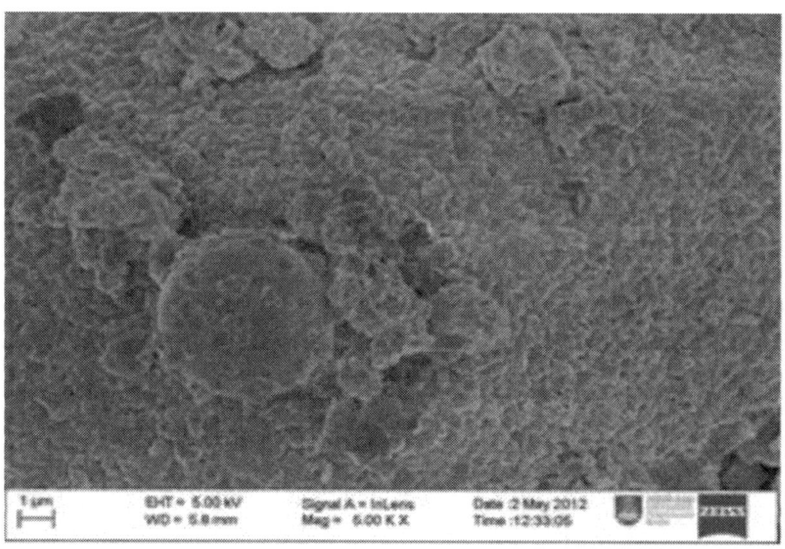

Figure 7. Micrographs of cement (x5000)

Table 2. Comparison of RECESS soft clay, cement, rubber chips and sand chemical composition (in percentage) by X-ray Fluorescence (XRF) test [7].

Chemical Composition	RECESS soft clay	Cement	Rubber chips	Sand
SiO2	59.1	18.30	20.50	96.00
Al2O3	27.5	4.68	16.20	3.44
Fe2O3	3.66	2.32	1.28	-
CaO	0.18	66.80	44.87	-
MgO	1.09	1.59	-	-
Na2O	0.18	0.28	-	-
K2O	1.96	0.57	-	-
SO3	5.25	5.03	6.89	-
TiO2	0.63	0.21	-	0.29

3. Experimental Methodology & Investigations

3.1 Field Emission Scanning Electron Microscope (FESEM)

FESEM is a microscope that differs from an optical microscope because it works with electrons (i.e. particles with a negative charge) instead of light. These electrons are liberated by a field emission source. The object is scanned by electrons in a random and a zigzag pattern. It can discern and help visualize very small topographic details on the surface or entire or fractioned objects as small as 1 nanometer (i.e. billionth of a millimeter).

The working principle of FESEM: A narrow scanning beam of electrons is bombarded on the surface of the sample. As a result, secondary electrons are released from each impact spot on the sample. A detector catches the secondary electrons and produces an electronic signal providing a video scan-image that can be seen on a monitor or as a digital image that can be saved and processed further. The detector used in this study was an In-lens detector.

Prior to microscopy observations, small but representative pieces were collected from one of UCS samples that replicates others and then oven dried for approximately 24 hours. A similar small piece of cement-rubber chips specimen was mounted on a copper specimen holder and then coated with a thin layer of gold coating about 15 nm thick to provide surface conductivity by using sputter coater.

When handling the small piece of clay with cement-rubber chips specimen, it was made certain that a flat surface (as the top side) was chosen as a specimen before coating. A flat surfaced specimen is needed to enhance the capture on the micrograph. The coated pieces were placed in FESEM Carl Zeiss (SUPRA 40 VP with GEMINI column) operating at 5 kV to observe the micro-structural changes in the matrix of the stabilised specimens and to visually examine the resulting products of hydration.

For fine particles such as cement-sand specimen, the steps for coating of the particles are not needed because of its fine size particles. The FESEM examination was performed on a similar small portion of the specimen (± 4 mm), but it is believed to be representative of the reaction process between rubber chips and cement matrix or sand and cement matrix.

4. Results and Discussions

According to past research, the collapsibility of loess decreases with the increasing clay particle content. However the results of the investigations on loess collected from Yucheng and Kelan of Shanxi province, Xi'an of Shanxi province in China (shown in Fig. 8), there does not appear to be an obviously marked relationship between collapsibility coefficient and clay particle content.

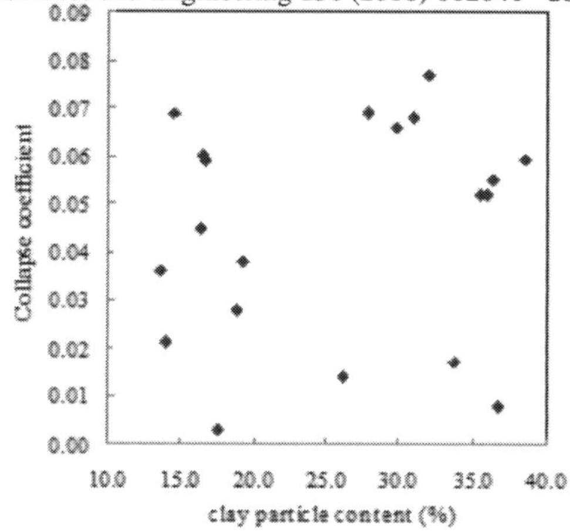

Figure 8. Research observations between collapsibility coefficient of Loess and clay particle content.

Coefficient of collapsibility is a parameter that quantifies the collapsibility of the loess. According to Chinese code for building construction in collapsible loess regions, GB 50025-2004 [8], the loess is called as collapsible loess when coefficient of its collapsibility is greater than 0.015.

On the whole, the correlation between the collapsibility of loess and the content of the clay particles is not good enough. The clay particle content of some samples with great collapsibility coefficient is greater than 20%, even more than 30%. It contradicts with the traditional common view that the loess with the clay particle content exceeding 15%~20% will not collapse. Hence the authors demonstrate that it is difficult to assess the collapsibility of loess from simple observation of its clay particle content alone. There lacks a one-to-one correspondence between the collapsibility and the clay particle content. However, the microstructure analysis indicates that the states of the clay particles existing in the loess have great influence on its properties.

Grain and aggregate are the main types of solid composites for loess soil. The aggregate is the group of grains or particles stuck together by cementing materials like clay particles; it acts like a single particle or grain. It is impossible to judge the collapsibility of the soil only depending on the type of solid composite. The connection types of solid composites include contact and cement. Even cement connection may be more stable than contact connection; further research is needed to judge the collapsibility of the loess soil only based on its type of connection of solid composites.

There are two types of voids, i.e. open void and embedded void, for the loess microstructure. A soil with open voids is loose, and its microstructure is not stable. If there are many open voids existing in the loess soil, it would collapse when the soil is immersed in the water and under the action of certain load. Therefore, the existence of the open voids is the definitive internal factor for collapsibility of the loess, while both loading and wetting are the definitive external factors responsible for collapse of loess.

Figs. 9 and 10 show the microstructure of the stabilized clays SC-5C-5R and SC-10C-5R after 28 days of curing. From these two figures, it is evident that the amount of calcium silicate hydrates (CSH) and calcium aluminate hydrates (CAH), which are two cement hydration products formed in the clay stabilized with cement is greater and its fabric appeared less porous than that of untreated clay. When comparing stabilized clay to the untreated clay, it can be seen clearly that a bonded structure exists, whereby the homogeneous texture of clay particles interact with cement hydration products. As a result of such inter-particle cementation, an increase in small-strain shear modulus can be expected [9].

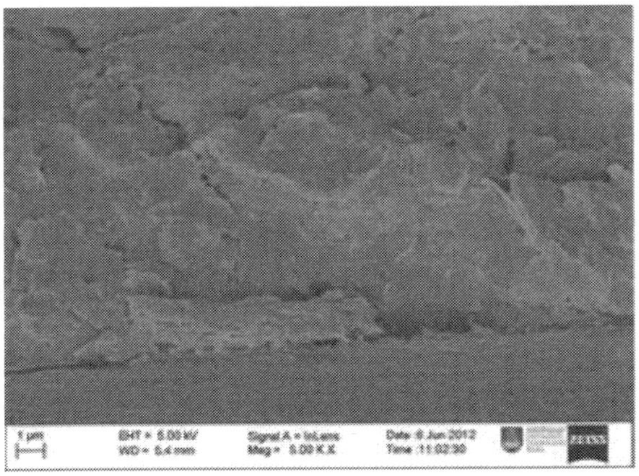

Figure 9. SEM micrographs for SC-5C-5R after 28days (x5000).

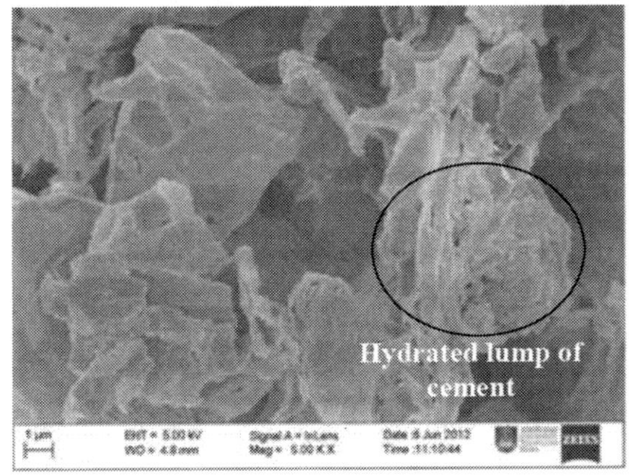

Figure 10. SEM micrographs for SC-10C-5R after 28 days (x5000).

SEM micrographs clearly indicate that cement interacts with the clay minerals to form new cementing materials, viz. calcium silicate hydrates (CSH) and aluminate hydrates (CAH). Interface bonding between cement-rubber chips and cement-sand can be observed within the stabilized clay. This consequently produces high stiffness and strength but often a low resistance to fracture, i.e. brittle behavior. These new materials functioned as additives that filled the inter-granular clay voids or sometimes coat the clay particles that result in decreased soil porosity and permeability.

The micrograph for the SC-5C-5R (sand cement sample with 5% cement and 5% rubber chips) (Fig. 9) seem to be more evenly distributed in the mix as compared to the SC-10C-5R mix. The micrograph (Fig. 10), for the latter sample shows that the matrix is more porous and the particles are agglomerated or aggregated. This can be due to the higher moisture content in the soft clay that reacts with the increased cement content (10 % C). The addition of cement to soft clay makes it more aggregated and flocculated, as compared with untreated soil. Stacks of clay particles can still be seen, indicating that not all soft clay particles are cemented. On the other hand at 10 % cement, the presence of cement seems obviously distributed throughout the mixture. These observations also complement and agree with the laboratory UCS test, where it was noted and is also known that the higher cement content will give higher strength and stiffness [7].

Fig. 11 shows a micrograph with a larger magnification (x10000) of a similar sample as that in Fig. 10. It is observed that larger pores represent a looser bonding between cement-rubber chips samples. Figs. 12 and 13 show micrographs for samples SC-5C-10S and SC-10C-10S after 28 days, respectively. As observed, both FESEM images look similar even though the cement content has been doubled in the case of Fig. 13. As compared to the cement-rubber chips samples, the cement-sand sample pores are well-knitted together and infilled with hydration products [10]. The processes of cement hydration; alite (C3S–Ca3SiO5) and belite (β-C2S–Ca2SiO4) are the main components in Portland cement, were virtually undetected in Fig. 13, confirming that primary hydration was completed within 28 days. Portlandite, naturally occurring form of calcium hydroxide (Ca(OH)2) or the peak of calcium aluminum oxide (Ca-Al-O) can also be seen in Fig. 12 with higher cement content.

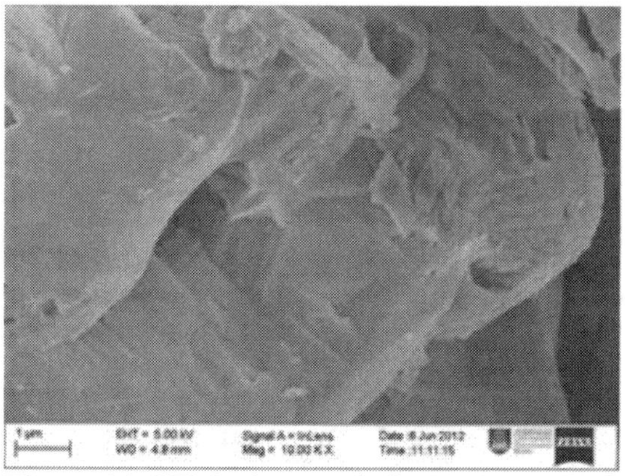

Figure 11. SEM micrograph for SC-10C-5R after 28 days (x10000).

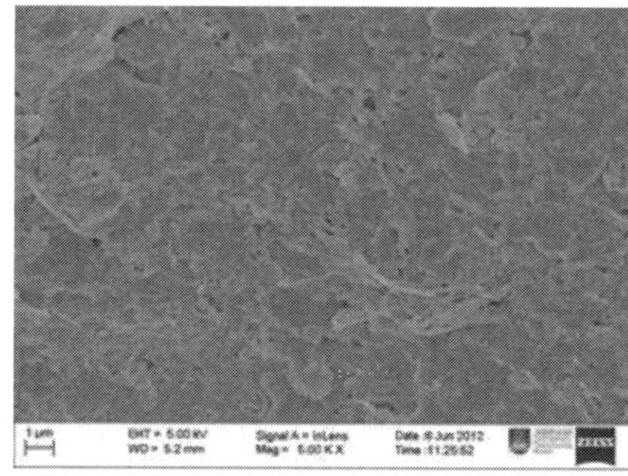

Figure 12. SEM micrograph for SC-5C-10S after 28 days (x 5000).

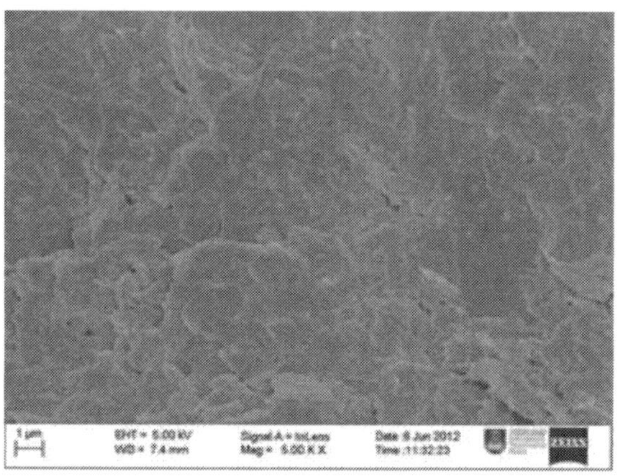

Figure 13. SEM micrographs for SC-10C-10S after 28 days.

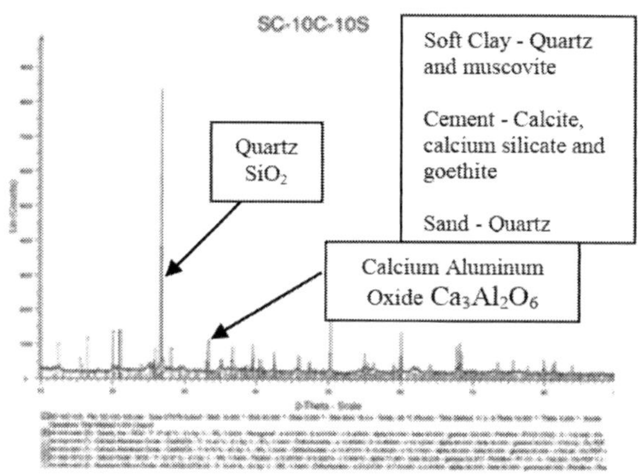

Figure 14. XRD (X-ray Diffraction) pattern of SC-10C-10S [7].

When more cement is added, particle agglomeration takes place with the cementation due to secondary particles. Similar observations were also reported by Hossain and Sakai [11]. The SEM images published by Emiroglu et al. [4] showed that the addition of waste tyre rubber in normal concrete does not affect the C-S-H formation in concrete (viz. C-S-H morphologies of normal and rubberized concrete are the same).

Some previous researcher [12] have discussed about the strength of the stabilised clay. Fig. 15 shows the compressive strength (qu) obtained from laboratory testing for untreated soft clay is 13.48 kPa. Also, SC-10C-5R sample shows no change or only a slight reduction in the strength for the field test result. Another observation is that the qu at 28 days for laboratory, semi-controlled and field test are very close to one another within 6 to 12 % of each other. A very promising field test result was observed for SC-10C-15S result that showed improvement in the properties. This shows that mixing sand into the stabilized clay gives a stiffer stabilized soil compared to the rubber chips addition. Both SC-10C-5R and SC-10C-15S showed an increase in strength as the curing days increased except in the field test result for cement-rubber chips after 28 days.

IOP Conf. Series: Materials Science and Engineering **136** (2016) 012041 doi:10.1088/1757-899X/136/1/012041

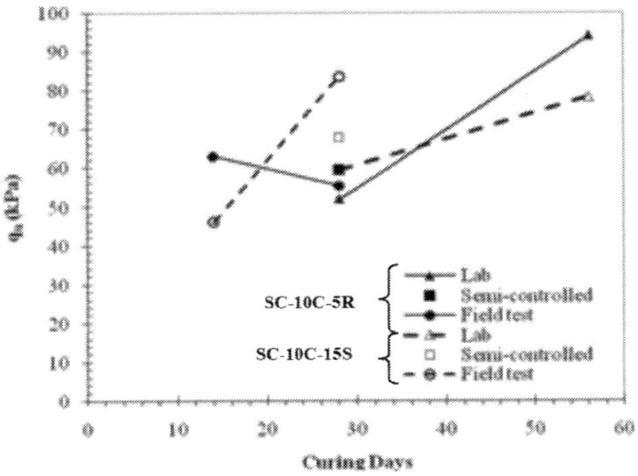

Figure 15. Undrained shear strength for SC-10C-5R and SC-10C-15S samples [12].

The results from SEM showed that a discontinuity can be observed in the rubber-matrix interface, indicating rubber adhesion to cement paste was poor, while for NaOH-treated rubber particles, an adhered joint was noticeably present. The powdered tyre rubber particles were surface treated with NaOH aqueous solution for 20 minutes, which enhanced the bonding of the rubber particles to the cement matrix. The mechanical behaviour of soil is strongly influenced by the shape, size and surface characteristics of the soil particles [14]. According to Lee and Lee [13], the soil microstructure is also a function of the individual particle properties, and strongly influences the soil behaviour. Such complementing details on particle properties can only be observed with the aid of a SEM. Prior to sample preparation the pore water in the soil needs to be removed through replacement or by freeze drying. This process of removal of pore water is very difficult to be done without disturbing the structure of the sample and some debate still exists as to what method is most suitable for different types of soils.

SEM result for bonding of kaolin-cement was that the addition of cement causes the kaolin to be less aggregated but more flocculated as compared with the untreated one [13]. They observed that 5 % or less of cement addition did not have significant effect on the mechanical properties of the stabilised soil, when observed through the SEM photomicrographs.

Hossain and Sakai [11] studied the formation of secondary particles that accompanied the cementing reaction after 7 days curing. They reached this conclusion through SEM observations of the formation of larger sized particles in the soil when 0.2, 0.4 and 0.6 % cement was added. This flocculation brought the clay particles together by cementing them to form a compound or secondarily agglomerated particle.

Horpibulsuk et al [10] also observed that in cement-stabilised silty clay, there was a strength gain with time. They observed that the large pores were filled with the cementitious products that resulted in a pore volume which was smaller and consequently the total pore volume decreased.

The terms aggregation/ agglomeration is used synonymously in this paper to refer to a process of particle collection into an unorganized entity of a larger size. Flocculation is used in the context of the process of individual particles of clay aggregates becoming clotted mass or precipitated into small lumps. Flocculation occurs as a result of chemical reaction between clay particles and salt water. Conversely, dispersion is the process of breakdown of grouped particles into individual particles and evenly distributed.

5. Conclusion

The analysis of the microstructure of loess soils from different regions in China shows that loess collapse happens when the soil mainly possesses open void which is the prime inner cause of the collapsibility of loess. Overall, the FESEM observations help to demonstrate bonding between cement, rubber chips/ sand and soft clay. Larger pores were seen in enlarged FESEM image that occurred for cement-rubber chips sample, which represent a looser bonding between cement-rubber chips samples as compared to the cement-sand sample. Whereas, the microstructure of cement-sand is more closely bonded compared to cement-rubber chips samples and therefore influences the geotechnical properties of the stabilized clay. Sand seems to give better result in unconfined compressive strength than the rubber chips in the semi-controlled and field test result. The potential use of rubber chips and sand in this research is not so obvious and the cement content needs to be increased in order to give greater improvement to the strength of the soft clay.

Acknowledgment

The authors thank Universiti Tun Hussein Onn Malaysia for supporting this research under the Postgraduate Incentive Research Grant. The research on Loess received financial support from NSRF in China and NSRF of Shanxi province.

References

[1] Assallay A M, Rogers C D F, Smalley I J, 1997 Formation and collapse of metastable particlepacking and open structures in loess deposits. *Engineering Geology*, **48** 101-115
[2] Xie D Y, 2001 Exploration of some new tendencies in research of loess soil mechanics. *Chinese Journal of Geotechnical Engineering*. **23(1)** 3-13.
[3] Zhang M H, Xie Y L, Liu B J, 2005 Analysis of Compression Deformation Characteristics of Loess during Moistening and De-moistening Process. *Journal of Hunan University of Science & Technology(Natural Science Edition)*, **26(9)** 50-55.
[4] Emiroglu M, Kelestemur M H and Yildiz S, 2007 An investigation on ITZ microstructure containing waste vehicle tire, *Proc. of 8th Int. Fracture Conf.*, 7-9 November 2007 453-459.
[5] Segre N and Joekes I, 2000 Use of tire rubber particles as addition to cement paste, *Cement and Concrete Research*. **30(9)** 1421-1425.
[6] Mitarai Y, Yasuhara K, Kikuchi Y and Karmor A K, 2006 Application of the cement treated clay with added tire chips to the sealing materials of coastal waste disposal site, *5th ICEG Environmental Geotechnics, Thomas Thelford, London*. 757-764.
[7] Ho M H, 2014 The potential of using cement-rubber chips and cement-sand as additives in stabilised soft clay, Universiti Tun Hussein Onn Malaysia: Ph.D. Thesis.
[8] Chinese code for building construction in collapsible regions, GB50025-2004. (in Chinese)
[9] Verástegui Flores R D, Di Emidio G and Van Impe W F, 2010 Small-strain shear modulus and strength increase of cement-treated clay, *Geotech. Testing J.* **33(1)** 62-71.
[10] Horpibulsuk S, Rachan R, Chinkulkijniwa A T, Raksachoon Y and Suddeepong A, 2010 Analysis of strength development in cement-stabilised silty clay from microstructural considerations, *Construction and Building Materials*. **24** 2011-2021.
[11] Hossain M Z and Sakai T, 2008 The effectiveness of nominal dosage of ordinary cement on strength and permeability of clayey soil, *Journal Japan Soc. Soil Pys*. **110** 25-35.
[12] Ho M H, Wijeyesekera D C and Bakar I, 2013 A comparative study of laboratory-controlled, semi field-controlled and field testing for cement stabilised clay using rubber chips and sand, *Soft Soil Engineering International Conference (SEIC)* Sept. 2013, Kuching, Sarawak.
[13] Lee K H and Lee S, 2002 Mechanical properties of weakly bonded cement stabilised kaolin, *KSCE J. Civil Eng*. **6(4)** 389-398.
[14] Mitchell J K, Fundamentals of soil behavior, 2nd Ed., John Wiley & Sons, Inc., New York, (1993).

IOP Conf. Series: Materials Science and Engineering **136** (2016) 012042 doi:10.1088/1757-899X/136/1/012042

Mitigating Settlement of Structures founded on Peat

D C Wijeyesekera[1,2], L Numbikannu[2,1], T N H T Ismail[2.1] and I Bakar[1,2]
[1] Research Center for Soft Soil, Universiti Tun Hussein Onn Malaysia, 86400 Batu Pahat Johor, MALAYSIA
[2] Faculty of Civil and Environmental Engineering, Universiti Tun Hussein Onn Malaysia, 86400 Batu Pahat Johor, MALAYSIA

E-mail: devapriya@uthm.edu.my

Abstract. Observations made of two common failures of structures founded on peat/organic soil in Johor, Malaysia is presented. Critical evaluation of current lightweight fill technology to mitigate such settlement is also discussed. Lightweight technology, such as Expanded Polystyrene (EPS), has been used in construction on soft yielding ground for decades. Regrettably, some published information of EPS failures to perform on construction sites are also cited in this paper. This paper outlines some concepts leading to the development of an alternative innovative lightweight fill is that the idealised cellular structure of the GCM permit free flow of water and complemented by the mat structure which evens out any differential settlement A further highlight of this paper is the monitoring of the field performance of this lightweight fill (GCM) as a feasible alternative to fill weight reduction on yielding ground.. Hence, a prime research objective was to compare the fill settlements observed with 1m high fill of surcharge loading on peat ground (comparison of the case of using a partial 0.6m high GCM and that of a total of 1m of conventional sand backfill).

Keywords: Soft soil, cellular mat, Settlement mitigation.

1. Introduction

Stability of civil engineering structures depends on the geotechnical properties of the foundation such as shear strength, permeability, and compressibility of the soil. The low strength of highly organic soil such as peat cause excessive settlement and failure of the civil engineering superstructure. Consequent cracking occur on the main superstructure or pavement when there is excessive settlement leading to super structure failure.

Figure 1 (a) shows an example of such settlement failure observed at Parit Nipah, Johor, Malaysia. Although, the main activity at Parit Nipah is farming and therefore the volume of even moderately heavy trucks are limited but the road surfaces shows signs of distress (longitudinal cracking, potholes and differential settlements giving rise to a bumpy road). This adverse problem is due to non-uniform and excessive consolidation settlement. Bumpy roads result from such differentially settling roads proved to be damaging, uncomfortable and hazardous to the threesome of vehicle, driver and passenger. The second example of failure is that which has occurred repeatedly at the bridge approach in Parit Yaani, Johor, Malaysia (Figure 1 (b)). This is a popular road used by local community to commute from Sri Garding to Parit Yaani and it is therefore heavily used. This road has been repaired several times with the placing of additional asphalt. However, sometime (6 month) after repair this approach begins to settle again. Such affect can be and have been modeled in the laboratory through physical and software modeling. Both sites are considered unfavourable as the local drift material on both sites is either highly organic clay or decomposed peat.

Such soft organic soils are prone to moisture induced ground movements which pose undesirable challenges to the field construction even during this modern age. These types of soil are simply not capable of bearing the load exerted by a structure due its extraordinarily high field moisture content that can be as much as 800% at times [1]. As a result, the structure sinks into the soft soil, somewhat similar to how a person standing in the mud sinks into soft and wet clay. Figure 2 shows the distribution of peat throughout Peninsular Malaysia and their occurrence seems to be confined to the coastal area. Hence, there is an urgently emerging construction issues for structures founded on such areas of yielding soil. Peat and organic soil have been identified as major group of problematic and challenging soils from a civil engineering perspective. Peat is found in many countries, including peninsular Malaysia, where 3.0 million hectares or 8% of the country land area is covered with tropical peat. Peat covers more than 4 million km^2 of the planet's surface and represents 50 – 70% of the total wetlands on the earth [2]. Peat soil is by nature complex and highly variable and consequently its properties cannot be determined with precision. Peat is an organic soil with more than 75% of organic content and therefore poses many problematic physical and mechanical issues such as: soil surface subsidence, high water table, loose soil structure, and high volumes of subsurface woody debris. It is noteworthy that contrarily, these properties are the most favourably desired properties for agricultural purpose; the very productive palm oil farms are popularly located on peat ground.

Figure 1. Road settlement failure; (a) roller coaster scenarios in different settling road at Parit Nipah, (b) bridge approach settlement in Parit Yaani

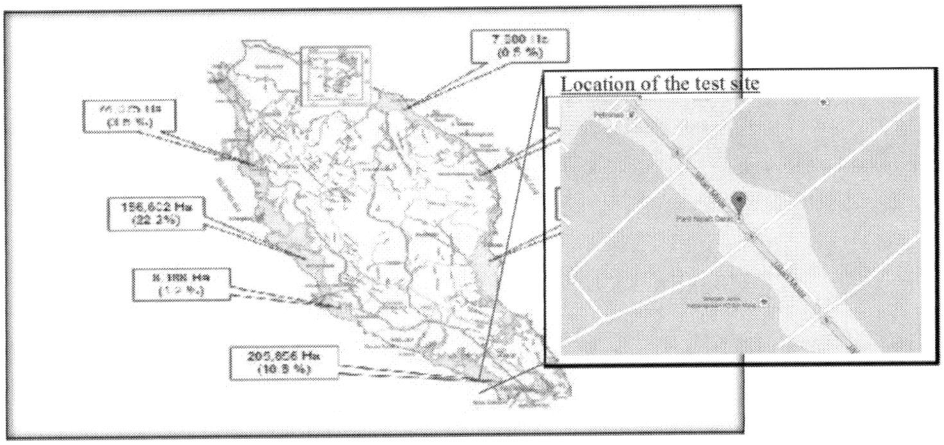

Figure 2. Extent of peat (highlighted in yellow) in Peninsular Malaysia and the location of the Parit Nipah test site. (Modified from [3])

Many researchers have attempted to provide a mitigating solution with the addition of stabilizing material such as cement, lime, fly-ash, pozzolanas and blast furnace slags [4]. However, the effect of this type of stabilisations is much localised and temporary because the distribution of this non-uniform peat is widely variable both in area and depth.

With decades of research on Forensic Geotechnical Engineering, many innovative approaches have been used to mitigate settlement of structures on peat. For example, Expanded Polystyrene (EPS) was adopted for retaining structures to provide a reduction in lateral pressure as well as preventing settlement and improving waterproofing. Regrettably there are documented instances where the EPS construction sites have experienced floatation failure situation associated with buoyancy forces arising from water level fluctuations. Two examples of such failures were published in Horvath 1999 [5]. The first case was a failure that occurred at Carousel Mall in Syracuse New York. The fill was under construction when a water main burst, filling the surrounding area of the EPS with water. The second case shown is from a roadway in Norway during an unusually high flood. Hence, this project focuses on the development of another innovative lightweight fill which could reduce settlement of structure on peat and eventually reduce the impact on environment.

Lightweight cellular mat structures are presented as a further feasible alternative of weight reducing material in this research study. The idealised cellular structure of the GCM permits free flow of water while the mat structure evens out any differential settlement. The use of cellular solids has been introduced widely to replace or substitute the use of conventional materials like soil and rock. Cellular solid mechanic is different from soil mechanics. Cellular solid is made up of a cell structure that help share the load, and soils are particulate (as illustrated in Figure 3) [6]. Figure 4 shows examples of natural materials with cellular structure as found in wood, cork, sponge, cancellous bone, skull and Plant Stems.

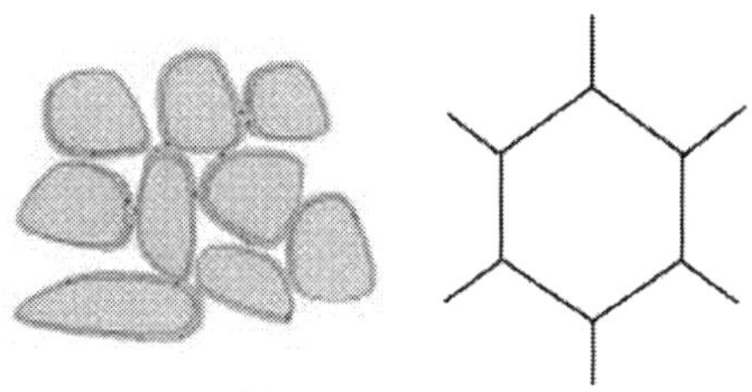

(a) (b)

Figure 3. Mechanics of material: (a) soil particle, (b) cell structure

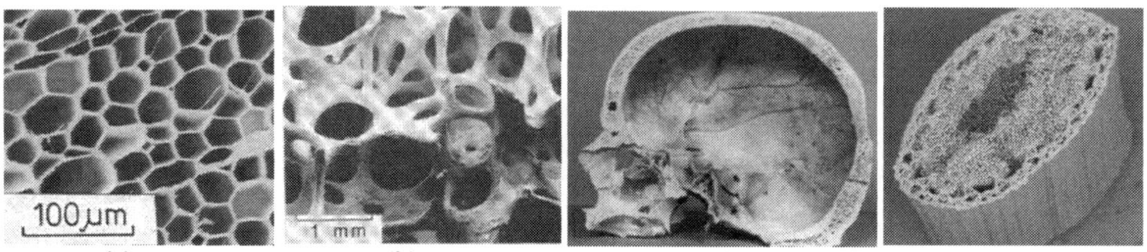

(a) (b) (c) (d)

Figure 4. Some example of nature cellular structure: (a) wood, (b) cancellous bone, (c) skull, (d) plant stems [7]

Mitigating settlement failure of embankments using cellular mat technology has been demonstrated through laboratory testing and physical modelling. This paper presents a field study case to demonstrate the capability and potential of the lightweight cellular mat structures on the peat ground and how it can be used to reduce the total and differential settlement through accelerating consolidation settlement.

2. Conceptual and Model Study of the Cellular Mat Structure

Physical modelling was carried out through laboratory testing to show the concept of failure by application of lightweight cellular mat structure (illustrated as a rigid foundation) compared to the conventional material (sand used to represent a flexible foundation). This study identified three types of settlement as shown in Figure 5. Figure 5(a) shows the uniform settlement that occurs with a rigid foundation. The tilt of superstructures in Figure 5(b) is a result of variable thickness of the foundation materials leading to differential settlement. The non-uniform settlement in Figure 5(c) is a result of the flexibility of the foundation structures as portrayed by the effect of the particulate material in the conventional field.

Figure 6 shows two different types of failures that occurred at the Palace of Fine Arts, Mexico City and Transcano Elevator, Canada, respectively. The Palace of Fine Arts (The Palacio de Bellas Artes) was built comprising a structural steel frame supported on a raft (or mat) foundation on soft clay. This structure settled in several stages. Based on the pattern of settlement shown in Figure 5 this is referred to as a case of uniform settlement. The second case study shows the non-uniform tilt settlement failure that occurred at Transcano Elevator, Canada. The foundation of the elevator consisted of a large rectangular raft resting on a plastic clay deposit [9]. The foundation settled as a rotational slip and the failure of the structure was due to the instability of the clay to support load placed.

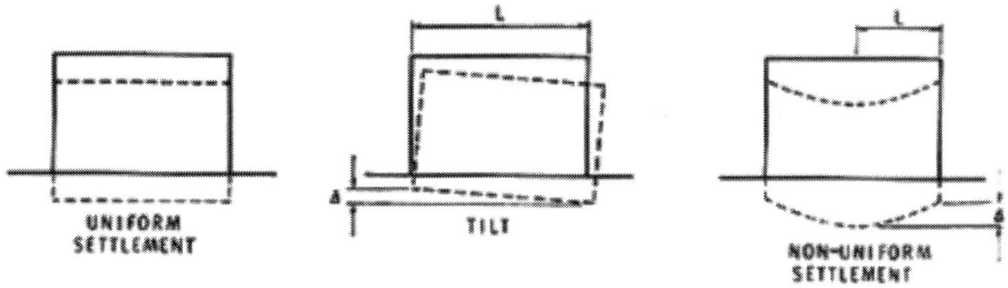

(a) (b) (c)

Figure 5. The diagram of settlement failure [8]

(a)

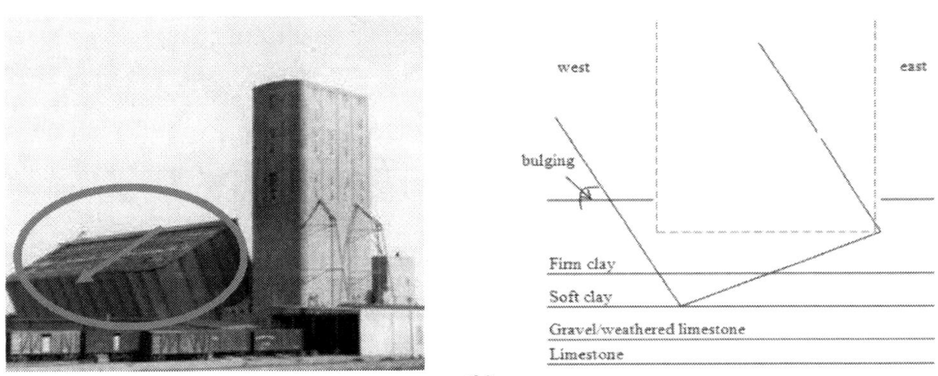

(b)

Figure 6. Settlement failure; (a) uniform settlement occur on The Palace of Fine Arts (b) non-uniform failure of the Transcano Elevator, Canada

3. Site soil condition

Peat soil is by nature complex and has wide varying engineering properties depending largely on the deposit's depth below the ground surface and its proximity to the water table. Table 1 shows a summary of the varying properties of peat in Johor by. Based on the vane shear tests conducted, the undrained shear strength of peat at Parit Nipah is around 8-11 kPa (as shown in Table 1). Natural moisture contents observed are commonly in the range of 670 and 808%. Moreover, the liquid limit and plastic limit are generally in the range of 119 – 410% and 123 - 243, displaying high sensitivity. Compression index as high as 3.76 – 5.30 has been reported by [10]. At this high compressibility, strain rate effects can be significant.

The effectiveness of the cellular mat structure to mitigate settlement was mainly conducted on Parit Nipah peat, which extends to a maximum depth of 6.5m. Underlying this layer is silty clay. The tests were divided into two models – model (1) contained no cellular mat structure and model (2) had cellular mat structure at 0.6m depth, in order to compare the settlement failure of lightweight technology and conventional fill material. These two models were setup in the soil box having a size of 1x1x1.2m on the same site but next to each other. Square cellular mats with size of 0.95m width, 0.95m length and 0.2m depth were fabricated. The first soil box was filled with sand (as a conventional fill material) up to 1m depth. Three layers of cellular mats were placed in the second soil box together with a 0.4m sand cover, amounting to a total height of 1m.

Table 1. Physical properties of typical hemic peat soil in Johore.

Parameter	Johore Hemic Peat	Parit Nipah Peat		
	[11]	[12]	[13]	[10]
Undrained shear strength (kPa)	-	-	-	8 - 14
Natural water content, W (%)	230 – 500	808	791	670 – 730
Liquid limit, LL (%)	220 – 250	-	119	250 – 410
Plastic Limit, PL (%)	-	-	-	123 – 243
Specific gravity (Gs)	1.48 – 1.8	1.27	1.88	1.52 – 1.59
Organic content (%)	80 – 96	95.44	78.76	73.5 – 83.5
Unit weight (kN/m^3)	7.5 – 10.2	-	-	-

4. Field Instruments for monitoring the performance

Considering the limitations of the site condition and cost, it was decided to adopt conventional geodetic surveying methods to monitor vertical displacement (or deformation). A special settlement gauge staff was developed to monitor the settlement of the soft ground. The uniqueness of this instrumentation is that it is lightweight hence not intrusive, easy to use, inexpensive and requires less human resources are required. Moreover, these instruments will be available on site until the completion of the testing period which is approximately about 6 months. Since these instruments will not be moved, more plausible results are expected to be obtained. There will be less soil disturbance compare to other conventional methods. Five settlement gauges were installed in each case and these included an observation at the centre of each soil box test. All settlement gauges were read through the digital auto level (with 0.001m accuracy) to monitor vertical settlement.

The current settlement readings obtained after 60 days of monitoring are shown in Table 2. These readings show evidence that cellular mat structure can be effectively used to reduce settlement and accelerate consolidation settlement. Model 1 with no ground improvement had 41% more subsurface settlement than model 2. The field monitoring of settlement is still ongoing to date.

Table 2. Summary of settlement observations.

Model	Description	Maximum vertical settlement for 60 days (mm)
Model 1	Conventional material only without cellular mats	63.12
Model 2	0.6m depth of cellular mats + 0.4m depth of sand	44.53

The in-situ settlement time plots observed at the centre point of each field soil box test are shown in Figure 6. These plots present the ground level settlement observations at the base of the fill loading, with Figure 7(a) plotted using Casagrande's log time method, and Figure 7(b) plotted using Taylor's square root of time method. The settlement data is more demonstrable with Casagrande's method than the Taylor's square root of time method. Both figures illustrate the primary settlement of the lightweight fill material to cease somewhere around 25mm (for the conventional fill loading around 33mm for Casagrande's method and 37mm using Taylor's method). This confirms that the lightweight fill material reduces and accelerates settlement that occurs on peat ground.

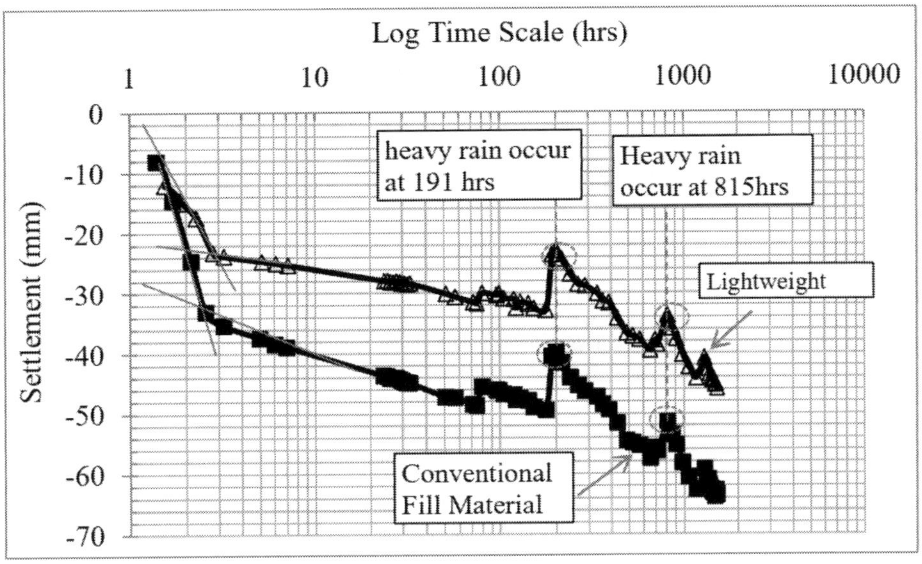

Figure 7(a). Measured settlement at the centre of the model (Casagrande's method)

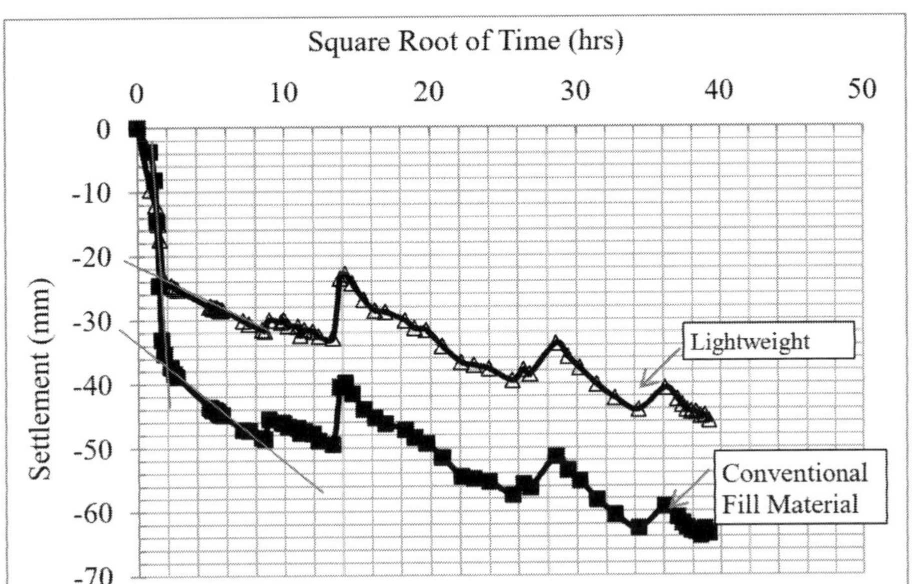

Figure 7(b). Measured settlement at the centre of the models (Taylor's method)

Both observations presented in Figure 7(a), show two sharp rises (indicating a temporary ground heave) in both setups occurring at a time of around 191 and 815 hours after the commencement of the construction phase. This might be due to the heavy rain that occurred at the test site on the night previous to the day of the said observation. The water table at the site recorded the highest level, corresponding to a depth of 0.358m and 0.470m from the surface respectively. The heavy rain fall caused the ground in the vicinity of the test site to be flooded. It was observed that at the covered the test site the water entered the sand fill box from the bottom and caused the ground to rise and heave (also made the sand moist and heavier). Even though so, the lightweight fill experienced lesser heave compared to the conventional fill. This confirms that the cellular structure of the innovative lightweight material functioned well in reducing the floatation by allowing the flow of water through it.

5. Applicability of Terzaghi Theory for Peat

The commonly adopted Terzaghi's one dimensional consolidation theory assumes that the soils are fully saturated, homogeneous, the soil compression and water flow occurs vertically, and to obey a generalized Darcy's law which takes into accounts the movement of solid particles. Moreover, the classical Terzaghi theory only considers small strain consolidation which is definitely contrary to the settlements experienced with the peat ground settlement.

Hence, for the field study on fibrous peat at Parit Nipah, it can be considered that some of the assumptions of Terzaghi theory is invalid and therefore the Terzaghi theory is not appropriate to predict settlement on peat soil. This is because fibrous peat is not homogeneous and the consolidation process takes a long time due to the high fibrous and high moisture content. However, hemic peat which is very highly decomposed maybe considered as a homogenous soil. Besides that, peat is a type of soft soil that is expected to behave as anisotropic. It is to be expected that if there is a dominant direction along which the fibres (old tree roots etc) are aligned, the peat will show anisotropic behaviour. The reaction of peat when loaded perpendicular to the main fibre direction will be different than it is loaded in the main fibre direction.

Furthermore, since peat is a soft soil, settlement predictions need to consider instead the large strain consolidation theory such as the Gibson theory. A further restriction to the use of small strain theory to peat settlement is that it also invalidates the assumption that the soil particles are incompressible and defining strain in an element of a fixed volume and ignoring the boundary movement as consolidation occurs.

6. Conclusions

This study has helped to obtain the field evidence for performance of the innovative lightweight material as a feasible and effective alternative lightweight fill.. It is concluded that the lightweight fill mat can help mitigate settlement in both reducing and accelerating settlement that occur on soft yielding ground like peat. Field performance observations showed that the GCM has reduced the total settlement by 41% compared with that observed using conventional fill material. Furthermore, by using GCM the primary consolidation only needs 25 hours to complete compared to the conventional fill material which required 44 hours. The promising performance as seen from the preliminary data obtained creates an opportunity for the material to be tested in the scale of a real trial embankment.

References

[1] Huat B B K, Asadi A 2013 Most Current Ideas and Latest Advances in Soil Improvement at University Putra Malaysia. *Proceeding of Soft Soil Engineering International Conferences (SEIC'13)*, Sarawak, Malaysia.

[2] Abdullah M, Huat B B K, Kamaruddin R, Ali A K and Duraisamy Y 2007 Design and Performance of EPS Footing for Lightweight Farm Structure on Peat Soil. *American Journal of Applied Sciences* **4**(7) 484-490.

[3] Ab Kadir M I 2009 Long Term Consolidation Study on the Tropical Peat at Pekan, Pahang, Malaysia. Universiti Malaysia Pahang: *Bachelor Degree Thesis*.
[4] Makusa G P Soil Stabilization Methods and Materials. Department of Civil, Environmental and Natural Resources Engineering. Lulea University of Technology, Sweden.
[5] Horvath J S 1999 Geofoam Fills and The Non-Issue Of Buoyancy. *Lessons Learned from Failure Involving Geofoam in Roads and Embankments: Report No.* CE/GE-99-1, Civil Engineering Manhattan College.
[6] Ajdari A 2008 Mechanical Behavior of Cellular Structures: A Finite Element Study. Northeastern University: Mechanical Engineering, *Master's Thesis*.
[7] Gibson L J 2005 Biomechanics of cellular solids. *Journal of Biomechanics* **38** 377–399.
[8] Information on http://www.gf.uns.ac.rs/-wus/wus07/web3.htm.
[9] Barnes G 2000 Soil Mechanics: Principle and Practice. 2nd ed. London: Macmillan Pres LDT.
[10] Bakar I and Wijeyesekera D C 2014 Soft Soil Engineering of Challenging Grounds. Fourth Progress Report MTUN COE Research Grant.
[11] Zainorabidin A and Wijeyesekera D C 2007 Geotechnical Challenges with Malaysia Peat. *Advances in Computing and Technology Conference* Universiti of East London.
[12] Kheong C S 2011 Analysis of Peat Fabric Response to Various Stress and Intensities. Universiti Tun Hussein Onn Malaysia: Bachelor Degree Thesis.
[13] Atemin H H 2012 A Study on Consolidation and Permeability Properties of Tropical Peat. Universiti Tun Hussein Onn Malaysia: Bachelor Degree Thesis.

Stabilization/Solidification Remediation Method for Contaminated Soil: A Review

S A A Tajudin[1,2], M A M Azmi[2,1] and A T A Nabila[2,1]
[1] Research Center for Soft Soil, Universiti Tun Hussein Onn Malaysia, 86400 Batu Pahat Johor, MALAYSIA
[2] Faculty of Civil and Environmental Engineering, Universiti Tun Hussein Onn Malaysia, 86400 Batu Pahat Johor, MALAYSIA

E-mail: saifulaz@uthm.edu.my

Abstract. Stabilization/Solidification (S/S) is typically a process that involves a mixing of waste with binders to reduce the volume of contaminant leachability by means of physical and chemical characteristics to convert waste in the environment that goes to landfill or others possibly channels. Stabilization is attempts to reduce the solubility or chemical reactivity of the waste by changing the physical and chemical properties. While, solidification attempt to convert the waste into easily handled solids with low hazardous level. These two processes are often discussed together since they have a similar purpose of improvement than containment of potential pollutants in treated wastes. The primary objective of this review is to investigate the materials used as a binder in Stabilization/Solidification (S/S) method as well as the ability of these binders to remediate the contaminated soils especially by heavy metals.

Keywords: Soft soil, stabilization/solidification, contaminated soils.

1. Introduction

Heavy metal contaminations in soil are causing a serious threat to the environment and human health. However, there are several technologies that have been developed to treat and remediate the contaminated soil. Remediation technologies can be classified according to immobilization or extraction (action that is applied to metals), in-situ or ex-situ (location that is applied to metals) and other types of technologies [1]. The treatments of heavy metal contaminated soils are limited to two main strategies which is immobilization and extraction. Immobilization approaches aim at stabilizing the metals by minimizing the leaching characteristics of the soil matrix and change the metals to less soluble, toxic or bioavailable form in the soil to reduce the risks of human health and the environment [2]. While, extraction approaches are referred to a process that separates the metals from the soil's composition, reduce the concentration of metals as well as reduce the volume of the entire contaminated medium. However, between these two strategies, immobilization approaches is identified to have advantages in terms of cost-effective and environmentally friendly.

Currently, there are several remediation methods that have been implemented using immobilization approaches. One of the promising methods is Stabilization/Solidification method. Hunce et., al. [3] defines the Stabilization/Solidification (S/S) method as a technique that aims in immobilizing contaminants by converting them into a less soluble form and encapsulating them with the creation of durable matrix. Other research that were done by Lasheen et., al. [4] stated that heavy metal wastes normally needs S/S method processes to reduce contaminant leaching prior to landfill disposal. Another research by Yao et., al. [5] mentioned that the Stabilization/Solidification (S/S) method is commonly used to reduce the mobilization of contaminants within a hardened mass (solidification) and chemical conversion of contaminants into less soluble form (stabilization). In the same way, Voglar and Lestan [6] stated that the S/S method has emerged as an efficient technique for the treatment of sites that are contaminated with potentially toxic metals. Therefore, this study were conducted to review the current knowledge on the materials used as a binders in the Stabilization/Solidification (S/S) remediation method focusing on the soil contaminated by heavy metals.

2. Stabilization/Solidification Remediation method

Cement-based S/S technology has been shown to be effective in immobilizing the heavy metals even without additional additives [7]. Du et., al. [8] has studied the leaching behavior of Pb contaminants by using OPC as a binder. This research concluded that at pH 2.0, this strongly acidic condition has resulted in substantial lowered leachate pH and significantly increased the amount of Pb leached. Contradictory to the condition, when OPC was added in S/S sample from 12% to 18%, it resulted in a decreased amount of Pb leached. In another study, Li et., al. [9] concluded that Pb concentration has been leached out from the solidified specimens using OPC as a binder at 109, 83 and 71 mg respectively with cement ratio of 0.2, 0.3 and 0.4. Another research by Wang et., al. [10] has showed an excellent capacity of OPC in remediating the contaminated soil at a 17 year-old site. This research found that the Toxicity Characteristic Leaching Procedure (TCLP) test sample containing Cu, Ni, Zn, Pb and Cd has satisfied the drinking water standard.

Malviya and Chaudhary [11] has also used the OPC to remediate soil that is contaminated by Pb, Zn, Cu, Fe and Mn. As a result, they argued that sample containing OPC at pH $\geq$ 12 leached less Pb. Then, it was observed that the concentrations of Zn, Cu, Fe and Mn were also decreased in alkaline conditions. This study concluded that the leachability of heavy metals studied are very pH dependent. Similar observation were reported by Voglar and Lestan [6] which showed that concentrations of Cd, Pb, Zn and Ni decreased in alkaline conditions on TCLP extraction and met the regulatory limit for heavy metals in soils.

Furthermore, a part from OPC, there is an interesting study using Calcium Aluminate cement (CAC). Voglar and Lestan [12] studied the use of CAC and pozzolanic cement (PC) as a binder in the S/S sample. The result indicates that there are significant reductions in soil leachability of Cd, Pb, Zn, Cu, Ni and As into deionized water below the limit of quantification. Similarly, Voglar and Lestan [13] stated that there are higher mechanical strength usage of up to 7.65 Nmm^{-2} when using CAC in S/S soil samples. They also argue that S/S with CAC effectively reduces the leachability of Zn, Pb, Cu, As, Cd and Ni. On the other hand, Navarro-Blasco et., al. [14] studied the sorbent capacities of CAC for toxic metals and found that the total uptake of Pb, Zn and Cu could be achieved of up to 3 wt %.

However, there were some researchers who were interested in substituting the cement content with other additives such as fly ash, lime and pozzolan materials. Jing et., al. [15] have studied the effects of Pb leachability in S/S method using OPC, lime and fly ash. As a result, samples containing 10% of lime showed a significant reduction of 43 mg/L Pb concentration compared to samples containing 10% OPC alone with 699 mg/L Pb concentration at 28 curing days. In the same study, the researcher mixed lime and fly ash and found it to have a better result as the reduction of Pb concentration was at it highest at 8 mg/L at 28 curing days. The usage of fly ash was also highlighted by Moon et., al. [16], Anastasiadou et., al. [17] and Xi et., al. [18]. The results indicated that by adding fly ash into the S/S samples, the strength could be enhanced but they argue that the strength of sample was mainly derived from cement. However, by partially replacing the cement with fly ash, it showed a significant reduction on the leachability. Additionally, concentrations of heavy metals were reduced to below the regulatory limit for TCLP test compared to sample with cement standalones.

Nonetheless, there were also other researchers who tried mixing the OPC with waste materials which are available free of charge. The main objective of using these waste materials was to investigate the ability in the soil remediation process and if it could help in preserving the environment towards a more sustainable circumstance. Yin et., al. [19] have used the admixture of OPC with Rice Husk ash (RHA) to remediate the Pb contaminated soil. The result shows that leachability of Pb from crushed block were reduced due to the incorporation of RHA into the binder system where the Pb concentration was reduced from 3.70 to 1.06 mg/L (deionized water), 0.66 to 0.08 mg/L (acetic acid) and 3.22 to 1.88 mg/l (sulfuric/nitric acid). Another study by Jain et., al. [20] showed the effects of Cr (VI) on solidification and hydration behavior of Ordinary Portland cement (OPC) and rice husk ash (RHA) blended (10%, 20%, and 30%) cement. The result indicates that the addition of RHA accelerates final settings as compared to control samples (OPC), while retardation in setting time has been observed when there is an increase in rice husk ash concentration (10%–30%). Additionally, the TCLP test shows that the retention capacity of OPC and RHA blended samples were in the range of 92% to 99% and the leached Cr (VI) concentration was under the allowable limit (5 mg/l) of U.S. Environmental Protecting Agency (EPA). Table 1 summarizes the overview of soil remediation by S/S method.

Table 1. Soil remediation by S/S method over 10 years from 2004 to 2014.

Binder	Heavy metals studied	Findings	Location	References
OPC, lime and fly ash	Pb	- TCLP indicated that Pb final pH divide into acid, neutral and alkaline condition. - GANC of S/S materials leached out all Pb concentration - In MEP, Pb was leached out in average concentration at 4.8 mg/L	USA	[15]
Ordinary Portland cement (OPC)	Pb, Zn, Cu, Fe, Mn	- Diffusion coefficient of all metals studied was above 11.5 indicating low mobility in cement matrix. - Sample containing OPC attained the required strength after 28 days of curing.	India	[11]
Quicklime, fly ash	Pb	- Test result in leaching indicated that S/S treatment was effective in immobilizing Pb and can be accepted according to leachability index (LX)	USA	[16]
OPC and Rice Husk ash (RHA)	Pb	- Results indicates that usage of OPC incorporated with RHA to remediate Pb concentration in S/S method is more favourable in reducing the leachability the usage the OPC standalone	Malaysia	[19]
OPC	Cd, Pb, Cu, Ni, As	- Through TCLP, concentration of Cd, Pb, Zn and Ni generally decreased, concentration of As remained unchanged while concentration of Cu increased	Slovenia	[6]

OPC, CAC and Pozzolanic cement (PC)	Cd, Pb, Zn, Cu, Ni, As	-	TCLP testing reduced soil leachability of Cd, Pb, Cu, Ni and As into deionized water below the limit and produce high value in mechanical strength up to 12 Nmm^{-2}	Slovenia	[12]
OPC and Rice Husk ash (RHA)	Cr	-	Solidification studies show that the compressive strength of controls and rice husk ash blended samples increases with increase in the curing period and maximum strength was observed with 20%RHA blended samples	India	[20]
Fly ash (FA), bottom ash (BA) and OPC	Cr, Fe, Ni, Cu, Cd, Ba	- - -	Cement based S/S exhibited a UCS of 0.55 to 16.12 MPa UCS was decreased as the percentage of cement reduced UCS of S/S decreased with partial additional of FA and BA	Greece	[17]
Calcium Aluminate cement (CAC) and Sulfate Resistant Portland cement (SRC)	Zn, Pb, Cu, As, Cd, Ni	- -	CAC produced S/S soil higher in mechanical strength up to 7.65 Nmm^{-2} Result of TCLP indicate that S/S with CAC reduced leachability more than SRC	Slovenia	[13]
Calcium Alite cement (CAC)	Pb, Zn, Cu	-	CAC showed good sorbent capacities for the toxic metals where metals were fully retained when added below 3% (w/w) by cement	Spain	[14]
OPC	Pb	-	Result shows that the acidic leachant (pH 2.0) altered the leaching behavior of the soil in semi dynamic leaching test	China	[8]
OPC	Pb	-	Lead was leach out from 109, 83 and 71 mg to 37, 30 and 25 mg respectively in semi dynamic leaching test	China	[9]
OPC	Cu, Ni, Zn, Pb, Cd	-	Result indicates that S/S treatment at 17 year service time satisfied the TCLP standard level where Ni was the most stable with around 40% remained in soil	UK	[10]
OPC, lime, montmorillonite , diatomite, sepiolite, and fly ash	Pb	- -	Result demonstrate that the additives with mass ratio cement: fly ash: quicklime were found to be the most effective ratio. Pb contaminated level of 10,000 mg/kg could be successful remediated	China	[18]

3. Conclusion
Soil contamination especially by hazardous heavy metals is pointed as a main contributor to environmental problems throughout the world. To ensure that these problems can be reduced and finally be solved, a systematic prevention and remediation methods should be taken. It is evident from this review that S/S remediation methods using various types of binders attempted of addressing these problems. S/S method using OPC are the most frequently studied for the remediation of heavy metals in soil. However, this review have revealed that S/S method using OPC incorporated with other potential additive such as fly ash, bottom ash and lime showed satisfactory results as well as help on reducing the remediation cost. This review also found that the use of recycle material from agricultural waste such as Rice Hush

Ash (RHA) capable to improve the quality of soil treatment as well as contributed towards a sustainability remediation method.

Acknowledgement
The authors would like to thank the Ministry of Higher Education and Universiti Tun Hussein Onn, Malaysia for supporting this research under Fundamental Research Grant Scheme (FRGS) Vot. 1455.

References

[1] Dermont G, Bergeron M and Mercier G, 2008 Metal-Contaminated Soils : Remediation Practices 188–209.
[2] Tantawy M A, El-Roudi A M and Salem A A, 2012 Immobilization of Cr(VI) in bagasse ash blended cement pastes, *Constr. Build. Mater* **30** 218–223.
[3] Hunce S Y, Akgul D, Demir G, and Mertoglu B, 2012 Solidification/stabilization of landfill leachate concentrate using different aggregate materials, *Waste Manag.*, **32**(7), 1394–400.
[4] Lasheen M R, Ashmawy A M, Ibrahim H S, and Moniem S M A, 2013 Pozzolanic-based materials for stabilization/solidification of contaminated sludge with hazardous heavy metal: case study, *Desalin. Water Treat*, **51** 2644–2655.
[5] Yao Z, Li J, Xie H, and Yu C 2012 Review on Remediation Technologies of Soil Contaminated by Heavy Metals, *Procedia Environ. Sci*, **16** 722–729.
[6] Voglar G E and Lestan D, 2010 Solidification/stabilisation of metals contaminated industrial soil from former Zn smelter in Celje, Slovenia, using cement as a hydraulic binder, *J. Hazard. Mater*, **178(1–3)** 926–33.
[7] Napia C, Sinsiri T, Jaturapitakkul C, and Chindaprasirt P, 2012 Leaching of heavy metals from solidified waste using Portland cement and zeolite as a binder, *Waste Manag*, **32(7)** 1459–1467.
[8] Du Y J, Wei M L, Reddy K R, Liu Z P and Jin F, 2014 Effect of acid rain pH on leaching behavior of cement stabilized lead-contaminated soil, *J. Hazard. Mater*, **271(2014)** 131–40.
[9] Li J S, Xue Q, Wang P, Li Z Z and Liu L, 2014 Effect of drying-wetting cycles on leaching behavior of cement solidified lead-contaminated soil, *Chemosphere*, **117** 10–13.
[10] Wang F, Wang H and Al-Tabbaa A, 2014 Leachability and heavy metal speciation of 17-year old stabilised/solidified contaminated site soils, *J. Hazard. Mater*, **278** 144–51.
[11] Malviya R and Chaudhary R, 2006 Factors affecting hazardous waste solidification/stabilization: a review, *J. Hazard. Mater*, **137(1)** 267–76.
[12] Voglar G E and Lestan D, 2011 Efficiency modeling of solidification/stabilization of multi-metal contaminated industrial soil using cement and additives, *J. Hazard. Mater*, **192(2)** 753–762.
[13] Voglar G E and Lestan D, 2013 Equilibrium leaching of toxic elements from cement stabilized soil, *J. Hazard. Mater*, **246–247** 18–25.
[14] Navarro-Blasco I, Duran A, Sirera R, J. Fernandez M and Alvarez J I, 2013 Solidification/stabilization of toxic metals in calcium aluminate cement matrices, *J. Hazard. Mater*, **260** 89–103.
[15] Jing C, Meng X, and Korfiatis G P, 2004 Lead leachability in stabilized/solidified soil

samples evaluated with different leaching tests, *J. Hazard. Mater*, **114(1–3)** 101–110.

[16] Moon D H and Dermatas D, 2006 An evaluation of lead leachability from stabilized/solidified soils under modified semi-dynamic leaching conditions, *Eng. Geol*, **85(1–2)** 67–74.

[17] Anastasiadou K, Christopoulos K, Mousios E, and Gidarakos E, 2012 Solidification/stabilization of fly and bottom ash from medical waste incineration facility, *J. Hazard. Mater*, **207–208** 165–170.

[18] Xi Y, Wu X and Xiong H, 2014 Solidification/Stabilization of Pb-contaminated Soils with Cement and Other Additives, *Soil Sediment Contam. An Int. J*, **23(8)** 887–898.

[19] Yin C Y, Mahmud H and Shaaban M G, 2006 Stabilization/solidification of lead-contaminated soil using cement and rice husk ash., *J. Hazard. Mater*, **137(3)** 1758–64.

[20] Jain N, 2011 Solidification and Leachability of Cr (VI) in Rice Husk Ash-Blended Cement, *ISRN Civ. Eng*, **2011(Vi)** 1–6.

Adsorption Study on Moringa Oleifera Seeds and Musa Cavendish as Natural Water Purification Agents for Removal of Lead, Nickel and Cadmium from Drinking Water

N A A Aziz[1], N Jayasuriya[1] and L Fan [1]
[1]School of Engineering, RMIT University, Melbourne, VIC 3001, AUSTRALIA

E-mail: s3412996@student.rmit.edu.au

Abstract: The effectiveness of plant based materials Moringa oleifera (Moringa) seeds and Musa cavendish (banana peel) for removing heavy metals namely lead (Pb), nickel (Ni) and cadmium (Cd) from contaminated groundwater was studied. Tests were carried out with individual and combined biomass at neutral pH condition on synthetic groundwater samples. The optimum biomass doses were determined as 200 mg/L for single biomass and 400 mg/L (in the ratio of 200 mg/L: 200 mg/L) for combined biomasses and used for adsorption isotherm studies with contact time of 30 minutes. Results showed that combined biomasses was able to met the Pb, Ni and Cd WHO standards from higher Pb, Ni and Cd initial concentrations which were up to 40 μg/L, 50 μg/L 9 μg/L, respectively compared to individual biomass of Moringa seed and banana peel. Moringa seeds exhibited the highest removal of Pb (81%) while the combined biomasses was most effective in removing Ni (74%) and Cd (97%) over wider their initial concentration ranges. The experimental data were linearized with Langmuir and Freundlich adsorption isotherm models. Freundlich model described the Pb adsorption better than the Langmuir model for all the tested biomasses. However, the Langmuir model fit better with the experimental data of Ni adsorption by Moringa seeds. Both models showed negligible differences in the coefficient of determination (R2) when applied for Ni and Cd adsorption on banana peel and combined biomasses, suggesting that there were multiple layers on the biomass interacting with the metals. Chemisorption is suggested to be involved in Pb adsorption for all tested biomasses as the value of nF calculated was lower than one. This type of adsorption could explain the phenomenon of different behavior of Pb removal and the higher Pb adsorption capacity (represented by KF values) compared to Ni and Cd. The study demonstrates that Moringa seeds, banana peel and their combination have the potential to be used as a natural alternative to the other water treatment agents for removing the Pb, Ni and Cd from drinking water.

Keywords: Adsorption, cadmium, coagulation, drinking water, freundlich isotherm, langmuir isotherm, lead, moringa oleifera, musa cavendish, nickel.

1. Introduction

Most people living in remote areas of developing countries are largely dependent on groundwater as a source of drinking water. However, the groundwater can be contaminated with various pollutants such as heavy metals. Heavy metals are present in groundwater due to poor land use practices including excessive use of agrochemicals. Average lead (Pb) levels of 287 μg/L and 330 μg/L have been recorded in groundwater samples collected from some areas in India and Pakistan, respectively [1, 2]. [2] reported that the average level of nickel (Ni) in groundwater in Pakistan was up to 960 μg/L. In Sri Lanka, Cd average levels in groundwater have been recorded between 5 to 23 μg/L [3]. The concentrations of the heavy metals far exceeded those recommended by WHO drinking water quality standards (10 μg/L, 20 μg/L and 3 μg/L for Pb, Ni and Cd, respectively). Therefore, it is of extremely importance to remove these harmful substances from the drinking water to prevent the potential adverse effects to human health.

Poor economic situations in some developing countries require low-cost drinking water treatment solutions for the affected communities for which conventional/advanced drinking water treatment is not viable. Application of plant-based materials for drinking water treatment is environmentally sustainable, cost-effective as they are locally available, non-toxic, easy to prepare and the residual generated is easy to dispose.

Various natural-based materials have been studied as potential low-cost coagulants for water purification including Moringa oleifera (Moringa, MO) seed [4, 5]. MO seed is one of the popular natural materials investigated as it has shown ability in removing various pollutants such as copper, nickel, zinc and color from water and wastewater [6, 7]. MO is a vegetable that can be found widely in Asian countries, growing well in any tropical areas including many in Africa. Besides, Musa cavendish (banana peel, MC) was also reported to be a promising plant-based material for removing As, Pb and Cd [8, 9]. The use of MC in water purification fits well with the concept of environmental sustainability since the agricultural waste are recycled to minimize the wastes in an environmentally friendly way [10, 11].

Adsorption was regarded as a key removal mechanism when these types of biomass material were used for water treatment [12]. Despite considerable attention has been paid by current researchers on the application of natural materials as water purification agents to remove various kind of pollutants, understanding of the mechanism involved in the treatment process is still lacking. The removal mechanism for one heavy metal could be different to one another. Therefore, it is important to understand the mechanism that contributing to the removal of a specific heavy metal from the aqueous solution using the different type of biomass rather than just its application. The aim of this study was to examine whether the existing adsorption isotherm models can best describe the behavior and treatment efficiency in removing Pb, Ni and Cd with Moringa seeds (MO), Musa cavendish/banana peel (MC) and their combination (MO+MC) from contaminated groundwater and to understand the adsorption mechanism of particular heavy metal described by the presented models.

2. Methodology

2.1 Preparation of synthetic groundwater samples

Synthetic groundwater samples containing individual Pb, Ni and Cd of known concentrations were prepared with deionized water [13]. All chemicals used in the study, including lead carbonate, nickel carbonate and cadmium nitrate tetrahydrate were of analytical grade and obtained from Sigma-Aldrich (Australia). The chemical stock solutions with a concentration of 1000 mg/L of each heavy metal were prepared and then diluted to the target concentrations as per need for the experiments. The pH of the water samples was adjusted using either 0.1 M H2SO4 and 0.1 M of NaOH. Table 1 presents the initial concentrations of the contaminants in the synthetic groundwater samples. The initial concentration values in this study were based on the mean groundwater quality values reported in the literature [1, 2, 3].

Table-1. Characteristics of synthetic groundwater samples.

Parameter	Unit	Initial concentration (μg/L)	WHO drinking water quality standard
Pb	μg/L	5, 10, 20, 40, 60, 80, 100	10
Ni	μg/L	10, 20, 30, 40, 50, 80, 100	20
Cd	μg/L	1, 3, 5, 7, 9, 10, 11	3
pH	-	7.0	6.5-8.5

2.2 Preparation of biomass

2.2.1 Moringa oleifera (moringa seeds)
Moringa seeds (MO) were bought from Pacific Blue Consulting Pty. Ltd. (The Moringa Shop, Australia). High-quality seeds were selected and cleaned from debris. Husks and external coats were removed [14]. The selected MO seeds were then rinsed with deionized water and dried in an oven at 100°C for 1 hour. The dried seeds were ground to a fine powder and sieved through a stainless steel mesh with an opening size of 300 μm to obtain MO particles of similar size. One gram of the MO seed powder was added to 1 L deionized water and stirred for 10 minutes using a magnetic stirrer to create active constituents to prepare an MO stock solution of 1000 mg/L.

2.2.2 Musa cavendish (banana peel)
Ripe bananas (MC) were purchased from a local market. The peels were gently separated from the fruits. The peels were cleaned and dried in an oven for 1 hour at 100°C. Then, the dried peels were cut into small pieces, ground and sieved through a stainless steel mesh with an opening size of 300 μm [15]. One gram of powdered banana peel was added to 1000 mL deionized water and stirred for 10 minutes to generate active constituents making it a stock solution with the MC concentration of 1000 mg/L.

2.3 Batch experiments
A standard jar tester (Lovibond, Phipps & Bird) was used for batch tests for determining the removal of the target contaminants by the selected natural materials. All experiments were performed at room temperature (20°C ± 1°C) and were carried out with 1 L water samples in acrylic jars. The pH of each sample was adjusted to 7.0 with 0.1 M NaOH or 0.1 M H_2SO_4 prior to the batch experiment. The water sample was then mixed with the biomass of a required dosage. Then the sample was subjected to a rapid mixing at 150 rpm for 3 min, followed by the slow mixing at 30 rpm for 30 min [5]. The agitated solutions were permitted to settle for 30 min before the supernatant was taken for sampling. The samples were filtered through 0.45 μm fiberglass membrane filters (Advantec) prior to the analysis to determine the removal efficiency of the target contaminants. Pb, Ni and Cd concentrations were determined using Agilent 7700 series Inductively Coupled Plasma Mass Spectrometry (ICP-MS). pH measurement was carried out using Seven Easy Mettler Toledo pH meter. In the investigations with the combined biomass (MO+MC), two different biomasses were added to the water sample concurrently. The experiments were conducted in triplicate (N=3) to ensure the reproducibility, and the mean values were reported.

In previous studies [13,16] evaluated the optimum biomass doses for removing arsenic (As), Pb, Ni, Cd, fluoride (F) and turbidity from single and multi-element synthetic groundwater samples. From the above studies, the optimum biomass concentrations which gave the best removal efficiencies were found to be 200 mg/L (for MO and MC, separately) and 200+200 mg/L (for combined MO and MC). Above biomass concentrations were therefore used in the adsorption isotherm studies.

2.4 Adsorption isotherms.
Adsorption isotherm studies were carried out by using batch experiments as described in the previous section. The adsorption data have been subjected to Langmuir [17] and Freundlich [18] isotherm models (Equations 1 & 2, respectively).

Langmuir theory assumed that the adsorption process takes place on a set of distinct localized adsorption sites within the natural material and no interactions between adsorbate molecules on adjacent sites. In linear form, it can be written as Equation 1.

$$\frac{C_e}{Q_e} = \frac{1}{Q_m} C_e + \frac{1}{K_L Q_m} \qquad (1)$$

where Q_m and K_L are the Langmuir constants. Q_m is the monolayer adsorption capacity (mg/g), K_L is adsorption constant (L/mg), Ce is the equilibrium concentration of the heavy metal (mg/L) and Q_e is the amount of heavy metal adsorbed (mg/g) (Equation 3). The Q_m and K_L can be determined from the gradient and the intercept of the linear graph between $\frac{C_e}{Q_e}$ and C_e. The Freundlich isotherm accounts for a multiple sites adsorption for heterogeneous surfaces (Equation 2).

$$ln\ (Q_e) = ln\ (K_F) + (\frac{1}{n_F})\ ln(C_e) \qquad (2)$$

where K_F and n_F are Freundlich adsorption constants. Q_e and C_e are as defined earlier for Equation 1. n_F (dimensionless) and K_F (L/g) can be determined from the gradient and the intercept of the linear graph between ln (Q_e) and ln (C_e).

$$Q_e = \frac{(C_0 - C_e)V}{m} \qquad (3)$$

where C_0 is the initial concentration of heavy metal in the solution (mg/L), V is the volume of the solution (L) and m is the mass of the biomass used (g). Percentage removal of heavy metal was calculated using Equation 4.

$$\%\ \text{Removal} = \frac{(C_0 - C_e)}{C_0} \times 100 \qquad (4)$$

where C_0 and C_e are as defined earlier in Equation 3 and 1, respectively.

3. Result and Discussions

3.1 Effect of initial concentration of heavy metals

The initial concentrations for Pb, Ni and Cd of the synthetic groundwater samples were varied as shown in Table 1. Figure 1a depicts the relationship between initial Pb concentration and removal efficiency at fixed dosages of 200 mg/L for MO and MC, and 200+200 mg/L for MO+MC. For MO, the increase of initial Pb concentration from 5 µg/L to 60 µg/L increased the Pb removal from 65% to 81%. However, Pb removal remained unchanged (around 80% to 81%) for concentrations between 60 to 100 µg/L. Similar to MO, the removal percentage increased from 32% to 58% with MC until the Pb concentration is 60 µg/L. The removal of Pb did not increase with the increase of initial Pb concentration beyond 60 µg/L. The combined biomasses gave similar removal rates to MO alone. MC showed the least removal efficiency compared to MO and MO+MC.

In contrast to Pb, removal efficiencies decrease with the increase of Ni and Cd concentrations. Figure 1b shows that the removal efficiency decreased rapidly from 98% to 42% with MO when Ni concentration was increased from 10 to 30 µg/L. The rate of decrease in removal efficiency was lower with MC and MO+MC when the initial Ni concentrations were gradually increased. As depicted in Figure 1b the removal efficiency of Ni was improved by combining MO and MC.

Similar to Ni, MO+MC showed a better removal efficiency of Cd compared to MO and MC alone. There was a sudden drop in removal efficiency when the Cd concentration increased from 3 to 5 µg/L and further reduction was shown when Cd concentration was beyond 9 µg/L for MC and MO+MC (Figure 1c). [19] attributed this behavior to the saturation of adsorption sites on biomass's surface. The % reduction of Cd with MO did not vary much (63% to 54%) when Cd initial concentration increased from 1 to 11 µg/L.

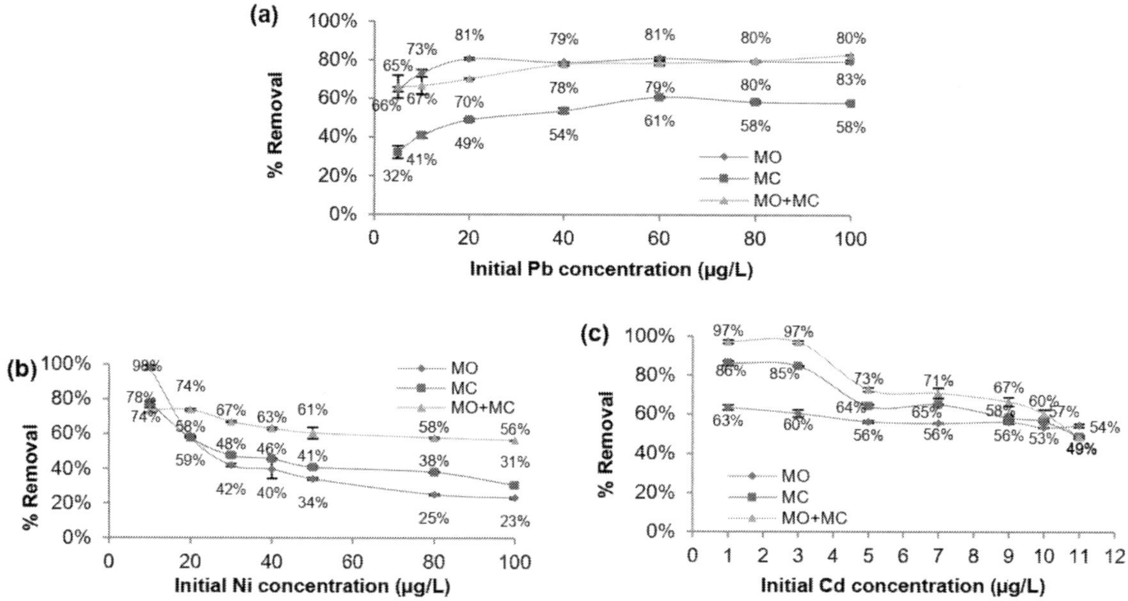

Figure 1. The relationship between removal efficiency of plant-based biomasses and initial concentrations of **(a)** Pb, **(b)** Ni and **(c)** Cd.

3.2 Adsorption Isotherms

Adsorption of Pb, Ni and Cd by MO, MC and MO+MC were modeled using the Langmuir (Equation 1) and Freundlich (Equation 2) isotherms with the quality of the fit assessed using the coefficient of determination (R^2).

The experimental plots of Pb adsorption by MO, MC and MO+MC (Figure 2), were best fitted with Freundlich model with R^2 values of 0.97, 0.99 and 0.99, respectively (Figure 2b). [12] also reported that Freundlich model showed a better fit for Pb adsorption experimental data on MO. This finding suggested that the Pb was interacting with multi-layer of the biomass's sites for adsorption. Negative slopes and intercepts obtained for Pb adsorption on all tested biomasses (Figure 2a), suggesting that the adsorption behavior do not obey the Langmuir's theory assumption (surface of biomass is energetically homogeneous). In nature, MO and MC may have heterogeneous available adsorption sites due to its abundant availability of functional groups, considering the origin of the materials used in this study being plant-based materials [11]. Therefore, in this case Pb might have chemical interactions with the specific functional groups present on the biomass's surface. Such interactions could lead to the generation of new chemical bonds that could improve the adsorption capacity of the biomass.

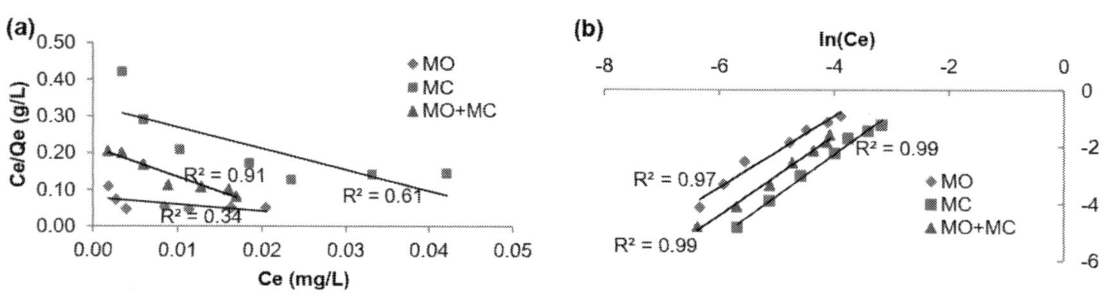

Figure 2. (a) Langmuir and **(b)** Freundlich isotherms of Pb on the MO, MC and MO+MC.

The experimental plots of Ni adsorption on the MC and MO+MC were best fitted with Freundlich model with R^2 values of 0.97 and 0.99, respectively (Figure 3b). On the other hand, MO exhibited a better fitness with the Langmuir model (R^2 = 0.96) as shown in Figure 3a than with the Freundlich model (R^2 = 0.74), indicating that greater monolayer adsorption occurred rather than multilayer adsorption.

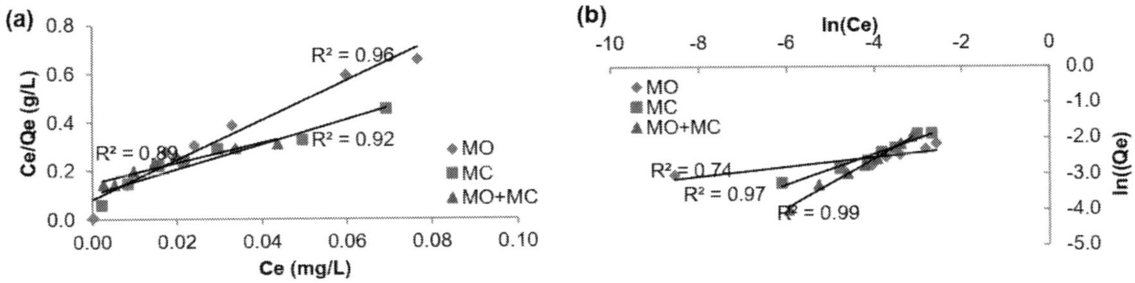

Figure 3. (a) Langmuir and **(b)** Freundlich isotherms of Ni on the MO, MC and MO+MC.

Based on Figure 4(a and b), the Freundlich model demonstrated a better fit than the Langmuir model for Cd adsorption by MO. Similar finding was reported by [20]. The difference between the R^2 values from both models were insignificant when adsorption by MC was considered, suggesting that there is a possibility that Cd adsorption onto these biomasses occurred in multilayers. This finding reinforces the conclusions made by [12] that the insignificant difference between R^2 values obtained for both models, indicating a possible existence of more than one type of adsorption site interacting with the metal.

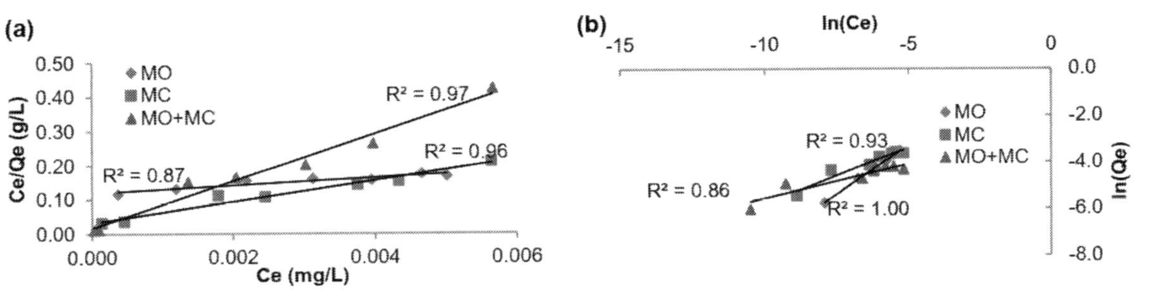

Figure 4. (a) Langmuir and **(b)** Freundlich isotherms of Cd on the MO, MC and MO+MC.

Table 2 presents the model parameters and the coefficient of determination (R^2) obtained by fitting Langmuir and Freundlich models (Equations 1 & 2) with the experimental data.

Negative values were obtained for Q_m (adsorption capacity) and K_L (binding energy) by fitting the Langmuir linearization model with MO, MC and MO+MC biomasses for Pb adsorption. This was due to the negative gradient of the linear graph as shown in Figure 2a. This result indicated that the adsorption behavior of the treatment did not follow the theory assumptions of Langmuir isotherm due to possibility to undertakes a specific kind of adsorption mechanism (i.e. chemisorption) as the behavior of Pb removal was different when compared to Ni and Cd (Figure 1a).

In the Langmuir linearization of MO, MC and MO+MC for Ni adsorption, it was verified that MO+MC had a higher adsorption capacity of Ni (Q_m) compared to MO and MC alone. However, MO had the highest binding energy (K_L) with Ni. The binding energy (K_L) of Cd by MO+MC was significantly higher in the Langmuir linearization compared with MO and MC. This was likely the reason that the Cd removal efficiency was better achieved by MO+MC compared to MO and MC alone (Figure 1c).

Table 2. Parameters of equilibrium isotherm of Langmuir and Freundlich models for Pb, Ni and Cd adsorption of MO, MC and MO+MC.

Heavy metal	Biomass	Langmuir isotherm model			Freundlich isotherm model		
		Q_m (mg/g)	K_L (L/mg)	R^2	K_F (L/g)	n_F	R^2
Pb	MO	-0.540	-23.422	0.34	59.626	0.799	0.97
	MC	-0.174	-17.563	0.61	33.991	0.691	0.99
	MO+MC	-0.125	-37.191	0.91	54.136	0.715	0.99
Ni	MO	0.122	101.832	0.96	0.133	7.628	0.74
	MC	0.195	49.712	0.92	0.477	2.342	0.97
	MO+MC	0.239	28.222	0.89	1.199	1.450	0.99
Cd	MO	0.090	93.294	0.87	2.625	1.180	1.00
	MC	0.032	974.905	0.96	0.368	2.093	0.93
	MO+MC	0.015	4168.667	0.97	0.073	3.396	0.86
	MC	0.368	2.093	0.93			
	MO+MC	0.073	3.396	0.86			

In the Freundlich linearization, the K_F constants (adsorption capacity) obtained for Pb by all biomasses were much higher than those of Ni and Cd. This result was consistent with the increase of Pb removal with the increase in initial Pb concentration (Figure 1a).

The n_F value in the Freundlich model represents the reactivity of biomass's active sites. According to [21], if a value of $n_F = 1$, the adsorption is linear, for $n_F < 1$, the adsorption is chemisorption, and for $n_F > 1$ the adsorption is a favorable physical adsorption. Thus, analyzing the n_F values (Table 2), adsorption of Pb onto MO, MC and MO+MC was much favored the chemisorption. In this case, adsorption of Pb on MO, MC and MO+MC was likely to generate new chemical bonds at the biomass's surface which could lead to the improvement of the biomass adsorption capacity. This phenomenon could explain the reason of the increase of removal efficiency with the increase of initial Pb concentration and the higher values of K_F (adsorption capacity) compared to Ni and Cd. On the other hand, physical adsorption was shown favorable for adsorption of Ni and Cd on MO, MC and MO+MC.

4. Conclusion

The adsorption of lead (Pb), nickel (Ni) and cadmium (Cd) by naturally occurring plant-based materials of *Moringa oleifera* (MO) seeds, *Musa cavendish* (MC) and their combination (MO+MC) was investigated and following conclusions are drawn from the study:

- MO showed the highest removal efficiency for Pb while MO+MC showed the highest removal for Ni and Cd compared to MO and MC alone.
- Freundlich model best described the Pb adsorption for all the tested biomasses. The experimental data with Pb did not obey the Langmuir's model adsorption theory assumptions.
- Adsorption of Pb on MO, MC and MO+MC favored the chemisorption type of adsorption. This phenomenon could explain the reason of different behavior of removal Pb and higher adsorption capacity were obtained for Pb compared to Ni and Cd.
- Data linearization demonstrated the best fit by the Langmuir model for Ni adsorption on MO. Negligible difference between R^2 from both isotherms for Ni and Cd adsorption by MC and MO+MC, suggests that there were multiple layers on the biomass interacting with the metals.
- MO, MC and MO+MC show potential to be used as alternative treatment material for purification of drinking waters polluted with heavy metals such as Pb, Ni and Cd as they are available and low-cost in the affected developing countries.

Acknowledgements

The authors would like to thank Ministry of Higher Education Malaysia and Universiti Tun Hussein Onn Malaysia for providing scholarship to the first author to conduct the research.

References

[1] Buragohain, M., Bhuyan, B., & Sarma, H. P. (2010). Seasonal variations of lead, arsenic, cadmium and aluminium contamination of groundwater in Dhemaji district, Assam, India. *Environmental Monitoring and Assessment*, *170*(1-4), 345–51.

[2] Midrar-Ul-Haq, Khattak, R. A., Puno, H. K., Saif, M. S., & Memon, K. S. (2005). Surface and groundwater contamination in NWFP and Sindh provinces with respect to trace elements. *Int J Agri Biol*, *7*, 214–217.

[3] Bandara, J. M. R. S., Wijewardena, H. V. P., Liyanege, J., Upul, M. A., & Bandara, J. M. U. A. (2010). Chronic renal failure in Sri Lanka caused by elevated dietary cadmium: Trojan horse of the green revolution. *Toxicology Letters*, *198*(1), 33–39.

[4] Ali, E. N., Muyibi, S. A., Salleh, H. M., Salleh, M. R. M., & Alam, M. Z. (2009). Moringa Oleifera Seeds As Natural Coagulant for Water Treatment. In *Thirteen International Water Technology Conference, IWTC, Hurghada, Egypt,* 13, 163–168.

[5] Ndabigengesere, A., Narasiah, K. S., & Talbot, B. G. (1995). Active agents and mechanism of coagulation of turbid waters using Moringa oleifera. *Water Research*, *29*(2), 703–710.

[6] Kalavathy, M. H., & Miranda, L. R. (2010). Moringa oleifera—A solid phase extractant for the removal of copper, nickel and zinc from aqueous solutions. *Chemical Engineering Journal*, *158*(2), 188–199.

[7] Prasad, R. K. (2009). Color removal from distillery spent wash through coagulation using Moringa oleifera seeds: Use of optimum response surface methodology. *Journal of Hazardous Materials*, *165*(1–3), 804–811.

[8] Anwar, J., Shafique, U., Waheed-uz-Zaman, Salman, M., Dar, A., & Anwar, S. (2010). Removal of Pb(II) and Cd(II) from water by adsorption on peels of banana. *Bioresource Technology*, *101*(6), 1752–5.

[9] Kamsonlian, S., Balomajumder, C., & Chand, S. (2012). A Potential of biosorbent derived from banana peel for removal of As (III) from contaminated water. *International Journal of Chemical Sciences and Applications*, *3*(2), 269–275.

[10] Nguyen, T. A. H., Ngo, H. H., Guo, W. S., Zhang, J., Liang, S., Yue, Q. Y., Li, Q., Nguyen, T. V. (2013). Applicability of agricultural waste and by-products for adsorptive removal of heavy metals from wastewater. *Bioresource Technology*, *148*, 574–85.

[11] Okoro, I. A., & Okoro, S. O. (2011). Agricultural by Products as Green Chemistry Absorbents for the Removal and Recovery of Metal Ions from Waste-Water Environments. *Continental J. Water, Air and Soil Pollution*, *2*(1), 15–22.

[12] Junior, A. C. G., Meneghel, A. P., Rubio, F., Strey, L., Dragunski, D. C., & Coelho, G. F. (2013). Applicability of *Moringa oleifera* Lam. pie as an adsorbent for removal of heavy metals from waters. *Revista Brasileira de Engenharia Agrícola E Ambiental*, *17*(1), 94–99.

[13] Aziz, N.A.A., Jayasuriya, N. & Fan, L. (2014). Effectiveness of plant-based indigenous materials for the removal of heavy metals and fluoride from drinking water. In *5th International Conference on Sustainable Built Environment Proceedings*. Kandy, Sri Lanka, 3, 34-41.

[14] Okuda, T., Baes, A. U., Nishijima, W., & Okada, M. (2001). Coagulation Mechanism of Salt Solution-Extracted Active Component in Moringa oleifera Seeds. *Water Research*, *35*(3), 830–834.

[15] Liu, C., Ngo, H. H., Guo, W., & Tung, K.-L. (2012). Optimal conditions for preparation of banana peels, sugarcane bagasse and watermelon rind in removing copper from water. *Bioresource Technology*, *119*, 349–54.

[16] Aziz, N.A.A., Jayasuriya, N. & Fan, L. (2015). Application of Moringa oleifera seeds and Musa cavendish as coagulants for lead, nickel and cadmium removal from drinking water. In: *Asia Pacific Confederation of Chemical Engineering Congress 2015: APCChE 2015, incorporating CHEMECA 2015*. Melbourne, Australia, 1774-1781.

[17] Langmuir, I. (1918). The adsorption of gases on plane surface of glass, mica and platinum. *Journal of the Chemical Society*, 40, 1361–1403.

[18] Freundlich, H. (1926). Colloid and capillary chemistry. *London: Methuen,* 397–414.

[19] Kumar, P. S., Ramalingam, S., Sathyaselvabala, V., Kirupha, S. D., Murugesan, A., & Sivanesan, S. (2012). Removal of Cd(II) from aqueous solution by agricultural waste cashew nut shell. *Korean J. Chem. Eng.*, (29), 756–768.

[20] Mataka, L. M., Sajidu, S. M. I., Masamba, W. R. L., & Mwatseteza, J. F. (2010). Cadmium sorption by Moringa stenopetala and Moringa oleifera seed powders : Batch, time, temperature, pH and

adsorption isotherm studies. *International Journal of Water Resources and Environmental Engineering*, 2(3), 50–59.

[21] Bhatt, A. S., Sakaria, P. L., Vasudevan, M., Pawar, R. R., Sudheesh, N., Bajaj, H. C., & Mody, H. M. (2012). Adsorption of an anionic dye from aqueous medium by organoclays: equilibrium modeling, kinetic and thermodynamic exploration. *RSC Advances*, *2*, 8663–8671.

Preparation and Characterization of Activated Cow Bone Powder for the Adsorption of Cadmium from Palm Oil Mill Effluent

A AbdulRahman[1], A A A Latiff[1], Z Daud[1], M B Ridzuan[1], N F M D[1], A.H Jagaba[1]
[1]Faculty of Civil and Environmental Engineering, Universiti Tun Hussein Onn Malaysia, 86400, Batu Pahat, Johor, MALAYSIA

E-mail: aziz@uthm.edu.my

Abstract: Several studies have been conducted on the removal of heavy metals from palm oil mill effluent. In this study, cow bones were developed as an adsorbent for the removal of cadmium II from POME. A batch experiment was conducted to investigate the effectiveness of the prepared activated cow bone powder for the sorption of cadmium II from raw POME. The experiment was carried out under fixed conditions using 100mg/L raw POME combined with different adsorbent dosage of CBP of 184.471 Ra(nm) surface roughness. The equilibrium adsorption capacity of the hydrophobic CBP of average contact angle 890 was determined from the relationship between the initial and equilibrium liquid phase concentrations of POME. The optimum adsorption of cadmium II on CBP was at 10g adsorbent dosage for sample 1 and 2 at 97.8% and 96.93% respectively. The least uptake was at 30g adsorbent weight for both samples at average of 95.1% for both samples. The effective removal of cadmium ion showed that CBP has a great potential for the treatment of heavy metal in POME.

Keywords: POME, cow bone, adsorption.

1. Introduction

Most Wastewater containing heavy metals is of utmost importance globally because of their potential treat to both aquatic ecosystem and the soil environment. According to World Health Organization, heavy metal is hazardous even at low concentration and causes various forms of diseases [1,2]. The presence of cadmium in water is very toxic for the aquatic ecosystem [3] the effect of the water consumption containing cadmium has been linked to cancerous growth, hormonal disorder, renal failure etc. Therefore, effective treatment of cadmium is essential for the preservation of aquatic ecosystem. Several technologies based research has been documented for the removal of heavy metals from aqueous solution.

Some of the methods are coagulation and flocculation [4], photocatalyst [5], membrane filtration process [6]. In the treatment of palm oil mill effluent (POME), researches have been conducted by Harsono et al. [7] using anaerobic treatment. Hazlan [8] also reported that evaporation process can be used for the treatment of POME and also electrocoagulation to reduce the pollutants [9]. However, the above treatment methods can only be applicable for high concentration of POME and also very expensive to operate.

In recent studies, there is growing interest in the removal of heavy metals by adsorption method. Using locally sourced natural materials prepared as adsorbent. Some of the materials include chitosan-clay composite, walnut shell [10], kaolinite and montmorillonite [11] and palm kernel shell charcoal [12]. The method of adsorption among researchers for the removal of heavy metals is due to the cost effectiveness, availability of the naturally sourced materials and efficiency in the treatment of high strength industrial wastewater with quality effluent [13,14]. The reversible nature of adsorption process implies that regeneration of adsorbents is possible through desorption processes [15]. The ability to reuse spent by regeneration suggests that adsorption technology is well suited for the preservation of the climate by reducing the necessity of burning and discharging of spent adsorbent media.

The use of adsorption for the treatment of high strength industrial wastewater such as POME has rarely been reported in literature. Hence, its inclusion will enhance the adsorption technology and will provide an alternative treatment of heavy of cadmium in POME. The production of POME especially in

Malaysia known to be one of the largest producers of palm oil recorded to have national production rate of 0.67 cubic meters per tonne of fresh fruit bunch (FFB) processed by mill is a major challenge to researchers [16]. Effective treatment of heavy metals such as cadmium from POME which has been proven by [3] to be very toxic to aquatic environment requires very efficient technology in facing the current challenges. The method of adsorption using prepared cow bones was adopted for this research to investigate the treatment efficiency of activated cow bone powder for the removal of cadmium from POME. Bone powder is obtained from ignition process of crushed bones initially formed as bone char, the process is achieved by heating at 500-700°C. Bone char is produced through the carbonization process of cattle bones, it contains 10% carbon and 90% of hydroxyapatite [17]. The bone charcoal also have the composition of tricalcium phosphate (70-76 wt %) and calcium carbonate (7-9wt %) [18]. Use of bone char has been applied in the study of [17] for the removal of fluouride from water solution. The uses of bone powder have also been applied in the study of [18] for the removal of lead in aqueous solution. However, the application of cow bone powder for the treatment of high strength industrial wastewater has rarely been reported in literature.

This study will provide an alternative treatment technology for agro based industrial wastewater especially Malaysia which is known to be the largest producer of palm oil in the world after Indonesia [5]. The main objective of our present study is to investigate the treatment efficiency of prepared activated carbon produced from cow bones at different adsorbent dosage of cow bone powder (CBP) for the adsorption of cadmium from POME.

2. Materials and Methods

2.1 Preparation and Characterization of Adsorbent

Fresh femur of cow bones was obtained at Bazaar Kedai Rakyat Johor. The bones were carefully washed with water severally, cleaned from meat and later washed again repeatedly to remove impurities on the surfaces. The bones were rinsed with de-ionized water and transferred to the oven at 80 °C to dry. The dried bones were crushed and ignited at 700°C for 5 hours. The loss of ignition (LOI) was evaluated by weighing the sample before and after ignition. The bone char produced were further reduced using the ceramic ball mill grinder. The solubility of the surface of the prepared media in contact with deionized water is illustrated in Figure 1 and was achieved using VCA optima.

Finally the average particle size of less than 150μm was used for the study and was obtained using the 150μm sieve size. The field emission scanning microscopy JEOL JSM-7600F was used to determine the surface morphology and the average particle sizes used for the study. The samples were coated with electric conductive gold film prior to the investigation. The X-ray fluourescence (XRF) S4 pioneer model was used to determine the chemical composition of the media.

2.2 Adsorbate

The POME sample used for the study was obtained from Kilang Sawit PPNJ Kahang Johor. The sample was stored in a 30 L plastic container, transported and stored in the laboratory according to standard methods for the preservation and storage of wastewater [19]. The characterization of the raw POME was achieved before preservation and storage. The result of the characterization is represented in Table 2.

2.3 Batch Study

The batch adsorption study was carried out at 30°C in a 250 ml conical flask containing 100 ml of the raw POME. The agitation of the mixture was performed at 200rpm and a predetermined period of 105 minutes contact time for all analysis. The pH 7 was adopted for the study, the adjustment was conducted using solution of 30%diluted sodium hydroxide and dilute sulphuric acid. The rate of adsorption of cadmium II on CBP was achieved at different adsorbent dosage to obtain the optimum dosage for the uptake of the heavy metal. A set of replicate was prepared for each of the adsorbent mixture. For both sample and replicate, the conical flasks were removed from the orbital shaker and filtered at the end of the contact time, the filtration of the sample was performed with 0.45μm glass fibre membrane and was analysed using inductively coupled plasma (ICP) mass spectrometer Elan 9000. The result presented as averages of three measurements. Relative standard deviation (RSDs) of less than 5% was obtained for the samples and replicates. The pattern of adsorption is presented in Figure 2. The rate of adsorption of cadmium on CBP was calculated as:

$$\%\,removal = \frac{C_o - C_e}{C_o} x100 \qquad (1)$$

Where;

Co	=	The initial liquid face concentration (mg/L)
Ce	=	The equilibrium liquid-phase concentrations (mg/L)

3. Result and Discussions

The behaviour of the adsorbent media in de-ionized water as observed through VCA optima as illustrated in Figure 1.

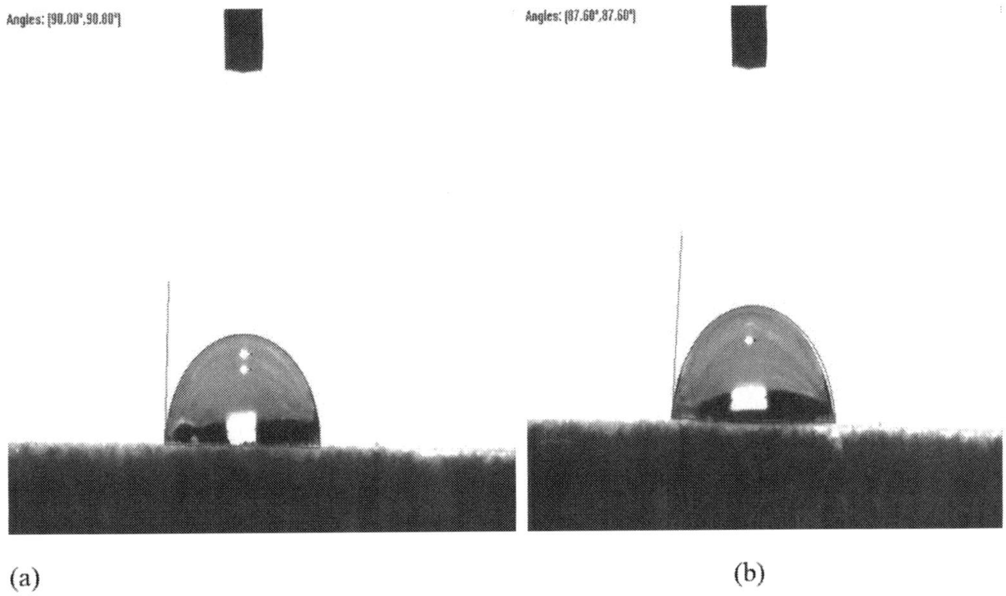

Figure 1. Microphotographs of sample with (a) 90.8o and (b) 87.60o contact angles using the water drop penetration time test.

According to Leelamanie [20], hydrophobic media using water drop penetration time have contact angles within the range of 88-93°C. The effectiveness of CBP for the removal of Cd by varying the weight of the media showed the adsorptive capacity of CBP at different dosage for the uptake of Cd^{2+} from raw POME. The percentage removal by weight of adsorbent after treatment is illustrated in Figure 3. The results of the ICP mass spectrometer Elan 9000 test for the adsorbent signified the effectiveness of the prepared media, the results are presented for the sample and the replicate. The optimum removal was observed at 10 g dosage with 97.8% uptake for sample 1 and 96.3 % uptake for the replicate. Further dosage showed a decrease in the rate of adsorption for both samples. The lowest removal of cadmium (II) was 0.819 μg/L (94.5%) and 0.65 μg/L (95.6%) at dosage of 30g for both samples. The rate of adsorption was partly constant for 20 and 30g dosage.

The adsorptive behaviour of CBP depends on the pore size arrangement of the particles, the hydrophobic surface and the chemical composition of the media. The high removal rate of cadmium (II) is reasonably possible as a result of the availability of active sites and larger pore spaces. The macro pores and meso pores provided readily available sites for the cadmium ions to react with the surfaces internally and externally. The active sites were responsible for >90% removal of Cd(II). The hydrophobic surface provided electrostatic interaction between sorption sites and the metal ions. Also, the significance of calcium in the chemical composition makes high capacity for exchangeable ions at neutral pH possible, this cause very strong electrostatic interaction between the positively charged sites of CaO and the anionic sites of POME [21], the carbon surfaces influenced sorption and the hydroxyapatite arrangement of the phosphate ions provided sites for the attraction of cadmium ions. The surface morphology and the particle distribution as shown in Figure 2. The atomic force microscope AFM XE-100 park system was used to determine the surface roughness of the adsorbent media. The sample was stored at room temperature without any noticeable form of bacterial contamination throughout the period of the study.

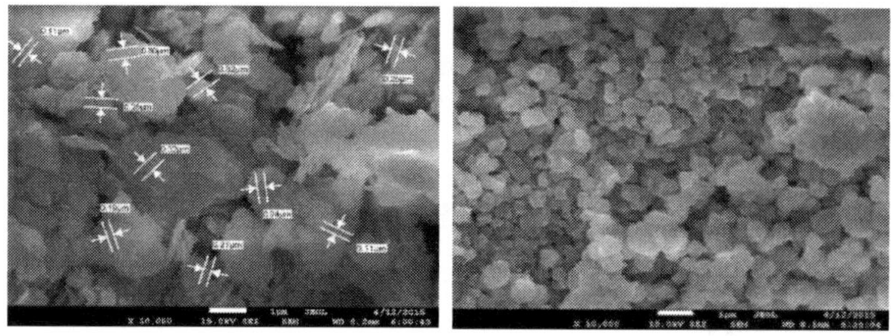

Figure 2. Particle size distribution of CBP.

Table 1. The result of the XRF analysis.

Formula	Concentration(%)
Orig-g	8
Added-g	2
CaO	49.80
P_2O_5	32.90
C	1.00
Na_2O	0.96
SiO_2	0.89
MgO	0.78
Al_2O_3	0.44
Cl	0.11
K	0<LLD
Fe	0<LLD
LOI	13.13

Table 2. The characteristics of the CBP and raw POME.

	CBP	Raw POME	
Parameter	Value	Parameter	Value
Bulk density(g/ml)	0.81	pH	4.5
Solid density (g/ml)	3.24	Suspended solids (SS) mg/L	10,333.3
Porosity	73.9%	Oil and grease (mg/L)	2233.3
Length(μm)	0.777	Ammonia Nitrogen (mg/L)	937
Roughness Ra(nm)	184.471	COD (mg/L)	53,374
		Cadmium at pH 4.5	15.6 μg/L

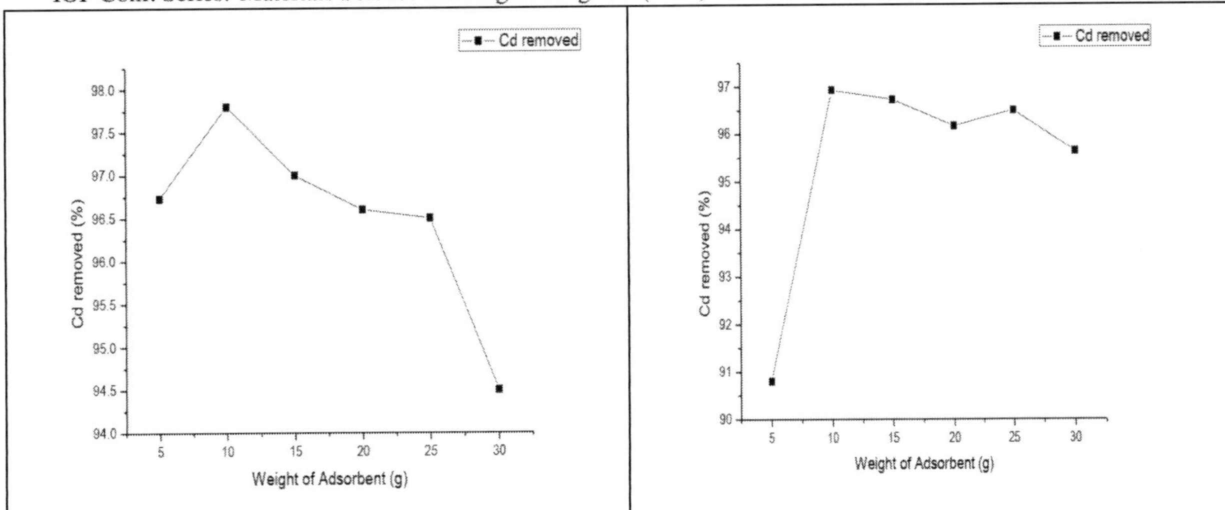

Sample 1	Sample 2

Figure 3. Effect of adsorbent dosage on the removal of Cadmium II from POME.

4. Conclusion

The purpose of this study was to investigate the adsorption efficiency of CBP used as alternative to activated carbon for cadmium removal from POME. The performance of the CBP was analysed in the batch study conducted using different adsorbent dosage of the media. The investigation showed that the prepared activated CBP offers an effective non-toxic alternative treatment for the uptake of cadmium ion from raw POME. The rate of adsorption is a function of the effectiveness of the adsorbent media and the concentration of the adsorbate. The results showed that CBP acheived >90% removal of cadmium ion throughout the study. Thererfore it can be widely applied as an effective media for the treatment of cadmium ion in raw POME.

References

[1] Lourie E, Patil V, Gjengedal E (2010). Efficient purification of heavymetal- contaminated water by microalgae-activated pine bark. Water Air Soil Pollut, 210, pp. 493--500.

[2] Beesley L., Inneh O.S., Norton O.S., Moreno-Jimenez E., Pardo T, Clemente R, Dawson J.C (2014). Assesing the influence of compost and biochar amendments on the mobility and toxicity of metals and arsenic in a naturally contaminated mine soil. Environmental Pollution 186, pp. 196--202.

[3] Johri, N., Jacquillet, G., & Unwin, R. (2010). Heavy metal poisoning: the effects of cadmium on the kidney. BioMetals, 783–792.

[4] Ahmad Y., Yaakub Z., Akhtar P., Sopian K (2015).Production of biogas and performance evaluation of existing treatment processes in palm oil mill effluent (POME). Renewable and sustainable energy 1260-1278.

[5] Chang S.H.(2014). An overview of empty fruit bunch from oil palm as feedstock for bio-oil production. Biomass Bioenergy, 62, pp. 174--181.

[6] Idris M.A., Jamal M.S., Muyibi S.A (2010). Tertiary treatment of biologically treated palm oil mill effluent (POME) using UF membrane system: effect of MWCO and transmembrane pressure. Int J. Chem Environ.Eng. 108-112.

[7] Harsono S.S., Grundmann P., Soebronto.S(2014) Anaerobic treatment of palm oil mill effluent: Potential contribution to net energy yield and reduction of greenhouse gas emissions from biodiesel production. Journal of Cleaner Production.619-627.

[8] Hazlan Bin Haris (2006) Treatment of Palm Oil Mill Effluent (POME) using memebrane bioreactor: A thesis submitted in fulfillment of the requirements for the award of the degree of Bachelor of Chemical Engineering. Faculty of Chemical and Natural Resources Engineering Technology.

[9] Daud Z., Abdul Latif A, Ab. Aziz N., Awang H (2013). Treatment of palm oil mill effluent by electrocoagulation with Alumina electrodes. Australian Journal of Basic and Applied Sciences. 452-463.

[10] Mokri G., Modrishala N., Behnajady M.A., Vahid B (2015) Adsorption of C.I Acid Red 97 dye from aqueous solution onto walnut shell. Kinetics, thermodynamics parameters isotherms. International Journal of Environmental Science and Technology. 1401-1408.

[11] Bhattacharyya K.G., Gupta S.S., Sarma G.K (2015). Kinetics, equilibrium isotherms and thermodynamics of adsorption of Congo red onto natural and acid treated kaolinite and montmorillonite. Desalination and water treatment. 530-542.

[12] Olayemi E.A., Adeyemi A.F., Olabanji I.O (2012) Removal of Pb2+ and Cd2+ ions from wastewaters using palm kernel shell charcoal. Research Journal in Engineering and Applied Sciences.308-313.

[13] Rizzo L., Fiorentino A., Grassi. M, Attanasio D., Guida, M (2015). Advanced treatment of urban wastewater by sand flitration and graphene adsorption for wastewater reuse: Effect on a mixture of pharmaceuticals and toxicity, Journal of environmental chemical engineering 3,pp. 122--128.

[14] Darael H., Mittal A., Noorisepehr M., Mittal J (2015). Separation of chromium from water samples using eggshell powder as a low-cost sorbent: Kinetic and thermodynamic studies.Desalination and Water Treatment, 53, pp. 214--220.

[15] Tresintsi S., Simeonnidis K., Katsikini M., Paloura E., Bantsis G., Mitrakas M (2014). A novel approach for arsenic adsorbents regeneration using MgO. Journal of Hazardous Materials, 265, pp. 21--225.

[16] Ng, F.Y., Yew, F.K., Basiron,Y, Sundram, K.(2011). A renewable future driven with Malaysian palm oil-based green technology.J.Oil Palm Environ.2, 1-7

[17] Medellin-Castillo N.A., Leyva- Ramos R., Padilla-Ortega., Ocampo Perez R., Flores-Cano J.V, Berber-Mendoza M.S (2014). Adsorption capacity of bone char for removing flouride from water solution . Role of hydroxyapatite content, adsorption mechanism and competing anions. Journal of industrial and Engineering Chemistry. 4014-4021.

[18] Cheung, C.W., Porter, J.F and Mckay G (2001). Sorption kinetics analysis for the removal of cadmium ions from effluents using bone char. Wat. Res, pp. 605--612.

[19] APHA, AWWA & WPCF. 2005. Standard Methods for the Examination of Water and Wastewater.

[20] Leelamanie D.A.L, Karube J, Yoshida A (2010). Characterizing water repellency indices: contact angle and water drop penetration time of hydrophobized sand. Soil Science and plant nutrition. Page 179-187.

[21] Debnath S, Kitinya J, Onyanyo M-S (2014)Removal of Congo red from aqueous solution by two variants of calcium and iron based mixed oxide nano-particle agglomerates. Journal of industrial and Engineering Chemistry (20) 2119-2129.

Treatment of Wastewater From Car Washes Using Natural Coagulation and Filtration System

A A Al-Gheethi[1], R M S R Mohamed[1*], M A A Rahman[1], M R Johari[1] and A H M Kassim[1]

[1]Faculty of Civil and Environmental Engineering, Universiti Tun Hussein Onn Malaysia, 86400 Batu Pahat, Johor, MALAYSIA

Email: maya@uthm.edu.my

ABSTRACT

Wastewater generated from carwash is one of the main wastewater resources, which contribute effectively in the increasing of environmental contamination due to the chemical characteristics of the car wastes. The present work aimed to develop an integrated treatment system for carwash wastewater based on coagulation and flocculation using *Moringa oleifera* and Ferrous Sulphate ($FeSO_4.7H_2O$) as well as natural filtration system. The carwash wastewater samples were collected from carwash station located at Parit Raja, Johor, Malaysia. The treatment system of car wash wastewater was designed in the lab scale in four stages included, aeration, coagulation and flocculation, sedimentation and filtration. The coagulation and flocculation unit was carried out using different dosage (35, 70, 105 and 140 mg L^{-1}) of *M. oleifera* and $FeSO_4.7H_2O$, respectively. The efficiency of the integrated treatment system to treat carwash wastewater and to meet Environmental Quality Act (EQA 1974) was evaluated based on the analysis of pH, dissolved oxygen (DO), chemical oxygen demand (COD) and turbidity (NTU). The integrated treatment system was efficient for treatment of raw carwash wastewater. The treated carwash wastewaters meet EQA 1974 regulation 2009 (Standards A) in the term of pH and DO while, turbidity and COD reduced in the wastewater to meet Standards B. The integrated treatment system designed here with natural coagulant (*M. oleifera*) and filtration unit were effective for primary treatment of carwash wastewater before the final disposal or to be reused again for carwash process.

Keywords: Coagulation, flocculation, filtration, sedimentation, moringa oleifera, ferrous sulphate.

1. Introduction

Among several types of wastewater, the carwash wastewater represents one of the heavily contaminated wastes with high impurities. It was due to presence of sand and particles, oil and grease, surfactants, detergent, phosphates and hydrofluoric acid [1]. Therefore, the direct disposal for wastewater into the drainage exacerbates the natural water pollution [2,3,4]. On the other hand, the discharge of the car washes wastewater directly into sewerage network might effect negatively on the efficiency of the treatment processes of sewage due to presence of heavy metals or at least represent a burden on the sewage treatment plant.

Further, the treatment of wastewater at the resource point might be easier and more effective than that if they have collected into one treatment plant. This is because, the individual treatment process will take sufficient time to remove most impurities and contaminants. Indeed, the treatment of small amount of wastewater does not require a complicated treatment system to produce high quality of treated wastewater. Moreover, the simple treatment system might be more efficiency to produce good quality of treated wastewater suitable for car wash again [5]. There are no quality regulations for the water, which are used for car wash. Therefore, the wastewater generated from car washes with high turbidity might be treated using coagulation and flocculation methods and reused again, both methods are efficient for turbidity and economic [6]. In order to improve the water quality characteristics, an integrated treatment system might be provided.

The coagulation and flocculation treatment are one of the important stage in the wastewater treatment processes. Several of chemical substances such as ferric sulphate, ferrous sulphate, ferric chloride and ferric chloride sulphate are used as coagulant and flocculants. However, the new direction is to use the natural substances. The application of these substances included free toxic by-products such as carcinogenic compounds and low cost. The utilisation of natural coagulants provides an alternative way to lower the coagulants and flocculants cost and to improve the water quality characteristics for safe utilization [7]. In previous study, [8] investigated the efficiency of *Moringa oleifera* and *Strychnospotatorum* seeds as natural coagulant compared with synthetic common coagulants for treating car wash wastewater. The study indicated that both *M. oleifera* and *S. potatorum* seeds contain coagulating substances capable of removing turbidity, *chemical oxygen demand* (COD), *Biochemical oxygen demand* (BOD) and *Total suspended solids* (TSS).

In the present study, a treatment process of car wash wastewater was designed based on the previous study [7,8]. The treatment process was enhanced with an integrated system comprises aeration, coagulation and flocculation, sedimentation and filtration unit to produce high quality of treated wastewater. The coagulant used was *M. oleifera* and ferrous sulphate ($FeSO_4.7H_2O$). The efficiency of the integrated treatment system to treat carwash wastewater and to meet Environmental Quality Act 1974 was evaluated based on the analysis of pH, dissolved oxygen (DO), chemical oxygen demand (COD) and turbidity (NTU).

2. Material and Methods

2.1 Collection of carwash wastewater samples

The carwash wastewater samples were collected from a Pusat Cuci Kereta Automatic Parit Raja. The car wash station is located at the Jalan Manis 1/2, 86400 Parit Raja, Johor, Malaysia (103°103'019) (Figure 1). The samples were collected weekly during the period from January to March 2015. The samples were collected between 9 and 11 am, where high number of vehicles were coming for the washing process based on the data collected from the station management. The samples were taken using grab sampling technique into the plastic bottles (2 L) and transported to the laboratory according to [9].

Figure 1. Location Map of Pusat Cuci Kereta Automatik Parit Raja (retrieved from Google map on November 3, 2014) Analysis of wastewater samples

The wastewater samples were tested for pH, Turbidity, Chemical Oxygen Demand (COD), and Dissolved Oxygen (DO). pH value of the sample was determined using a calibrated pH meter (EC Probe Meter). Turbidity of the wastewater sample was tested using the Attenuated Radiation Method (Method 3750, HACH DR 6000). COD were determined using Reactor Digestion Method (Method 8000, HACH DR 6000), finally, DO was tested using DO meter (EC Probe Meter). All the analysis was carried out according to [9].

2.2 Preparation of Moringa oleifera and ferrous sulphate

Moringa oleifera seeds (Figure 2) were purchased from a local market and was dried under the sunlight to improve level of polyelectrolyte presented in the kernels [10]. Then, the seeds were grounded into fine particles (0-0.20 mm) and dried at 100°C for 24 hrs. The dried seeds were milled into fine powder by using domestic blender and then passed through 50-mesh sieve according to coagulant-method [11]. Ferrous sulphate ($FeSO_4.7H_2O$) salts (CAS:7782-63-0, Fisher Scientific) was supplied from Tai`an Ming Chen Chemical Import and Export Co. Ltd, China.

Figure 2. *Moringa oleifera* seeds

2.3 Design of integrated treatment system

The treatment system used in this study is presented in Figure 3. It consists of four stages included aeration, coagulation and flocculation unit, sedimentation and filtration unit.

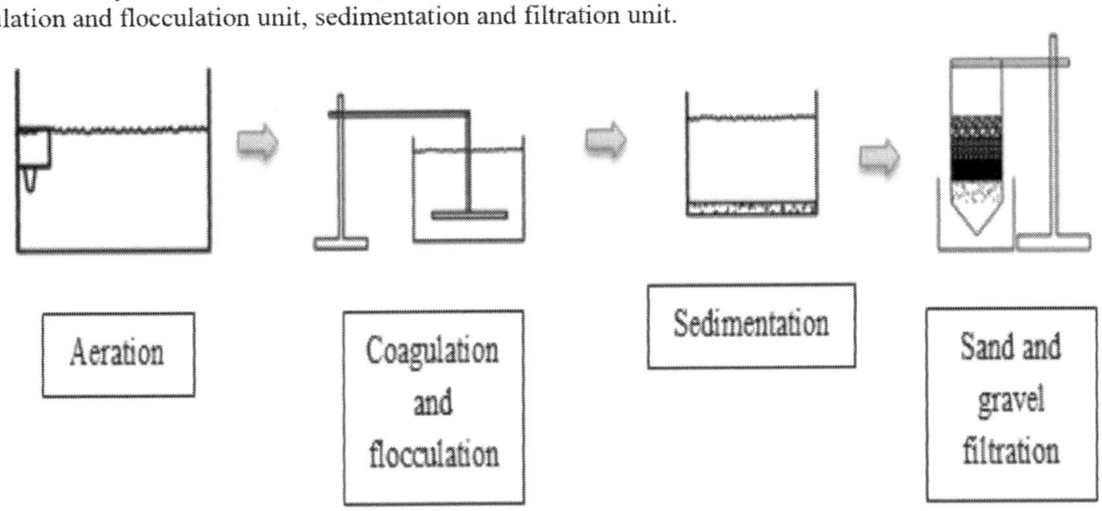

Figure 3. Schematic diagram of the integrated flocculation carwash wastewater system.

In this system, a volume of one litre of wastewater sample was transported manually into the aeration tank (25cm×17cm×18cm), and aerated for 30 min. The wastewater was pumped into the flocculation and coagulation tank using a water pump (Shanda), 120V. The flocculation and coagulation tank was made of *High-density polyethylene* (HDPE) transparent bottle (14cm×20cm×14cm) and Direct Current motor (DC Geared Motor, DC12V) with 1.1 watt, 103 rpm, 410 mA, 127.4 mN and 8 mm of shaft diameter. A weight of 35 mg of *M. oleiferra* was added into the flocculation and coagulation tank. The wastewater sample was mixed for one hour. Thereafter, the supernatant was pumped slowly into sedimentation tank and allowed to settle for one hour. The supernatant was then filtrated using a natural filtration system consisted of sand and gravel.

The treatment process was repeated with different doses of *M. oleiferra* and $FeSO_4.7\ H_2O$ (70,105 and 140 mg, respectively) According to [7,8]. The treated wastewater was analysed for pH, turbidity, DO and COD parameters.

3. Results and discussion

3.1 Characterization of raw carwash wastewater

The mean of pH, turbidity, DO and COD of raw car wash wastewater are illustrated in Table 1.

Table 1. Parameters of raw carwash wastewater samples.

Parameter	Raw car washes wastewater	Effluent Standard by Environmental Quality Act 1974
pH	6.96	6-9
Turbidity (NTU)	275.1	<5
Dissolved Oxygen (DO) (mg/L)	2.55	>7
Chemical Oxygen Demand (COD) (mg/L)	220	<50

Its appeared that the turbidity, DO and COD were higher than the standards limits of Environmental Quality Act 1974. While, pH was within the standards limits. These results are in agreement with those reported by [12] in Malaysia and [13] in India. However, they were differed from that reported in previous study [8]. Turbidity in carwash wastewater in this study was more than that in the last study. In contrast, COD concentrations was less than that recorded in the previous study. These differences would be related to the source carwash wastewater and type of vehicles.

The high value of turbidity is due to the presence dirt, mud and brake particles in wastewater which are washed off from vehicles and are comparatively large molecule size of pollutants [12]. The increasing the turbidity of wastewater indicated that the wastewater has high concentrations of total suspended solid (TSS) [1]. However, TSS is not investigated in the current study. The high concentration of COD might be related to the presence of detergents utilized during the washing process, which cause oxidation process of organic compound and then increasing COD [14]. It can be indicated that the parameters of raw carwash wastewaters are higher than the standards limits required for disposal process. Therefore, these wastes should be treated to improve their characteristics and prevent contamination of the environment.

3.2 Efficiency of integrated treatment system

The efficiency of the integrated treatment system design here for treating of carwash wastewater was evaluated based on pH, turbidity, DO and COD as described in section materials and method. The characteristics of wastewater treated with different doses of *M. oleiferra* and $FeSO_4.7\ H_2O$ are presented in Table 2.

Table 2. Characteristics of carwash wastewater treated with different doses of *M. oleifera* and $FeSO_4.7H_2O$.

Dose ($mg\ L^{-1}$)	pH		Turbidity (NTU)		DO ($mg\ L^{-1}$)		COD ($mg\ L^{-1}$)	
	M. oleifera	$FeSO_4.7H_2O$	*M. oleifera*	$FeSO_4.7H_2O$	*M. oleifera*	$FeSO_4.7H_2O$	*M. oleifera*	$FeSO_4.7H_2O$
0	6.96		275.1		2.55		220	
35	6.89	5.65	39.7	47.7	4.19	3.6	215	217
70	7.03	4.83	23.5	30.7	5.89	4.91	200	213
105	6.29	4.13	14.3	18.8	7.22	6.78	160	180
140	6.17	3.93	8.42	18.3	7.13	6.38	143	169
Standards (A)	6-9		<5		>7		50	
Standards (B)	5.5-9		50		5 – 7		100	

It can be noted that pH, and DO of treated wastewater reduced to meet EQA 1974, Regulation 2009 [15]. The maximum value of pH was recorded with 70 mg L^{-1} of *M. oleifera*. In contrast, pH was decreased with increasing of $FeSO_4.H_2O$ to pH 3.93 with 140 mg L^{-1}. These results indicated that the natural coagulant is more efficient to adjust pH value of wastewater than chemical coagulant ones. These findings were similar to that recorded by [8]. However, in that study the author reported that the alum and $FeSO_4$ was better as coagulant because little or no further addition of chemical would be required to correct the finished water pH. Conversely, this study demonstrated that *M. oleifera* is the best as coagulant because the pH value still within the standards limits 6-9 even at high concentrations, while it decreased to pH 5.65 at the low concentrations of $FeSO_4$.

The coagulation using *M. oleifera* increased of wastewater DO from 2.55 to 7.22 mg L^{-1} with 105 mg L^{-1}, while $FeSO_4.7H_2O$ increased it to 6.78 at same concentrations. Thus, in the term of DO, only carwash wastewater treated with *M. oleifera* has met EQA 1974, Regulation 2009 (Standard A).

Both coagulants were effective for reducing turbidity from 275.1 NTU to 8.42 and 18.3 NTU respectively at 140 mg L^{-1} of concentrations. The turbidity concentration still more than the EQA 1974, Regulation 2009 (Standard A). However, the wastewater treated with *M. oleifera* was almost near to meet the standards limits (<5 NTU), the removal percentage was >97%. [16] also showed that the coagulation of wastewater with *M. oleifera* reduced turbidity by more than 99%.

COD exhibited less response for both coagulant substrates; it reduced from 220 mg L^{-1} to 143 and 169 mg L^{-1} with 140 mg L^{-1} of *M. oleifera* and $FeSO_4 7H_2O$ respectively. However, the wastewater meet the regulations required for Standards B. The maximum reduction of COD was 35 and 23.1% recorded at 140 mg L^{-1} of *M. oleifera* and $FeSO_4$, respectively. This removal percentage is less than that reported by [7] where 60% of reduction for COD was observed at 60 mg L^{-1} of *M. oleifera*. These differences might be due to the turbidity concentrations of raw carwash wastewater, 275.1 NTU in this work comparing to 39-173 NTU in the study conducted by [7]. Therefore, in the current study the interaction between turbidity and COD removal was analyzed at $p<0.05$. It was noted that the turbidity concentration and COD removal correlated significantly with $R^2=0.96$.

In general, *M. oleifera* was more effective in the treatment of raw carwash of pollutants than $FeSO_4.7H_2O$ at different concentrations investigated in this study. These differences might be related to nature of *M. oleifera* as adsorbent for H^+ ions and TSS and has the ability to neutralization of solution [17]. Thus, the turbidity in the wastewater treated with *M. oleifera* was less than that treated with $FeSO_4.7H_2O$.

In addition to efficiency of *M. oleifera* compared to $FeSO_4.7H_2O$, it is a natural product, low cost and eco-friendly since no chemical additive is needed or toxic by-product is generated from the coagulation process.

4. Conclusion

It can be concluded that the integrated treatment system designed here with natural coagulant (*M. oleifera*) and filtration unit was effective for primary treatment of carwash wastewater. *M. oleifera* was effective as coagulant for treatment of raw car washes wastewater. The wastewater treated with low concentrations of *M. oleifera* meet the regulation required Standard A in the term of pH and DO. The reduction of turbidity and COD reduced to meet EQA 1974 regulation 2009 (Standards B). Moreover, the parameters of treated car washes wastewater is appropriate to be reused again for car washing processes.

Acknowledgments

The authors gratefully acknowledge the Ministry of Science Technology and Innovation (MOSTI) for the research project financial support under FRGS vot Grant No. 1453.

References

[1] Mandac, RS, Bogunović, B., Žagar, D., Faganeli, J. (2014) Riverine impact on the thermohaline properties, turbidity and suspended solids in a shallow bay (Bay of Koper, northern Adriatic Sea). *ActaAdriatica*, 55(2): 195 – 212.

[2] Al-Gheethi A.A.S., Norli, I.; Lalung, J.; Azieda, T. and Ab. Kadir, M. O. (2013) Reduction of faecal indicators and elimination of pathogens from sewage treated effluents by heat treatment. Caspian Journal of Applied Sciences Research, 2 (2): 29-45.

[3] Al-Gheethi, AA, Radin Maya Saphira RM, Efaq AN, Amir HK. (2015) Reduction of microbial risk associated with greywater utilized for irrigation. A review. Water and health Journal (IWA, ISI, IF 1.45). In press.

[4] Jangkorn, S., Kuhakaew, S., Theantanoo, S., Klinlaor, H., Sriwiriyarat, T. (2011) Evaluation of reusing alum sludge for the coagulation of industrial wastewater containing mixed anionic surfactants. *J. Environ. Sci.,* 23 (4), 587-594.

[5] Khamis N, (2010) Degradation of Lead From Car Wash Waste Water Sample Using Ultrasonic Assisted Extraction Process. Faculty of Chemical & Natural Resources Engineering, UMP, Malaysia.

[6] Firdaus, S. (2013) Car wash industry in Malaysia: Treatment of car wash effluent using ultrafiltration and nanofiltration membranes. *Sep. Purif. Technol.,* 104, 26–31.

[7] Mohamed RMSR., Abdul Rahman, N., MohdKassim AH. (2014a) *Moringa Oleifera* and *StrychnosPotatorum* Seeds as natural coagulant compared with synthetic common coagulants in treating car wash wastewater: Case Study 1. *Asian J. Apple. Sci.*, 2 (5), 693-700.

[8] Mohamed, RMSR., Mariam, A, Kutty, I., Kassim AH. (2014b) Efficiency of using commercial and natural coagulants in treating car wash wastewater treatment, *Aust. J. Basic Appl. Sci.,* 8 (16): 227-234

[9] APHA (2005) American Public Health Association, Inc. Standard methods for the examination of water and wastewater (21st Ed.). New York. USA, 2005.

[10] Nirmala, R., Jadhav, M.V. (2012) Enhancing filtrate quality of turbid water incorporating seeds of *Strychnospotatorum*, pads of cactus opuntia and mucilage extracted from the fruits of cocciniaindica as coagulants. *J. Environ. Res. Dev.* 7(2):668-674.

[11] Suhartini, S., NurHidayat, Rosaliana, E. (2013) Influence of powdered *Moringa oleifera* seeds and natural filter media on the characteristics of tapioca starch wastewater. *Int. Rec. Org. Waste Agri*. 2: 1-11.

[12] Lau, W. J., Ismail, A. F., Firdaus, S. (2013) Car wash industry in Malaysia: Treatment of car wash effluent using ultrafiltration and nanofiltration membranes. *Sep. Purif. Technol., 104*, 26–31

[13] Mazumder, D., Mukherjee, S. (2011) Treatment of automobile service station wastewater by coagulation and activated sludge process. *Int. J. Environ. Sci. Dev.*, 2(1), 64–69.

[14] Boussu, K., Kindts, K., Vandecasteele, C., Van der Bruggen, B. (2007) Applicability of nanofiltration in the carwash industry. *Sep. Purif. Technol.*, 54:139–46.

[15] EQA 1974, (2011) Environmental Quality (Sewage) Regulations 2009. International Law Book Science, Kuala Lumpur.

[16] Sanghi, R., Dixit, A. and Guha, S. (2006) Sequential batch culture studies for the decolorization of reactive dye by Coriolusversicolor. *Bioresour. Technol.,* 97, 396–400.

[17] Mangale, S.M., Sonal, C.G. and Raut, P.D. (2012) Use Moringa Oleifera (drumstik) seed as natural absorbent and an antimicrobial agent for groundwater treatment. *Res. J. Recent Sci.* 1 (3), pp.31-40, 2012.

Solid Waste Composition and Quantification at Taman Melewar, Parit Raja, Batu Pahat

A A Kadir[1], S S S Z Abidin[1]
[1]Faculty of Civil and Environmental Engineering, Universiti Tun Hussein Onn Malaysia, 86400 Batu Pahat, Johor, MALAYSIA

E-mail: aeslina@uthm.edu.my

Abstract. The poor management of solid waste is noticeable through the increasing of the solid waste each year and the difficulties in disposing the waste in the current available landfill. This study was undertaken to analyze the quantity and composition of waste generation in Taman melewar. Taman Melewar is a student residential area and this study is focusing on student's daily waste composition. The objective of this study was to identify the amount of solid waste generation, analyze and classify the composition of solid waste in Taman Melewar. The waste collection was conducted for 50 houses on a daily basis for two weeks. The average household waste generation rate was 0.082 kg/person/day. Organic waste was the major constituent of waste production. The average of organic waste represents about 72.4% followed by paper (9%), plastics film (5.5%), plastics rigid (4.7%), napkins (3.8%), tetrapek (1.3%), glass (1.1%), household hazardous waste (0.85%), textiles (0.52%), metal (0.51%) and rubber (0.34%). The moisture content was ranging from 27.67% to 28.68%. An evaluation was made based on student's behavior towards waste production and recycling. In conclusion, the results revealed that organic waste is the highest waste generated and recycling habits is also poor in Taman Melewar.

Keywords: Waste generation, waste composition, student residential area.

1. Introduction

It is to be estimated that the waste generation rate in Malaysia is about 9,092,000 tons/year as in 2020 correspondence with the increasing in population [1]. Meanwhile, based on previous study [2] the average solid waste generated per person per day in Malaysia is ranging from 0.5 kg to 0.8 kg in rural area whiles at the urban area, the average of solid waste generated per person per day at urban area is 1.9 kg. Hence the waste generation rates are increasing each year especially in the urban area. This will eventually lead to insufficient landfill and difficulties in managing the waste. In managing solid waste, it is highly important to consider the characteristic of the solid waste generated by knowing the waste composition, generation rates and sources [3]. The source of the waste generated comes from different sectors. This also may include waste from residential, commercial, industries and institutional. Statistics had showed that nearly 50% of the waste generated in Malaysia is household waste [2].

Meanwhile, according to Dinie et. al., [1] 36.73% of waste generated are from household followed by 28.34% from industrial and construction wastes and 34.93% from other sources. In summary, household waste contribute the most on the waste production. Moreover, the pattern of the consumption of household waste is changing correspondence with the changing lifestyle of the Malaysian over the years.

In planning a good waste management, it is highly important to identify the waste composition on the household waste. Waste composition consists of variable type of materials or substances and waste composition is different on each country. Therefore, the availability of reliable data is also different and it is highly dependable on the country waste generation. Basically, household waste is a type of waste based on the normal household production. The waste production is based on day to day operations of households [4]. In Malaysia, the composition of households waste are consists of organic waste, paper, plastic, napkins, tetrapek, glass, metal, household hazardous waste, rubber, textiles, leather, garden and others waste. Furthermore, waste composition can be divided into two categories which is organic and inorganic waste. The quantity of organic and inorganic waste is depending on the country itself. The data on the waste composition is limited especially in local area. Therefore, reliable data on waste composition is highly needed.

For the past years, Parit Raja's population increased and the economy becomes more stable due to the existence of Universiti Tun Hussein Onn Malaysia (UTHM), which clearly had increased the waste generation in the area. In order to implement proper planning on Household Solid Waste (HSW) management, a study on the local needs to be done. A reliable data on the waste generation rate is difficult to achieve especially on small town or city like Parit Raja.

According to Ojeda-Benitez et.al. [5] by analyzing the household waste the waste generation rate and any potential use for recyclable wastes can be identified. Meanwhile Suthar et. al. [6] mentioned that the local resident's composition of HSW reflects their way of their living and attitude Nevertheless, the composition of waste generated are able to determine the possible treatment method and through proper management practice, the waste composition rate will be changed. The composition of the waste generated is inconsistent due to seasonal, standard of living, legislation factor and culture which makes it is more difficult in defining and measuring the waste composition rate [7]. In addition, the students are among the community living in Parit Raja. Due to that, the consumption of HSW pattern might be different.

Parit Raja is lacking of disposing facilities for recyclable items. Additionally, slum area is usually are not included in the waste collection routes by the local authorities. Therefore, the waste generated by the household is fully under the responsibility of the resident. In addition, previous paper by [8] stated that the waste generation rate in slum areas are expected to increase if the urbanization is increasing at the rate of which faster than the local authorities can handle. Without a proper guidance, this will only contribute to the household waste generation rate and eventually will worsen the household waste management. Other than that, lack of concern or knowledge among the local society especially students will eventually bring out a negative impact to the composition of the household waste. The practice of 3R concept can be determined by the composition of waste generated by the house resident. The aim of this study is to determine the waste composition of Taman Melewar, Parit Raja, Batu Pahat. This study were able to reveal the local attitude on the waste management throughout the waste composition that was collected during the investigation.

2. Materials and methods

2.1 Sampling area

The study was carried out in one of the UTHM's residential college which is in Taman Melewar. The selected area consists of student residential area. Taman Melewar is one of the UTHM's residential colleges in which accommodate about 433 students and staffs. Basically, this residential area comprised of 77 single storey terrace house. However, only 50 houses were occupied and willing to give full commitments. As for the household size, it can reach up to 8 residents per household.

2.2 Survey
A survey on the selected house needs to be done in order to identify the waste retain, the number of household members and also to seek for their permission to participate in the study. During the process of surveying, the participants were briefed regarding the survey and the procedures on the waste collection from their house. Each house that agreed to participate in this study was labeled with a reminder notes on not to discard their wastes out of the house.

2.3 Sampling method
This study was conducted according to Malaysian Standard MS 2505:2012. Since the purpose of this study is to identify the solid waste generation, the selection of sampling point will be waste as-generated. Sampling waste as-generated known as the waste that will contribute to the total amount of waste generated from the household. The waste also including both the waste kept at home for recycling or reuses purpose and the unwanted waste place at the collection point. The targeted households were provided with two plastic bags with labeled. Each plastic bag was labeled Monday to Sunday and the plastic bags also had been labeled as organic and inorganic. The waste collection process was done by collecting wastes directly from house to house. The plastic bags were collected daily for a week. The collected plastic bags (organic and inorganic) for each house were straightly weighted at the location and the data will be recorded daily. In addition, the house number of the collected waste also had been recorded for reference purpose. The organic wastes were straightly weighed and recorded. Meanwhile, the collected inorganic wastes were weighed and brought to the laboratory for sorting purposes. This procedure was repeated for one week until the minimum requirement for waste collection obtained (minimum of 50 houses or 200 kg). The collection of wastes was done twice in order to get more accurate data. The inorganic wastes were sorted accordingly into category and weighed in the laboratory. The sorting process was done daily. The collected wastes were sorted accordingly into 14 categories as listed in the Malaysian Standard MS 2505:2012 guidelines. Weighing was carried out three times and the average value was recorded in the data sheet. The weight value recorded as wet weight with a weight scale. During this process, it is important to use the personal protective equipment (PPE) to avoid any unwanted incident and hygiene purpose.

3. Results and discussion

3.1 Analysis on waste generation rate
Figure-1 represent about the total of waste collected by day on week 1. The data demonstrated on day 6 (Saturday) and day 7 (Sunday) have the least of total waste by 27.46 kg and 28.65 kg. This might be due to the outdoor activities involved by the students during weekend. The highest number of waste collected is on day 1 (Monday) with 38.87 kg. The highest inorganic waste being produced was also on day 1(Monday) with 12.44 kg, meanwhile, organic waste was highly produced on day 3 (Wednesday) with 27.5 kg. The lowest organic being produced was on day 6 (Saturday) with 17.67 kg and as for inorganic was on day 2 (Tuesday) with 7.57 kg. The overall waste collected for the total 50 houses in 7 days was 227.91 kg. The total organic waste was 160.81 kg. Followed by the inorganic waste collected for the week was 67.1 kg. In average, the waste collected for 7 days was found to be 32.56 kg per day. In conclusion, the total waste generation rate was 0.0841 kg/person/day.

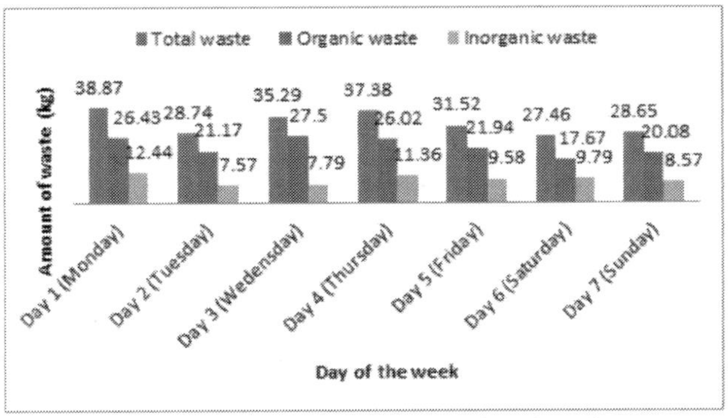

Figure 1. Data of week collection (week 1).

On the other hand, Figure-2 represents the total of waste collected by day on week 2. The waste collected for week 2 also showed the same characteristic as the sample for week 1 in which day 6 (Saturday) and day 7 (Sunday) has the least waste by 25.32 kg and 30.2 kg. The highest number of waste collected was on day 1 (Monday) by 36 kg. As for the organic consumption pattern, the highest organic waste recorded was on day 1 by 29.19 kg and the lowest was on day 6 by only 19.53 kg. Inorganic waste was the highest in day 5 (Friday) by 10.39 kg and the lowest in day 6 by only 5.79 kg. In conclusion, the overall data of waste collected for 7 days was 220.84 kg. The highest organic and inorganic waste for the week recorded was 164.47 kg and 56.17 kg. The average value of waste in 7 days period was 31.55 kg and the total waste generation rate for week 2 was 0.0815 kg/person/day.

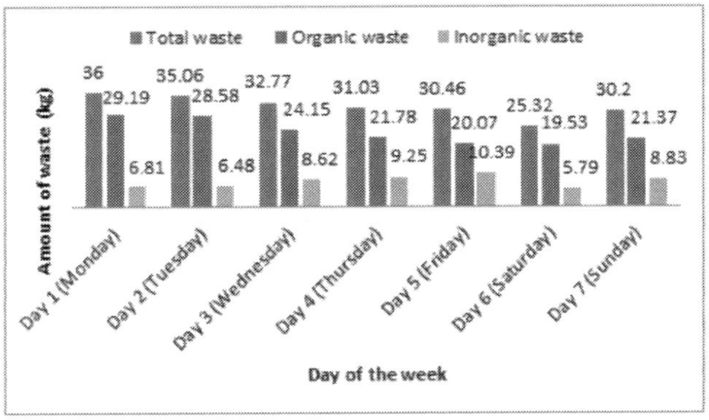

Figure 2. Data of week collection (week 2).

3.2 Waste composition

The waste composition was divided into 15 categories namely, food organic, paper, tetrapek, plastics film, plastics rigid, napkins, textiles, rubber, leather, wood, garden, glass, metal, household hazardous waste and others. Figure-3 represents the waste composition for week 1 and Figure-4 represent the waste composition for week 2. Based on the pie chart on Figure-3, the highest weight of waste collected was food waste which is 71% (160.14 kg) then followed by paper with 10% (21.03 kg). The paper waste may include boxes, newspaper, magazines and others paper. Next is plastic film and plastic rigid with 6% (14.26 kg) and 5% (11.56 kg). Plastic films consist of polystyrene, foam and the most common is plastic bags. Meanwhile plastic rigid may consists of waste from bottles, pipes, plastic toys, hard plastic, spectacles and crates. Napkins may include diapers, facial cottons and tissues. The reading for napkins was 3% (7.36 kg). Other than that, the lowest reading values of wastes were tetrapek with 1% (2.51 kg), metal with 1% (1.3 kg), textiles with 1% (1.79 kg) and glass with 1% (3 kg). Household hazardous waste recorded to be 1% (2.35 kg) and it may consist of battery and mostly medicines.

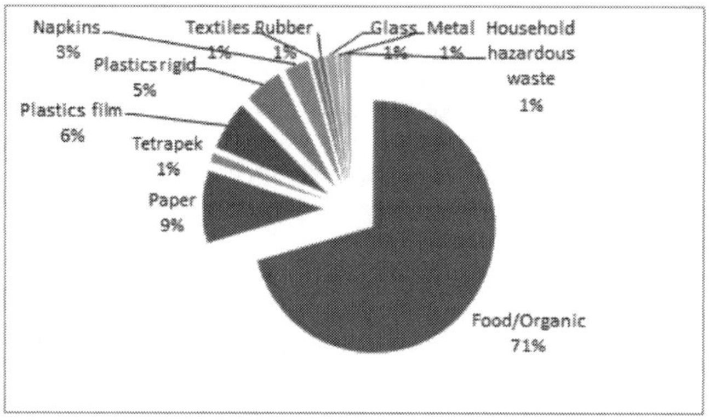

Figure 3. Waste composition for week 1.

Based on Figure-4, the highest amount of waste collected was still food waste with 74% (162.37kg) and the second highest waste collected was paper with 9% (18.99 kg). Plastic films and plastic rigids also one of the highest wastes recorded in this study which was 5% (10.25 kg) and 4% (9.5 kg). Next, napkins weight recorded was the same as plastics rigid which was 4% (9.35 kg). Tetrapek contributes to only 2% (3.27 kg) of the total waste collected. As for glass and household solid waste, it contributes 1% of total waste production. Glass value was 1.8 kg and household solid waste was 1.45 kg. Meanwhile, the lowest value of waste collected in a week is textiles with 0.5 kg followed by metal with 0.96 kg. The glass consists of mirror and glass while household hazardous waste consists of medicines, liquid fluid, light bulbs and pesticides.

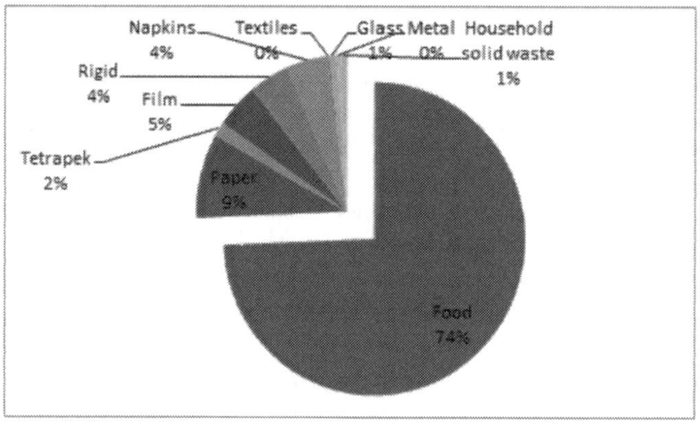

Figure 4. Waste composition for week 2.

Based on the above data, the differences of number of waste production between the food waste and other categories varied significantly as for both samples. Taman Melewar is one of the UTHM's residential colleges. Therefore, the residents of Melewar were not allowed to cook in their house. Despite the limitation, the food waste contributed most to the waste production in Taman Melewar. The high rate of food waste production might be due to the habit of the residents. This finding is also similar with the previous study that claimed production of food waste is higher mainly because food is the essential needs for people [9]. In addition, the society preferably to throw this type of waste rather than recycling it. This explained the high production of this type of waste.

3.3 Waste characterization

The moisture content for each category was determined by drying the samples for 24 hours in an oven with temperature of 105°C. 500g of each waste were taken representing the overall weight of the waste. The value of dry mass was then calculated to get the moisture content representing week 1 and week 2. From the data, organic waste had the highest moisture content for 34%. Meanwhile, the recyclable items like tetrapek had the highest moisture content in compared with all the others categories by 37.5%. Tetrapek was used as a liquid container. Hence, this explained the higher moisture content value of the waste. Besides tetrapek and organic waste, the moisture content for napkins was 22%. The value was considerably high. Plastics film, rubber, paper and glass wastes had lower moisture content level. This might be due to their chemical compositions that do not absorb the moisture. The overall weight of waste after sorting for week 1 was 226.8 kg and as for week 2 was 218.44 kg. In summary, the value of moisture content for week 1 was 27.67% and week 2 was 28.68%. To compare with previous study, moisture content in Taman Melewar was considerably low for both weeks. Based on previous study, the moisture content for 113 households was approximately 50% [10]. Owing to the fact that the wastes collected in this study were not exposed to rain as the wastes were collected directly from the house and it is sorted properly by day.

4. Conclusion and recommendation

It can be concluded that the average household waste generation for both weeks was 0.082 kg/person/day of which food waste was dominating the waste generation rate in Taman Melewar. The waste generation rate in this study was low compared with previous study. The result was indirectly proved that the bigger the households are, the lower the daily waste generation rate.

The average food waste accounted for the highest proportion by 72.4% (161.26 kg) and the least was rubber by 0.34% (0.75 kg). Among the recyclable items, paper fraction accounted for the highest proportion about 9% (20.01 kg) followed by plastics film and plastics rigid by 5.5% (12.26 kg) and 4.7% (10.53 kg). Other than that, the average napkins production was 3.8% (8.36 kg). In addition, tetrapek, household hazardous waste and textiles average value was 1.3% (2.89 kg), 0.85% (1.9 kg) and 0.51% (1.15 kg). Meanwhile, glass and metal were comparatively low. Glass was about 1.1% (2.4 kg) and metal was only 0.51% (1.13 kg). The study also revealed that the moisture content was ranging from 27.67% to 28.68%. These values were representing from two weeks of sampling.

The results had proven that the residents of Taman Melewar have a poor practice in recycling. In addition, the high rate of the production of food waste needs to be addressed carefully. The high in number of recycling waste like paper and plastics reveals the most of the residents do not practice the 3R concept which is reuse, reduce and recycle. Improper waste facilities might be one of the reasons to the poor waste segregation. Proper waste facilities need to be provided in order to increase the recycling waste. Without a doubt, waste segregation is highly important. By segregation, not only it will help the waste sorters, it will also help to improve the selection of recyclable items. There are many parts of the wastes that are useful in terms of economic and environment. The future research is highly needed in order to gain more reliable information regarding household waste composition and generation rate.

References

[1] Richardson, O. P., (2005), Household Behaviour Towards Waste Recycling In Ampang Jaya and Subang Jaya, Selangor
[2] Abas, A., Wee, S. T. (2014), Municipal Solid Waste Management: An Insight Towards Sustainability
[3] Dinie, M., Samsudin, M., & Don, M. M. (2013). Jurnal Teknologi Full paper Municipal Solid Waste Management in Malaysia : Current Practices , Challenges and Prospect, *1*, 95–101.
[4] Pakpour, A. H., Zeidi, I. M., Emamjomeh, M. M., Asefzadeh, S., & Pearson, H. (2014). Household waste behaviours among a community sample in Iran: an application of the theory of planned behaviour. Waste Management (New York, N.Y.), 34(6), 980–6.
[5] Ojedabenitez, S., Armijodevega, C., & Ramirezbarreto, M. (2003). Characterization and quantification of household solid wastes in a Mexican city. Resources, Conservation and Recycling, 39(3), 211–222.
[6] Suthar, S., & Singh, P. (2015). Household solid waste generation and composition in different

family size and socio-economic groups: A case study. Sustainable Cities and Society, 14, 56–63.

[7] Yusuf, M. S., Maya, R., Radin, S., & Wurochekke, A. A. (n.d.). Municipal Solid Waste Composition in Urban Municipality: Case Study of Parit Raja.

[8] Lindell, A. (2012). Achievement of sustainable solid waste management in developing countries.

[9] Armi, M., Samah, A., Manaf, L. A., Ahsan, A., Nor, W., Sulaiman, A. Silva, J. L. D. (2013). Household Solid Waste Composition in Balakong City , Malaysia : Trend and Management, 22(6), 1807–1816.

[10] Qu, X., Li, Z., Xie, X., Sui, Y., Yang, L., & Chen, Y. (2009). Survey of composition and generation rate of household wastes in Beijing, China. Waste Management (New York, N.Y.), 29(10), 2618–24.

Solid Waste Composition Study at Taman Universiti, Parit Raja, Batu Pahat

A A Kadir[1], M S A M Sani[2]
[1]Faculty of Civil Engineering and Environmental,University of Tun Hussein Onn Malaysia, 86400 Batu Pahat, Johor, MALAYSIA

E-mail: aeslina@uthm.edu.my

Abstract. Solid waste management is recognised as one of the most challenging issues confronted by both the developed and developing countries. The problems rise due to growing population in current years which results in increased generation of waste with various compositions. The aim of this study was to determine the waste compositions at Taman Universiti. Taman Universiti is a mix residential and commercial area which a preferred residential location amongst students and lecturers due to its proximate location to UTHM main campus. The waste collection was carried out for 50 houses on a daily basis. The collection and sorting out method was conducted according to Malaysian Standard MS 2505:2012 and the data was collected and recorded The result showed that the average generation rate of household waste at Taman Universiti was 0.16kg/person/day and the moisture content was approximately ranging from 61%-68%. Household wastes collected were categorized and it consisted of food and organic, paper, rigid plastics, plastics film, baby diapers, glass, tetra pak, household hazardous waste, metal, rubber, textiles, garden waste and leather. The proportion of each wastes were approximately 64.67%, 9.36%, 9.22%, 5.33%, 3.51%, 2.53%, 1.37%, 1.05%, 0.84%, 0.85%, 0.80%, 0.27%, and 0.23%, respectively. Results from the analyses indicated that the food and organic waste are the major composition of household waste at Taman Universiti followed by the paper, rigid plastics, and plastic film. Meanwhile, the proportion of baby diapers, glass, tetra pak, household hazardous waste, metal, rubber, textiles, and garden decreasing accordingly. In addition, leather was recognized as the least category that contributed to the household waste.

Keywords: Waste composition, household waste, waste generation.

1. Introduction

Malaysia as a developing country has experiencing various urban transformations and rapid growth economy in parallel to extension of latest technology over the last decade. As a result, the waste generation in Malaysia has been increased gradually since year 2000 according to a report by the Ministry of Housing and Local Government. The source of the wastes may come from residential, institutional, or commercial sources which also known as Municipal Solid Waste (MSW).

According to Law of Malaysia [1] solid waste is defined as any scrap material or rejected product due to any application of any process, any substance that needed to be disposed of as being worn out, broken, contaminated or anything that is required by the authorities to be disposed of, and exclude wastes, sewage and radioactive materials. Municipal solid waste commonly disposed through landfilling method or incineration which only about 2% of the waste generation is recycled or treated by biological composting [2].

The household waste (HW) is an important part of the municipal solid waste (MSW) stream [3]. Since household waste is the main source of MSW, therefore reliable information related to HW generation and composition is necessary to develope a 3Rs approach. In many part of developing country, the composition of household waste also reveal the trends of waste reuse and recycling habit which is in practice as informal act. In addition, the socio-economical conditions of the certain households and urban society also may be measured regarding to household waste capacity. Increment of the income will effect the consumption pattern of household which resulted the changes quantities and composition of household waste.

Municipal solid waste is refering to garbage generated by household and commercial sources. There are three sources of municipal solid waste in Malaysia including domestic, commercial and industrial solid waste. The wastes that come from households and public areas, including waste collected from residential buildings, litter bins, streets, marine areas and country parks are known as domestic solid waste. Meanwhile, commercial solid waste is the waste which comes from shops, restaurants, hotels, offices and markets in private housing estates and the other waste which is generated by all industries, but does not include construction and demolition waste, chemical waste or other special waste is called as industrial solid waste.

Initially, the trends of MSW generation in Malaysia have been studied since the early 1980s. Due to rapid population and economic growth, generation of Malaysia MSW has increased by more than 91% over the past 10 years [4]. The 36.73% of wastes are household waste, 28.34% industrial and construction wastes, and 34.93% of waste comes from other sources based on collection in central and southern regions of Malaysia [5]. A bulk density above 200 kg/m^3 and a high moisture content proves that Malaysian solid waste contains a very high concentration of organic waste. However, based on a waste characterization study, found that the major components of Malaysia MSW are foods, papers and plastics where it covers 80% of overall weight [6]. Therefore, continuous effort is essential to identify a prudent management approach to minimize the burden of existing MSW disposal systems.

Household waste is normally defined as waste that is produced by normal household activities [7]. Household wastes are consisted of food/organic, paper, plastics rigid, plastics film, baby diapers glass, tetra pak, household hazardous waste, metal, rubber, textiles, garden, and leather. Municipal waste management have to allocate most cost to household waste since it is one of the major sources of municipal solid waste [8]. Increasing amount and complexity of waste has become one of numerous challenges faced by the waste management sector globally. Therefore, the environment will be effected in term of its quality if household waste and supermarket waste are not disposed or recycled properly.

Besides, information on waste composition is essential in accessing possible choice for sustainable waste disposals, reuse and recycling. Due to the unknown pathways of waste streams in the environment especially with all the existence of various waste system and management as well as limited survey possibilities, the detailed information on the Malaysian waste composition is rather under developed and doubtful accuracy. Knowledge of solid waste necessary knowledges vital for an effective management of urban solid waste [9] [10]. However based on the previous study, waste composition in Malaysia is known to be dominated by putrescible wastes which insert processed kitchen waste and food waste are included mainly in rural landfills, followed by paper waste and plastic-based waste. The recyclable items represent about 60% of total waste volume without the retrieval activity which is obviously becoming the larger amount of wastes [11]. Therefore, in this study organic waste such as food waste are estimated to be the dominant in the wastes category.

IOP Conf. Series: Materials Science and Engineering **136** (2016) 012048 doi:10.1088/1757-899X/136/1/012048

2. Materials and methods

The sampling boundary area, size required (number of premises), period and method of sampling, categorization and laboratory analysis was carried out according to Malaysian Standard MS 2505:2012 Guidelines for sampling of household solid waste-composition and characterisation analysis prepared by Department of Standards Malaysia. The solid waste sample was analysed in the Environmental Laboratory of Faculty of Civil and Environmental Engineering University Tun Hussein Onn Malaysia. This study was conducted by covering 41 unit of single-storey houses at Taman Universiti. A target amount of 200 kg of wastes were collected daily from the 41 houses (Figure-1).

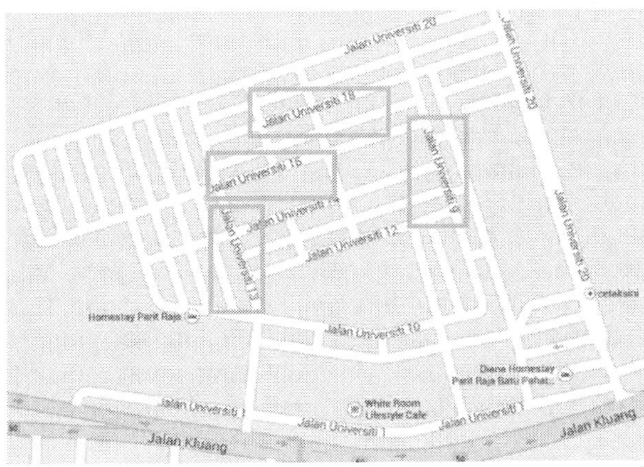

Figure-1. Location of selected single-storey houses at Taman Universiti.
(Source: https://www.google.com.my/maps)

The aim of this study were focusing on waste discarded and waste retained. The survey was conducted by distributing questionnaires to the houses owner and the number of surveys conducted were 100 % of the total number of houses to be sampled (completed survey). In addition, the survey was within the sampling boundary. As a result, the study gathered the details on household waste generation and recycling behavior, at the local level.

During the sample collection, the residents were provided with two plastic bags with label recyclable and non-recyclable on daily basis to put all their waste into these bags accordingly. The composition of solid waste disposal was identified by separating the recyclable and non-recyclable item. The collection of recycling materials occured on the same day with other waste collection. However, the organic waste and recycling materials was ensured that they were not mixed together and there was no spillage occured. Then, the sample waste was sorted and classified into waste composition accordingly. Goggle and face mask were used for protection purposes during sorting out.

The total weight of the waste composition was determined by weighing all the categories of household wastes. In order to have a precise reading, the waste was weighed for three times and an average value was taken. These procedures, were followed throughout the experiment. The weight of waste compositions were recorded on the worksheet classification of waste compositions and was illustrated in terms of percentage. As a result, the most abundant household waste in Taman Universiti area were determined based on the weight of the recorded wastes. The moisture sample for each categories of waste was taken by weighed it into 500 g each and recorded as wet weight of sample. Then, the wet sample was dried to a constant weight at a temperature not exceeding 105°C using oven at RECESS laboratory in Universiti Tun Huessein Onn. The sample allowed to be cooled and then weighed again to achieve the last reading. Then, the sample was recorded as dry weight of sample. As for the last step, the moisture content of each categories were determined.

3. Results and discussion

3.1 Survey

A questionnaire that consist of two sections was distributed within the sample boundary of Taman Universiti. For the first part, consists of respondents' basic informations and as for the second part, two questions regarding the household solid wastes were given. The distribution of questionnaires cover 41 number of single storey terrace houses which most of the respondents mainly consist of students and local residents. Throughout the survey, essential informations such as total number of family members living in the house, amount of recyclable items produced and recycling method practiced by the residents had been identified. Average total numbers of people living in the house ranged between 4-7 person per house.

Analysis from the questionnaire shown that kitchen waste dominates the amount of waste produced. For week one collection, kitchen waste was estimated 76kg followed by others such as papers, magazine and plastic rigid which are 10.4kg, 10.2kg, and 9.91kg respectively. The domination factor is due to the daily cooking activites by the residents. The amount of weight for other items such as newspapers (9.6kg), drinking plastics (7kg), steel can (5.2kg), others (5kg), aluminium can (4.4kg), glass bottle (4.3kg) and plastic bag (3.71kg) are decreased accordingly. Meanwhile, papers had the highest number of recyclables produced for the survey conducted during the first week of collection sample by 197 numbers followed by drinking plastics (189), kitchen waste (177), plastic rigid (119), and napkins (100) based on the survey on the estimation quantity of recyclables produced by number. The highest number of other papers produced for week one may contributed by the students who discarded their books, learning modules and used tissue papers.

Survey on the preference of recycling method by the residents in Taman Universiti indicate that most of them do not practice any of the recycling method and discarded all the wastes generated to the dustbin as comingled wastes. Some residents prefer to bring the waste such as magazine (4 houses), newspaper (5 houses), plastic drinking (5 houses), and plastic rigid (4 houses) to the recycling centre which made this method as second most popular compared to others.

Furthermore, the feedback on solid waste management and recycling in community given by the resident also stated that 12% of them agreed that the solid waste management in Taman Universiti is at satisfying level meanwhile 10% of them gave compliment for the very good service provided. However 2% from the number of residents are unhappy for the bad service and 5% of them stated that the service is not too good and need to be improved. One of the ideas suggested by the respondent in the way to improve the system is to perform daily waste collection in order to prevent any unpleasant smell coming from the wastes which will attract uninvited animals such as rat, snakes and cockroaches to their neighborhood. However, all the data collected from the survey mostly are inconsistent due to non-cooperative behavior from the respondents and also difficulties in estimating the waste produced.

3.2 Generation rates and composition

All materials discarded, whether or not they are later recycled or disposed in a landfill was considered as waste generated. The total waste collected for Taman Universiti for the first week is 231.47 kg, meanwhile for the second week is 226.07 kg which made it 228.77 kg as average. Then, this amount was calculated to a daily per capita generation rate of 0.1613 kg/person/day and 0.1575 kg/person/day for both weeks. Based on the waste generation by day, Day 1 (Monday) hold the highest number of waste being produced compared to the other day. Nevertheless, the waste productions continue to drop until day 5. Day 6 and day 7 can be considered as the lowest waste production of the week. This might be due to the weekend in which most of the residents going out which contribute to the less waste production.

The number of waste collected were classified into 15 categories throughout the week. The pie chart shows the percentage of waste by its category. Based on Figure-2, it can be concluded that the highest percentage for the first week waste collection at Taman Universiti was food waste which is 68.3%. The second highest waste being produced was plastics rigid and paper. with 8% respectively.

Next, napkins and plastics film which is 4% for each category meanwhile glass and household hazardous waste contribute to 2% of the total waste produce in Taman Universiti. Rubber, textiles, metals, glass and household hazardous waste are all considered as lowest waste production. Rubber, textiles, tetra pak and metals each contribute to only 1% of waste.

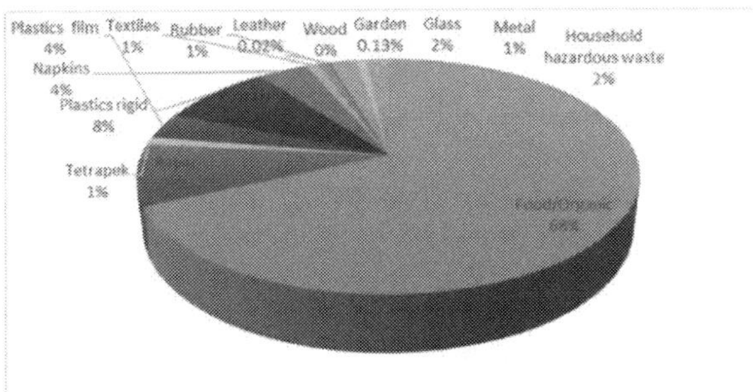

Figure-2. Composition distribution of household waste (week 1)

On the other hand, Figure-2 shows the classification of waste in Taman Universiti according to its categeory by percentage for the second week. The pattern of waste production is still the same compared with the first week data collection. Similar to week 1, food waste is still the highest waste being produced in Taman Universiti which is 61% in total for 7 days then followed by paper and plastics rigid by 10% for each category. Next is plastics film by 7%. The rest of the categories hold the lowest waste production. Babies diapers and glass hold by 3%, tetrapek 2% while textiles, leather, rubber and metals by 1%.

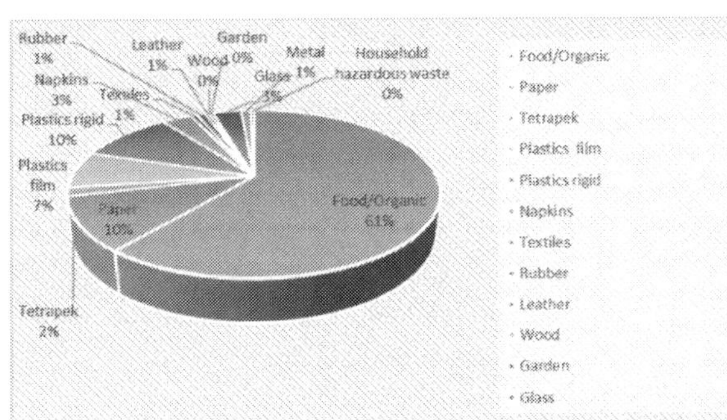

Figure-3. Composition distribution of household waste (week 2)

During two weeks collection period, a total of 228.33kg and 223.92kg of waste was sorted involving fifteen categories of household waste with average 226.13kg (Figure-3). The pattern of waste production is still the same for both sampling. Food waste still contribute most of the waste production and followed by paper, plastics film and plastics rigid. From the previous study, organic waste is also the main composition in Kuala Lumpur, Malaysia with 74% from total solid waste generation [12]. In addition, 61% from solid waste generation in Sangammer City, India is also organic waste [13]. In Pulau Pinang, Malaysia shown by United Nation Development Programme (UNDP) under administration of local council, Majlis Perbandaran Seberang Perai (MPSP) determined 63% organic waste from municipal solid waste collected under their authority area (United Nation Development Programme, Malaysia, 2008). Food wastes became main contributor of waste composition with the average from two weeks sampling time of 146.24kg.

The values of the moisture content for total waste collection in first week are 28.99% and 21.24% for the second week. Generally, the determination of moisture content of wastes may assist in obtaining water content of wastes and thus leachate from landfill not only contributed by waste moisture but also from rain and groundwater. Futhermore, moisture content also a vital controlling parameter during the treatment process of waste incineration and composition.

4. Conclusion and recommendation

This study had successfully determined that average total rate of waste generation for two weeks was 0.16kg/person/day, of which food waste accounted for the highest proportion (approximately 61%-68%) meanwhile wood and garden accounted as the lowest. The total moisture content of solid wastes collected for the first week was 28.99% and 21.24% for the second week. The result from the waste collection showed some similarities with the result obtained by the survey which organic wastes become the highest composition followed by other papers, magazine and plastic rigid accordingly. High amount of organic waste produced in Taman Universiti need to be taken seriously, and require major concern regarding on the problem that may occur from it.

References

[1] United Nation Development Programme, Malaysia. (2008) . Malaysia Developing a Solid Waste Management: Model for Penang. Kuala Lumpur, Malaysia.

[2] Chen, C. C. (2005). An Evaluation of Optimal Application of Government Subsidies on Recycling of Recyclable Waste. Polish Journal of Environmental Studies. 14(2), pp.137-144

[3] Dangi, M. B., Pretz, C. R., Urynowicz, M. A., Gerow, K. G., and Reddy, J. M. (2011). Municipalsolid waste generation in Kathmandu, Nepal. Journal of Environmental Management, Vol 92, pp.240–249.

[4[Periathamby, A., Hamid, F. S., and Khidzir, K. (2009). Evolution of Solid Waste Management in Malaysia: Impacts and Implications of The Solid Waste. Journal of Material Cycles and Waste Management. 2009. Bill, 2007. 11 pp.96–103.

[5] Abdullah, A. R. (1995). Environmental pollution in Malaysia: trends and prospects. Trends in Analytical Chemistry, 14, pp.191–198.

[6] Kathirvale S., Muhd Yunus, M. N. Sopian K., and Samsuddin A. H. (2003). Energy potential from municipal solid waste in Malaysia. Journal of Renewable Energy, 29, pp.559-567.

[7] Mbande, C. (2003). Appropriate approach in measuring waste generation, composition and density in developing areas. Journal of the South African Institution of Civil Engineering, 45, pp.2–10.

[8] Karak, T., Bhagat, R.M. and Bhattacharyya, P. (2012). Critical Reviews in Environmental Science and Technology. Municipal solid waste generation, composition, and management: The world scenario, 42, pp.1509–1630.

[9] McDougal, F., White, P., Franke, M. and Hindle, P. (2002). Integrated Solid Waste Management: A Life Cycle Inventory. Second ed. Oxford: BlackwellScience UK.

[10] Zhang, H., Zhao, Y. and Qi, J. (2007). Study on use of MSWI fly ash in ceramic tile, pp.106-114.

[11] Osman, M. S., Nasir, M. H. and Mujeebu, M. A. (2009). Assessment of municipal solid waste generation and recyclable materials potential in Kuala Lumpur, Malaysia. Waste Management, 29(7): 2209–13.

[12] Budhiarta, I., Siwar, C. and Basri, H. (2012). Current Status of Municipal Solid Waste Generation in Malaysia. International Journal on Advanced ScienceEngineering Information Technology. 2(2), pp.2088-5334

[13] Thitame, S. N., Pondhe, G. M. and Meshram, D. C (2009). Characterisation and composition of Municipal Solid Waste (MSW) generated in Sangamer City, District Ahmednagar, Maharashtra, India. Environ Monit Assess, 170, pp.1-5.

Performance of *Elaeis Guineensis* Leaves Compost in Filter Media for Stormwater Treament Through Column Study

H Takaijudin [1,2*], A A Ghani [1], N A Zakaria [1] and L L Tze[1]
[1] River Engineering and Urban Drainage Centre, Engineering Campus, Universiti Sains Malaysia, Seri Ampangan, 14300 Nibong Tebal, Penang, Malaysia.
[2] Department of Civil Engineering, Faculty of Engineering, Universiti Teknologi Petronas, 31750, Tronoh, Perak, Malaysia.

E-mail: *husna_takaijudin@petronas.com.my

Abstract: Compost based materials arv e widely used in filter media for improving soil capability and plant growth. The aim of this paper is to evaluate different types of compost materials used in engineered soil media through soil column investigation. Three (3) column, namely C1 (control), C2 and C3 had different types compost (10%) which were, commercial compost namely PEATGRO, Compost A and Compost B were prepared with 60 % medium sand and 30% of topsoil. The diluted stormwater runoff was flushed to the columns and it was run for six (6) hour experiment. The influent and effluent samples were collected and tested for Water Quality Index (WQI) parameters. The results deduced that C3 with *Elaeis Guineensis* leaves compost (Compost B) achieved 90.45 (Class II) better than control condition which accomplished 84 (Class II) based on WQI Classification. C3 with Compost A (*African Mahogany* Leaves Compost) obtained only 59.39 (Class III). C3 with the composition of Compost B effectively removed most pollutants, including Chemical Oxygen Demand (COD, Ammoniacal Nitrogen (NH_3-N), were reduced by 89±4% and 96.6±0.9%, respectively. The result concluded that *Elaeis Guineensis* leaves compost is recommended to be used as part of engineered soil media due to its capabilities in eliminating stormwater pollutants.

Keywords: Elaeis Guineensis leaves compost, stormwater pollutants, soil column study, filter media.

1. Introduction

Compost based material has been introduced as part of engineered soil media composition in bioretention to alleviate the problem associated with stormwater contaminant. It was believed that compost material capable to provide microbial populations which allow more microbial activities and supply carbon source, nutrients and moisture [1]. Typically, engineered soil media composition has 10-20% of compost, 50-60% sand and 20-30%topsoil in the detention system. Compost consists of large amounts of organic matter offer significant functions in enhancing soil capabilities and vegetation growth.

However, the use of compost must need to take into consideration because it led to nutrient leaching out of their system. Compost properties also assist on remaining greater hydraulic conductivity as compared to topsoil in long term duration due to having a porous structure [2].

Compost is usually applied in an urban garden for energizing the soil. However, it always depends on the type and quality of compost which then effect on metal solubility [3]. Besides, it has potential to discharge contaminants which originally exist in the compost which resulted leachate flows to the surface runoff or groundwater [4].

Organic leaves are one of the potential materials that can be used in the composting process. Brown materials such as old leaves, woodchips and tree bark are good materials which act as a carbon source in compost while green material such as chicken manure or cow dung supplied nutrient and high organic content. Alcala et al. [1] found that compost product was effectively treated nitrate from surface water.

Oil Palm (*Elaeis Guineensis)* leaves is usually will be dumped as a waste nearby or being used for livestock feed especially for cattle and goats [5]. These waste would not constantly used effectively which lead to land filling and open burning to dispose the oil palm residues [6]. Sidik et al. [7] reported that Oil Palm leaves had potential in removing oil spill which had maximum adsorption capacity of 1176±12.92mg/g at 303 K. However, the use of Oil Palm leaves as compost type for infiltration purpose was less documented. For this reason, this is an opportunity to make these types of leaves as part of the organic compost. Hence, this paper present to carry out the potential of Oil Palm (*Elaeis Guineensis)* leaves compost together with typical soil composition in filter media which then enhancing stormwater quality.

2. Material and Methods

2.1 Compost Preparation

Three compost materials were chosen based on product availability and quality as shown in Figure 1. PEATGRO compost was obtained from local supplier as typical compost product. Another 2 (two) compost was prepared in the lab by grinding the brown materials (old leaves) into small pieces. Compost A consisted of 1:1 ratio of green material (cow grass) and brown material (African Mahogany (*Khaya senegalensis*) leaves), 1% of chicken manures and water. Compost B was prepared by 1:1 ratio of green material (cow grass) and brown material (oil palm (*Elaeis Guineensis*) leaves), 1% (by weight) of chicken manure and water. This compost preparation method was recommended by [14].

The main characteristics of compost mainly moisture content, conductivity,and pH measurement were tested. Dry mass determination was generated to compute moisture content [1]. A 10g of the compost sample was dried for 24 hours in 105-110°C. Equation 1 is used to calculate moisture content. For pH and conductivity measurement, 5g dried sample was weighted and adding 25 ml of deioinzed water [1]. The solution was stirred for 2 minutes and leave to stable for another 5 minutes. Then, the samples were tested using YSI Professional Plus Multiparameter Instrument. The instrument was calibrated for pH measurement before measuring compost samples. Calibration was attained using two (2) buffer solutions of pH 4 and 7.

$$M_n = \frac{w_w - w_d}{w_w} x100 \qquad (1)$$

Where;

M_n	=	Moisture content of compost (%),
W_w	=	Soil weight during wet condition,
W_d	=	Soil weight after 24 hours dry

PEATGRO **COMPOST A** **COMPOST B**

Figure 1. Three (3) different types of compost materials.

2.2 Soil column preparation

Three acrylic soil columns (160L, 0.16 m^2) were designated for the evaluation of various types of compost in engineered soil media. Four layers were consisted of ponding layer, engineered soil media layer, transition layer (coarse sand with 0.6 - 2mm diameter of grain size) and underdrain layer (gravel with 6-9 mm diameter of grain size) to mimic a typical impermeable system of bio retention as illustrated in Figure 1.

Various materials of compost (10 % only) were mixed with 60% of medium sand and 30 % topsoil. Medium sand and topsoil were dried at 105°C within 24 hours and sieved with a particle size range of 0.2-0.6 mm (BSI, 1990) and less than 2 mm of grain size [8]. Then, all materials were mixed uniformly with suitable composition using mechanical mixing drum. Once completed, the mixtures were filled in acrylic columns with 27 blows using fabricated light hammer at every 100 mm depth of engineered soil media to have uniform compaction since it also influence the hydraulic conductivity.

2.3 Stormwater collection

Approximately 500 L of raw stormwater runoff samples was collected at existing drain nearby Parit Buntar, Perak. The area is in urban areas with mixed development which has school, hotels, shop lots, restaurants and others. Then, the initial TSS concentration was tested to obtain the specific range of TSS before experimental work is carried out. In this study, the initial of mean TSS concentration was 1712 mg/L. A diluted runoff was prepared in a pollutant source tank with 0.8 times the average mean TSS concentration in Parit Buntar area.

2.4 Sampling and chemical analysis

All soil columns were flushed with tap water with 1-2 day to obtain saturated condition. Then, it was dried for at least one day to ensure tap water was released into the system. Then, six (6) hour experiment was taking place with stormwater pollutant where a volume of 500 L stormwater runoff was discharged to the columns. The purpose of short duration experiment was to simulate a short duration storm event and following treatment of small amount of runoff [1].The flow was adjusted until constant ponding level is achieved. The influent was sampled at the beginning of experiment. The effluent samples were collected in every 10 minutes for the first 30 minutes and every 30 minutes thereafter [9]. There were six (6) WQI parameters were tested mainly, pH, DO, BOD, COD, TSS, and NH_3-N. pH and DO were measured during sampling work using YSI Profesional Plus Multiparameter Instrument. BOD was tested using 5-Day BOD techniques. Each sample was tested and analyzed for nitrogen and phosphorus concentration. The stormwater sample collected was from a drain outlet between a parking lot and an open field. The nitrogen concentration was tested using the Nessler Method where the unit is in mg/L NH_3-N.

Figure 2. A soil column experiment in treating stormwater pollutants.

2.5 Statistical Analysis
The results of water samples were analyzed using one-way Analysis of Variance (ANOVA) and Tukey's Test Method with $p<0.05$ using Minitab 16. The results were compared among the varying types of compost materials in engineered soil media.

3. Result and Discussion

3.1Comparison of Compost Products' Characteristics
Table 1 lists the physical components in compost materials. From the data, it can be concluded that PEATGRO and Compost B have similar moisture % content, 41±1% and 47±1% respectively. Compost A had the highest moisture content which was 68±4%. This presumably because of those higher substances for more quickly breakdown ''green'' proteinaceous material [1]. The conductivity among three compost was significantly different through non parametric Kruskal Wallis test with median 757, 157.1 and 40.1 μs/cm, respectively. Greater pH was obtained in Compost B which indicated alkaline condition. Similarly to Alcala et al. (2009) reported that the pH ranged between 6.5 to 8while Murray et al. (2011) obtained pH of compost at 7.3.

3.1 Stormwater Treatment Performance
Average treated stormwater runoff (Effluent) concentration for each different system were lower than raw stormwater runoff (Influent) excluding DO as presented in Table 2. It was found that DO levels were increased after the treatment. Based on Table 2, DO saturation was significantly different among the compost. C3 obtained higher DO (6.5±1.8mg/L) than control condition (C1) which has 5.3±1.7mg/L. DO in C2 was the lowest with only 2.6±0.6 mg/L. Higher DO indicated that the effluent samples were treated well in engineered soil media.

Table 1. Physical properties of compost materials (Values in mean + standard deviation (SD)).

Physical properties	Compost		
	PEATGRO	Compost A	Compost B
Moisture content (%)	41±1	68±4	47±1
Conductivity (μs/cm)	746±26	41.2±4	157.4±2.2
pH	7.2±0.2	7.0±0.0	8.2±0.6

Conversely, there was no statistically difference between TSS concentrations of treated storm water. In addition, all filter media satisfy the Water Quality Index (WQI) which fell under Class I (DOE, 2010).All the systems capable to filter physical sediments excellently regardless on the types of compost materials used in the system. This may be explained that the physical sediments were easily trapped at the top layer of filter media regardless the surface zone. However, higher volume of sediment trapped in the system might tend to clog pore sizes which affected hydraulic conductivity as highlighted by [10].

COD concentration of each treatment was comparable. Based on Tukey's test method, C1 and C3 are not statistically different because the confidence interval includes 0. However, there was a statistically different between the mean COD concentration of C1 and C2 and C2 and C3. This concluded that C3 has capabilities slightly similar to C1 as a control condition. In the early stage of the experiment, the effluent concentration of COD was greater than influent in column C2 which indicate negative removal efficiency. After half an hour running, the concentration, reduced and slightly constant until at the end of the experiment. It might be explained by COD concentration in the soil composition of C2 was higher and the filter media system yet to stabilize.

Treated stormwater from C1 and C3 were achieved Class II of WQI which the concentration of COD ranged between 10-25 mg/L (DOE, 2010). The lowest treatment was C2 which only felt under Class IV with value ranged 50-100 mg/L. Those progresses about COD demonstrated those effects from claiming influent mode of the carbon source distribution. COD could be firstly corrupted by vigorous microorganisms controlled by dissolved oxygen. [11].

Table 2. Water quality in raw stormwater (Influent) and treated stormwater (Effluent) (mean value ± standard deviation, n= 12 (Influent), n=36 per contaminants).

Parameters	Influent	Effluent		
		C1	C2	C3
pH	6.6±0.1	6.4±0.3	6.4± 0.1	6.5±0.2
NH_3-N (mg/L)	5.9±1.3	0.4±0.1	0.8±0.2	0.2±0.0
COD (mg/L)	123±4.2	14.9±3.5	70.2±6.9	13.5±4.9
TSS (mg/L)	1332±132	9.6±5.6	18.6±14.6	5.3±3.7
DO (mg/L)	1.5±1.3	5.3±1.7	2.6±0.6	6.5±1.8
DO_{sat} (%)	19.4±17	70.1±21	34.3±8.4	88.3±21
BOD (mg/L)	7.1±0.8	0.4±0.3	2.8±1.5	2.0±1.0
Overall Class	V	II	III	II
WQI	29.4	84.0	59.4	90.5

3.2 Treatment Efficiency

Urban runoff consisted of organic matters can be filtered, adsorbed and preserved for oxidative degradation by microorganisms [12] using biofiltration system. Hence, the capabilities of compost mixtures in filter media have been demonstrated in Figure 3. COD treatment efficiency was slightly similar in C1 (87.84±2.86%) and C3 (88.98±4.02%) which had no statistically different among both treatment. C2 which consisted of African Mahogany leaves compost obtained the lowest treatment efficiency with only 42.86±55.98%. Greater standard deviation (SD) is due to there was a negative (-) removal efficiency at the beginning of experimental work. This may be explained by filter media system in C2 yet to stabilize due to contain higher amount of organic matter in the system itself which insufficiently degraded by microorganisms.

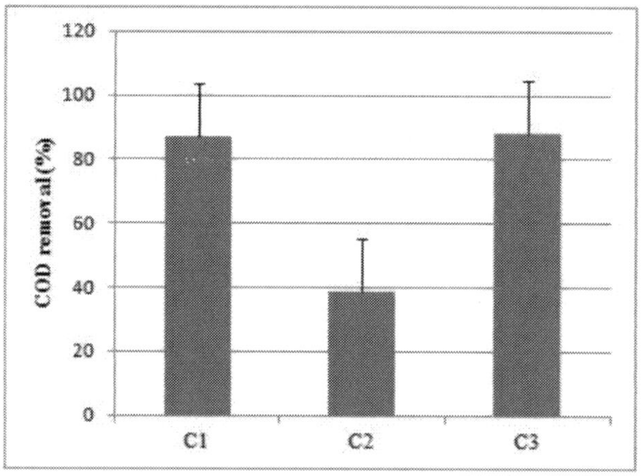

Figure 3. Treatment efficiency of COD among different types of compost materials.

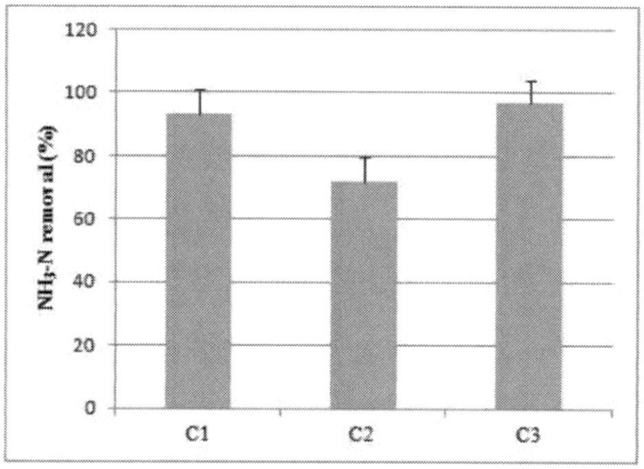

Figure 4. Treatment efficiency of NH_3-N among different types of compost materials.

Figure 4 illustrates the efficiency of the filter media system in treating NH_3-N. By comparing using Tukey Method, C2 and C3 had significantly different between these two treatments. C2 had 83.48±8.36% which was lower than C3 with 96.62±0.85%. However, there was no significant difference between C1 and C2 and C1 and C3. Based on this result, it was summarized that all media capable to remove NH_3-N excellently and C1 was successfully depleting NH_3-N concentration which slightly higher than C1(92.95±1.91%) as a control condition. This might be affected by the presence of an active community of ammonia oxidizing bacteria in the filter media system [13].

4. Conclusions

The comparison between the type of compost materials in the filter media system were obtained through soil column experimental work. Water quality analysis was carried out through standard laboratory testing and classified using WQI and removal efficiency.

C3 with existing oil palm leaves, compost in the engineered soil composition achieved successful performance, which fell under Class II in 90.5 of WQI indices which slightly higher than control condition (C1) that consisted of manufactured compost (PEATGRO) which only obtained 84.0. The mixture African Mahogany leaves compost, sand and topsoil had the lowest performance which only achieved Class III with 59.4 indices. The compost itself might contain higher organic matter which then contributes to the less performance in the treatment system.

Only COD and NH_3-N removal efficiency were discussed in this paper. COD in urban runoff was effectively captured by C1 and C3 with more than 85% removal. C2 probably had compost leaching problem in the early stage of experimental work which indicate by negative (-) removal. COD was treated less than 50% pollutant removal in C2. Conversely, NH_3-N removal performed better than COD. All treatment systems capable to reduce NH_3-N concentration very well with more than 80%. However, C1 and C3 excellently removed NH_3-N with more than 90% removal rates. Hence, oil palm leaves compost is recommended as a potential organic compost to be used as part of the component in the filter media system.

References

[1] Alcala, M., Jr., Jones, K. D., Ren, J., and Andreassen, T. E. (2009). Compost product optimization for surface water nitrate treatment in biofiltration applications. Bioresource Technology, 100(17), pp. 3991-3996.

[2] Lim, H. S., Lim, W., Hu, J. Y., Ziegler, A., and Ong, S. L. (2015). Comparison of filter media materials for heavy metal removal from urban stormwater runoff using biofiltration systems. Journal of Environmental Management, 147(0), pp.24-33.

[3] Murray, H., Pinchin, T. A., and Macfie, S. M. (2011). Compost application affects metal uptake in plants grown in urban garden soils and potential human health risk. Journal of Soils and Sediments, 11(5), pp.815-829.

[4] Iqbal, H., Garcia-Perez, M., and Flury, M. (2015). Effect of biochar on leaching of organic carbon, nitrogen, and phosphorus from compost in bioretention systems. Science of The Total Environment, 521–522(0), pp.37-45.

[5] Arun, N., Wan Ismail, W.H., and Iskandar, N.A. (2011). Comparison of composting potential of oil palm leaves using indigeneous microorganism and red wiggler worm (Eisenia Fetida). Research Report. Research Management Institute, Universiti Teknologi MARA (UiTM).

[6] Hashim, R., Saari, N., Sulaiman, O., Sugimoto, T., Hiziroglu, S., Sato, M., Tanaka, R.(2011, Effect of particle geometry on the properties of binderless particleboard manufactured from oil palm trunk, Mater. Des., 32, pp.246-254.

[7] Sidik, S.M., Jalil, A.A., Triwahyono, S., Adam S.H., Satar, M.A.H., Hameed, B.H., (2012). Modified oil palm leaves adsorbent with enhanced hydrophobicity for crude oil removal. Chemical Engineering Journal,203, pp. 9-18.

[8] Takaijudin,H., Ghani., A. Ab., Zakaria, N.A., & Azamathulla, H.Md.. (2014). Preliminary Study on the Impacts of Variation Engineered Soil Composition in Bioretention on Hydraulic Performance. Proceedings on 13th International Conference on Urban Drainage, Sarawak, Malaysia, 7-12 September 2014 1-8.

[9] Good, J. F., O'Sullivan, A. D., Wicke, D., and Cochrane, T. A. (2012). Contaminant removal and hydraulic conductivity of laboratory rain garden systems for stormwater treatment. Water Sci Technol, 65(12), pp. 2154-2161.

[10] Takaijudin,H., Ghani, A. Ab., Zakaria, N.A., and Lau T.L. (2015). The Influence of Filter Depths in Capturing Nutrient Contaminants for Non-Vegetated Bioretention Column: A Perliminary Study. Paper presented at the 36th IAHR World Congress, The Hague, Netherlands.

[11] Wang, Y., Song, X., Ding, Y., Niu, R., Zhao, X., and Yan, D. (2013). The impact of influent mode on nitrogen removal in horizontal subsurface flow constructed wetlands: A simple analysis of hydraulic efficiency and nutrient distribution. Ecological Engineering, 60(0), pp.271-275.

[12] Hou,L., Liu, F., Feng, C., and Wan, L. (2014). Removal of chemical oxygen demand and dissolved nutrients by a sunken lawn infiltration system during intermittent storm events. Water Sci Technol, 69(2), pp.398-406.

[13] Tyrrel, S. F., Seymour, I., and Harris, J. A. (2008). Bioremediation of leachate from a green waste composting facility using waste-derived filter media. Bioresource Technology, 99(16), pp.7657-7664.

[14] Ministry of Domestic Trade Co-operatives and Consumerism of Malaysia, (2008). Membuat Baja Kompos. Buletin Pengguna: Amalan Penggunaan Lestari, pp.10. Kuala Lumpur.

[15] British Standard Institution (BSI) (1990). Methods of Test for Soils for Civil Engineering Purposes Part 2:Classification Tests, pp. 27-59. London.

[16] Department of Environment (DOE) of Malaysia, (2010). Malaysia Environment Quality Report 2010, pp. 76-77. Kuala Lumpur: Department of Environment Malaysia.

IOP Conf. Series: Materials Science and Engineering **136** (2016) 012051 doi:10.1088/1757-899X/136/1/012051

A Potential Waste to be Selected as Media for Metal and Nutrient Removal

N Zayadi[1], N Othman[1], R Hamdan[1]
[1,2,3] Faculty of Civil and Environmental Engineering, Universiti Tun Hussein Onn Malaysia, Beg Berkunci 101, 86400, Parit Raja, Batu Pahat, Johor, Malaysia.

E-mail: nabilahzayadi@gmail.com

Abstract: This study describes the potential of application of cassava peel, banana peel, coconut shell, and coconut coir to be selected as metal removal while limestone and steel slag for nutrient removal. The media were characterized by X-Ray Fluorescence (XRF), Fourier Transform Infrared (FTIR), Field Emission Scanning Electron Microscopy-Energy Dispersive X-Ray (FESEM-EDX), and X-Ray Powder Diffraction (XRD). The results of XRF analysis medias show the present of calcium oxide, CaO which confirm the high efficiency in adsorbing metal ions and nutrient which is in agreement with the result of XRD. The characteristics of medias by FTIR analysis also confirmed the involvement of alcohol, carboxylic, alkanes, amines and ethers which play important role to reduce ions while FESEM-EDX indicates the porous structures of study medias. The characterization analysis highlight that cassava peel and steel slag were selected as a potential media in this study.

Keywords: Heavy metal, nutrient, cassava peel, steel slag.

1. Introduction

The discharged of large amount of wastewater with heavy metals and nutrients in water body leads to environmental problem. Nutrient compounds are becoming increasingly significant in water and wastewater management because the nutrients discharge of nitrogen and phosphorus into river and lakes can cause adverse influences on our environment and life. An excessive increase in the quantities of these nutrients in the aquatic surroundings disturbs the ecological balance, resulting in severe damage to environment including of eutrophication. For heavy metals once they enter the food chain, high concentration of heavy metals may accumulate in the human body through biomagnification process. Excess nutrients lead to alga bloom at the water body and finally deplete oxygen transfer and light penetration from atmosphere, owing to dying of aquatic life [1][2].

The discharged of large amount of wastewater with heavy metals and nutrients in water body leads to environmental problem. Nutrient compounds are becoming increasingly significant in water and wastewater management because the nutrients discharge of nitrogen and phosphorus into river and lakes can cause adverse influences on our environment and life. An excessive increase in the quantities of these nutrients in the aquatic surroundings disturbs the ecological balance, resulting in severe damage to environment including of eutrophication. For heavy metals once they enter the food chain, high concentration of heavy metals may accumulate in the human body through biomagnification process. Excess nutrients lead to alga bloom at the water body and finally deplete oxygen transfer and light penetration from atmosphere, owing to dying of aquatic life [1][2].

Due to the problems and situation arise of heavy metals and nutrients in water body, interest toward potential of natural and industrial wastes has gain attention among the researchers in order to sequester these pollutants. Current literature indicates different types of pollutants were treated with different types of media.

Natural waste most often used to treat heavy metal while industrial waste were focused on removal of nutrients and only few studies on removal of heavy metals [3][4][5].

Natural wastes such as agriculture wastes are cheap, unlimited, easily disposed by incineration, and even reusable after being rejuvenated shell [6]. Agricultural wastes such as cassava peel, banana peel, coconut coir, coconut shell have metal-binding capacities and are able to remove unwanted heavy metals from contaminated water at low cost [7][8][9][10].

Previous research also shows that steel slag and limestone show a high affinity for nutrient binding. Steel slag and limestone which mostly co-product of industrial waste were an interesting potential market for its use as filter substrate to treat wastewater [11][12][13].

Among the media mention previously, conditions for potential wastes will be selected based on largely available, cheap, rich in functional groups which potentially able to bind heavy metals, the presence of rich basic cations (Calcium and Magnesium) to facilitate nutrient removal, and its unique properties such as large surface area, highly porous and adsorptive [14][15][16].

In order to meet these criteria, this paper were focused on characterization of media through XRF, FTIR, FESEM-EDX, XRD and TG analysis. In this study, agricultural wastes namely cassava peel, banana peel, coconut coir, and coconut shell were compared to be selected as heavy metals removal while industrial wastes such as limestome and steel slag for nutrient removal. Once selected, the most potential agricultural waste were converted to activated carbon to improved their life cycle. Limited research has been reported on using combination media to treat heavy metals and nutrient. Therefore, this research will explore the potential used of media through characteristic studies.

2. Materials and Methods

2.1Preparation of media

Cassava peel, banana peel, coconut coir and coconut shell were collected at Kilang Kerepek Ahmad Shah, Parit Raja Darat, Johor. Raw fruit peels were soaked together with distilled water for further clean up for a minimum of 24 hours and dried in an oven overnight at 60 °C [17].

Meanwhile, Electric Arc Furnace Slag (EAFS) were collected at Antara Steel Mills, Pasir Gudang, Johor, while limestone (LS) were collected from Perwaja Steel Sdn. Bhd., located in Telok Kelong Industrial Area, Kemaman, Terengganu. EAFS and LS were washed with tap water two times and then followed by distilled water once. This process was done in order to remove any dirt or particulate that adheres to the media before being used for batch study. The washed media were oven dried at 105 °C for 24 hours and sieved to obtain 10 – 20 mm of particle size [18].

2.2 X-Ray Fluorescence (XRF)

XRF is particularly used in the analysis of chemical compounds of specimen. XRF analysis were conducted in Analytical Environmental Laboratory, Faculty of Civil and Environmental Engineering, UTHM [19]. Agricultural wastes and industrial waste were prepared as pressed powder pellet XRF, meanwhile coconut coir was performed by loose technique.

2.3 Fourier Transform Infrared (FTIR)

FTIR were used to determine functional groups of medias. FTIR analysis were conducted in Statics and Dynamics Laboratory, Faculty of Mechanical and Manufacturing Engineering, UTHM [20].

2.4 Field Emission Scanning Electron Microscopy (FE-SEM) coupled with Energy Dispersive X-ray spectroscopy (EDX spectra)

FE-SEM coupled with EDX analysis were used to identify the size, particle shape, and surface microstructures. Analysis were performed at Microelectronics and Nanotechnology-Shamsuddin Research Centre (MINT-SRC), UTHM. The FE-SEM images are recorded using JEOL JSM-7600F Field Emission Scanning Electron Microscope (USA).

2.5 X-ray diffraction (XRD)

XRD were used to characterize the precipitates (covering the surface of media). The samples were analyzed by Cu K radiation with a scanning rate of 0.05° per second 40 Kv/20A, at 10°≥2≤90°at Material Science Laboratory, Faculty of Mechanical and Manufacturing Engineering, UTHM.

3. Result And Discussion

3.1 XRF

The results from Table 1 and 2 show XRF analysis of agricultural wastes. Based on the results, calcium oxide, (CaO) of cassava peel demonstrates as the highest chemical compound compared to other agricultural wastes. Meanwhile, Table 3 also shows the presence of CaO which is highest in limestone compare to EAFS. However, the presence of iron oxide, silicon oxide, aluminium oxide and magnesium oxide is also important factors to be considered in nutrient removal as it also proved to provide adsorption sites for anions like As, Cr, and P [21]. XRF analysis of industrial wastes show that most of the important chemical compounds were highest in EAFS compare to limestone. Therefore, result of XRF analysis for both media indicates that cassava peel and EAFS has high potential in heavy metals nutrients removal, respectively.

3.2 XRD

The XRD pattern of agricultural wastes (see Figure 1) obtained reveal that, the major diffraction peaks are around 27° to 33° and Table 4 demonstrates most of agricultural wastes contain calcium compounds in their media. Meanwhile, Figure 2 show XRD pattern of industrial wastes of limestone and EAFS. From Figure 2, most of major diffraction peaks of limestone and EAFS are around 28° to 31° and 21° to 57°, respectively. Table 5 indicates most of calcium compounds were also found in EAFS and limestone. Similar trends were also observed from previous studies [22]. This result of XRD is in agreement with the result of XRD obtained. Therefore, the present work suggests the possibility of using cassava peel and EAFS since most of the significant compound were found at major diffraction peaks.

3.3 FTIR

FTIR groups of agricultural wastes of cassava peel, banana peel, coconut shell, and coconut coir is shown in Table 6. The adsorption band at 3330.4 cm^{-1}, 3305.3 cm^{-1}, 3343.9 cm^{-1}, and 3336.2 cm^{-1} of cassava peel banana peel, coconut shell, and coconut coir is attributable to the alcohol group, respectively. Similar functional group were also observed from previous studies of this medias [7][8][10].

Table 1. XRF analysis of agricultural wastes.

Formula	Media		
	Cassava peel (%)	Banana peel (%)	Coconut shell (%)
C	10	10	10
CaO	4.6	2.45	0.16
K_2O	3.43	13.70	1.03
SiO_2	1.73	0.94	-
P_2O_5	0.36	0.62	-
Al_2O_3	1.11	-	-
Fe_2O_3	0.90	-	0.76
SO_3	0.55	1.05	-
CI	0.50	4.60	0.28
MgO	-	0.19	-
Br	-	0.15	-
MoO3	-	-	0.20

Table 2. XRF analysis of coconut coir (loose technique).

Formula	Concentration
Al	0.011
Ca	0.1551
CH2	98.6
Cl	0.207
Cr	0.0151
Cu	0.0012
Fe	0.07006
K	0.5478
Mg	0.0355
Mn	0.0018
Ni	0.0044
P	0.0779
S	0.00463
Si	0.186
Zn	0.0011

Table 3. XRF analysis of industrial waste.

Formula	Media	
	EAFS (%)	Limestone (%)
Fe_2O_3	29.20	0.48
CaO	21.80	70.10
SiO_2	16.90	1.43
Al_2O_3	7.10	0.46
MnO	5.59	-
MgO	3.70	0.29
Cr2O3	1.46	-
C	1.00	-
TiO2	0.72	-
BaO	0.25	-
P_2O_5	0.34	-
Na2O	0.21	-
K2O5	-	0.15
Zn	0<LLD	-
S	0<LLD	-
Mn	-	0<LLD

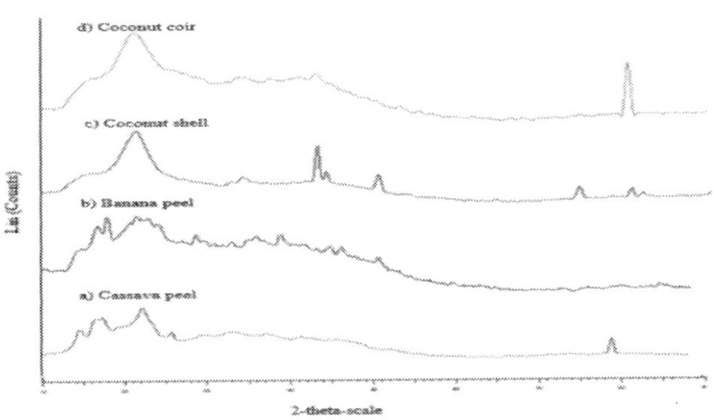

Figure 1. XRD pattern of agricultural wastes.

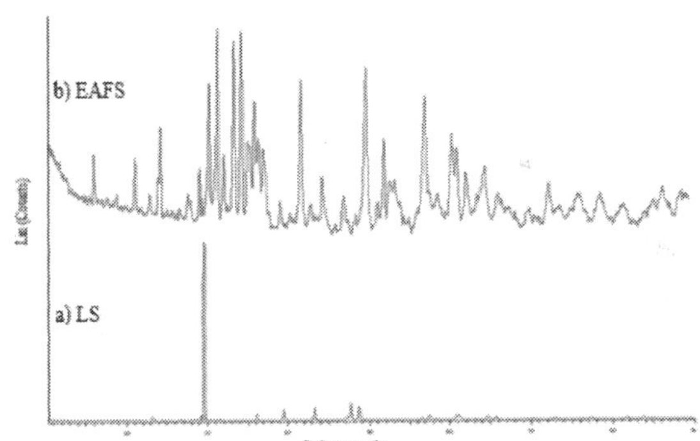

Figure 2. XRD pattern of industrial wastes.

IOP Conf. Series: Materials Science and Engineering **136** (2016) 012051 doi:10.1088/1757-899X/136/1/012051

Table 4. Identified patterns list of agricultural wastes.

Media	Peak	Compound	Media	Peak	Compound
a) Cassava peel	26	Carbon	b) Banana peel	21	Adipic acid
	17	L-Lactide-poly(ethylene glycol)		18	Magnesium adipate tetrahydrate
	15	Iron Chloride Hydrate		32	Calcium Silicon
	17	4-Chlorophenyl sulfone			
	27	Diclone			
	17/23	Starch (maize)			
	27	Silicon Oxide			
	17	Diosgenin			
c) Coconut shell	23	Decamethylene glycol	d) Coconut coir	27	Carbon
	22/34	Sulfur Oxide Graphite		27	Potassium Magnesium Silicate
	30	Mucochloric acid		27	Silicon Oxide
	19/23	n-Docosane		27	Magnesium Chromium Oxide
	30	Ascorbic acid		26	Potassium Chromium Oxide
	27	Carbon		27	Calcium Hydrogen Phosphate
	32	Calcium Hydride		27	Magnesium Phosphate
	33	Calcium Carbide		27	Potassium ethyl malonate
	27	Potassium Iron Sulfate			
	16	p-Hydroxybenzaldehyde			
	28	Potassium Calcium Sulfate			
	21	Adipic acid			
	29	Calcium Chlorate Hydrate			
	36	Potassium Hydrogen Tartrate			
	28	Calcium Sulfite Hydrate			

The characteristic band region at 2919.7 cm^{-1} were found similar for agricultural wastes which suggests the presence of alkanes groups. Alcohol and alkane groups are the potential sites for biosorption and the uptake of metal depends on various factors such as abundance of sites, their accessibility, chemical state and affinity between the adsorption site and metal. This finding is comparable with the previous study whereas FTIR analysis confirmed the involvement of these groups [7][8][10]. The involvement of these groups in the media might involved in adsorption of metal ions.

Table 5. Identified patterns list of industrial wastes.

Media	Peak	Compound	Diffraction
a) LS	29	Calcium Carbonate	104
	28	Calcium Aluminum Silicate	004
	31	Calcium Aluminum Silicate	211
	28	Potassium Aluminum Oxalate Hydrate	231
	11	Silicon Oxide	100
	27	Silicon Oxide	101
	27	Calcium Oxide	
	28	Silicon	111
	28	Calcium Aluminum Silicate	004
b) EAFS	21/30	Potassium Carbide	111/113
	34/50/57	Calcium Iron Silicate	211/222/260
	24	Nickel benzenesulfonate hexahydrate	660
	28	Iron di(hydroxybenzenesulfonate) octahydrate	-216
	31	Calcium Aluminum Silicate	211
	28	Calcium	111
	28	Calcium Aluminum Silicate	004
	34/50	Calcium Iron Silicate	211/222

3.4 SEM

The SEM is one of the useful techniques applied in the examination of the surface morphology of biosorbents. The surface morphology of the agricultural and industrial wastes was studied with the use of FESEM (Figures 3 to 8). The pictures indicate that the media has a rough, irregular and porous surface thus providing large area for metal–surface interaction and biosorption. Meanwhile, a microporous structure was observed for EAFS and limestone.

Tables 7 and 8 shows EDX results of agricultural and industrial wastes and most of the element were also found in XRF and XRD results. Si was found to be highest in cassava peel among others agricultural wastes while Fe were found to be highest in EAFS compare to limestone. These elements were considered as important factor in heavy metals and nutrient removal [15].

Table 6. FTIR groups of agricultural waste.

Sample	Wave number (cm^{-1})	Classification	Bond	Mode
Cassava peel	3330.4	alcohols	OH	stretching
		amines	NH	stretching
	2919.7	alkanes	CH	symmetric stretching
Banana peel	3305.3	alcohols	OH	stretching
		amines	NH	stretching
	2919.7	alkanes	CH	symmetric stretching
Coconut shell	3343.9	alcohols	OH	stretching
		amides	NH	antisymmetric stretching
	2919.7	alkanes	CH	symmetric stretching
Coconut coir	3336.2	alcohols	OH	stretching
		amines	NH	stretching
	2919.7	alkanes	CH	symmetric stretching

Figure 3. Cassava peel.

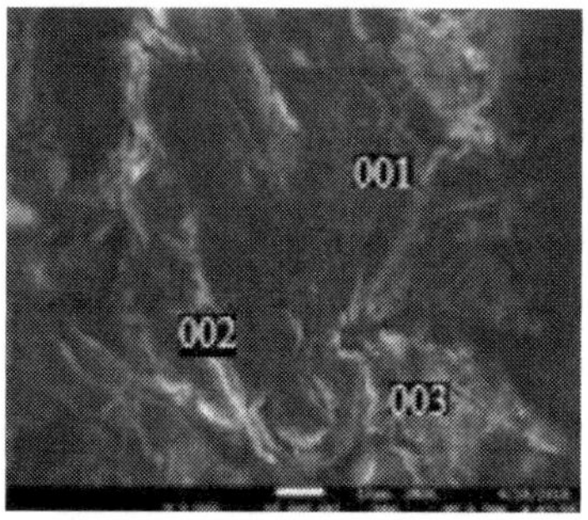

Figure 4. Banana peel.

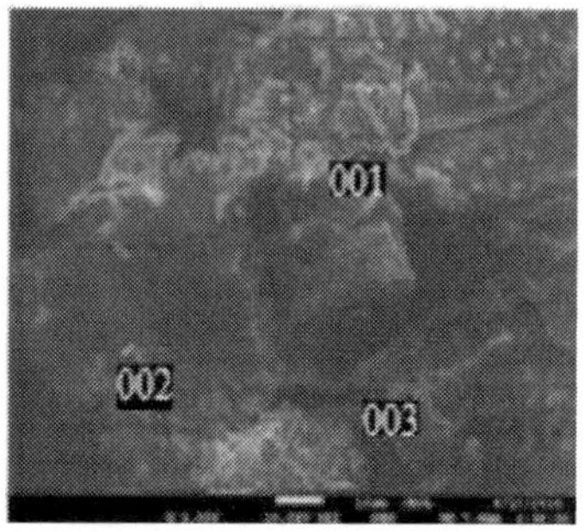

Figure 5. Coconut shell.

Figure 6. Coconut coir.

Figure 7. Limestone.

Figure 8. EAFS.

Table 7. EDX results of agricultural waste

Element	Weight (%)			
	Cassava peel	Banana peel	Coconut shell	Coconut coir
C	52.37	52.84	59.52	58.99
O	42.26	47.01	40.00	39.67
Al	1.85	-	-	-
Si	2.98	0.15	0.05	0.21
Ca	0.10	-	0.10	0.36
Fe	0.44	-	0.19	0.19
Cl	-	-	0.02	0.12
K	-	-	0.12	0.32
Mg	-	-	-	0.14
Totals	100.00	100.00	100.00	100.00

Table 8. EDX results of industrial wastes

Element	Weight (%)	
	Limestone	EAFS
C	13.58	12.18
O	53.75	44.78
Mg	0.46	0.83
Al	3.12	1.35
Si	2.68	1.81
Ca	25.52	17.66
Fe	0.89	18.20
Na	-	0.94
Totals	100.00	100.00

4. Conclusion

Characterization study XRF, XRD, FTIR, and FESEM-EDX indicates that cassava peel and EAFS as potential wastes to be selected for heavy metals and nutrient removal. The presence of Si, Ca, Mg, Fe, rich in functional groups and surface porosity of media were significant factors which attribute to facilitate heavy metals and nutrient removal. In future work, to increase the performances of cassava peels several modifications will be apply in term of the preparation. Cassava peel and EAFS can effectively be used to remove heavy metals and nutrient in polluted water.

References

[1] Wiesmann, U., Choi, I. S. & Dombrowski, E-M. (2007). *Fundamentals of Biological Wastewater Treatment*. Weinheim GER: WILEY-VCH.

[2] Johnson, M. L. & Mara D. D. (2007). Ammonia removal from facultative pond effluents in a constructed wetland and an aerated rock filter: performance comparison in winter and summer. *Water Environment Research,* 79(5), 567–570.

[3] Tudor, H. E. A., Gryte, C. C. & Harris, C. C. (2006). *Seashells: Detoxifying Agents for Metal-Contaminated Waters. Water, Air, and Soil Pollution,* 173, 209–242.

[4] Zayadi, N., & Othman, N. (2013). Characterization and optimization of heavy metals biosorption by fish scale, *Advanced Materials Research*, 795, 260-265.

[5] Claveau-Mallet, D., Wallace, S. & Comeau, Y. (2013). Removal of phosphorus, fluoride and metals from a gypsum mining leachate using steel slag filters. *Water Research*, 47, 1512-1520.

[6] Murthy, C.V.R., Ramesh, P. and Ramesh, A. (2012). Study of Biosorption of Cu (II) from Aqueous Solutions by Coconut Shell Powder. *Chemical Sciences Journal*, 1-9.

[7] Memon, J. R., Memon, S. Q., Bhanger, M.I., Memon, G. Z., El-Turki, A. & Geoffrey, C. A. (2008). Characterization of banana peel by scanning electron microscopy and FT-IR spectroscopy and its use for cadmium removal. *Colloids and Surfaces B: Biointerfaces*, 66, 260–265.

[8] Kosasih, A. N., Febrianto, J., Sunarso, J., Ju, Yi-Hsu., Indraswati, N. & Ismadji, S. (2010). Sequestering of Cu(II) from aqueous solution using cassava peel (*Manihot esculenta*). *Journal of Hazardous Materials*, 180, 366–374.

[9] Gonzalez, M. H., Arau´jo, G. C. L., Pelizaro, C. B., Menezes, E. A., Lemos, S. G., Sousa, G. B. & Nogueira, A. R. A. (2008). Coconut coir as biosorbent for Cr(VI) removal from laboratory wastewater. *Journal of Hazardous Materials,* 159, 252–256.

[10] Bledzki, A. K., Mamun, A. A. & Volk, J. (2010). Barley husk and coconut shell reinforced polypropylene composites: The effect of fibre physical, chemical and surface properties. *Composites Science and Technology*, 70, 840–846.

[11] Chazarenc, F., Kacem, M., Gérente, C. & Andrès, Y. (2008). 'Active' filters: a mini-reviewon the use of industrial by-products for upgrading phosphorus removal from treatment wetlands. *In: Proceedings of the 11th International Conference on Wetland Systems for Water Pollution Control, Indore, India,* 1–7 November.

[12] Camargo-Valero, M.A., Johnson, M., Mather, T. & Mara, D. D. (2009). Enhanced phosphorus removal in a waste stabilization pond system with blast furnace slag filters. *Desalination Water Treatment*, 4, 122-127.

[13] Claveau-Mallet, D., Wallace, S. & Comeau, Y. (2013). Removal of phosphorus, fluoride and metals from a gypsum mining leachate using steel slag filters. *Water Research*, 47, 1512-1520.

[14] Pratt, C., Parsons, S. A., Soares, A. & Martin, B. D. (2012). Biologically and chemically mediated adsorption and precipitation of phosphorus from wastewater. *Current Opinion in Biotechnology*, 23, 890-896.

[15] Babel, S. & Kurniawan, T. A. (2003). Low-cost adsorbents for heavy metals uptake from contaminated water: a review. *Journal of hazardous materials*, 97(1-3), 219–243.

[16] Rocha, C. G., Zaia, D. A. M., Alfaya, R. V. D. S. & Alfaya, A. A. D. S. (2009). Use of rice straw as biosorbent for removal of Cu(II), Zn(II), Cd(II) and Hg(II) ions in industrial effluents. *Journal of hazardous materials*, 166(1), 383–388.

[17] Achak, M., Hafidi, A., Ouazzani, N., Sayadi, S. & Mandi, L. (2009). Low cost biosorbent "banana peel" for the removal of phenolic compounds from olive mill wastewater: kinetic and equilibrium studies. *Journal of hazardous materials*, 166(1), 117–125.

[18] Wan-Mohamed, W-A., Hamdan, R. & Othman, N. (2015). Study of the pH Effects on the Phosphorus Removal Mechanism in Lab-Scale Electric Arc Furnace Slag and Limestone Filters in Synthetic Wastewater. Applied Mechanics and Materials, (752-753), 277-282.

[19] Bruker AXS. (2007). S4 EXPLORER / S4 PIONEER X-ray Spectrometer. Karlsruhe, GRMN.

[20] Perkin Elmer. (2005). Spectrum 100 Series User's Guide. UK.

[21] Oh, C., Rhee, S., Oh, M. & Park, J. (2012). Removal characteristics of As(III) and As(V) from acidic aqueous solution by steel making slag. *Journal of Hazardous Materials, 213-214.*

[22] Madakson, P.B., Yawas, D.S. & Apasi, A. (2012). Characterization of Coconut Shell Ash for Potential Utilization in Metal Matrix Composites for Automotive Applications. *International Journal of Engineering Science and Technology (IJEST),* 4(3).

Biodegradation of PAHs in Soil: Influence of Initial PAHs Concentration

N A F M Kamil[1] and S A Talib[2]
[1] Faculty of Civil and Environmental Engineering, Universiti Tun Hussein Onn Malaysia, 86400 Batu Pahat, Johor, MALAYSIA
[2] Institute for Infrastructure Engineering and Sustanable Management, Faculty of Civil Engineering, Universiti Teknologi MARA, 40450 Shah Alam, Selangor, MALAYSIA

E-mail: noramani@uthm.edu.my

Abstract: Most studies on biodegradation of Polycyclic Aromatic Hydrocarbons (PAHs) evaluate the effect of initial PAHs concentration in liquid medium. There are limited studies on evaluation in solid medium such as contaminated soil. This study investigated the potential of the bacteria, Corynebacterium urealyticum isolated from municipal sludge in degrading phenanthrene contaminated soil in different phenanthrene concentration. Batch experiments were conducted over 20 days in reactors containing artificially contaminated phenanthrene soil at different concentration inoculated with a bacterial culture. This study established the optimum condition for phenanthrene degradation by the bacteria under non-indigenous condition at 500 mg/kg of initial phenanthrene concentration. High initial concentration required longer duration for biodegradation process compared to low initial concentration. The bacteria can survive for three days for all initial phenanthrene concentrations.

Keywords: Biodegradation, contaminated soil, corynebacterium urealyticum, phenanthrene.

1. Introduction

PAHs are highly toxic, mutagenic and carcinogenic that can be linked to other health problems which had attracted increasing interest in the research community (Li et al., [6]; Zhao et al., [10]). Accidental discharge of high concentration PAHs due to human activities has led to serious concerns on human health and ecological system well-being [2]. The concern for human health arises from the bioaccumulation of PAHs through plants or animals exposed to contaminated soil. PAHs can accumulate in food chains due to their affinity for fatty tissue [9]. PAHs was tested on laboratory animals and results had shown that it was it was responsible for tumor, cancer, reproductive problems, damage to skin and immune system [3]. In addition, the pregnant mice fed with PAHs also showed other harmful effects, such as birth defects and decreased body weight. Studies in animals have also shown that PAHs can cause harmful effects on the body system for fighting disease after exposure with PAHs.

The use of bacteria in PAHs bioremediation can be affected by different factors. These factors may lead the bioremediation process to become either more or less efficient. Thus, optimization of these factors is required to achieve high degradation efficiency in the bioremediation process. In addition, the optimization process can enhance the bacterial activity in soil.

Organic contaminants are utilized by bacteria for anabolism and metabolism [1]. High variation of substrate concentration is influenced either by natural or anthropogenic activities. Zhao et al. [10] reported that degradation rate was reduced with the higher phenanthrene concentration by inoculated Tistrella sp and Sphingomonas sp. The study reported the highest degradation was achieved at minimum PAHs concentrations, i.e., 250 ppm. Results from Zhao et al. [10] were consistent with the study reported by Janbandhu and Fulekar [5], where the highest degradation of phenanthrene was achieved at lowest concentration, i.e., 100 ppm. High contaminant concentrations may have reduced the sorption behavior in microorganism's cells. Therefore, higher concentrations had increased toxicity of contaminants to microorganisms and subsequently reduce the bacterial activity.

Established studies show that high organic contaminant concentrations had reduced the degradation rates. Mostly established studies were conducted in liquid medium. However, there is limited report in solid medium such as contaminated soil. The specific aim of this present work is to evaluate the potential of Corynebacterium urealyticum in degrading phenanthrene in different initial phenanthrene concentration. This study also evaluated the growth rate of the bacteria at different initial phenanthrene concentrations.

2. Methodology

2.1 Materials
All chemicals used for extraction, preparation of minimal media, and bacteria culture were of analytical grade and supplied by Merck, Germany. The phenanthrene standard for the gas chromatography mass spectrometer (GCMS) analyses was also obtained from Merck, Germany. Ultra-pure water (UPW) used in this study was produced from Alga Purelab Ultra (18.2 MΩ, United Kingdom).

2.2 Sample Preparation

2.2.1 Preparation of contaminated sand
Silica sand was sampled and washed three times in a sand:water mixture of 1:2 (w:v), air dried in the dark, and passed through a 1 mm sieve. The sand samples were stored in a 500 ml glass jar and dried at 60°C until a constant mass was recorded. Phenanthrene was dissolved in hexane to produce phenanthrene/hexane solution. The prepared sand samples were autoclaved (Hirayama, HVE-50, Japan) for 20 minutes at 121°C and spiked with the phenanthrene/hexane solution. Hexane was evaporated under continuous mixing to ensure homogenous distribution of phenanthrene in the sand samples. The samples were stored at 4^0C until to be used in the biodegradation studies. The initial concentration of phenanthrene was verified in triplicates before being used for biodegradation study.

2.2.2 Preparation of minimal media
Minimal media solution was prepared by dissolving 8.5 g Na2HPO4, 3.0 g KH2PO4, 0.5 g NaCl, 1.0 g NH4Cl, 0.5 g MgSO4.7H2O, 0.0147 g CaCl2, 0.0004 g CuSO4, 0.001 g KI, 0.004 g MnSO4.H2O, 0.004 g ZnSO4, 0.005 g H3BO3, and 0.002 g FeCl3 in 1L UPW. The prepared minimal media were autoclaved (Hirayama, HVE-50, Japan) for 20 minutes at 121°C.

2.2.3 Preparation of bacteria strain
The strain, Sphingobacterium spiritovorum, was isolated from municipal sludge by Othman [7] preserved at -800C in microbeads (microbankTM, Round Rock, Texas). The strain was revived by transferring a few of the frozen beads into universal bottles containing 20 ml of Nutrient Broth that was incubated at 30°C for three days. A series of dilution streaking was performed and the strain was sub-cultured three times to attain the active bacterium before it was used in biodegradation studies.

2.3 Growth Curve Test
The plate count method was used to establish the growth curve after the bacteria was inoculated in contaminated sand. The bacteria number in the sand samples were quantified by mixing 1 g of sand with 9 mL of sterile phosphate buffered saline and homogenized at high speed for 1 minute using a vortex mixer. Successive 1/10 dilutions were made by adding 1 mL of the sand suspension to 9 mL of phosphate buffered saline. An aliquot (0.1 mL) of each dilution was transferred to nutrient agar on Petri dish. The dishes were incubated at 30°C at an inverted position. After 4 days, the numbers of bacterial colonies were counted using a plate counter. Plates with different dilutions were prepared and those with colonies in the range of 30 to 300 were used to estimate the number of bacteria. This number of colonies was then multiplied by the dilution factor to find the total number of bacteria per 1g of the sand. The numbers of colonies was expressed as colony-forming units per gram of sand (CFU/g). All tests were conducted in triplicate.

IOP Conf. Series: Materials Science and Engineering **136** (2016) 012052 doi:10.1088/1757-899X/136/1/012052

2.4 Biodegradation Study

Batch experiments were conducted in a 250 ml Erlenmeyer flask as the reactor. The reactor contained 20 g autoclaved sand, 10 mg/kg, 100 mg/kg, 250 mg/kg, 500 mg/kg and 1000 mg/kg initial phenanthrene concentration, 5.6 ml minimal media, and 1.4 ml bacterial inoculums. 5.6 ml of minimal media constituted 40 percent of the total volume. This volume was selected based on the quantity of inorganic nutrient required for bioremediation of PAHs in soil [6]. 1.4 ml of bacterial inoculum occupied 10 percent of the total volume and this quantity was selected based on an optimization study conducted by Othman [7]. The bacteria inoculated into the samples were collected at the middle of the exponential growth phase based on established growth curves, i.e., on day 4. This experiment was repeated three times and the average bacteria concentration on day 4 was $1\pm1.2\times107$ cells/g soil. Then, the strain was inoculated into contaminated soil. After inoculation, all flasks were shaken in an incubator shaker at 150 rpm in the dark at 30°C. Sterile water was supplied at 2 percent remaining weight every day. All samples were analyzed in triplicate. Control samples consisted of 20 g autoclaved sand, 10 mg/kg, 100 mg/kg, 250 mg/kg, 500 mg/kg and 1000 mg/kg phenanthrene, and 5.6 ml minimal media without bacterial inoculums. The plate count method was performed on the control samples and no colony was found.

2.5 Extraction and Analysis

For sample analysis, 500 mg of the contaminated sand sample was dissolved in 25 ml of n-hexane and acetone 7:3 (v/v). The extractions were performed using the pressurized microwave extraction system (MAE) Multiwave 3000 (Rotor 8XF100 SOLV and solvent safety system; Graz, Austria). All samples that were placed in the MAE were extracted for 40 minutes under a pressure of 10 bars. When the extraction period was completed, the equipment was allowed to cool down to room temperature (20 minutes). Subsequently, the samples were filtered with a Whatman fiber filter with a pore size of 11 µm and stored in a 25 ml universal bottle. The samples were concentrated by means of a rotary evaporator to 1 ml. The extraction method had been pre-tested for recovery efficiency for phenanthrene and demonstrated an efficiency of 99 percent. Phenanthrene concentrations were analysed using gas chromatography mass spectrometer (Perkin Elmer Clarus 600; Shelton, Connecticut), equipped with Elite Column 5MS of 30 m long X 0.25 mm internal dimension X 0.25 µm thickness. The injector was operated at 250°C in the splitless mode with a 3 minute splitless period. Helium was used as the carrier gas with 1 ml/min constant flow rate. The column temperature was initially set at 50°C for 1 minute, increased to 250°C at a rate of 25°C/min, kept constant at 1 minute, and held constant until the end of the 22 minute total run time.

3. Results and Discussions

Phenanthrene degradation by the bacteria was evaluated at different initial phenanthrene concentrations, namely, 10 mg/kg, 100 mg/kg, 200 mg/kg, 500 mg/kg and 1000 mg/kg. Figure 1 and Figure 2 show the phenanthrene degradation for low concentrations that is, 200 mg/kg, 100 mg/kg and 10 mg/kg and higher concentrations, that is, 1000 mg/kg and 500 mg/kg, respectively.

Degradation trend can be divided into two phases, phase 1: rapid degradation and phase 2: very slow degradation. Experiment durations in phase 1 were different for each concentration. The durations in phase 1 were observed on day 4, day 5 and day 6 for low concentrations, that is, 10 mg/kg, 100 mg/kg and 200 mg/kg, respectively. On the other hand, for high concentrations, the durations in phase 1 were observed on day 9 and day 7 for 1000 mg/kg and 500 mg/kg, respectively. This result shows that the experiment durations were shorter for low concentrations compared to high concentrations, i.e., 10 days and 15 days for low and high concentrations, respectively.

Based on degradation rates, the optimum initial concentration for phenanthrene degradation was observed at 500 mg/kg. As initial phenanthrene concentration increased, the degradation rate was also increased except for 1000 mg/kg of initial phenanthrene concentration. For samples at high of phenanthrene concentration, such as 500 mg/kg, the bacteria have more carbon sources that can be metabolized to bring more energy.

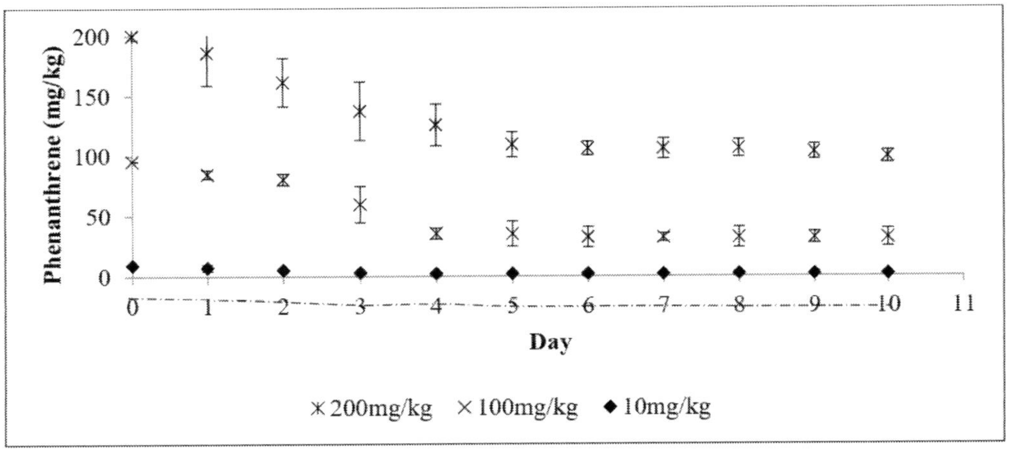

Figure1. Phenanthrene degradation by the bacteria for low initial concentrations, i.e., 200mg/kg, 100 mg/kg and 10 mg/kg.

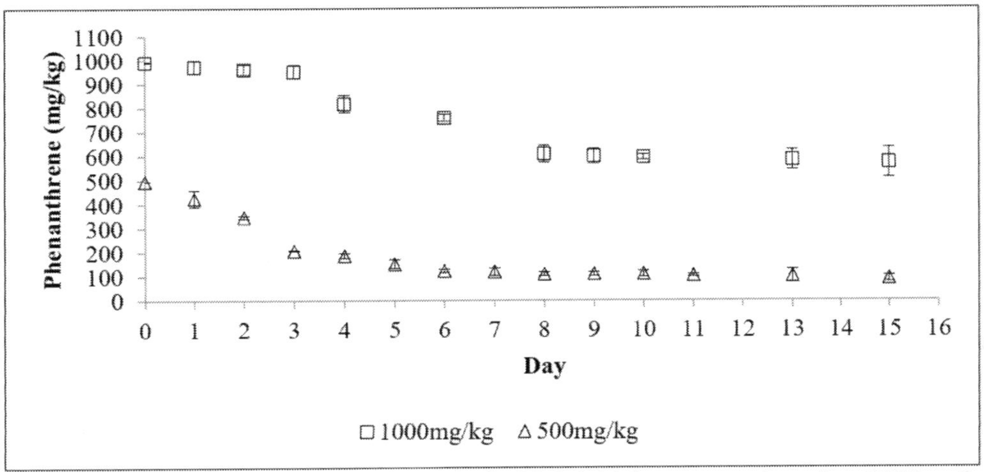

Figure 2. Phenanthrene degradation by the bacteria for high initial concentrations, i.e., 1000mg/kg and 500 mg/kg.

However, very high phenanthrene concentration, such as 1000 mg/kg, had saturated the bacteria cell with the substrate. Therefore, the performance was reduced at 1000 mg/kg of initial phenanthrene concentration. Performance of the bacteria in degrading at different phenanthrene concentrations in this study was not consistent with the results from a study conducted by Othman [7]. The bacteria was previously used by Othman [7] to degrade phenanthrene with initial concentration of 1 ppm, 5 ppm, 10 ppm, 100 ppm and 500 ppm in minimal media. The study reported that the bacteria achieved a highest degradation rate for initial concentration of 100 ppm. Then, degradation rate was reduced at 500 ppm of initial phenanthrene concentration. In contrast, higher degradation rate was observed for initial concentration of 500 mg/kg in this study. Therefore, at high phenanthrene concentration, the bacteria performed better in a contaminated soil compared to minimal media.

Phenanthrene degradation by the bacteria increased with initial concentration, suggests a great potential in application of the bacteria in degradation of contaminated soil that is containing PAHs at high concentration. In chemical treatment, high contaminant concentration requires high of the chemical agent to remediate the contaminant [4]. This high concentration of chemical agent can result in toxicity and produces other problems such as health hazard [8].

Statistical analysis using ANOVA was conducted to establish the correlation between initial phenanthrene concentrations and degradation rates. From ANOVA, at the 5% level of significance, it was found that initial phenanthrene concentrations have significant effect to the degradation rates ($p<0.05$). This analysis shows that the degradation rates were influenced by initial phenanthrene concentrations.

Figures 3 and Figure 4 show the growth of the bacteria at different initial phenanthrene concentrations, which indirectly indicates the strains survival. Bacteria numbers were increased until day 3 for all initial phenanthrene concentrations. After day 3, bacteria numbers were reduced until at the end of the experiment. This result indicates that the bacteria can survive for three days for all initial phenanthrene concentrations. As initial phenanthrene concentration increased, the growth rates were also increased. The growth rates for initial phenanthrene concentration of 10 mg/kg, 100 mg/kg, 200 mg/kg, 500 mg/kg and 1000 mg/kg were observed to be 3.91×107 cfu/g soil/day, 2.06×108 cfu/g soil/day, 3.04×108 cfu/g soil/day, 4.15×108 cfu/g soil/day and 3.31×108 cfu/g soil/day, respectively.

As previously discussed in this section, the bacteria have high carbon sources to bring more energy in high initial phenanthrene concentration. This high energy resulted in better growth. Therefore, the bacteria grew better at high concentrations compared to low concentrations. For low concentrations, phenanthrene can be limited and retard the growth of the strain.

Statistical analysis using ANOVA was conducted to establish the correlation between phenanthrene concentrations and growth rates. At the 5% level of significance, it was found that initial phenanthrene concentrations have significant effect to the growth rates ($p<0.05$). This analysis shows that the growth rates were influenced by initial phenanthrene concentrations.

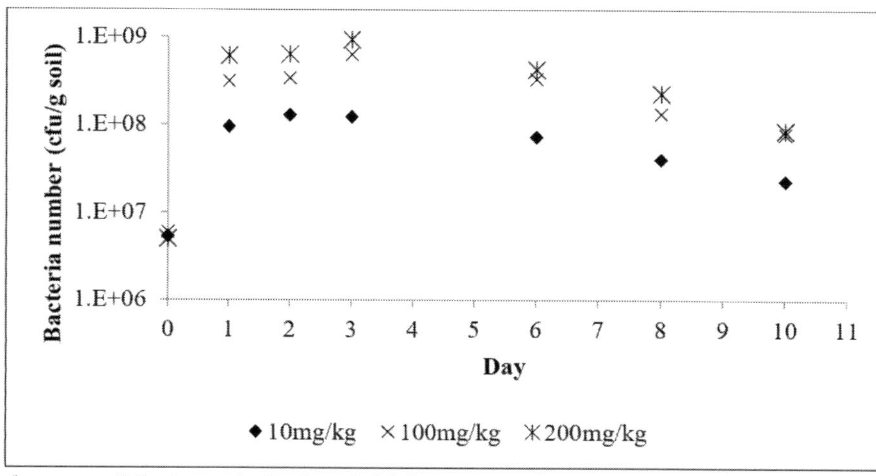

Figure3. Growth curve of the bacteria for low initial concentrations, i.e., 200mg/kg, 100 mg/kg and 10 mg/kg.

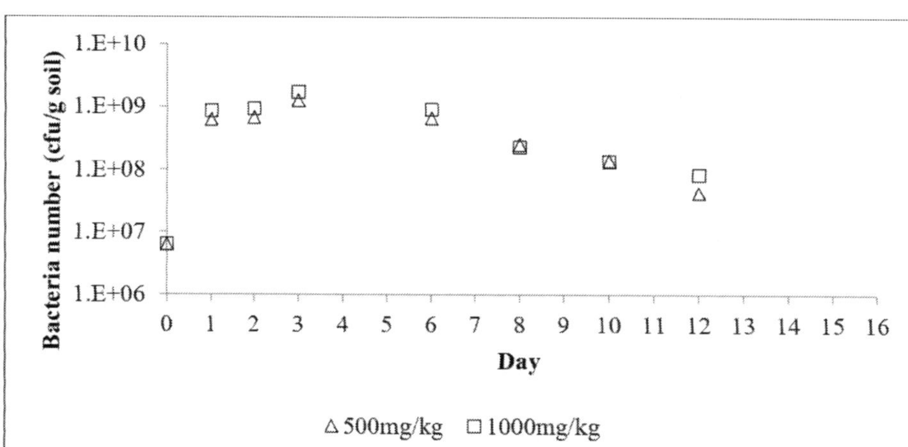

Figure 4. Growth curve of the bacteria for high initial concentrations, i.e., 1000mg/kg and 500 mg/kg.

4. Conclusions

Optimization studies provide optimum environmental condition for PAHs degradation and bacterial growth. This condition can be used in designing the soil bioremediation with a focus on PAHs as soil contaminants. This study established the optimum condition for phenanthrene degradation by the bacteria at 500 mg/kg of initial phenanthrene concentration. The growth optimum also obtained at 500 mg/kg of initial phenanthrene concentration. Optimization of PAHs biodegradation should also be conducted on different factors, e.g., soil types, salinity, moisture and organic matter. This is important to evaluate the effectiveness of potential bacteria as PAHs degrading bacterias for in-situ remediation.

References

[1] Boopathy, R. (2000). Factors limiting bioremediation technologies. Bioresource Technology, 74, 63-67.
[2] Chaineau, C. H., Rougeux, G., Yepremian, C. & Oudot J. (2005). Effects of nutrient concentration on the biodegradation of crude oil and associated microbial populations in the soil. Soil Biology and Biochemistry, 37:1490–1497.
[3] Faust, R. A. (1993). Toxicity Summary for Phenanthrene. Oak Ridge National Laboratory, Chemical Hazard Evaluation Group.
[4] Gan, S., Laua, E.V. & Ng, H.K. (2009). Remediation of soils contaminated with polycyclic aromatic hydrocarbons (PAHs). Journal of Hazardous Materials, 172: 532–549.
[5] Janbandhu, A. & Fulekar, M.H. (2011). Biodegradation of phenanthrene using adapted microbial consortium isolated from petrochemical contaminated environment. Journal of Hazardous Materials, 187,333–340.
[6] Li, X. J., Li, P. J., Lin, X., Zhang, C. G., Li, Q. & Gong, Z. Q. (2008). Biodegradation of aged polycyclic aromatic hydrocarbons (PAHs) by microbial consortia in soil and slurry phases. Journal of Hazardous Materials, 150,21-26.
[7] Othman, N. (2010). Municipal sludge characteristics and biodegradation of selected PAHs by indigenous bacteria strain, Ph.D thesis, Faculty of Civil Engineering, Universiti Teknologi MARA (Unpublish).
[8] Scullion, J., (2006). Remediating polluted soils. Naturwissenschaften, 93:51–65.
[9] Van der Oost, R., Beyer, J. & Vermeulen, N.P.E. (2003). Fish bioaccumulation and biomarkers in environmental risk assessment: A review. Environmental Toxicology and Pharmacology, 13,57–149.
[10] Zhao, H. P., Wang, L., Ren, J. R., Li, Z., Li, M. & Gao, H. W. (2008). Isolation and characterization of phenanthrene-degrading strains Sphingomonas sp ZP1 and Tistrella sp ZP5. Journal of Hazardous Materials.152:1293-1300.

Drinking Water Quality of Water Vending Machines in Parit Raja, Batu Pahat, Johor

N H Hashim[1] and H M Yusop[2]
[1]Department of Civil Engineering Technology, Faculty of Engineering Technlogy Universiti Tun Hussein Onn Malaysia, 86400 Batu Pahat, Johor, MALAYSIA
[2]Faculty of Civil and Environmental Engineering Universiti Tun Hussein Onn Malaysia, 86400 Batu Pahat, Johor, MALAYSIA.

E-mail: haslina@uthm.edu.my

Abstract. An increased in demand from the consumer due to their perceptions on tap water quality is identified as one of the major factor on why they are mentally prepared to pay for the price of the better quality drinking water. The thought that filtered water quality including that are commercially available in the market such as mineral and bottled drinking water and from the drinking water vending machine makes they highly confident on the level of hygiene, safety and the mineral content of this type of drinking water. This study was investigated the vended water quality from the drinking water vending machine in eight locations in Parit Raja are in terms of pH, total dissolve solids (TDS), turbidity, mineral content (chromium, arsenic, cadmium, lead and nickel), total organic carbon (TOC), pH, total colony-forming units (CFU) and total coliform. All experiments were conducted in one month duration in triplicate samples for each sampling event. The results indicated the TDS and all heavy metals in eight vended water machines in Parit Raja area were found to be below the Food Act 1983, Regulation 360C (Standard for Packaged Drinking Water and Vended water, 2012) and Malaysian Drinking Water Quality, Ministry of Health 1983. No coliform was presence in any of the vended water samples. pH was found to be slightly excess the limit provided while turbidity was found to be 45 to 95 times more higher than 0.1 NTU as required by the Malaysian Food Act Regulation. The data obtained in this study would suggest the important of routine maintenance and inspection of vended water provider in order to maintain a good quality, hygienic and safety level of vended water.

Keywords: water quality, vending machine, drinking water.

1. Introduction

Water is essential in maintaining life in all organisms. Although it is abundantly available in most of the country however, according to WHO [1], 663 million people lack in access to improved drinking-water and 1.8 billion people use faecally contaminated drinking-water in their daily life. Drinking water must be well-treated and must be safe for consumption without any harm based on drinking water quality standard stated by WHO and health body in each country. In Malaysia, potable water sources that are mainly sourced from surface water is treated, distributed and extensively monitored by local authorities. For example, Syarikat Air Johor (SAJ) is a private company that responsible with the treatment and distributions of clean water supply in Johor.

Daily needs for water increases from the early infancy (about ~0.6 L) to 3.3 L for man with modest physical activity and can be up to 4,5 L for active man. In the other hand, women are generally exhibit lower (about 0.5 to 1 L less) than man [2]. In fact, drinking a lot of water has not only to avoid water dehydration and maintaining good body function but it also become a common practice in weight control [3]. Therefore, the water purity and and health implication of drinking water must be a priority to ensure there is no adverse effect for human due to drinking water consumption.

Issues on water resource and drinking water have grown worldwidely and public concern have been raised over years on the water constituents even they are present in a trace concentration down to ng/L. In Malaysia, several studies have been recorded on the presence of organochlorine pesticides and plasticizers in a river basin [4], pharmaceuticals in surface water [5], heavy metals in freshwater [6], ammonia and manganese in drinking water treatment [7] and minerals content in drinking water [8].

Malaysian consumes several types of drinking water such as unfiltered tap water, filtered water, bottled water, mineral water and vending machines water. As conducted by Lee [9], boiled and unfiltered tap water was consumed by 52.3% of population compared to boiled and filtered water (25.6%) while 6.2% for bottled water. Although study by [8] shows that the mineral compositions in the tap water supplied were below the standard limits prescribed by the Malaysia Drinking Water Quality Standard 2010 [10] and WHO Guidelines [11] there are some perceptions of Malaysian tap water consumer regarding the unacceptable taste and unpleasant appearance of tap water [12]. In addition, the consumer also believed that the use of tap water could invite more health risk. As a result, many of them changed to filtered, bottled water (drinking and mineral water) and vending machines or water dispenser machine (usually .claimed as reverse osmosis (RO) water).

There are several types of RO water machine available in Malaysia either for home used, or for commercial used. For home used RO comprises a simple machine or even a few filters together, joint between the pipe while for the commercial used, it commonly looks larger than the home used and exists in the form of vending machine type as it consists several filter systems inside the machine. Instead of the consumer perceptions and experience towards the tap water, the low cost of vending machine water as low as 20 cents per liter seems as a good choice for the consumer those who are not afford to buy the home used RO one, compared to bottle water which can cost more than RM 2 for the same amount [13]. However, reported by Aruna & Camoens [13], precautions have to be taken by the consumers when using this vending machine water as six out of 14 samples randomly picked from several places in Kuala Lumpur were found to have either coliform or E.coli and did not contain free chlorine for disinfection purpose. The similar cases were also reported in Los Angelas [14] and Dubai [15]. All incidents were concluded to be related to the poor maintenance and improper hygienic conditions of water vending machine.

Due to the local demand and requirement, the common placement of water vending machines is located in public are in university campus and hostel area. Therefore, in this study, samples from eight locations in Parit Raja area were collected in order to evaluate the quality of provided RO water vending machines in terms of physical, chemical and biological paramenters. The parameters then were compared to Food Act [16], Regulation 360C (Standard for Packaged Drinking Water and Vended water, 2012) and Malaysian Drinking Water Quality, Ministry of Health 1983 to ensure their quality and safety This information is important as it may reveal the level of quality and safety of vended water in Parit Raja to surrounded consumer.

2. Materials and methods

Parit Raja is a small town in Batu Pahat district, located about 95 km from Johor Baharu city, Johor, Malaysia. Parit Raja has undergone rapid development and substantial growth due to the expansion of Universiti Tun Hussien Onn Malaysia (UTHM) which accommodates approximately 15, 300 students in varies education levels [17]. This study was investigated the quality of drinking water supplied by drinking water vending machines in Parit Raja area. A total of 64 samples of vended water were collected from eight drinking water vending machines (as in Figure 1) in Parit Raja area. From eight selected locations, four of the vending machines were located in UTHM residential colleges, while another four were located at Parit Raja town (PR). Samples were collected twice a week for one month monitoring (April to June 2015) with triplicate samples for each sampling. All vending machine has claimed to use reverse osmosis (RO) system for their water drinking treatment.

Samples were collected in a 1 L plastic container which was preserved in a 1:1 nitric acid solution for 2 days and were rinsed with milli-Q ultrapure water (Millipore, US) prior to use. The vended water samples were collected in the morning in the container and were sealed and transported to the laboratory in a cold container with temperature around 4 to 6°C. Samples were filtered through 0.45 μm pore diameter membrane filters and undergone pH adjustment with nitric acid to pH 2 using HACH pH meter. No preservatives were added to any water samples and all testing were conducted within five days from the the day of collection.

Figure 1. Drinking water vending machine.

Samples were analyzed for heavy metals content which are cadmium (Cd), arsenic (As), lead (Pb), chromium (Cr) and nickel (Ni) using Inductive Couple Plasma – Mass Spectrometer ELAN 900 (Perkin Elmer, US). Total organic carbon was determined using Total Organic Carbon Analyzer TOC-VCSH (Shimadzu, Japan), turbidity with turbidity meter HACH 2010 (HACH, US) and total dissolved solid (TDS) by SenION 5 (HACH, US). Total colony forming unit (CFU) was determined at 37°C in 24 hours [18]. Colilert Reagent was added to the vended water samples in vessels prior to 35°C and 24 hours incubation for the coliforms and E.coli presence or absence test.

3. Results and Discussions

Concentrations of selected parameters were tabulated as in Table 1 to Table 4. Result was reported as a mean of triplicate samples. The concentrations of all determined parameters were compared with Food Act 1983, Regulation 360C (Vended water), Malaysian Food Regulation and Malaysian Drinking Water Standard. Table 1 shows the pH and TDS concentration for four consecutive weeks. All samples collected from PR were found to have higher pH (5.5 – 6.4) than the value stated in Malaysian Food regulation 1983 and Malaysian Drinking Water Standard during that month. Values given by vended water samples in UTHM somehow were slightly above both regulations. On the other hand, all TDS concentrations were in good agreement with Malaysian Drinking Water Standard as no value provided for TDS in Malaysian Food Regulation.

Table 1. pH and TDS concentration.

	pH				TDS (mg/L)			
Location	**W1**	**W2**	**W3**	**W4**	**W1**	**W2**	**W3**	**W4**
UTHM 1	6.1	5.7	6.7	7.1	109.0	110.6	107.2	110.1
UTHM 2	7.1	6.0	6.4	6.4	9.9	9.9	11.0	11.4
UTHM 3	6.7	6.4	6.6	6.4	4.3	4.2	4.9	5.0
UTHM 4	6.5	6.4	6.2	6.6	4.9	6.0	5.0	5.2
PR 1	5.9	6.2	5.9	6.4	14.2	19.4	21.2	22.2
PR 2	5.5	6.0	5.9	6.0	95.9	103.9	103.4	101.9
PR 3	5.6	5.9	6.0	6.1	35.4	44.5	33.4	37.3
PR 4	5.8	6.1	5.9	5.9	99.6	101.6	103.3	102.3

Malaysian Food Regulation 1983; pH: 6.5 – 8.5
Malaysian Drinking Water Standard: pH 6.5 – 9.0, TDS: 1000 mg/L

Turbidity values given by all samples as tabulated in Table 2 were in the range of 4.5 to 9.5 mg/L. Based on both standards, these values were 45 to 95 times higher than the value set by Malaysian Food Regulation which is 0.1 NTU and almost doubled from the Malaysian Drinking Water Standard. Turbidity was set to be 0.1 NTU in vended water and 5 NTU in drinking water since turbidity was found to be associated with the presence of suspended matter such as clay, silt, organic or inorganic matter and might also contain other microorganisms [19]. As commercial RO membrane system was design to remove particle efficiently from at least 0.1 nm depending on the filter type, it is expected that this vended water has only containing the pure water and turbidity concentration less than raw water which is 5 NTU. This might suggest the inefficient filtration capacity of the RO system in the vending machine due to the originality of supplied RO membrane or the poor maintenance of the vending machine.

In TOC observation, none of the standards were regulated on TOC value. However, it can be presence in drinking water sample from less than 100 μg/L to up 25,000 μg/L [20] and indicates the presence of organic matter in the water samples. As indicated in the table, TOC concentrations were observed in relatively low level in all vended water samples.

Table 2. Turbidity and total organic carbon concentration.

	Turbidity (NTU)				TOC (mg/L)			
Location	**W1**	**W2**	**W3**	**W4**	**W1**	**W2**	**W3**	**W4**
UTHM 1	5.0	5.0	4.5	4.5	2.7	3.1	1.8	1.2
UTHM 2	5.0	9.0	5.0	5.0	2.6	3.7	2.5	1.5
UTHM 3	8.0	9.5	3.5	5.0	2.8	3.8	2.5	1.5
UTHM 4	6.5	7.5	5.0	3.5	2.1	3.3	2.3	1.3
PR 1	3.5	4.0	5.0	4.5	3.4	2.9	2.1	1.1
PR 2	5.5	4.5	5.0	5.5	2.3	2.5	2.0	1.1
PR 3	5.0	6.0	4.5	4.5	1.7	2.9	1.9	1.1
PR 4	4.0	3.5	3.5	5.0	2.2	2.8	1.5	1.4

Malaysian Food Regulation 1983; Turbidity: 0.1 NTU
Malaysian Drinking Water Standard: Turbidity: 5 NTU

Table 3 shows the total colony forming unit in the samples. However, no testing was done on the presence of coliform in the water samples although it was set as one of the most important requirement in both standards. Though, the numbers given by Table 3 indicated some types of microorganism were presence in the vended water samples and thus suggested water must be boiled first prior to used. Coliforms and E.coli presence and absence tests were performed using colilert powder indicated negative results when none of the vended sample water showing any colors of fluorescence. This result indicates that the vended water quality in Parit Raja area are still safe to be consumed in terms of total coliform presence as 6 samples over 49 samples were detected to contain total group members in dispenser machines in Ajman, UAE [15]. Similar observation was also revealed the presence of total coliforms in 13 samples out of 40 drinking water vending machines in Los Angeles due improper scheduled maintenance [14].

Table 3. Total Colony Forming Unit.

	CFU (cell/ml)			
Location	**W1**	**W2**	**W3**	**W4**
UTHM 1	62.5	64.5	67.5	73.5
UTHM 2	34	58	54	60
UTHM 3	51.5	68	68	79.5
UTHM 4	36	63	40	45.5
PR 1	76	91	89.5	96.5
PR 2	63	84	68	88
PR 3	49	58.5	60.5	67.5
PR 4	82	91	88	82.5

All heavy metals were analyzed simultaneously and determinations were based on six point calibration curve as tabulated in Table 4. Cd in all vended water samples were found to be in the range of 0.03 to 0.15 μg/L. While As, Pb, Cr and Ni were determined as low as 0.04 μg/L and can be up to 0.68 μg/L, 0.17 μg/L to 1.53 μg/L, 0.12 μg/L to 0.78 μg/L and 0.19 μg/L to 0.82 μg/L, respectively. Malaysian Food Regulation has regulated more stringent value about five-fold increase for Cd, Pb, Cr and Ni, and 10-fold increase for As compared to Malaysian Drinking Water Standard. However, results have revealed that all samples were below the minimum concentration level recommended by Malaysian Food Regulation.

Based on the location, vended water at four stations in UTHM have recorded 0.03 to 0.10 μg/L of Cd, 0.04 to 0.68 μg/L of As, 0.17 to 1.43 μg/L of Pb, 0.12 to 0.57 μg/L of Pb and 0.12 to 0.57 μg/L of Cr. While in PR, four stations of vended water were recorded to have Cd in the range of 0.06 to 0.15 μg/L, 0.08 to 0.60 μg/L of As, 0.21 to 1.53 μg/L of Pb, 0.19 to 0.78 μg/L of Cr and 0.19 to 0.57 of Ni. In general, both locations have almost the similar values of the concentration range suggested that the performance of drinking water vending machines are in the satisfactory level for investigated heavy metals.

Table 5 shows the mean concentrations of Cd, As, Pb, Cr and Ni in all eight drinking water vending machines in both locations. From the mean concentrations, vended water from PR 2 drinking water vending machine has the highest concentration of Cd, Pb and Cr compared to other stations. Cd was recorded to be as high as 0.12 μg/L, 0.94 μg/L of Pb and 0.71 μg/L of Cr. While the highest As was observed at UTHM 1 with 0.62 μg/L and Ni at UTHM 2 with 0.55 μg/L. It may suggest that PR 2 vending machines was having poor maintenance and less inspection by the provider during the monitoring month.

Table 4. Cd, As, Cr, Pb and Ni concentration in vended water samples.

	Cd				As				Pb			
Location	**W1**	**W2**	**W3**	**W4**	**W1**	**W2**	**W3**	**W4**	**W1**	**W2**	**W3**	**W4**
UTHM 1	0.05	0.06	0.04	0.10	0.50	0.71	0.68	0.52	0.17	0.71	0.43	0.96
UTHM 2	0.05	0.03	0.06	0.07	0.07	0.14	0.13	0.09	0.62	0.93	1.34	0.59
UTHM 3	0.06	0.04	0.08	0.08	0.07	0.08	0.10	0.07	0.49	0.63	1.00	0.20
UTHM 4	0.06	0.07	0.04	0.10	0.04	0.07	0.04	0.06	0.17	0.59	0.52	0.38
PR 1	0.07	0.06	0.06	0.11	0.10	0.09	0.09	0.08	0.22	0.54	0.58	0.36
PR 2	0.08	0.15	0.06	0.16	0.36	0.37	0.42	0.34	0.46	1.53	0.78	0.69
PR 3	0.10	0.05	0.11	0.07	0.20	0.26	0.25	0.18	0.21	0.30	1.27	0.25
PR 4	0.09	0.08	0.09	0.15	0.53	0.55	0.60	0.50	0.44	0.79	0.87	0.51

	Cr				Ni			
Location	**W1**	**W2**	**W3**	**W4**	**W1**	**W2**	**W3**	**W4**
UTHM 1	0.57	0.49	0.61	0.48	0.55	0.43	0.45	0.51
UTHM 2	0.44	0.48	0.54	0.42	0.38	0.51	0.82	0.40
UTHM 3	0.16	0.17	0.15	0.23	0.41	0.26	0.27	0.38
UTHM 4	0.12	0.20	0.12	0.15	0.39	0.55	0.21	0.59
PR 1	0.22	0.30	0.34	0.19	0.54	0.37	0.22	0.52
PR 2	0.64	0.67	0.78	0.69	0.52	0.52	0.39	0.57
PR 3	0.35	0.36	0.41	0.36	0.20	0.19	0.20	0.26
PR 4	0.40	0.41	0.59	0.40	0.57	0.51	0.55	0.57

Malaysian Food Regulation 1983; Cd: 0.6 μg/L; As: 1 μg/L; Pb: 2 μg/L; Cr: 10 μg/L; Ni: 4 μg/L
Malaysian Drinking Water Standard: Cd: 3 μg/L; As: 10 μg/L; Pb: 10 μg/L; 50 μg/L; Ni: 20 μg/L

Table 5. The mean concentrations of heavy metals at UTHM and PR for the whole month.

Location	Cd	As	Pb	Cr	Ni
UTHM 1	0.06±0.04	0.62±0.10	0.65±0.43	0.53±0.14	0.47±0.05
UTHM 2	0.05±0.03	0.11±0.03	0.92±0.34	0.48±0.11	0.55±0.33
UTHM 3	0.06±0.03	0.08±0.01	0.60±0.43	0.18±0.07	0.32±0.09
UTHM 4	0.07±0.03	0.06±0.01	0.46±0.28	0.15±0.04	0.44±0.18
PR 1	0.08±0.03	0.09±0.01	0.47±0.19	0.27±0.09	0.38±0.15
PR 2	0.12±0.05	0.37±0.04	0.94±0.51	0.71±0.09	0.50±0.08
PR 3	0.08±0.03	0.23±0.04	0.57±0.54	0.37±0.05	0.22±0.04
PR 4	0.11±0.03	0.55±0.04	0.69±0.29	0.46±0.09	0.54±0.05

Drinking water standard supplied by water provider in Malaysia has to follow Malaysian Drinking Water Standard provided by Department of Environment. There were several studies in Malaysia investigated the presence of heavy metals in Malaysia drinking water. Reviewed by Hafiza et al. [21], Pb was detected in Malaysia drinking water due to the corrosion and plumbing system in water distribution system. Stagnation of water in a piping system caused the leached Pb to be in contact with stagnant water. The presence of heavy metals was also related to the source of the water supply such as groundwater and related activity occurred surrounding the water sources. After water treatment process, treated water supposedly reached the standards provided by Malaysian Drinking

Water Standard. Water supply normally was boiled prior to use by household however, was filtered using self-use filter system, or paid filter system such as vending machine was expected to remove more trace metals and microorganism left in water supply. Once the maintenance and inspection are poor and improper, the hygienic aspect, the safety level and the filtration performance will reduce [14]. The reduction in vended water quality thus will cause infection, spreading of disease and toxic effect to consumer [8,14,15]. Furthermore, as reviewed by [21], heavy metals are now becoming a world concern due to their absorption and accumulation ability in human body, displacement of essential minerals and interruption on cell activity.

Instead of the maintenance and inspection by the provider, the installation site of the drinking water vending machine must be taken into account. Study by Al Moosa et al. [15] the placement of water dispenser in a playing area or being exposed to direct sunlight was observed to be high in total coliform as it increases with increases of ambient temperature.

4. Conclusion

This study determined the concentration of pH, turbidity, total dissolved solid (TDS), total organic carbon (TOC), total colony forming unit, and heavy metals (Pb, As, Cd, Cr and Ni) from eight drinking water vending machines in Parit Raja, Batu Pahat, Johor, Malaysia area. Presence and absence test of coliform was also conducted to all vended water samples. In conclusion, TDS and all heavy metals in eight vended water machines in Parit Raja area were found to be below the Food Act 1983, Regulation 360C (Standard for Packaged Drinking Water and Vended water, 2012) and Malaysian Drinking Water Quality, Ministry of Health 1983. No coliform was presence in any of the vended water samples. pH was found to be slightly excess the limit provided while turbidity was found to be 45 to 95 times more higher than 0.1 NTU as required by the Malaysian Food Act Regulation. The findings of this study suggest that regular and proper maintenance of drinking water vending machines by the provider must be conducted to ensure the level of hygiene, safety and satisfactory among consumer are always in the best place.

References

[1] WHO (2015). 25 years Progress on Sanitation and Drinking Water: 2015 Update and MDG Assessment. World Health Organization. Available at: http://www.who.int/about/licensing/copyright_form/en/index.html.

[2] Sawka, M.N., Cheuvront, S.N. & Carter, R. (2005). Human Water Needs. Nutrition Reviews, 63(6), pp.30–39.

[3] Muckelbauer, R. et al. (2013). Association between water consumption and body weight outcomes : a systematic review. Am J Cli Nutr, pp.1–18.

[4] Santhi, V.A. & Mustafa, A.M. (2013). Assessment of organochlorine pesticides and plasticisers in the Selangor River basin and possible pollution sources. Environmental Monitoring and Assessment, 185(2), pp.1541–1554.

[5] Al-Odaini, N.A. et al. (2011). The occurrence of human pharmaceuticals in wastewater effluents and surface water of Langat River and its tributaries, Malaysia. International Journal of Environmental Analytical Chemistry, 93(3), pp.245–264.

[6] Mohammad, E. & Idris, M. (2008). Heavy metal concentrations in water and sediments in Tasik Chini, a freshwater lake, Malaysia. Environmental Monitoring and Assessment, 141(1), pp.297–307.

[7] Hasan, H.A. et al. (2011). Problems of Ammonia and Managanese in Malaysian Drinking Water Treatments. World Appi;ied Sciences Journal, 12(10), pp.1818–4952.

[8] Azlan, A. et al. (2012). Evaluation of Minerals Content of Drinking Water in Malaysia. The Scientific World Journal, 2012, pp.1–10.

[9] Lee, C. (2005). Water Tariff and Development, University of Malaysa, Selangor, Malaysia.

[10] Ministry of Health Malaysia (2010). Drinking Water Quality Surveillance Programme: Drinking Water Quality Standard.

[11] WHO (2011). Guidelines for Drinking-water Quality: 4th Ed. World Health Organization.
[12] Khattab, T.H. & Abdul Wahid, N. (2015). Penang User's Perception of Domestic Water Quality, Health Risk Concern and Willingness to Pay: A Pilot Study. Jurnal Teknologi (UTM), 74(11), pp.93–107.
[13] Aruna, P. & Camoens, A. (2014). Bacteria in Your Drinking Water. The Star Online.
[14] Schillinger, J. & Du Vall Knorr, S. (2004). Drinking-Water Quality and Issues Associated with Water Vending Machines in City of Los Angeles. Journal of Environmental Health, 66(6), pp.25–31.
[15] Al Moosa, M.E. et al. (2015). Microbiological Quality of Drinking Water from Water Dispenser Machines. International Journal of Environmental Science and Development, 6(9), pp.710–713.
[16] Ministry of Health (1983). Food Act 1983, Malaysia.
[17] Executive Information System (EIS) (2016), Total Campus Integrated System (TCIS), Universiti Tun Hussein Onn Malaysia.
[18] Bartram, J. and Balance, R. (1996), Water Quality Monitoring - A Practical Guide to the Design and Implementation of Freshwater Quality Studies and Monitoring, United Nations Environment Programme and the World Health Organization
[19] LeChevallier, M.W., Evans, T.M. & Seidler, R.J. (1981). Effect of turbidity on chlorination efficiency and bacterial persistence in drinking water. Applied and Environmental Microbiology, 42(1), pp.159–167.
[20] APHA (2005). Standard Methods for the Examination of Water and Wastewater 21st ed., American Public Health Association, US.
[21] Hafiza, N. et al. (2015). Drinking water studies : A review on heavy metal , application of biomarker and health risk assessment (a special focus in Malaysia). Journal of Epidemiology and Global Health.

Optimization of low ring polycylic aromatic biodegradation

N Othman[1] , S Abdul-Talib[2], C C Tay[3]
[1] Micropollutant Research Centre, Faculty of Civil and Environmental Engineering
University Tun Hussein Onn, Malaysia
[2].Faculty of Civil Engineering, University Teknologi Mara, Shah Alam Selangor
[3]. Faculty of Applied Science, University Teknologi Mara, Arau. Perlis

E-Mail : norzila@uthm.edu.my

Abstract. Polycyclic aromatic hydrocarbons (PAHs) are recalcitrance and persistence that finally turn into problematic environmental contaminants. Microbial degradation is considered to be the primary mechanism of PAHs removal from the environment due to its organic criteria. This study is carried out to optimize degradation process of low ring PAHs. Bacteria used in this study was isolated from sludge collected from Kolej Mawar, Universiti Teknologi MARA, Shah Alam, Selangor. Working condition namely, substrate concentration, bacteria concentration, pH and temperature were optimized. PAHs in the liquid sample was extracted by using solid phase microextractio equipped with a 7μm polydimethylsiloxane (PDMS) SPME fibr . Removal of PAHs were assessed by measuring PAHs concentration using GC-FID. Results from the optimization study of biodegradation indicated that maximum rate of PAHs removal occurred at 100 mgL^{-1} of PAHs, 10% bacteria concentration, pH 7.0 and 30°C. These working condition had proved the effectiveness of using bacteria in biodegradation process of PAHs.

Keywords: bioremediation, persistent organic pollutant, degradation percentage

1. Introduction

PAHs represent a large family of organic compounds that are considered environmental pollutants. These compounds are widespread in environment and it is believed as the most abundant organic compounds in the universe that might pose a threat to the environment and mankind [1,2,3]. Significant levels are detected in air, water, soils and sediments [1,4]. PAHs impact the aquatic environment by targeting the bottom feeder such as polycaetes, nematode, bivalves and crustaceans [5], thus creating a disturbance in the ecological balance of the aquatic systems. PAHs may be released into the environment when the soils or sediments are disturbed.

Nowadays, PAHs can also be found in many household products such as mothballs, cosmetics and hair dyes. In addition, some food products such as roasted coffee, peanut and barbecued meat also constitute other source of PAHs in domestic setting. These products at the end of their used would eventually end up in the wastewater treatment plants or landfill thus contributing to PAHs presence.in water, sludge and soil.

Microbial degradation is among favarobale method in removing PAHs from the contaminated sites. Previous studies had proved successful process of PAHs degradation using several type of microbes isolated from different sources namely, riverine sediment, sludge, petrochemical sludge and soil [6,7,8,9].

Information on optimization condition is serve as an important parameter for ensuring complete biodegradation and bioremediation of PAHs on site. This study is carried out to optimize degradation process of low ring PAHs. Parameter such as substrate concentration, bacteria concentration, pH and temperature were optimized.

2. Method

2.1 Bacteria Strain
Strain of bacteria *Corynebacterium uroalyticum* was isolated from domestic municipal sludge. The sludge was sampled by placing a long hose tube at the bottom of the clarifier and pumping it out to the drying bed. Sludge sample was collected using the scoop method and stored in an air tight container. Samples for bacterial isolation was stored in air tight plastic container.and was stored in a refrigerator at 4° C at the Microbiology Laboratory, Faculty of Applied Science, UiTM to inhibit microbial activities. Two different sources of PAHs, single PAHs and mix PAHs were used as carbon source during isolation process.

2.2 Preparation of Bacterial Strain and PAH Degradation
A liquid inoculum was prepared from a pure culture of bacteria. A sterile swab was used to transfer bacterial colonies from minimal agar to sterile nutrient broth. Then the inoculum was incubated at 30^{0}C until the broth became turbid. The incubation time used depended on the turbidity test.

2.3 Design Experiment for Optimization study
Four parameters including the concentration of bacteria, concentration of PAH, pH and temperature were selected to determine their effects on the degradation of PAHs by the isolated bacteria strain Experiments were conducted with 50 ml sterile minimal media subjected to specific parameters to be investigated. The flasks were placed in an incubated shaker at 150 rpm for two weeks. Summary of the variables on effects of abiotic factors are presented in Table 1 [7,9].

Table 1. Summary of abiotic factors

Abiotic factors	Range of variables
Bacteria concentration	1 % to 12 %
Substrate concentration	1 ppm, 10 ppm, 50 ppm, 100 ppm and 500 ppm
pH	5, 6, 7, 7.5 and 8
Temperature	20^{0}C, 25^{0}C, 30^{0}C, 35^{0}C and 40^{0}C.

2.4 PAHs Extraction
An innovative solid phase microextraction (SPME) method optimized by [10] was adopted to extract PAHs. 20 ml of sample was kept in 25 ml glass bottle with a septum cap and placed in an ultrasonic water bath. The sample volume and vial size were kept constant in order to reduce effects of external factors on extraction of the analytes. Besides that, to ensure consistency and uniformity in adsorption and desorption, the period and depth of immersion, time between immersion and exposure to the GC and GC desorption parameters including temperature and time were kept constant [11,12].

2.5 Analytical Instruments for PAHs
Gas chromatography was performed on a Perkin Elmer Clarus 500 system coupled to flame ionization detection (FID) system. In order to analyze naphthalene, acenapthylene, acenapthene, fluorene, anthracene and phenanthrene, Elite Column 5MS with 30m long 0.25mm internal dimension and 0.25μm thickness was used to separate the compounds. The injector was operated at 250°C in the splitless mode with a two minute splitless period. Helium was used as the carrier gas with 1ml/min constant flow rate. The column temperature was initially set at 50°C for 1 min, increased to 150°C at a rate of 15°C/min and held at 1 min, and finally ramped at 5°C/min to 300°C and held constant until the end of the 35 min total run time. Identification of analytes in the chromatograms was based on retention time.

2.6 *Abbreviations*
A few abbreviation were used in the results to name the PAHs and mix of PAHs. N-napthalene, AC-acenapthylene, A-acenapthene, F-fluorene, P-phenantherene. M1-mix pahs 1, M2-mix PAHs 2 and M3-mix PAHs 3.

3. Results and discussion
Four parameters including the initial concentration of bacteria, pH, temperature, concentration of substrate were tested for investigating their effects on the degradation of PAHs using the isolated bacteria strains.

The data in optimization study was analyzed using a general linear analysis of variance (ANOVA) model. For multiple comparisons of data means, Turkey's honestly significant difference test with an $\alpha=0.05$ was performed.

The study on the optimal growth conditions for PAH oxidation by isolated bacteria species could increase the removal rate of PAHs from sludge, soil and other matrices. Apart from that these experiments will provide knowledge for further investigations.

3.1 Effect of Initial Bacteria Concentration on PAHs degradation
Degradation of PAHs by *Corynebacterium uroalyticum* (P1) will be used. PAHs degradation was investigated with 50 ml of aqueous media, two weeks (14 days) contact time at pH 7 and with bacteria concentration varied from 1% to 12% at 30^0C. A typical graph on the effect of bacteria concentration on phenanthrene degradation is shown in Figure 1.

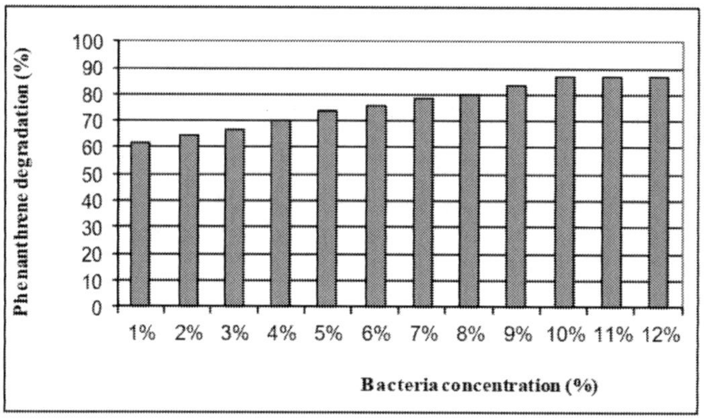

Figure 1. Phenanthrene degradation at different concentration of *Corynebacterium uroalyticum* (P1) after 14 days of incubation

Figure 1 shows that after 14 days incubation, the degradation of phenanthrene increase as the initial concentration of bacteria increase. The PAHs degradation reaches the optimum level at 10% of bacteria concentration. The rate for 1% bacteria concentration is 3.9 ppm per day. The rate increased to 6.2 ppm per day, for 10% bacteria concentration and remains unchanged at 12% bacteria concentration. Therefore, the concentration of bacteria was selected as 10% for further experiments.

More enzymes could be synthesized with the increase in bacteria concentration. Each enzyme has a certain area on its surface that is known as active sites. The active site is the region at which the enzyme forms a loose association with PAHs as its substrate. Therefore, the more enzymes are secreted the more active sites will be provided for PAHs. Enzymes remain unchanged while they speed up degradation of PAHs.

The percentage of degradation increased as the concentration of bacteria increased and finally reached a steady state known as chemical equilibrium. At this point the enzyme's active site or binding site is working at full capacity.[13,7] reported the same range of bacteria concentration used for biodegradation of PAHs.

For the purpose of comparison among the different pair of bacteria strains and PAHs, a normalized graph on PAHs degradation for different bacteria concentrations were constructed as shown in Figure-2 for single PAH and Figure-3 for mixed PAHs.

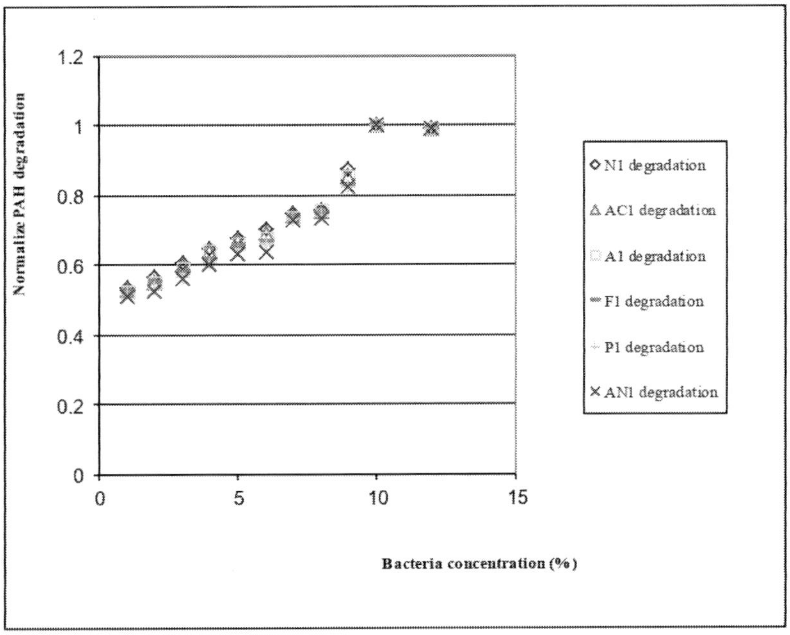

Figure 2. Normalize PAHs degradation at different bacteria concentration (single PAHs substrate)

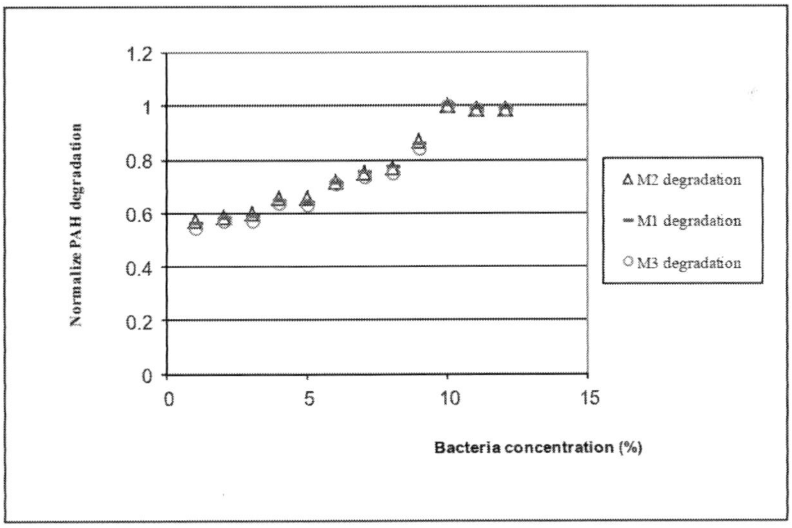

Figure 3. Normalize PAHs degradation at different bacteria concentration (mixed PAHs substrate)

Generally, it can be observed that all strains of PAHs degrading bacteria show increasing trend as the concentration of bacteria increase (as shown in Figures 2 and 3). Degradation of phenanthrene remain unchanged at 10% bacteria concentration.

Analysis of variance (ANOVA) performed on the results showed that at 5% level of significance, the degradation of phenanthrene was significantly affected by concentration of bacteria. Hence, higher concentration of bacteria result in faster degradation process.

3.1 Effect of pH on PAHs degradation

Degradation of phenanthrene using *Corynebacterium uroalyticum* (P1) was used for the purpose of discussion effect on pH. pH plays an important role in PAHs biodegradation. Effect of pH on PAHs degradation was evaluated at pH 5, 6, 7, 7.5 and 8. Like other proteins, enzymes in bacteria are affected by extreme pH such as strong acid or alkali. Thus, this study evaluated the effect of pH on PAHs degradation at pH value that is mildly acid and mildly alkali. Typical graph on effect of pH on PAHs biodegradation by *Corynebacterium uroalyticum* is shown in Figure-4. Based on Figure-4, degradation of phenanthrene by *Corynebacterium uroalyticum* shows increasing trend as pH increase. The degradation reaches optimum level at pH 7. At pH 7.5 and 8 degradation processes decrease.

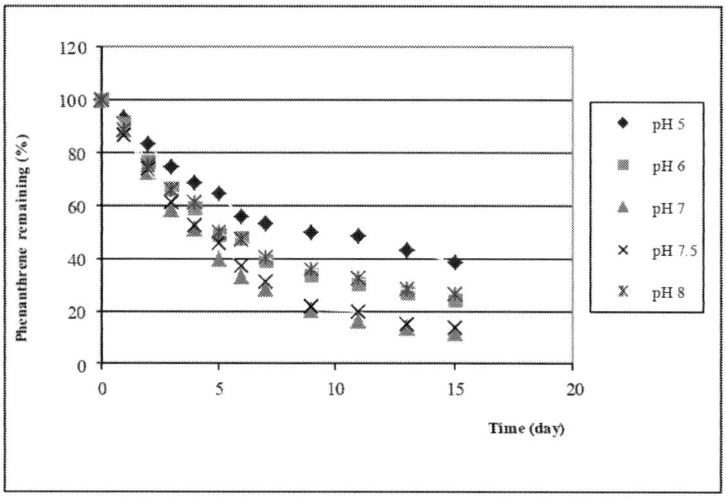

Figure 4. Effects of pH on phenanthrene degradation by *Corynebacterium uroalyticum* (P1)

At equilibrium, biodegradation of PAHs on *Corynebacterium uroalyticum* as a function of pH is shown in Figure-4, which indicated that at pH 5 and pH 6, slower biodegradation were observed with the range of phenanthrene degradation 61% to 75% respectively. While at pH 7, 88% of phenanthrene was degraded. For the pH value higher than 7, biodegradation process is slower than optimum level. The result shows that the optimum degradation was accomplished at pH 7.

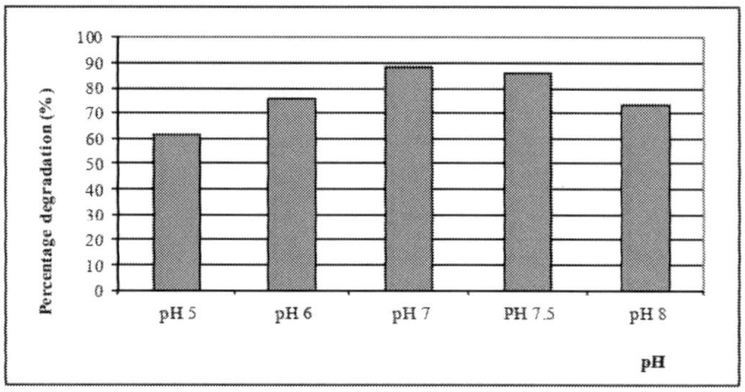

Figure 5. Phenanthrene degradation by *Corynebacterium uroalyticum* (P1)

Changes in pH can alter the electrical charge on various chemical groups in enzymes molecules. Changes in electrical charge probably alter the enzyme's ability to bind its substrate and catalyze a reaction. Imbalance of the electrical charges in very acidic and alkali condition can disrupt hydrogen bonds and other weak forces that maintain enzyme structure. Such disruption of enzyme structure is called denaturation. Thus, this phenomenon resulted in poor biodegradation process. This explanation was also offered by [7]. They also had pointed out that microbial enzymes have an optimum pH at which they function most effectively, thus it is related to an organism normal environment.

The results from this study suggest that the biodegradation of PAHs strongly depends on pH, as it affects the ionization state of the main functional group in PAHs which is benzene on bacterial cell [14]. Several studies reported by [15,7] indicated that optimum pH for PAHs biodegradation normally occurred at pH 7. In order to compare strength among pair of bacteria strains and PAHs at different pH, normalized graphs were constructed as shown in Figures 6 and 7.

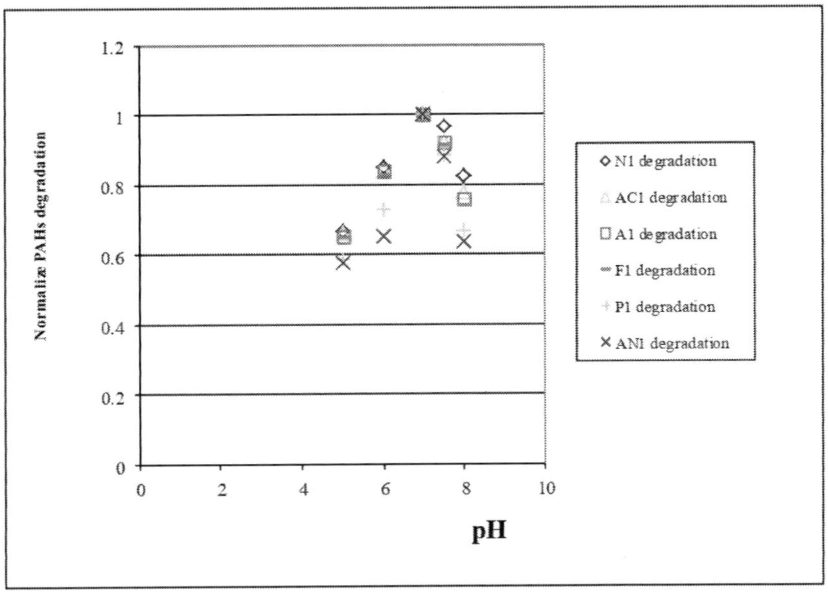

Figure 6. Normalize PAHs degradation for pH value (single PAHs substrate)

Figures 6 and 7, show that all strains of PAHs degrading bacteria show increasing trend as pH values increase from pH 5 to pH 7. Optimum pH for all strains was established at pH 7. The graphs show decreasing trend after the optimum pH. This trend indicates that the activity of PAHs degrading bacteria is maximum under neutral condition.

Statistical analysis (ANOVA) performed on the results at 5% level of significance indicate that the degradation of phenanthrene was significantlly affected by pH. Better degradation process was obtained with higher pH of substrate.

3.2 *Effect of temperature on PAHs degradation*

Degradation of phenanthrene using *Corynebacterium uroalyticum* as biodegrader is carried out at different initial temperature. Degradation of phenanthrene using *Corynebacterium uroalyticum* (P1) was used for the purpose of discussion effect of temperature. The temperatures chosen for temperature dependence study were 20^{0}C, 25^{0}C, 30^{0}C, 35^{0}C and 40^{0}C. Figure-8 shows a typical graph on biodegradation of phenanthrene as a function of temperature.

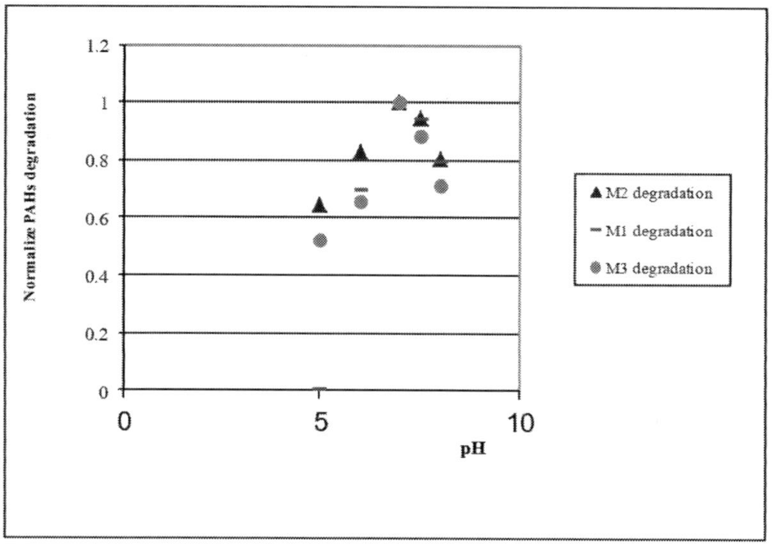

Figure 7. Normalize PAHs degradation for pH value (mixed PAHs substrate)

Figure 8 shows biodegradation of PAHs by *Corynebacterium uroalyticum* at equilibrium condition as a function of temperature. It is shown that at low temperature such as 20^{0}C low degradation process is observed, while at 25^{0}C the biodegradation process start to increase. The results show that the optimum temperature was attained at 30^{0}C. The equilibrium PAHs biodegradation is 87.2% at this temperature value.

Low degradation process at low temperature suggests that this temperature is not sufficient for the enzyme to speed up the reaction rate. Therefore, increased in temperature will result in faster reaction rate in degrading PAHs. In general enzymes as catalysts have certain temperature that it withstands and normally at optimum temperature will catalyze reaction most rapidly. Microbial enzymes likewise function best at optimum temperature which related to an organism's normal environment. Thus, reaction such as degradation process is normally optimum at mild temperature in microbial cell. Above 40^{0}C, however enzyme is rapidly denatured, and its activity decreased accordingly.

The results from this study suggest that the degradation of PAHs strongly depend on temperature. It affects the reaction rate to cleave binding in benzene as the main functional group. These observations were also reported by previous studies that indicated the optimum temperature for PAHs biodegradation occurred at 30^{0}C.

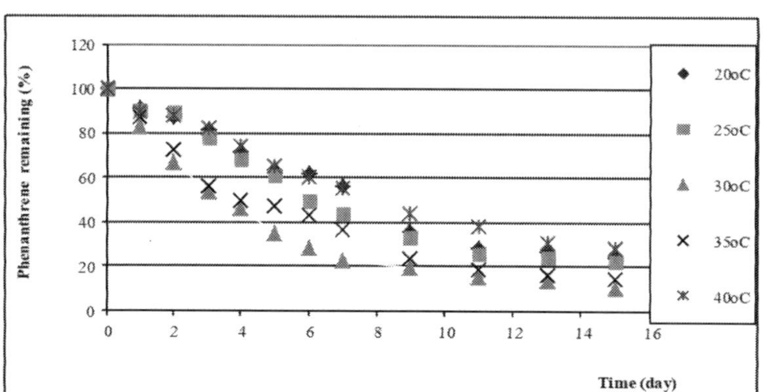

Figure-8. Effects of temperature on phenanthrene degradation by *Corynebacterium uroalyticum* (P1)

A normalized graph for each pair of bacteria and PAHs were drawn at different temperatures as shown in Figure 9 and 10. Figures 9 and 10 indicate that all strains of PAHs degrading bacteria show increasing trend as temperature increased from 20°C to 30°C. Nevertheless, after the optimum temperature at 30°C, the graphs decrease. Statistical analysis (ANOVA) was conducted on the results for degradation of phenanthrene at different temperature. At 5% level of significance it was found that the degradation of phenanthrene was significantly affected by temperature. Better degradation was obtained with higher temperature.

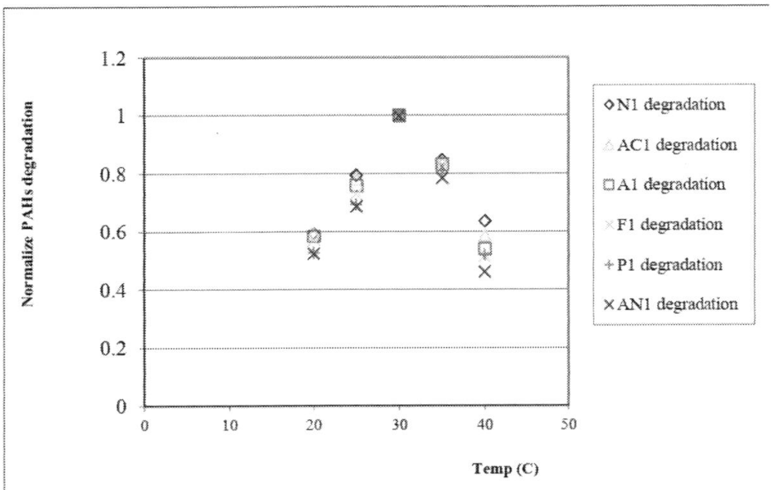

Figure 9. Normalize PAHs degradation for temperature (single PAHs substrate)

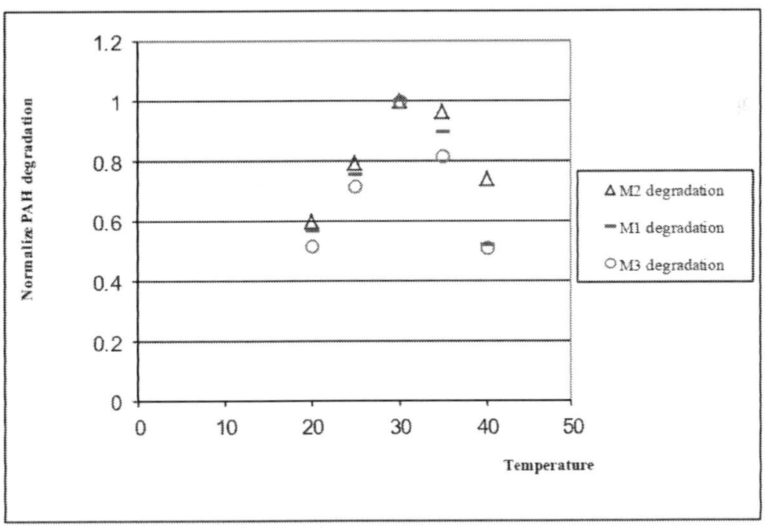

Figure 10. Normalize PAHs degradation for temperature (mixed PAHs substrate)

3.3 *Effects of Initial substrate concentration on PAHs degradation*

The experiment on effect of initial substrate was conducted at optimized parameters, 10% bacteria concentration, pH 7.0 and temperature at 30°C. The initial PAHs concentrations were concentration adopted from characteristic study, 10ppm, 50ppm, 100ppm and 500ppm. The lowest concentration was chosen based on PAHs concentration in sludge sample analyzed. Other PAHs concentrations were chosen based on previous studies as reported by [13,14]. PAHs degradation by *Corynebacterium uroalyticum* is affected by the change in phenanthrene concentration as shown in Figure-11. The percentages of degradation increase as concentration of substrate increase up to 100ppm of substrate concentration (Figure-11). Thus, degradation process is optimum at 100ppm of substrate or is known as a steady state. At this point the PAHs degrading bacteria have sufficient carbon sources for energy and growth.

There was about 87.3% phenanthrene degradation for cultures with the addition of 100 ppm PAHs but only 80.5% phenanthrene degradation at 500 ppm. The average degradation rate of phenanthrene is 6.00 ppm/day at 100ppm, 3.28 ppm/day at 500ppm, 0.98 ppm/day for 50pm, 0.45 ppm/day at 10 ppm, 0.12 ppm/day for 5ppm and 0.01 ppm/day at 1ppm (as shown in Figure-11). 100 ppm phenanthrene shows a higher efficiency to remove phenanthrene compare to 500 ppm phenanthrene culture. This may be due to the fact that too high the concentration of phenanthrene would saturate the bacteria cell with the substrate. Therefore, the optimum degradation on phenanthrene at 86.3% was obtained at phenanthrene concentration of 100ppm.

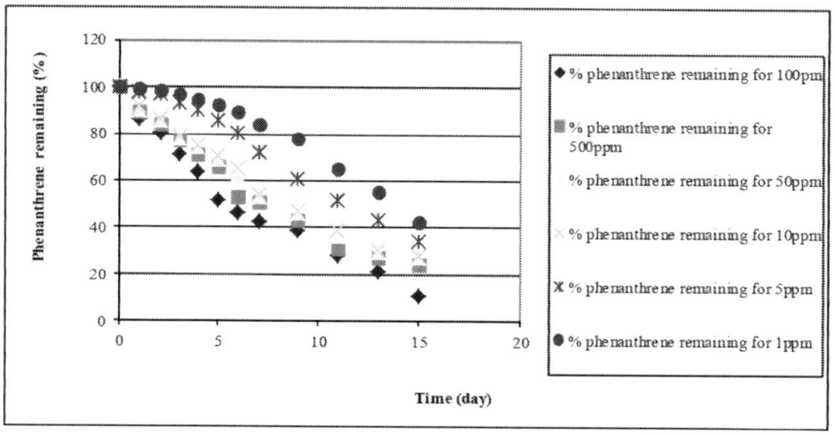

Figure 11. Effects of substrate concentration on phenanthrene degradation by *Corynebacterium uroalyticum* (P1)

Degradation of phenanthrene by *Corynebacterium uroalyticum* increased with substrate concentration indicates a great potential in application of *Corynebacterium uroalyticum* as a biodegrader to the treatment of sludge containing PAHs at high concentration.

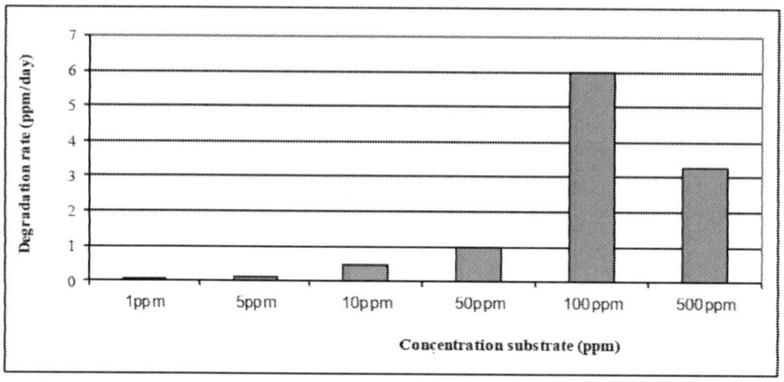

Figure 12. Average degradation rate for different concentration

Other researchers also reported that high concentration of PAHs were beneficial to the growth of isolated bacteria because PAHs could acts as a sole carbon source for bacterial growth. Therefore, low concentrations of PAHs would become a limiting factor for the bacterial culture to grow.

The degradation of PAHs at higher PAH concentration level shows an increasing trend because the increase of PAHs will increase the affinity of substrate toward formation of product. When neither substrate nor products are removed from the system, the reaction will ultimately reach a steady state known as chemical equilibrium. At equilibrium no net change was observed in the concentration of substrate or product. After the steady state, the substrate saturated and degradation process start decreasing and resulted in lower rate of reaction as shown in Figure-12 for 500ppm phenanthrene concentration. Similar observation had been put forward by [15,13]. In order to compare strength of strains at different substrate concentration a normalized graph are constructed as shown in Figure 13 and 14 for single and mixed PAH respectively. Figures 13 and 14 indicate that all PAHs show increasing trend as concentration of substrate increased. This figures also show that at 100ppm of substrate concentration degradation of PAHs is more effective compared to 500ppm.

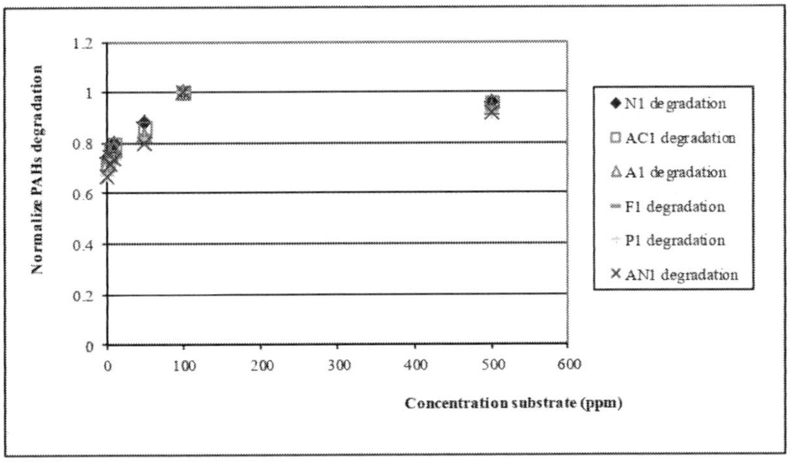

Figure 13. Normalize PAHs degradation for substrate concentration (single PAHs substrate)

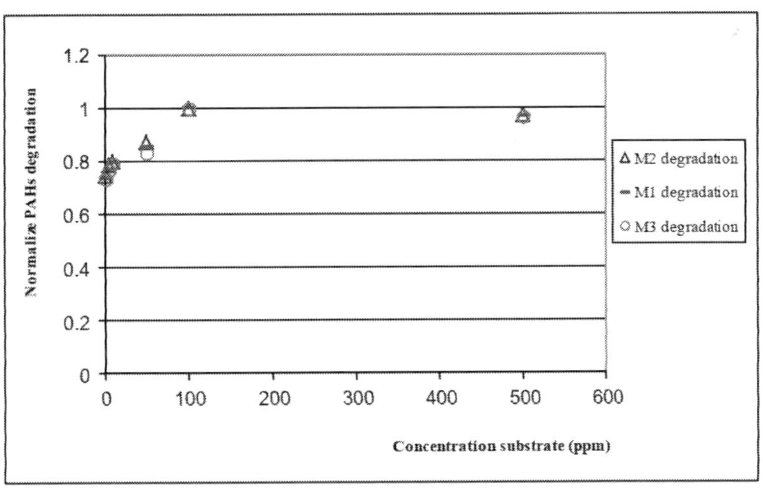

Figure 14. Normalize PAHs degradation for substrate concentration (mixed PAHs substrate)

In this optimization study PAHs act as a sole source of carbon and energy for bacterial growth. Thus, high concentration of substrate is valuable to cell development of the culture. The results from this study and previous studies also showed that too high concentration of substrate would become a limiting aspect for culture expansion.

Statistical analysis using ANOVA conducted on the data at a 5% level of significance showed that a significance different between concentration substrate 1ppm, 5ppm, 10ppm, 100ppm and 500ppm exist at 5% level of significance. Hence, higher concentration results in faster degradation process. level of significance. Hence, higher concentration results in faster degradation process.

4. Conclusion

Results from the optimization study of biodegradation indicated that maximum rate of PAHs removal occurred at 100 mgL^{-1} of PAHs, 10% bacteria concentration, pH 7.0 and 30°C. The bacteria isolated from single source of PAHs performed as well as bacteria isolated from mix PAHs.

References

[1] Harvey, R. G. (1991). *Polycyclic aromatic hydrocarbons: Chemistry and carcinogenicity*. Cambridge University Press. Cambridge, England.

[2] Salama, F. &. Allamandola, L. J. (1995). Polycyclic aromatic hydrocarbon ions and the diffuse interstellar bands. *Advances in Space Research*, 15(3)

[3] Mrozik, A., Piotrowska-Seget, Z. & Labuzek, S. (2003). Bacterial degradation and bioremediation of polycyclic aromatic hydrocarbons. *Polish Journal of Environmental Studies*. 12(1), pp.15-25.

[4] Yang, Y. & Hildebrand, F. (2006). Phenanthrene degradation in subcritical water. *Analytica Chimica Acta*. 555, pp.364-369.

[5] Moreno, A., Albertelli, A. & Fabiano, M. (2009). Nematode response to metal, PAHs and organic enrichment in mediterranean sea. *Marine Pollution Bulletin*, 58(8), pp.1192-1201

[6] Churchill, S. A., Harper, J. P., & Churhill, P. F. (1998). Isolation and characterization of *Mycobacterium* species capable of degrading three and four ring aromatic and aliphatic hydrocarbon. *Applied and Environmental Microbiology*, 65, pp.549-552.

[7] Yuan, S.Y., Wei, S.H., & Chang, B.V. (2000). Biodegradation of polycyclic aromatic hydrocarbons by a mixed culture. *Chemosphere*, 41, pp.1463-1468.

[8] Gaskin, S. & Bentham, R. (2005). Comparison of enrichment methods for the isolation of pyrene degrading bacteria. *International Biodeterioration and Biodegradation*, 56, pp.80-85.

[9] Yu, S.H., Ke, L., Wong, Y.S. & Tam, N.F.Y. (2005). Degradation of polycyclic aromatic hydrocarbons (PAHS) by a bacterial consortium enriched from mangrove sediments. *Environment International*, 31(2), pp.149-154.

[10] Othman, N., Hussain, N., Abdul- Talib, S. (2008). Extraction technique for determination of polycyclic aromatic hydrocarbons in municipal sludge. *Proceeding of Symposium on Urban Engineering and Sustainability*. Daejon, Korea. 30 March-1 April 2008, pp.357-364.

[11] Wan, H. B., Chi, H., Wong, M. K & Mok, C. Y. (1994). Solidphase microextraction using pencil lead as sorbent for analysis of organic pollutants in water. *Analytica Chimica Acta*, 298, pp.219-223

[12] King, A. J., Readman, J. W. & Zhou, J. L. (2001). The application of solid phase microextraction (SPME) to the analysis of polycylic aromatic hydrocarbons (PAHs). *Environmental Geochemistry and Health*, 25, pp.69-75.

[13] Romero, M. C., Cazau, M. C., Giorgieri, S. & Arambarri, A. M. (1998). Phenanthrene degradation by microorganism isolated from a contaminated stream. *Environmental Pollution*, 101, pp.355-359.

[14] Wong, J.W.C., Lai, K.M., Wan, C.K. Ma, K.K. & Fang, M. (2002). Isolation and optimization of PAH degradative bacteria from contaminated soil for PAHs bioremediation. *Water, Air and Soil Pollution*, 139, pp.1-13.

[15] Bouchez, M., Blanchet, D. & Vandecasteele, J. P. (1995). Degradation of polycyclic hydrocarbons by pure strains and by defined strains associations: Inhibition phenomena and cometabolism. *Applied Microbial Biotechnol*, 43, pp.156-164.

The Utilization of Banana Peel in the Fermentation Liquid in Food Waste Composting

A A Kadir[1], N A Rahman[1] and N W Azhari[1]
[1]Faculty of Civil and Environmental Engineering, Universiti Tun Hussein Onn Malaysia, 86400 Batu Pahat, Johor, MALAYSIA

E-mail: aeslina@uthm.edu.my

Abstract: Municipal solid waste in Malaysia contains a high amount of organic matters, particularly food waste. Food waste represents almost 60% from the total municipal solid waste disposed in the landfill. Food waste can be converted into useful materials such as compost. However, source separation of food waste for recycling is not commonly practiced in Malaysia due to various constraints. These constraints include low awareness among the waste generators and low demand of the products produced from the food waste such as composts. Composting is one of the alternatives that can be used in food waste disposal from Makanan Ringan Mas. The aim of the study is to convert food waste generated from Makanan Ringan Mas which is a medium sale industry located at Parit Kuari Darat, Batu Pahat by using composting method. The parameters which include temperature, pH value, NPK (Nitrogen, Phosphorus, Potassium) values has been examined. Banana peel is being used as the fermentation liquid whilst soil and coconut husk were used as the composting medium. Based on the results during the composting process, most of the pH value in each reactor is above 5 and approximately at neutral. This shown that the microbial respiration in the well controlled composting reactor was inhibited and had approached the mature phase. On the other hand, during the period of composting, the overall temperature range from 25 ° C to 47 ° C which shown the active phase for composting will occoured. As for NPK content Nitrogen value range is 35325 mg/L to 78775 mg/L, Phosphorus, 195.83 mg/L to 471 mg/L and potassium is 422.3 mg/L to 2046 mg/L which is sufficient to use for agricultural purpose. The comparison was made with available organic compost in the market and only showed slightly difference. Nevertheless, in comparison with common fertilizer, the NPK value of organic compost are considerably very low.

Keywords: Composting, food waste, fermentation liquid, banana peel.

1. Introduction

Composting practice is an alternative sustainable waste management practice to transform organic waste into valuable commodity. Composting benefits the environment because manure nutrients are converted to more stable forms and are less likely to reach groundwater or move in surface runoff [1].

Compost is valued for its organic matter content, and is typically used as a soil amendment to enhance the chemical, physical and biological properties of soil. Composting practice is an alternative sustainable waste management practice to transform organic waste into valuable commodity. Composting benefits the environment because manure nutrients are converted to more stable forms and are less likely to reach groundwater or move in surface runoff [1]. Compost is valued for its organic matter content, and is typically used as a soil amendment to enhance the chemical, physical and biological properties of soil.

Medium scale industry is specifically on food play an important role in developing the economy to establish, expand and strengthen the basic structure of industrial development of the country. There are more than 9000 food processing factories in Malaysia, of which 95% are classified as small-scale industry [2]. Many types of small scale industry such as frozen food and processed food are focusing on food production and generate food waste. One of the small scale industry in Parit Raja is Makanan Ringan Mas which is located at Parit Kuari Darat. Makanan Ringan Mas is a small scale industry that produced food products that come from the raw materials which are coconut, banana and tapioca peel. This small scale industry are facing the same problem and generated high amount of food waste for example coconut husk, tapioca peel, banana peel and others. Nevertheless, all the food waste management is not in a proper way because there are no facilities provided to manage the waste properly. At Makanan Ringan Mas, they generates very high level of processed and raw food waste daily and most of the waste were thrown in the river or by burning through open burning which leads to environmental pollution.

The goal of this study was to compost the food waste generated at Makanan Ringan Mas and determine the physical parameters which are temperature, pH and chemical parameters which are Nitrogen (N), Phosphorous (P) and Potassium (K). This study also to compare between NPK concentration produce from the composting process using banana peel as fermentation liquid with the chemical fertilizers.

2. Materials and Methods

2.1 Preparation of the starch fruit (banana peel) fermentation liquid

In order to fully utilize the food waste generated at Makanan Ringan Mas, the preparation for fermentation liquid is by using banana peel to replace fermented food and coconut candy to replace sugary water. Banana peel is an organic waste that highly rich in nutrient especially K, that could support the microbial growth in fermentation phase [3]. Using banana peel as fermentation liquid or waste in composting is considered as good as it stimulates soil microbial growth and activity with the subsequent mineralization of plant nutrients [4].

As replacement of brown sugar in common fermentation liquid, uses of candy which serves as food source for the microbes. Three (3) liters of water was mixed with 250 gram candy and 250 gram banana peel from Makanan Ringan Mas. Fermentation was occurred when bubbles starts to form, which normally occurs on the second day. Ideally, fermentation should take no longer than 7 days. Fermentation liquid was completed when the banana peel floats and the liquid settles at the bottom. During the fermentation process, the lid of the bottle should be loosened once a day to remove the gas that trapped in the bottle. The sweet and sour smell produced indicates that the fermentation liquid is successfully conducted.

2.2 Preparation of composting medium

Six (6) reactors were set up, one reactor act as control and other reactors were consisted of different types of food waste Each reactor was filled with 250 g of food waste daily according to the type of waste on the reactors as shown in Table 1.shows the flow research work and the phases of the research.

Table 1. Reactors with different types of food waste.

Reactor	Classification food waste	Types of food waste
x	Control	-
A	Processed food waste	Candy + chips
B	Raw food waste	Banana peel + tapioca peel + coconut husk
A+B	Processed food waste + Raw food waste	Candy + chips + tapioca peel + banana peel + coconut husk
C	Processed food waste + Raw food waste (Product A)	Candy + coconut husk
D	Processed food waste +Raw food waste (product B)`	Chips + tapioca peel + banana peel

2.3 Parameter Measurement

DR 6000 Spectrophotometer was used to analyze the concentration value of N and P, while Atomic Absorption Spectrophotometer (AAS) was used to get the concentration of K, by acoording to method in APHA Water and Wastewater. pH meter was used to measure the pH value of the reactor every week whether it is acid or alkali and thermometer was used to measure temperature every day.

3. Results and Discussions

3.1 Temperature

The temperatures of each reactor were measured before disposing any food waste into the reactor. The changes of temperature during the first month of the composting process. The highest temperature averages between 45°C to 47°C. The increase of temperature during the composting procedure is due to the high temperature generated by the microorganisms through respiration activities and decomposition of organic materials. The temperature of composted matter determines the rate of many biological processes and plays a basic role in microorganism succession understood as a change in the quantitative and qualitative composition of microorganism population [5].

The temperature pattern showed that there was a rapid rise from the initial mesophilic phase to the thermophilic phase for all reactors in the first three months, which lead to a high proportion of readily degradable substances. In that respect is no common definition of mesophilic and thermophilic, but more frequently than not refers to temperature up 40°C, and thermophilic temperatures are from 45° C up to 70°C. When the temperature reaches 40 °C to 50 °C, both phase microorganism takes place [6]. It is the active phase of composting.Most of the reactors reach the second phase in those three months of the composting process which the rates of temperature between 40 °C to 47 °C.

The most active reactor between these six months was reactor A that contain process of food waste (candy and chips) and reactor D (chips + tapioca peel +banana peel). This explains the findings that the microorganisms can withstand one extreme environmental factor, or which is high temperature [7].
Up till the final month of composting, most of the time, Reactor A and Reactor D always shows an increasing rate of temperature.

After the thermophilic phase which corresponds to a peak of degradation of fresh organic matter, the microbial activity decreases, as does the temperature especially in the Reactor B whereas less than control. This is called the cooling phase. This is supported by drop of the temperature below 40°C as the number of fungi in the composted mass increased. The compost maturation phase begins when the compost temperature falls to that of the ambient air.

3.2 pH value

The initial pH during week one among the six reactors is 5.10 to 7.32 as shown in Figure 1. In general, in the composting process, the pH falls below neutral in the beginning due to the formation of organic acid and later rise above neutral because the acids are consumed [8]. Most of the pH value in each reactor is above 5. The reason for this was the microbial respiration in the well-controlled composting reactor that was inhibited [8]. The highest pH obtained between six reactors was reactor B, when the composting process entered its thermophilic phase. The pH increase from week one followed by week two and three, however, by the end of week four, the value of pH decrease to 4.99. The decrease of pH value at the end of composting process may also obtained due to the increase of carbon dioxide (CO2) emissions from organic acids [8, 9].

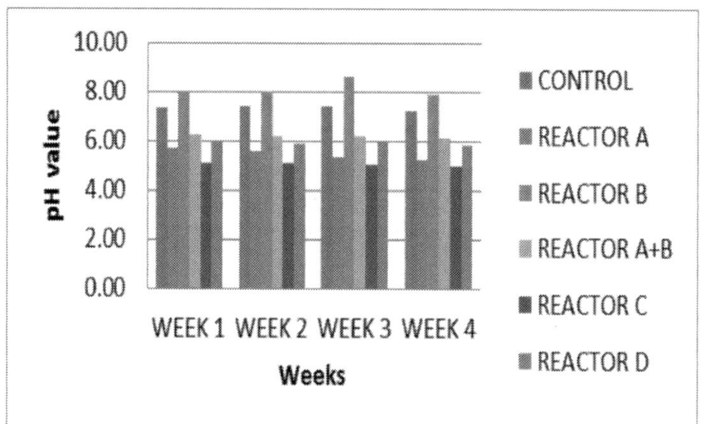

Figure 1. Changes in pH of composting mixtures against time.

3.3 Nitrogen (N) Concentration

Most of the reactors were not reached the matured phase of composting by week four especially for reactor A, reactor B and reactor D because the degradation of wastes is still taking place as shown in Figure 2. The mature phase which is the organic material continue to decompose and are converted to biologically stable humic substance In addition, the increase in the N value at the end of the composting period of all reactors might occur due to the usage of N by microorganism to build up cells, thus reducing the N, and some of the organisms will eventually die, which is recycled as N and thus contribute to increase the value [10].

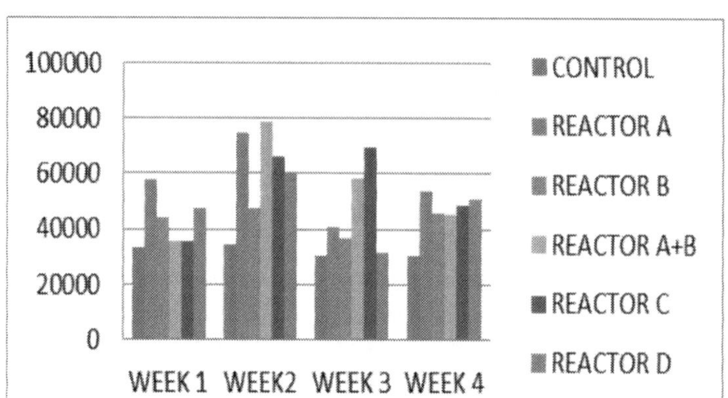

Figure 2. Nitrogen content in composting mixtures against time.

3.4 Phosphorus (P) concentration

In general, the trends of P concentration in the reactors are slightly increase within week four. P content in Reactor B gradually increased during the composting process as shown in Figure 3. The highest concentration was obtained in Reactor B and the lowest concentration was demonstrated in Reactor A. According to previous researcher, concentration of phosphorous increased during the composting process, and the water solubility of phosphorous decrease with humification so that the phosphorous solubility during the decomposition was subjected to further immobilization factors [11,12].

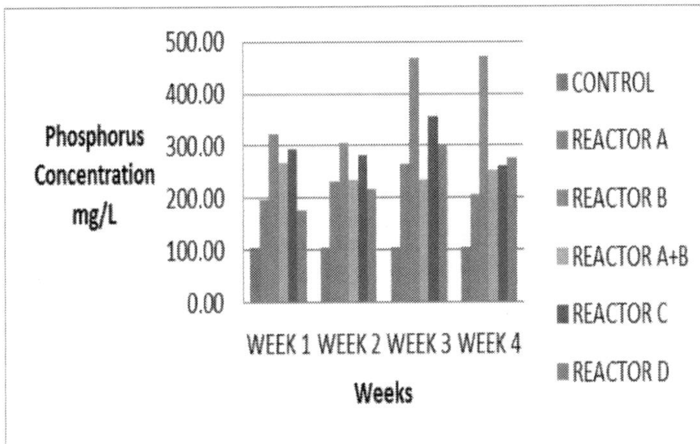

Figure 3. Phosphorus value in composting mixtures with weeks.

3.5 Potassium (K) concentration

Reactor A decrease from week one to week two but slightly similar from week three until week four. The same condition was demonstrated by Reactor A+B, Reactor C and Reactor D as shown in Figure 4. The concentration of K in Reactor B is slightly increased from week one until week four.This might due to high inherent content in banana peels [13].

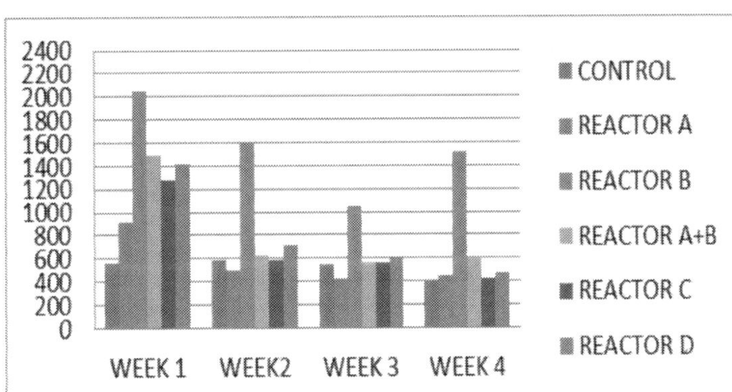

Figure 4. Potassium value in composting mixtures against time.

3.6 Comparison between organic compost and chemical fertilizer

Table 2 shows the comparison been NPK value from organic compost with chemical fertilizer. On the other hand, Table 3 shows the comparison of the organic compost from this study with the valueable organic fertilizer in the market. Nutrient compositions in chemical fertilizer are high to ensure the amount is suffiecient to assist plant to grow well. Chemical fertilizers are produced synthetically from inorganic materials. Since these chemical fertilizers are prepared from inorganic materials artificially, it may have some harmful acid that could affect the plant growth. As for the food waste organic fertilizer, the values shown potential as this value will be increased once all the food waste are fully degrade in the compost and matured enough to become fertilizer. Nevertheless. The comparison between organic compost and fertilizer shows similar results.

Table 2. Percentage of N-P-K value of the food waste composting.

Sample	N (%)	P (%)	K (%)
Reactor A	3.2	0.02	0.07
Reactor B	5.7	0.04	0.16
Reactor A+B	4.3	0.02	0.08
Reactor C	5.5	0.03	0.07
Reactor D	0.5	0.02	0.08
Chemical Fertilizer [14]	20	40	25

Table 3. Percentage of N-P-K values for organic compost.

Organic materials	N(%)	P(%)	K(%)
Cow dung	1.30	0.58	2.15
Cassava peeling compost	1.70	0.86	1.50
Rabbits dropping	1.04	0.99	2.05

Source: [15]

4. Conclusions

In this composting study, temperature and pH value during the composting process was found i reach the level of the proposed range of 40°C to 47°C, 5.1 to 8.6 and surpassed 90%. N content ranges from 35325 mg/L to 78775 mg/L, P from 195.83 mg/L to 471 mg/L and K from 422.3 mg/L to 2046 mg/l. Reactor B showed the most effective composting process in this composting study. Reactor B which consists of banana peel waste is easily decomposed and the nutrient produced in compost is much faster and higher than others. This is probably due to the oxygen and supplementary nutrient supply that could promote specific growth of microorganism essential for mineralization of organic substance. High K and N content in banana peel compost reactor is to assist in obtaining high potential K and N nutrient source in the soil. The concentration of N showed increased slightly between each reactor by week. On the other hand, the concentration of P and K are more focus on reactor B that are greater than other reactor. The different concentration indicated that presence of food waste added will differ amount of NPK due to microbial activity that produce microbes. There is a potential for this organic food waste from this composting study to be used as an organic fertilizer. Eventhough the NPK value is considerably low in comparison to chemical fertilizer, the NPK value from this study were found to be comparable with organic fertilizer. The results indicate that the value will be increased steadily as the compost process is fully matured.

References

[1] Larney F. J., Sullivan D. M., Buckley K. E., and Eghball B., (2006). The role of composting in recycling manure nutrients. vol 1.

[2] Senik G (2006). Small-Scale Food Processing Enterprises In Malaysia. pp. 1–9.

[3] Essien J. P., Akpan E. J., and Essien E. P.. (2005). Studies on mould growth and biomass production using waste banana peel. Bioresour. Technol., 96(13), pp. 1451–1456.

[4] Cercioglu M., Delibacak S., and Riza A.. (2014). Changes in physical conditions of a coarse textured soil by addition of organic wastes. vol. 3.

[5] Wolna-Maruwka, J. Dach, and a Sawicka. (2009). Effect of temperature on the number of selected microorganism groups and enzymatic activity of sewage sludge composted with different additions in cybernetic, Agron. Res., 7(2) pp. 875–890.

[6] Peigné J. and Girardin P. (2004). Environmental Impacts of Farm-Scale Composting Practices. Water, Air, Soil Pollut., 153(1–4), pp. 45–68.

[8] Sundberg C., Smårs S., and Jönsson H. (2004). Low pH as an inhibiting factor in the transition from mesophilic to thermophilic phase in composting. Bioresour. Technol., 95(2), pp. 145–150.

IOP Conf. Series: Materials Science and Engineering **136** (2016) 012055 doi:10.1088/1757-899X/136/1/012055

[7] Sundberg C., Yu D., Franke-Whittle I., Kauppi S., Smårs S., Insam H., Romantschuk M., and Jönsson H. (2013). Effects of pH and microbial composition on odour in food waste composting, Waste Manag, 33(1), pp. 204–211.

[9] Saithep, N., Dheeranupatana, S., Sumrit, P., Jeerat, S., Boonchalearmkit, S., Wongsanoon, J., and Jatisatienr, C., (2009). Composting of Tobacco Plant Waste by Manual Turning a Forced Aeration System. Journal of Science and Technology. 3(02): 248-260.

[10] Jusoh M. L. C., Manaf L. A., and Latiff P. A.. (2013). Composting of rice straw with effective microorganisms (EM) and its influence on compost quality. Iranian J. Environ. Health Sci. Eng., 10(1), p.p 17.

[11] Shymala, D.C., Belagali, S.L., (2012). Studies on Variations in Physico-Chemical and Biological Characteristics at Different Maturity Stages of Municipal Solid Waste Compost. International Journal of Environmental Sciences. 2(4): 1984-1997.

[12] Janakiam, T., and Sridev, K., (2010). Conversion of Waste into Wealth: A Study in Solid Waste Management. E-Journal of Chemistry. 7(4): 1340-1345.

[13] Kalemelawa F., Nishihara E., Endo T., Ahmad Z., Yeasmin R., Tenywa M. M., and Yamamoto S. (2012), An evaluation of aerobic and anaerobic composting of banana peels treated with different inoculums for soil nutrient replenishment. Bioresour. Technol., 126(375–82).

[14] Massri M. and Labban L. (2014). Comparison of Different Types of Fertilizers on Growth, Yield and Quality Properties of Watermelon (Citrllus lanatus). pp. 475–482.

[15] Adeniyan O. N., Ojo A. O., and Adediran J. A. (2011). Comparative study of different organic manures and NPK fertilizer for improvement of soil chemical properties and dry matter yield of maize in two different soils. 2(1). pp. 9–13.

Particulate Matter Levels in Ambient Air Adjacent to Industrial Area

R M S R Mohamed[1], N M S Nizam[1], A A Al-Gheethi[1], A Lajis[2] and A H M Kassim[1]
[1]Faculty of Civil and Environmental Engineering, Universiti Tun Hussein Onn Malaysia, 86400 Batu Pahat, Johor, MALAYSIA.
[2]Faculty of Mechanical and Manufacturing Engineering, Universiti Tun Hussein Onn Malaysia, 86400 Batu Pahat, Johor , MALAYSIA

E-mail: maya@uthm.edu.my

Abstract: Air quality in the residential areas adjacent to the industrial regions is of great concern due to the association with human health risks. In this work, the concentrations of particulate matter (PM_{10}) in the ambient air of UTHM campus was investigated tostudy the air qualityand their compliance to the Malaysian Ambient Air Quality Guidelines (AAQG). The PM_{10} samples were taken over 24 hours from the most significant area at UTHM including Stadium, KolejKediamanTunDr. Ismail (KKTDI) and MakmalBahan. The meteorological parameters; temperature, relative humidity, wind speed and wind direction as well as particulate matterwere estimated by using E-Sampler Particulate Matter (PM_{10}) Collector. The highest concentrations of PM_{10} (55.56 µg/m^3) was recorded at MakmalBahan during the working and weekend days. However, these concentrations are less than 150 µg/m^3. It can be concluded that although UTHM is surrounded by the industrial area, the air quality in the campus still within the standards limits.

Keywords: Air pollution, particulate matter.

1. Introduction

The economic growth in Malaysia depends on the manufacturing industries, especially electronics, chemicals and rubber. The developments in industrial sector has increased from 38.5% in 1980 to 44.5% in 2002. This developments has contributed in increasing of the gross domestic product (GDP) from 19.64% in 1980s to 31.56% in 2004 [1]. The electrical and electronics manufacturing sector has enhanced the Malaysian economicby more than 60% during the period from 1991 to 1996 [2]. However, most of industrial activities including the large combustion installations burning fuel oil or coal; cement factories; waste incinerators and road traffic, wood industries and others are correlated with the increasing of air pollution [3].

According to Tiwari et al. [4], the air quality in the developing countries has decreased due to the industrial growth, fossil fuel combustion, rapid urbanization and population growth that lead to the adverse effects on the climate. There are many of industrial areas around the world with a poor air quality due to coal combustion method in their industrial processes. For instance, Turkey is one of the most industrialized area with high PM_{10} concentration ranged from 102.3µg/m^3 during winter to 59.9µg/m^3 during summer [5]. In India, the PM_{10} concentrations is varied depended on the season. Certain areas such as Delhi has particulate concentrations as high as 435 µg/m^3 [6].

In order to protect the public health from the adverse effects of air pollutants, several countries have adopted regulations for air quality standards based on Particulate Matter (PM_{10}). In Malaysia, the Ambient Air Quality Guidelines (AAQG) has recommended that the PM_{10}, should be less than 150 µg/m^3/ 24 hours or less than 50 µg/m^3/ 12 months. Some countries such as Australia recommended more stringent standards where,

PM_{10} should be less than 50 μg/m^3/ 24 hours and 0 / 12 month. In Japan, PM_{10} should be between 100-200 μg/m^3/ 24 hours and 0 / 12 months (Table 1).

In Malaysia, the highest concentrations of PM_{10} was recorded in Penang near industrial areas. The annual average PM_{10} concentrations were 61.73 μg/m^3 in 2001, 75.03 μg/m^3 in 2002, 80.13 μg/m^3 in 2003, 92.31 μg/m^3 in 2004, 78.99 μg/m^3 in 2005, 49.81 μg/m^3 in 2006 and 45.45 μg/m^3 in 2007 [7]. However, those concentrations are still within AAQG standards.

The high concentrations of PM_{10} leads to several health concerns. Therefore, the current work aimed to quantify the PM_{10} in Universiti Tun Hussein Onn Malaysia (UTHM) campus to study the air quality and their compliance to AAQG.

Table 1. Ambient Air Quality Standard (AAQS).

Country Name	**AAQS Particulate Matter (PM_{10})**	
	Average time	**Guidelines (μg/m^3)**
Malaysia	24 Hour	150
	1 Year	50
Australia	24 Hour	50
	1 Year	None
China	24 Hour	150
	1 Year	None
European Union	24 Hour	50
	1 Year	35
Hong Kong	24 Hour	100
	1 Year	9
Japan	24 Hour	100 or 200
	1 Year	None
United States	24 Hour	150
	1 Year	1

2. Materials and Methods

2.1 Study area

University Tun Hussein Onn Malaysia (UTHM) is a public university established on 2007 and located at Parit Raja, Batu Pahat, about 100 km from Johor Baru, Malaysia (see Figure 1.)

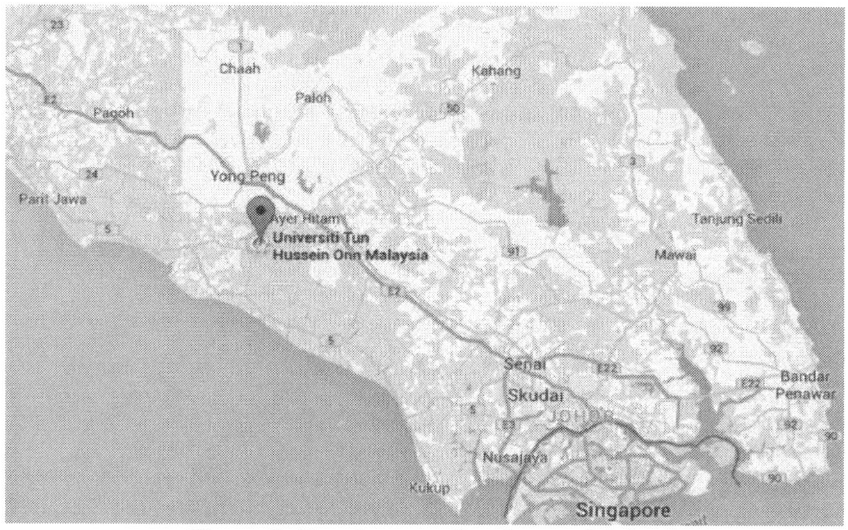

Figure 1. Google map of study area Malaysia.

Three monitoring stations included, the Stadium (Station 1), Kolej Kediaman Tun Dr. Ismail (KKTDI) (Station 2), and Makmal Bahan (Station 3) were selected for measuring the particulate matter, PM10 in UTHM campus. These locations were selected because they represent recreational areas, residential colleges and the laboratories which are significant points for estimating air quality in UTHM. Figure 2. shows the UTHM locations and sampling stations distance from the Parit Raja industrial areas.

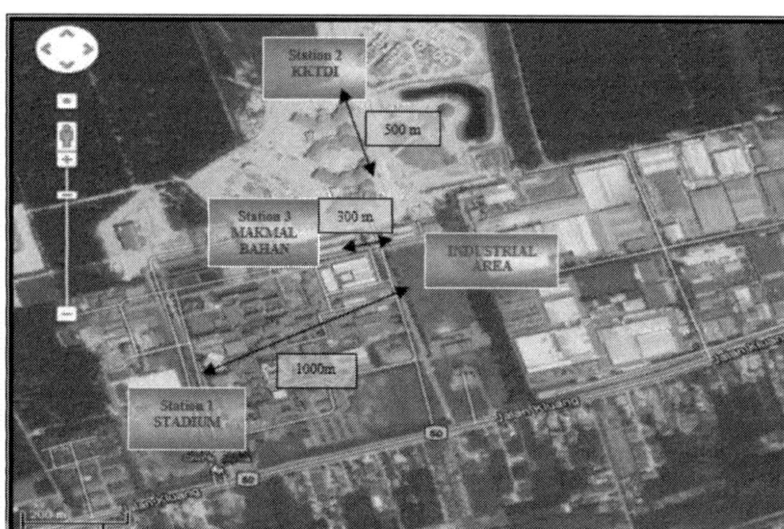

Figure 2. The study area (Observed on (December, 2015) googlemap.com).

2.2 Estimation of PM_{10}

The E-Sampler Particulate Matter (PM_{10}) Collector was used in this study (see Figure 3.). It was placed in open areas away from trees and buildings. Humidity sensor and a combined set for wind speed and wind direction was directed to the sky. The data of particles in the air, humidity, wind direction and wind speed as well as the ambient temperature from sampling area was recorded by E-Sampler. The filter paper was used for taking the particulate matter samples from ambient air. Filter paper was weighing before and after the sampling. The data was recorded over 24 hours. The concentration PM_{10} from E-Sampler was compared with data taken from the filter paper. The PM_{10} was estimated during working days and weekend day for three weeks between Jun to December 2011. The sampling was performed at least once a week at each stations.

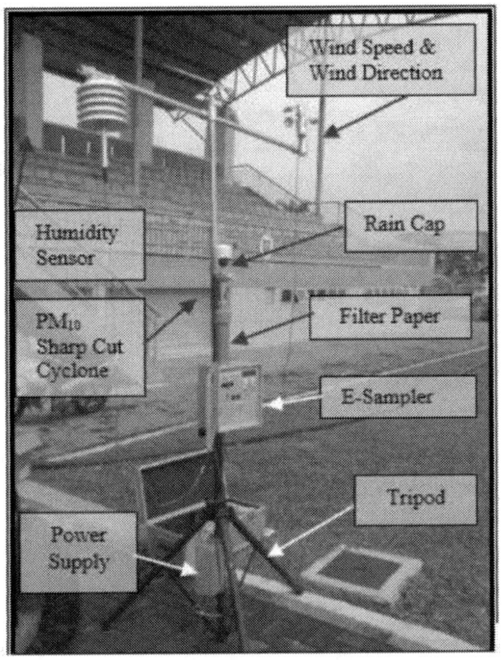

Figure 3. E-Sampler Particulate Matter (PM_{10}) Collector.

3. Results and Discussions

Air quality in urban areas is of great concern due to the association with human health risks [6]. Knowledge of the distribution and sources of $PM_{10}$is essential in understanding and establishment of a control strategy. According to AAQG six parameters namely carbon monoxide, nitrogen dioxide, ozone, sulphur dioxide and PM_{10} are the criteria for the air quality and should be determined to estimate the air pollutants. However, PM_{10} is the most important ambient air polluters which harmfully affect human health [8]. PM_{10} is used as a primary indication of air pollution because It is associated with the exacerbation of respiratory and cardiovascular conditions [9,10]. Therefore, this study focused on determination of PM_{10} concentrations. The increasing of PM_{10} to more than the standards limits recommended by AAQG should be followed by measuring the above criteria. In the present work, the mean of PM_{10} concentrations in the selected locations during the working days are presented in Figure 4.

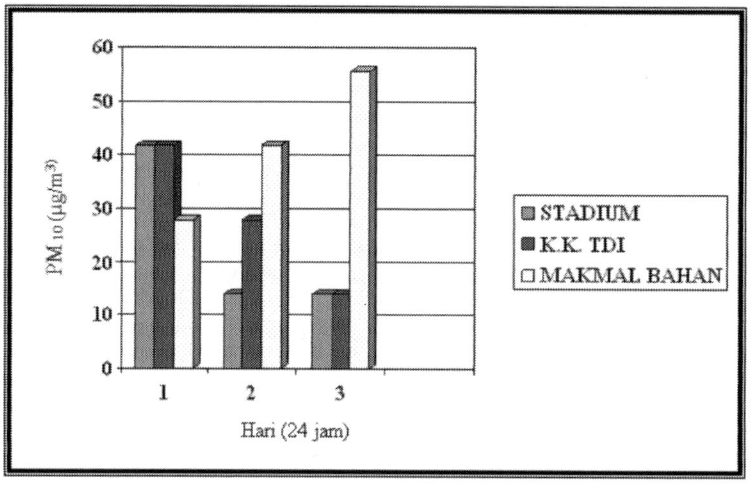

Figure 4. Concentration of PM_{10} in the selected locations during working days.

It can be noted that the highest PM_{10} concentrations were recorded at Makmal Bahan (55.56 µg/m³), while the lowest PM_{10} concentrations were observed at Stadium during the second and third week (12.03 µg/m³). In the first week, the estimation of air pollution at Stadium and KKTDI stations recorded higher PM_{10} concentrations than Makmal Bahan, 42 vs. 28 µg/m³.Air quality at KKTDI station exhibited the lowest PM_{10} concentrations during the third week (12.03 µg/m³), while were 28µg/m³ in the second week.

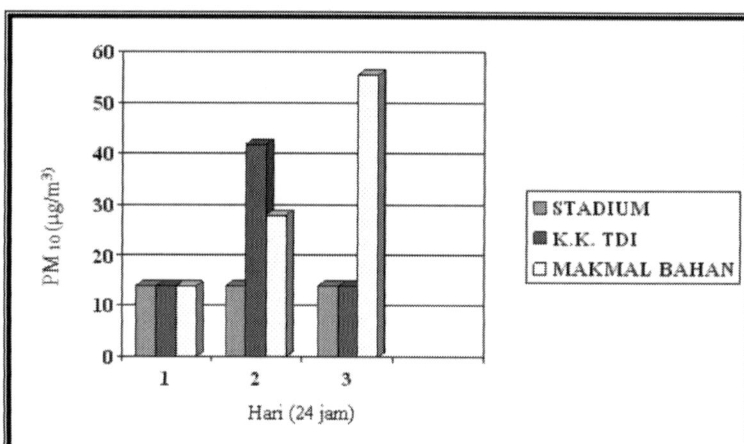

Figure 5. Concentration of PM_{10} in the selected locations during weekend days.

However, KKTDI station occurred the highest PM_{10} during the second week (42µg/m³). PM_{10} did not differ significantly during the first week among three stations (15µg/m³). In the comparison with AAQG, these concentrations are within the standard limits. Besides, the PM_{10} concentrations were quite less than the annual mean PM_{10} concentrations in India [6], Egypt [11], Eastern Mediterranean and Africa [12].

The differences in PM_{10} among the three stations might be due to the distance from the industrial area. Makmal Bahan is the nearest station of industrial area (300 m), while Stadium station is located 1000m from the industrial area. On the other hand, the differences between PM_{10}concentrations during the working and weekend days would be related to the meteorological factors. Therefore, the correlation between meteorological factors including temperature, humidity, wind direction and wind speed was analysed based on Pearson correlation coefficient ($p<0.05$).

At Stadium station, temperature and wind speed correlated significantly with PM_{10} concentrations during the working and weekend days. At Makmal Bahan station wind speed also correlated significantly with PM_{10} concentration during the working and weekend days, while temperature and humidity correlated significantly with PM_{10} concentrations during the weekend days only. Humidity correlated significantly with PM_{10} concentration at KKTDI station during the working days, in the contrast, wind direction correlated with PM_{10} concentration during the weekend days. The results demonstrated effect of meteorological factors on PM_{10} concentrations. These findings are consistent with Barmpadimos et al. [13] who indicated that there is a correlation between meteorological parameters and PM_{10} concentrations.

4. Conclusion

It can be concluded that although UTHM is surrounded by the industrial area, the air quality within the campus still within the standards limits. The meteorological factors play important roles in the PM_{10} concentrations. However, sampling station nearest to the industrial area, which was located at Makmal Bahan has shown a highest value of $PM_{10.}$

Acknowledgments
The authors gratefully acknowledge the UniversitiTun Hussein Onn Malaysia for the research project financial support under UTHM contract grant Vot. U396.

References

[1] Kamaruddin, R., Masnon, T. A. (2010) sources of growth in the manufacturing sector in Malaysia: Evidence from Ardl and Structural Decomposition Analysis. Asian.
[2] Asid, R. (2010) The technical efficiency analyses of manufacturing sector in Malaysia: Evidence from the First Industrial Master Plan (1986-1995). *Asian Social Science*, 6 (2).99-107.
[3] Jaecker-Voirol, A. Pelt, P. (2000) PM_{10} emission inventory in Ile de France for transport and industrial sources: PM_{10} re-suspension, a key factor for air quality. *Environmental Modelling & amp; Software* 15(6-7), 575-581.
[4] Tiwari, S., Chate, D. M., Srivastava, A. K., Bisht, D. S., Padmanabhamurty, B. (2012) Assessments of PM1, PM2.5 and PM10 Concentration in Delhi at Different Mean Cycles. Geofizika, 29, 125- 141.
[5] Pekey, B., Bozkurt, Z. B., Pekey, H., Doğan, G., Zararsiz, A., Efe, N., Tuncel, G. (2009) Indoor/outdoor concentrations and elemental composition of $PM_{10}/PM_{2.5}$ in urban/industrial areas of Kocaeli City, Turkey. *Indoor Air* 20(2), 112–125.
[6] Ravindra, K., Mittal, A.K., Grieken RV. (2001) Health Risk Assessment of Urban Suspended Particulate Matter with Special Reference to Polycyclic Aromatic Hydrocarbons: A Review. *Review on Environmental Health*, 16, 169-189.
[7] Ul-Saufie, A.Z., Yahaya, A. S, Ramli, N. A., Hamid, H.A. (2012) Future PM_{10} concentration prediction using quantile regression models. *2nd International Conference on Environmental and Agriculture Engineering IPCBEE,* 37, 15-19. IACSIT Press, Singapore
[8] Koelemeijer, R. B. A., Homan, C. D., &Matthijesen, J. (2006) Comparison of special and temporal variations of aerosol optical thickness and particulate matter over Europe. *Atmospheric Environment,* 40(27), 5304–5315.
[9] Kappos, A. D., Bruckmann, P., Eikman, T., Englert, N., Heinrich, U., Hoppe, P. (2004). Health effects of particles in ambient air. *International Journal of Hygiene and Environment Health*, 207(4), 399–407.
[10] WHO (2006) Air Quality Guidelines Global Update 2005: Particulate matter, ozone, nitrogen dioxide and sulfur dioxide. Geneva: World Health Organization.
[11] Safar Z.S., Labib M.W. (2010) Assessment of particulate matter and lead levels in the Greater Cairo area for the period 1998–2007. Journal of Advanced Research, 1, 53–63.
[12] WHO (2014) Public health, environmental and social determinants of health (PHE), Ambient (outdoor) air pollution in cities database 2014.
[13] Barmpadimos, I., Hueglin, C., Keller, J., Henne, S., Prevot A. H. S. (2011) Influence of meteorology on PM10 trends and variability in Switzerland from 1991 to 2008. *Atmos. Chem. Phys.*, 11, 1813-1835.

Food Waste Composting Study from Makanan Ringan Mas

A A Kadir[1], S N M Ismail[1] and S N Jamaludin[1]
[1]Faculty of Civil and Environmental Engineering,University Tun Hussein Onn Malaysia, 86400 Batu Pahat, Johor, MALAYSIA

E-mail: aeslina@uthm.edu.my

Abstract: The poor management of municipal solid waste in Malaysia has worsened over the years especially on food waste. Food waste represents almost 60% of the total municipal solid waste disposed in the landfill. Composting is one of low cost alternative method to dispose the food waste. This study is conducted to compost the food waste generation in Makanan Ringan Mas, which is a medium scale industry in Parit Kuari Darat due to the lack knowledge and exposure of food waste recycling practice. The aim of this study is to identify the physical and chemical parameters of composting food waste from Makanan Ringan Mas. The physical parameters were tested for temperature and pH value and the chemical parameter are Nitrogen, Phosphorus and Potassium. In this study, backyard composting was conducted with 6 reactors. Tapioca peel was used as fermentation liquid and soil and coconut grated were used as the fermentation bed. Backyard composting was conducted with six reactors. The overall results from the study showed that the temperature of the reactors were within the range which are from 30° to 50°C. The result of this study revealed that all the reactors which contain processed food waste tend to produce pH value within the range of 5 to 6 which can be categorized as slightly acidic. Meanwhile, the reactors which contained raw food waste tend to produce pH value within the range of 7 to 8 which can be categorized as neutral. The highest NPK obtained is from Reactor B that process only raw food waste. The average value of Nitrogen is 48540 mg/L, Phosphorus is 410 mg/L and Potassium is 1550 mg/L. From the comparison with common chemical fertilizer, it shows that NPK value from the composting are much lower than NPK of the common chemical fertilizer. However, comparison with NPK of organic fertilizer shown only slightly difference value in NPK.

Keywords: Food waste, backyard composting, npk, organic compost, tapioca peel.

1. Introduction

According to the Solid Waste Management Act 2007, Act 672[1], solid waste is defined as any unwanted substance; material or product that is broken, contaminated or worn out that required by the authority to be disposed. In Malaysia, landfill is the most common method of solid waste disposal and up to 50% of landfill is containing with food waste. Inarguably, the generation of Municipal Solid Waste is highly dependable on the food waste generation. Based on research by Saeed, et.al [2], the food waste generation rate is up to 60% from MSW. Basically food waste is generated from many different sources. Agriculture, household, commercial, institutional, industrial have all contributes to the food waste production.

Essentially, all of the mentioned sources are producing food waste but in different quantity. Therefore it is highly important to also include MSW management plan for all these different sources. There are many options in disposing food waste. There are by composting, sanitary landfills and incineration. However, this study is only focusing on composting.

Composting is a natural process that turns food waste into compost that is rich with nutrients for the soil. Composting the food wastes is able to reduce the food waste being disposed in landfill and also help the food industries in managing their wastes. As a result, the new organic fertilizer can be made. Li et.al.,[3] stated that composting helps to improve the soil structure and fertility. According to Chang and Hsu [4], the chemical changing process and the complex metabolic processes of various microorganisms vary with the composition of the material composting. Particularly, the outcome of the composting is depending on the waste characteristic.
Composting is an easy, economical and natural biodegradation process that takes organic waste usually food waste and turns into nutrient rich food for plants. Composting is a biochemical process converting various components in organic waste into relatively stable humus-like substances that can be used as a soil amendment or organic fertilizer [3].

Composting usually used for organic farming, occurs by allowing organic waste to sit in one place for months until microbes decompose it. Composting is one of the best method of waste disposal as it can turn unsafe organic into safe compost. Nevertheless, disadvantages of composting are it is a slow process and takes a lot of space. According to Hanc, Szakova, & Ochecova [5], the advantages of composting are that it minimizes damage to the environment and produces economically valuable products from the bio-waste.
Moreover, as an alternative waste disposal method, composting can divert waste from landfill, mitigate groundwater contamination, reduce air pollution and greenhouse gas (GHG) emissions, and generate useful products. Therefore, the studies related to composting processes have attracted more attentions of many researchers.

In this study, Makanan Ringan Mas industry have been chosen as sampling area because the area is far away from the main road, so the solid waste management in the area needs to be done own their own without collection and facilities provided. Solid waste generation from Makanan Ringan Mas Industry is mostly food waste. The food waste were burned and dumped into the river by the workers and that affect our environmental sanitation thus causing air and water pollution.

Due to the improper waste management at Makanan Ringan Mas Industry, investigation on food waste composting as an alternative disposal method was in need to be conducted to obtain a good practice of food waste management.

2. Materials and Methods

2.1 Collection of food waste

Food waste samples were collected from Makanan Ringan Mas, a small and medium industry producing food waste such as chips, coconut candy, coconut grated and fruit peels. It is located approximately 13.5 km away from Universiti Tun Hussein Onn Malaysia (UTHM). The collected food wastes were tapioca peel, banana peel, banana chips, coconut candies and grated coconut. The food wastes were collected twice a week in 6 months period throughout the study and the collected food wastes were weighed accordingly.

2.2 Preparation for fermentation liquid

For this study, tapioca peels were used as fermented food as a substitute for fermented soybeans product that are normally used and coconut candy were used to replace brown sugar. The tapioca peel were mixed with 250 gram coconut candy with 3 liter of water for 5 to 7 days to produce the fermentation liquid. The container was closed loosely because of the gases that produced during the fermentation process. The sweet and sour smells produced indicate that the fermentation liquid is a success.

2.3 Preparation for fermentation bed

The material that was used for the fermentation bed was grated coconut waste from the industry. Makanan Ringan Mas small scale industry has always producing a lot of grated coconut. To fully utilize the waste produce at Makanan Ringan Mas, the grated coconuts were used as the fermented bed material. To make the fermented bed, 2 kg soil and 1 kg grated coconut were mixed. The grated coconuts were used to substitute the rice husks as nitrogen source and soil as the carbon source in the fermented bed.

2.4 Preparation of the reactors

Baskets were used as reactors to carry out the process of composting. The baskets were covered with carpet to prevent the compost spilled and insect infestations. The reactors were then covered with breathable fabric for ventilation. Each reactor was labelled according to the type of food waste as shown in Table 1.

Table 1. Classification and types of food waste

Reactor	Classification of food waste	Types of food waste
CONTROL	Control	-
A	Processed food waste	Chips + candy
B	Raw food waste	Banana peel + tapioca peel + grated coconut
A+B	Processed food waste + raw food waste	Candy + chips + tapioca peel + banana peel + grated coconut
C	Processed food waste + raw food waste (Product A)	Candy + grated coconut
D	Processed food waste + raw food waste (Product B)	Chips + tapioca peel + banana peel

Each reactor was filled with 3 kg fermented beds that have been mixed with fermentation liquid. The reactors were covered with breathable fabrics for 3 to 5 days under room temperature (27°C). In addition, the reactors were placed indoor at Micropollutant Research Center (MPRC) to avoid any water from outside from entering the reactors.

After 5 days, food wastes were added into the reactors. The food wastes were cut into smaller sizes to make it easier to compost. Afterwards, the mixtures were stirred once a day to allow air to improve the composting process. When the fermented bed started to dry, water was added to ensure the right moisture for the bacteria in the reactor. The compost process would take up 3 to 6 months to become matured and could be used as organic fertilizer.

2.5 Physical and chemical parameters

All the compost samples from each reactor in this study were tested for physical and chemical parametes every week with two replicate of samples for every reactors.

In physical parameter, the pH value, and temperature from each reactor was measured. The pH values of the compost were determined by mixing the compost into distilled water. Ten gram of waste was weighed and 25 mL distilled was measured and put in a beaker. The samples were stirred for 10 minutes. After 3 minutes or when the samples have settled, the samples were tested with pH meter to get the reading.

The temperature was taken every day for each reactor to record the changes in temperature during composting. Each reactor temperature was taken three times at different places to have the average value of each of the reactors.

The chemical parameters were carried out according to standard APHA method. The chemical parameter including Nitrogen, Phosphorus and Potassium were conducted in this study. Nitrogen and phosphorus concentration was analyzing using DR6000 Spectrophotometer, while potassium concentration was analyzing using Atomic Absorption Spectroscopy (AAS).

3. Results and Discussions

3.1 Analysis on pH value

In this study, the pH values were measured during 5 month of composting process. It is to be highlighted from the Figure 1, the reactor A, C and D was below pH 7 which indicating that the waste samples were acidic. Acid pH values indicate a lack of maturity due to short composting time or occurrence of anaerobic conditions. The waste in reactor A contained processed food waste which are chips and candy. Reactor C contained processed and raw waste which are chips, candy, tapioca peel, banana peel and grated coconut waste. Meanwhile, Reactor D contained waste candy and grated coconut. Based on the result, the pH value for reactor control is neutral between 7 and 8. It is because there is no waste insert in the reactor. As for Reactor B, the pH value was in range of 7 and 9. This is because the wastes are raw food wastes which are only tapioca peel and banana peel.

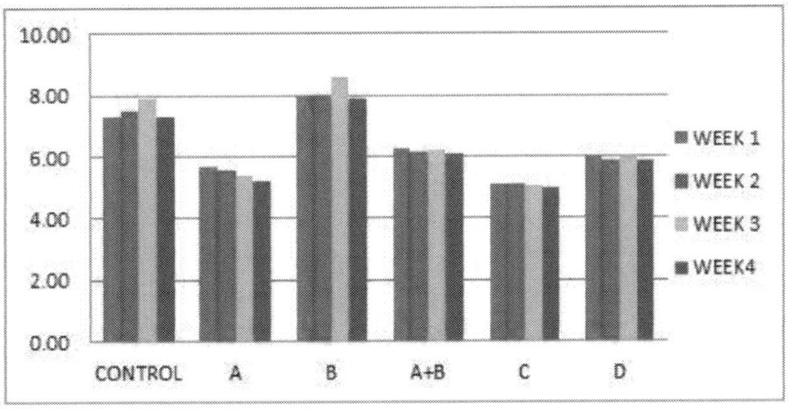

Figure 1. Analysis on pH value of every reactor.

3.2 Analysis on temperature

The temperature of composting for different reactors for the first month of the study increase significantly besides reactor control. At 20th November, the temperature recorded was the highest with 48°C. This indicates that the microorganisms such as bacteria and fungi are responsible on heat generation during composting cause in lost moisture content [6]. However, then the temperature began to fall slightly. As for the reactor control, the temperature did not have major changes because there is no waste added.

The temperatures were observed when Tapioca peel is utilized in the fermentation liquid from December 2014 to February 2015. Based on previous study by Tweib et al, [6], the initial temperature was 26.5°C and increased up to 58°C. Above all, all the temperature showed the same pattern. Based on these figures, the average temperature of the compost in all the reactors was between 30°C and 52°C from December 2014 until February 2015. There were certain weeks that the recorded temperature increases dramatically because each reactor was sprayed with water once a week. The water was added to the waste to attain the desired moisture level. Apparently, the waste did not produce enough moisture in order for the microbial activity to take place. The food waste are being disposed and mixed every day.

The temperature continued to show the same pattern for all the months except in March, 2015 where the temperature was ranging from 26°C to 35°C. In March, all the activity including waste inserted, water added and compost mixture were stopped to allow the compost in each reactor to mature and age. The drop in temperature also was mentioned in a study by Fatin et. al, [7] which stated that the temperature decrease as the activity of the microorganisms slowed, indicating that the compost was nearing its mature phase.

3.3 Nitrogen

Figure 2 shows the result of nitrogen for four weeks at the 5th month of composting. The graph shows the comparison of Nitrogen between the food waste reactors with the control. Reactor A was inserted with processed food waste from Makanan Ringan Mas such as chips and coconut candy. It showed that the nitrogen decreased steadily until it rises up at week four with 56000 mg/L. Meanwhile, Reactor B was inserted with raw food waste which is banana and tapioca peel. This reactor showed that the nitrogen value increased slowly and then significantly increased at the fourth week. Reactor B has the highest Nitrogen value by 96200 mg/L compared to other reactors. Reactors A+B was inserted with mixture of all food waste. The results showed that the nitrogen value fluctuated but as for the fourth week, the reactor continued to increase with the recorded value of 71400 mg/L. As for Reactor C, the inserted wastes were based on all types of coconut waste from Makanan Ringan Mas Industry. The graph was skewed-left. The highest Nitrogen recorded was at the third week which is 57200 mg/L. Reactor D was inserted with banana and tapioca wastes. Based on the graph, it showed that the result pattern was increased significantly until week four. The highest nitrogen value recorded was 57000mg/L.

As a conclusion, the pattern of nitrogen value on all of the reactors showed the highest value of nitrogen in week four except for reactor C. Reactor C obtained the highest nitrogen value was in week three. According to Haung et. al 2004, the reactors that have highest Nitrogen at the last week because, the Nitrogen fixing bacteria might also have contributed to the increase in Nitrogen in the later stage of composting.

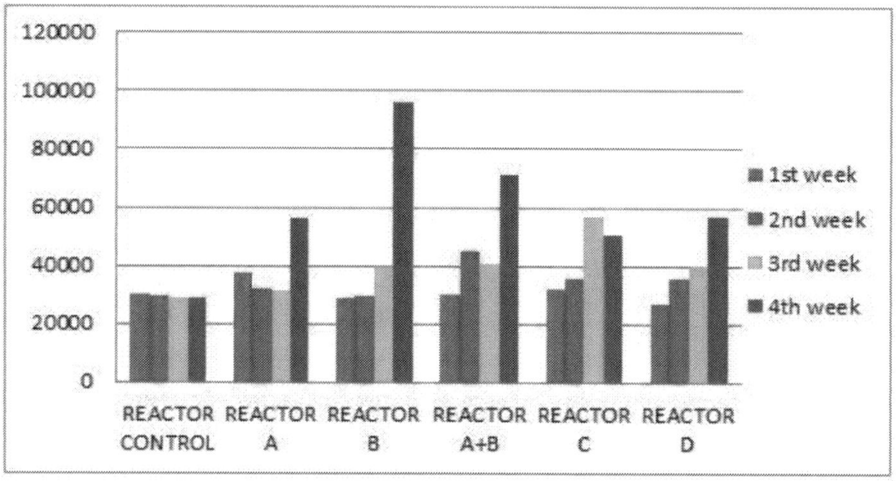

Figure 2. Nitrogen result.

3.4 Phosphorus

The result from the study indicate that the Reactors A and C shows a high level of phosphorus at the third week. However, for the whole four weeks, reactor B produced the highest phosphorus compared to other reactors. Reactor A+B did not show any big difference for the whole four weeks. On the other hand, reactors D results shows that the Phosphorus value increase steadily for all weeks.

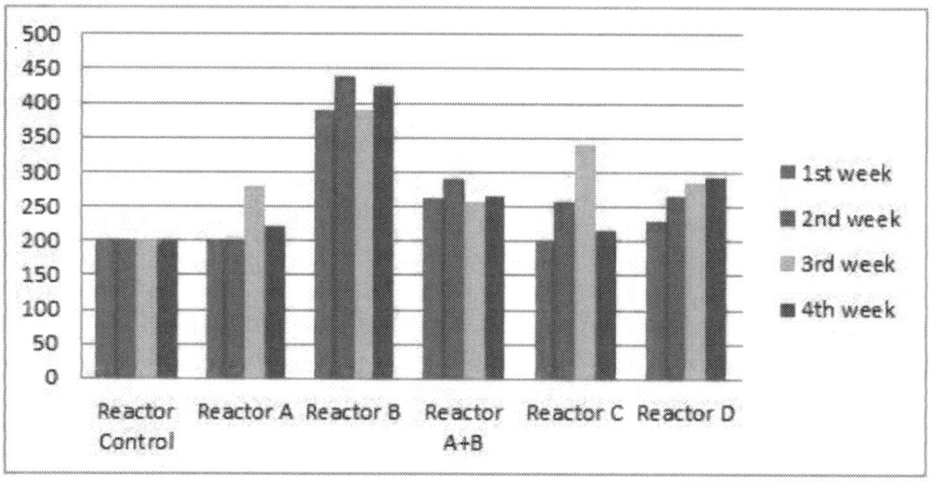

Figure 3. Phosphorus result.

Figure 4 shows potassium result of food waste reactors compared to control. Reactor A did not show any drastic changes from control reactor. Reactor B shows the highest potassium recorded for the first week with 1997 mg/L but then it drops to 1328 mg/L at the fourth week. Reactor A+B shows decreasing but then increasing steadily and recorded highest potassium at the week four. However, reactor C did not shows any drastic changes for three weeks until it increase strongly at week four. On the other hand, reactor D recorded the highest potassium at week 2 with 861 mg/L and then decrease to 652 mg/L at week four.

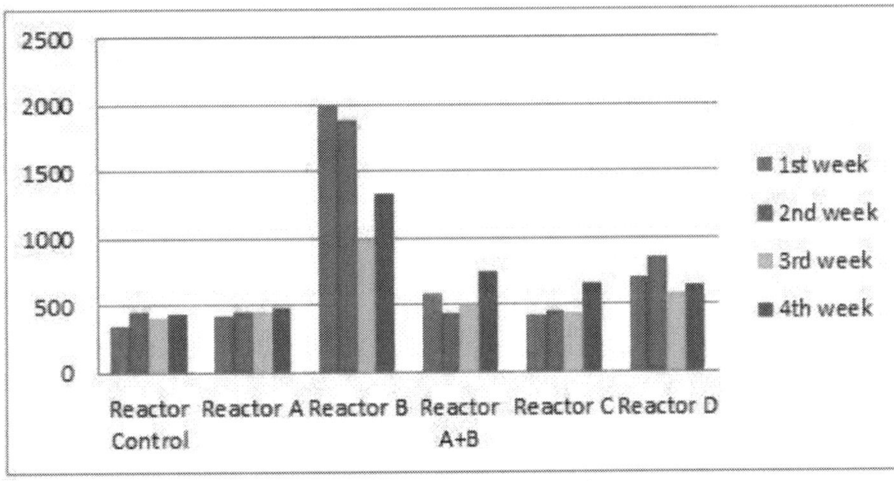

Figure 4. Potassium result.

3.5 Comparison of Nitrogen, Phosphorus and Potassium with chemical fertilizer

Table 2 shows the comparison of NPK between the chemical fertilizers with compost from the food waste composting. From the table it shows that the NPK value of chemical fertilizer is higher than NPK produced from the compost. Based on previous study by [7], Table 3 shows the nutrient from several of organic compost. From the tables it shows the NPK percentage is different between compost. However the comparison with other studies on organic compost only shown slightly difference in NPK compared with chemical fertilizer. The nutrient that being produced varied according to the type or organic compost and wastes that was used. From the table 2 it shows that reactor B with raw type of waste have produced the highest NPK than the rest of the reactors. Even though chemical fertilizers have high percentage in NPK, organic fertilizers are still the best fertilizers because of their micronutrient content and their ability to be both fertilizer and regenerate poor soil condition.

Table 2. Comparison of Nitrogen, Phosphorus and Potassium with chemical fertilizer.

Sample	Nitrogen, N (%)	Phosphorus,P (%)	Potassium, K (%)
Chemical fertilizer [8]	20	40	25
Reactor Control	2.93	0.020	0.041
Reactor A	3.92	0.023	0.045
Reactor B	4.85	0.041	0.155
Reactor A+B	4.68	0.027	0.057
Reactor C	4.39	0.025	0.050
Reactor D	4.00	0.027	0.070

Table 3. Chemical analysis for organic compost.

Sample [7]	Nitrogen, N (%)	Phosphorus, P (%)	Potassium, K (%)
Cow dung	1.3	0.58	2.15
Poultry manure	2.21	2.98	2.05
Cassava peelings compost	1.7	0.86	1.50
Rabbit droppings	1.04	0.99	2.05
Cane rat droppings	1.95	2.06	3.30

4. Conclusion

This study shows that composting process is one of the alternative ways to reduce food waste generation at Makanan Ringan Mas Industry. It can be concluded that from the results that obtained, all parameters have different value due to different wastes that were being used in each reactors. Primarily, every reactor showed different value in temperature. This can be assumed that, each reactor had a different temperature range due to the different in food wastes used. Higher in temperature revealed that there was a high microbial activity. Nevertheless, the temperature must be within the range of 30°C to 50°C. The overall results from the study showed that all of the reactors were within the range.

As for the pH value, the result of this study revealed that the reactors which contain processed food waste tend to produce pH value within the range of 5 to 6 which can be categorized as slightly acidic. Meanwhile, the reactors which contained raw food waste tend to produce pH value within the range of 7 to 8 which can be categorized as neutral. As for Reactor B pH value is 7 to 8 nearest to neutral which is the optimum value for composting. It can be concluded that different food wastes will results in different pH value.

The results also indicate that each waste produce different value of NPK. Moreover, the value of NPK in composting reactor did not showed significantly difference when it is compared with NPK in reactor control. From this study, it showed that tapioca peel can be used as the fermentation liquid in food waste composting, although it shows different value of NPK from the reactors. The highest NPK obtained is from reactor B that process only using raw food waste. The average value of Nitrogen is 48540 mg/L, Phosphorus is 410 mg/L and Potassium is 1550 mg/L. From the study it shows that the NPK compost only increases slightly when compared with reactor control. Longer observation is recommended to monitor the NPK value and more time was given for the compost to mature. The comparison with common chemical fertilizer shows that food waste composting from Makanan Ringan Mas determine lower NPK value. However, the comparison with available organic compost only shown slightly difference value in NPK. Nevertheless organic compost is much better because it is one of the alternative ways to treat waste that is more environmentally friendly. In addition composting can improve soil structure and soil fertility.

References

[1] Waste, S., & Management, P. C. (2007). Laws of Malaysia Act 672.

[2] Saeed, M. O., Hassan, M. N., & Mujeebu, M. A. (2000). Development of Municipal Solid Waste Generation and Recyclable Components Rate of Kuala Lumpur : Perspective Study, 1–7.

[3] Li, Z., Lu, H., Ren, L., & He, L. (2013). Chemosphere Experimental and modeling approaches for food waste composting : A review, 93, 1247–1257. doi:10.1016/j.chemosphere.2013.06.064

[4] Chang, T-E.Hsu , (2009) Effects of compositions on food waste composting, , Bioresource Technology 99: 8068-8074.

[5] Hanc, A., Szakova, J., & Ochecova, P. (2014). Differences in the mobility of Cd, Cu, Pb and Zn during composting of two types of household bio-waste collected in four seasons. Bioresource Technology, 168, 204–13. doi:10.1016/j.biortech.2014.01.125

[6] Tweib, S. A., Rahman, R. A., & Khalil, M. S. (2011). Composting of Solid Waste from Wet Market of Bandar Baru Bangi Malaysia, *5*(5), 975–983.

[7] Fatin, N., Saad, M., Nadrah, N., Zain, S., Ezlin, N., Basri, A., & Sofia, N. (2013). Jurnal Teknologi Full paper Composting of Mixed Yard and Food Wastes with Effective Microbes, 2, 89–95.

[8] Adeniyan, O. N., Ojo, A. O., & Adediran, J. A. (2011). Comparative study of different organic manures and NPK fertilizer for improvement of soil chemical properties and dry matter yield of maize in two different soils, 2(1), 9–13.

Water Quality Characteristics of Sembrong Dam Reservoir, Johor, Malaysia

S Mohd-Asharuddin[1], N Zayadi[1], W Rasit[1] and N Othman[1]
[1]Faculty of Civil and Environmental Engineering, Universiti Tun Hussein Onn Malaysia, 86400 Batu PPahat, Johor, MALAYSIA

E-mail: hf140120@siswa.uthm.edu.my

Abstract. A study of water quality and heavy metal content in Sembrong Dam water was conducted from April – August 2015. A total of 12 water quality parameters and 6 heavy metals were measured and classified based on the Interim National Water Quality Standard of Malaysia (INWQS). The measured and analyzed parameter variables were divided into three main categories which include physical, chemical and heavy metal contents. Physical and chemical parameter variables were temperature, dissolved oxygen (DO), biochemical oxygen demand (BOD), chemical oxygen demand (COD), total suspended solid (TSS), turbidity, pH, nitrate, phosphate, ammonium, conductivity and salinity. The heavy metals measured were copper (Cu), lead (Pb), aluminium (Al), chromium (Cr), ferum (Fe) and zinc (Zn). According to INWQS, the water salinity, conductivity, BOD, TSS and nitrate level fall under Class I, while the Ph, DO and turbidity lie under Class IIA. Furthermore, values of COD and ammonium were classified under Class III. The result also indicates that the Sembrong Dam water are not polluted with heavy metals since all heavy metal readings recorded were falls far below Class I.

Keywords: water quality, heavy metal, dam water.

1. Introduction

Surface water resources have played an important function as basic national water resource. The quality of water is affected by a wide spectrum of natural and human influences. The major natural influences are geological, hydrological and climatic, since these affect the quantity and the quality of water available [1]. In general, sources of pollutant in water are known as point and non-point sources. Nowadays, water quality monitoring has become a major concern to determine current condition and long term pattern for future management of any dam or reservoir [2].

According to UNEP (2003), dams serve to function as a visible tool for supplying freshwater resources which contributing to the development of socio-economic through supplying drinking water [3]. In Malaysia, example of dam namely Tasik Chini Dam (Pahang), Sembrong Dam (Johor) and Bakun Dam (Sarawak). Water pollution is becoming a serious problem in Malaysia and post a negative impacts on the sustainability of water resources plants and organisms, people's health and the country's economy. Therefore, quality of dam water should be monitored to the acceptable standard to ensure safe water that enter water treatment intake.

Sembrong dam is flood-control dam which is tapped by Syarikat Air Johor for water supply purpose since 1984. The reservoir provides a clean water supply to 240 000 consumers in the district of Kluang [4]. Due to indiscriminate farming, agricultural activities and the planting of oil palm trees which covers at least 87% of the dam's 130 km^2 catchment area, this water reservoir is slowly endangered due to eutrophication or better known as algae bloom. The purpose of the study is to examine the water quality status through several physic-chemical parameters as well as heavy metals concentration.

2. Methodology

2.1 Study Area

Sembrong dam is located about 10km from Ayer Hitam, Johor. The dam lies at the coordinate of 2°01'35"N - 1°58'29"N latitude and 103°09'32"E - 103°12'57"N longitude [4]. The reservoir surface area is 8.5 km^2, with capacity of 18 million m^3 along with a catchment area of 130 km^2 and dam height of 11 m. This reservoir is surrounded by agricultural farms and animal husbandry as pollutant sources and there is no buffer zone between these and the water body [5]. There are two major tributaries flowing into the reservoir which are Sembrong River and Marpo River [2].

2.2 Sampling and preservation

Surface water was collected in 500 ml HDPE bottle. The water were collected about 10cm below the water surface. The sample were stored in an icebox and transported back to the lab for analysis on the same day [6].

2.3 In-situ analysis

In-situ parameters studied were temperature, dissolved oxygen (DO), pH, and turbidity. These parameters were measured using YSI 650 Multiparameter Water Quality Monitoring Unit.

2.4 Ex-situ analysis

Metal analysis was carried out using Inductive Coupled Plasma Mass Spectrometry (ICP-MS) (Perkin Elmer, USA). Physico-chemical characteristics were measured.

3. Results and discussion

3.1 Water Quality Status

Results of the mean values of the physic-chemical parameters and heavy metal concentrations determined in the water of Sembrong dam are listed in Table 1. Values of water temperature recorded for Sembrong dam were ranged from 28.2 – 29.3°C with an average of 29.13°C. The temperature values did not indicate any spatial change but showing temperal variation. All temperature readings recorded was lower than INWQS normal level [7]. Based on Table 3, the mean temperature obtained in this study was slightly lower than a study conducted in 2011, which is 29.51°C [2] but in the range of temperature reported by [3]. Increase in temperature resulting in increase of metabolic activity of organism which will reduce the DO [8].

The pH values were recorded in the range of 6.73 – 7.82. All samples had neutral pH and lies in Class I according to the INWQS. In unpolluted waters, pH is controlled by the ionic balance between the carbon dioxide, carbonate and bicarbonate ions. The daily variations of pH value may due to photosynthesis and respiration cycles of algae in the waters [1].

The DO concentration ranged from 5.64 – 6.05 mg/L with average concentration of 5.86 mg/L. The average value obtained falls under Class IIA of INWQS. High value of DO recorded is possibly due to the photosynthesis of algae during blooms. Moreover, since algae having short life, thus decomposition consumes oxygen levels [5].

The water turbidity readings were varied from 23.8 to 28.8 NTU. The average value was 26.8 NTU which lies into Class I of INWQS for Malaysian surface water. Water turbidity reading indicates the scattering or absorption of incident light by the particles [9]. In this study, the turbidity of the water might be due to suspended solids which consist of particles from the eroded soil transported through surface runoff from the watershed upstream.

Table 1. The range, reading and mean values of water quality parameters recorded at different sampling time.

Parameter	In-situ study	Ex-situ study					INWQS Class
		Sample					
	April 2015 - August 2015	1 (4/8/2015)	2 (11/8/2015)	3 (18/8/2015)	4 (24/8/2015)	Mean	
Temperature (°C)	28.62 - 30.92	28.20	29.40	29.60	29.30	29.13	—
Salinity (%)	—	0.10	0.10	0.09	0.09	0.10	I
pH	7.79 - 8.71	6.72	7.17	7.82	7.50	7.30	IIA
Turbidity (NTU)	5.99 - 17.56	28.80	23.80	26.90	27.70	26.80	IIA
Conductivity (μs/cm)	—	196.80	195.50	194.00	194.40	195.18	I
DO (mg/L)	5.02 - 7.94	5.64	6.05	5.88	5.90	5.86	IIA
COD (mg/L)	—	60.87	43.86	52.63	61.40	54.69	III
BOD (mg/L)	—	2.74	2.00	2.43	3.05	2.55	I
TSS (mg/L)	—	19.50	21.20	21.30	20.62	20.65	I
Nitrate (mg/L)	—	0.62	0.18	0.20	0.25	0.31	I
Ammonia (mg/L)	—	1.12	1.18	1.12	1.12	1.14	III
Total phosphate (mg/L)	—	0.84	1.60	0.92	0.86	1.05	—

Table 2. Interim National Water Quality Standards for Malaysia (INWQS), (DOE, 2005).

Parameters	CLASSES					
	I	IIA	IIB	III	IV	V
Ammonical Nitrogen (mg/L)	0.1	0.3	0.3	0.9	2.7	>2.7
BOD (mg/L)	1	3	3	6	12	>12
COD (mg/L)	10	25	25	50	100	>100
DO (mg/L)	7	5-7	5-7	3-5	<3	<1
pH	6.5-6.8	6-9	6-9	5-9	5-9	—
Electric Conductivity (μm/cm)	1000	1000	—	—	6000	—
Salinity (%)	0.5	1	—	—	2	—
Total Dissolved Solids (mg/L)	500	1000	—	—	4000	—
Total Suspended Solids (mg/L)	25	50	50	150	300	300
Temperature (°C)	—	Normal +2°C		Normal +2°C		—
Turbidity (NTU)	5	50	50	150	300	300
Nitrate	Natural levels or absent	7	7	-	5	Levels above IV
Phosphorous		0.2	0.2	0.2	-	

Table 3. Water quality parameters in 3 different dams in Malaysia.

Parameter	Tasik Chini Dam [3]	St 12, Bakun Dam [9]	Sembrong Dam [2]
Sampling period	Oct 2004 - May 2005	6 Jan - 8 Jan 2012	22 Nov - 25 Nov 2011
Temperature (°C)	28.66 - 31.39	22.94	29.51
pH	6.52 - 6.61	6.41	7.04
Turbidity (NTU)	0 - 38.29	93.69	
DO (mg/L)	5.10 - 6.54	9.34	7.25
COD (mg/L)	12.90 - 22.76	—	—
BOD (mg/L)	1.01 - 1.89	—	3.73
TSS (mg/L)	6.49 - 11.36	—	17.2
Nitrate (mg/L)	0.032 - 0.706	—	
Total phosphate (mg/L)	0.051 - 0.083	—	0.18

Conductivity of the water samples fall between 194-196.8 µs/cm. The mean of water conductivity was 195.18 µs/cm and classified in Class I. The average conductivity value obtained was higher compared to the average conductivity reported by [2].

The total suspended solid (TSS) of the water samples collected were recorded in high value which ranged from 19.50 – 21.30 mg/L. The average concentration recorded was 20.65 mg/L and falls in Class I. The INWQS standard for acceptable range of TSS is 25-50 mg/L while the INWQS threshold level of TSS for supporting aquatic life in fresh water ecosystems is 150 mg/L [3].

The average ammonia value in the Sembrong water samples was 1.14 mg/L. The highest value was 1.18 mg/L while the lowest value was 1.12 mg/L. The ammonia level falls under Class III of INWQS. The high ammonia value in the water could be due to high TSS level content due to backflow from Sembrong River during wet season which might generating more NH3 this stimulating the water ammonia content [3].

The mean recorded for nitrate concentration was 0.31 mg/L. The concentration lies in the range of 0.18– 0.62 mg/L. The nitrate ion is usually originated from anthropogenic activities including agricultural works, domestic sewage and other effluents containing nitrogeneous compounds [3]. Based on INWQS classification, the nitrate concentration in the Sembrong dam falls into Class I.

The range of total phosphate were recorded between 0.84 – 1.60 mg/L. Total phosphate represented by organic and inorganic dissolved phosphorus and phosphate which is bounded to particulate matter [8]. High total phosphate value was due to agricultural discharge brough by the runoff in major rivers [3].

The average value of BOD in the water samples was 2.55 mg/L ranging from 2.00 – 3.05 mg/L. Improper sanitation systems at the nearby village and camp could result in extra biological loading which affecting the value of BOD [8]. The INWQS level for Malaysia surface water is 6 mg/L which indicates the BOD levels of the study are within the acceptable standard of INWQS.

The COD level of the water samples were measured to be varies from 43.86 – 61.40 mg/L with average value of 54.69 mg/L which falls under Class III of INWQS. Metal concentration of water is a indicators of degree of contamination. Figure 1 shows the concentration of heavy metals in water of the area under investigation. In comparison with Interim National Water Quality Standards Malaysian (INWQS) in Table 4, the data shows that the heavy metals concentration in Sembrong dam water falls far below class I which shows that the study area is not polluted with heavy metals.

IOP Conf. Series: Materials Science and Engineering **136** (2016) 012058 doi:10.1088/1757-899X/136/1/012058

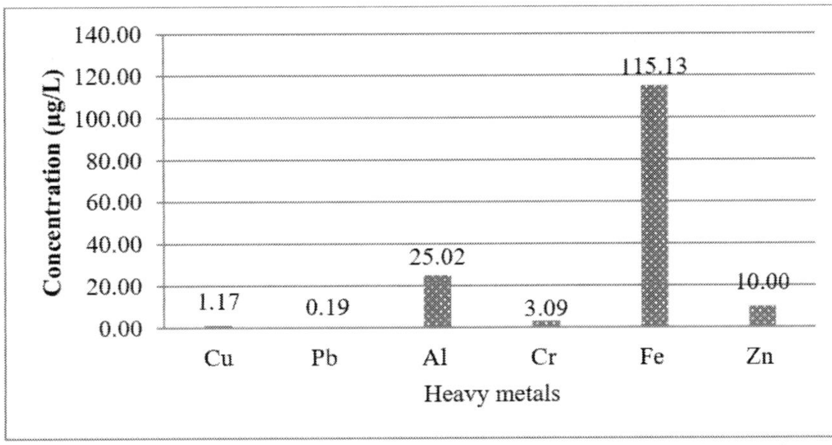

Figure 1. The average value of heavy metals concentration in Sembrong dam water.

Table 4. Interim National Water Quality Standards for Malaysia (INWQS) for heavy metals.

Parameters	Unit	CLASSES				
		I	IIA/ IIB	III	IV	V
Al	mg/l		—	- (0.06)	0.5	
As	mg/l		0.05	0.4(0.05)	0.1	
Ba	mg/l		1	—	—	
Cd	mg/l		0.01	0.01* (0.001)	0.01	
Cr (IV)	mg/l	N	0.05	1.4(0.05)	0.1	L
Cu	mg/l	A	0.02	—	0.2	E
Ca	mg/l	T	—	—	—	V
K	mg/l	R	—	—	—	L
Fe	mg/l	A	1	—	1(leaf) 5 (others)	S
Pb	mg/l	L	0.05	0.02* (0.01)	5	
Mn	mg/l		0.1	0.1	0.2	A
Hg	mg/l	L	0.001	0.004 (0.0001)	0.002	B
Ni	mg/l	E	0.05	0.9*	0.2	O
Se	mg/l	V	0.01	0.25 (0.04)	0.02	V
Ag	mg/l	E	0.05	0.0002	—	E
Sn	mg/l	L	—	0.004	—	
U	mg/l	S	—	—	—	IV
Zn	mg/l		5	0.4*	—	
Ba	mg/l		1	- (3.4)	0.8	
Cl	mg/l		200	—	80	
NO_3	mg/l		7	—	5	

4. Conclusion

The preliminary results obtained in this study show that some of the water quality parameters are quite high and fall in Class III according to INWQS. Therefore more research needs to be carried out to access pollution impact of the study area for the innovation steps to be taken to improve current water treatment system for sustainable clean water supply.

References

[1] Ashraf, M.A., Mohd. Jamil Maah, M.J and Yusoff, I. (2010). Study of Water Quality and Heavy Metals in Soil & Water of Ex-Mining Area Bestari Jaya, Peninsular Malaysia. International Journal of Basic & Applied Sciences IJBAS-IJENS, 10(3). pp. 7 – 23.

[2] Baharim, N.B., Yusop, Z., Askari, M., Yusoff, I., Wan Muhd Tahir, W.Z., Othman, Z. and Zainal Abidin, M.R. (2012). Preliminary results of stratification study in Sembrong reservoir, Peninsular Malaysia. 2ndinternational Conference On Water Resources.

[3] Shuhaimi-Othman, M., Ahmad, A. and Lim, E.C. (2008). Seasonal influence on water quality and heavy metals concentration in Tasik Chini, Peninsular Malaysia. Proceedings of Taal 2007: The 12th World Lake Conference. pp. 300 – 303.

[4] Awang, H., Daud, Z., Tan, L.W. and Mohd Hatta, M.Z. (2015). Hydrology Properties at Sembrong Dam Reservoir in Johor. Applied Mechanics and Materials, 773-774.pp. 1370-1374

[5] Sardool Singh, G.K., Kuppan, P., Goto, M.,Sugiura, N., Megat Mohd Noor, M.J. and Ujang, Z. (2013). Physical Water Quality and Algal Density for Remediation of Algal Blooms in Tropical Shallow Eutrophic Reservoir. Journal of Novel Carbon Resource Sciences, 7, pp. 33-41

[6] APHA, (1998). Standard Methods of Water and Waste Water Analysis. 20 Ed. Washington, D.C:American Public Health Association.

[7] Department of Environment (DOE) (2005). Interim National Water Quality Standards (INWQS) for Malaysia. http://www.doe.gov.my Accessed 28 Feb., 2009

[8] Sujaul Islam,M., Ismail, B.S., Muhammad Barzani, G., Sahibin, A.R. and Mohd Ekhwan, T. (2012). Hydrological Assessment and Water Quality Characteristics of Chini Lake, Pahang, Malaysia. American-Eurasian J. Agric. & Environ. Sci., 12(6), 737-749.

[9] Lee, N., Ling,T.Y. and Grinang, J. (2012). Physico-chemical Characteristics in the Filling Phase of Bakun Hydroelectric Reservoir, Sarawak, Malaysia. International Journal of Applied Science and Technology. 2(6). pp. 91-101.

Performance of Spent Mushroom Farming Waste (SMFW) Activated Carbon for Ni (II) Removal

N S Md Desa[1,2], Z Ab Ghani[1], S Abdul Talib[2] and C C Tay[1]
[1]Faculty of Applied Sciences, Universiti Teknologi MARA, Arau, Perlis, 02600 MALAYSIA
[2]Faculty of Applied Sciences, Universiti Teknologi MARA, Shah Alam, Selangor, 40450 MALAYSIA

E-mail: taychiay@perlis.uitm.edu.my

Abstract. The feasibility of a low cost agricultural waste of spent mushroom farming waste (SMFW) activated carbon for Ni(II) removal was investigated. The batch adsorption experiments of adsorbent dosage, pH, contact time, metal concentration, and temperature were determined. The samples were shaken at 125 rpm, filtered and analyzed using ICP-OES. The fifty percent of Ni(II) removal was obtained at 0.63 g of adsorbent dosage, pH 5-6 (unadjusted), 60 min contact time, 50 mg/L Ni(II) concentration and 25 °C temperature. The evaluated SMFW activated carbon showed the highest performance on Ni(II) removal compared to commercial Amberlite IRC86 resin and zeolite NK3. The result indicated that SMFW activated carbon is a high potential cation exchange adsorbent and suitable for adsorption process for metal removal. The obtained results contribute toward application of developed SMFW activated carbon in industrial pilot study.

Keywords: Activated carbon, adsorption, Ni(II) removal, spent mushroom farming waste (SMFW).

1. Introduction

Increasing industrialization has led to increase environmental problems due to the presence of metal in wastewater. Ni(II) was selected for treatment due to its widespread applications in industrial processes such as mining, electroplating and nickel-cadmium batteries. Ni(II) is a trace metal and is beneficial to human at low concentration. However, it accumulates become high concentration. A long-term exposure time, it causes chronic effects to human health such as lung cancer, renal disorder and skin dermatitis [1].

Most of the conventional methods in metal removal such as precipitation, membrane filtration, solvent extraction and electrodialysis are found to be inadequate [2]. This is due to high cost and inefficiency in low concentration of metal [2]. Conversely, adsorption technique is a treatment technology that gains attention of researches. Adsorption is a physico-chemical process where contaminants are bound to the carbon surface by intermolecular attraction using adsorbent. It offers advantages of low cost and simple operation [3].

Recently, the development of cost effective, highly efficient as well as eco-friendly materials as adsorbents is of great concern [2]. Activated carbon (AC) is one of the most widely used adsorbents for the adsorption of pollutants [4]. AC from SMFW is a low cost adsorbent consisting of sawdust and rice husk. Previous study by [1] in removing Ni(II) using raw SMC has shown that SMC has a potential in application of metal removal. Many previous studies by [5],[6], [7], [8] and [9] have carbonized sawdust and rice husk to generate high porosity of AC. This shows that both components of sawdust and rice husk have a great potential to produce high porosity of carbonaceous AC and suitable for application on adsorption of metal removal. Therefore, AC from SMFW has been selected in this study as there is lack of sufficient literature review on utilization of SMFW in AC preparation and application on metal removal.

This study aims to evaluate Ni(II) removal by spent mushroom farming waste (SMFW) activated carbon. The objectives of this study include optimization concerning performance on Ni(II) removal of the following parameters: adsorbent concentration, pH, contact time, Ni(II) concentration and temperature. Comparison with the commercial resin Amberlite IRC86 and zeolite NK3 were used to evaluate SMFW activated carbon performance on Ni(II) removal as they are widely used in local market.

2. Materials and methods

2.1 Materials

Spent mushroom farming waste (SMFW) used for preparation of activated carbon was provided by C&C Mushroom Cultivation Farm, Johor, Malaysia.

2.2 Activated carbon preparation

The preparation of AC from SMFW was accomplished using potassium hydroxide (KOH) as activating agent. The KOH is responsible for the enhancement of pores formation due to the intercalation of metallic potassium (K) in the carbon structure. The SMFW was impregnated with 3 M KOH for 24 hours respectively. The slurry was then dried at 80°C for 24 hours. The impregnated precursor was carbonized in furnace at 500°C with heating rate of 10°C/min. The produced char was washed with hydrochloric acid and distilled water to remove residual KOH and dried at 80 °C. Finally, the prepared AC was sieved to size of 150 micrometer and kept in drying cabinet prior to use. The AC preparation was indexed according to the procedure by [10].

2.3 Batch adsorption experiment

Batch adsorption experiments were carried out by weighted SMFW activated carbon adsorbent from 0.1 – 0.8 g. 50 ml of 50 mg/L Ni (II) solution was added into 250 mL Erlenmeyer flask and was agitated in an incubator shaker at 125 rpm for 60 minutes at 25 °C. The solutions were filtered and the filtrates were analyzed using ICP-OEM (7300DV, Perkin Elmer, USA). The parameter study includes initial pH in the range of 2 to 6, contact time of 1 min, temperature of 5 °C to 45 °C and Ni (II) concentration in the range of 10 to 60 mg/L. The sample was prepared in duplicated and the percentage of Ni (II) removal was calculated as described as equation (1)

$$percentage\ of\ metals\ removal = \frac{C_o - C_f}{C_o} x\,100\% \qquad (1)$$

where;

C_o= initial Ni (II) concentration (mg/L)
C_f= (final) Ni (II) concentration in solution (mg/L).

2.4 Evaluation of prepared adsorbent

The evaluation of SMFW activated carbon was carried out corresponding to the optimum condition prepared in batch adsorption experiment. 0.63 g of SMFW activated carbon was weighed in 250 mL Erlenmeyer flasks. The samples were agitated at 125 rpm for 60 minutes at 25°C in an incubator shaker. The solutions were filtered and the filtrates were analyzed using or ICP-OES (7300DV, Perkin Elmer, USA). Experiment was repeated using Amberlite IRC86 and zeolite NK3.

3. Result and discussion

3.1 Adsorbent amount

Figure1 shows the effect of increasing adsorbent concentration on the percentage removal of Ni(II). Ni(II) percentage removal was increased from 14.26% to 55.03% by increasing the adsorbent concentration at 0.1 g to 0.8 g. This is due to the availability of more active binding sites of surface functional groups and higher surface area [11]. The trend is identical for Ni(II) removal by almond husk activated carbon and doum seed activated carbon by [12] and [13] respectively. According to study by [1], half-saturation constant of fifty percent Ni(II) removal was applied since the experimental data did not show a clearly defined saturation stage. For this study, the half saturation constant of 0.63 g is determined using the sigmoidal curve fitting with r^2 of 0.9918.

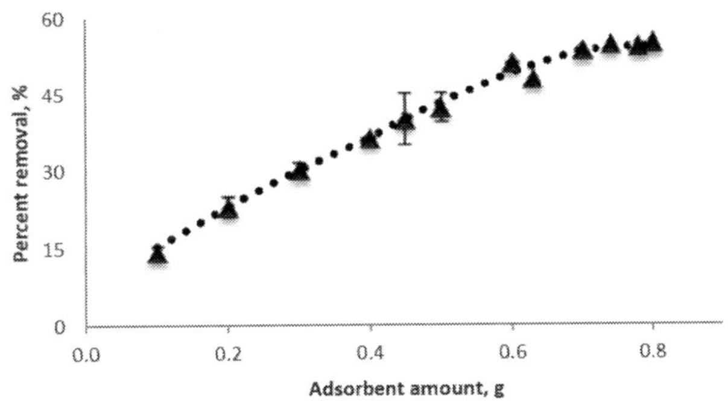

Figure 1. Effect of adsorbent amount on Ni(II) percent removal.

3.2 pH

Figure 2 illustrates the effect of pH ranging from 1 to 6 on percentage removal of Ni(II). The percentage removal of Ni(II) has gradually increased up to pH 4. An increase in metal ion removal as pH increased can be explained on the basis of a decrease in competition between proton (H^+) and positively charged metal ion at the active binding sites of surface functional group. Meanwhile, a decrease in positive charge of proton (H^+) has resulted in a lower repulsion of the adsorbing metal ion [7]. At saturation range pH 4-6, no further increase in Ni(II) removal was observed due to saturation of active binding sites of surface functional group by the cations. At alkaline pH, lower binding sites were attributed to metal hydroxide precipitation of $Ni(OH)_2$ [11], [13], [14], [3] and [2] described a similar trend of pH using olive stone activated carbon and sugarcane baggase pith activated carbon respectively. Table 1 shows the comparison of optimum range for Ni(II) removal using various types of activated carbon. This developed SMFW activated carbon has a comparable optimum pH with other recent studies. The pH 5-6 has been chosen in this study due to its highest performance of ion exchange between cation in the solution and surface sites as well as minimal use of chemical on pH adjustment.

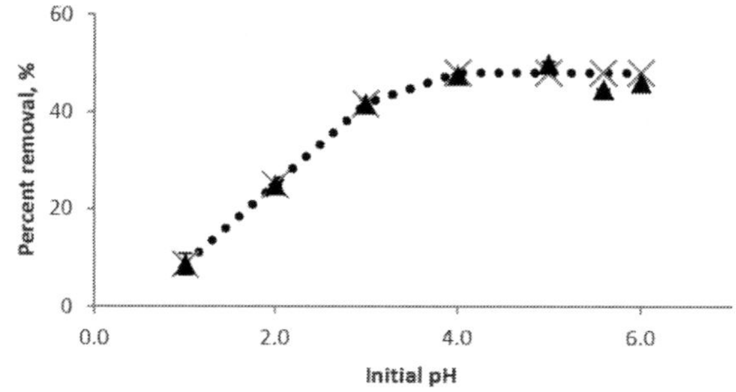

Figure 2. Effect of pH on Ni(II) percent removal.

Table 1. Comparison for optimum pH with other adsorbent using activated carbon

Types of activated carbon	Optimum pH	References
Doum seed coat	7.0	(Foo and Hameed, 2012)
Sugarcane bagasse pith	6.5	(Krishnan *et al.*, 2011)
Enteroporpha prolifera	5.95	Gueye *et al.*, 2014)
Rosa-Canina-Lfruits	5.0	(Erdogan *et al.*, 2005)
Spent mushroom farming waste	5.59	This study

3.3 Contact Time

The effect of contact time on Ni(II) percentage removal is illustrated in Figure 3. The percentage removal of Ni(II) increased at the beginning and gradually achieved equilibrium at 60 min for a 46.22% Ni(II) removal. Rapid Ni(II) removal at the beginning of the reaction was due to the availability of active binding sites of surface functional group that became saturated at equilibrium stage. A similar trend of observation was obtained from Ni(II) removal using Khulays activated bentonite and activated carbon from lignin of papermaking black liquor by [15].

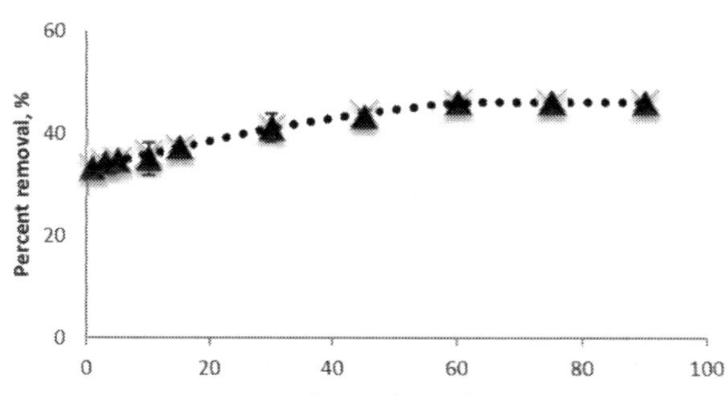

Figure 3. Effect of contact time on Ni(II) percent removal.

3.4 Ni(II) concentration

Figure 4 shows the effect of metal concentration on Ni(II) percentage removal. An increase of Ni(II) concentration from 10 ppm to 60 ppm caused a decrease in Ni(II) removal from 83.53 % to 48.59 %. At low Ni(II) concentration, more active binding sites of surface functional group were available. But, as the concentration increased, the number of ions competing for available active binding sites of surface functional group increased [13] thus, decreased in Ni(II) percentage removal. The Ni(II) concentration provides necessary driving force to overcome the resistance to the mass transfer between the aqueous and adsorbent [14]. This trend is consistent with the study by [13] and [2]using doum seed coat activated carbon and sugarcane bagasse pith activated carbon, respectively

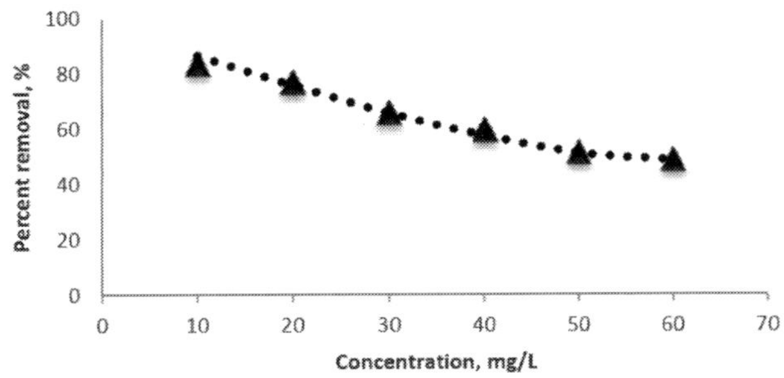

Figure 4. Effect of metal concentration on Ni(II) percent removal

3.5 Temperature

Figure 5 illustrates the effect of temperature on Ni(II) percentage removal. Percentage removal of Ni(II) increased from 39.15 % to 58.25 % with the increased of temperature from 5 °C to 45 °C. This observation revealed that at high temperature, diffusion rate increased and the adsorption process was enhanced. This trend of observation also revealed that Ni(II) adsorption was an endothermic reaction. Similar trend of temperature effect was observed by [12] using the almond husk activated carbon.

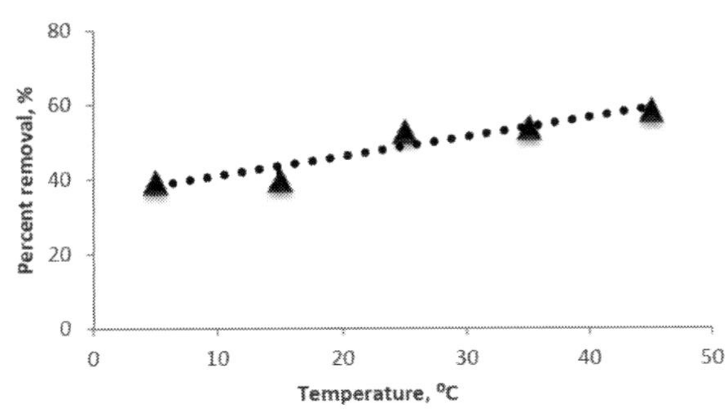

Figure 5. Effect of temperature on Ni(II) percent removal.

3.6 Evaluation of SMFW activated carbon

The SMFW activated carbon performance was evaluated with commercial materials of resin and zeolite under similar experimental design. The SMFW activated carbon showed the highest Ni(II) removal of 50.35% compared to commercial Amberlite IRC86 resin (39.31%) and NK3 zeolite (34.35%). The low Ni(II) removal of commercial Amberlite IRC86 resin was attributed to the weak acid resin having only carboxyl group as functional group. This is supported by [1] in removing of Ni(II) (46.61%) using Amberlite IRC86. Meanwhile, structure of zeolite NK3 consists of three dimensional tetrahedral frameworks of SiO_4 and AlO_4. The net negative charge replacement of Si^{4+} by Al^{3+} is balanced by exchangeable cation of potassium, sodium, calcium and Ni(II) ion in the solution [15]. The reaction could mainly be attributed to ion exchange to the microporous of zeolite sample and decreased the Ni(II) removal. This was supported by [16] in removing of Ni(II) (28%) using zeolite. This result also implied that the SMFW activated carbon has several typesof functional groups that contribute to adsorption process simultaneously. Thus, the SMFW activated carbon is the most outstanding adsorbent for Ni(II) removal.

4. Conclusion

The Ni(II) removal performance is evaluated using low cost agricultural waste of SMFW activated carbon. The fifty percent of Ni(II) removal was found at 0.63 g of adsorbent amount, 5-6 pH solution (unadjusted), 60 min contact time, 50 mg/L metal concentration and 25°C temperature. The effect of Ni(II) concentration and temperature on adsorption study were defined as inverse and direct relationship, respectively. The evaluation of Ni(II) removal showed that the SMFW activated carbon is a great adsorbent that showed highest performance compared to commercial resin and zeolite. This indicated that SMFW activated carbon has a high potential as cation exchange adsorbent for adsorption process.

Acknowledgements

The authors acknowledge the research grant RAGS, MOHE (RAGS/2013/UiTM/SG01/4) for funding, C & C Mushroom Cultivation Farm Sdn. Bhd for free sample and also UiTM for supporting the facility.

Reference

[1] Tay C C, Liew H H, Redzwan G, Yong S K, Surif S, and Abdul Talib S 2011 Pleurotus ostreatus spent mushroom compost as green biosorbent for nickel (II) biosorption, Water Science and Technology, 64(12) pp.2425–32.

[2] Krishnan K A, Sreejalekshmi K G and Baiju R S 2011 Nickel(II) adsorption onto biomass based activated carbon obtained from sugarcane bagasse pith, Bioresource Technology, 102(22) pp.10239–47.

[3] Bohli T, Ouederni A, Fiol N and Villaescusa I 2012 Uptake of Cd^{2+} and Ni^{2+} metal ions from aqueous solutions by activated carbons derived from waste olive stones, International Journal of Chemical Engineering and Applications, 3(4) pp.232–236.

[4] Duman O and Ayranci E 2010 Attachment of benzo-crown ethers onto activated carbon cloth to enhance the removal of chromium, cobalt and nickel ions from aqueous solutions by adsorption, Journal of Hazardous Materials, 176(1-3) pp.231–8.

[5] Foo K Y and Hameed B H 2012 Mesoporous activated carbon from wood sawdust by K_2CO_3 activation using microwave heating, Bioresource Technology, 111 pp.425–32.

[6] Moodley K, Singh R, Musapatika E T, Onyango M S and Ochieng A 2011 Removal of nickel from wastewater using an agricultural adsorbent, Water SA, 37(1) pp.41–46.

[7] Khan N, Yahaya E M, Faizal M, Mohamed P, Abustan I and Azmier M 2010 Effect of Preparation Conditions of Activated Carbon Prepared from Rice Husk by $ZnCl_2$ Activation for Removal of Cu (II) from Aqueous Solution, International Journal of Engineering and Technology, 10(06) pp.1–5.

[8] Chen Y, Zhu Y, Wang Z, Li Y, Wang L, Ding L and Guo Y 2011 Application studies of activated carbon derived from rice husks produced by chemical-thermal process a review, Advances in Colloid and Interface Science, 163(1) pp.39–52.

[9] Kumagai S, Sato M and Tashima D 2013 Electrical double-layer capacitance of micro- and mesoporous activated carbon prepared from rice husk and beet sugar, Electrochimica Acta, 114 pp.617–626.

[10] Gueye M, Richardson Y, Kafack F T and Blin J 2014 High efficiency activated carbons from African biomass residues for the removal of chromium(VI) from wastewater, Journal of Environmental Chemical Engineering, 2(1) pp.273–281.

[11] Kadirvelu K, Senthilkumar P, Thamaraiselv, K and Subburam V 2002 Activated carbon prepared from biomass as adsorbent: elimination of Ni(II) from aqueous solution, Bioresource Technology, 81(1) pp.87–90.

[12] Hasar H 2003 Adsorption of nickel(II) from aqueous solution onto activated carbon prepared from almond husk, Journal of Hazardous Materials, 97(1-3) pp.49–57.

[13] El-sadaawy M and Abdelwahab O 2014 Adsorptive removal of nickel from aqueous solutions by activated carbons from doum seed (Hyphaenethebaica) coat, Alexandria Engineering Journal, 53(2) pp.399–408.

[14] Erdoğan S, Önal Y, Akmil-Başar C, Bilmez-Erdemoğlu S, Sarıcı-Özdemir Ç, Köseoğlu E and İçduygu G 2005 Optimization of nickel adsorption from aqueous solution by using activated carbon prepared from waste apricot by chemical activation, Applied Surface Science, pp.1324-1331.

[15] Gao Y, Yue Q, Gao B, Sun Y, Wang W, Li Q and Wang Y 2013 Preparation of high surface area-activated carbon from lignin of paper making black liquor by KOH activation for Ni(II) adsorption, Chemical Engineering Journal, 217 pp.345–353.

[16] Shaheen S M, Derbalah A S and Moghanm F S 2012 Removal of Heavy Metals from Aqueous Solution by Zeolite in Competitive Sorption System, International Journal of Environmental Science and Development, 3(4) pp.362–367.

Study of The Maximum Uptake Capacity on Various Sizes of Electric Arc Furnace Slag in Phosphorus Aqueous Solutions

W M W Afnizan[1], R Hamdan[2] and N Othman[3]
[1,3] Faculty of Civil and Environmental Engineering Universiti Tun Hussein Onn Malaysia (UTHM), 86400 Parit Raja, Batu Pahat, Johor, Malaysia.
[2] Department of Engineering Technology, Faculty of Engineering Technology, Universiti Tun Hussein Onn Malaysia (UTHM), 86400 Parit Raja, Batu Pahat, Johor.

E-mail: afnizan@uthm.edu.my

Abstract. The high content of uncontrolled phosphorus concentration in wastewater has emerged as a major problem recently. The excessive amount of phosphorus that is originated from domestic waste, unproper treated waste from septic tanks, as well as agricultural activities have led to the eutrophication problem. Therefore, a laboratory experiment was initiated to evaluate the potential of the Electric Arc Furnace Slag (EAFS), a by-product waste from steel making industry in removing phosphorus concentrations in aqueous solutions. In this work several particle sizes ranging from (9.5-12.4 mm, 12.5-15.9 mm, 16.0-19.9 mm, 20.0-24.9 mm, 25-37.4 mm) with a known weight (20±0.28 g, 40±0.27 g, 60±0.30 g, 80±0.29 g and 100±0.38 g) were used to study the effect of different particle sizes towards phosphorus removal. Each particle size of EAFS was shaken in synthetic phosphorus solutions (10 mg/l, 20 mg/l, 30 mg/l, 40 mg/l and 50 mg/l) at a contact time of 2 hours. Final concentrations of phosphorus were sampled and the measurement was made using WESTCO Discrete Analyzer equipment. Results showed that the highest of the maximum uptake capacity of each EAFS particle size distribution achieved at 0.287, 0.313, 0.266, 0.241 and 0.25 mg/g as particle size range was varied from 9.5-12.4 mm to 25-37.4 mm. In conclusion, the maximum uptake capacity of each EAFS mostly was determined to occur at adsorbent weight of 20 to 40 g in most conditions.

Keywords: Steel slag, phosphorus concentration, particle size.

1. Introduction

The excessive of phosphorus in the aquatic environment create a serious hazard to the environment and negative health effects. According to Camargo et al. [1] phosphorus pollution in aquatic environments can be attributed to three major sources: industry, agriculture and wastewater treatment works. Wastewater from industrial, commercial, municipal, and individual households and runoff from residential, agricultural, and municipal are contributors to the phosphorous loading issue. Besides that, phosphates used in food processing, detergents and agriculture have resulted in the depletion of the essential plant nutrient. The presence of phosphorus in industrial wastes and its persistence in wastewater treatment plant effluents have led to the increase concentrations into water bodies and eventually has resulted a widespread concern over its environmental impact.

Recently several materials have been investigated due to their potential use as a filter media in a wastewater for phosphorus removal treatment system which include piles of natural, industrial by-products and man-made materials. The use of industrial by-products for phosphorus removal was started in Japan in late 1980s [2]. Among the industrial by-product material tested, steel slag appeared as an excellent adsorbent for phosphorus removal study. The most thoroughly investigated steel slag for their potential use in removing phosphorus were blast furnace slag (BFS) [1,2,3,4], electric arc steel furnace slag and melter slag [5,6,7]. Amongst the investigated slag, BFS has undergone the most numerous study for its phosphorus adsorption properties, with the result demonstrate that almost 60 % removal achieved in the field in case of both filter boxes and Constructed Wetland System (CWS). Therefore, a motivation arises for the author to further studies on the Electric Arc Furnace Slag (EAFS) particularly on the effect of various sizes towards the maximum uptake capacity of the adsorbent..

2. Materials and methods

2.1 Media preparation

The collected Electric Arc Furnace Slag (EAFS) media from Perwaja Steel Sdn. Bhd., located in Telok Kelong Industrial Area, Kemaman, Terengganu, Malaysia were sieved using test sieve shaker Endecotts Lombard Rd. London, model Sw193BR, England with British Standard sieve BS410/1986 into a uniform size of 9.5-12.4 mm, 12.5-15.9 mm, 16.0-19.9 mm, 20.0-24.9 mm, 25-37.4 mm for a period of 15 minutes. The retained EAFS media on each sieve were collected and washed with tap water twice followed by distilled water once. This process was done in order to remove any dirt or particulate that adheres to the media before being used for batch study. The washed media were oven dried at 105 °C for 24 hours and kept in a heavy duty air-tight plastic bag.

2.2 Synthetic phosphorus solutions

Phosphorus concentration in synthetic solutions was measured in the dissolved reactive inorganic phosphorus (orthophosphate) form. This species is one of the major biologically-available nutrient that is responsible in contributing and accelerating eutrophication process [8].

Synthetic stock phosphorus solution was prepared by dissolving 0.4394 g of potassium phosphate monobasic, KH_2PO_4 which was oven dried previously inside the electrical oven at 105 °C for 1 hour and cooled down inside dessicator. The above amount KH_2PO_4 was then dissolved in 900 mL of reagent water and diluted to 1 L and this procedure gave phosphorus stock solution of 100 mg/l. The preparation of other diluted solutions were using ultrapure laboratory water. The desired synthetic phosphorus concentrations of 10, 20, 30 , 40 and 50 mg P/L were prepared through the following relationship (Equation1)

$$C_1V_1 = C_2V_2 \tag{1}$$

which;

C1 is the phosphorus concentration of stock solution (mg/L) before dilution,
C2 is the desired phosphorus concentration (mg/L) to be prepared after dilution,
V1 is the stock solution volume taken from the volumetric flask (mL) before dilution, and
V2 is the filled volume of ultrapure laboratory water (UPW) (mL)

to the mark of the diluted volumetric flask after a known volume V_1 was added into it . For example, 10 mg P/L of the desired concentration can be prepared inside 1 L of the diluted volumetric flask by taking 10 mL of stock solution from 1 L volumetric flask (before dilution) and placed it inside 1 L of another empty volumetric flask and diluted with UPW until reached the mark. The same procedures were implemented to prepare the other desired solutions used for the synthetic wastewater.

Moreover the maximum uptake capacity of EAFS was calculated using Equation 2.

$$q_e = \frac{C_o - C_e}{W} xV \tag{2}$$

which;
qe is the uptake capacity of P adsorbed by EAFS (mg P/g),
Co is the initial concentration of phosphorus (mg/l),
Ce is the final concentration of phosphorus (mg/l),
W is the weight of EAFS (g) and
V is the solution volume (l).

Meanwhile removal of phosphorus concentration was observed using EAFS at particle size range from 9.5 – 37.4 mm and the result was expressed in percentage of removal based on Equation 3.

$$RP = \frac{C_o - C_e}{C_o} x100 \quad (3)$$

which;
RP is the removal percentage of phosphorus taken up by EAFS adsorbent.

2.3 Study condition design
All experiment conditions that were set during the study is summarized in Table-1.

Table 1. Study conditions on the maximum uptake capacity of EAFS adsorbent.

Parameter	Value
Particle size (mm)	9.5-12.4, 12.5-15.9, 16.0-19.9, 20.0-24.9, 25-37.4
Initial concentration (mg P/L)	10, 20, 30, 40, 50
Contact time (hours)	2
Adsorbent dosage (g)	20, 40, 60, 80, 100
pH	Original (4 – 6)
Temperature (°C)	Room (22 – 27)
Solution volume (ml)	200
Shaking speed (rpm)	150

According to Table-1, laboratory experimental work was conducted using five different particle size distributions with the aim to determine the maximum uptake capacity of EAFS for each size that could offer towards phosphorus removal study.

Meanwhile, initial phosphorus concentration was selected between 10 - 50 mg/L based on the typical residential untreated wastewater for total phosphorus concentration (5 - 20 mg/L). The maximum value of phosphorus concentration for this study was set up to 50 mg/L to make sure that when equilibrium state is achieved there should be solute concentration is remaining in solution after the adsorption takes place. If this is not happened the adsorption maximum uptake capacity could not be established because the sites on matrix surfaces is still not saturated with the attach adsorbate and consequently the equilibrium state is not attained. Such condition will occur when running the experiment at low concentration with more adsorbent dosage added to the solution.

Furthermore, the required volume of phosphorus solution for this study was determined at 200 mL (using 500 mL conical flask) since the largest particle size range (25.0 - 37.4 mm) in average diameter of both adsorbents will be completely submerged at this volume when it is added to the solution. This is very crucial because during shaking process, the whole surface of adsorbent should be covered by the solution so that the phosphate ions can attach to it and hence achive the actual of the maximum uptake capacity state.

3.0 Result and Discussion

Maximum uptake capacity of EAFS and removal efficiency of phosphorus concentration Figure-1 to Figure-6 show the relationship of different EAFS weight on the maximum P uptake capacity (qe) and P removal percentage (RP) efficiency. In general, it was observed that as the weight of EAFS adsorbent increases, the maximum P uptake capacity decreases. Similar trend was shown for all initial concentrations (10 – 50 mg/l) and all EAFS particle sizes (9.5 – 12.4 mm to 25.0 – 37.4 mm) – Figure-1, Figure-3, and Figure-5. According to Figure-1, as EAFS adsorbens were shaken in 2 hours contact time, the highest qe were recorded at 0.016 to 0.287 mg/g when initial concentrations were adjusted from 20 – 50 mg/l.

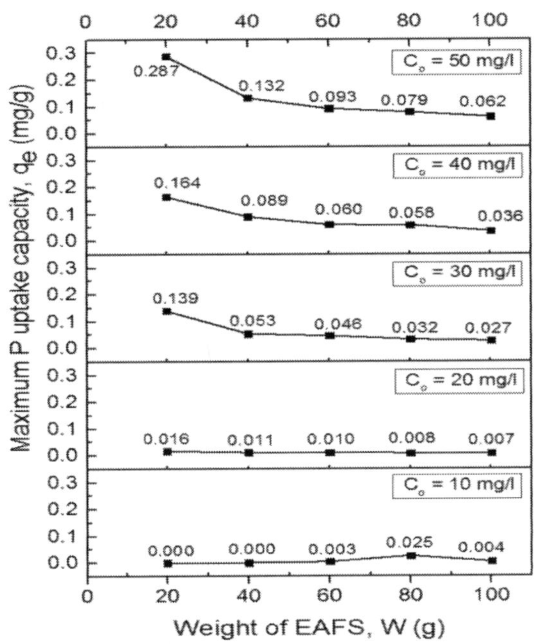

Figure 1. The effect of initial phosphorus concentrations on the maximum P uptake capacity (EAFS particle size: 9.5 – 12.4 mm, Co = 10 – 50 mg/l).

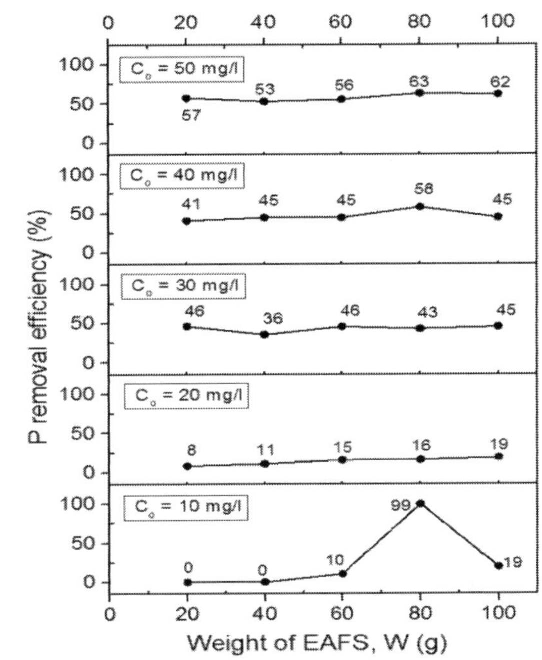

Figure 2. The percentage of P removal efficiency on various initial phosphorus concentrations (EAFS particle size: 9.5 – 12.4 mm, Co = 10 – 50 mg/l).

The highest RP of phosphous concentrations were found to occur as the weight of EAFS adsorbents used increases (Figure-2). Value of RP appeared higher at lower initial P concentration (10 mg/l) with the RP achieved almost 100%. However, the pattern for other initial concentrations (30, 40 and 50 mg/l) demonstrated that the RP values did not much different when the adsorbent weight increases up to 100 g. As such 20 g of EAFS adsorbent is adequate to remove about 45% - 60% of initial phosphorus concentrations from the solutions for 2 hours contact time.

Similar result was also obtained for particle size of 16.0 – 19.9 mm. Values of qe were found to decrease as the weight of EAFS adsorbent increases. The highest values of qe were occurred at lower adsorbent weight (20g) with 0.060, 0.117, 0.204 and 0.266 mg/g as the initial concentration of P was elevated from 20 – 50 mg/l.

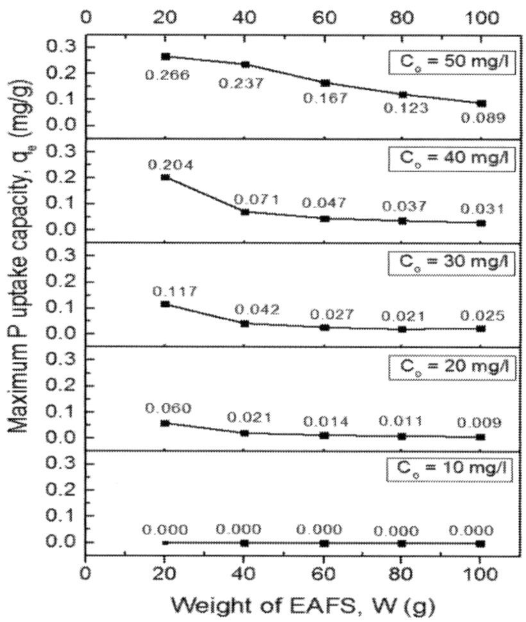

Figure 3. The effect of initial phosphorus concentrations on the maximum P uptake capacity (EAFS particle size: 16.0 – 19.9 mm, Co = 10 – 50 mg/l).

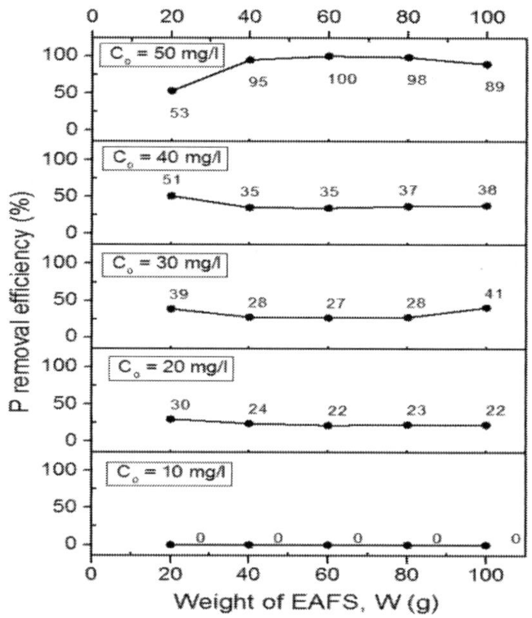

Figure 4. The percentage of P removal efficiency on various initial phosphorus concentrations (EAFS particle size: 16.0 – 19.9 mm, Co = 10 – 50 mg/l).

However, it was noted that bigger particle sizes produce higher RP of phosphorus tremendously especially at the highest initial concentration (50 mg/l). An amount of 40 g EAFS was enough to produce RP close to 100% at 2 hours contact time.

Experiment was also conducted at the biggest selected particle size range (25.0 – 37.4 mm) under the same control condition. Weight of 20 g was not obtained in this set of experiment because on average, a single of EAFS adsorbent for this size range produce the weight of more than 25 g. Again, the same pattern was observed when more adsorbent weight was shaken inside the initial phosphorus solution (Figure-5 and Figure-6). Values of qe was reached at its peak when lower weight of adsorbent was used. An amount of 40 g gave the highest value of qe (0.073, 0.079 and 0.250 mg/g) for each intial concentrations (30, 40 and 50 mg/l). Furthermore, an amount of 40 g was found capable to produce completely RP of phosphorus (100%) particularly at the highest intial concentration (50 mg/l). Nevertheless, the lowest concentration (10 mg/l) was revealed to produce a very poor removal of P which was zero removal (0%).

A comparison of the percentage removal between the present study and that of Bird and Drizo (2010) shows that EAFS was significantly capable in removing the dissolved reactive phosphorus (DRB). In the researcher's work, two different particle sizes of EAFS (5 –10 mm and 10 – 20 mm) were utilized where the latter size was used at the inlet of the filter as a purpose to reduce clogging problem. These sizes were employed inside two 0.242 m3 polypropylene drums which connected in series. They found that the use of the above mentioned sizes were capable to provide removal efficiency up to 70%. Compared with the present study it was found that bigger size range (25.0 – 37.4 mm) was determined could provide higher removal efficiency (almost 100%) at a high phosphorus concentration (50 mg/l).

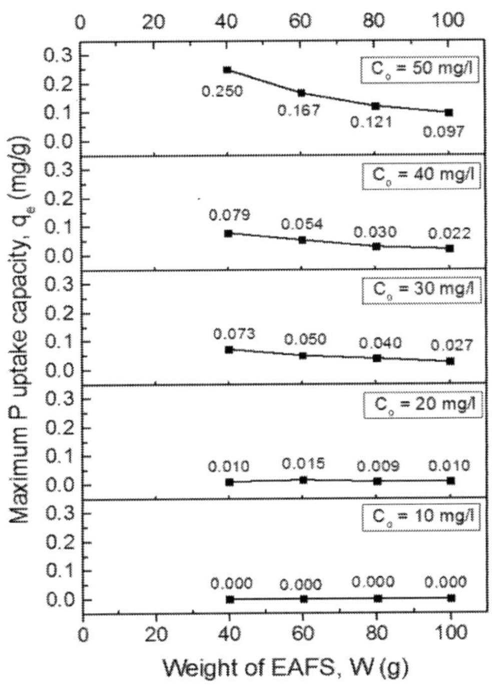

Figure 5. The effect of initial phosphorus concentrations on the maximum P uptake capacity (EAFS particle size: 25.0 – 37.4 mm, Co = 10 – 50 mg/l).

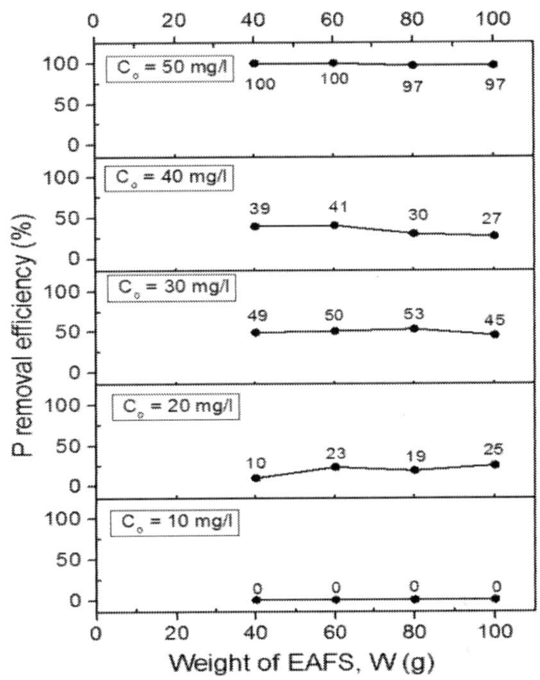

Figure 6. The percentage of P removal efficiency on various initial phosphorus concentrations(EAFS particle size: 25.0 – 37.4 mm, Co = 10 – 50 mg/l).

3.1 The highest of the maximum uptake capacity of EAFS and removal efficiency of phosphorus concentration

The optimum weight of EAFS that produce the highest of the maximum uptake capacity and RP efficiency for each initial concentration were then analyzed and shown in Figure-7 and Figure-8. Apparently, with an amount of 20 – 40 g of EAFS utilized, the qe value was increased as the intial phosphorus concentrations were raised from 10 – 50 mg/l. Values of qe were found at the highest when Co = 50 mg/l was used for all particles size distribution (9.5-12.4 mm, 12.5-15.9 mm, 16.0-19.9 mm, 20.0-24.9 mm, 25-37.4 mm) and reported as 0.287, 0.313, 0.266, 0.241 and 0.25 mg/g respectively (Figure-7).

The highest RP efficiency of phosphorus was determined to increase as the particle size increases, however, on contrary for smaller size (9.5 – 12.5 mm) the efficiency portrayed a fluctuating value. This could be attributed to the adsorbent was saturated with phosphate ion when it was in contact with high phosphorus concentration thus creating the potential of desorption to occur. Particle size distribution of 25 – 37 mm demonstrated the best removal efficiency (100%) of phosphorus concentration especially at higher phosphorus concentration (50 mg/l) (Figure-8).

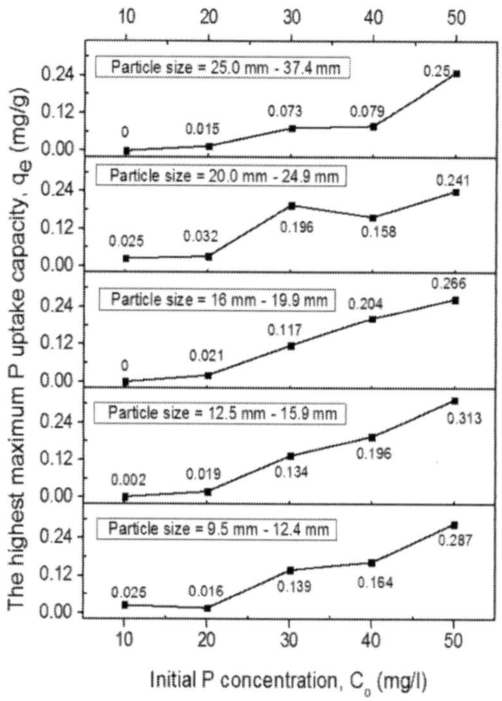

Figure 7. The highest of the maximum P uptake capacity on various particle sizes of EAFS (EAFS particle size: 9.5 – 12.4 mm to 25.0 – 37.4 mm).

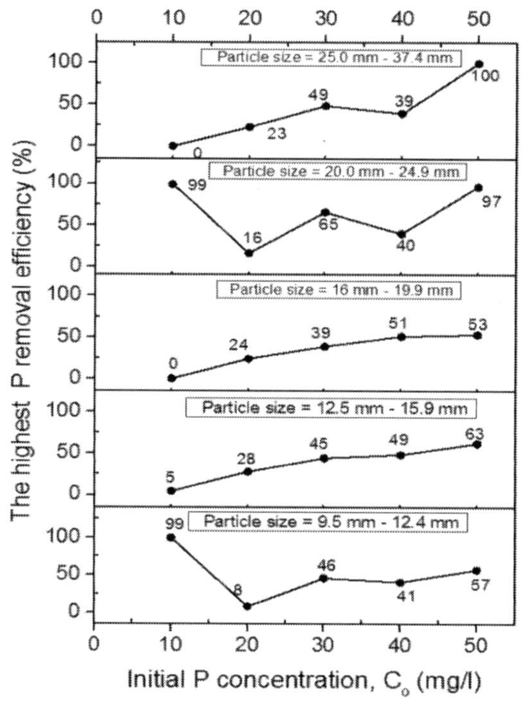

Figure 8. The highest percentage of P removal efficiency on various particle sizes of EAF (EAFS particle size: 9.5 – 12.4 mm to 25.0 – 37.4 mm).

4.0 Conclusions

In conclusion, the maximum uptake capacity of EAFS was excellent at lower adsorbent weight. Moreover, bigger particle sizes showed a tremendous phosphorus removal as initial concentration increases. It is suggested that future study should be conducted on the equilibrium state of the adsorbate. This is vital because equilibrium studies on adsorption provide information about the capacity of the sorbent or the amount required to remove a unit mass of pollution.

Acknowledgment

We are extremely indebted to all our funders: especially to Malaysian Ministry of Education under Exploratory Research Grant Scheme (Vot. E031), Universiti Tun Hussein Onn Malaysia and Perwaja Steel Sdn. Bhd. for providing us Electric Arc Furnace Slag (EAFS) samples during this research

References

[1] Camargo Valero, M. A., Johnson, M. L., Mather, T. and Mara, D.D. (2009). Enhanced Phosphorus Removal In a WSP System With Blast Furnace Slag Filters. Desalination and Water Treatment 4, pp.122–127.

[2] Yamada, H., Kayama, M., Saito, K. and Hara, M. (1986). A fundamental research on phosphate removal by using slag. Water research, 20(5), pp.547 – 557.

[3] Mann, R. A. (1997). Phosphorus adsorption and desorption characteristics of constructed wetland gravels and steelworks by-products. Australian Journal of Soil Research 35, pp.375–384.

[4] Agyeia, N. M., Strydomb, C. A. and Potgieter J.H. (2002). The removal of phosphate ions from aqueous solution by fly ash, slag, ordinary Portland cement and related blends. Cement and Concrete Research 7 32(12), pp.1889-189.

[5] Drizo, A., Forget, C., Chapuis, R.P. and Comeau, Y. (2002). Phosphorus removal by EAF steel slag - A parameter for the estimation of the longevity of constructed wetland systems. Environmental Science and Technology 36, pp.4642-4648.
[6] Lu, S., Bai, S. and Sahn, H. (2008). Mechanisms of phosphate removal from aqueous solution by blast furnace slag and steel furnace slag. Journal of Zhejiang University – Science A 9 (1), pp.125–132.
[7] Bird, S. C. and Drizo, A. (2010). EAF Steel Slag Filters for Phosphorus Removal from Milk Parlor Effluent: The Effects of Solids Loading, Alternate Feeding Regimes and In-Series Design. Water 2, pp.484-499.
[8] Reynolds, C.S. and Davies, P.S. (2001). Sources and bioavailability of phosphorus fraction in freshwaters : A British perspective. Biol. Rev., 76, pp.27 - 64.

Effect of Phosphoric Acid Concentration on the Characteristics of Sugarcane Bagasse Activated Carbon

M R M Adib[1], W M S W Suraya[1], H Rafidah[2], A R M Amirza[1], M H M N Attahirah[1], M S N Q Hani[1] and M S Adnan[1]
[1] Faculty of Civil and Environmental Engineering, Universiti Tun Hussein Onn Malaysia (UTHM), 86400 Batu Pahat, Johor, MALAYSIA
[2] Faculty of Engineering Technology, Universiti Tun Hussein Onn Malaysia (UTHM), 86400 Batu Pahat, Johor, MALAYSIA

E-mail: adib@uthm.edu.my

Abstract. Impregnation method is one of the crucial steps involved in producing activated carbon using chemical activation process. Chemicals employed in this step is effective at decomposing the structure of material and forming micropores that helps in adsorption of contaminants. This paper explains thorough procedures that have been involved in producing sugarcane bagasse activated carbon (SBAC) by using 5%, 10%, 20%, 30% phosphoric acid (H3PO4) during the impregnation step. Concentration of H3PO4 used in the process of producing SBAC was optimized through several tests including bulk density, ash content, iodine adsorption and pore size diameter and the charactesristic of optimum SBAC produced has been compared with commercial activated carbon (CAC). Batch study has been carried out by using the SBAC produced from optimum condition to investigate the performance of SBAC in removal of turbidity and chemical oxygen demand (COD) from textile wastewater. From characteristic study, SBAC with 30% H3PO4 has shown the optimum value of bulk density, ash content, iodine adsorption and pore size diameter of 0.3023 g cm-3, 4.35%, 974.96 mg/g and 0.21-0.41 μm, respectively. These values are comparable to the characteristics of CAC. Experimental result from the batch study has been concluded that the SBAC has a promising potential in removing turbidity and COD of 75.5% and 66.3%, respectively which was a slightly lower than CAC which were able to remove 82.8% of turbidity and 70% of COD. As a conclusion, the SBAC is comparable with CAC in terms of their characteristics and the capability of removing contaminants from textile wastewater. Therefore, it has a commercial value to be used as an alternative of low-cost material in producing CAC.

Keywords: Sugarcane bagasse, activated carbon, impregnation method.

1. Introduction

Activated carbon is one of the adsorption agents that have best efficiency in removing contaminants from wastewater [1]. However, due to the non-affordable market price associated with intensive regeneration process, activated carbon is not favourable to manufacturer. This difficulty has led to search for the use of cheap and efficient alternative materials such as rice husk, bamboo, sugarcane bagasse and tamarind kernel powder [2]. Therefore, this study has been designed focusing on the use of sugarcane bagasse to produce activated carbon, as a means of replacing the expensive conventional activated carbon.

Impregnation method is one of the steps involved in chemical activation where raw materials are immersed fully or partially in certain acid. Chemicals employed in chemical activation ($ZnCl_2$, H_3PO_4, and H_2SO_4) are effective at decomposing the structure of the raw material and forming microspores [3]. Figure-1 shows the pores characterization of an activated carbon.

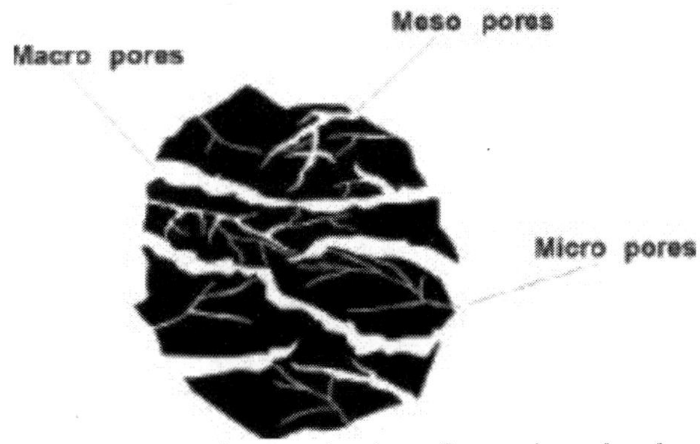

Figure 1. Pores characterization of an activated carbon.

Even though activated carbon from sugarcane bagasse has been investigated for removal of contaminants from wastewater by previous researchers, there are a few studies on treatment efficiency using different concentration of acid in producing activated carbon. Thus, the concentration of phosphoric acid used in impregnation method and its effect towards the performance of SBAC in removing turbidity and COD is highlighted in this study.

Activated carbon from sugarcane bagasse emerges as one of the alternatives that has been introduced to replace conventional activated carbon. Besides of its low cost, it is also easily available [4]. Crushed sugarcane produces 3 tonnes of bagasse for every 10 tonnes of it. This will consume one part of the landfill area. As landfill scarcity is one of the big issues that Malaysia confronting today, thus reuse of bagasse as raw material for producing filter media help in reducing production and disposal of organic waste at landfill.

2. Materials and Methodology

2.1 Material preparation

Sugarcane bagasse is taken from small-medium industries around Parit Raja, Johor. Raw bagasse was washed with tap water to remove contaminants and excess sugar and dried in dry oven at 105 °C overnight. Dried bagasse was impregnated with 5%, 10%, 20%, 30% phosphoric acid (H_3PO_4) for another 24 hours. In impregnation process, raw bagasse was fully immersed in acid with the ratio of 1g bagasse to 300 ml acid. After impregnated for 24 hours, impregnated bagasse is then carbonized at 500 °C in furnace for two hours. Produced activated carbon was washed with pure water until the pH of activated carbon reached 3-7. Activated carbon produced was then sieved to 60 μm to 2.00 mm by sieve analysis testing, referring to ASTM C136-06 Standard Test Method for Sieve Analysis of Fine and Course Aggregates [5]. A workflow of study is shown in Figure-2. Figure-3 shows activated carbon produced from sugarcane bagasse.

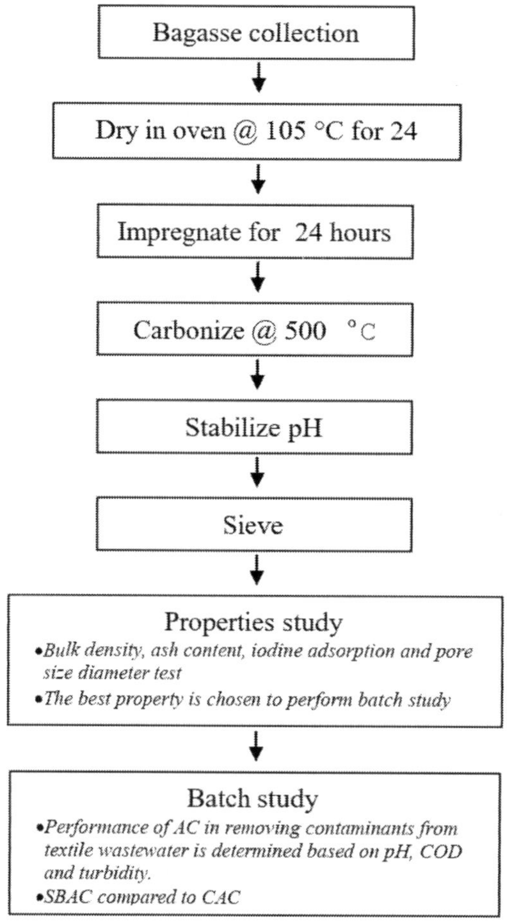

Figure 2. Workflow of study.

Figure 3. Sugarcane bagasse activated carbon.

2.2 Material characterization

All activated carbon produced from various concentrations of acid were undergone several tests to determine their bulk density, ash content, adsorption capacity (iodine), pore size diameter and morphology of the activated carbon produced. SBAC with the best properties was chosen to be continued in batch study.

2.3 Bulk density

Bulk density (expressed as g cm-3) is defined as the mass of a unit volume of the sample in air, including both the pore system and the voids among the particles [3]. Generally, an adsorbent with bulk density need to be generated frequently because it can hold adsorbate per unit weight [2]. In this test, 10 ml measuring cylinder was dried in over at 110 °C for 30 minutes. Powdered activated carbon was filled into cylinder with three layers and tapped about 300 times for each layer until its fully compacted and reweighed [5,6].

$$Bulk\ density, Db = 1000\frac{m}{v} \qquad (1)$$

Where,

Db	=	bulk density in g cm-3
m	=	mass of sample in g
v	=	volume of sample in measuring cylinder

2.4 Ash content

Ash content is a test that provide the total ash content although in some cases, a more detailed analysis of the ash is needed. It is the residue that remains when carbonaceous portion is burned off [7]. This test is carried out using method from ASTM [8]. Empty crucible is ignited in murfle furnace at 500°C for 15 minutes Allow crucible to cool to room temperature in dessicator and weigh to nearest 0.1 mg. 1.0 g of activated carbon is transferred into crucible and ignite for another 1 hour. Crucible is allow to cool in dessicator to room temperatureand reweighed again [5,6].

$$A_c = \frac{F-G}{B-G}\ x\ 100 \qquad (2)$$

Where,

Ac	=	ash content in %
F	=	Mass of crucible and ashed sample in g
B	=	Mass of crucible and dried sample in g
G	=	Mass of empty crucible in g

2.5 Iodine adsorption

Iodine adsorption indicates the internal surface area of carbon where in many cases, the result is close to Brunauer-Emmett-Teller (BET) surface area [3]. It has been established that, the iodine number in mg/g gives an estimate of the surface area in m2/g [9], and measures the porosity of pores with dimensions between 1.0-1.5nm [10]. According to method from AWWA [11,12], 10 ml of 5% hydrochloric acid is boiled with 1g of activated carbon. 100 ml of 0.10N iodine solution is added and shaken vigorously for 30 secondes. After filtration, filtrate is titrated with 0.10N sodium thiosulphate solution with starch as indicator.

$$I_n = \frac{X}{M}A \qquad (3)$$

Where,

In	=	iodine adsorption in mg/g

X	=	mg of iodine adsorbed by activated carbon,
N1	=	normality of iodine solution
N2	=	normality of sodium thiosulphate solution
V	=	volume of sodium thiosulphate solution in ml
M	=	mass of activated carbon in g
A	=	correction factor

$$I\,n = \frac{v-x}{y}\,x\left(\frac{v}{w}\right) x\; M(126.9) \qquad (4)$$

Where,

In	=	iodine adsorption in mg/g
v	=	volume of iodine solution used for titration (ml)
x	=	Volume of sodium thiosulphate used for carbon free aliquot (ml)
y	=	Volume of sodium thiosulphate used for blank solution (ml)
w	=	mass of activated carbon in g
M	=	molarity of iodine solution used

2.6 Pore size diameter and morphological view

Morphological view and pore size is measured using Field Emission Scanning Electron Microscopy (FESEM). FESEM is microscope that works with electrons instead of light. FESEM was used to visualize very small topographic details on the surface are of SBAC. In this study, a range of pore size from selected surface area were used to observe whether the pore is well-developed or ruptured.

2.7 Batch study

Batch study was conducted to measure performance of activated carbon produced. Here, 12.5g of activated carbon with the best properties was chosen to neutralize pH, COD and turbidity from 250 ml textile wastewater with contact time of 24 hours. A comparable study is conducted towards CAC. Figure-4 shows detail on the batch study.

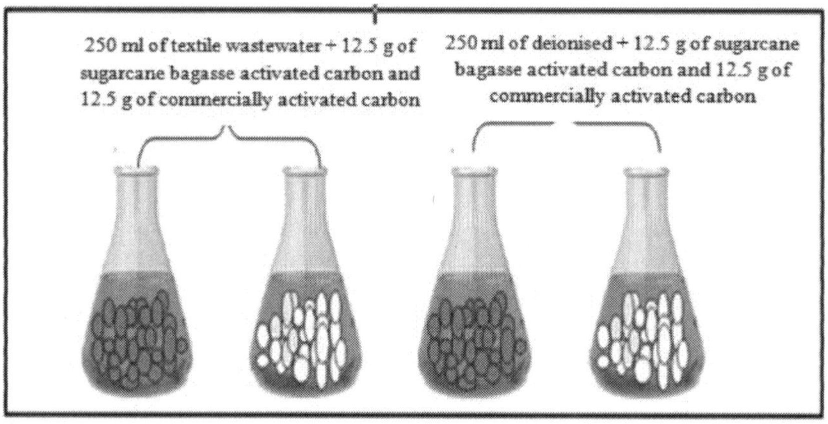

Figure 4. Batch study.

3. Results and Discussion

3.1 Bulk density

Determination of bulk density depends on the shape, size and density of individual particles and are useful in estimating tank or packing volume [3]. Figure-5 shows bulk density of activated carbon for impregnation using 5%, 10%, 20%, 30% of H3PO4. The trend shows that bulk density was decreased with the increasing of acid concentration. According to Yakout and El-Deen[13] the trend indicates the increasing of carbon particle in activated carbon along with increasing of acid concentration. As the lowest bulk density indicates a good adsorbant, thus it can be said that 30% H3PO4 has the lowest bulk density from the others with 0.3023 g cm-3.

3.2 Ash content

Ash content indicates the quality of activated carbon in terms of mechanical strength. Having a large number of ash in activated carbon, indicates the poor strength of carbon produced. Ash content is also related to carbon yield percentage. The increment of ash content in activated carbon will decrease the carbon yield. Figure-6 shows ash content from SBAC. From figure-6, 30% H_3PO_4 has the lowest ash contents with 4.35%, hence having the best mechanical strength.

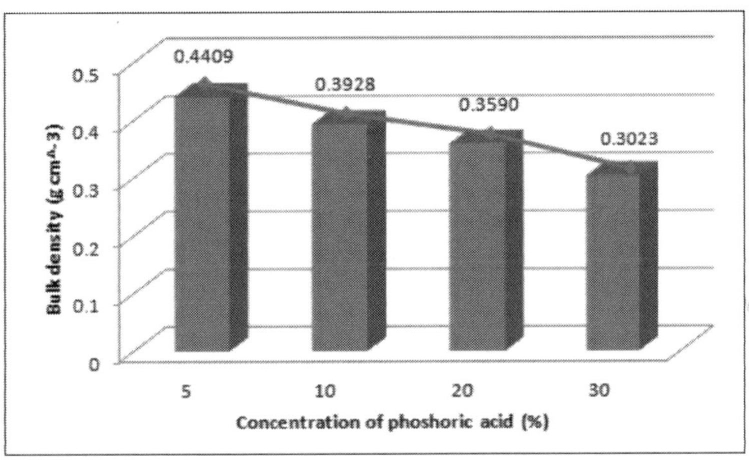

Figure 5. Bulk density of SBAC.

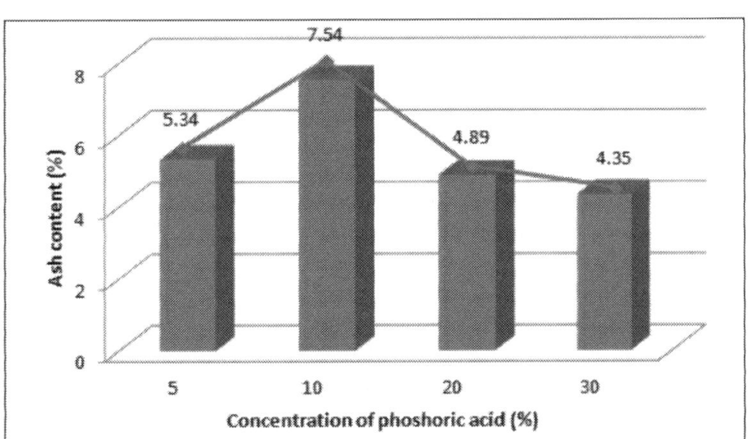

Figure 6. Ash content of SBAC.

3.3 Iodine adsorption
Practically, iodine number is the amount of iodine in milligrams, adsorbed per gram of carbon when the equilibrium concentration (Ce) of iodine is 0.02M. Figure-7 illustrates iodine number for various concentration of H_3PO_4. According to AWWA [12], activated carbon recommended for water treatment must have iodine number ranging from 600 to 1100 mg/g. From the figure-7, it shows that all produced AC comply with the standard set by AWWA [12]. Moreover, iodine number of SBAC from 30% H3PO4 is comparable to commercial activated carbon.

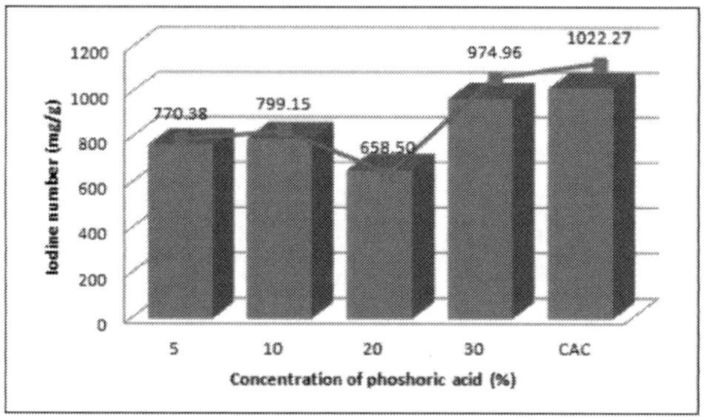

Figure 7. Iodine number of SBAC and CAC.

3.4 Pore size diameter and morphological view
Generally, observing pore size diameter is conducted to see how well the chemicals employed in impregnation method decompose raw material and develop micropores. Porosity is one of the important properties in activated carbon. The more micropores form in activated carbon, the bigger the surface area, hence the performance of activated carbon in absorbing contaminants from wastewater is more efficient. Table 1 shows morphological view and pore size diameter of SBAC and CAC. It can be concluded that all SBAC produced having a comparable pore size diameter with CAC (0.14-0.92 mm).

4. Conclusions
From this study, SBAC with 30% H3PO4 has shown the optimum value of bulk density, ash content, iodine adsorption and pore size diameter of 0.3023 g cm-3, 4.35%, 974.96 mg/g and 0.21-0.41 μm, respectively. It was chosen to have the best performance in removing contaminants from textile wastewater. This study proves that SBAC performance is comparable to CAC. This is shown when SBAC is able to achieve 75.5% and 66.3% in removal of turbidity and COD, respectively which was a slightly lower than CAC which able to remove 82.8% of turbidity and 70% of COD. This study also will allow a reappraisal of optimizing acid used in impregnation method of sugarcane bagasse activated carbon production.

The production of activated carbon by using agricultural waste is more economical compared to commercially activated carbon. This research will also assist nation guidelines and educational drives to alleviate the water pollution awareness. Indirectly, this study will immerse a degree of awareness of the "waste to wealth" concept as well as pleasant drainage could invite eco-tourism in Malaysia.

Table 1. Morphological view and pore size diameter.

Morphological view	Pore size diameter (μm)	Morphological view	Pore size diameter (μm)
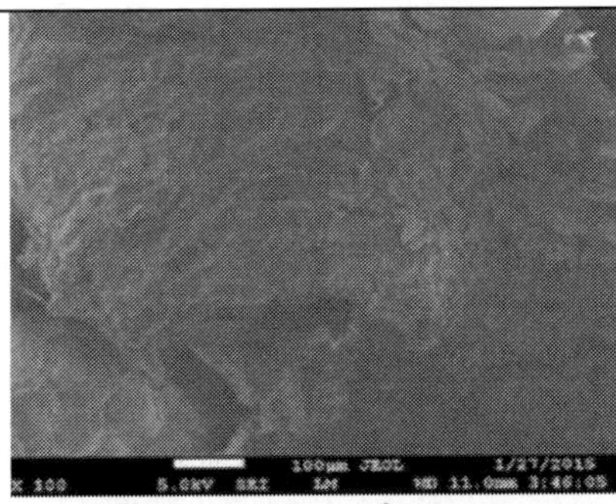 Raw sugarcane bagasse	no pores was detected	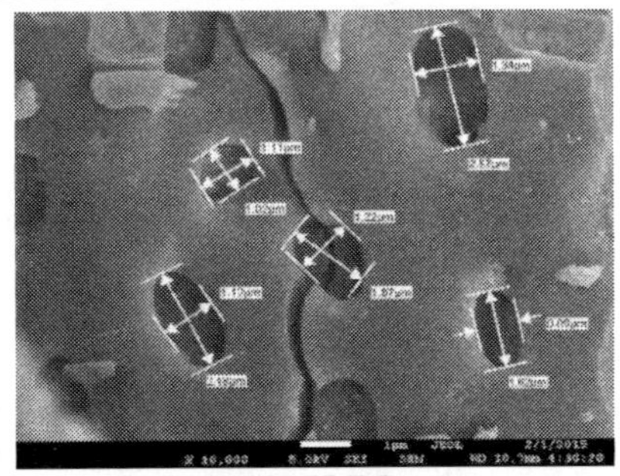 10% H_3PO_4	0.15 – 0.66
 5% H_3PO_4	0.28 – 0.92	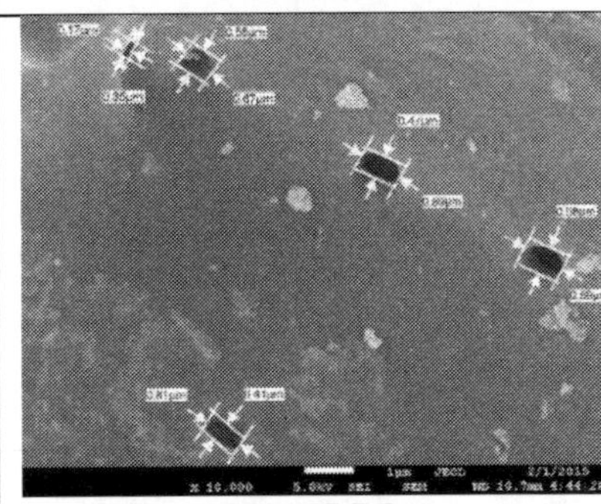 20% H_3PO_4	0.34 - 0.85
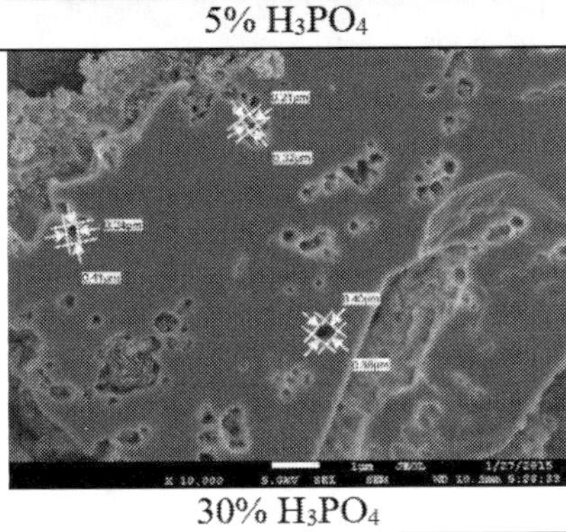 30% H_3PO_4	0.21 – 0.41	 CAC	0.14 – 0.92

Acknowledgment

The author wouold like to express gratitude to the Ministry of Higher Education Malaysia under Fundaamental Research Grant Scheme (FRGS: Vot 1456), Department of Environment Malaysia (DOE) and Universiti Tun Hussein Onn Malaysia for the support in preparing this paper. Grateful acknowledgement also goes to all that are involved directly and indirectly in completing this paper.

IOP Conf. Series: Materials Science and Engineering **136** (2016) 012061 doi:10.1088/1757-899X/136/1/012061

References

[1] Ferhan C. and Ozgur A. (2011). Activated Carbon for Water and Wastewater Treatment: Integration of Adsorption and Biological Treatment. First Edition. Weinheim: WILEY-VCH Verlag GmbH & Co. KGaA.
[2] Baseri J. R, Palanisamy P.N & Sivakumar P. (2012). Preparation and characterization of activated carbon from thevetia peruviana for the removal of dyes from textile waste water. Advances in Applied Science Research, 3(1), 377-383.
[3] Marsh H. & Reinoso F.R. Activated carbon. 1st Edition. Great Britain: Elsevier Ltd. 2006.
[4] Siti Khadijah C.O., Siti Fatimah C.O., N. Aina Misnon, and F. Hanim K. (2012). Utilization of Sugarcane Bagasse in the Production of Activated Carbon for Groundwater Treatment. International Journal of Engineering and Applied Sciences. ISSN: 2305-8269: Vol 1(2). December 2012.
[5] ASTM C136 / C136M-14, Standard Test Method for Sieve Analysis of Fine and Coarse Aggregates, ASTM International, West Conshohocken, PA, 2014, www.astm.org.
[6] ASTM D2854-09(2014), Standard Test Method for Apparent Density of Activated Carbon, ASTM International, West Conshohocken, PA, 2014, www.astm.org.
[7] Verla A.W., M. Horsfall, E.N. Verla, A.I. Spiff, O.A. Ekpete. (2012). Preparation and characterization of activated carbon from fluted pumpkin (Telfairia occidentalis hook f.) seed shell. Asian Journal of Natural & Applied Sciences. ISSN:2186-8476, Vol.1. No.3. September 2012.
[8] ASTM D2866-11, Standard Test Method for Total Ash Content of Activated Carbon, ASTM International, West Conshohocken, PA, 2011, www.astm.org
[9] Gergova K., Petrov N., & Eser S. (1994). Adsorption properties and microstructure of pyrolysis. Carbon, 32(4), 693-702.
[10] Collin G. J., Fauziah A.A., Hasnul F.M.Z, & Siti F.D. (2006). Treatment of landfill leachate in Kayu Madang, Sabah: Porosit and adsorption studies (Part 2). Asian Chemistry Letters, 10 (3-4), 89-94.
[11] American Water Works Association (AWWA). (1978). Standards for activated carbon. American Water Works Association, ANSI/AWWAB600-78 Denver Co.
[12] American Water Works Association (AWWA). (1991). Standards for granular activated carbons. American Water Works Association, ANSI/AWWAB604-90 Denver Co.
[13] Yakout S.M & El-Deen G.S. (2012). Characterization of activated carbon prepared by phosphoric activation of olives stones. Arabian Journal of Chemistry. Doi 10.1016/j.arabjc.2011.12.002

The Characterization of Chlorophyll-A and Microalgae Isolation Process of Wastewater Collected at Sembrong Dam

R Wellson[1], N Othman[1] and H M Matias-Peralta[2]
[1]Faculty of Civil and Environmental Engineering, University Tun Hussein Onn Malaysia, 86400 Batu Pahat, Johor, MALAYSIA
[2]Faculty of Science, Technology and Human Development, University Tun Hussein Onn Malaysia, 86400 Batu Pahat, Johor, MALAYSIA

E-mail: Wellsonrichter@gmail.com

Abstract. Recently, there has been an increasing number of river water quality deterioration that has brought into water quality disruptions that entering dams including in Johor and one of them is occurred in Sembrong Dam in Johor. Sembrong Dam is a major water source for some 120,000 people in the districts of Kluang and parts of Batu Pahat. The quality of water in Sembrong should be well-monitored in ensuring the continuous distribution of clean and safe water supply to peoples. Based on the news reported by The Star news dated on 11 May 2015, the water bodies in Sembrong Dam are polluted by the algae blooms which has started to cause problems in treating water phase by clogging up the filters and causing the production to be reduced and finally resulting in frequent water disruptions to residents. Therefore, there is a need to study the water quality of the dam water prior to further water treatment. One of important characterizations is by measuring chlorophyll-a and the isolation of the dominant microalgae species in the water body in which they are able to indicate the level of water pollution. This paper presents the determination of chlorophyll-a and the isolation of microalgae strains collected from Sembrong Dam. Chlorophyll-a is a photosynthetic pigment present in all species of phytoplankton, including algae and in some photosynthetic bacteria, known as cyanobacteria. The method used in measuring the chlorophyll-a is based on the standard method of IS0 10 260. The average chlorophyll-a concentration measured at Sembrong Dam is 175.9 $\mu g\ L^{-1}$ and it is responsible for the appearance of green color in the sample and it is categorized into hypereutrophic state which is highly polluted. The technique used for isolation of microalgae strains is traditional method which is by spreading the sample on agar. The pure isolate indicated that the genus Botryococcus is the dominant algae species which is characterized morphologically. Both chlorophyll-a and microalgae isolation are good biological indicator that indicate the pollution of Sembrong Dam. The pure culture is very important that it can be used for further studies with series of different tests to understand its properties and character for sustainability approach towards environmentally friendly as well as for microalgae removal formula.

Keywords: Microalgae, isolation, chlorophyll-a.

1. Introduction

Sembrong Dam was previously built for flood mitigation in 1984 and due to the demand for clean water supply to peoples, it has been providing water for human consumption since 1990. It covers up to 775ha and supplies 55 million litres of treated water via its Sembrong Barat water treatment plant daily [1]. Since that, the water quality control in Sembrong Dam has always been monitored to prevent from water quality deterioration that can affect the treatment water process as well as to provide clean water to peoples.

Unfortunately the water bodies in Sembrong Dam has been degraded by two general type of pollution which is point sources and nonpoint sources. Point sources is a single, identifiable source of pollution while the nonpoint source is those inputs and impacts which occur over a wide area and are not easily attributed to a single source. Based on the study by [2], the nonpoint sources that contribute to the increasing pollution are coming from indiscriminate farming activities around the dam, agriculture activities, husbandry and planting of oil palm trees which cover at least 87% of the dam's 130 sq km catchment area.

The Star news dated on 11 May 2015, in this year had reported that the dam is slowly dying due to the eutrophication or better known as algae bloom and had caused 244 disruptions at the Sembrong water treatment. This phenomena has turned the dam water into green water which indicates the present of high concentration of nutrients in the dam water such as nitrogen and phosphorus [3]. The source of the nutrients, mainly due to the excessive fertilizer used in oil palm plantation, animal waste and discharges from agriculture and sewerage. The Star news described that the algae bloom has started to cause problems when treating water, as they clog up the filters, causing the production to be reduced or halted and finally resulting in frequent water disruptions to residents.

The pollution by microalgae blooms is concerned to affect the process of water treatment in Sembrong Dam by carrying high loading of microalgae that would make the treatment process more costly. Normally, high cost will be needed in the primary separation of alum flocculation to reduce the sediment of microalgae and more chlorine is required to fully kill the microalgae which would cause the high concentration of Trihalomethanes (THMs). THMs constitutes the major category of chlorinated disinfection byproducts, formed by the reaction between residual chlorine and natural organic matter in water such as algae, river weeds, and decaying leaves. They are tasteless and odourless, but harmful and potentially toxic [4]. Lifetime exposure to THMs may pose risks to human health. Several epidemiologic studies have suggested a link between THMs exposure and risk of bladder, colon, and rectum cancers [5]. However, knowledge on characterization of the water body in Sembrong Dam will help in monitoring the level of eutrophication by measuring the concentration of chloropyll-a in the water. According to Robinson et al [6], chlorophyll-a is probably the most-often used estimator in North America of algal biomass in lakes and streams by providing an estimate for measuring algal weight and volume. More than that, chlorophyll-a able to relate the relationship between nutrient concentration and other biological phenomena in aquatic ecosystem as well as effective measure for trophic status for pollution classification [7]. The results of chlorophyll-a with average reading $175.9\ \mu g\ L^{-1}$ in this study has indicated that the dam water is polluted and classified under hypereurotrophic state.

Algae is a good indicators of water quality and many lakes are characterized based on their dominant phytoplankton groups. The microalgae species found in the Sembrong dam can be further studied in enhancing the water treatment process as well as can be utilized in many field such us in medicine, foods, agriculture, chemical, drugs and treatment for wastewater. More than that, the disadvantages of the microalage is when excessive blooms occurs in which they will produce scums and lead to the discoloration of water and when left untreated, these alage could suffocate fish and affect the entire marine environment [8]. The high loading of microalgae also will cause the cost of treatment water to be increased due to the extra amount of chemical and additional treatment system needed to remove the alage.

The most dominant microalgae species can be identified by observing the samples under the microscope on Phytoplankton chamber and by isolation process. The overall chlorophyll-a concentration comes from the pigment of dominant species in which it becomes important indicator of pollution as well will be useful if studied further in reducing them in Sembrong Dam and the isolated alage can be studied further for the optimization for biodiesel production.

2. Materials & Method

2.1 Field collection

Water samples with visible microalgal population were collected from River sembrong dam located about 10km from Air Hitam in the state of Johor on the Air Hitam-Kluang road. The sampling was conducted during dry season for ensuring undiluted concentration of contaminant in the water. Fourth times of sampling were conducted within 4 consecutive weeks in which one sampling for microalgae isolation and Chlorophyl-a and the other for chlorophyll- a. All the samples locations were taken at the same location which is at the water intake of water treatment plant. The samples were collected by using two of 4 L bottle sampling for each time sampling. The bottles containing sample were wrapped with aluminium foil and placed into a polystyrene box containing ice cubes to be carried to laboratory for about 15 minutes travel time using a car. The standard method used for sampling is based on [9].

2.2 The Determination of Chlorophyll-a

In determining the chlorophyll-a ESS Method 150.1 [10] and IS0 10260 are adapted [11]. This experiment is conducted immediately after the samples arrived in the laboratory. In this experiment, samples of 100 m and 2.5 mL of $MgCO_3$ were directly filtered through a 47 mm Whatman GF/C filter. After that, the filter paper then was folded by using tweezer and placed into a graduated boiling tube and pushed down. Then, 90 % of methanol is added into the graduated boiling tubes up to 10 mL mark and capped very loosely and the tubes are placed into the water bath with 60 ° C for 2 minutes. The resultant solutions are filtered through a 25mm GF/C filter and the filtrate was collected. The absorbance of both extracts was measured at 665 and 750 nm. The concentration of chlorophyll -a was determined according to [11]. The steps were repeated for another replicates.

2.3 Microalgae isolation

Alga isolation in this study involved the combination of dilution techniques and with standard plating agar methods to separate algal populations. Once the samples arrived in Lab, 200 ml of sample was centrifuged at 4000 rpm with 28 °C and washed by the addition of autoclaved distilled water for three times centrifugation to remove the mucilage or other contaminant. Right after the sample was rinsed, a simple dilution was conducted by using four 10 mL test tubes with the addition of 1mL of every test tube to aid the isolation process. Sterilized plastic petri dishes (100 × 15 mm) were used for the preparation of agar plates. The media growth used is the modified bold's basal medium that is used for many of the green algae. The 50 agar plates were made by mixing the 1000 mL of distilled water, 20 mL of BBM, and 23 g of pure agar powder altogether in the beaker and autoclaved for 121o C for 15 minutes. After that, two agar plating techniques have been applied which are by using spreading and streaking method. In spreading procedure, a drop of sample from the test tube is put on the agar plate by using a disposal Pasteur pipettes and the glass hockey stick is used in spreading a drop of sample on agar thoroughly until the spreading became sticky. After that, the petri dish is covered with its lid and wrapped it with parafilm to avoid from contamination. The sample is stacked upside down inside the incubator and left for 14 days and steps were repeated for the other three diluted samples. In streaking, the wire loop is sterilized by putting the loop in flame using Bunsen burner inside the fume cabinet to sterilize it and the sterilized wire loop is cooled down for few seconds and dipped into the sample. The wire loop then pulled out and the sample was streaked back and forth across the surface of the agar. After that, the petri dish is covered with its lid and wrapped it with parafilm to avoid from contamination. The sample is stacked upside down inside the incubator and left for 14 days and the steps were Repeated for the other three diluted sample.

On the 14th day, there were number of algal colonies in form of green dots observed inside the agar plates. Once the algae colonies formed, the next step to proceed is to culture the colonies into the conical flasks. In transferring the algal colonies, a green dot is taken out by using sterilized wire loop and placed it into the conical flask containing BBM. There were 10 green dots of algae colonies transferred and cultured into the 10 conical flasks. Culturing the algae is based on the natural condition by covering the conical flask's mouth with cotton wool, have it shaken twice per day and exposed it with the sunlight. After the 14 days passed, the algae cultures turned green and ready to be checked under the inverted microscope. By observing the 10 algae cultures from different conical flasks the result showed that there were still two and three species of algae living in the culture. As the pure culture is still have not yet obtained, the streaking and spreading have been conducted once again as well as the transferring the colonies into 12 conical flasks. Finally, the pure culture is obtained. Figure-1(a) is the colonies formed on the surface of the agar and Figure-1(b) is the culturing of algae colonies into the conical flasks..

Figure 1(a). Colonies formed.

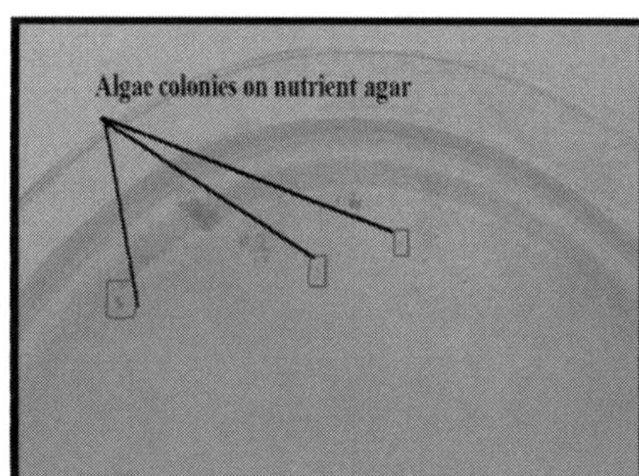

Figure 1(b). Culturing algae colonies.

3. Results and discussion

3.1 Concentration of Chlorophyll-a

One of the easy and efficient ways to monitor the quality of water in Sembrong Dam is by measuring the level of chlorophyll-a. As all phytoplankton have chlorophyll-a, then chlorophyll-a is an important parameter to be monitored. In addition, it can be used for continuous or long-term monitoring. Lakes are commonly classified according to their trophic state, a term that describes how "green" the lake is as measured by the amount of algae biomass in the water. Three trophic state categories that are used

IOP Conf. Series: Materials Science and Engineering **136** (2016) 012062 doi:10.1088/1757-899X/136/1/012062

to describe lakes as they grow progressively greener are oligotrophic, mesotrophic, and eutrophic. Lakes with extreme trophic indices may also be considered hyperoligotrophic or hypereutrophic.

Since the the classification of trophic based on chlorophyll-a is still not established by NWQS, the classification of the trophic status was conducted based on Us Trophic Status Index [12]. This Index was developed as a part of of The National Eutrophication Survey (NES), comparing the work of some investigator on chlorophyll-a vs tropic state as show in in Table-1 [13]. More than that, the comparison with the Carlson Trophic State index [14] was made as in Table-2.

The average value of chlorophyll-a for four weeks of Sembrong Dam is 162.78 µgL-1 as shown in Figure 2 . By comparing the result with Table 1, the trophic status of Sembrong Dam is belong to eutrophic state which is following the range given by Academy, Dobson and EPA-NES but is not valid for Sakamoto. Therefore, when the result is compared with Table 2 which stated that the chlorophyll-a ranged from 56 – 155+ and secchi depth ranged from 0.25 - 0.5 is belong to hypereutrophic state. By comparison, the average chlorophyll-a of sembrong dam is 162.78 µgL-1 and the depth of sembrong dam ranged from 0m to 7m have classified Sembrong Dam as in hypereutrophic state [2].

The result also shows that the river quality is in the range considered problematic by the New Hampshire Department of Environmental Services which provides the chlorophyll guidelines for river quality. Chlorophyll-a measurement below 7 µg/l is within a desirable range. 7-15 µg/l is less than desirable, while over 15 µg/l is considered problematic. Furthermore, Based on the study of 5 lakes conducted by [15] one of the lakes named Feldberger Haus with the depth 6m has the highest measurement of chlorophyll-a which is 175.9 µg L−1 and with that value he categorized the lake is belong to hypereutrophic state.

According to Dietrich [16] hypereutrophic lakes are very nutrient-rich lakes characterized by frequent and severe nuisance algal blooms and low transparency. More than that, hypereutrophic lakes have a visibility depth of less than 3 feet, they have greater than 40 micrograms/litre total chlorophyll and greater than 100 micrograms/liter phosphorus.

Table 1. Trophic state vs. chlorophyll-a [13].

Trophic Condition	Chlorophyll-a (µgL^{-1})			
	Sakamoto	Academy	Dobson	EPA-NES
Oligotrophic	0.3-2.5	0-4	0-4.3	<7
Mesotrophic	1-15	4-10	4.3-8.8	7-12
Eutrophic	5-140	>10	8.8	> 12

Table 2. Trophic status [14].

Secchi Depth, SD (Metres)	Chlorophyll-a	Tropic class µgL^{-1}
>8 - 4	0—2.6	Oligotrophic
4 - 2	2.6—20	Mesotrophic
2 – 0.5	20—56	Eutrophic
0.5 -0.25	56—155+	Hypereutrophic

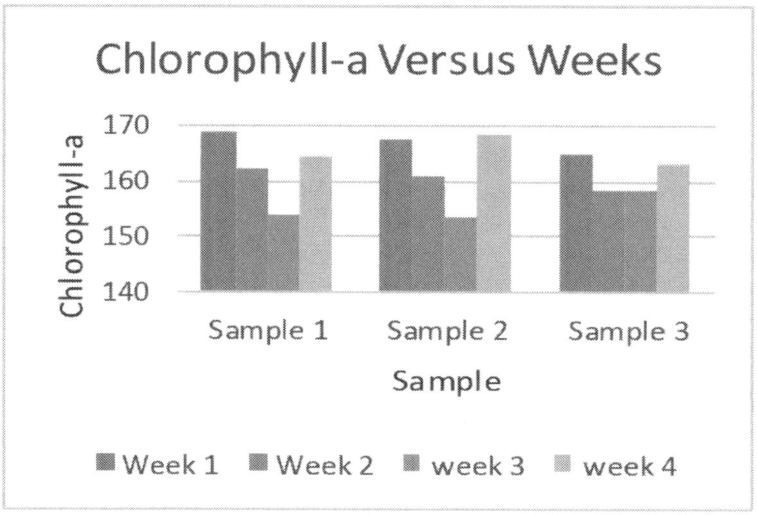

Figure 2. Chlorophyll-a sample reading within 4 consecutive weeks.

3.2 Morphological Identification

The water from sembrong dam was sampled onto agar plates. Green microalgae clones were picked up after a few cycles of agar plate spreading. This general classification method was only used to distinguish isolates on the most basic level. Identification of these isolates to the genus level was based on the morphology of the individual cells following microscopic examination. The colony characteristics and morphological features of the Sembrong Dam water isolate have demonstrated it is close similarity with the genus Botryococcus. The individual cells of the colonies were in the range of 3- 11μm. Cells are spherical in shape and the variation in colonial size is depending upon the daughter colonies which remain attached to one another. More than that, cells are generally green to yellowish and some of them are in dark green, they also show orange yellow colouration as well.

Based on the study by Dayananda et al. [17] the Bortyococcus brunaii were collected from freshwater ponds in Mahabalipuram, Tamil Nadu, India. Specimens were isolated and examined for morphological features using microscopic and was found to be genus Botryococcus and after further after 18S rRNA sequence analysis was carried out the analysis has shown more than 90% similarities with genus Botryococcus and in particular to the species B. braunii. In the morphological, the species of the genus Botrococcus were mainly distinguished based on colony size, colour and the cell shape. Based on Dayananda morphological result, it almost same with the algae isolate from Sembrong Dam and the colonies shape, colour and size is observed and distinguished as shown in Figure-2(a) and Figure-2(b). By comparison, both are spherical and oval-shaped cells. More than that, both isolates also structured irregularly and the cells are attached to each other that links two and more distinct clumps of cell. The colour of both algae isolate also the same which are generally green to yellowish green, some of the cells are dark green colour. The size of individual cell for both isolate are in the range between 3-11 μm.

According to Metzger and Largeau [18], the Botryococcus braunii cells are green color, and together by a lipid matrix under microscopic, whereas some of the strains have dark green color with irregular colonies consisting of hundreds of elliptical cells interconnected by strand of though mucilage. More than that, sometimes Botryococcus. braunii cells are attached to each other by a refringent material that sometimes links two or more distinct clumps of cell. It has been reported that, Botryococcus braunii exists in the form of blooms in fresh water bodies like ponds, lakes and reservoirs and the existence of Botryococcus brunaii. in USA, Ivory Coast, Portugal, Bolivia, Morocco, India, Philippines, Thailand, France and West Indies has confirmed its wide distribution [18].

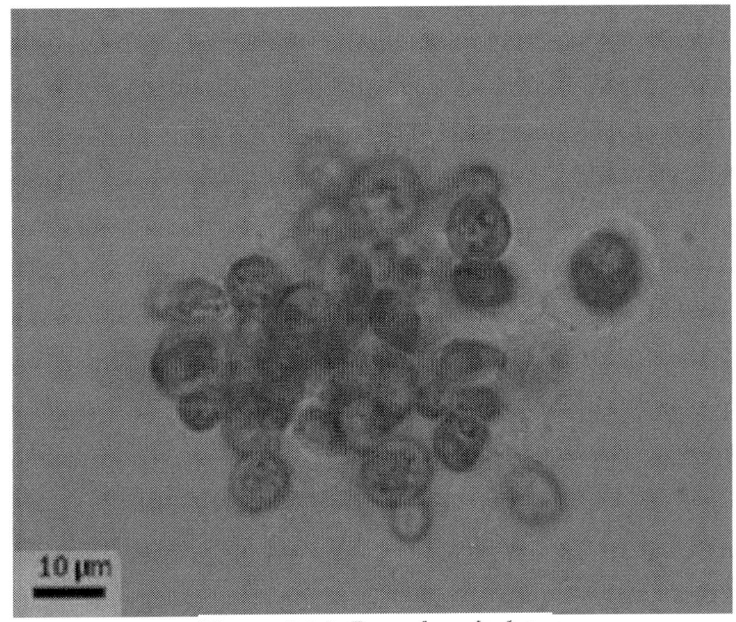

Figure 2 (a). Pure algae isolate.

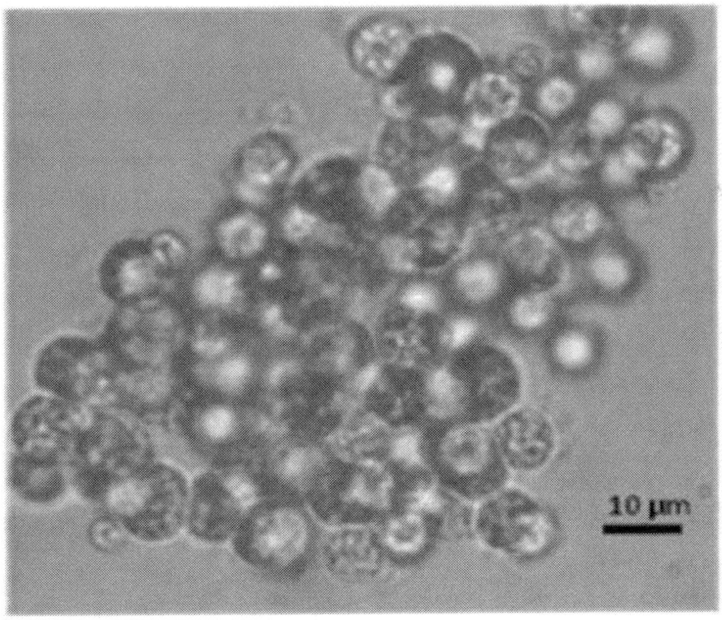

Figure 2 (b). Botryococcus Braunii [17].

4. Conclusion

As conclusion, the concentration of chlorophyll-a of Sembrong Dam has proved that the water body is in hypereutrophic state which is heavily polluted by the microalgae blooms and one of the dominant species of microalgae is botryococcus braunii. The further study can be done is to use the alage as the indicator for pollution and the alage also can be studied for the potential of biodisesl production.

References

[1] Sharip, Z. & Jusoh, J., 2010. Integrated lake basin management and its importance for Lake Chini and other lakes in Malaysia. Research and Management, (15), pp.41– 51.

[2] Daud, Z., Awang, H., Zainuri, M and Hatta, M. (2015). Hydrology Properties and Water Quality Assessment of the Sembrong Dam, Johor, Malaysia. Social and Behavioral Sciences 195, pp.2868 – 2873.

[3] Chislock, M. F., Doster, E., Zitomer, R. A. & Wilson, A. E. (2013) Eutrophication: Causes, Consequences, and Controls in Aquatic Ecosystems. Nature Education Knowledge 4(4), pp.10.

[4] Uza, M., R. LeCraw, L. Fortin, H. Broomer, and A. Edmonds. 1997. "Optimization of WTPs for the Control of Trihalomethanes." Ontario Water and Wastewater Division Conference.

[5] Chowdhury, M. J. Rodriguez, and R. Sadiq. (2011). "Disinfection byproducts in Canadian provinces: associated cancer risks and medical expenses," Journal of Hazardous Materials, vol. 187, no. 1–3, pp. 574–584.

[6] Robinson, K.W., S.M. Flanagan, J.D. Ayotte, K.W. Campo, A. Chalmers, J. F. Coles, and T.F. Cuffney. 2004. Water Quality in the New England Coastal Basins: Maine, New Hampshire, Massachusetts, and Rhode Island, 1999–2001. U.S. Geological Survey Circular 1226, pp.38.

[7] Wickliff, J.L and Aronoff, S.(2015). Quantitative Measurement of Leaf Chlorophylls by Spectrophotometry of Their Pheophytins in Aqueous Alcoholic Extracts. Department of Biochemistry & Biophysics, & the Institute for Atomic Research, Iowa State University, Ames.

[8] Duong, V.T., Yan, L., Nowak, E and Peer, M.S. (2012). Microalgae Isolation and selection for prospective Biodiesel Production. Energies 2012, (5), pp.1835-1849

[9] APHA (2005). Standard Methods for the Examination of Water and Wastewater 21st ed., American Public Health Association, US.

[10] ESS Method 150.1:Chlorophyll-a Spectrophotometric, Environmental Sciences Section, Inorganic Chemistry Unit, Wisconsin State Lab of Hygiene, 465 Henry Mall, Madison, WI 53706. Equation for chlorophyll a from Jeffrey and Humphrey (1975).

[11] ISO 10260:1992. Water quality measurement of biochemical parameters spectrometric determination of the chlorophyll-a concentration. International Organization for Standardization, Geneva, Switzerland, 1992.

[12] Trophic state index". Wikipedia: The Free Encuclopedia. Wikimedia Foundatio, Inc., 29 Sept. 2015. <https://en.wikipedia.org/wiki/Trophic/state/index>.

[13] Premazzi and Chiaudani. (1992). Ecological Quality of Surface Waters. Quality Assessment Schemes for European Community Lakes Commission of the European Communities, Joint Research Centre, Ispra (Varese).

[14] Carlson R.E. and J. Simpson (1996) A Coordinator's Guide to Volunteer Lake Monitoring Methods. North American Lake Management Society.pp.96.

[15] Peter, K., Judit, P., Riner, K., Koschel, L. K. and Frank, G.(2008) Chlorophyll a concentration across a trophic gradient of lakes: An estimator of phytoplankton biomass?.Ecology and Management of Inland Waters, (38), pp.327-338.

[16] Dictrich, U.(1980).Stability and multiple steady states of hypereutrophic ecosystem Development in hydrobiology. Hypertrophic Ecosystem, (2), pp.235-247.

[17] Dayananda, C., Kumudha, A., Sarada, R. and Ravishankar, G. A. (2010). Isolation, characterization and outdoor cultivation of green microalgae Botryococcus sp. Scientific Research and Essays. 5(17), pp. 2497-2505.

[18] Metzger P, Largeau C (2005). Botryococcus braunii: a rich source for hydrocarbons and related ether lipids. Apply Microbiol Biotechnol 66, pp. 486-496.

BBD Optimization of K-ZnO Catalyst Modification Process for Heterogeneous Transesterification of Rice Bran Oil to Biodiesel

K S Kabo[1], A R Yacob[1], W A W A Bakar[1], N A Buang[1], A M Bello[1] and A Ruskam[1]

[1]Department of Chemistry, Faculty of Science, Universiti Teknologi Malaysia, 81310 UTM Johor Bahru, Johor, Malaysia

E-mail: kamalskabo@yahoo.com

Abstract. Environmentally benign zinc oxide (ZnO) was modified with 0-15% (wt.) potassium through wet impregnation and used in transesterification of rice bran oil (RBO) to form biodiesel. The catalyst was characterized by X-Ray powder Diffraction (XRD), its basic sites determined by back titration and Response Surface Methodology (RSM) Box-Behnken Design (BBD) was used to optimize the modification process variables on the basic sites of the catalyst. The transesterification product, biodiesel was analyzed by Nuclear Magnetic Resonance (NMR) spectroscopy. The result reveals K-modified ZnO with highly increased basic sites. Quadratic model with high regression $R^2 = 0.9995$ was obtained from the ANOVA of modification process, optimization at maximum basic sites criterion gave optimum modification conditions of K-loading = 8.5 % (wt.), calcination temperature = 480 °C and time = 4 hours with response and basic sites = 8.14 mmol/g which is in close agreement with the experimental value of 7.64 mmol/g. The catalyst was used and a value of 95.53 % biodiesel conversion was obtained and effect of potassium leaching was not significant in the process

Keywords: K-modified-zinc oxide, basic sites optimization, biodiesel, catalyst leaching.

1. Introduction

There is increasing concern for environmental safety and need for sustainable energy resources for use as green transportation fuels. Biodiesel is among the best alternative energy sources as doesn't increase carbon to the atmosphere and can be used in its pure form without major modifications to current diesel engines. It is obtained from renewable sources and so becomes part of the solution to the global energy crisis due to increased consumption, leading to rapid depletion of existing fossil fuel reserves, high cost and increased greenhouse emissions [1]. Unfortunately, commercialization of biodiesel is challenged by a number of problems including availability and cost of feedstock and effective catalyst formulation.

Zinc oxide has very low toxicity and solubility levels which are good properties needed in heterogeneous catalyst as there is less danger of pollution and enable ease of phase separation after product formation [2]. Alone, it has low activity in transesterification reaction and so in order to use it at lower temperature, its properties need to be modified [2,3,4,5]. But the catalysts preparation in the reported literature involves the use of solvent for co-precipitation or harsher reaction conditions are used during the transesterification, both of which can pose environmental hazard or raise the cost of biodiesel production. Potassium hydroxid (KOH) alone is very good homogeneous catalyst with excellent yield in transesterification reactions for biodiesel, but there are so many problems of phase, catalyst and glycerol separation associated with homogeneous catalysts. The use of KOH supported on some substances for heterogeneous application was also investigated and it was reported to show remarkable improvement on the catalytic activity of substances used [6,7]. RBO oil is reported to be a good substrate for biodiesel due to the availability [8,9,10] but it is largely wasted and remain unexploited especially in Africa. This idea can be used to prevent wastage, boost economic value of rice, improving its production for food and utilizing the bran for alternative energy generation. Since ZnO does not dissolve in biodiesel [2], then its modification with more active substance to prepare a modified catalyst will help to produce non-leaching, recyclable catalyst that can be used for commercial biodiesel production.

An equally important aspect of biodiesel production is the optimization of the process to achieve minimum cost for competitive commercialization of the product. Many studies were reported using various mathematical models and tools to optimize reaction conditions for biodiesel production using modified heterogeneous catalysts [6, 11,12,13]. However, optimization of modification process of heterogeneous catalysts prior to application is not reported in the previous studies and that can help in the reduction of excessive use of active substances (main cause of leaching), high energy and time used in calcination all of which are important in cost and environmental saving. Response surface experimental Box–Behnken Design (BBD) was used because it covers entire design space at minimum number of experimental runs [14,15].

In this research, zinc oxide was modified with K to prepare active catalyst with high biodiesel yield at atmospheric pressure, low catalyst loading, temperature and methanol to oil ratio. BBD was used to optimize the effect of K-loading, calcination temperature and time on the basic sites response. The best catalyst prepared was then used for the transesterification of rice bran oil to biodiesel. The potassium leaching was also evaluated.

2. Materials and methods

2.1 Materials

The analytical grade reagents used were supplied by QRëC™; ZnO, KOH, NaOH, HCl, and methanol while 99.8% deuterated chloroform for NMR was obtained from Merck, Germany.

2.2 Catalyst Preparation

Basic Sites Modification of the Catalyst was carried out in order to improve its basic sites through wet impregnation method. In the preparation of a catalyst, 10g of zinc oxide was measured and transferred into 200 ml beaker and 25 ml distilled water added. 3% KOH (w) accurately measured into the beaker and the set-up mounted onto hot plate with magnetic stirrer. The mixture was maintained at 90°C stirred with gentle evaporation until the water dried off and then transferred to oven and kept at 105°C for 8 hours for complete drying. The sample was then transferred to furnace and calcined at required temperature and time. This is the basic sites modified catalyst used in transesterification for biodiesel. All ZnO-XK modified catalysts were prepared in similar way. X is equivalent to 3, 6, 9, 12 and 15% (wt.) KOH used. The use of KOH on nano-crystalline CaO prepared by wet impregnation as reported by [16] have shown an excellent biodiesel conversion and recyclable catalyst without any significant potassium leaching.

2.3 Characterization of the Catalyst

The biodiesel was prepared by using 250 ml double necked round bottomed flask is fitted with a condenser and thermometer, then mounted onto retort stand. In one of the reactions, appropriate molar ratio of methanol is measured into the round flask followed by needed weight percent of the catalyst relative to oil. The mixture was then vigorously stirred for 20 min with the aid of magnetic stirrer to form a uniform alkoxide mixture. 22 ml (20g) of rice bran oil was then added and the set up heated in paraffin oil at 65°C temperature under magnetic stirring for 3 hours. The product was cooled, allowed to stand overnight and centrifuged and three layers were distinctly observed. The small upper layer of unreacted methanol was removed while the middle layer of biodiesel was carefully decanted from the lower layer which contains glycerol and the settled solid catalyst. The transesterification reaction is given in Scheme 1.

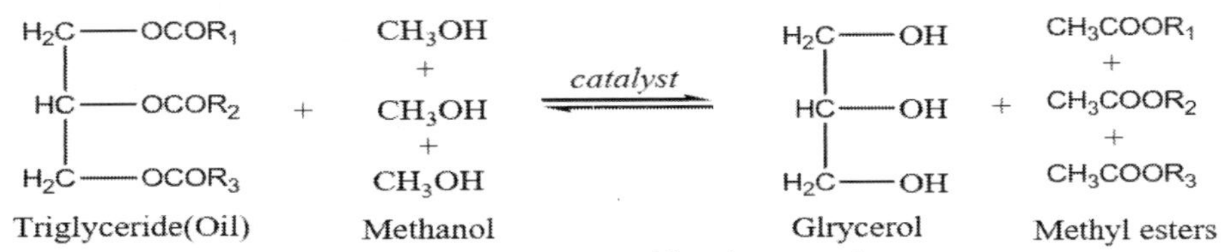

Scheme 1. Transesterification reaction.

2.4 NMR Analysis of the Transesterification Products

The substances obtained after the transesterification were analyzed by ^{1}HNMR spectroscopy for both confirmation of the functional groups and percentage yield determination. The NMR spectra of biodiesel samples were obtained using Bruker 400.

2.5 Design of Experiment

Optimization was performed to evaluate the optimum conditions needed to prepare the catalyst with high amount of basic sites needed. In the modification process, experiments were performed randomly in order to prevent systematic error. Response Surface Methodology Box–Behnken Design (RSM-BBD) was used to study the effects of the variables on response because it covers entire design space at minimum number of experimental runs [14,15].

The amount of basic sites was evaluated to analyze the effect of three variables; potassium loading, X1 (0- 15 % (wt)), calcination temperature, X2 (300-800 °C) and time, X3 (3-6 hours) at three numeric factors levels on the response, basic sites. Based on the design, a total number of 17 experiments with 5 center points were obtained. The experiments were then carried-out at random and the yield was compared against the predicted.

2.6 Statistical Analysis and Model Fitting

Statistical software Design Expert 7.1 was used. Second order full factorial equation was used in order to optimize the effect of the parameters on the response. For predicting the optimal point after performing experiments, a second-order polynomial quadratic equation (Equation 1) was fitted to correlate the relationship between independent variables and responses and interactive effect of the process variables. Regression coefficients of the quadratic model were evaluated by analysis of variance (ANOVA). All the terms in the model were tested by student's F-test and significance of the F-values at probability levels ($p \leq 0.05$) were analysed. The developed mathematical models were used for the construction of three dimensional (3D) response surface plots in order to predict the relationships between independent and dependent variables. The experimental data were evaluated with determination coefficient (R^2), adjusted determination of coefficient (R^2 adj.), predicted determination of coefficient (R^2 pred.). Verification experiments were performed under the optimal conditions and the value of the experiments was compared with the predicted values of the developed model equations.

$$Y = \beta_o + \sum \beta_i x_i + \sum \beta_{ii} x_i^2 + \sum \beta_{ij} x_i x_j + \mathcal{E} \quad (1)$$

Where,

Y is the response factor (molar yield of "biodiesel" reaction),

x_i the ith independent factor;

β_o the intercept;
β_i the first order coefficient of the model;
β_{ii} the quadratic coefficient of i factor;
β_{ij} the linear coefficients of the model for interaction between i and j factors; and
$\mathcal{E}$ the experimental error [24].

3. Results and Discussions

In this research, zinc oxide was modified with KOH by wet impregnation, the catalysts prepared were characterized by XRD and basic back titration method. BBD experimental design was used to evaluate the effect of catalyst modification process variables on the response basic sites. The transesterification was carried out and the results are presented as follows.

3.1 Basic Strength Sites Back Titration Analysis

The results of basic sites analysis obtained from back titration showed moderate basic sites compared to modified catalysts using the same method, 9.80 mmol/g from was obtained from Na-modified metal oxides [17]. All KOH-modified ZnO sample have higher basic sites compared to the pure calcined ZnO samples. The combination of KOH with calcination temperature increases the basic strength of the catalysts. This occurs as a result of interaction of potassium with zinc oxide at high temperature which became adsorbed to form K-ZnO single phase and improve its basic strength. In contrast, calcination at low temperatures does not significantly improve the basic sites due to inability of the potassium to strongly adsorb on ZnO surfaces. The result also showed that, at optimum conditions, the use of KOH above 9 % did not cause significant increase in basic sites. This indicates that, the basic sites came from the modified catalyst and not from the KOH itself. This is in agreement with the results of KOH modified CaO [16] and K/MgO [18], which reported neither increased basic sites nor catalytic performance and higher amount of KOH and better K-Mg surface interactions at optimum conditions respectively. Also, at temperatures above 600°C, decomposition of K_2O takes place and this leads to the deterioration of the resultant basic sites in the catalysts. The summary of basic sites result is given in Table 1.

Table 1. Basic sites of the catalysts prepared under optimized conditions.

Catalyst	0K-ZnO	3K-ZnO	6K-ZnO	9K-ZnO	12K-ZnO	15K-ZnO
Basic Sites (mmol/g)	3.61	4.37	6.74	7.93	8.05	8.13

3.2 X-Ray Powder Diffraction (XRD) Analysis

The XRD analysis was carried out to investigate the presence of K2O in the ZnO crystal phase. The spectra in Figure 1 indicates hexagonal wurtzite zinc oxide, zincite with reference number 01-075-1526. The three most intense peaks appeared at the $2\theta = 36.53$ (101), 32.07 (100) and 34.46 (002) respectively. Others are $2\theta =$ 47.79 (102), 57.16 (110), 63.10 (103), 67.07 (200), 68.49 (112) and 69.77o (201) [19]. The K2O peaks could be observed at $2\theta = 28$, 31, 33, and 51o [7,16,20]. In the XRD spectra obtained from 8.5% K- modified ZnO, the presence of K_2O could be observed. The intensity of K_2O peaks is weak due to low concentration of K on the surface and does not increase with the amount of KOH used, suggesting fine deposition of K on to the crystalline lattices of the ZnO. It also indicate interaction between K and Zn which may weaken the Zn^{2+}-O^{2-} bond thereby enabling better interaction with methanol and thus increasing its catalytic activity and yield. In all the KOH-modified ZnO catalysts, ZnO peak intensities remain approximately the same indicating that there was no increase in crystallite size when the K is attached to the ZnO crystalline lattice.

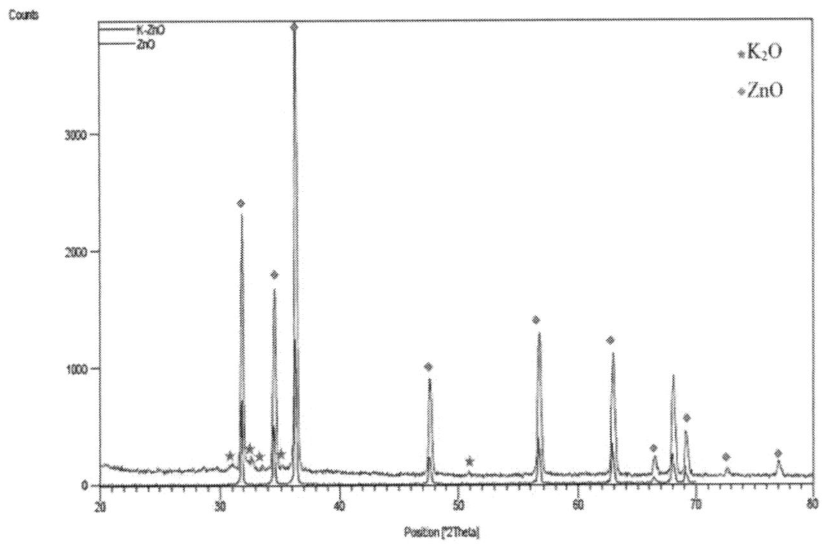

Figure 1. XRD spectra of pure and K-modified ZnO.

3.3 Result of Basic Sites Experimental Design and Statistical Analysis

For basic sites evaluation, three factors at three levels BBD was used in the optimization of the process variable; KOH loading (X_1), calcination temperature (X_2)and time(X_3) on the response basic sites. A total of 17 experiments were randomly performed to avoid systematic error and the experimental results were compared with the predicted responses as presented in Table 2 and Figure 2. The final quadratic Equation 2 that explain the mathematical relation of the independent variables; X_1, X_2 and X_3 with the response, basic sites was obtained as

$$Y = 7.70 + 2.40X_1 - 0.73X_2 + 0.37X_3 - 0.79X_1 X_2 - 0.70X_1X_3 + 0.37X_2X_3 - 2.10X_1^2 - 1.94X_2^2 - 1.84X_3^2 \quad (2)$$

Table 2. BBD with corresponding experimental results for the basic sites.

Run	KOH Load (X_1)	*Calc. Temp (X_2)	*Calc. Time (X_3)	*Exp. Basic Sites (mmol/g)	*Pred. Basic Sites (mmol/g)
1	0.00	550.00	6.00	1.66	1.68
2	0.00	550.00	3.00	1.04	1.03
3	15.00	800.00	4.50	4.56	4.53
4	7.50	550.00	4.50	7.57	7.70
5	7.50	300.00	6.00	3.96	3.90
6	0.00	300.00	4.50	1.16	1.19
7	15.00	300.00	4.50	7.53	7.57
8	7.50	550.00	4.50	7.63	7.70
9	7.50	550.00	4.50	7.85	7.70
10	15.00	550.00	3.00	7.24	7.21
11	7.50	550.00	4.50	7.72	7.70
12	7.50	800.00	3.00	3.12	3.18
13	7.50	300.00	3.00	5.40	5.38
14	15.00	550.00	6.00	5.08	5.09
15	0.00	800.00	4.50	1.36	1.32
16	7.50	550.00	4.50	7.73	7.70
17	7.50	800.00	6.00	3.17	3.19

*Calc.= calcination *Exp.= experimental *Pred.= predicted

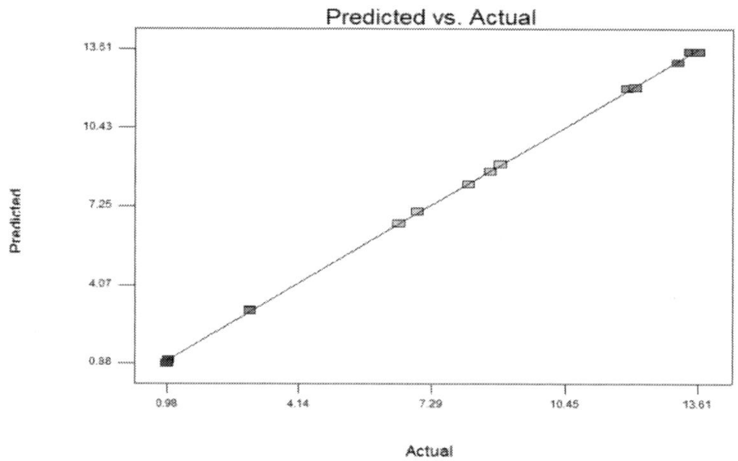

Figure 2. Diagnostic plot of comparison between experimental and predicted basic sites in modification.

In order to validate the model, analysis of variance was used and its summary presented in Table 3. The model F-value of 1438.50 implies that the model is significant and that there is only a 0.01% chance that a Model F-Value this value could occur due to noise. The P-values of < 0.001 obtained indicates that all the model terms are significant. The "lack of fit" F-Values of 0.42 also implies that this factor is not significant. The R^2-value 0.9995 obtained for the model is very high. There is also a good agreement between the predicted R^2-value of 0.9988 and the adjusted R^2-value of 0.9973. Thus, the above analysis indicates that this predicted quadratic model is in good agreement with the experimental responses and it can be adopted.

Table 3. Analysis of variance (ANOVA) for the basic sites modification process.

Source	SS	DF	MS	F	P	Comment
Model	110.92	9	12.32	1438.50	< 0.0001	Significant
X_1	46.03	1	46.03	5372.64	< 0.0001	Significant
X_2	4.26	1	4.26	497.58	< 0.0001	Significant
X_3	1.07	1	1.07	125.25	< 0.0001	Significant
X_1X_2	2.51	1	2.51	293.22	< 0.0001	Significant
X_1X_3	1.93	1	1.93	225.51	< 0.0001	Significant
X_2X_3	0.56	1	0.56	64.78	< 0.0001	Significant
X_1^2	18.61	1	18.61	2172.38	< 0.0001	Significant
X_2^2	15.93	1	15.93	1859.10	< 0.0001	Significant
X_3^2	14.29	1	14.29	1668.32	< 0.0001	Significant
Residual	0.060	7	0.008568			
Lack of fit	0.014	3	0.004792	0.42	0.7488	Not Significant
Pure error	0.046	4	0.011			
Total	110.98	16				

SS: sum of squares; DF: degree of freedom; F: f-value; P: p-value

3.4 *Effect of KOH loading*

KOH loading is the single most important factor in this analysis and it has a positive linear effect on the response basic sites in the modification process as can be observed in Equation 3 and Table 3. From the 3D graphical plot Figures 3 (a and b), it can be seen the KOH loading has slight negative interaction effect with both higher calcination temperature and time. This is due higher temperature and prolonged heating causes aggregation of the particles and decreased surface area and active sites.

It also shows negative intermediate quadratic effect in the general process. In both interactions, the KOH loading leads to higher response, though its increase peaks at about 7.5 % KOH loading at 550°C and 5 hours and starts decreasing as the temperature and time are increased to higher than their mean values. It shows that the optimum amount needed for the modification should not exceed 11.25 %, above which it become waste.

3.5 *Effect of Calcination Temperature*
Temperature is the second important factor in this modification process as can be observed in Equation 3 and Table 3. The heat is necessary to provide the energy needed to properly adsorb and disperses potassium on the ZnO crystalline structures. From the 3D plot in Figures 3 (b and c), it can be observed that higher temperature has a negative quadratic effect on the modification process. There was increase in the response basic sites as the calcination temperature is raised up to 550°C. Above that, the quadratic effect become negative causing a significant decrease of the response, basic sites. This is due to the decomposition of active substance, K_2O at higher temperatures and possibility of annealing the particles to become large with decreased active sites.

3.6 *Effect of Calcination Time*
In this modification process, calcination time is the third important factor as can be observed in Equation 3 and Table 3. Adequate time is needed for effective modification, less time does not enable the active substance to adsorb and when time is prolonged there is possibility of its decomposition or annealing to occur. From the 3D plot in Figures 3 (a and c), it can be observed that it has negative interaction with both KOH loading and temperature at higher values. Its quadratic effect on the modification process is also negative. There was increase in the response basic sites as the calcination time is raised up to 4.50 hours. Above that, the quadratic effect become negative causing a significant decrease of the response, basic sites. This is due to the decomposition of active substance, K_2O as calcination time become longer. The residence calcination time can also cause annealing the particles to become larger with decreased active sites.

3.7 *Optimization of Basic Sites Modification*
Optimal values of the modification variables were obtained by solving the quadratic equation at maximum basic sites criterion. Considering these parameters ($X_1 = 8.68$, $X_2 = 479.71$, $X_3 = 4.23$ and Basic sites = 8.14) repeated experiments were performed under similar conditions. Experimental value of 7.64 mmol/g was obtained which in good agreement with the predicted value of 8.14 mmol/g and this KOH modified ZnO catalyst chosen in the of transesterification process.

IOP Conf. Series: Materials Science and Engineering **136** (2016) 012063 doi:10.1088/1757-899X/136/1/012063

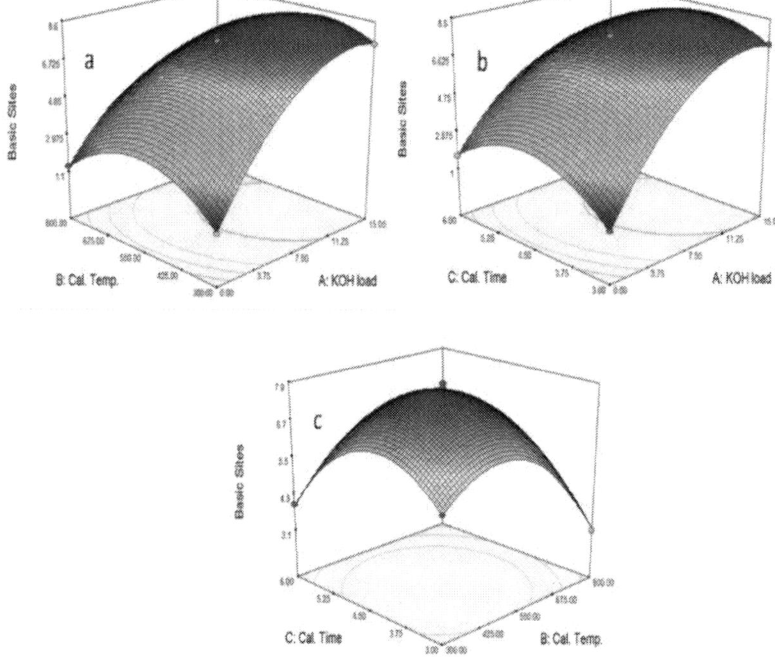

Figure 3. 3D plots of basic sites as a response of (a) KOH loading and calcination time, (b) KOH loading and calcination temperature and (c) calcination temperature and time

3.8 Results of HNMR Biodiesel Analysis

Biodiesel was prepared from ZnO and KOH-modified ZnO at 65°C, 3 hrs and 9:1 methanol to oil ratio. The samples were characterized by ^{1}HMR for both qualitative and quantitative analysis. The disappearance of methoxy triglyceride protons peak at $\delta = 4.3$ ppm and formation of fatty acid ester methoxy protons peak at $\delta = 3.67$ ppm as shown in Figure 4 indicated the formation of biodiesel while the ratio of the $\delta = 3.67$ ppm methoxy protons of biodiesel with $\delta = 2.3$ ppm methylinic protons of oil is used for quantitative determination. The spectra obtained were analyzed with the Bruker TOPSPIN 2.1 software and the yield was calculated using Equation 3 from ratio of the peak area of methylene protons of the triglycerides (A_1) which appears at $\delta = 2.3$ ppm with that of methoxy protons present in the fatty acid methyl esters of the biodiesel which appears at $\delta = 3.67$ppm (A_2) [21,22]. A percent conversion of 95.53 % was observed, which shows a very good conversion compared to other heterogeneous catalysts.

$$\%\,Yied = \frac{A_1/_3}{A_2/_2} \times 100\% \qquad (3)$$

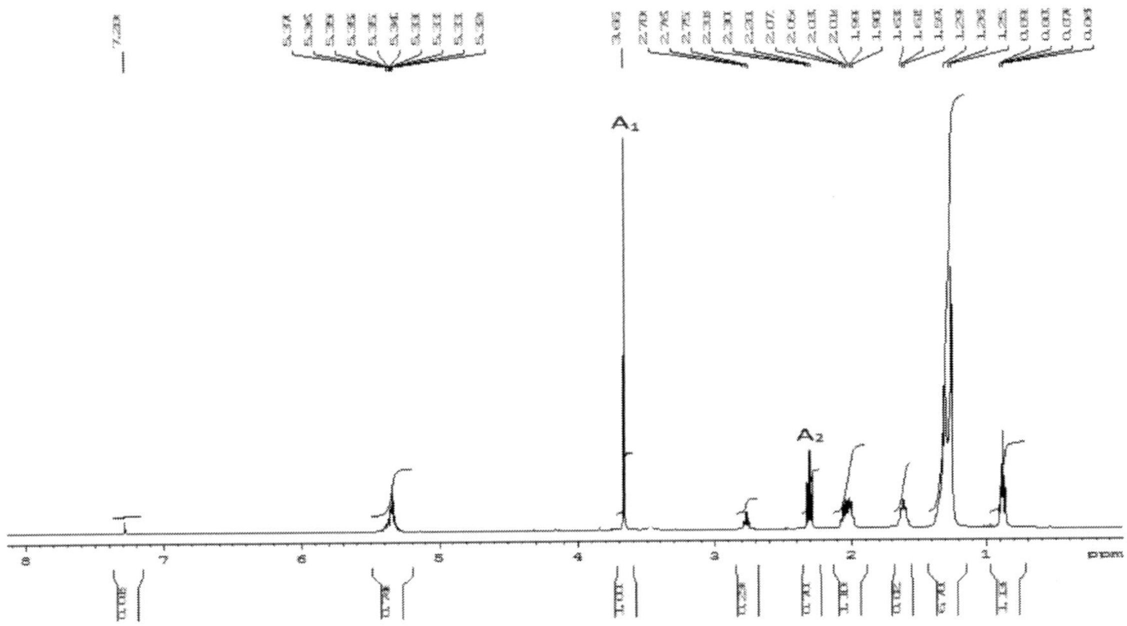

Figure 4. NMR Spectra of 95.53% biodiesel prepared under optimized conditions of catalyst loading = 5%, temperature = 65 °C, time = 3 hrs mins and MOR = 9.

3.9 Catalyst Leaching

Leaching is the major cause of catalyst deactivation and also the possibility of homogeneous reaction which can have negative effect on the overall process. In order to screen the possibility of potassium leaching into biodiesel, a simple test as adopted by [16,23]. The optimized catalyst and methanol were mixed in the same experimental conditions as carried out in the transesterification process, but in the absence of RBO oil. The reaction was allowed to proceed for 3 h and the catalyst finally recovered by filtration. The methanol filtrate was used in transesterification reaction of RBO. The reaction was carried out for 3 hours at 65 °C and only about 4.30% of FAME yield of about was observed. This is showing that the catalyst leaching only slightly contributed in the process and the reaction was generally catalysed by the modified catalyst, supporting XRD result that the modification process produced a highly dispersed K-ZnO catalyst.

4. Conclusion

In order to conserve energy and improve sustainability, a model was developed for optimum conditions of catalyst modification process which provided stable catalyst with improved basic sites. Rice bran oil which is mainly wasted in the farms was used in the transesterification and high conversion to biodiesel was obtained. The catalyst leaching is not significant and so the process is heterogeneously catalyzed and environmental friendly.

References

[1] Khan, T.M.Y., Atabani, A.E., Badruddin, I.A., Badarudin, A., Khayoon, M.S., and Triwahyono, S. (2014) Recent scenario and technologies to utilize non-edible oils for biodiesel production. Renewable and Sustainable Energy Reviews, 37, 840–851. Elsevier.

[2] Yan, S., Mohan, S., Dimaggio, C., Kim, M., Ng, K.Y.S., and Salley, S.O. (2010) Long term activity of modified ZnO nanoparticles for transesterification. Fuel, 89, 2844–2852. Elsevier Ltd.

[3] Yoo, S.J., Lee, H.-S., Veriansyah, B., Kim, J., Kim, J.-D., and Lee, Y.-W. (2010) Synthesis of biodiesel from rapeseed oil using supercritical methanol with metal oxide catalysts. Bioresource technology, 101, 8686–9. Elsevier Ltd.

[4] Wan, L., Liu, H., and Skala, D. (2014) Environmental Biodiesel production from soybean oil in subcritical methanol using MnCO 3 / ZnO as catalyst. "Applied Catalysis B, Environmental," 152-153, 352–359. Elsevier B.V.

[5] Baskar, G. and Aiswarya, R. (2015) Biodiesel production from waste cooking oil using copper doped zinc oxide nanocomposite as heterogeneous catalyst. BIORESOURCE TECHNOLOGY. Elsevier Ltd.

[6] Liao, C.-C. and Chung, T.-W. (2013) Optimization of process conditions using response surface methodology for the microwave-assisted transesterification of Jatropha oil with KOH impregnated CaO as catalyst. Chemical Engineering Research and Design, 91, 2457–2464. Institution of Chemical Engineers.

[7] Takase, M., Zhang, M., Feng, W., Chen, Y., Zhao, T., Cobbina, S.J., Yang, L., and Wu, X. (2014) Application of zirconia modified with KOH as heterogeneous solid base catalyst to new non-edible oil for biodiesel. ENERGY CONVERSION AND MANAGEMENT, 80, 117–125. Elsevier Ltd.

[8] Patil, P.D., Gude, V.G., and Deng, S. (2009) Biodiesel Production from Jatropha Curcas, Waste Cooking, and Camelina Sativa Oils. Industrial & Engineering Chemistry Research, 48, 10850–10856.

[9] Gunawan, S., Maulana, S., Anwar, K., and Widjaja, T. (2011) Rice bran, a potential source of biodiesel production in Indonesia. Industrial Crops and Products, 33, 624–628. Elsevier B.V.

[10] Sun, Y., Ponnusamy, S., Muppaneni, T., Reddy, H.K., Patil, P.D., Li, C., Jiang, L., and Deng, S. (2014) Optimization of high-energy density biodiesel production from camelina sativa oil under supercritical 1-butanol conditions. Fuel, 135, 522–529. Elsevier Ltd.

[11] Mahdavi, V. and Monajemi, A. (2014) Optimization of operational conditions for biodiesel production from cottonseed oil on CaO–MgO/Al2O3 solid base catalysts. Journal of the Taiwan Institute of Chemical Engineers, 45, 2286–2292. Taiwan Institute of Chemical Engineers.

[12] Sánchez, M., Navas, M., Ruggera, J.F., Casella, M.L., Aracil, J., and Martínez, M. (2014) Biodiesel production optimization using γAl2O3 based catalysts. Energy, 73, 661–669.

[13] Veiga, P.M., Luna, A.S., de Figueiredo Portilho, M., de Oliveira Veloso, C., and Henriques, C.A. (2014) Zn,Al-catalysts for heterogeneous biodiesel production: Basicity and process optimization. Energy, 75, 453–462.

[14] Nazwanie, W., Abdullah, W., Azelee, W., Abu, W., Wan Abdullah, W.N., Wan Abu Bakar, W.A., Ali, R., and Embong, Z. (2014) Oxidative desulfurization of commercial diesel catalyzed by tert-butyl hydroperoxide polymolybdate on alumina: optimization by Box-Behnken design. Clean Technologies and Environmental Policy, 433–441.

[15] Witek-Krowiak, A., Chojnacka, K., Podstawczyk, D., Dawiec, A., and Pokomeda, K. (2014) Application of response surface methodology and artificial neural network methods in modelling and optimization of biosorption process. Bioresource technology, 160, 150–60. Elsevier Ltd.

[16] Kumar, D. and Ali, A. (2012) Nanocrystalline K–CaO for the transesterification of a variety of feedstocks: Structure, kinetics and catalytic properties. Biomass and Bioenergy, 46, 459–468. Elsevier Ltd.

[17] Singh, A.K. and Fernando, S.D. (2009) Preparation and Reaction Kinetics Studies of Na-based Mixed Metal Oxide for Transesterification. Energy & Fuels, 9, 5160–5164.

[18] Querini, C.., Cornaglia, L.., Ulla, M.., and Miró, E.. (1999) Catalytic combustion of diesel soot on Co,K/MgO catalysts. Effect of the potassium loading on activity and stability. Applied Catalysis B: Environmental, 20, 165–177.

[19] Hariharan, C. (2006) Photocatalytic degradation of organic contaminants in water by ZnO nanoparticles: Revisited. Applied Catalysis A: General, 304, 55–61.

[20] Mutreja, V., Singh, S., and Ali, A. (2014) Potassium impregnated nanocrystalline mixed oxides of La and Mg as heterogeneous catalysts for transesterification. Renewable Energy, 62, 226–233. Elsevier Ltd.

[21] Guzmán-vargas, A., Santos-gutiérrez, T., Lima, E., Flores-moreno, J.L., Oliver-tolentino, M.A., and Martínez-ortiz, M.D.J. (2015) Efficient KF loaded on MgCaAl hydrotalcite-like compounds in the transesterification of Jatropha curcas oil. Journal of Alloys and Compounds, 1–6.

[22] Kaur, N. and Ali, A. (2015) Lithium zirconate as solid catalyst for simultaneous esterification and transesterification of low quality triglycerides. "Applied Catalysis A, General," 489, 193–202. Elsevier B.V.

[23] Chen, J., Jia, L., Guo, X., Xiang, L., and Lou, S. (2014) Production of novel biodiesel from transesterification over KF-modified Ca–Al hydrotalcite catalyst. RSC Adv., 4, 60025–60033.

[24] Thirugnanasambandham, K., Sivakumar, V., Maran, J.P., and Kandasamy, S. (2014) Chitosan based grey wastewater treatment — A statistical design approach. Carbohydrate Polymers, 99, 593–600. Elsevier Ltd.

Application of Geographic Information Systems (GIS) in Analysing Rainfall Distribution Patterns in Batu Pahat District

A A Kadir[1*], M Kaamin[1], N S Azizan[2], S Sahat[1], S M Bukari[2], M Mokhtar[1], N Ngadiman[1] and N B Hamid[1]
[1]Center of Diploma Studies, Universiti Tun Hussein Onn Malaysia, 86400 Batu Pahat, Johor, MALAYSIA
[2]Faculty of Civil and Environmental Engineering, Universiti Tun Hussein Onn Malaysia, 86400 Batu Pahat, Johor, MALAYSIA

E-mail: aeslina@uthm.edu.my

Abstract: Rainfall forecasting reports are crucial to provide information and warnings to the population in a particular location. The Malaysian Meteorology Department (MMD) is a department that plays an important role in monitoring the situation and issued the statement of changes in weather and provides services such as weather advisories and gives warnings when the situation requires. Uncertain weather situations normally have created panic situation, especially in big cities because of flash floods due to poor drainage management. Usually, local authorities provided rainfall data in tables, and it is difficult to analyse to acquire the rainfall trend. Therefore, Geographic Information System (GIS) applications are commonly used to generate rainfall patterns in visual formation with a combination of characteristics of rainfall data and then can be used by stakeholders to facilitate the process of analysis and forecasting rainfall. The objective of this study is to determine the pattern of rainfall distribution using GIS applications in Batu Pahat district to assist interested parties to understand and easy to analyse the rainfall data in visual form or mapping form. Rainfall data for a period of 10 years (2004-2013) and monthly data (Dec 2006 – Feb 2007) are provided by the Department of Irrigation and Drainage (DID) for 12 stations in the district of Batu Pahat, and rainfall maps in each year was obtained using the interpolation Inverse Distance Weighted (IDW) method was used in this research. The rainfall map was then analyzed to identify the highest rainfall that was received during the period of study. For the conclusion, this study has proved that rainfall analysis using GIS application is efficient to be used in gaining information of rainfall patterns as the results show that the highest rainfall occurred in 2006 and 2007, and it were the years of major floods occurrence in Batu Pahat district.

Keywords: Geographic information system (GIS), inverse distance weighted (idw), rainfall distribution.

1. Introduction

Malaysian climate condition has a uniform temperature, high humidity and abundant of rainfall. Therefore, climate forecasts can help certain parties in future planning involving agriculture industry, fishing industry, flood prediction and many more [1]. In addition, rain forecast may also help to identify specific locations, areas or regions in a country that receives the highest annual rainfall. Furthermore, weather forecasting plays an important role to give information about the weather and can provide early warning of the possibility of severe weather phenomena, high winds and rough seas. Together with these warnings, information will be disseminating through the mass media. It is intend to inform the public as soon as possible about the adverse weather conditions and sea conditions that were forecasted [2].

Traditionally, rainfall data were obtained from local authorities at meteorological stations (rain gauges data collection). Using Thiessen Polygons or Inverse Distance Weighting (IDW) data in each rain gauge was measured to gain its total volume of rainfall on the area. How accurate the rain gauge in predicting rainfall volume is depends on how consistent the gauge estimates where it is actually represent the rainfall distribution between each gauge. It is hardly to get the accurate results using rain gauge alone, therefore it was not surprising that many hydrologic analysis frequently exhibit large uncertainties [3]. Thus, the usage of GIS applications to produces the rainfall patterns in the form of visual and combined with the attribute data of rainfall can be used by stakeholders to facilitate the process of analysis and forecasting rain. Consequently, would help to solve the problems associated with rainfall data. IDW function should be used when a set of points is dense enough to capture the extent of surface area to which area that need to be analysed. This method gives more accuracy in interpolation results because it gives values close to the minimum and maximum values of the sample data. In this study, IDW function was chosen because the set of rainfall stations is dense enough to capture all the surface in Batu Pahat district.

This study focused on the district of Batu Pahat. This study aims to determine the pattern of 10 years rainfall distribution using GIS applications between 2004 and 2013 which includes 12 rainfall stations. The main data used were the rainfall data. In this study, the rainfall data was obtained from the Department of Irrigation and Drainage (DID). The map of the area in Batu Pahat district was also used in this study. The aim of this study was to determine the pattern of rainfall in the area of Batu Pahat in the form of mapping based on information and criteria required in the use of GIS applications. GIS application effectiveness is determined by using the analysis result of the rainfall distribution pattern based on cause and effect of the highest rainfall identified during the review period.

1.1. Rainfall Intensities

Rainfall is a form of precipitation that fell to earth as one of the processes to complete the cycle of the earth. According to [4], rain is one of the precipitation in liquid form where the diameter particulars is more than 0.5 mm. Effectiveness or impact of the rain on a system is dependent on the intensity. Rainfall intensity may be influenced by a number of internal factors such as the saturation of water molecules, latent heat of fusion, the rate of evaporation, cloud formation process, the initial coverage rate and external factors such as the position of latitude, the movement of the monsoons and the diverse landscape of earth [5].

In general, monthly and annual rainfall received throughout Malaysia is greatly influenced by the monsoon. The influence of monsoon forming a cycle of rain based on the northeast monsoon, the southwest monsoon and two transition periods [6]. Previous studies show that at the beginning of the season northeast monsoon, normally throughout the whole of Peninsular Malaysia was sluggish, particularly in the east coast. However, at the end of that season, many areas began to experience dry conditions [7].

1.2. GIS Application and IDW

A GIS is a computer system for capturing, storing, checking, and displaying data related to positions on Earth's surface. GIS can shows many different kinds of data on one map. This enables people to more easily see, analyse, and understand patterns and relationships. With GIS technology, people can compare the locations of different things in order to discover how they relate to each other [8].

Interpolation is the procedure used to predict cell value for location that lack sample points. The method of IDW should be used when the set of points is dense enough to capture the extent of local surface variation needed for surface. IDW determines cell values using a linear-weighted combination set of sample points. The weight assigned is a function of the distance of an output cell location. The greater the distance, the less influence the cell has on the output value [9].

2. Methodology

In this study, the Batu Pahat rainfall data, map of Batu Pahat District and GPS coordinates for rainfall stations were used (Figure 1). Analysis of rainfall data was based on the analysis of space and time. For the analysis of space, location of rainfall stations was plotted by using GPS, then the amount of annual rainfall was plotted by each rainfall station. After that, the isohyet maps that represented the annual rainfall for each year was predicted by using the interpolation of IDW method in GIS software. This interpolation is the procedure used to predict cell value for location that lack of sample points. The aim of isohyet map development using GIS was to identify the distribution of rainfall patterns and to compare the rainfall distribution between years. In addition, the changing patterns of rainfall from year to year can also be analysed. Other than that, the analysis of changes in rainfall per year was analysed by using the graph changes in rainfall for each station. The analysis was then performed to identify the highest rainfall received during the review period and then the causes and effects of the highest rainfall were identified. From this analysis, the average rainfall during this past decade can be determined. Furthermore, extreme phenomena that occurred during the ten years can also been identified.

3. Result and Discussion

The analysis has also been made based on rainfall stations where the rainfall patterns has shown in graphical form between 2004 and 2013 (Figure 2). It indicates that the darker the colour show in the map, the higher the rainfall received on that area. From Figure 2, it shows that in year 2005 and 2006, the rainfall occurred was heavier than usual, which are exceeding 2500 mm per year.

Figure 3 shows that the highest rainfalls occurred in 2007 recorded at Ladang Sri Gading station was 4444 mm. In addition, the stations at Pintu Kawalan Parit Saidi Barat, Pintu Kawalan Parit Bintang, Pintu Kawalan Sembrong, Pintu Kawalan Separap and Ladang Union in Yong Peng has also recorded that the highest rainfall occurred at each station over the past 10 years were in 2007. In the same figure clearly shows that Rumah Pengawas Parit Raja, Pusat Kawalan Parit Bintang, Pintu Kawalan Sembrong, Ladang Union in Yong Peng and Pintu Kawalan Tampok have similar rainfall reading start from 2004 until 2013 with average 1823 mm. Furthermore, the average rainfall for each year was also been calculated and the results show that the highest average per year over 10 years was occurred in year 2007 followed by 2006, 2012, 2011, 2004, 2005, 2013, 2008, 2009 and 2010. From the analysis, it can be shown that the presence of heavy rains were occurred in 2006 and 2007.

This study further proved by an analysis of the pattern of rainfall in December 2006 until February 2007. This analysis has created by using cumulative monthly rainfall data for 3 months data starting from December 2006 until February 2007. From the analysis, it shows that heavy rainfall has certainly occurred in 2006 and 2007. The analysis of rainfall patterns using GIS application has identified almost all areas in Batu Pahat received the rainfall exceeds 1000 mm within 3 months. The total rainfall in Batu Pahat in this period was the highest rainfall in the whole of Peninsular Malaysia. The results of rainfall pattern using GIS applications were compared with isohyet map produced by the DID. Isohyet map from DID shows that the highest rainfall also occurred in Batu Pahat District in year 2006 and 2007 where it shows that Batu Pahat district received rainfall more than 1000 mm. From the comparison, it proved that a heavy rain was absolutely occurred in the area of Batu Pahat in December 2006 – February 2007. Figure 4 shows the comparison between rainfall pattern using GIS and isohyet map from DID where both of the isohyet map show the same result which are both of them received rainfall more than 1000 mm within 3 month. It means that the analysis of using GIS has proven the effectiveness in analysing the rainfall data in the mapping form or visual form.

The results of this study then were used to identify the causes and effects that occur by consequences of heavy rains in 2006 and 2007. According to meteorological expert [10], floods in Johor in year 2006 and 2007 was due to extraordinary rains caused by element tropical climate phenomenon called the Madden Julian Oscillation (MJO). MJO originates from the Indian Ocean and across 4 times a year. When MJO meet Northeast Monsoon season in Peninsular Malaysia by the end of the year, it will rain more often than usual.

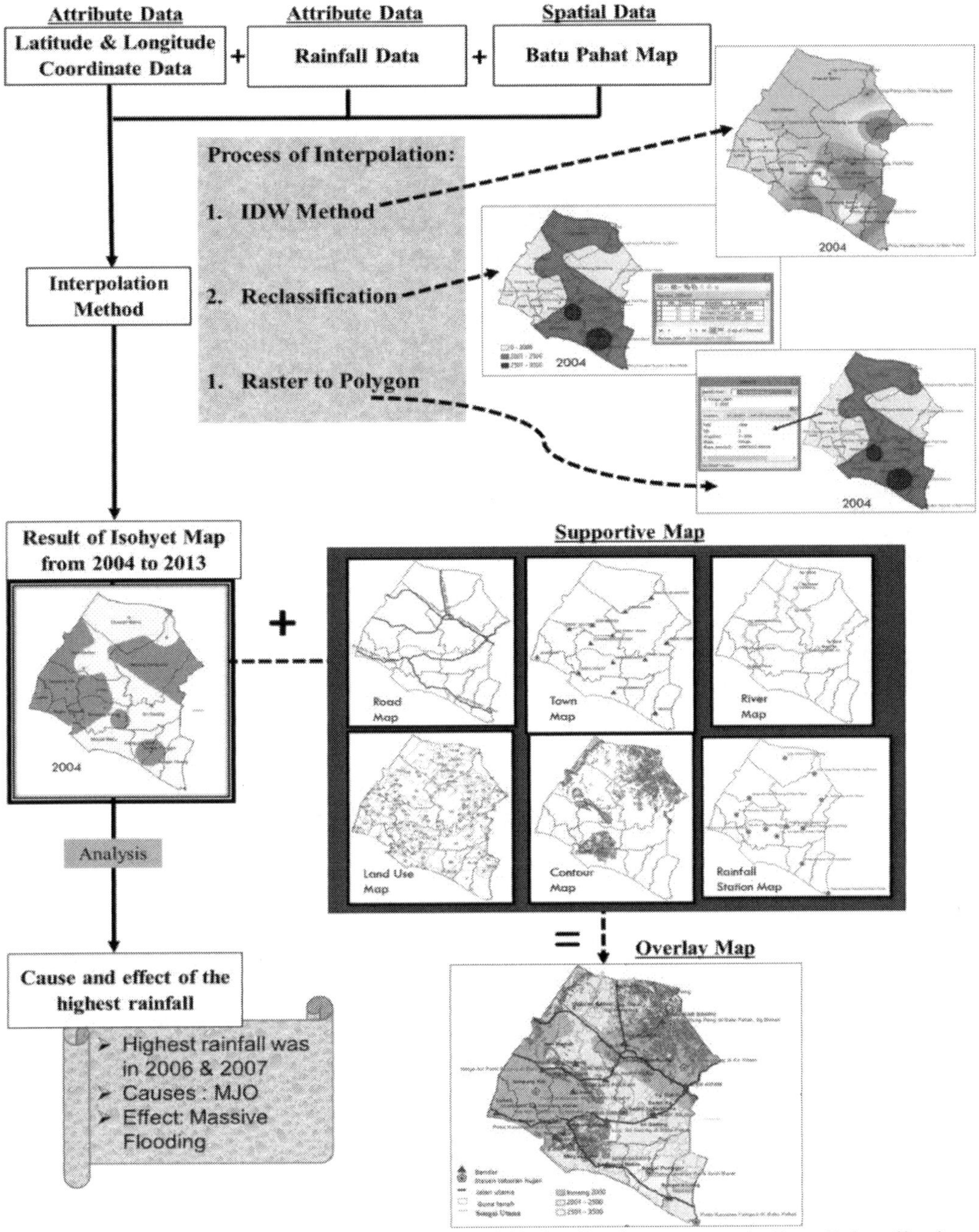

Figure 1. Methodology of analyzing the rainfall distribution patterns in Batu Pahat district.

Figure 2. Distribution patterns of rainfall in Batu Pahat area year 2004 – 2013.

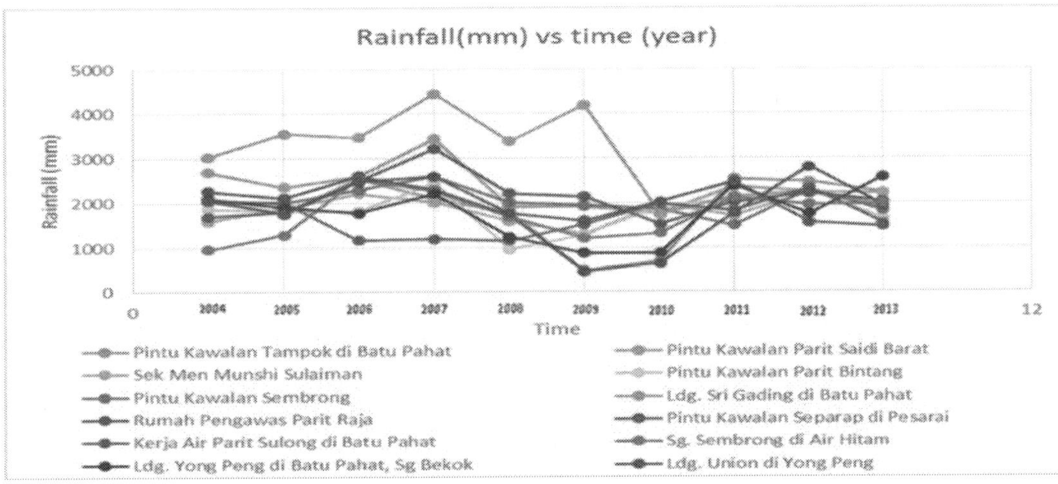

Figure 3. Graph of rainfall (mm) versus time (year).

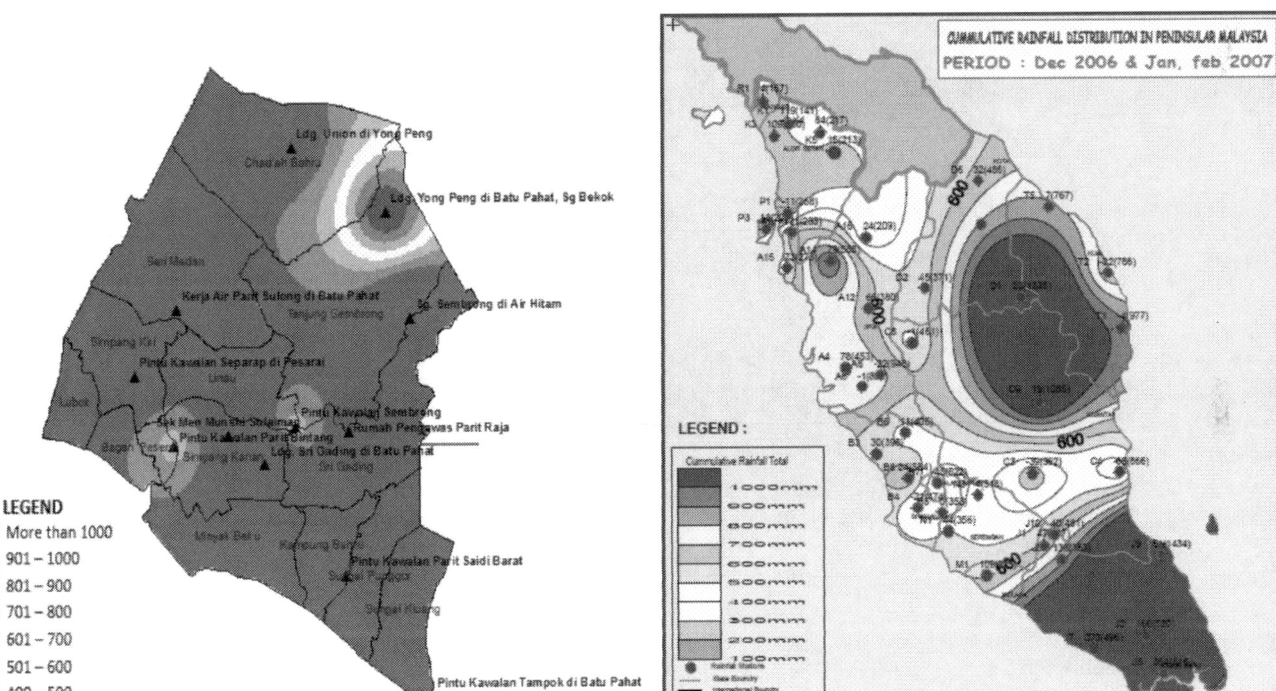

Figure 4. Comparison between rainfall pattern using GIS and isohyet map from DID.

MJO increase the amount of the northeast wind that brings excess water and encourage extraordinary monsoon rains when the monsoon came across to Peninsular Malaysia. Although the MJO across 4 times a year, but rainstorms only occur when it reaches the other climate phenomena such as monsoon. While the impact of heavy rains have proved that Johor was hit by severe flooding on 19 to 31 December 2006 (first flood) and on 12 to 17 January 2007 (second flood) involved eight districts including Johor Bahru, Kota Tinggi, Batu Pahat, Muar, Segamat, Kluang, Pontian and Mersing. The phenomenons of extraordinary heavy rains have caused severe damages to this all eight districts due to extreme flood. As the result more than 103,000 were ordered to move in evacuation centers. In addition, the floods have also killed at least 17 peoples during the flood when it happens where people are beginning to move back home after the first floods began to recede. Furthermore, many schools, the administration of local authorities, hospitals and business activities also disrupted due to flooding [11]. Figure 5 indicates the flood image in 2006 and 2007, where the flood will normally happens at the northeast of Peninsular Malaysia. However in 2006 and 2007, flooding occurred in the southern of Peninsular Malaysia which are normally should experience a dry period. This phenomena occurred due to the increasing amount of northeast winds and bring more heavy rains to the southern of Peninsular Malaysia.

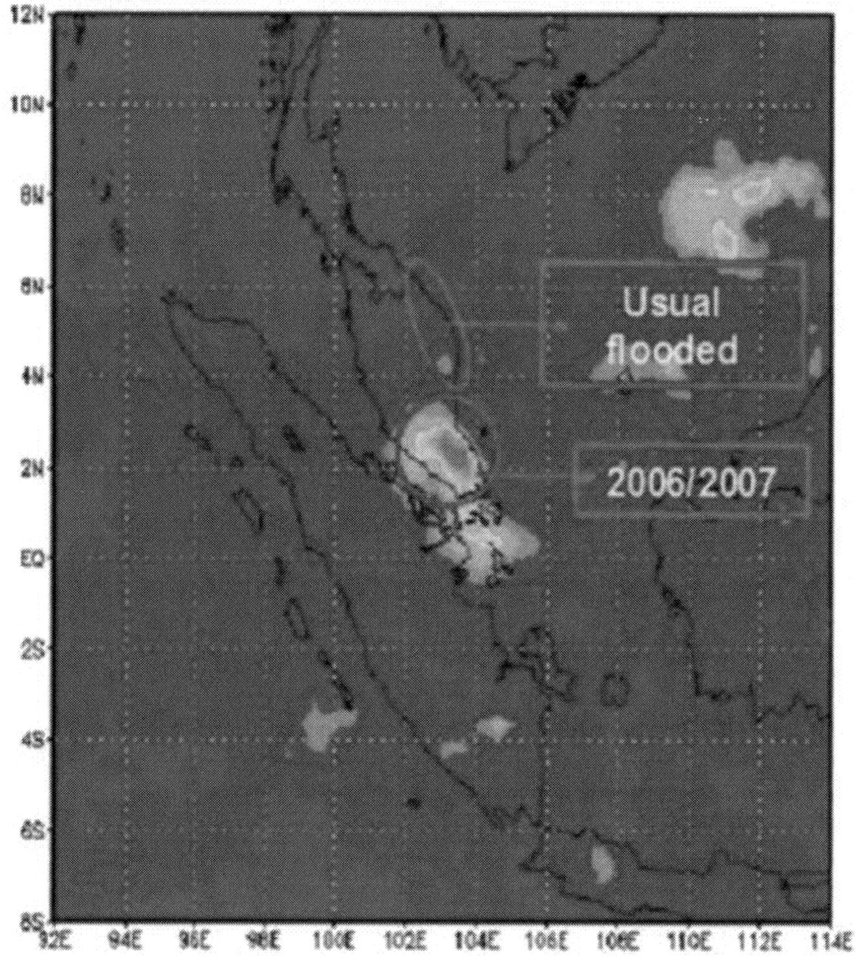

Figure-5. Flood image (2006 and 2007).

4. Conclusions

In conclusion, application of Geographic Information Systems (GIS) in determining the distribution patterns of rainfall in Batu Pahat area have proved their effectiveness where the analysis have shown that 2006 and 2007 were the years of the heavy rainfall occurrence and the application has proved the existence of massive flooding event caused by heavy rains that year.

Overall, the application of GIS is one of the appropriate tools in handling information and involved operation such as planning, observation, collection, storage, management and analysis of data to produce information that can used in the decision-making process. GIS applications are also able to create a better performance in term of hydrological processes rather than using rain gauges alone.

Acknowledgements

Thanks to UTHM for allowing this study to be carried out and help to simplify the process of getting data and also thank to the Department of Irrigation and Drainage (DID) to provide rainfall data in carrying out the study of application Geographic Information Systems (GIS) in determining the distribution patterns of rainfall in Batu Pahat area.

References

[1] Roudier, P., Muller, B., d'Aquino, P., Roncoli, C., Soumaré, M. A., Batté, L., & Sultan, B. (2014). The role of climate forecasts in smallholder agriculture: Lessons from Participatory Research in Two Communities in Senegal. Climate Risk Management, 2, 42-55.

[2] Bahari, M., Dollah R.B., Sap M. N., Bakri A. (2006). Study of Clustering Technique in Data Mining For Climate Data. Universiti Teknologi Malaysia.

[3] Hoblit, Brian C. and David C. Curtis. (2001). GIS Highlights Importance of High-Resolution Radar Rainfall Data. Proceedings of the Twenty-First Annual ESRI International User Conference. San Diego, California.

[4] Wan Ruslan Ismail (1994) Pengantar Hidrologi. Kuala Lumpur: Dewan Bahasa Dan Pustaka.

[5] Mohd Sani, S.F. & Rindam, M. (2011). Analisis Taburan Hujan dan Impaknya Kepada Sumber Air di Pulau Pinang. Universiti Sains Malaysia: Tesis Sarjana Muda.

[6] Ahmad, N. (2003). Pengkelasan Data Kajicuaca Di Dalam Peramalan Hujan Dari Pelbagai Sumber Menggunakan Rangkaian Neural. Universiti Teknologi Malaysia: Tesis Sarjana Sains.

[7] Ahmad, S. & Yassen, M.E. (2005). Urban Climate Research in Malaysia. IAUC Newsletter.

[8] Mohd Zambri, M. R. (1999). Sistem Maklumat Geografi Untuk Aplikasi Kejuruteraan Tentera. Buletin GIS, Jabatan Ukur dan Pemetaan, pp. 2-5

[9] Colin Childs. Interpolating Surfaces in ArcGIS Spatial Analyst. Esri Education Services. 2004.

[10] F. T. Tangang, L. Juneng, E. Salimun, Kwan M. S., Loh J. L. & H. Muhamad (2012), Climate Change and Variability over Malaysia: Gaps in Science and Research Information. Sains Malaysiana 41(11) UKM, 1355–1366

[11] Nasir, A.S. (2008). Kajian Banjir di Sungai Muar. Universiti Teknologi Malaysia: Tesis Sarjana Muda.

[12] Gao, M., He, X., Luo, D., Jiang, J. and Teng, Q. (2013). Object tracking using firefly algorithm. IET Computer Vision, 7(4), pp.227--237.

Hydrogeochemistry Characteristics in Kampong Salang, Tioman Island, Pahang, Malaysia

A N Idris[1,2], A Z Aris[1], S. M. Praveena[3], S. Suratman[2], I Tawnie[2], M K N Samsuddin[2] and A Sefei[2]
[1]Environmental Forensics Research Centre, Faculty of Environmental Studies, 43400 UPM, Serdang, Selangor, MALAYSIA
[2]Hydrogeology Research Centre, National Hydraulic Research Institute of Malaysia 43300 Seri Kembangan, Selangor, MALAYSIA
[3]Department of Environmental and Occupational Health, Faculty of Medical and Health Science, 43400 UPM, Serdang, SELANGOR

E-mail: azrul@nahrim.gov.my

Abstract: The constituents of major ions of groundwater samples has been used to identify the chemical characteristics and the type of groundwater in the low lying area in Kampong Salang located at north-west of Tioman Island in Pahang state. This study aimed to identify the groundwater quality with some geochemical process and to understand groundwater characteristics in the low-lying aquifer of the study area. From that, fourty two (42) groundwater samples were collected and analysed for various parameters. Physical and chemical parameters of groundwater such as pH, electrical conductivity (EC), total dissolved solids (TDS), Ca, Na, Mg, K, HCO_3, Cl, SO_4, NO_3, CO_3, were determined. TDS did not exceed 1,000 mg/l with a maximum of 359 mg/l for sampling activities. Generally, groundwater in the study area is fresh, based on the analytical results. Hydrochemical facies of groundwater in the study area are dominated by Ca-HCO_3 and Ca-Mg-HCO_3 water type. A software known as PHREEQC was used to calculate the saturation indices for the main mineral phases with respect to carbonate and sulphate minerals. Most of the minerals were saturated to undersaturated with respect to carbonate minerals and undersaturated with respect to sulphate minerals composition. Maximum ionic strength in the study area is 0.009 which was found 90 m from the sea. Mostly minerals from calcite and dolomite precipitation are from bedrock of pyroxene hornfel.

Keywords: Hydrogeochemistry, PHREEQC, saturation indexes, undersaturated ionic strength.

1. Introduction

Groundwater is utilised as one of the sources beside surface water for regular water supply in most of the small island. Seawater intrusion is often become a major constraint to optimal utilisation of fresh groundwater from island aquifers. Excessive groundwater abstraction, has led to large-scale lowering of groundwater tables and changing the direction and magnitude of groundwater gradients [1]. Understanding the change of groundwater chemistry is important to appreciate the characteristic of water facies in the tropical islands [2]. Coastal aquifers, particularly tropical islands, fragile in nature and the alteration of the equilibrium state between freshwater and saltwater, thereby affecting socioeconomic development and damaging the entire ecosystem. This is because coastal aquifers are usually in direct contact with the sea, creating a state of equilibrium between the freshwater and saltwater. Distortion of the equilibrium usually occurs through either an increase or a decrease in the flow of freshwater, leading to alteration of the freshwater/saltwater equilibrium interface [3].

It is pertinent to understand the hydraulic properties and hydrochemical characteristics of the groundwater as a movement below ground surface is very slow. In general, groundwater flow below ground surface along its path will increase the concentration of the chemical species [4]. The interrelationship between groundwater flow, hydrogeologic properties and hydrochemistry is related to chemical quality and lithology of the area of study.

Chemistry of groundwater depends very much on the various factor such as geology, degree of chemical weathering, rock type, quality of recharge water and inputs from sources other than water-rock interaction [5]. The interactions will result in a different groundwater quality.

This study aimed to identify the groundwater quality with some geochemical process and to understand groundwater characteristics in the low-lying aquifer of the study area. The concentrations of major and minor ions of groundwater are essential in understanding geochemical processes affecting water quality. The results of this study may contribute to the optimisation of water resources management.

2. Study Area

Kampong Salang, Tioman Island (Figure 1) with a landmass of approximately 1.5 km^2, most of the surface terrain is hilly with a maximum elevation of 350 m above sea level The island is under the control of Tioman Development Authority and the study area located between 2°52.580' to 2°52.582'N latitude and 104°09.291' to 104°09.275'E longitude. It is one of the main area of tourists' congregation and well known to the divers as one of the best coral reef locations in Malaysia. About 40% of Kampong Salang is a low lying area and about 60% of it's covered by forest. The climate of the island is tropical, receiving over 2,500 mm of rainfall annually, most of it during the monsoon season between November and February, with a temperature range between 28 and 31 °C and humidity of about 80%. Groundwater levels tend to form a surface that is sub-parallel to the topography. Areas of low topography relief therefore tend to have very low groundwater gradients compared to areas with higher relief. This factor needs to be considered in relation to movement of water away from the location where it is recharged artificially. The river in the northern part of Tioman Island especially in Kampong Salang is influenced by weather condition. Salang River water level very low during the dry season where river flows into the ground as subsurface river or groundwater while during the monsoon season there are a lot of streams appearing and water flows into the main river.

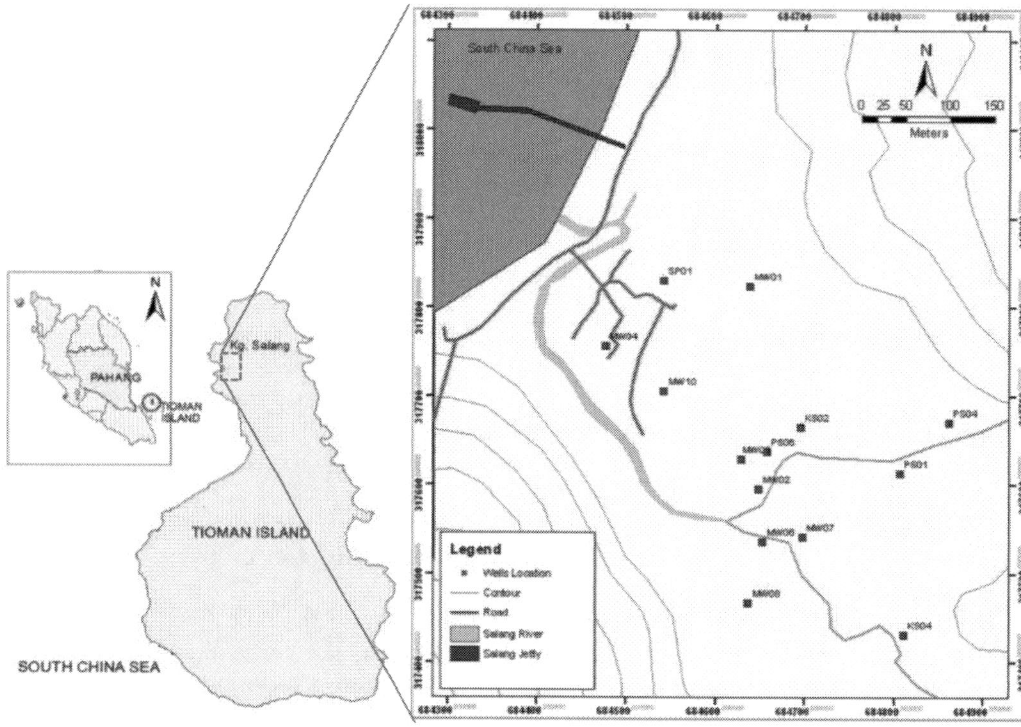

Figure 1. Kampong Salang, Tioman Island.

3. Materials and Methods

Fourty two groundwater samples were collected from 14 wells at Kampong Salang from April, June and August 2013. The locations of sampling points are shown in Figure 1. Groundwater sampled using polyethylene bottles based on the methods described in [6]. All samples were vacuum-filtered in the field for further analyses of cations using Whatman cellulose Nylon-type Millipore filters of 0.45-um pore size. In-situ analyses were conducted in the field using multiparameter sonde YSI 6600 for total dissolved solid (TDS), electrical conductivity (EC), pH, dissolve oxygen (DO), redox potential and temperature. Chloride and bicarbonate were determined using argentometric and titration methods in the laboratory respectively. Major cations (Ca^{2+}, Na^{+}, Mg^{2+} and K^{+}) were determined in the laboratory using inductively coupled plasma mass spectrometry (ICP-MS) model Nexion 300X, made by PerkinElmer from United State of America (USA) according to the Environmental Protection Agency (EPA). Sulfate and Nitrate were detected using HACH DR2800 Portable Spectrophotometer. All the analyses were performed in the water quality laboratory and follow the procedures referred to [6].

The statistical analysed for descriptive statistics water quality parameter presented by mean, standard deviation, minimum and maximum. The saturation parameters for carbonate (calcite, aragonite and dolomite) of hydrogeochemical components of groundwater were set up using numerical model program PHREEQC [7]. PHREEQC for windows software was used to simulate different scenarios which is based on an ion-association aqueous model.

4. Results and Discussion

4.1. Statistical Analysis

The groundwater results obtained from geochemical process occurring interaction between water and geological material which groundwater flow movement in subsurface [8]. The water quality analyses included all physical and chemical parameters. pH and electrical conductivity value ranging from 6.55 to 8.01 and 0.07 to 0.55 mS/cm, respectively. Groundwater in the study area falls under fresh type of water (TDS < 1,000). TDS value ranges from 47 to 359 mg/l. Redox potential and salinity of the study area ranging from -138 to 177 mV and 0.03 to 8.30 ppt, respectively (Table 1).

Concentration of calcium which is derived from calcium rich mineral like pyroxenes and feldspar. The concentration ranging from 4.75 to 95.97 mg/l which may also came from corals and shells. The sources of magnesium are due to the ion exchange of minerals in rocks and soils by water. Mg and HCO_3 ions in the groundwater samples ranged from 1.08 to 23.80 and 46.36 to 336.72 mg/l, respectively (Table 2). Bicarbonate in the study area indicates the biodegradation of the organic matter in the sub surface environment [9] Na and K groundwater concentration in the study area range from 6.19 to 53.55 and 0.32 to 26.99 mg/l. Chloride and sulphate concentrations in groundwater increase from recharge to discharge area range from 11.9 to 40.7 and 0 to 52 mg/l. This is due from the intrusion of sea water. Higher nitrate concentration in groundwater was found near to sewage tank of chalet which is 15.97 mg/l. Total hardness as $CaCO_3$ is ranging from 70 to 200 mg/l is considering from moderately to hard water

4.2. Saturation Index (SI)

Saturation index are used to evaluate the degree of equilibrium between water and minerals. Calcite, dolomite, gypsum and anhydrite saturation indices were calculated to test the possibility that SO_{4-} reduction and dolomitisation processes are taking place. The saturation index of a mineral is obtained from equation (1):

$$SI = \log(IAP/Kt) \tag{1}$$

where IAP is the ion activity product of the chemical species in solution, Kt is the equilibrium solubility product for the chemical involved at sample temperature. SI which is less than zero indicate groundwater is undersaturated which respect to the composition of the minerals. SI with greater than zero indicate the groundwater is supersaturated with respect to the particular mineral.

Figure 4 shows the plots of SI of some minerals against TDS for all the investigated groundwater. Nearly all the water samples were slightly oversaturated to undersaturate with respect to calcite, and dolomite and all samples undersaturated with respect to gypsum and anhydrite, suggesting that these carbonate mineral phases.

The saturation indices of some common carbonate (calcite and dolomite) and sulphate (gypsum and anhydrite) minerals are shown in Figure 2. Most of the groundwater samples from April, June and August 2013 are slightly to moderately undersaturated with respect to the carbonate minerals (calcite and dolomite) indicating their dissolve in the aquifer matrix. Few samples are slightly oversaturated with respect to carbonate minerals indicating that these minerals can precipitate by the aggressive groundwater. The groundwater are also moderately undersaturated with respect to sulphate minerals (gypsum and anhydrite) indicating that more sulphate minerals will dissolve to increase their saturation states. Zero concentration of sulphate in samples resulting of no saturation index of both gypsum and anhydrite in April and June 2013. In the Gibbs plot, the rock is the dominant processes controlling major ion composition of groundwater that shows in Figure 3 may have influenced the chemical composition of the study area. Table 3 shows in Ca-HCO_3 water type the mean values of $SI_{calcite}$ and $SI_{dolomite}$ are -0.04 and -0.71 respectively. In Ca-SO_4 water type the mean values of SI_{gypsum} and $SI_{anhydrite}$ are -3.01 and -3.22 respectively.

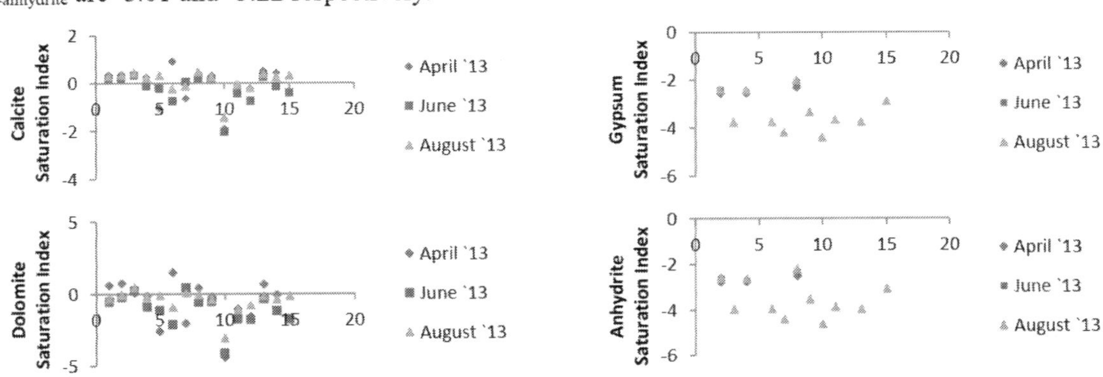

Figure 1. Saturation indices of the common carbonate and sulphate minerals in the collected groundwater samples.

Seawater intrusion will increased the ionic strength and soil redox condition. Figure 5 shows the decreasing of ionic strength from north-west of the study area to south-east where the downstream is at the south-east of the study area. In this study there is a slight intrusion of seawater but not obvious significant. Calcium bicarbonate groundwater has an ionic strength of about 0.02, whereas seawater has an ionic strength of 0.70 [10][8]. Maximum ionic strength in the area is 0.009 (value is 90 m from sea).

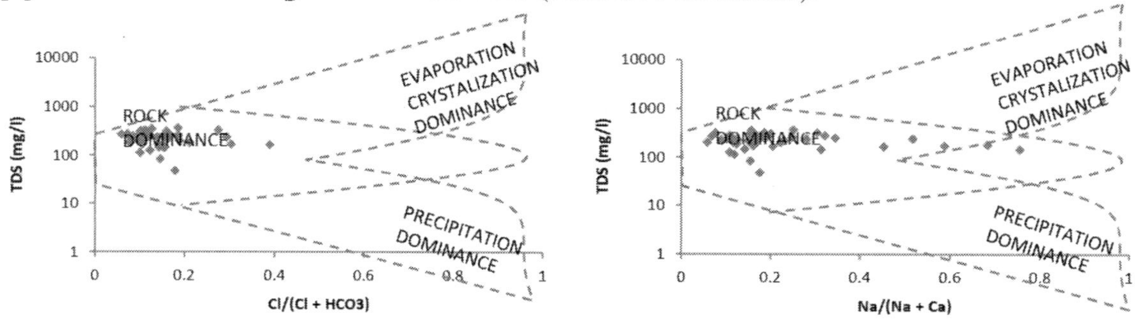

Figure 2. Mechanism governing groundwater chemistry (after [11]).

4.3. *Hydrochemical Facies*

A Piper diagram (Figure 6) was created for the low-lying of Kg Salang area using the analytical data obtained from the hydrochemical analysis. In general, the sample points can be classified in the piper diagram into 3 fields. They are i) Ca-HCO_3 type, ii) Ca-Mg- HCO_3 type and iii) Ca-Na- HCO_3 type. However, in the present study water types were confined to the first type. In the study area, major groundwater type is Ca-HCO_3 and Ca-Mg-HCO_3. Evaluation of the water types using piper plot suggests that there is a clear indication of the contribution from the weathering of pyroxenes in the hard rocks. In Figure 7 shows that all groundwater samples fall under the subdivision of alkaline which had been proposed by [12]. The diagram shows geochemical classification and hydrochemical parameters of groundwater. The sample was taken on April, June and August 2013 in this period dissolution of the minerals are the major processes occurring in the groundwater environment. Since the flow is sluggish there will be much time for precipitation. Dominance of Ca and Mg in the groundwater samples collected suggested an inverse ion exchange process. During this process Ca from the aquifer matrix will be exchanged by Na from the groundwater. Sluggish flow in this relatively low lying area enables sufficient rock-water interactions.

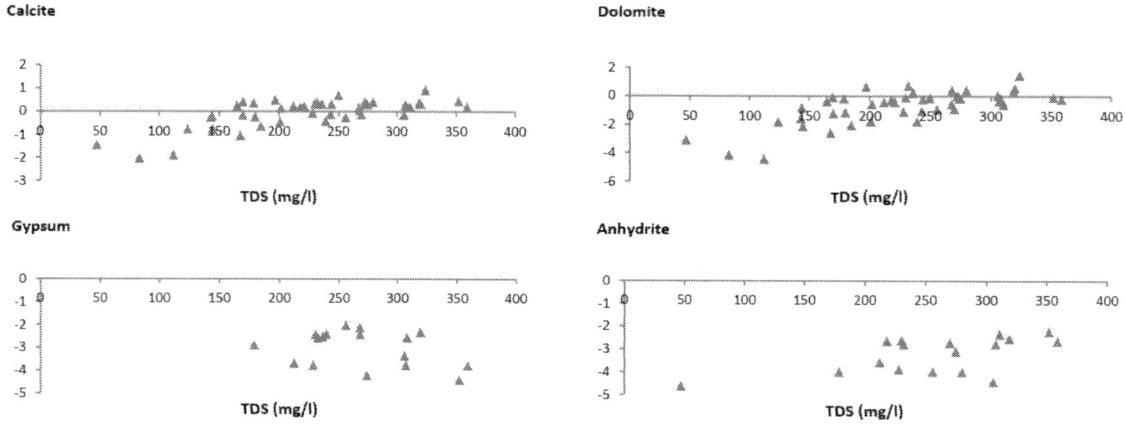

Figure 3. Plots of saturation indexes with respect to some carbonate and sulphate minerals against total dissolved solids (TDS).

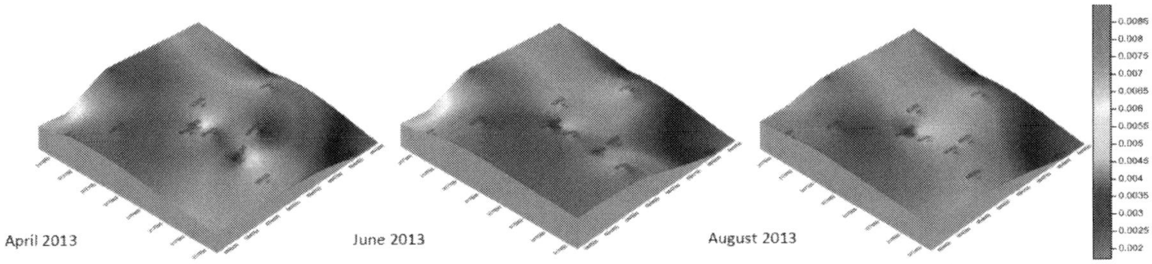

Figure 4. Contour map of ionic strength for April, June and August 2013.

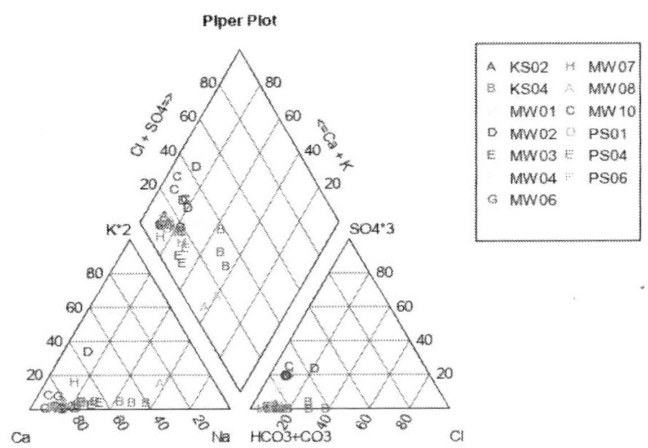

Figure 5. Piper diagram of groundwater chemical facies.

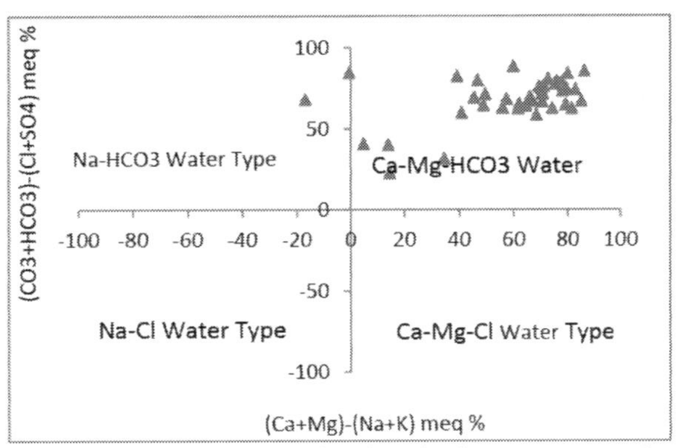

Figure 6. Diagram of geochemical classification and hydrochemical parameters of groundwater (after [12]).

5. Conclusion

The geological of the study area confirm the interaction of rock and water which precipitate calcite and dolomite minerals. These minerals are from the pyroxene hornfel which is found in the upstream of the study area. The problem of groundwater in Kg. Salang, Tioman Island is rather complicated; it represent the tidal effect of monsoon season and infiltrate into low lying aquifer area. Based on the results obtained from PHREEQC, the fresh groundwater is yet not contaminated by the seawater as there were no excessive pumping activities in the village of the island. Results of SI show that nearly all of groundwater samples were saturated to under saturated with respect to carbonate minerals and undersaturated with respect to sulphate minerals. The shielding of ion lowers the activity of dissolved species that occur around charged. Interpretation of hydrochemistry analysis indicate that groundwater in the study area is fresh. The dominant hydrochemical facieses of groundwater is Ca-HCO_3 and Ca-Mg-HCO_3. The groundwater is considering as moderately hard to hard water that contains dissolved hardness minerals (calcium and magnesium).

References

[1] Melloul, A.J. and Goldberg, L.C. (1998). Early indicator Signals of Groundwater Contamination: The Case of seawater Encroahment. Environmental Geology, 33(4): pp.279-288.
[2] Aris, A.Z., Abdullah, M.H. and Kim, K.W. (2007). Hydrogeochemistry of Groundwater in Manukan Island, Sabah. The Malaysian Journal of Analytical Sciences, 11(2): pp.407-413.
[3] Rey, J., Martínez, J., Barberá, G.G., García-Aróstegui, J.L., García-Pintado, J, and Martínez-Vicente, D. (2013). Geophysical characterization of the complex dynamics of groundwater and seawater exchange in a highly stressed aquifer system linked to a coastal lagoon (SE Spain). Environ Earth Sci, 70(5):pp.2271– 2282
[4] Kortatsi, B.K. (2007). Hydrochemical Framework of Groundwater in The Ankobra Basin, Ghana. Aquatic Geochemistry, 13(1): pp.41-74.
[5] Aghazadeh, N. and Mogaddam A.A. (2010). Assessment of Groundwater Quality and its Suitability for Drinking and Agricultural Uses in the Oshnavieh Area, Northwest of Iran. Scientific Research. Journal of Environmental Protection, 1: pp.30-40.
[6] APHA. (2005). Standard methods for examination of water and wastewater, 21st ed. American Public Health Association, Washington.
[7] Parkhurst, D.L. Appelo, C.A.J. (2005). User's Cuide to PHREEQC (Version 2) - A Computer Program for Speciation, Batch-Reaction, One-Dimensional Transport, and Inverse Geochemical Calculations. U.S. Geological Survey Water-Resources Investigations Report 99-4259, pp. 310.
[8] Appelo, C.A.J. and Postma, D. (2005). Geochemistry, Groundwater and Pollution. 2nd ed. Rotterdam, Balkema.
[9] Idris, A.N., Aris, A. Z., Suratman, S. and Tawnie, I. (2013). International Conference on Environmental Forensics iENFORCES 2013, 12. In From Sources to Solution, 7: pp.121-126.
[10] Deutsch, W.J. (1997). Groundwater Geochemistry; Fundamentals and Applications to Contamination. Boca Raton: Lewis Publishers.
[11] Gibbs, R.J. (1970). Mechanisms Controlling World Water Chemistry, Science, 17: pp.1088–1090.
[12] Chadha, D.K. (1999). A Proposed New Diagram for Geochemical Classification of Natural Water and Interpretation of Chemical Data. Hydrogeology Journal, 7: pp.431–439.
[13] Appelo, C.A.J. and Postma, D. (1996). Geochemistry, Groundwater and Pollution. Rotterdam, Balkema, 536p.
[14] Hounslow, A.W. (1995). Water Quality Data - Analysis and Interpretation. CRC Press LLC, 85p.
[15] Panthi, S.R., (2003). Carbonate Chemistry and Calcium Carbonate Saturation State of Rural Water Supply Projects in Nepal. Seventh International Water Technology Conference Egypt, pp.545-560.
[16] Sawyer, G.N. and McCartly, D.L., 1967. Chemistry of Sanitary Engineers, 2nd ed. McGraw Hill, New York: 518p.

Appendixes

Table 1. Mean, standard deviation, minimum and maximum value of physical parameters for April, June and August 2013 of groundwater samples from 14 wells.

Parameter		**April `13**	**June`13**	**August `13**
Temp	Mean	26.63	26.65	26.87
	Standard Deviation	0.68	0.70	0.73
	Minimum	25.64	25.84	26.04
	Maximum	28.01	28.20	28.48
EC (mS/cm)	Mean	0.36	0.31	0.38
	Standard Deviation	0.11	0.10	0.12
	Minimum	0.17	0.13	0.07
	Maximum	0.50	0.48	0.55
TDS (mg/l)	Mean	231.29	200.86	244.67
	Standard Deviation	70.81	62.01	80.64
	Minimum	112.00	83.00	47.00
	Maximum	324.00	311.00	359.00
pH	Mean	7.40	7.16	7.49
	Standard Deviation	0.48	0.33	0.23
	Minimum	6.55	6.55	7.11
	Maximum	8.01	7.63	7.86
Redox (mV)	Mean	-80.14	-22.21	-63.02
	Standard Deviation	85.34	91.32	59.16
	Minimum	-205.00	-138.00	-141.00
	Maximum	149.00	177.00	35.30
Salinity (ppt)	Mean	0.75	0.15	0.18

IOP Conf. Series: Materials Science and Engineering **136** (2016) 012065 doi:10.1088/1757-899X/136/1/012065

Standard Deviation	2.18	0.05	0.06
Minimum	0.08	0.06	0.03
Maximum	8.30	0.23	0.26

Table 2. Value of chemical parameters of groundwater samples

Parameters	Mean	Median	Standard Deviation	Minimum	Maximum
K	2.60	1.10	4.54	0.32	26.99
Na	13.54	11.20	9.63	6.19	53.55
Mg	5.53	4.02	4.74	1.08	23.80
Ca	48.43	49.20	22.31	4.75	95.97
HCO_3	203.58	215.01	73.73	46.36	336.72
Cl	16.90	16.04	4.56	11.90	40.70
CO_3	2.19	1.20	2.49	0.00	15.00
NO_3	0.51	0.30	0.60	0.00	15.97
SO_4	4.51	0.02	9.54	0.00	52.00

Table 3. Statistical summary of the saturation index minerals in Kampong Salang.

	$SI_{Calcite}$	$SI_{Dolomite}$	SI_{Gypsum}	$SI_{Anhydrite}$
Mean	-0.04	-0.71	-3.01	-3.22
Standard Deviation	0.62	1.20	0.77	0.78
Minimum	-2.03	-4.42	-4.40	-4.62
Maximum	0.92	1.48	-2.02	-2.23

Centerline Bed Elevation Profile of Sand Bed Channel due to Bar Formation

D A Tholibon[1], J. Ariffin[1], J Abdullah[1] and N S Muhamad[2]
[1] Faculty of Civil Engineering, Universiti Teknologi MARA, Shah Alam, Selangor, MALAYSIA
[2] Faculty of Civil Engineering, Universiti Teknologi MARA, Jengka, Pahang, MALAYSIA

E-mail: duratulaintholibon@gmail.com

Abstract: Numerous data on bar formation have been accumulated yet the methods to predict bar geometry especially bar height are still insufficient. Objectives of this study to determine the trend in term of a significant difference of centreline bed elevation profile along the longitudinal distance. This can be investigate by carried out an experimental work in an erodible sand bed channel using a large-scale physical river model. The study included the various hydraulic characteristics with steady flow rates and sediment supply. An experimental work consists of four matrices of flow rate and channel width with other variables namely grains size and bed slope were kept constant. Analysis have included the discussion on a significant difference of centreline bed elevation profile along the longitudinal distance. As a conclusion the higher velocity in the smaller channel width have induced erosion of the banks that resulted in elevation increase while the larger flow rates have contributed to higher elevation.

Keywords: Bar formation, bar profile, physical river model.

1. Introduction

The flow in a straight channel with erodible bed is unstable and large-scale migrating bedforms may develop. These bed forms usually called alternate bars that height and wavelength are scale to the flow depth and channel width respectively [1][2]. Basically there are three general types of bars include mid-channel bars also called braid bars, point bars and mouth bars. The mid-channel bars usually common in braided rivers, which point bars are normally formed in meandering rivers and mouth bars are usually found in river deltas. The locations of bars are greatly influenced by the geometry of the river and the flow rate [2]. Point bars formed on the inside of meander bends in meandering river because the shallow flow and low shear stresses reduce the amount of material that can be carried away. The excess material falls out of transport and bars are eventually formed. The aims of this paper is to discuss the trend in term of a significant difference of centreline bed elevation profile along the longitudinal distance.

2. Literature Reviews

For the purpose of explanation of stream channel processes, many experiments have already been implemented as indicated in Table 1. The type and characteristics for each experimental condition is being stated clearly. Most of the study used rectangular cross section and straight channel for their initial experimental condition. [3][4] carried out experimental work in the largest physical model under various conditions in order to clarify the fluvial processes of alluvial river channels in macro scale. The results indicate that channel under complicated conditions have common aspects to those under simple conditions.

Table 1. Studies on bar formation using physical simulation by selected investigators.

No.	Researcher (year)	Flume size (m)			Channel cross section	Channel type	Channel size (m)			Flow rate (L/s)	slope	Max time (hr)	Grain size (mm)	Sediment supply
		width	length	depth			width	length	Depth					
1	Hong and Davies (1979)	2	5	0.15	rectangular	straight	0.1	5	0.02	0.003 to 0.018	0.04 to 0.1	50	0.17	0.0015 to 0.05 kg/min
2	Fujita and Muramoto (1981)	7.5	243	1.5	trapezoidal	straight	0.25 to 1	42.3 to 130	0.05 to 0.4	5 to 20	0.005 to 0.002	100	0.61	
3	Fujita and Muramoto (1982)	1.24	12	0.55	rectangular	straight	0.25 to 0.55	8	0.55	1 to 3	0.01 to 0.005	19	0.39	0 to 3.65 cm/s
4	Ikeda (1984)	0.5	15		rectangular	straight	0.32	15		1.01 to 5.24			1.3	
5	Jaeggi (1984)	0.3	25		rectangular	straight				0.5 to 10	0.002 to 0.012		0.52, 1.8, and 4	
6	Fujita and muramoto (1985)	0.7	18	0.55	rectangular	straight and expanding	0 to 0.55	18		1.39 to 3	0.01 to 0.005	8	0.99	
		7.5	43	0.55	rectangular	straight	2	43		43.78	0.005 to 0.003	9	0.88	
		0.5	20.5	0.55	rectangular	straight		18.4	0.08	1.9 to 5.6	0.01 to 0.005	5	0.99	
7	Ashmore (1990)	2	10		trapezoidal	straight	2	10	0.07	12 to 45	0.01 to 0.015	2	1.16	
8	Lanzoni (2000a)	1.5	55	1	rectangular	straight	1.5	55	1	20 to 47	0.00162 to 0.00517	816	0.48	
9	Federici And Paola (2003)	2m	9.5m		trapezoidal	straight				0.35	0.014	until the bar occur	0.5	100g/min
10	Bertoldi and Tubino (2005)	0.6	12		trapezoidal	straight	0.08 to 0.12	12	0.15 to 0.2	0.117 to 0.5	0.01 to 0.02	2	0.5 to 1.3	
11	Egozi and Ashmore (2009)	3	18	0.15	trapezoidal	straight	0.5	18	0.015	1 to 3	0.005 to 0.025	70	1.2	
12	Bertoldi et al., 2009a)	2.9	25		rectangular	straight	2.9	25	0.15 to 0.2	0.3 to 1.5	0.007 to 0.017	84	0.63	
13	An et. al. (2013)	1.3	3	0.3	rectangular	straight	1.3	3	0.2	0.107 to 0.231	0.05	10	0.28	10 to 40 kg/hr

[5] concluded that in undivided channel, a short submerged central bar is deposited during a high flow. Because of some local conditions not all transported through this particular reach and some accumulate in the centre of the channel. According to [5] the bar formation will initiate braiding as the bed shear stress slightly above the critical value. However, finding by [6] indicates that at a higher bed shear stress the narrow channel bed begins to deform by a more obvious alternating bar which leads to braiding by chute cutoff mechanism. [7] found that the central bar formation had only occur at the restricted width or it will depend on the presence of bed particles large enough (relative to flow depth) to form a nucleus for bar development. [8] studied the mechanisms of evolution of bifurcations within a single channel. They observed that flow divergence invariably leads to the formation of central bars for any Froude number and aspect ratio (width/depth) of the incoming stream. The relatively high value of shields stress ($\tau^* > 0.15$) contributes to stable bifurcation and for low values of shields stress ($\tau^* < 0.15$), a bifurcation display an oscillatory instability. [9] used the Hurst index by [10] to examine the relationship between the length and width of bars. The results indicate that river with larger width promotes river bank constraint on bar reduction. With complete development of river width, river tends to eliminate bank constraint and leads to a wide shape bar in a large braided river.

3. Methodology

The experimental setup consists of three sets of experimental case with four different flow rates. Three channel widths, 20 cm, 30 cm and 40 cm were used. The initial experimental channel was designed with a 45° bend at the upstream to facilitate flow. The initial conditions for each experiment were set as shown in Table 2.

Table 2. Initial conditions for all experiments.

Variables	Range
Water flow rate, Q_w	4.97 m^3/hr, 6.64 m^3/hr, 8.62 m^3/hr and 10.91 m^3/hr
Channel width	20 cm, 30 cm and 40 cm
Sediment supply, Qs	0.021 m^3/hr
Bed slope, S	0.006
Grain sizes, d_s	1.15 mm

V-notch weir was located upstream of the physical river model to regulate flow. Calibration of the flow depth to flow rate was carried out prior to the experiment graduations. Graduations scales on the left indicates the flow depth and on the right indicates flow rate. The graduations were placed slanted along the V-notch face. During the experiment run, if a flow of 4.97m^3/hr is achieved, one can simply read from the graduation on the right. The amount of flow required and the corresponding flow depth can be read from the graduation on the left of the V-notch face. Dimensions of the V-notch weir is shown in Figure 1.

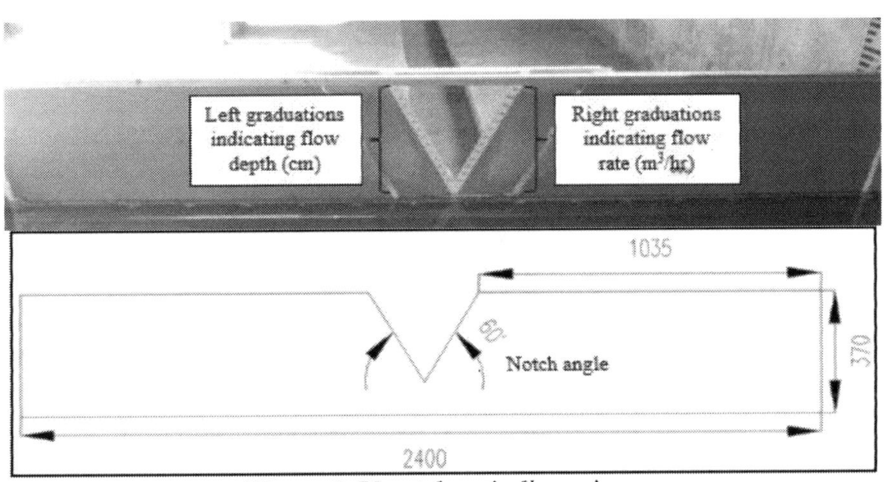

Figure 1. V-notch weir dimensions.

The flow rate was can be calculated using the following relationship;

$$Q = 4.28C\tan\left(\frac{\theta}{2}\right)(h+k)^{\left(\frac{5}{2}\right)} \quad (1)$$

Where Q is discharge, C is discharge coefficient (0.607), θ is notch angle equal to 60°, h is height of water above the V-notch.

Velocity was measured using a Nixon Streamflo Velocity Meter model 430 as shown in Figure 2. Calibration of velocity meter was conducted with the aid of an Acoustic Doppler Velocimetry (ADV). The ADV uses the Acoustic Doppler Effect to measure velocity by measuring the velocity of small particles in water and has been widely used to measure the velocity of water in a wide variety of hydraulic research. The calibration was also carried out by using the same model of Nixon Streamflo to check on the accuracy of the readings as shown in Figure 3.

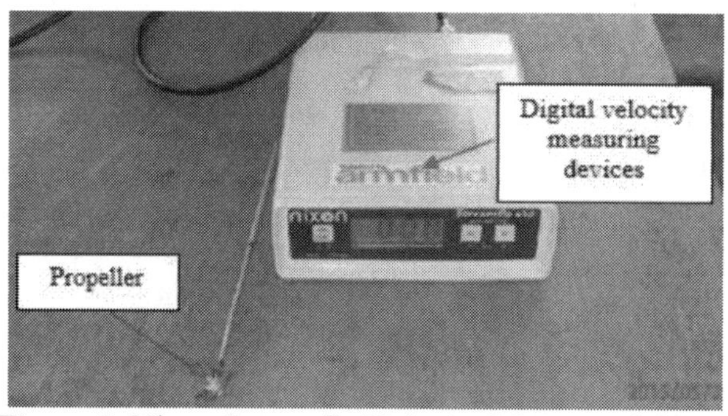

Figure 2. Nixon Streamflo Velocity Meter model 403.

Figure 3. Calibration of Nixon Streamflo Velocity Meter.

A miniature flow sensing probe was used to measure velocities as low as 5.0 cm/s with the channel depth at least 2 cm. The probe was lowered into the channel with the propeller placed perpendicular to the flow to avoid readings error as depicted in Figure 4. The measured velocities were recorded using a digital velocity measuring device.

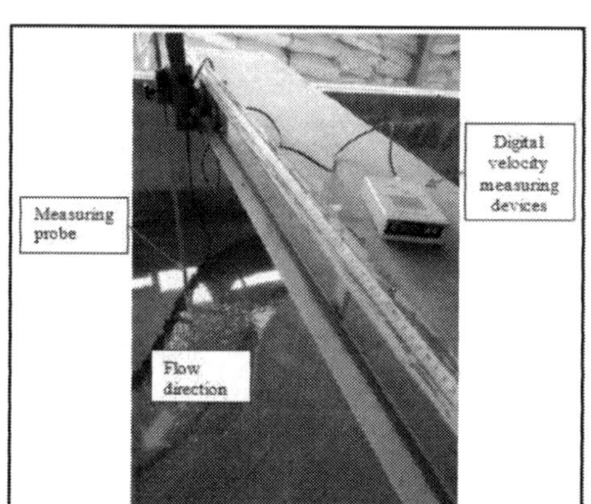

Figure 4. Digital velocity measuring devices.

4. Results and Discussions

Data acquisition during the test included flow depth and bed elevation. Flow depth was measured at the channel centreline using a ruler attached to the adjustable profiler as shown in Figure 5. The bed profile was surveyed periodically using the laser distance meter device attached to the gauge on a regular grid spacing 5cm in the transverse direction and 20 cm in the longitudinal direction. The equipment was moved manually along the platform for a study reach of 10 m beginning 5 m downstream of the entrance weir as to minimize any turbulence flow entrance and sediment feed effects on the channel pattern. This 10 m length is more than 10 times the average wetted width of the braided channels which provides sufficient length to average out local variations and sampling effects on braided pattern and braiding intensity measurements.

Data on channel pattern development were collected along the study reach and was stopped at the equilibrium condition when bar features have shown nearly constant values. Bar profiles development for the different matrices of flow rates and channel width were generated using Surfer, a software used for 3-Dimensional elevation plots. The experimental run being conducted continuously until the bar had developed and remained stable at a certain location. It is observed that for lower flow rates took the longer duration of the bar formation process that is until day sixth. Each experimental run followed the same method and sequence of measurement, until bar formation occurred. The initial dry bed elevation was measured and recorded as the datum for the channel. Flow was imposed into the channel and the sediment was fed into the channel at the weir front. Initial depth of water and velocity were recorded at 1m interval. Measurement was recorded daily and stopped as the formation of mid bar became stable. This is indicated by zero water flow on top of the bars as the bar top were exposed. Longer durations were recorded for lower flow rate with the experiments stopped when the channel had reached the side walls of the model.

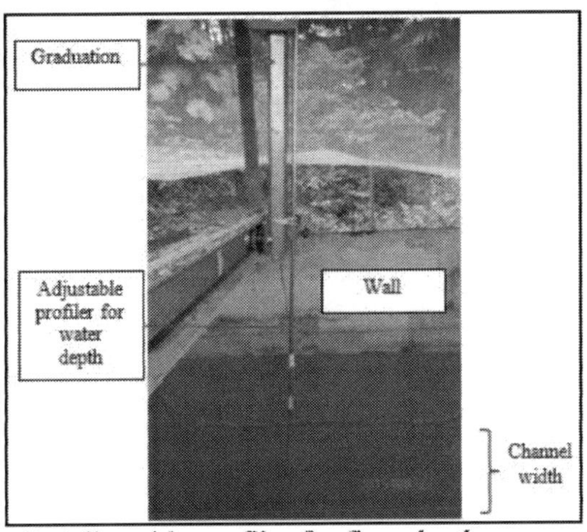

Figure 5. Adjustable profiler for flow depth measurement.

4.1 The Elevation of Centreline Profiles along Longitudinal Distance

The centreline profile of the channel were plotted from the first day until the end of experiment. Figure-6 (a) through Figure-6 (c) depicts the changes of bed elevation along the longitudinal distance. The datum of the initial bed taken as 0 cm and the initial bank height was 10 cm. The centreline axis was assumed to represent the bed elevation changes along the channel. The average elevation different between the initial and the final elevation was summarized in Table 3.

Table 3. The average elevation different between the initial and the final profile.

Flow rates Q (m^3/hr)	Different in average elevation (cm)		
	Channel width B = 20 cm	Channel width B = 30 cm	Channel width B = 40 cm
Q1 = 4.97	2.74	5.50	6.80
Q2 = 6.64	2.60	4.35	5.70
Q3 = 8.62	1.80	3.35	3.50
Q4 = 10.91	0.80	2.30	2.80

For lower flow rates, Q1 = 4.97 m^3/hr and Q2 = 6.64 m^3/hr depict the decreasing plotting line trends towards the downstream section. Flowrates Q3 = 8.62 m^3/hr and Q4 = 10.91 m^3/hr indicate more complex configuration as it shows more fluctuating plotting line compare to Q1 = 4.97 m^3/hr and Q2 = 6.64 m^3/hr. Centreline profile along the longitudinal distance depicts that the average elevation for 20 cm channel width yield the highest average elevation. The difference in average elevation indicates the same trend for the four flow matrices. This evidence by larger differences from day 1 until the end of experiment for the smaller flow rates while smaller difference for the larger flow rates.

The average elevation difference between the initial and the final profile indicates higher different for 40 cm channel width as compared to 20 cm channel width. The higher velocity in the 20 cm channel width has induced erosion of the banks that resulted in the rapid elevation increase. This leads to small changes in average elevation from first day to final day of the experiment. Erosion in 40 cm channel width occurs gradually, thus lead to higher changes of elevation as compared to 20 cm channel width.

It can also be noted that, changes in the average elevation for higher flow rates are smaller as compared to lower flow rate. The above can be attributed to the initial bank collapse as a result of the high flow. The eroded material is transported downstream by the high flow and subsequently deposited along the test length of the channel.

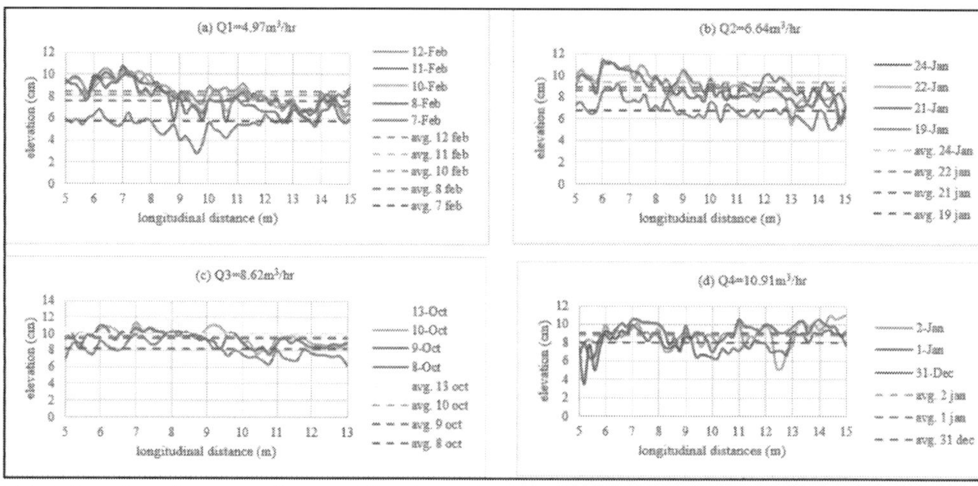

Figure 6(a). Elevation for centreline profile along the longitudinal distance for channel 20 cm width for (a) 4.97 m^3/hr (b) 6.64 m^3/hr (c) 8.62 m^3/hr and (d) 10.91 m^3/hr.

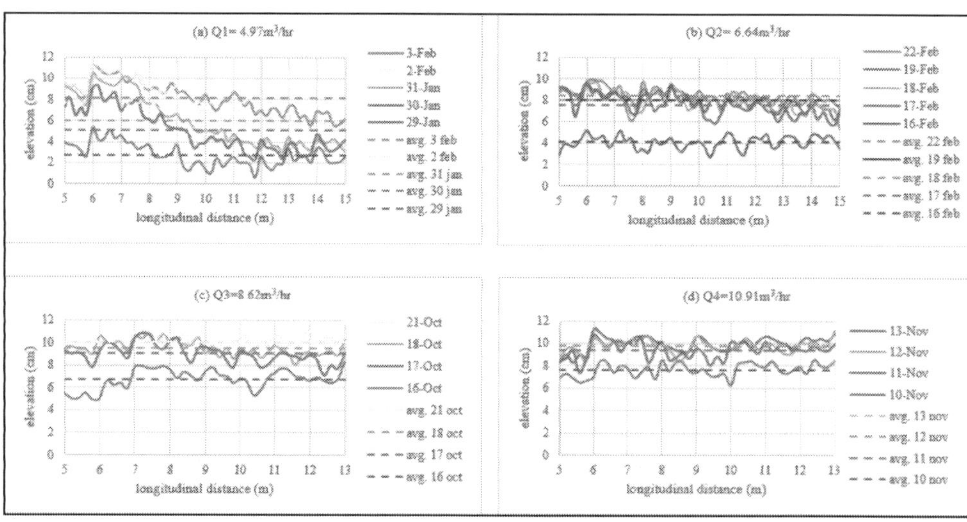

Figure 6(b). Elevation for centreline profile along the longitudinal distance for channel 30 cm width for (a) 4.97 m^3/hr (b) 6.64 m^3/hr (c) 8.62 m^3/hr and (d) 10.91 m^3/hr.

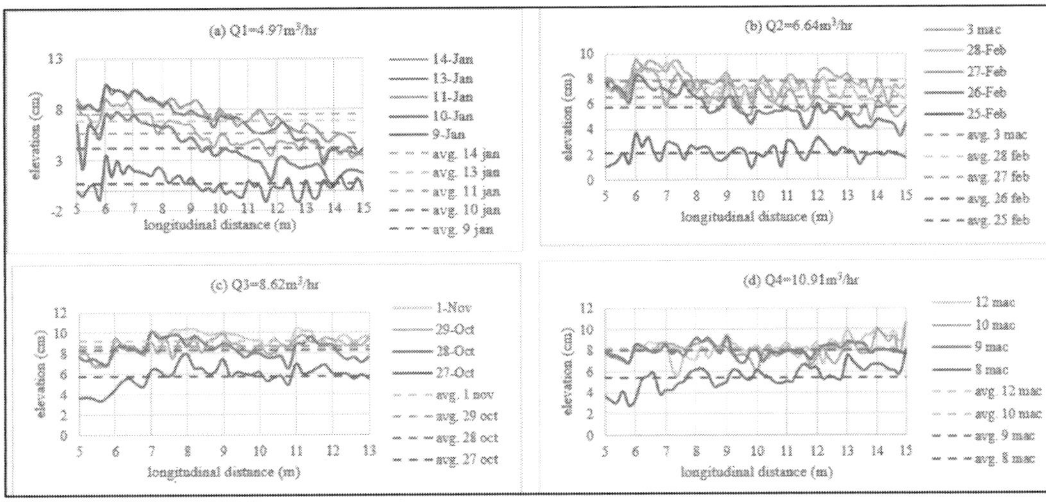

Figure 6(c). Elevation for centreline profile along the longitudinal distance for channel 40 cm width for (a) 4.97 m^3/hr (b) 6.64 m^3/hr (c) 8.62 m^3/hr and (d) 10.91 m^3/hr.

5. Conclusions

Analysis on the bed configuration profiles for Q3 = 8.62 m^3/hr and Q4 = 10.91 m^3/hr indicates more complex configuration compare to Q1 = 4.97 m^3/hr and Q2 = 6.64 m^3/hr. This can be attributed to the higher flow rate transporting sediment from the upper reach. Centreline profile along the longitudinal distance depicts that the average elevation for 20 cm channel width yield the highest average elevation. The higher velocity in the smaller channel width have induced erosion of the banks that resulted in elevation increase. The larger flow rates have contributed to higher elevation.

Acknowledgements

The authors gratefully acknowledge the support provided by the Ministry of Higher Education Malaysia (600- RMI/RAGS 5/3 (63/2013)) and Research Management Institute (RMI) of Universiti Teknologi MARA, Shah Alam, Malaysia.

References

[1] Tubino, M., Repetto, R., & Zolezzi, G. (1999). Free bars in rivers. Journal of Hydraulic Research, 37. 759–775.
[2] Hooke, J. M., & Yorke, L. (2011). Channel bar dynamics on multi-decadal timescales in an active meandering river. Earth Surface Processes and Landforms, 36, 1910–1928.
[3] Fujita, Y., & Muramoto, Y. (1981) Experimental Studies on stream channel process in alluvial rivers. Bulletin of the Disaster Prevention Research Institute, 32,49-96.
[4] Fujita, Y., & Muramoto, Y. (1982) The widening process of straight stream channels in alluvial rivers. Bulletin of the Disaster Prevention Research Institute, 32,115-141.
[5] Leopold, L. B., & Wolman, M. G. (1957). River channels patterns; braided, meandering and straight. Washington.
[6] Ashmore, P. (1990). How do gravel–bed rivers braid. Canada Journal Earth Science, 28, 326-341.
[7] Hong, L. E. B. A., & Davies, T. R. H. (1979). A studyof stream braiding. Geological Society of America Bulletineo, 90, 1839–1859.
[8] Federici, B., & Paola, C. (2003). Dynamics of channel bifurcations in noncohesive Sediments. Water Resources Research, 39, 1-15.
[9] An H.S., Chen, S., Chan, H., & Hsu, Y., (2013). Dimension and frequency of bar formation in a braided river. International Journal of Sediment Research, 28, 358-367.
[10] Walsh, J. & Hicks, D.M. (2002). Braided channels: Self-similar or self-affine? Water Resource Research, 38, 1801-1806.
[11] Bertoldi, W. & Tubino, M. (2005). River Bifurcation:Bed and bank evolution of bifurcating channels. Water Resources Research, 41, 1-12

[12] Bertoldi, W., Zanoni, L., & Tubino, M. (2009a). Planform dynamics of braided streams. Earth Surface Processes And Landforms, 34, 547–557.

[13] Egozi, R. & Ashmore, P. (2009). Experimental analysis of braided channel pattern response to increased discharge. Journal of Geophysical Research, 114, 1-15.

[14] Fujita, Y., & Muramoto, Y. (1985). Studies on the Process of Development of Alternate Bars. Bulletin of the Disaster Prevention Research Institute, 35, 55-86.

[15] Ikeda, S. (1984). Prediction of alternate bar wavelength and height. Journal Hydraulic Engineering, 110, 371-386.

[16] Jaeggi, M.N.R. (1984). Formation and effects of alternate bars. Journal Hydraulics Engineer, 110, 142–156.

[17] Lanzoni, S. (2000a). Experiments on bar formation in a straight flume: 1. Uniform sediment. Water Resources Research, 36, 3337–3349.

Monitoring on The Quality and Quantity of DIY Rainwater Harvesting System

H Kasmin[1], N H Bakar[1] and M M Zubir
[1] Faculty of Civil and Environmental Engineering, Universiti Tun Hussein Onn Malaysia, 86400 Batu Pahat Johor, MALAYSIA

E-Mail: hartini@uthm.edu.my

Abstract. Rainwater harvesting is an alternative sources of water supply and can be used for potable and non-potable uses. It could helps to store treated rainwater for more beneficial use and also for flood mitigation. Sustainable approach for flooding problem reduction in urban areas is by slowing down the rate of surface runoff flows at source by providing more storage area/tank. In order to understand the performance of a rainwater harvesting system (RWH), a preliminary monitoring on a 'do it yourself' (DIY) RWH model with additional first-flush strategy for water quality treatment was done. The main concept behind first flush diversion is to prevent initial polluted rainwater from entering the storage tank. Based on seven rainfall events observed in Parit Raja, both quality and quantity of the rainfalls were analysed. For rainwater quality, the samples from first flush diverter and storage tank were taken to understand their performance based on pH, dissolved oxygen (DO), turbidity, total dissolved solid (TDS), total suspended solid (TSS), chemical oxygen demand (COD) and biochemical oxygen demand (BOD) parameters. While for rainwater quantity, hydrograph analysis were done based on the performance of total rainfall and runoff, peak flow of rainfall and runoff; and delayed time parameters. Based on Interim National Water Quality Standard (INWQS) and National Drinking Water Quality Standard (NDWQS), first flush diverter apparently helps on water quality improvement in storage tanks when pH, DO, TDS, TSS and turbidity were classified as Class I (INWQS) and is allowable for drinking; but BOD and COD parameters were classified as Class III (INWQS). Hence, it has potential to be used as potable usage but will need extensive treatment to reduce its poor microbial quality. Based on the maximum observed rainfall event which had total volume of 3195.5 liter, had peakflow reduction from 0.00071 m^3/s to 0.00034 m^3/s and delayed runoff between 5 and 10 minutes after rainfall started. It concludes that the performance of water retention could be due to total rainfall and the tank capacity. Therefore, RWH has a potential to be used as potable use and at the same time it also has a potential to reduce local urban flooding.

Keywords: Rainwater quality, rainwater quantity, first flush strategy, portable use, water supply.

1. Introduction

Malaysia is a country with an equatorial climate that characterized by hot and humid throughout the year of two monsoon seasons. With regions that consist variety of topographical combinations such as floodplains, highlands, lowlands and coastal areas, Malaysia is typically receives around 2000 mm to 4000 mm rainfall anually [1,2]. However, with continuous and frequent intense rainfalls occurence may causes flooding problems especially in urban areas. Various approaches have been implemented by authorities to reduce and prevent flooding problems such as drainage and irrigation system improvement based on Sustainable Urban

Drainage System (SUDS). Rainwater harvesting (RWH) system is one of the SUDS structure that use for home/building rainwater reservoirs that provides more storage at source, hence surface runoff flow can be reduced. Apart from water control at source, RWH also has ability to be retrofitted around the existing housing areas and buildings at any time needed.

According to [3], the increase in impervious surfaces, including rooftops and pavement such as roads, driveways, parking lots and many more, the decreases the amount of water that soaks into the ground, or infiltrates and increases the amount of surface runoff. The excessive runoff will carries contaminants directly into water bodies because of less infiltration, peak flows of storm water runoff becomes larger and will arrive earlier, hence the magnitude of urban floods will increase [3] On the contrary, when Malaysia land is blessed with higher intense rainfall, without proper management to handle the plentiful water, rainwater will be wasted, downstream area will be flooded but ironically, clean water supply is scarce.

Growth of population and expansion in urbanization, industrialization and irrigated agricultural is imposing growing demand and pressure on water resource [4]. A new development of water resource from rainwater is very important to make sure there is no water shortage in the future [4]. The RWH system is a simple collection of water from the roof areas where the rain falls and the utilization of rainwater may involves for domestic and the agriculture purpose [5]. With estimation of average annual rainfall at 3000 mm or 990 billion m^3 over Malaysian land, 566 billion m^3 is estimated to becomes surface runoff. Hence, before water shortage becomes threat and crisis for country, further undertanding for promoting and encouraging public to implement the system need to be done.

2. Material And Methods

2.1 Components of DIY rainwater harvesting system

Rainwater harvesting components are commonly consists of roof, gutter, downspout, first-flush diverter, filter and storage tank [6] (Figure 1). The 7 m × 5 m roof is a catchment area for RWH system but roof also could collect dust and other debris such as leaves, bird dropping and roof corrosion on them during dry days. The quality of drinking water can be much improved if this debris is not allowed to enter the storage tank [7]. Second component is a 7 m gutter which has function to catch the rain from the roof catchment surface and transport it to the cistern [8]. Third component of rainwater harvesting system is downspouts which brings rainwater from a gutter. The downspouts using 80 mm of polyvinyl chloride (PVC) pipelines were used to carry rainwater from the catchment area to storage tanks (Figure 1).

Figure 1. Full components of RWH from the catchment area.

Fourth component is a first flush diverter which use to improve the quality of roof runoff prior to storage. Two bottles are holed at the bottom and being glued together and it is equivalent to 1800 ml of volume when its full of water (Figure 2). The DIY glued bottles acts as first-flush diverter which use to collect the first flush of contaminated rainwater especially after long period of dry days. As the water rises in the diverter, the plastic ball will stop the next rainwater from roof from entering the first flush diverter. The following rainwater will continue to flow to the last component which is storage tank. Storage tank will collect and store the clean rainwater from roof. Storage can be aboveground or underground, and consists of barrels, tanks, cisterns, or bladders. In Malaysia, a standard size of tank is 1 m^3 for every 100 m^2 roof catchment area [6]. Therefore, a tank with 0.385 m^3 in size is suitable for 38.5 m^2 of roof area. However, in this study, a storage box with 0.045 m^3 in size was used as a tank. The size of the tank used was just for this study only with assumption that if the small storage capacity of water will results in good reduction and delaying runoff, hence the larger tank storage would do more effectively.

Figure 2. Left: A DIY first flush diverter; Right: A DIY storage tank.

Table 1 shows the material used to build a DIY rainwater harvesting with first flush diverter. All material is available to be found in market.

Table 1. Material and quantity used in DIY rainwater harvesting system.

No	Material	Quantity
1	Gutter	1
2	80 mm diameter PVC	1
3	20 mm diameter PVC	1
4	Bracket	3
5	PVC T '20 mm'	1
6	Glue	1
7	Bottle	2
8	Storage box	1

3. Data Collection

For rainfall data collection, a manual DIY raingauge was used. Point rainfalls were measured using this raingauge by measuring the cumulative rainfalls in this bottle, and was installed adjacent to the roof catchment. Whilst for storage runoff data, a DIY weir at side of tank was made (Figure 3) and the water level of overflow rainwater were measured. Total volume of retained rainwater in tank before it started to overflow through the weir is 15.48 liter. The outflow discharge, Q is calculated based on the equation of [9]:

$$Q = \frac{2}{3} C_d \sqrt{2g} L H_1^{3/2} \qquad (1)$$

Where C_d is coefficient of discharge, L is spanning length of rectangular weir and H_1 is the depth of flow above the weir crest. The variation of Cd for rectangular sharp-crested weir is given by Rehbock formula [9]:

$$C_d = 0.611 + 0.08 \frac{H_1}{P} \qquad (2)$$

Where P is height of the weir and the formula is valid only for $H_1/P = \leq 5.0$. During rainfall events, both data rainfall and runoff were manually recorded at 5 minutes interval.

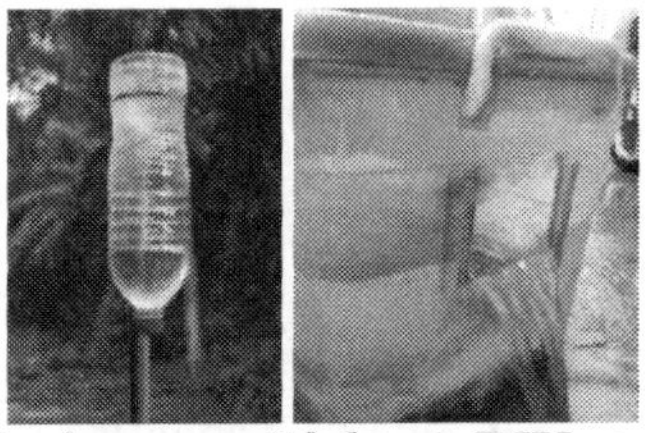

Figure 3. Left: A DIY raingauge; Right: A DIY storage tank and its weir.

For rainwater quality, during rainfall events, samples from the first flush diverter and storage tank were taken to be tested on the following parameters: pH, dissolved oxygen (DO), turbidity, total dissolved solid (TDS), total suspended solid (TSS), chemical oxygen demand (COD) and biochemical oxygen demand (BOD) (Table 2). The analysed parameters were then compared to Interim National Water Quality Standard [10] and National Drinking Water Quality Standard [11]. The RWH system is installed in housing area at Jalan Parit Hj Rais, Parit Raja Batu Pahat. The housing area is located near to oil palm plantations and factories. In agricultural areas, rainwater is expected to have higher concentration of nitrates due to fertilizer residue in the atmosphere [12]. However, limited parameters were tested due to limitation of reagent available.

Table 2. List of equipments used for each parameter tested.

Parameter	Equipment
pH	pH Meter
Dissolved oxygen	DO meter
Turbidity test	DR/4000 1-Inch Cell Adapter
Total Dissolved Solid (TDS)	TDS meter
Total Suspended Solid (TSS)	DR 6000
Chemical Oxygen Demand (COD)	DRB200 reactor
Biochemical Oxygen Demand (BOD)	DO meter

4. Results And Discussions

Both monitoring were not always available at the same time frame due to limitation on the laboratory usage. Water samples for rainwater quality were taken on 19/3/15, 22/3/15, 31/3/15, 1/4/15 and 12/4/15; while for rainwater quantity observation were taken on 29/3/15, 31/3/15, 1/4/15, 4/4/15 and 12/4/15 rainfall events.

4.1 Monitoring on rainwater quality

For each rainfall events, the average from three sample readings were taken. From observations, there were 10 days of dry days before rainfall event on 19/3/15 started, and it takes 3 days of dry days prior to rainfall events on 22/3/15. The next rainfalls were a continuous rainfall events from 30/3/15 to 31/3/15. After 8 days of dry days, another rainfall event occured on 12/4/15 (Table 3). The categorization of total rainfall (in mm) are as follows; < 25.4 mm (small storms), 25.4-76.2 mm (medium storms) and > 76.2 mm (large storms).

Generally, all pH readings in first flush samples are slightly acidic than storage tank (Table 3 & Table 4). However, all samples were fall between 6.5 and 8.5 which is in neutral condition and were classified as Class 1 by INQWS and acceptable values for drinking water. Sample 4 proved that with frequent rainfall, rainwater will flushed away all the air pollutant as stated by [13], the pH of rainfall would be 7.0 if there were nothing else in the air.

Table-3 Date and type of storm events

Sample	Date of sampling	Type of storms
1	19/3/15	Small
2	22/3/15	Small
3	29/3/15	Small
4	31/3/15	Medium
5	1/4/15	Large
6	4/4/15	Small
7	12/4/15	Small

The concentration readings for DO for both components also shows that the rainwater have an adequate supply of dissolved oxigen to be classified as Class I. For turbidity, 5 NTU is a standard for drinking water quality, but for our readings in first flush exposed that, it is still acceptable for raw water quality which are less than 1000 NTU. After treatment, the turbidity concentrations are better in storage tank and its even better if it has larger rainfalls. TDS and TSS also shows good improvement after treatment to be classified then as Class I for samples in storage tank and acceptable for drinking water standard even for TDS and TSS in first flush samples.

Table 4. Comparison from average reading of first flush samples.

Test	First Flush Sample					INWQS
(pH)	6.5	6.54	6.56	6.90	6.51	Class I
DO (mg/L)	7.31	7.92	7.68	7.90	7.85	Class I
Turbidity (NTU)	44.0	22.5	39.3	11.2	35.2	Class IIA
TDS (mg/L)	138	136	105	58.7	104	Class I
TSS (mg/L)	48.0	45.0	30.3	15.0	31.3	Class IIA
COD (mg/L)	191	175	125	68.6	103	Class V
BOD (mg/L)	16	14	14.3	13.4	13.6	Class V

However, the concentrations of COD and BOD from the samples are quite high and could not be accepted as raw water quality which supposed to have acceptable values around 10 mg/L and 6 mg/L, respectively. Hence, based on INWQS, the concentrations of both COD and BOD can be classified as Class V for samples in first flush diverter, and improved to Class III in storage tank samples.

The monitoring on rainwater quality exposed the information that, for RWH system can be used as drinking water, but extensive treatment for COD and BOD are needed.

Table 5 Comparison from average reading of storage tank samples.

Test	Storage Tank Sample					INWQS
(pH)	6.54	6.61	6.87	7.21	6.73	Class I
DO (mg/L)	7.05	7.60	7.45	7.76	7.72	Class I
Turbidity (NTU)	3.1	2.5	2.4	0.4	2.2	Class I
TDS (mg/L)	44.5	48.2	36.3	6.60	23.7	Class I
TSS (mg/L)	2.0	1.0	1	0.7	1	Class I
COD (mg/L)	50	34.2	46	25.3	37.3	Class III
BOD (mg/L)	5.86	2.07	5.44	1.67	4.3	Class III

4.2 Monitoring on rainwater quantity

For each rainfall events, the quantity of rainfall were recorded when every rainfall started until it stopped, in order to observe the rise in water level in the storage tank. All the water will flows into gutter, downspout and into the 1800 ml of first flush diverter, prior to fullfill the 15.48 liter tank capacity. Overall, total volume of 17.28 liter of rainwater will be retained for every rainfall occurance, if the tank can be emptied at every time after rainfall. Hence, for this DIY system, from the observation of each rainfall (Figure 4), it is expected that for medium and larger storm, it will typically takes between 5 and 10 minutes to fullfill the 17.28 liter; at least between 15 and 20 minutes for medium rainfall that weighted almost close to larger values (Figure 4(b) and Figure 4 (c)). It can be varied for medium and small rainfalls to fullfill the 17.28 liter where it will depends on the characteristis of the storm; the types and durations. For example small rainfall on 29/3/15 has 10 mm of rainfall (duration of 170 minutes) takes about 115 minutes to fill in the 17.28 liters, while for rainfall on 4/4/15 with 17 mm (duration of 140 minutes) needs only 50 minutes to fullfill the 17.28 liters. However, small rainfall on 12/4/15 which has 20 mm of total rainfall within duration of 115 minutes, only needs between 15-20 minutes to fills in the storage tank (Figure 4(e)).

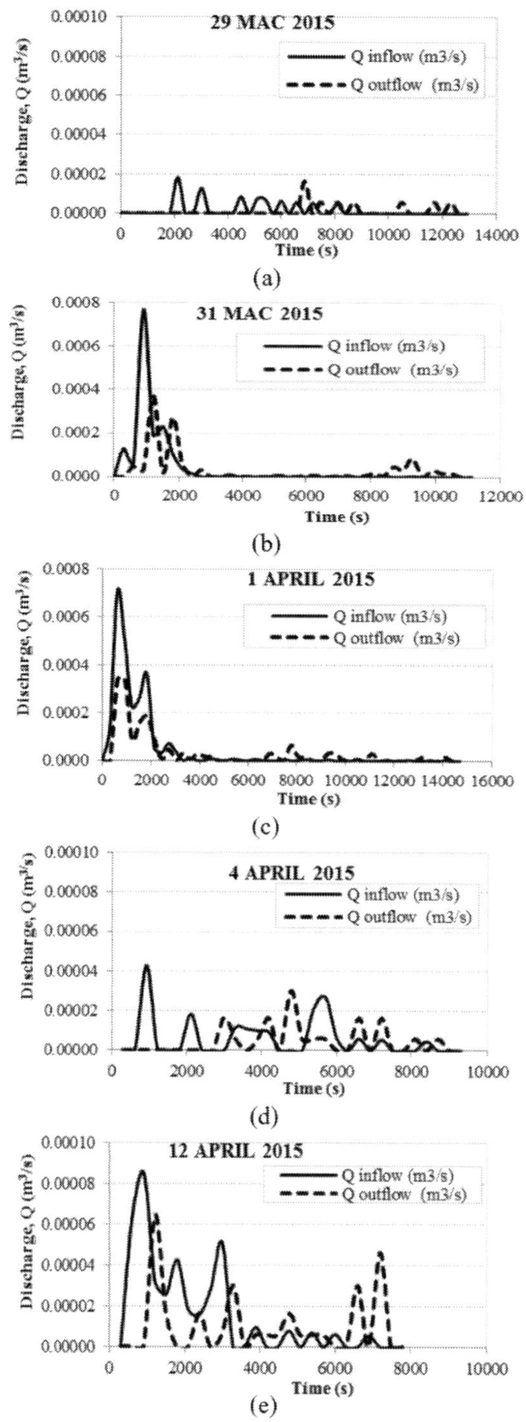

Figure4. (a) to (e): The inflow and outflow of each monitored rainfall for rainwater quantity.

The values of inflow discharge, Q is measured by rainfall data collected using the DIY raingauge. Based on the area of roof catchment, the point rainfalls were then converted into volume in m^3. At the same time, water levels of overflowed rainwater via rectangular weir are measured and then being converted into the outflow discharge using Equation (1).

Based on the hydrographs analyses (Figure 4), it can be seen that the total volume of the outflow discharges were reduced and retained, and it is obviously known that total retained volume will be the same volume provided by the first flush diverter and tank capacity which is 17.28 liter. It applicable for all types of rainfall events including rainfall on 1/4/15 that produced 3195.5 liter of total rainfall volume. However, for this large rainfall type, with small tank capacity like the DIY RWH, could delays runoff to start after rainfall started between 5-10 minutes, and reduced the peakflow from 0.00071 m^3/s to 0.00034 m^3/s would gives significant impacts on the flow at the dowstream drainage area. Hence, with bigger tank capacity, the delays and the volume of outflow discharge could perform better than the DIY RWH.

5. Conclusions

Based on simple preliminary monitoring on the quality and quantity of 'do it yourself' rainwater harvesting system, it is observed that RWH system can be performed as the best storage structure to control water at source for tropical country with high average rainfalls like Malaysia. Although RWH concept has been implemented since few thousand years ago in many countries like USA, China, Japan [4] and few more including Malaysia, but the use of RWH system at most of the time typically highlighted for water supply purposes only. Therefore understanding on rainwater quality is essential for water supply especially for drinking water. However, in this study, the concentration of some parameters like water hardness (dissolved calcium and dissolved magnesium), iron, nitrates, chlorides, sulphates and conductivity [14], parameters that originate from atmospheric sources that impact the waterbodies by rain [15] such as sulfur compound, nitrogen compounds, pesticides, mercury and other toxics are not extensively measured; considering that the location of RWH system is in the agricultural and industrial area. Still, this study highlights that the rainwater is in neutral condition, however the rainwater contained higher concentration of COD and BOD represents the quality of rainwater in the area are having inadequate oxygen available and having microbial activities on the roof catchment site, respectively; and the variety results of parameters concentration could depends on the weather, geography location and activities around the location area.

RWH system also has a role as retention and detention structure although it could not stand alone without any supporting structures to mitigate flood. Still, if one single small storage tank like DIY RWH system can retained, reduced and delayed small portion of the large storm at one particular house; with another i.e 10 houses and their 10 small storage tanks, more volumes can be retained, detained and delayed for a residential area. Larger storage of tank capacity would do better and less surface runoff would be produced at downstream areas of urban areas. The retained/stored water would then be used as an alternative source for water supply for each house, and indirectly could reduce the utility bills for the houses.

References

[1] Jamaludin Suhaila & Abdul Aziz Jemain, (2007). Fitting daily rainfall amount in Malaysia using the normal transform distribution. *Journal of Applied Sciences*, 7(14), 1880-1886.

[2] Hartini K., Virginia S. & Simon De-V., (2014). An evaluation of green roof hydrological performance in a Malaysian context. *13th International Conference on Urban Drainage, Sarawak, Malaysia.* 7 – 12 September 2014.

[3] Donaldson, S., (2004). The Effects of Urbanization on the Water Cycle. *Presentation for NEMO Project. USA: The University of Nevada.*

[4] Che-Ani, A. I., Shaari, N., A. Sairi, M.F.M. Zain and M.M. Tahir., (2009). Rainwater Harvesting as an Alternative Water Supply in the Future, *European Journal of Scientific Research*, Vol.34 No 1, 132-140

[5] Mohamed, S.F., & Mohd Rais, N.F., (2012). Potential of Rainwater harvesting implementation from multiple economic impacts measure. *OIDA International Journal of Sustainable Development.* 4 (8). pp. 65-70. ISSN 1923-6654.

[6] MSMA, (2012). Urban stormwater management manual for Malaysia. 2nd Edition. *Department of Irrigation and Drainage. Retrieved at* http://water.selangor.gov.my/web_ver2/bahagian/bsma/index.php

[7] Mosley, L., (2005). Water Quality of Rainwater Harvesting Systems. *Sopact Miscellaneous Report 579. SOPAC Water Quality Officer*. (February, 2005).

[8] Texas Water Development Board (TWDB), (2005). The Texas manual on rainwater harvesting system. *Third Edition, Austin, Texas. Retrieved at* http://www.ecy.wa.gov/programs/wr/hq/pdf/texas_rw_harvestmanual_3rdedition.pdf

[9] Subramanya, K., (2000). Flow in open channel. *Tata McGraw-Hill Publishing Company Limited, New Delhi.*

[10] INWQS, Interim National Water Quality Standard, (2015). *Retrieved at* http://www.gunungganang.com.my/pdf/Malaysian-Policies-Standards-guidelines/Standards/National%20Water%20Quality%20Standards.pdf

[11] NDWQS, National Drinking Water Quality Standard, (2015). *Retrieved at* http://kmam.moh.gov.my/public-user/drinking-water-quality-standard.html
[12] Thomas, P.R. & G.R. Greene, (1993). Rainwater quality from different roof catchment. *Water Science Technology*. 28 (3-5) pp. 291-297.
[13] Sehgal, J. D., (2005). A Guide to rainwater harvesting in Malaysia. *Rotary Club of Johor Bahru, District 3319 of Rotary International.*
[14] Narashimha Rao C., Dorairaju S.V., Bujagendra Raju M. & Chalapathi P.V., (2011). Statistical analysis of drinking water quality and its impact on human health in Chandragiri, near Tirupati, India. *ECO Services International. Retrieved at* http://www.eco-web.com/edi/111219.html
[15] EPA, Environmental Protection Agency, (2015). Air pollution and water quality. *Retrieved at* http://water.epa.gov/lawsregs/lawsguidance/cwa/tmdl/airdeposition_index.cfm

Seasonal Hypoxia Occurrence At Terengganu Estuary, Malaysia And its Potential Formation Mechanisms

H L Lee[1,2], F Tangang[1], M H Wahap[2] and S YAng[2]

[1] The National University of Malaysia, Faculty of Science and Technology
[2] National Hydraulic Research Institute of Malaysia (NAHRIM)

E-mail: hllee@nahrim.gov.my

Abstract. Hypoxia is commonly defined as oxygen depletion when the dissolved oxygen (DO) levels fall below 2 mg L^{-1}. In this paper, physical factors contributed to hypoxic condition in the diurnal tide, Terengganu Estuary has been experimentally and analytically studies. Hourly profiles on DO, current speed and nutrient were measured during the wet and dry seasons for continuous period approximately 30 hours, which cover both high and low tide at two fixed stations, located at the river mouth and 9.50 km upstream from the river mouth, respectively. It was observed that the entire estuary was free from hypoxia during the wet season. However, hypoxia was developed at the river mouth during the dry season with lowest reading of the DO level was 0.78 mg L^{-1}. It was demonstrated that the combination of high contamination of nutrient, low current speed, the geometrical formation and water stratification are recognized as important role in formatting the hypoxia at the Terengganu river mouth.

Keywords: hypoxia, dissolve oxygen (DO), estuary dynamics, river discharge, nutrient.

1. Introduction

The oxygen depletion in the estuarine environment is a global problem and reported increasing in severity and frequency [1,2,3]. Dissolved oxygen levels in the water columns are maintained through the interaction of numerous physical and biological factors, i.e., water column stratification, exchanges between the atmosphere and sea and its transport, biological oxygen production and consumption processes [4, 5]. Imbalances among these factors may lead to the development of hypoxia in the estuary. However, it was reported that since 1960s, excessive of anthropogenic inputs nutrients into the marine water has dramatically increased cases of hypoxia occurrences in the estuarine and coastal areas [6, 4, 7].

Dissolved Oxygen (DO) is an important water quality metric because it affects the wellbeing of marine organisms. Low DO can lead to catastrophic consequences for marine ecosystems and local economies that rely on commercial and recreational fisheries resources [8, 2]. Hypoxia is commonly defined as occurring when DO levels fall below 2 mg L^{-1} [9, 8, 2, 4, 10]. However, it was recognized that detecting hypoxia is difficult because DO value is extremely variable over short time scales, i.e., less than a day [8] or few days in estuaries [11].

Previous study on the DO dynamics and factors affecting the incidence of hypoxia in the estuary off the tropical area is limited [6, 1]. The main objective of this paper is to investigate the status of hypoxic event at Terengganu Estuary and to examine factors that contributed to its occurrences. The Terengganu Estuary, which is located in the east coast of the Peninsular Malaysia, is considered as a rain forest estuary of salt-wedge type that facing southern region of South China Sea.

IOP Conf. Series: Materials Science and Engineering **136** (2016) 012068 doi:10.1088/1757-899X/136/1/012068

2.Material And Methods

2.1 Study area

The Terengganu River basin is located at 5.34°N and 103.15°E on the east coast of Peninsular Malaysia (Fig. 1a). The catchment area of the basin is about 4,600 km^2, consists of two main tributaries that drained into the Terengganu Estuary, namely the Terengganu River and Nerus River (Fig. 1b). At the river mouth, a breakwater has been constructed with an opening width of 200 m with the water depths for navigation channel is 8 to 10 m depth. The Terengganu River is the main stream in the catchment area, where its length is approximately 65 km, it drains from the Kenyir Dam that located at mountainous area with elevation of 100 m above the local mean sea level datum and flows eastwards (Fig. 1a), draining into South China Sea

2.2 Data Sampling

The dissolved oxygen (DO) was measured at fixed stations, at the river mouth (the YB station with water depth approximately 7 m) and at upstream reaches 9.50 km from the river mouth, the Manir station with water depth approximately 5 m (Fig. 1b). Data campaign was done during the wet (5 to 6 March 2012) and dry (15 to 16 July 2012) seasons for duration of approximately 30 hours. Both data campaigns were carried up during end stage of neap tide, or early stage of spring tide. The measurements were made by two teams with fibre boats that anchored at the fixed stations. Hourly DO profiles were measured using YSI Professional Plus instrument with appropriate weight at the bottom, where measurements were made at 0.5 m below the water surface and at subsequent depth intervals of 1.0 m till it reached the bed layer. The YSI equipment was calibrated and verified before being deployed at the site.

In additional, the total organic carbon (TOC) was also been measured during the data campaign. Replicate samples of TOC were collected at the sub-surface (0.5 m below the water surface) and mid-depth of water depth. The measurements were carried out in hourly and continuous for 30 hours durations, which cover high and low water for wet and dry seasons. The Horizontal Van Dorn water sampler was used to collect water samples while high density polyethylene (HDPE) plastic bottles (1000 ml) were used as a water sample container. All the samples were preserved in Coleman ice-container box temperature below 4°C at the time of collection and were sent immediately to laboratory for TOC analysis. The determination of TOC concentration was based on [12], which related to method 2540C and 2540D in APHA (2005).

Two units of Nortek Aquadopp current profiler with a transducer head of 2 MHz were used to measure profiles current at the YB and Manir stations. Measurements were made during the wet (5 to 6 March 2012) and dry (15 to 16 July 2012). The instrument was installed within a stainless steel frame and deployed at the bed layer. These equipments were calibrated and verified upon being used at the site.

2.3 Oxygen concentration

The effects of low DO in the water columns were of particular concern for all the aquatic and marine habitats. Under this study, five categories of oxygen concentration were assigned [13]:

i) hypoxic or severe biological stress (0 to 2 mg L^{-1});
ii) moderate biological stress (> 2 to 4 mg L^{-1});
iii) mild biological stress (> 4 to 6 mg L^{-1});
iv) normoxic (> 6 to 9 mg L^{-1}); and
v) supersaturated (> 9 or 10 mg L^{-1})

3. Results

3.1 Seasonal changes in dissolved oxygen (DO) within Terengganu Estuary

It was observed that the depletion of the DO was severe during the dry season compared to the wet season (Fig. 2a). Moreover, water column was critical and in the hypoxic level at the river mouth (the YB station) compared to the upstream (the Manir station). The temporal and spatial DO concentrations at both stations (Fig. 2b), illustrated the DO levels were fairly homogenous in the entire estuary during the wet season; with DO concentration in the range of approximately 6 mg L^{-1} from the upstream to river mouth and almost consistence from the surface layers to bottom layers. On the other hand, during the dry season, the DO levels at these two stations were markedly different with average concentration of 3.76 ± 0.56 mg L^{-1} (n=159) and 1.81 ± 0.62 mg L^{-1} (n=260) for the Manir and YB station, respectively. In addition, it was observed that 93 % throughout the measurement period the bottom layer of DO levels at the YB station (river mouth) were below the hypoxic threshold (2 mg L^{-1}), with the minimum value was recorded 0.78 mg L^{-1} that occurred during the high water slack associated with local time from 6 to 10 o'clock in the morning.

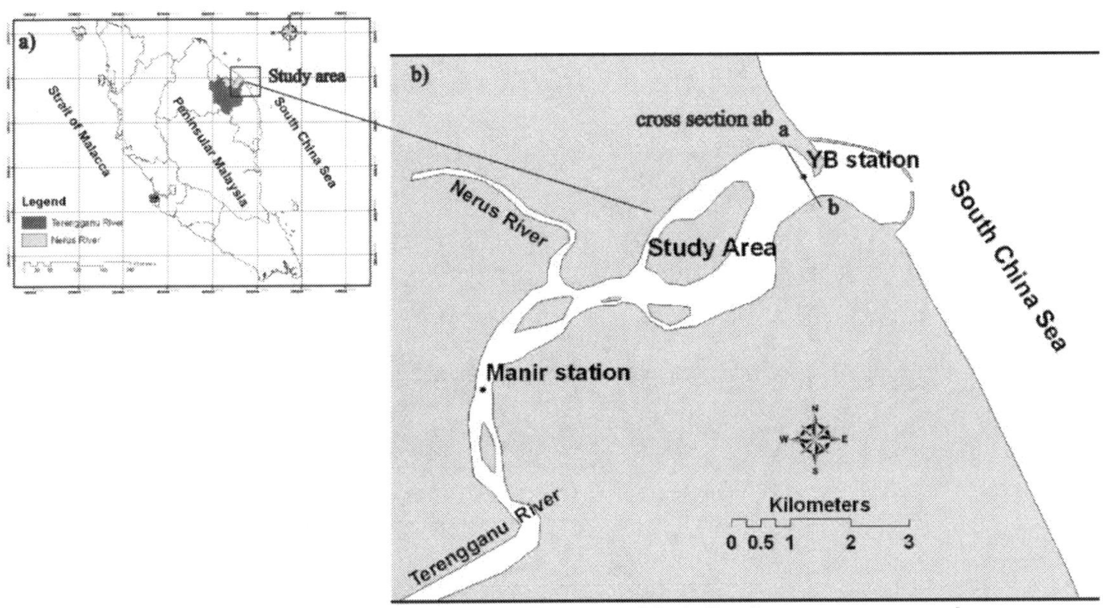

Figure 1. Location of the Terengganu Estuary with defined measurement stations.

Generally it can conclude that entire estuary is free from hypoxia during the wet season and hypoxia is only occurred at the river mouth during the dry season. It was observed that the DO concentration at the upstream reach of estuary (Manir station) was absolved from hypoxic condition regardless of season. However, the hypoxic conditions occurred at the river mouth (YB station) might be due to seasonality factors controlling the hypoxia at this station. Thus, the remainder of this paper focuses on understanding various roles of physical factors that may contribute to hypoxic condition at only the YB station.

3.2 The Formation Mechanism Of Hypoxia At The YB Station

3.2.1 Seasonal variation of hydrodynamic and DO

The temporal evolutions of current for the wet and dry season are displayed in Figures 3a and 3b respectively. It indicated the estuary is forced by diurnal tides, with the main tidal forcing is due to the K1 tidal constituent, which has a tidal period of approximately 24 hours. Analysis on average seasonal tidal period for both seasons, which illustrated that approximately 30% (7 hours) of the tidal period was driven by the baroclinic gradient during flood tide while the remaining 70% (17 hours) was driven by barotropic pressure during ebb tide, thus the estuary is experiencing longer durations downstream discharge compared to upstream flow.

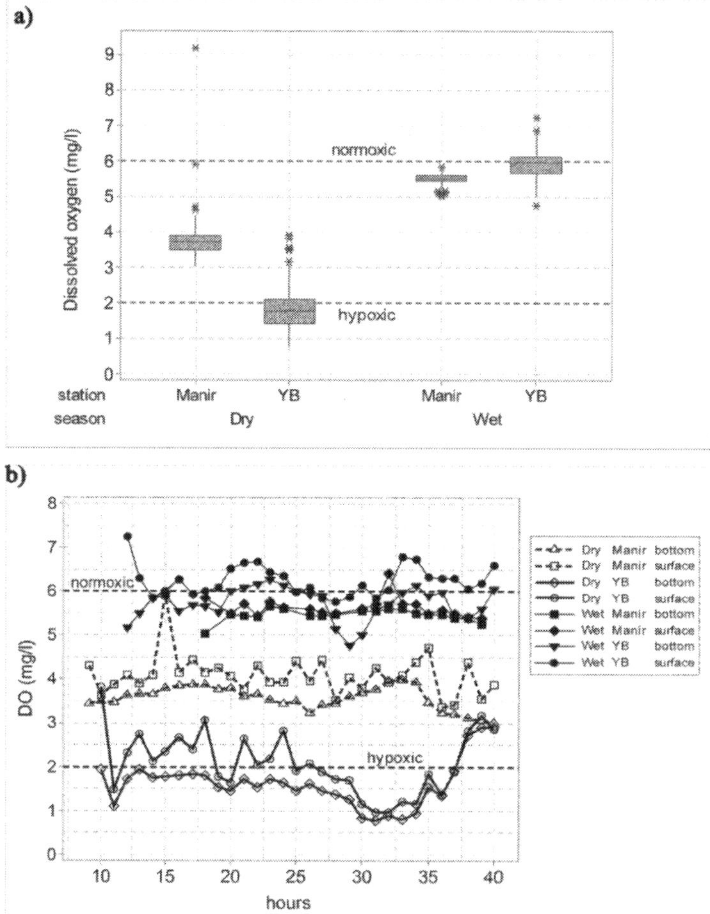

Figure 2. Seasonal changes of DO in the estuary where the Manir station located at upstream, the YB station at the river mouth.

Homogenous current profiles flow was observed during the wet season (Fig. 3a) compared to the dry season (Fig. 3b), higher freshwater discharge during the wet season (approximately 300 m^3s^{-1}, recorded at river gauging station at Kpg Tanggol, about 30 km upstream from river mouth) compared to the dry season (approximately 180 m^3s^{-1}) perhaps contributed to this phenomenon. [14]reported most of the Asia regions estuaries experiences big volume of freshwater flow during the monsoon season or the wet season and it makes the estuary behaviour almost like a river. Both seasons illustrated the magnitude of the ebb current was higher than the flood current, with the median flow magnitude of the ebb and flood tide were 0.32 ms^{-1} and 0.11 ms^{-1} respectively for both seasons. [15] finding on the Maipo River in Chile, which is also subjected to diurnal tides, recorded almost similarly magnitude of tidal currents of approximately 0.2 ms^{-1} and 0.1 ms^{-1} during the ebb and flood tide respectively.

The DO profiles as shown in Figure 4 illustrated the DO magnitude during the wet and dry season vary from 4.76 to 7.25 mg L^{-1} and 0.78 to 3.89 mg L^{-1} respectively. Low DO in the hypoxic condition was observed during the dry season, particularly two meters from the bottom layers (Fig. 4b). Critical extent of hypoxic events with the magnitude lowest then 1.00 mg L^{-1} was observed during the high water slack, where the entire water column was in the hypoxic level. Relationship between current flow and hypoxia in the water columns (Fig. 3b and 4b) during the dry season illustrated that hypoxia was developed associated with current flow less than 20 cms^{-1}. Finding by [16], reported the hypoxia is developed under the weak gravitation circulation mode of less than 5 cms^{-1}.

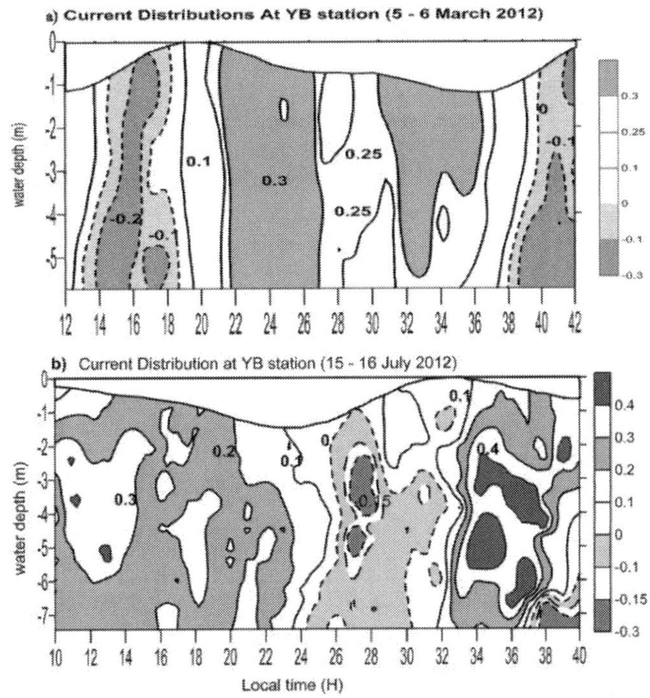

Figure 3. Comparison on current flow magnitude a) the wet season and b) the dry season. *negative current magnitude presents current flow towards upstream.

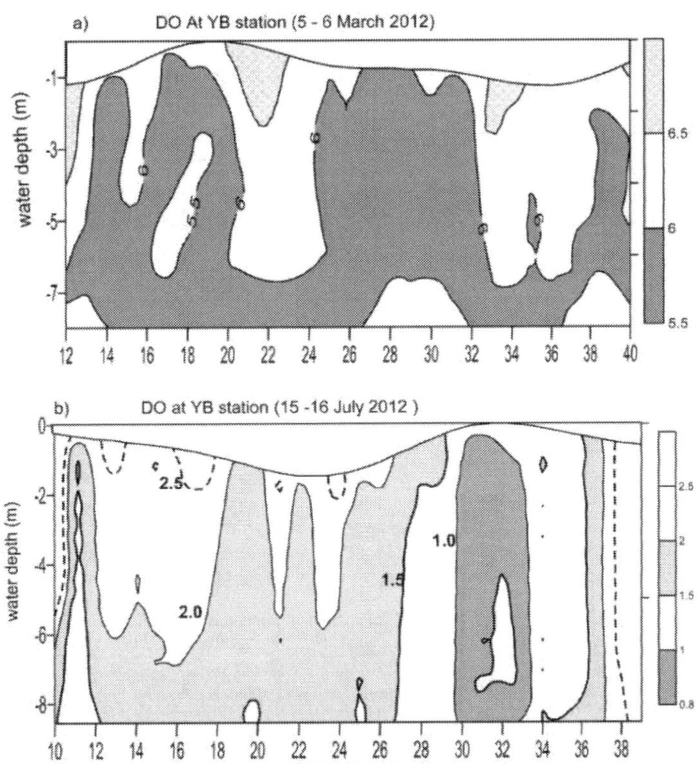

Figure 4. Comparison on the DO profiles for the YB station during a) the wet season and b) the dry season.

3.3 Water column profiles stratification during dry season

Analysis on the vertical distribution of the salinity and DO profile during the high water slack (Fig. 5a) and low water slack (Fig. 5b), illustrated during the high water slack, 80% of water column in the YB station was intruded by saline water concentrations of 31 ppt., apparently entire station at this particular moment was categorized as hypoxic stage as 100% of DO level was lesser than 2 mg L^{-1}(Fig. 5a). Meanwhile, during the low water slack, evident of sharp halocline that separated fresh and saline water was observed at the YB station (Fig. 5b), thus it strengthen the stratification and weaker the movement of vertical water column across the halocline [17, 16] developed significant depletion of the DO within 2 m water depth from bed level, where magnitude of the DO was 2.2 mg L^{-1} at the surface and 1.74 mg L^{-1} at the bed layer.

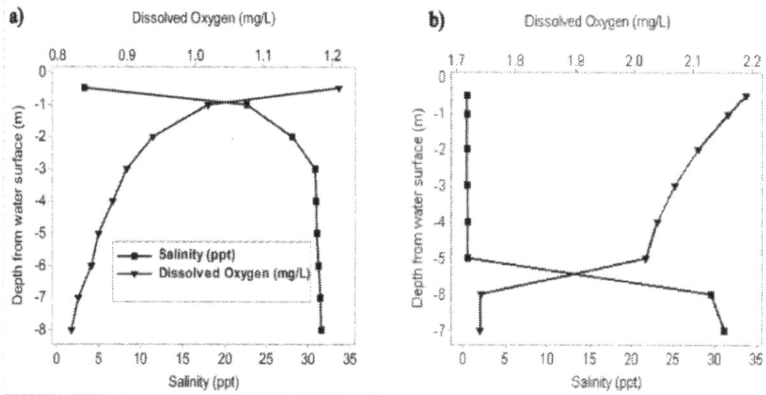

Figure 5. Water profiles stratification during high water slack a) and low water slack b) at the YB station.

3.4 Seasonal nutrient conditions

Analysis on the temporal and spatial nutrient effluents namely total organic carbon (TOC) in the water column at the YB station (Fig. 6), indicates the TOC level was approximately 167 ± 40 μM and 377.5±187.7 μM during the wet and dry season respectively. Similarly, the high contamination of TOC (> 300 μM) also been reported by [18] at the study site in the Gold Coast, Australia. It was reported the high ratio of TOC contents generally the key factors in enhancing biochemical process that required higher demand of oxygen for respiration and organic breakdown processes that causes the depletion of oxygen in the water column [4; 19].

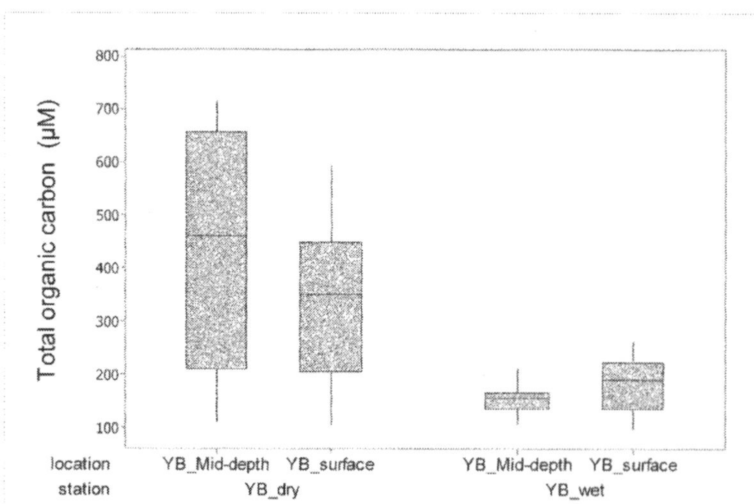

Figure 6. Comparisons on the levels of TOC (μM) during the dry and wet season at the YB station.

4. Discussion and Conclusion

The temporal analysis on the distribution of the DO in the study area indicated the DO level during the wet season was higher compared to the dry season. Large fresh water inputs from the upstream and high density of rains during the wet season enhanced advection pressure in the estuary (Fig. 2), which has transported out most of the contamination in the estuary into near-shore area, resultant to high level of oxygen concentration in the water column during the wet season. However in the dry season, low freshwater released from Kenyir Dam associated with lesser rains [20], developed shallow water column at the study site that weaken the advection process and enhance the residence time in the estuary [21, 22] When this scenario happen coupled with warmer water temperature and water column stratification (Fig. 5) during the dry season enhancing microbial productivity through biochemical process, thus it consumes more oxygen for respiration and decomposition of organic matter [21, 8, 23] which perhaps contributed to oxygen depletion. Similar finding was reported by [9] on tropical region estuary, northern Gulf of Mexico estuary, where the study shows that vertical diffusive oxygen exchange and advective oxygen transport are slow during dry season leading to significant extent of hypoxic events during summer.

The YB station located at the river mouth situated in town center and undergoing rapid developments in the industry sector and urbanization [24], where from morning to noon is the peak hours in daily activities, effluents input from the excessive organic matter loaded from adjacent main wet market[20] into water column could lead to accumulation of nutrients in the water column. Moreover if this situation associated with flood tide during the morning peak hours from 6 to 10 am (Fig. 3b) that minimized the outflow of contaminated blackish water with high concentration of nutrients (Fig. 6) into nearshore, associated with the low current flow magnitude (less than 10 cms^{-1}, Fig. 3b) and strong stratification in the water column (Fig. 5a), which has enhanced the biochemical process and activated the respiration processes, result in significant oxygen depletion took place (Fig. 4b).

Occurrence of the hypoxia is subjected to variations, such as resident time particularly in hydrodynamic flow, contamination of nutrient concentrations in the water column, stratification levels in water column, geometry condition at the study site and others. It was observed that at the study area, hypoxia was developed through combination of low hydrodynamic flow associated with high concentration of nutrient. During the wet season, even though the current flow was low, which was less than 20 cms^{-1} (Fig. 3a) but due to low level of nutrient (Fig. 6), it could not trigger the formation of hypoxia at the study area (Fig. 4a). However, during the dry season, high concentration of nutrient (Fig. 6) coupled with low current magnitude that was lesser than 20 cms^{-1} (Fig. 3b), it developed hypoxia in the water column (Fig. 4b). Thus, perhaps hydrodynamic is not the main cause of hypoxia at the study area, but have to be combined with others variation to generate hypoxic condition. In addition, geometrical conditions too contribute to main consequence on leading the hypoxia [23, 19]. Position of the YB station was located at the deep navigation channel and semi enclosed breakwater's arms (Fig. 1b & 7), where the deep trough and semi enclosed arms is favorable for sedimentation trapping of fine sediments from upper stream [23],it would isolate the bottom water in the navigation from exchange with surrounding oxygen rich water, through advection process due to stratification between fresh and saline water (Fig. 5a &b), making the bottom water inside the trough to be rather stable, leading to the hypoxic events.

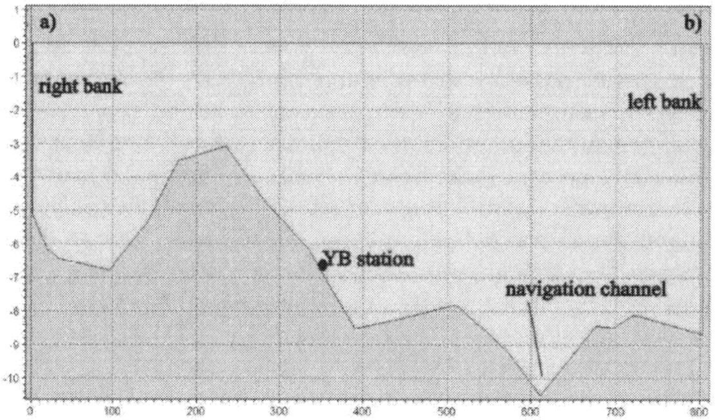

Figure 7. Cross section of navigation channel and position of the YB station (cross-section at ab refered to Fig. 1b).

In conclusion, hypoxia was developed during dry season at the study area particularly at the river mouth. It has demonstrated the linkage between the nutrients contamination, hydrodynamics, stratification and geometrical formation were the major controlling factors in developing hypoxic levels in the Terengganu river mouth.

Acknowledgments
The authos wish to thank Dr. Zelina Zaiton Ibrahim and Prof Mahendran Shitan from UPM on their guiden for the data collection campaign. We also wish to thank members of Research Centre for Coastal Management, NAHRIM for their help in data collections.

References

[1] Chen, C.C., Gong, G. C., Shiah, F.K. (2007). Hypoxia in the east China Sea: one of the largest coastal low-oxygen areas in the world. *Marine Environmental Research* 64, pp. 399-408
[2] DiMarco, S.F., Chapman, P., Walker, N., Hetkand, R. D. (2010). Does local topography control hypoxia on the easter Texas-louisiana shelf?. *Journal of Marine System* 80, pp. 25-35
[3] Tinh, T.T.K., Hai, N.X., Quang, P.V., Nguyen, T.Y. (2014). Variation of soil, water quality and impacts on biodiversity in Xuan Thuy Ramsar sie. ARPN Journal of Agricultural and Biological Science. (9), No 3
[4] Zhang, H., Li, S. (2010). Effects of physical and biochemical process on the dissolved oxygen budget for Pearl River Estuary during summer. *Journal of Marine Systems* 79, pp. 65-88
[5] Brown, C.A. and Power, J.P. (2011). Historic and recent patterns of dissolved oxygen in the Yaquina Estuary (Oregon, USA): Importance of anthropogenic activities and ocean conditions. *Estuarine, Coastal and Shelf Science* 92, pp. 446-455
[6] Dai, M., Guo, X., Zhai, W., Yuan, L., Wang, B., Wang, L., Cai, P., Tang, T., Cai, W.J. (2006). Oxygen depletion in the upper reach of the Pearl River estuary during a winter drought. *Marine Chemistry* 102, pp. 159-169
[7] Kountoura, K., Zacharias, I. (2014). Annual hypoxia dynamics in a semi-enclosed Mediterranean gulf. *Journal of Marine Systems* 139, pp. 320-331
[8] Nezlin, P.N., Kamer, k., Hyde, J., Stein, E.D. (2009). Dissolved oxygen dynamics in a eutophic estuary, upper Newport Bay, California. *Estuarine, Coastal and Shelf Science* 82, pp. 139-151
[9] Hagy, J.D., Murrell, M.C. (2007). Susceptibility of a northern Gulf of Mexico estuary to hypoxia: an analysis using box models. *Estuarine, Coastal and Shelf Science 74, pp.* 239-253

[11] Gao, X., Song, J. (2008). Dissolved oxygen and O_2 Flux across the water air interface of the Changjiang Estuary in May 2003. *Journal of Marine System* 74, pp. 343-350

[12] Suratman, S., Weston, K., Jickells, T. and Fernand, L. (2009). Spatial and seasonal changes of dissolved and particulate organic C in the North Sea. *Hydrobiologia*, pp. 628, 13-25.

[13] Roegner, G.C., Needoba, J.A., Baptista, A.M. (2011). Coastal upwelling supplies oxygen-depleted water to the Columbia River estuary" *PlosOne* 6, issue 4 e18672

[14] Vijith, V., Sundar, D. and Shetye. (2009). Time-dependence of salinity in monsoonal estuaries. *Estuarine, Coastal and Shelf Science* 85, pp. 601-608

[15] Piñones, A., Levinson, A., Narváez, D.A., Vargas, C.A., Navarrete, S.A., Yuras, G., Castilla, J.C., (2005). Wind induced diurnal variability in river plume motion. *Estuarine, Coastal and Shelf Science* 65, pp. 513-525.

[16] Lin, J., Xie, L., Pietrafesa, L.J., Shen, J., Mallin, M.A., Durako, M.j. (2006). Dissolved oxygen stratification in two micro-tidal partially mixed estuary. *Estuarine, Coastal and Shelf Science* 70, pp. 423-437

[17] Yin, K., Lin, Z., Ke, Z. (2004). Temporal and spatial distribution of dissolved oxygen in the Pearl River Estuary and adjacent coastal waters. *Continental Shelf Research* 24, pp. 1935-1948

[18] Parvez, M., Goonetilleke, A., Ayoko, G.A., Egodawatta, P. (2011). Effects of climate change on the wash-off of volatile organic compounds from urban roads. Science of the Total Environment 409, 3934-3942

[19] Testa, J. T., Li, Y., Lee, Y.J., Li, M., Brady, D.C., Tor o, D.M., Kemp, W.M., Fitzpatrick, J.J. (2014). Quantifying the effects of nutrient loading on dissolved O2 cycling and hypoxia in Chesapeake Bay using a coupled hydrodynamic-biogeochemical model. *Journal of Marine Systems*139, pp.139-158

[20] DID (Department of Irrigation and Drainage, Terengganu, Malaysia) 2010. Terengganu River integrated river basin management study: DID

[21] Wei, H., He, Y., Li, Q., Liu, Z., Wang, H. (2007). Summer hypoxia adjacent to Changjiang Estuary. *Journal of Marine Systems* 67, pp. 292-303

[22] Nijs, M.A.J.D, Pietrzak, J.D. (2011). Advection of salt wedge and evolution of the internal flow structure in the Rotterdam Waterway." *Journal of Physical Oceanography,* volume 41

[23] Wang, B. (2009). Hydromorphological mechanisms leading to hypoxia off the Chanhjiang Estuary. *Marine Environmental Research* 67, pp. 53-58

[24] Sultan, K., Shazili. N. A. (2010). Geochemical baselines of major, minor and trace elements in the tropical sediments of the Terengganu River basin, Malaysia. *International Journal Of Sediment Research* 24, pp. 340-354.

Geophysical and Hydrochemical Characteristics of Groundwater at Kerian Irrigation Scheme

M A A Nazri[1], L W Tan[1] , H Kasmin[1] , S Syafalni[2] and I Abustan[3]
[1]Faculty of Civil and Environmental Engineering, Universiti Tun Hussein Onn Malaysia, 86400 Batu Pahat, Johor, MALAYSIA.
[2] Binus University, Kebon Jeruk, 11530 Jakarta Barat, INDONESIA
[3] School of Civil Engineering, Universiti Sains Malaysia, 14300 NibongTebal, Pulau Pinang, MALAYSIA

E-mail: ariffn@uthm.edu.my

Abstract. .A study was conducted to determine the potential of groundwater uses at the Kerian irrigation scheme, specifically the Selinsing irrigation area. Resistivity image profiling method was employed at four sampling locations within the study area to identify the depth and variation of aquifer layer and the possibility of hard layer presence. Investigations conducted revealed that the thickness of aquifer within the study area varies between 5 m and 10 m, which is located at a depth of 30 m to 70 m below the ground surface. The thickness of clay layer is between 1 m and 80 m which dominates the upper subsurface layer. Seven samples of groundwater were used in determining their hydrochemical and physicochemical characteristics, ionic composition, and the quality suitability for irrigation use. A preliminary characterisation based on the piper diagram provides the hydrofacies classification while Stiff diagram is used to exhibit the ionic relationship. The degree of correlation between cations and anions has been estimated in order to assess their mutual relationships. Strong positive correlation exists for Ca^{2+}–Mg^{2+}, Ca^{2+}–Na^{+}, Na^{+}–Mg^{2+}, Na^{+}–K^{+}, Ca^{2+}–K^{+}, Mg^{2+}–K^{+}, Mg^{2+}–Cl^{-}, K^{+}–HCO_3^{-}, and Na^{+}–Cl^{-}. Profusion of elements also reflects the composition of aquifer and the climatic conditions, which possibly contribute to the genetic relationship.

Keywords: Geophysical, groundwater, hydrochemical.

1. Introduction

Groundwater is an integral part of water resources. It is a main source of water supply and plays a vital role in the eco-environmental maintenance and socio-economic development in arid and semi-arid regions of the world [1]. Sustainable water management is vital for maintaining the ecosystem's balance and economic development [2] and [3].

Geophysics is the application of physics to study the earth such as by taking measurements at or near the surface of the earth. As these surfaces were influenced by the internal distribution of physical structures and properties, the variation of subsurface structure which possesses different physical attributes is able to reflect the condition of the earth structure [4].

Non-intrusive geophysical techniques such as electrical resistivity survey has been applied for many decades to determine the thickness and resistivity of the layered media as well as to map the geological environment of existing aquifer. This method has been greatly favored due to the contrast between the electrical properties of water and soil [5]. In other words, the subsurface material is represented by unique bulk electrical conductivity value. The measure of ground conductivity shows the strength of the secondary magnetic field detected by a current-carrying transmitter coil [6]. In the past, Schlumberger sounding is commonly used in determining the resistivity variation with depth [7]. However, it is very difficult to perform resistivity soundings everywhere without prior information. These classical geophysical techniques determine the physical properties of the subsurface, which are only indirect indicators of water content and hydraulic conductivity. Due to rapid development of technology, geophysical hardware and software tools have provide the opportunity to obtain high resolution quality of geophysical data. They are widely applied to supplement conventional borehole sampling which is expensive and provide information based on the effects at the sampling site [8]. They are also chosen due to the simplicity of the techniques, easy interpretation and rugged nature of the associated instrumentation.

Meanwhile, hydrochemistry is the subdivision of hydrology which deals with chemical characteristics of water. The sustainable development and effective management of the valuable groundwater resouces of an area requires better understanding of hydrochemical characteristics of the groundwater and its evolution under natural water circulation processes, as well as the origin and residence time of the groundwater [9]. The quality of groundwater is equally important in identifying suitable uses of water. Water quality investigation is a significant agenda in groundwater studies. The ever increasing new knowledge on the hydrochemical processes governing groundwater quality has led to more exploration of groundwater use. Hydrochemistry studies provide insight into the interaction between groundwater and its environment and contribute to better groundwater resource management and planning [10]. Variation of groundwater quality of an area is a function of physicochemical parameters that are greatly influenced by geological formations and anthropogenic activities [11].

Due to the importance of hydrochemical characteristics and its influence on water quality, the aim of this study is to assess the spatial distribution of hydrochemical constituents of groundwater. The proposed method relates the assessed constituents for better understanding of hydrogeological and its suitability for agriculture use, with an application to Kerian irrigation scheme.

1.1 Study area

Currently, approximately 20% of the land in Malaysia is utilised for agricultural purposes. More than two thirds are for industrial crops, mainly rubber and oil palm plantations while paddy cultivation covers 660,000 ha (2% of total land area)[12].

Selinsing irrigation scheme located in the southern part of Kerian district (4°56′10″N - 4°57′34″N and 100°38′27″E - 100°39′21″E) is a typical irrigation scheme and is one of the granary areas in Malaysia.

Selinsing scheme experiences a humid tropical climate with high average annual rainfall of about 2,420 mm and temperature between 27°C and 28°C. The scheme lies mostly along the coastal fringe where flat marine alluvial soil provides favorable environment for the agricultural activities such as paddy growing [13].

Selinsing canal originating from Bukit Merah reservoir is designed to supply water for agricultural activities in compartments G and H (Figure 1), covering an area of almost 12,166 ac. (49.23 km^2).

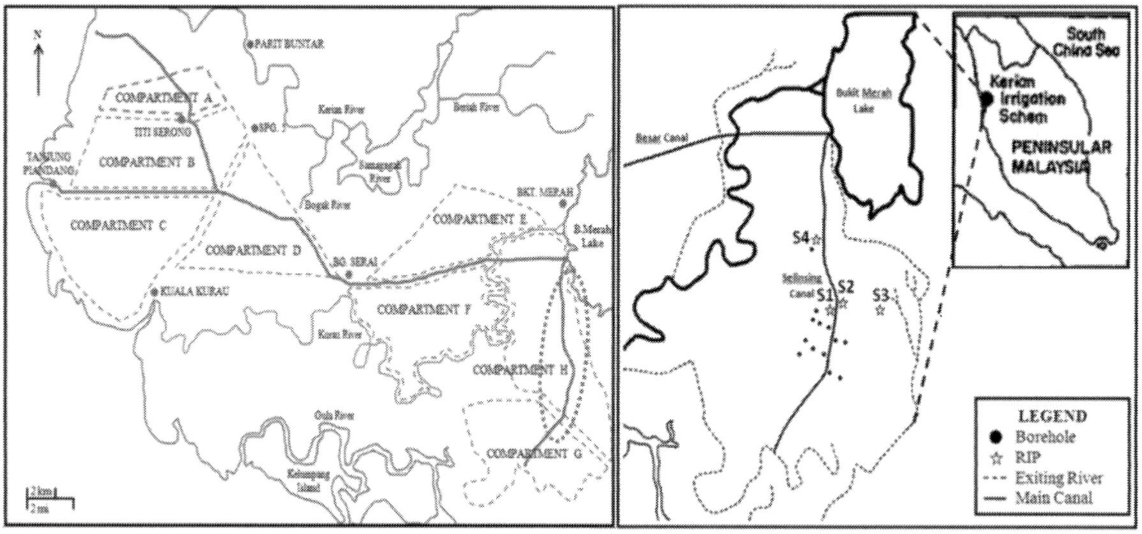

Figure 1. Location map of study area and resistivity image profiling (RIP) points.

2. Methods

2.1 Geophysical

Resistivity image profiling (RIP) survey was conducted on field using Terrameter SAS 4000 system, LUND ES464 electrode selector system, multi-core cables with total layout length of 800 m span with 10 m spacing between each electrodes. It is a fully automated technique that uses a number of electrodes connected by a multicore cable. Since measurements are done automatically, settings of electrode array and others have to be configured according to the tasks by taking account into the depths of interest, data density, and time spent.

The resistivity meter controls the positions of current and potential electrodes through a microprocessor controlled electronic module [14].

The RES2DINV resistivity inversion software [15] was used to automatically invert the apparent resistivity data from field and forward it into the two-dimensional models. RES2DINV is a least square inversion that uses smoothing technique [16] and [17]. By applying the appropriate survey and data interpretation techniques, 2D and 3D visualisations of the subsurface profile can be obtained. The smoothness-constrained least-square inversion method is used to produce the subsurface resistivity models. This inversion method minimises the square of the differences between the measured and calculated apparent resistivity values and produces earth resistivity models with gradual transitions across zones of different resistivities [18]. The smooth inversion method gives better results for areas where the subsurface resistivity changes in a gradual manner, while the blocky inversion method gives significantly better results where there are sharp boundaries [19].

However, the model with the lowest possible RMS error occasionally shows large and unrealistic variations in the resistivity values. Most of the inversion models produce very small RMS error, i.e. less than 3% [14]. Generally, the most prudent approach is to choose the model at the iteration after which the RMS error does not change significantly. This usually occurs between the 3rd and 5th iterations (Loke, 2004). Resistivities are adjusted iteratively until an acceptable agreement between the input data and the model responses is reached [20].

2.2 Hydrochemical

Groundwater samples from 7 tube wells were analysed between April and July 2009. The samples were extracted at depths ranging between 5 m and 45 m. Water samples were collected using suction pumps after purging at least three well volumes. Samples were stored in pre-sterilised polyethylene bottles without any preservative to avoid contamination.

In order to eliminate any potential contaminant, the sample bottles were triple-rinsed with MilliQ water (18 mΩ-cm) and then filled with 50%, v/v solution of reagent-grade HCl and MilliQ water prior to sampling [21]. Before collecting the sample, the bottle was rinsed 3 times with the sample water for conditioning purposes. Only ultra-pure water and deionised water were used for preparing solutions for analysis.

Concentrations of unstable parameters such as total dissolved solids (TDS), pH, temperature, and electrical conductivity were determined on site with Hanna HI9828 multiparameter probe to minimise the contact with air. The water samples were preserved and brought to the laboratory in minimum period of time for further analyses, as determined by [22] standard methods.

For analysing major cations (Na^+, Mg^{2+}, K^+, Ca^{2+}) and anions (Cl^-, SO_4^{2-}, NO_3^-), the groundwater was filtered with a hand-held syringe using 0.45 μm membrane filter. The cations and anions were analysed by Dionex ICS-1000 with AS-DV50 auto sampler. Calibration for anion and cation analyses was performed using appropriately diluted standards which are issued by the ICS supplier periodically. Alkalinity was determined by the Gran titration method using 0.1 N HCl.

3. Results and discussion

3.1 Resistivity image profiling (RIP) survey

The 2D resistivity imaging survey was conducted along the Selinsing canal. Electrical signals were used to identify the resistivity value of the subsurface soil layers, and the data is inversed into the 2D resistivity model profiles by the RES2DINV software. The *x*-axis of the model signifies the distance along the cross section of the ground while the *y*-axis represents the vertical depth of the ground from the surface. The intensity of the resistivity is differentiated by various colors, thus the layer of acquifer can be identified. The resistivity of aquifer layer which contains groundwater ranges between 10 Ωm and 100 Ωm [23] and [24].

For Selinsing 1 area (Figure 2a), layer of clay was found at the earth surface, extending to a depth of 20 m below. The aquifer layer was found at a depth of about 30 m below the ground. Wide thin layer of low resistivity (blue colored zone) having 10 Ωm to 75 Ωm is observed to have depth varying between 20 m and 50 m below ground surface. This resistivity is either influenced by the clay layer or due to the infiltration from the nearest waterway.

Similar to S1 profile, Selinsing 2 (Figure 2b) and Selinsing 3 (Figure 2c) also show the same characteristics of the low resistivity layer.

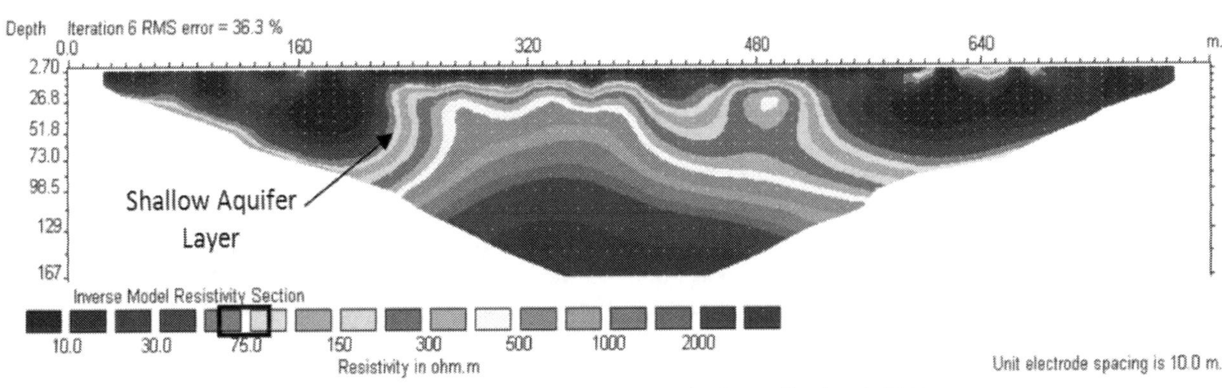

Figure 2a. Two-dimensional resistivity profile for Selinsing 1 (S1).

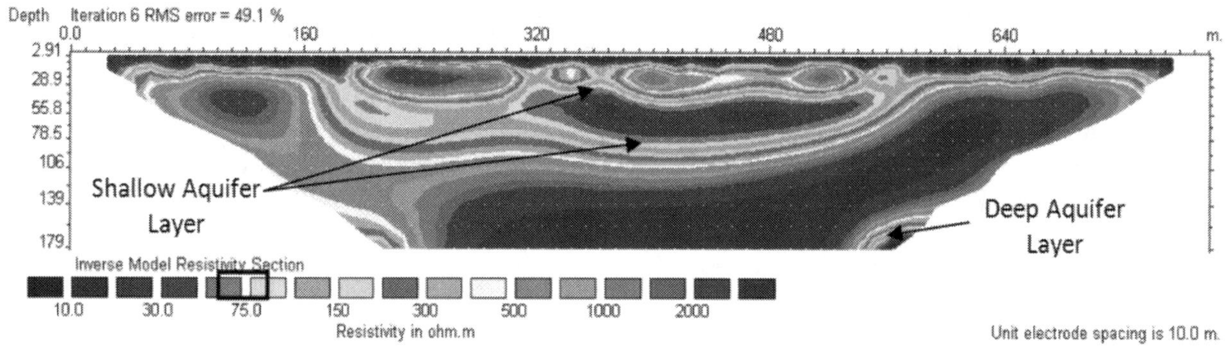

Figure 2b. Two-dimensional resistivity profile for Selinsing 2 (S2).

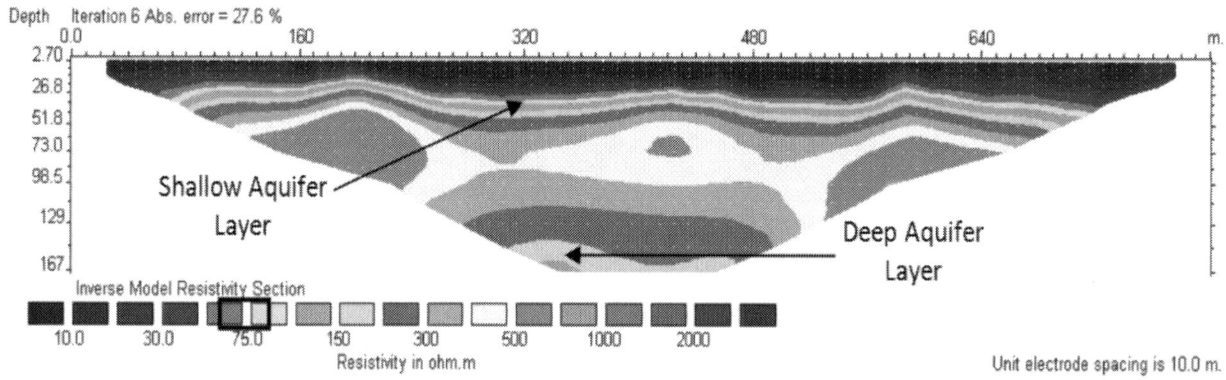

Figure 2c. Two-dimensional resistivity profile for Selinsing 3 (S3).

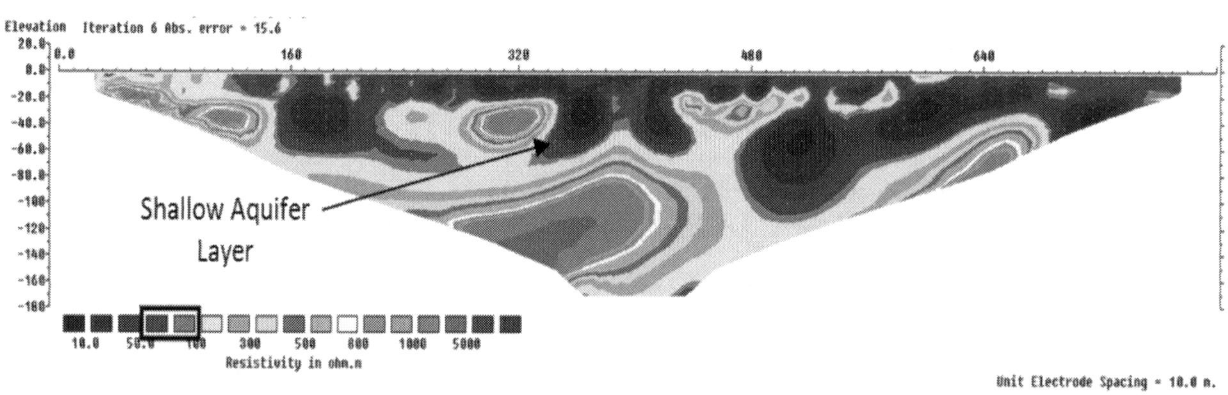

Figure 2d. Two-dimensional resistivity profile for Selinsing 4 (S4).

The thickness of aquifer layer is found to be varying between 5 m and 10 m. The aquifer layer is located at the depth of 30 m to 70 m below the ground surface, and even up to 140 m in some exceptional case. The main feature of the inverted resistivity profiles is that the aquifer occur in layers, which can be considered as consolidated porous media. The observation at Selinsing 4 (Figure 2d) showed that the aquifer layer is encountered at the depth of 40 m to 80 m, with moderate aquifer thickness of about 20 m.

IOP Conf. Series: Materials Science and Engineering **136** (2016) 012070 doi:10.1088/1757-899X/136/1/012070

3.2 Hydrochemical analyses

Table 1 shows the concentraiton of physicochemical parameters in the groundwater along Selinsing canal. Correlation matrix among 7 groundwater quality parameters of Selinsing canal is shown in Table 2. Meanwhile, the sodium concentration (NA%), SAR, total hardness (TH), and RSC of the groundwater are provided in Table 3.

Table 1. Ionic variation in the groundwater along Selinsing canal (April-July 2009).

Well	pH	EC	TDS	Ca^{2+}	Mg^{2+}	Na^+	K^+	CO_3^-	HCO_3^-	Cl^-	SO_4^{2-}	NO_3^-
BH1	7.10	590	45	7.73	0.94	1.82	1.42	0	135.6	19.89	4.66	0.12
BH2	6.85	495	74	8.09	1.08	4.97	3.10	0	119.8	15.32	46.18	5.51
BH3	6.17	129	46	7.83	0.58	1.79	2.69	0	164.7	3.61	17.81	2.88
BH4	7.00	516	122	8.34	2.65	39.96	2.28	18	125.7	54.44	11.61	0.44
BH5	7.06	629	101	6.14	2.95	21.60	1.84	0	122.0	33.50	28.40	0.59
BH8	5.36	721	30	3.00	0.61	7.38	1.20	10	115.3	44.98	3.71	2.93
BH12	7.50	732	62	22.74	0.19	6.88	2.22	13	36.79	2.43	5.68	0.66

Note: All values are in mg/L except pH and EC

Table 2. Correlation of groundwater physicochemical parameters along Selinsing canal (April-July 2009).

	Ca^{2+}	Mg^{2+}	Na^+	K^+	CO_3^-	HCO_3^-	Cl^-	SO_4^{2-}
Ca^{2+}								
Mg^{2+}	0.955							
Na^+	0.965	0.971						
K^+	0.942	0.906	0.973					
CO_3^-	0.445	0.178	0.229	0.317				
HCO_3^-	0.851	0.866	0.909	0.905	0.624			
Cl^-	0.872	0.918	0.945	0.871	0.389	0.885		
SO_4^{2-}	0.044	0.125	0.028	−0.006	0.495	0.144	0.034	
NO_3^-	−0.374	−0.379	−0.369	−0.278	−0.133	−0.191	−0.400	0.520

Table 3. Total hardness (TH), RSC, SAR, and Na concentration of groundwater along Selinsing canal.

Borehole ID	TH	RSC	SAR	Na%
BH1	23.18	1.80	0.16	27.20
BH2	24.65	1.50	0.44	46.84
BH3	21.94	2.30	0.17	34.76
BH4	31.74	2.00	3.09	79.35
BH5	27.48	1.50	1.79	72.01
BH8	10.00	2.00	1.02	70.39
BH12	57.57	-0.11	0.39	28.41

The groundwater is considered suitable for irrigation purposes if the TDS concentration is less than 1000 mg/L. TDS concentration between 1000 mg/L and 2000 mg/L shows fair quality while the quality is inferior when the salinity exceeds 2000 mg/L. All 7 groundwater wells being sampled were considered suitable for irrigation use (Table 1).

Table 4. Major ions distribution in the shallow groundwater along the Selinsing canal.

Elements	Min. (mg/L)	Max. (mg/L)	Mean (mg/L)
Na^+	1.79	39.96	12.06
K^+	1.20	3.10	2.11
Ca^{2+}	3.00	22.74	9.12
Mg^{2+}	0.19	2.95	1.29
Cl^-	2.43	54.44	24.88
HCO_3^-	36.79	164.70	117.13
CO_3^-	0.00	18.00	5.86
SO_4^{2-}	3.71	46.18	16.86
NO_3^-	0.12	5.51	1.88

Sodium is the most abundant element found in the groundwater at the study area. Its concentration ranges from 1.79 mg/L to 39.96 mg/L with a mean of 12.06 mg/L (Table 4). Quite the opposite, the concentration of potassium is consistent, with value of 1.20 mg/L to 3.10 mg/L and an average of 2.11 mg/L. Both sodium and potassium have good solubility and mobility, and will not precipitate at any pH. The possible source of contamination is from the seawater.

As shown in Figure 3a, the mutual relation between Na and Cl is a linear correlation which indicates genetic relation with meteoric water [25]. This is proven by the correlation between the ions as shown in Table 2, where Na^+–K^+ is 0.973, Na^+–Cl^- is 0.945, and K^+–Cl^- is 0.871.

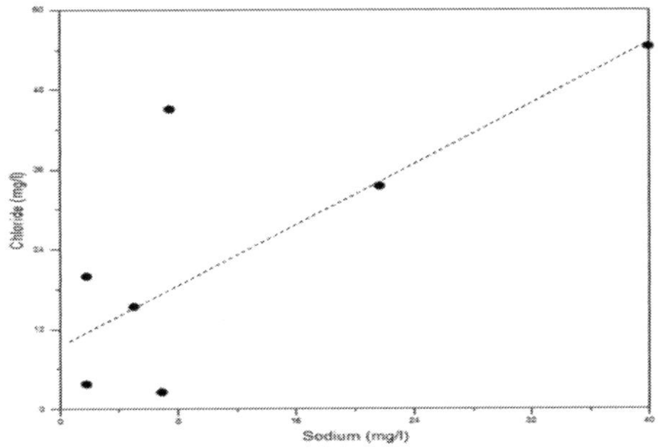

Figure 3a. Correlation between Na^{+} and Cl^{-} elements.

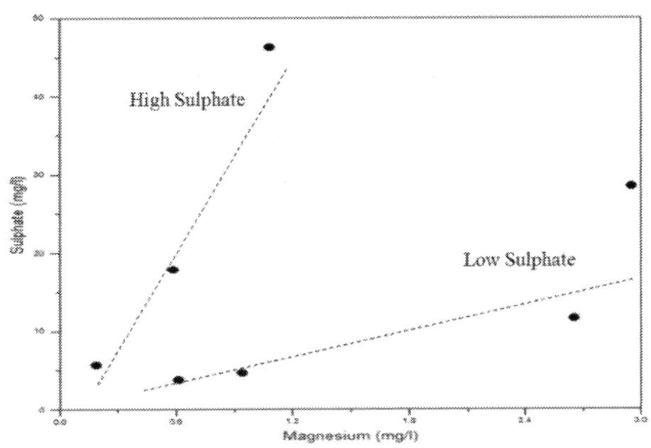

Figure 3b. Correlation between Mg^{2+} and SO_4^{2-} elements.

The Mg^{2+} and SO_4^{2-} relationship shows dual characteristics (Figure 3b). Sulphate SO_4^{2-} concentration varies from 3.71 mg/L to 46.18 mg/L with an average of 16.86 mg/L, whereas the average concentration of magnesium Mg^{2+} is 1.29 mg/L with minimum and maximum of 0.19 mg/L and 2.95 mg/L, respectively. The maximum limit for sulphate and magnesium as prescribed by WHO for usage are 200 mg/L and 50 mg/L, respectively. The sampled groundwater has Mg^{2+} and SO_4^{2-} concentrations below the limits.

The correlation coefficient r between the 7 groundwater quality parameters (in Table 2) shows the statistical measure of the interdependence of two or more parameters. The closer correlation value to 1, the relationship between the physicochemical parameters of underground water becomes stronger, which is given as

$$r = \frac{N\Sigma xy - (\Sigma x)(\Sigma y)}{\sqrt{\left[N\Sigma x^2 - (\Sigma x)^2\right]\left[N\Sigma y^2 - (\Sigma y)^2\right]}} \quad (1)$$

where,
x, y = two different parameters,
N = size of the sample

Distribution of the major ions shows compositional variations in the shallow groundwater samples (Table 3). In general, the trend of concentration distribution among cations shows $Na^{+} > Ca^{2+}$

> K^+ > Mg^{2+}. Anions exhibit similar distribution pattern, where HCO_3^- > Cl^- > SO_4^{2-} > CO_3^-. The dispersion of ions in the groundwater at the Kerian irrigation scheme probably indicates similar geochemical environment, climatic conditions and pH, except for few anomalies.

pH of the groundwater varies between 5.36 and 7.50, with the mean of 6.72, indicating slight acidity. According to WHO standard, the desirable limit of pH for drinking water is between 6.5 and 8.5. Only BH3 and BH8 samples are not within the desired limit.

Electrical conductivity (EC) of groundwater is due to the presence of various dissolved salts. EC obtained from the groundwater samples recorded values between 129 μS/cm and 732 μS/cm, with an average of 544.6 μS/cm. WHO prescribed the maximum limit of EC for drinking water as 1500 μS/cm. Thus, all the groundwater samples do not exceed the maximum permissible limit. However, for irrigation purposes, water with conductivity less than 2250 μS/cm has been used without problem for a considerable time. The U.S. Salinity Laboratory Staff (1954) has categorised groundwater based on EC (Table 5). According to this categorisation, the sample of BH3 is in the excellent category while the other six samples, i.e. BH1, BH2, BH4, BH5, BH8, and BH12 is in the good category.

Table 5. Water quality classification (U.S. Salinity Laboratory Staff, 1954).

Quality of water	Electrical conductivity (μS/cm)	Sodium adsorption ratio (epm)
Excellent	Up to 250 (BH3)	Up to 10 (all samples)
Good	250 – 750 (excluding BH3)	10 – 18 (nil)
Fair	750 – 2250 (nil)	18 – 26 (nil)
Poor	> 2250 (nil)	> 26 (nil)

*nil – not in list

Generally, most of the groundwater samples have hardness due to the presence of calcium and magnesium. The total hardness (TH) is given as

$$TH = 2.497\,Ca^{2+} + 4.115\,Mg^{2+} \quad (2)$$

This study has found that the total hardness of the groundwater samples ranges between 10 mg/L and 57.57 mg/L. The permissible limit of total hardness is 100 mg/L for drinking water as designated by WHO international standard. It has been shown that none of the groundwater samples exceed the maximum permissible limit. It also indicates that most of the groundwater samples fall in the soft water category.

According to [26], the residue sodium carbonate (RSC) has maximum influence on the aptness of water for irrigation purposes. Thus, SAR value acts as a significant tool to evaluate the aptness of irrigation water. It is important to ensure that all the samples have low RSC. The index quantifies the proportion of Na to Ca and Mg ions in a sample. RSC is the excess sum of carbonate and bicarbonate in groundwater over the sum of calcium and magnesium which is given as

$$RSC = \left(CO^{2-} + HCO_3^-\right) - \left(Ca^{2+} + Mg^{2+}\right) \quad (3)$$

where, the concentration is in meq/L.

The classification of irrigation water according to RSC is provided in Table 6. All the groundwater samples are categorised as good for irrigation.

Table 6. Groundwater quality based on the residual sodium carbonate (RSC).

Quality of water	RSC
Good	< 1.25 (BH12)
Doubtful	1.25 – 2.5 (BH1, BH2, BH3, BH4, BH5, & BH8)
Unsuitable	> 2.5 (nil)

*nil – not in list

The calculated SAR of the samples (Table 3) is between 0.16 and 3.09 with a mean of 1.01. SAR of less than 10 indicates that the groundwater has low sodium hazard, according to Todd (1980).

EC and SAR are plotted on the US salinity diagram, as shown in Figure 4. EC is taken as salinity hazard and SAR is the alkalinity hazard. Groundwater samples from BH1, BH2, BH4, BH5, BH8, and BH12 fall in the C_2S_1 quality, i.e. with medium salinity hazard and low sodium hazard. Sample of BH3 has the C_1S_1 quality, i.e. low salinity hazard and low sodium hazard.

This indicates that the groundwater of Selinsing canal can be used for irrigation since it is low in SAR with medium salinity hazard. Table 5 summarises the groundwater quality classification of salinity hazard and sodium hazard.

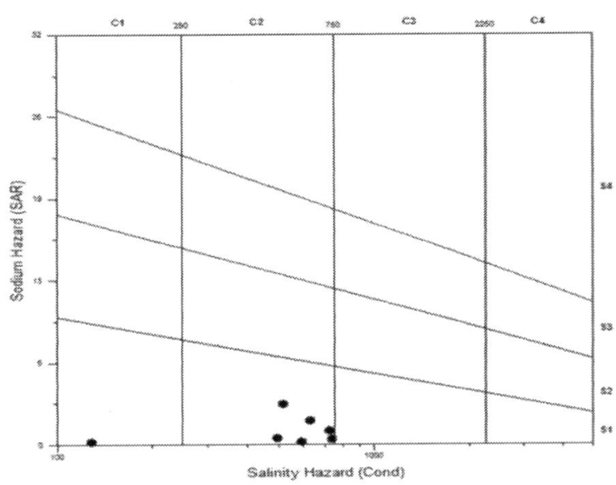

Figure 4. Suitability of groundwater for irrigation purposes based on the USSL diagram.

Sodium percentage obtained from the groundwater samples varies between 27.20% to 79.35%, with an average of 51.28%. Na% reflects that the classification of groundwater, i.e. either excellent, good, permissible, doubtful, and unsuitable [27]. Table 7 shows that four out of the seven samples are in the good and permissible categories, while the others are doubtful for irrigation. Sodium ions have the tendency to be absorbed by clay particles, displacing Mg^{2+} and Ca^{2+} ions if the concentration of sodium is high in irrigation water. This chemical reaction process is soil resulted in the reduction of permeability and internal drainage.

Table 7. Groundwater quality based on the Na%.

Quality of water	Na%
Excellent	< 20% (nil)
Good	20 – 40% (BH1, BH3 & BH12)
Permissible	40 – 60% (BH2)
Doubtful	60 – 80% (BH4, BH5 & BH8)
Unsuitable	> 80% (nil)

*nil – not in list

The hydrochemical facies analysis reflects the chemical processes in certain lithological environment and specific geochemical condition. The general classification and trend of variation of the groundwater samples are shown in the Hill-Piper diagram in Figure 5 [28], [29] and [30]. The hydrochemical pattern diagram helps in hydrogeochemical facies classification [31]. The diagram has been classified into four hydrochemical facies based on the dominance of particular cations and anions, i.e. facies 1: Ca^{2+}–Mg^{2+}–HCO_3^- type I; facies 2: Na^+–K^+–HCO_3^- type II; facies 3: Na^+–K^+–Cl^-–SO_4^{2-} type III and facies 4: Ca^{2+}–Mg^{2+}–Cl^-–SO_4^{2-} type IV. Most of the groundwater samples fall in type II facies, i.e. BH4, BH5, and BH8 and type IV facies, i.e. BH1, BH2, BH3, and BH12.

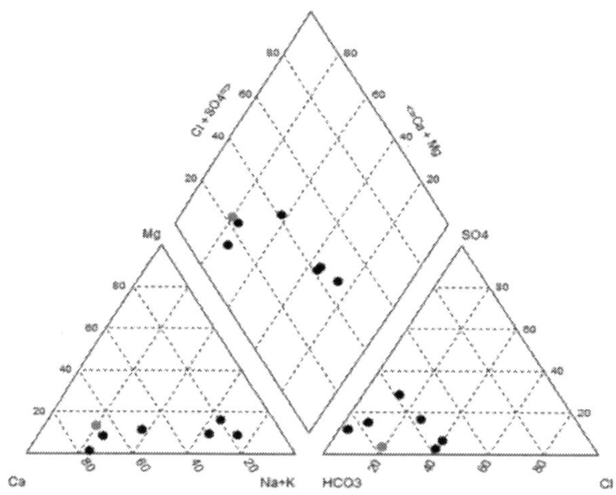

Figure 5. Piper diagram for the chemical composition of the groundwater samples of Selinsing canal.

The Ca^{2+}/Mg^{2+} and Na^+/Cl^- ratios are valuable in the determination of salinisation [25]. All the samples have Ca^{2+}/Mg^{2+} ratio of extremely higher than 1.0, which indicates the exposure of carbonate rocks in the study area (Table 8).

Table 8. Salinisation based on Ca^{2+}/Mg^{2+} and Na^{+}/Cl^{-} ratios.

Borehole ID	Ca^{2+}/Mg^{2+} ratio	Na^{+}/Cl^{-} ratio
BH1	8.223	0.692
BH2	7.491	1.475
BH3	13.500	0.496
BH4	3.147	0.734
BH5	2.081	0.645
BH8	4.918	0.975
BH12	119.68	2.831

However, the Na^{+}/Cl^{-} ratio is below 1.0, indicating fresh water existence. The Na^{+}/Cl^{-} ratio of the collected groundwater samples is between 0.496 and 2.831 with a mean of 1.121. The samples with high Na^{+}/Cl^{-} ratio of more than 1.0 shows evaporation and concentration [32].

4. Conclusion

From the resistivity image profiling and lithological data constructed from the deep boring method, the presence of shallow groundwater has been detected along the Selinsing canal. The thickness of the aquifer varies between 5 m and 10 m, located at depths ranging from 30 m to 70 m below the mostly-clayey ground surface. The top ground surface consists of clay with thickness between 1 m and 80 m. Resistivities of clay varies between 3 Ωm and 100 Ωm [23] and [24] whereas aquifer resistivities ranges from 10 Ωm to 100 Ωm.

The major ionic composition showed wide disparities in the hydrochemical properties of the groundwater at Kerian irrigation scheme. The degree of correlation between cations and anions is determined to assess their relationship. Profusion of elements was interpreted in relations to the composition of aquifer and the climatic conditions. The significant hydrogeophysical parameters have been estimated for reasonable assessment of groundwater quality for irrigation purposes. Relationships between various elements were identified from correlation coefficient, which reflects genetic association. Strong positive correlation exists for Ca^{2+}–Mg^{2+}, Ca^{2+}–Na^{+}, Na^{+}–Mg^{2+}, Na^{+}–K^{+}, Ca^{2+}–K^{+}, Mg^{2+}–K^{+}, Mg^{2+}–Cl^{-}, K^{+}–HCO_3^{-}, and Na^{+}–Cl^{-}.

Acknowledgements

The study is supported by a grant from the Malaysia Ministry of Science, Technology and Innovation and Universiti Sains Malaysia. Special thanks is extended to the Malaysian Nuclear Agency team for their cooperation.

References

[1] Zhang G, Deng W, Yang Y S and Salama R B 2007 Evolution study of a regional groundwater system using hydrochemistry and stable isotopes in Songnen Plain, northeast China, Hydrological Processes, 21, pp.1055--1065.

[2] Narasimhan T N 2005 Hydrogeology in North America: past and future. Hydrogeology Journal, 13(1), pp.7--24.

[3] Esteller M V and Andreu J M 2005 Anthropic effects on hydrochemical characteristics of the Valle de Toluca aquifer (central Mexico), Hydrogeology Journal, 13(2), pp.378—390.
[4] Kearey P, Brooks M and Hill I 2002 An Introduction to Geophysical Exploration, 3rd ed. Oxford: Blackwell Science Ltd.
[5] Panthulu T V, Krishnaiah C and Shirke J M 2001 Detection of seepage paths in earth dams using self-potential and electrical resistivity methods, Engineering Geology, 59(3-4), pp.281--295.
[6] Obikoya I B and Bennell J D 2012 Geophysical investigation of the fresh-saline water interface in the coastal area of Abergwyngregyn. Journal of Environmental Protection, 3, pp.1039--1046.
[7] Zohdy A A R and Jackson D B 1969 Application of deep electrical soundings for groundwater exploration in Hawaii, Geophysics, 34(4), pp.584--600.
[8] Granato G E and Smith K P 1999 Robowell: An automated process for monitoring ground water quality using established sampling protocols. Ground Water Monitoring and Remediation, 19(4), pp.81--89.
[9] Su Y H, Zhu G F, Feng Q and Li Z Z 2009 Environmental isotropic and hydrochemical study of groundwater in the Ejina Basin, northwest China, Environmental Geology, 58(3), pp.601--614.
[10] Adams S, Titus R, Pietersen K., Tredoux G and Harris C 2001 Hydrochemical characteristics of aquifers near Sutherland in the Western Karoo, South Africa, Journal of Hydrology, 241, pp. 91--103.
[11] Tank D K and Chandel C P S 2010 A hydrochemical elucidation of the groundwater composition under domestic and irrigated land in Jaipur City, Environmental Monitoring and Assessment, 166(1), pp.69--77.
[12] Ghazalli M A 1998 Modernisation of Irrigation and Drainage Management for Agricultural Production, Ampang, Kuala Lumpur: Department of Irrigation and Drainage Malaysia.
[13] Valera A and Mohd Desa M N 1991 Design-Management Interactions of Malaysia's Kerian Irrigation Scheme, Colombo, Sri Lanka: International Irrigation Management Institute.
[14] Dutta S, Krishnamurthy N S, Arora T, Rao V A, Ahmed S and Baltassat J M 2005 Localization of water bearing fractured zones in a hard rock area using integrated geophysical techniques in Andhra Pradesh, India, Hydrogeology Journal, 14, pp.760--766.
[15] Loke M H and Dahlin T 2002 A comparison of the Gauss-Newton and quasi-Newton methods in resistivity imaging inversion, Journal of Applied Geophysics, 49, pp.149--162.
[16] deGroot-Hedlin C and Constable S 1990 Occam's inversion to generate smooth, two-dimensional models from magnetotelluric data, Geophysics, 55(12), pp.1613--1624.
[17] Ellis R G and Oldenburg D W 1994 Applied geophysical inversion, Geophysical Journal International, 116, pp.5--11.
[18] Seaton W J and Burbey T J 2002 Evaluation of two-dimensional resistivity methods in a fractured crystalline-rock terrane, Journal of Applied Geophysics, 51(1), pp.21--41.
[19] Loke M H, Acworth I and Dahlin T 2003 A comparison of smooth and blocky inversion methods in 2D electrical imaging surveys, Exploration Geophysics, 34, pp.182--187.
[20] Dahlin T 2001 The development of DC resistivity imaging techniques. Computers and Geosciences, 27, pp.1019--1029.
[21] Johannesson K H, Cortés A and Kilroy K C 2004 Reconnaissence isotopic and hydrochemical study of Cuatro Ciénegas groundwater, Coahuila, México, Journal of South America Earth Sciences, 17, pp.171--180.
[22] APHA 2012 Standard Methods for the Examination of Water and Wastewater, 22nd ed. Washington, DC, American Public Health Association, American Water Works Association and Water Environment Federation.
[23] Keller G V and Frischknecht F C 1966 Electrical methods in geophysical prospecting. Monographs on Electromagnetic Waves, 10, Oxford: Pergamon Press.

[24] Daniels F and Alberty R A 1961 Physical Chemistry, 2nd ed. New York: John Wiley.
[25] Al-Ahmadi M E and Aljahdali N S 1999 Groundwater quality assessment in Wadi Tharad area, southern Saudi Arabia, Journal of King Abdulaziz University: Earth Sci*ences*, 11, pp.15--32.
[26] Bokhari A Y and Khan M Z A 1992 Deterministic modeling of Al-Madina Al-Munwarah groundwater quality using lumped parameter approach, Journal of King Abdulaziz University: Earth Sciences, 5, pp.89--107.
[27] Wilcox L V 1955 Classification and Use of Irrigation Waters, Washington DC, Department of Agriculture.
[28] Hill R A 1940 Geochemical patterns in Coachella Valley, California, Transactions-American Geophysical Union, 21, pp.46--53.
[29] Piper A M 1944 A graphic procedure in the geochemical interpretation of water-analyses. Transactions-American Geophysical Union, 25, pp.914--923.
[30] Piper A M 1953 A graphic procedure in the geochemical interpretation of water analyses. USGS Ground Water Note No. 12. USGS.
[31] Back W and Hanshaw B 1965 Chemical geohydrology. In V. T. Chow (Ed.). Advances in Hydroscience, New York: Academic Press, Inc.,pp.49--109.
[32] Naseem S, Nargis Y and Sheikh S A 2003 Geochemistry and quality of groundwater of Karachi area, Pakistan, Hamdard Medicus, XLVI, pp.113--122.

Time Variations of Scour Below Submerged Skewed Pipelines

M A M Yusoff[1], H M Azamathulla[2] and A A Ghani[3]
[1]Faculty of Civil and Environmental Engineering, U niversiti Tun Hussein Onn Malaysia, 86400 Batu Pahat Johor, MALAYSIA
[2]Faculty of Engineering, Universiti Of Tabuk, SAUDI ARABIA
[3]River Engineering and Urban Drainage Research Centre (REDAC), Engineering Campus, Universiti Sains Malaysia, Penang, MALAYSIA

E-mail: azlany@uthm.edu.my

Abstract. The presence of pipe across river initiates the piping effect combined with the stagnation eddy and vortex system in the vicinity of the pipeline. The main objective of the research is to investigate the physics of scour below skewed pipeline in river crossing as well as the time variations of the scour development. In this study, the experiments were conducted for four different angles of pipe (30°, 45°, 60° and 90°) across a channel and placed on the sediment bed with $e/D = 0$. The scour development for flow shallowness $y/D = 3$ and $y/D = 4$, initiated at downstream side of the pipe, where the bed sediment appeared to be ejected from the bed due to the piping effect process. At the initial stage, the scour process for 150mm flow depth enlarged rapidly. Whilst, the scour process for 200 mm flow depth slowly developed and after certain time, the sediment bed scoured rapidly. The scour depth increased considerably at development stage and the suspended load near the bed especially below the pipe decreased significantly compared to the initial stage. The rate of sediment eroded from the sediment bed decreased at the stabilization phase. The equilibrium phase of the scour depth considered achieved as the dimensions of the scour hole do not change significantly.

Keywords: Pipeline scour, piping, scour mechanism, skewed pipeline, flow mechanism.

1. Introduction

The presence of the pipeline in the river or stream obstructed the flow pattern within the structure may cause erosion or scour that can be divided into general scour and local scour. The changes in flow characteristics lead to a change of sediment transport capacity, hence affecting the equilibrium and the stability of the actual sediment transport capacity. The hydraulic conditions eventually are adjusted to a new state of equilibrium through scour process [1]. The impact of the structure on the flow in the river directly creates the local scour that is relatively has a shorter time scale than the general scour. The onset scour phenomenon is attributed to the seepage flow underneath the pipeline, which is caused by a large jump in the pressure or a different relative pressure coefficient (Cp) between the upstream and downstream sides [2]. Dey and Singh[3] concluded that the maximum scour depth below the pipelines with upward seepage through the bed was smaller than that without seepage. The three-dimensional flow field around the pipeline is extremely complicated due to the separation and multiple vortices that occur around the pipes. The difference of the pressure causes tunnel erosion which is affected by the seepage process. The increasing length of the seepage flow along the streamlines can reduce the pressure gradient as well as the seepage effect. The piping effect occurs when the pressure gradient exceeds the flotation gradient of the bed sediment and the associated upward flow. Tunnel erosion does not occur when the ratio of the low depth to the pipe diameter exceeds 3.5 and when the embedded ratio is too high [4].

Previous research confirmed that the scour hole below submerged pipelines could be regarded as a function of the Froude number and of a dimensionless clearance between the pipe and the undisturbed bed [5]. Kjeldsen et al. [6] found that scour depth, ds is affected by the approach flow velocity, U and pipe diameter, D but the effect of flow depth, y and the sediment size d50 is not included. Mao [7] identified that there are two cases of scour process, (a) jet period, which decides the maximum scour depth, and (b) wake period, which decides the location of scour depth. He also observed that for ds/D<1, the scour depth, ds is a weak function of the flow Shields parameter. Sumer et al. [8] compared the effective Shields parameter with the time-averaged value to assess the influence of lee wake on the scour profile downstream of the pipelines. The onset of scour below pipelines occur due to the piping effect and no scour takes place if the ratio of the approach flow depth to the pipe diameter exceed 3.5, and flow shallowness, y/D<3.5 for the pipelines that laid on the sediment bed.

Chiew [4] investigated the influence of the gap discharge, qg in estimation of the scour depth and proposed an iterative method with an empirical function that relate, y/D with the gap discharge, qg. The investigation of the scour characteristics of submerged pipelines continues to be a concern for hydraulic engineers. Most of the common practices for pipeline cross-river in Malaysia are transverse crossing modes (Figure 1). None of the aforementioned studies examined the influence of skew pipelines cross-river under clear-water scour condition. Deng et al. [9] analyzed and presented some pertinent protective measures based on the principle of hazard mitigation. They also presented the relation between pipe axis and river flow direction. The cross-river pipeline construction projects are the key projects in long-distance pipeline construction. Based on years of geological hazard investigation and washout protection in Zhongxian-Wuhan Pipeline and Lanzhou-Zhengzhou-Changsha Oil Pipeline, scouring and erosion as well as change in the river`s course are the main dynamic factors in river washout [9].

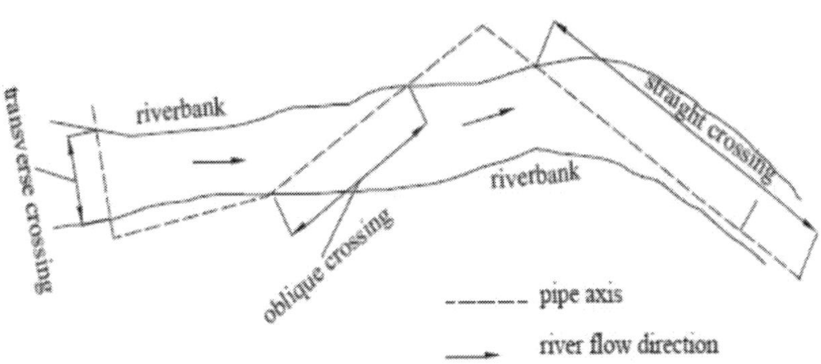

Figure 1. Relation between the pipeline axis and river flow direction [10]

Boyd et al. [11] suggested that the designed of pipelines across Musselshell River need to avoid meander bends and braided channels as well as buried the pipeline as deep as possible under the river bed. If the river bed does not have sufficient natural armoring, carefully construct a graded rock apron that extends far enough upstream and downstream from the buried siphon to assure the channel bed over the siphon fits the overall river channel gradients. Therefore, the main objective of this research is to conduct experimental and simulations of scour caused by submerged skew pipelines across the river.

2. Materials and methods

The experiments will use a 12 m long concrete (mortar finish) rectangular flume with a width of 1.5 m and a depth of 0.9 m. A PVC pipe with a 50 mm diameter is used in the experiment. The pipe will be placed crossing the flume at 30°, 45°, 60° and 90° and on the sediment bed as shown in Figure 2. A sand bed will be used in these experiments. The 12 meter test flume was divided into three sections which were the false floor, sediment recess and sediment basin. The false floor was constructed to maintain the same bed level of the sediment recess in the channel. After the false floor, the sediment recess section is filled with 0.3 m long, 1.5 m wide and 0.3 m deep of gravel. The gravel functioned as scour protection for the bed after the false floor. A sediment recess was 7m long, 1.5 m wide and 0.3 m deep, and located after the false floor. The sediment basin was located at the downstream part to cater the sediment that was transported through the channel. The sediment recess is raised 0.3 m from the bottom floor with a slope, S = 1:1000. The sediment particle that is used in the experiment can be classified into very coarse sand with mean sediment size, d_{50} = 0.58 mm. The water from the three interconnected tanks was supplied to the channel through the centrifugal pump. After entering the channel through the sump section, the water was allowed to flow into the main sump again where the water was recirculated. Water was allowed to stay over the bed for some time to remove the air voids and fill it with water.

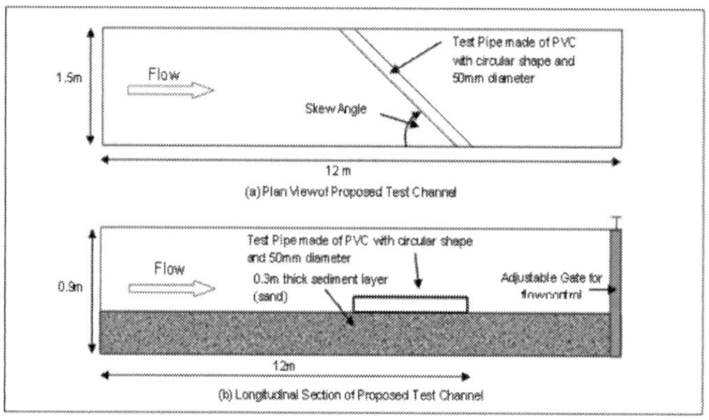

Figure 2. Test channel dimension and setup.

The pipeline model particularly PVC circular pipe was installed carefully in the middle of sediment recess and will be adjusted according to the desire angle of attack which are 30°, 45°, 60° and 90°. The water was allowed to flow through the test section and the scouring process was observed from the side walls. The arrangement was made such that if the experiment stopped due to any reasons, the water remained in the channel. This arrangement helped in resuming the experiment from the same point, where it was stopped without any disturbance to the scour hole developed. Water was allowed to flow at a specific flow depth for specific time duration. The experiments were conducted for initial flow depths of 150 mm and 200 mm. At the end of each experiment, the water was drained out of the test section as well as scour holed. Scour profile was measured for depth, width and length of scour hole. The equilibrium scour depth achieved when the scour hole did not change for quite some of time. The maximum scour profile achieved at that particular point.

3. Results and discussion

The maximum scour depth and width for the 30° and 60° of pipe installation are lower than 90° of pipe installation. From the physical observations, the scour tendency for the submerged skewed pipe is near the left bank (Figure 3). The hydrodynamic forces which consisted of drag force (in line with the flow direction) and lift force (normal to the flow direction) have been affected by the presence of the skew pipe. The skew pipe has changed the force direction as well as shifted the location of scour hole. There were also different rate of scour process for flow shallowness, y/D = 3 and y/D = 4. It can be seen that there was a rapid change of scour development for y/D = 3 compared to y/D = 4. The time-variation scour depth showed that the rate of scour for y/D = 3 was higher than y/D = 4.

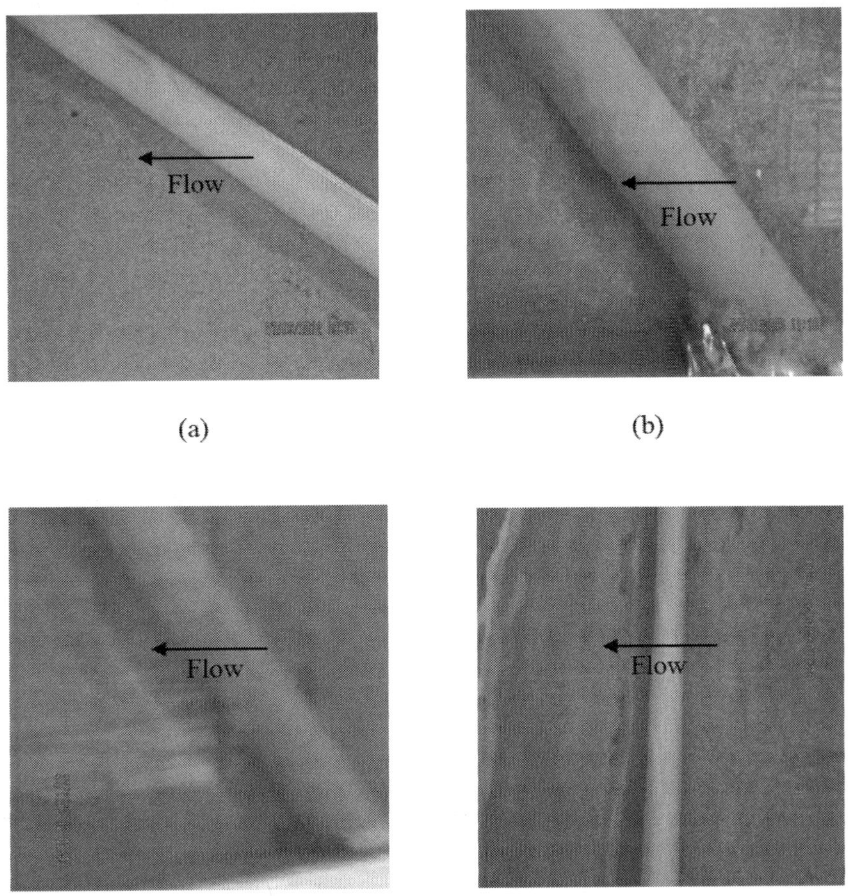

(a) (b) (c) (d)

Figure 3. Scour depth location below pipe; (a) 30° (b) 45°(c) 60°, and (d) 90°.

The difference pressure between the upstream and downstream the pipe created a small gap below the pipe. At this stage, the jetting flow started to eroded and transported the sediment away from the pipe. The tunnel erosion process continued to erode the sediment until certain stage where the combination of luff erosion and lee erosion started to dominate. The luff erosion occurs at the upstream side which was caused by the eddy formation in front of the pipe while the lee erosion at the downstream of the pipe, caused by the reemergence of main flow and turbulent wake (vortices system) downstream the pipe.

3.1 Time variation of scour depth

The temporal development of the scour hole below submerged skewed pipeline on a flat plane showed the tunnel scour for both flow shallowness, y/D = 3 and y/D = 4 initiated at the downstream side of the pipe. The bed sediment appeared to be ejected from the bed, showing that piping could be the cause of failure. Once the sand particles at the downstream side of the pipe were lifted off the bed, they were transported out by the flow. At the same time, the stagnation eddy at the upstream side of the pipe created a small depression, and the sand barrier was quickly breached leading to the formation of a tunnel under the pipe. In the tunnel, the sediment particles were entrained, and vigorous sediment transport occurred in the early stages. The sand that eroded from the tunnel formed a bar downstream of the scour hole, and with time, the sand bar propagated downstream and lee erosion, caused by the turbulent wake and reattachment of the main flow over the pipe began to dominate. Scouring continued until equilibrium was reached, when the temporal shear stress and turbulent agitation near the bed were no longer able to transport bed material from the scour hole. Figure 4 show the scatter plot for the effect of water depth on the scour profile and it can be observed that the scour process for flow depth 150 mm enlarged rapidly with time duration at the initial stage compared to scour development for 200 mm flow depth.

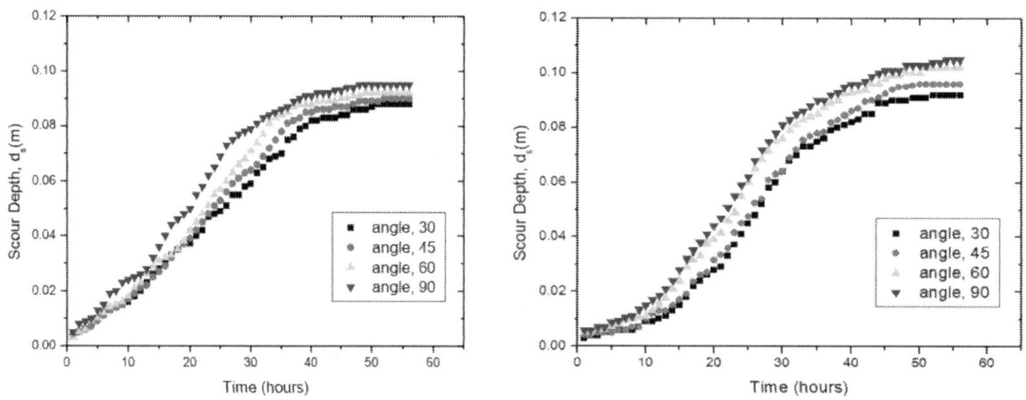

(a) (b)

Figure 4. Time variations of scour depth for (a) 150 mm water depth and (b) 200 mm water depth.

The strong pressurized flow between the upstream and downstream side of the pipeline caused the tunnel erosion effect to occur underneath the pipeline. The upward forces due to the ground-water flow under the pipe inducing instability to the particles and moved away the sediment particles. The vortices due to the pipe obstacles and seepage flow leads to the formation of small opening under the pipe and increased the flow concentration in the gap which augmented the shear stress on the sediment bed below the pipe. At the initial stage, the scour process for 150mm flow depth enlarged rapidly. Whilst, the scour process for 200 mm flow depth slowly developed and after certain time, the sediment bed scoured rapidly. The scour hole enlarged rapidly for 150 mm flow depth due to the jetting of flow underneath the pipeline which creates the lift forces for the sediment to move from the scour hole.

Eventually, a compact secondary armor layer from the non-uniform sediment offering resistance to the smaller sediment particles that inhibited the progress of further scouring. After the armor layer collapsed, lee wake and tunnel erosions prevailed. When the gap below the pipe considerably large at the end of the tunnel erosion, the vortex shedding began and swept the sediment below the pipe as the convection of flow towards downstream. This case mainly prevailed for the flow depth equal to 200 mm as the ratio y/D exceed 3.5 because the flow not adequate to induce force on the armor layer thus, retarded the scour process at early stage. The evolution of scour phase was divided into four phase: an initial phase, a development phase, a stabilization phase and an equilibrium phase, which was distinguished by previous researchers [12].

3.2 Effect of approach flow depth relative to pipe diameter on scour depth

Figure 5 show the relationships between the normalized scour depth relative to pipe diameter and the dimensionless time variation of pipeline scour. The pattern of the time variation seems to be in a logarithmic scale as suggested by several researchers. The figures showed that the scour depth (ds/D) increases with the flow shallowness (y/D). The final scour depth at high t/te values, may considered as the equilibrium scour depth increases with the approach flow velocities which creates higher flow depth. The approach flow depth considered as the dominant factor for the experiments as the scour depth keep increases with the flow shallowness. Evolution of the scour depth can be divided into three stages; (a) initial stage which the scouring rate increases sharply with time, (b) development stage where the scour process increases steadily with time and (c) equilibrium stage where the scour depth remains unchanged with time relatively. At this point, the scour depth and time are equilibrium, respectively. Alternatively, there was a possible tendency of d_s/D to be independent of y/D as the ratio y/D > 4. Dey and Singh [3] mentioned that the data plot for larger pipe diameter sizes are scattered and tends to be independent for y/D> 4. This aspect can be justified that the mechanism of scour depth below pipes governed by the three different erosion; (a) luff erosion caused by the eddy formation upstream the pipe (b) lee erosion caused by the vortices system downstream the pipe, and (c) tunnel erosion caused by the increase of flow concentration below the pipe and jetting flow. But, there is little argument on the scour potentials characterized by the near bed or lower flow zone as the approach flow depth beyond five times the pipe diameter [13].

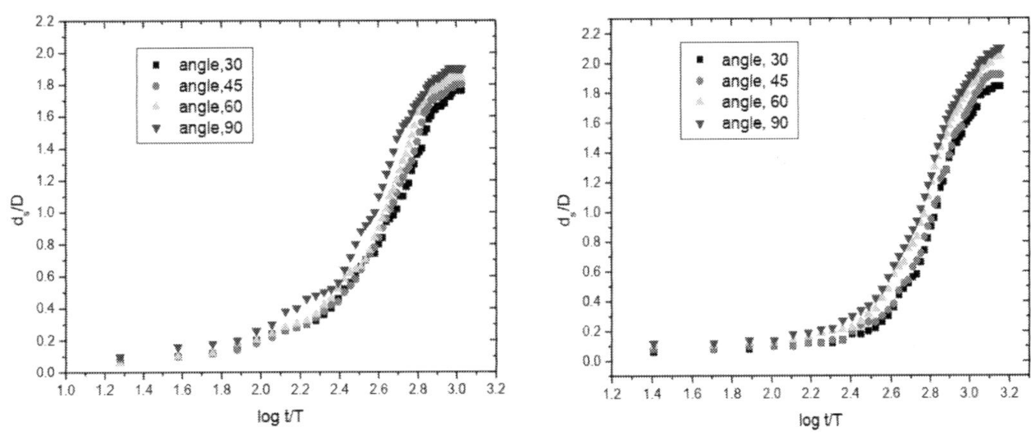

(a) (b)

Figure 5. Evaluation of scour depth (d_s/D) with non-dimensional time dependent, log (t/T) for (a) y/D = 3, and (b) y/D = 4.

4. Conclusion and recommendation

The change of flow direction affected by the skewed pipe obstacle created difference pressure between the upstream and downstream side of the pipe. The presence of the seepage effect created the uplift forces for the sediment underneath the pipe and sediment particles forced to transport downstream. The scour hole enlarged rapidly at the initial stage for flow depth 150 mm due to the tunnel erosion effect while the scour process for 200 mm flow depth increase gradually till reach the equilibrium stages. The hydrodynamic forces which consisted of drag force (in line with the flow direction) and lift force (normal to the flow direction) have been affected by the presence of the skew pipe. The skew pipe has changed the force direction as well as shifted the location of scour hole. There were also different rate of scour process for flow shallowness, y/D = 3 and y/D = 4. It can be seen that there was a rapid change of scour development for y/D = 3 compared to y/D = 4. The time- variation scour depth showed that the rate of scour for y/D = 3 was higher than y/D = 4. The difference pressure between the upstream and downstream the pipe created a small gap below the pipe. At this stage, the jetting flow started to eroded and transported the sediment away from the pipe. The tunnel erosion process continued to erode the sediment until certain stage where the combination of luff erosion and lee erosion started to dominate. The luff erosion occurs at the upstream side which was caused by the eddy formation in front of the pipe while the leee rosion at the downstream of the pipe, caused by the reemergence of main flow and turbulent wake (vortices system) downstream the pipe. The skew pipe also was affected the time scale of scour process where the shorter time of scour process for skew pipe more than 45° (60° and 90°). The time scale of scour process for y/D = 4 was shorter than flow shallowness, y/D = 3, although at the initial stage, it can be seen that the scour hole for y/D = 3 enlarged rapidly. The study found that the initiation of scour below submerged skewed pipeline is predominantly caused by piping as a result of the pressure gradient across the pipeline. Therefore, it is possible to inhibit this failure, thereby eradicating tunnel scour below submerged skewed pipeline completely, if the pressure gradient across the pipe can be kept reduced by increasing the flotation gradient of the bed sediment. Further investigation on others parameters such as grains size, high approach flow and differet pipe sizes need to be carried out to ensure the possibility of placing a pipe across channel and minimize the cost of operations.

Acknowledgements
The authors take the opportunity to thank the Director of River Engineering and Urban Drainage Centre (REDAC) for giving the chance to conduct the experiments at their facilities. Authors are grateful to University Sains Malaysia for providing intensive grant.

References
[1] Hoffmans, G., J., C., M. and Verheij, H., J. (1997) Scour manual.
[2] Mao, Y. (1988) Seabed scour under pipelines. *Proc. of 7th Int, Symp. on Offshore Mech and. Arctic Engrg.,* Houston, Texas, 33-38.
[3] Dey, S., and Singh, N. P. (2008). " Clear-water scour below underwater pipelines under steady flow". *J. Hydraul. Eng.,* 134 (5), 588-600.
[4] Chiew, Y.M., (1991). Prediction of maximum scour depth at submarine pipelines. *J. Hydraulic Engineering* , 117 (4), 452-466.
[5] Maza, J. A. (1987) Introduction to river engineering. Advanced course on water resources management, Uniersita Italiana per Stranieri, Perugia, Italy.
[6] Kjeldsen, S. P., Gjørsvik, O., Bringaker, K. G., and Jacobsen, J. (1973). "Local scour near offshore pipelines." *Proc., 2nd Int. Conf. on Port and Ocean Engineering under Arctic Conditions*, Univ. of Iceland, 308–331.
[7] Mao, Y. (1986) The interaction between a pipeline and an erodible bed. *Ser. Paper 39,* Institute of Hdrodynamics and Hydraulic Engineering, Technical University of Denmark, Lyngby, Denmark.
[8] Sumer, B. M., Jensen, H. R., Mao, Y., and Fredsøe, J. (1988) Effect of lee-wave on scour below pipelines in currents." *J., Wtrwy., Port, Coast. and Oc. Engrg.*, ASCE, 114(5), 599-614.
[9] Deng, Q., Li, L., Zhang, Y., (2013) An Analysis of River Washout Influence and Protective Measures of Long-distance Gas and Oil Pipelines, *International Conference on Pipelines and Trenchless Technology for Construction, Oil and Gas, Engineering* (ICPTT 2012), Wuhan, China, October 19–22, 2012, ASCE, pp. 1280–1293
[10] Qin, S.Y. and Zhao, M.H., (2009) Protection measures for pipeline crossing river in mountain region. *Journal of Soil Engineering and Foundation.* 85-87.
[11] Boyd, K., Kellogg, W., and Pick, T. (2013) Musselshell Best Management Practices: Pipeline Crossings, Lower Musselshell Conservation District, December, 2013.
[12] Breusers, H.N.C. and Raudkivi, A.J. (1991) Scouring. A.A Balkema, Rotterdam, Netherland, 1991.
[13] Bijker, E. W., and Leeuwestein, W. (1984). Interaction between pipelines and the seabed under influence of waves and currents." *Proc.,Symp. on Int. Union of Theoretical* .

Sediment Transport Dynamic in a Meandering Fluvial System: Case Study of Chini River

M H M Nazir[1], S Awang[1], A J Shaaban[1], N K E M Yahaya[1], A M Jusoh[1], M A R M A Arumugam[1] and A A Ghani[2]

[1] River Basin Research Centre, National Hydraulic Research Institute of Malaysia, 43300, Seri Kembangan, Selangor, MALAYSIA

[2] River Engineering and Urban Drainage Research Centre, Universiti Sains Malaysia, Engineering Campus, 14300 Nibong Tebal, Pulau Pinang, MALAYSIA

E-mail: hafifinazir@gmail.com

Abstract. Sedimentation in river reduces the flood carrying capacity which lead to the increasing of inundation area in the river basin. Basic sediment transport can predict the fluvial processes in natural rivers and stream through modeling approaches. However, the sediment transport dynamic in a small meandering and low-lying fluvial system is considered scarce in Malaysia. The aim of this study was to analyze the current riverbed erosion and sedimentation scenarios along the Chini River, Pekan, Pahang. The present study revealed that silt and clay has potentially been eroded several parts of the river. Sinuosity index (1.98) indicates that Chini River is very unstable and continuous erosion process in waterways has increase the riverbank instability due to the meandering factors. The riverbed erosional and depositional process in the Chini River is a sluggish process since the lake reduces the flow velocity and causes the deposited particles into the silt and clay soil at the bed of the lake. Besides, the bed layer of the lake comprised of cohesive silt and clayey composition that tend to attach the larger grain size of sediment. The present study estimated the total sediment accumulated along the Chini River is 1.72 ton. The HEC-RAS was employed in the simulations and in general the model performed well, once all parameters were set within their effective ranges.

Keywords: Sediment transport, meandering fluvial system.

1. Introduction

Chini River catchment has unique features as it was represented by a low lying lake as a detention area for a large flood event during the monsoon season. With seven mains feeder-rivers as its tributaries that allowing water coming from uncontrolled man-induced land development, sedimentation problem becomes vulnerable to more severe impact such as eutrophication and flood carrying capacity reduction [1].

Sediment transport dynamic within a catchment is reliant on its physical and hydrological characteristics [2,3] while these factors are intensely inter-related [4]. In describing the physical factors, a term of "meandering" of a river channel play a vital role in sedimentation dynamic [5,6,7,8,9], hydrological and ecological processes [10,11].

Meandering is a primary property of many natural rivers and meander bends are among the most recognizable and common features of the alluvial landscape. River meanders are of fundamental interest to many branches of pure and applied science, including geomorphology, geology, physical geography, and engineering. Meanders form across a variety of spatial scales produce characteristic depositional sequences within modern and ancient alluvium and recognition of these deposits lies at the heart of the development of predictive alluvial facies models. The movement of meandering rivers through lateral migration can be an important societal concern because such movement often threatens adjacent property and infrastructure.

Even with an lavishness of previous effort on sinuous waterways, our acquaintance of the fluid and sediment dynamics of complex meander curves in small low-lying undulating rivers, such as extended composite meander loops, is surprisingly sparse. As a consequence, our capacity to model and predict accurately the evolution behavior of complex bends in this type of rivers is still far from complete.

The present study attempts a step-change in our understanding of the sediment transport dynamic of a small meander river by exploring the interactions between one-dimensional flow structure, sediment transport and riverbed morphology within several complex meander loops along the low-lying catchment of Chini River.

Figure 1. The reach of Chini River and established sampling locations.

2. Methodology

The meandering characteristics of Chini River was analyzed according to an equation derived by [12].

$$K = Lc / Lv \qquad (1)$$

where K = Sinuosity Index, Lc = length of the valley axis, Lv= straight line distance between the two extreme points of the valley

Collection of sediment transport data and hydraulic data were made according to [13]. Analysis of particle size distribution of bed material (**Table 1**) was carried out using sieve analysis and hydrometry according to BS 1377-2:1990 (Methods of test for soils for civil engineering purposes:

Classification test. Analysis of sediment size is the measurement and information of size and particles range for a set of material. These collected dataset were later used as input for HEC RAS Sediment transport modelling. The Laursen-Copeland equation was chosen for the sediment transport modeling since it is the most suitable for silt and clay particles [14].

$$C_m = 0.01\gamma\left(\frac{d_s}{D}\right)^{7/6}\left(\frac{\tau_o'}{\tau_c}-1\right)f\left(\frac{u_*}{\omega}\right) \quad (2)$$

where Cm = sediment discharge concentration (lb ft3),
γ = specific weight of water (lb ft3),
ds = mean grain size (ft),
D = effective flow depth (ft),
τc = critical shear stress (lb ft-2),
τo' = bed stress acting on bed grains (lb ft-2), u*= shear velocity (ft s-1),
ω = fall velocity (ft s-1), and
f = function of the ratio of the latter two variables as defined by a figure in that has been incorporated into HEC-RAS.

Table 1. Bed material sediment results.

Locations	**Coarse Sand (%)**	**Fine sand (%)**	**Total non-cohesive (%)**	**Silt (%)**	**Clay (%)**	**Total cohesive (%)**	**D_{50} (mm)**
Lake confluence CH4000	3	22	25	58	17	75	0.0175
CH3200	0	3	3	61	36	97	0.018
Bridge CH950	5	43	48	35	17	52	0.08
Weir CH425	1	24	25	62	13	75	0.018
Confluence Chini River	2	4	6	69	25	94	0.0075

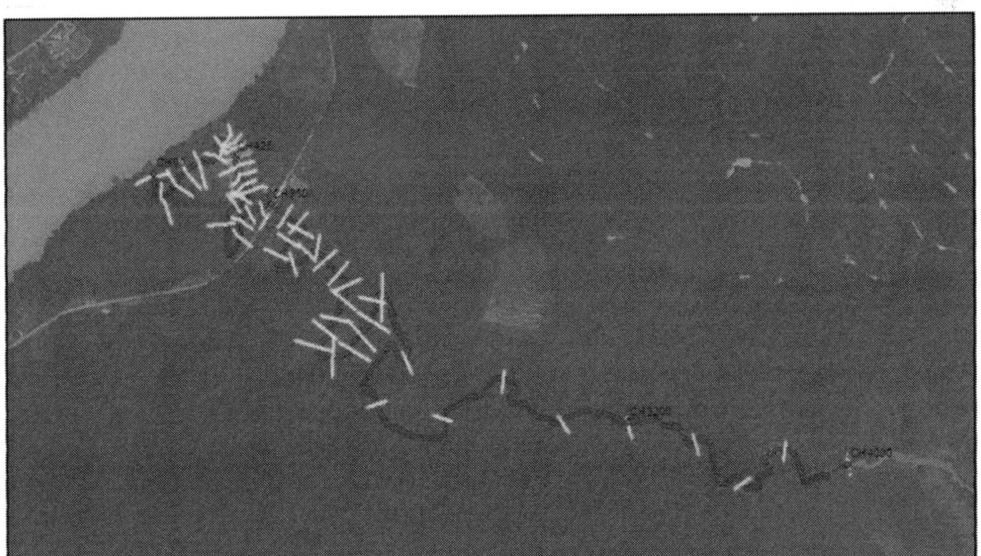

Figure 2. Particle size distribution curve for analysed bed material sediment samples at 5 established locations.

HEC-RAS is applied to develop one-dimensional Chini River sediment transport model. The model was calibrated by obtaining high goodness of fit between observed and predicted water level of hydraulic simulation run. With having the high accuracy of hydraulic model, sediment transport analysis then simulated for the current discharge and sediment supply. [15] produced a curve diagram plotting the tractive velocity flow between the incipient movement of grains and grain size. Thus, it was found that there was a correlation which clearly indicates that the coarse grain size is proportional to the velocity of the moving water. Hjulstrom curve gives a details interpretation about the flow velocity required to move and transport the various grain sizes in the composition of the deposit.

3. Results and discussion

3.1 Morphological - Meandering characteristics of Sg Chini

The sinuosity index, (K) for Chini River is 1.98 and categorized as a very unstable condition [16]. Chini River has greater percentage of suspended load with finer grained sediments dominated at all river sections. The range from flow discharge was between 1.30 m^3/s to 1.97 m^3/s. Most part of the riverbanks are found to be cohesive with little coarse sediment is observed.

3.2 Particle size distribution analysis

Analysis of sediment size is the measurement and information of size and particles range for a set of material. The analysis conducted through two types of laboratory analysis known as sieve analysis and hydrometer test. The size of bed material is the most important physical characteristics affecting the process of sediment transportation [17,18,19].

From the bed material sediment sampling analysis as shown in Table-1 and Figure-2, it was found that the silt particles dominates 35-69% of bed particles within the Chini River compared to sand particles. Sediment mixtures with a fraction of clay particles (D_{50}< 0.063 mm) which about 13-36% have cohesive properties because electro-statistical forces are higher than the gravity force acting between the particles.

Consequently, the sediment particles do not behave as individual particles but then to stick together forming a floc whose size and settling velocity are much larger than those of the individual particles. The D_{50} for bed particles and suspended particles has been interpolated to yield the spatial distribution within the Chini River water body.

Consolidation rate is an important factor in determining the level of soil erodibility. Fresh mud has a very loose texture in which the mud flocs has a low density. The wet bulk density of such cohesive deposit is within the range of 1050 to 1100 kg/m^3 with estimated of 95% or more consists of water and in this phase, the cohesive forces be present in the deposits are still very low and tend to have effortless erosional effect [20].

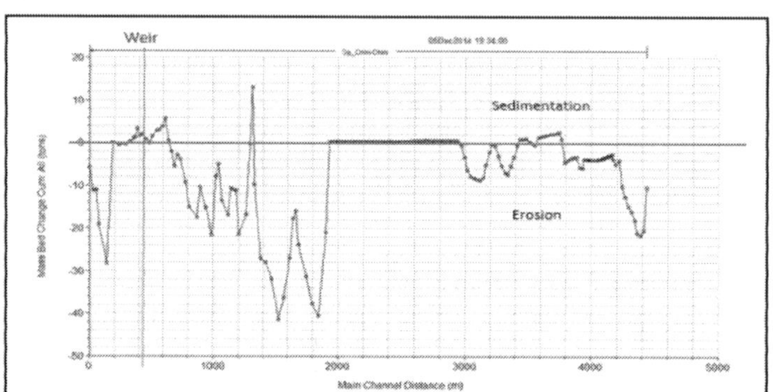

Figure 3. Cumulative weight of the sediments in the end of simulation period (5 Dec 2014) at all cross section for the current condition.

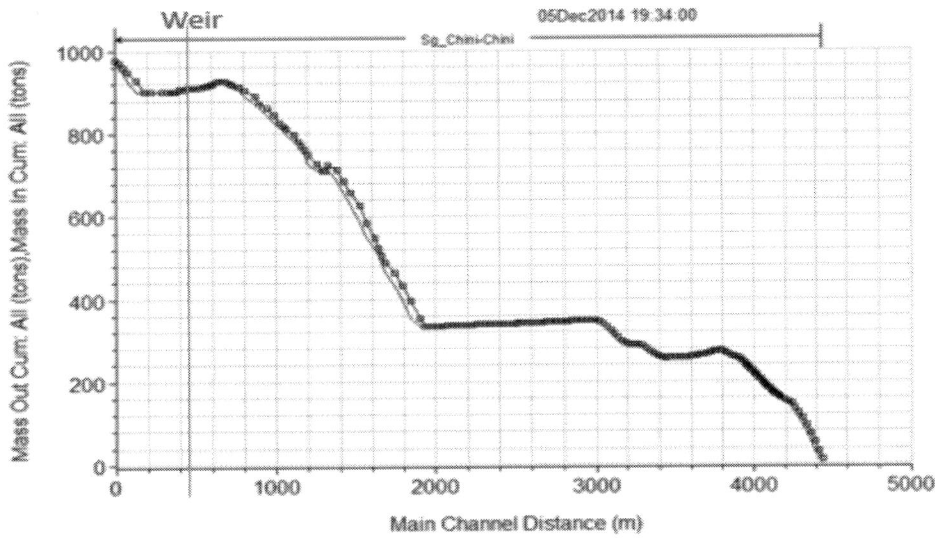

Figure 4. The accumulative weight of inflow and outflow sediments in all cross sections at the end of simulation period (5 Dec 2014).

3.3 Riverbed erosion and sedimentation dynamic of Chini River with the current condition

The simulation of flow sedimentation are presented as the transverse and longitudinal profile variation of the river, shear stress variation, sediment transport capacity and the volume of outflow sediment that were studied at different locations. The change in riverbed geometry during the model run can also be evaluated by comparing the longitudinal cumulative mass change in tons of sediment. Starting at the upstream boundary, this output sums to the mass change through every cross-section (Figure-3). At a steep slope sections, the observation indicated that is is an aggradation (positive slope) or degradation condition(negative slope).

Figure-4 gives the cumulative weight of inflow and outflow sediments values based on Laursen-Copeland sediment transport equation. As shown by the interpolation of the HEC-RAS model, the erosional activities along the river occurred at the upstream and middle stream of the river and at the downstream.

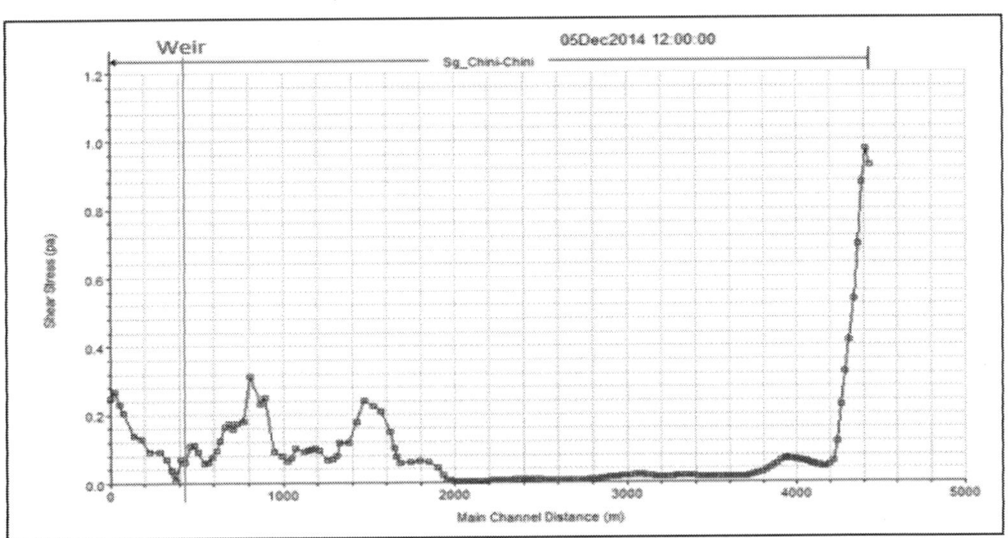

Figure 5. The shear stress variation during the simulation period (5 Dec 2014) at all river sections.

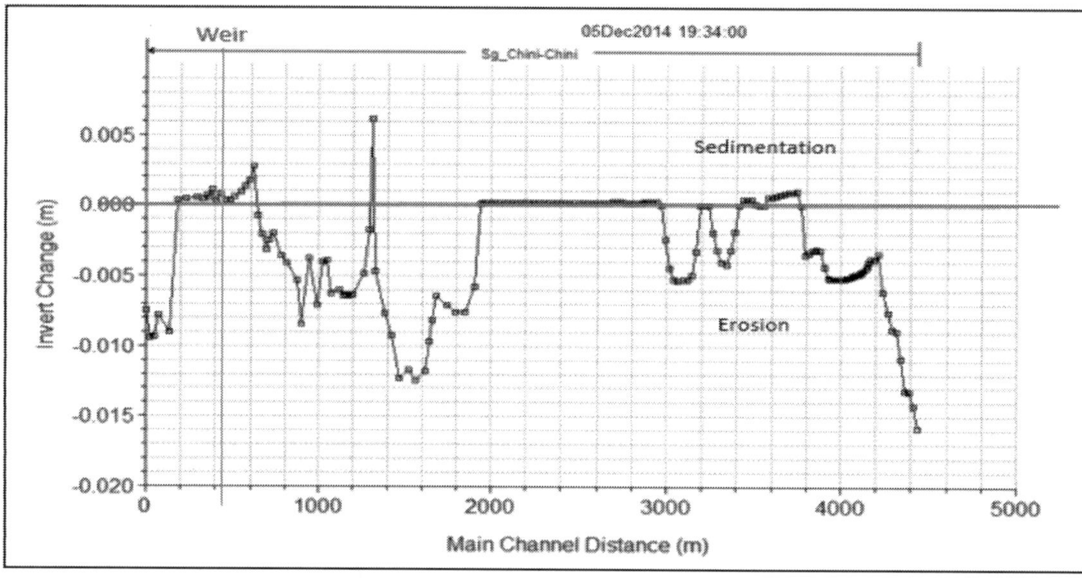

Figure 6. Invert change (m) of Sg Chini at all river sections after end of simulation (5 Dec 2014).

For example, the abrupt changes of slope between CH1800 and CH1600 cause the significant erosional effect. The accumulative weight of inflow and outflow reveal the extent of erosion and sedimentation in different sections. The total mass in during the simulation period of 18 November to 5 December is 907.97 tonnes and the mass out is 906.25 tonnes. Hence the total deposition of sediment along the river section is 1.72 tonnes.

The shear stress variation during the simulation period across the Chini River at each section has been shown in Figure-5. The highest value of shear stress is equal to 0.95 Pa in the section of CH4000 and the lowest is 0.00013 at CH2000. The shear stress variation when approaching the weir seem to decrease significantly especially from the CH800 to CH425 (Weir) due to the diminishing of velocity and drope slope. The shear stress at the weir is 0.07 Pa. The lower shear stress indicates the lower transport rate of coarse sediment at approaching the weir but increase in the deposition of coarse sediment but incessantly transporting the fine sediment.

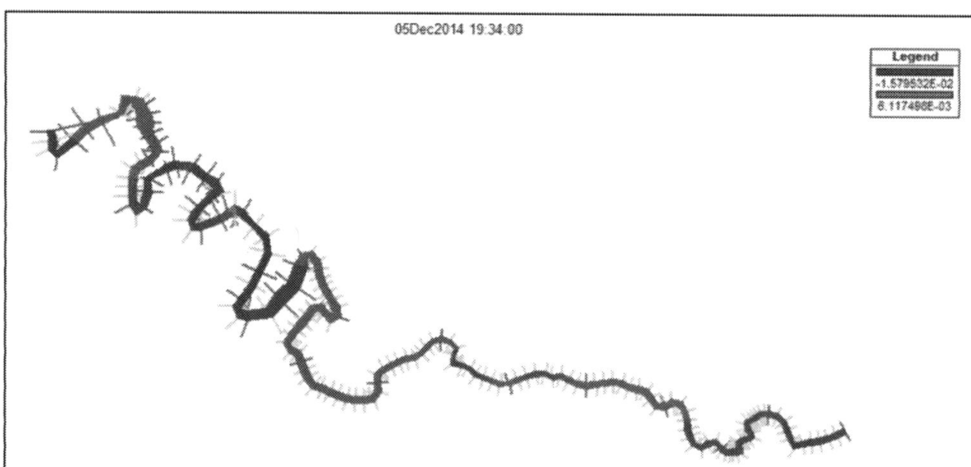

Figure 7: Schematic view of bed elevation change (m) of Sg Chini at all river sections at the end of simulation (5 Dec 2014).

IOP Conf. Series: Materials Science and Engineering **136** (2016) 012072 doi:10.1088/1757-899X/136/1/012072

The longitudinal profiles are shown in Figure-6 of Chini River along with the existing structures as well as sedimentation and erosion in different areas in the form of invert changes. It was observed that, riverbed erosion occurred at many locations along the river due to the high proportions of silt and clay (fine sediment) which can be transitionally transported as suspended sediment. These types of sediment is effortlessly can be moved from the riverbed.

Meanwhile, the bed elevation change along the Chini River represented by the pictographic view as in Figure-7. The sedimentations that occurred along the Sg Chini are symbolized by the red zone which having the positive value of bed elevation of the channel (increment of bed elevation). The blue zone represented the erosional zone which having the negative values indicating the decrease of bed elevation of the riverbed. The highest change in riverbed level due to depositional of sediment is approximately 0.0061 m. While the highest change in riverbed level due to erosion is approximately 0.016 m. The riverbed elevation change at the weir is 0.0025 m.

The expression of relative magnitude of the sediment transport capacity significantly related to flow velocity and interpreted through the Hjulstrom Curve [15]. From that curve, it is possible to identify ranges of flow velocities that will mobilize varying sediment sizes. Figure-8 illustrates potentials to erode on different particle sizes along the Chini River. I is expected that sand will only erode at the upstream of the Chini River in view of the fact that ample velocity to strain the sand particle from the riverbed. While the silt and clay has enough velocity to erode for any of the section in Chini River.

Anticipated sediment deposition form an important part of this study. Its characteristics may be observed from Figure-9 which illustrates potentials to deposit on different particle sizes along the Chini River in which the silt, clay, sand and gravel deposition area is alienated based on the velocity computed from the model simulation at each locations. As all-encompassing, it is expected that gravels will not be transported and will be deposited at the upstream of the river (CH4400-CH4250,CH1600-CH1400, CH925-CH750, CH100-CH00) because of the limited transport capacity due to deficient velocity. However it is noted that, finer materials are not likely to deposit in this reach because of it has sufficient velocity to transport these type of materials in the form of suspended sediment along the reach.

3.4 Area of active of riverbed erosion and deposition

The erosional and depositional process in the Chini River is a slow process since the lake impoundment cause the flow velocity to decrease. The sediment component on any sizes will deposit into the silt and clayey condition of the bottom bed of the lake. Hence, there is no significant changes of erosion and deposition part of the Chini River.

The area of deposition is significant at the weir of the Chini River and at the upstream part especially between CH2000 and CH3200. Whereas, the active erosion along the Chini River are located between CH800 and CH1800 due to high velocity compared to the upstream area.

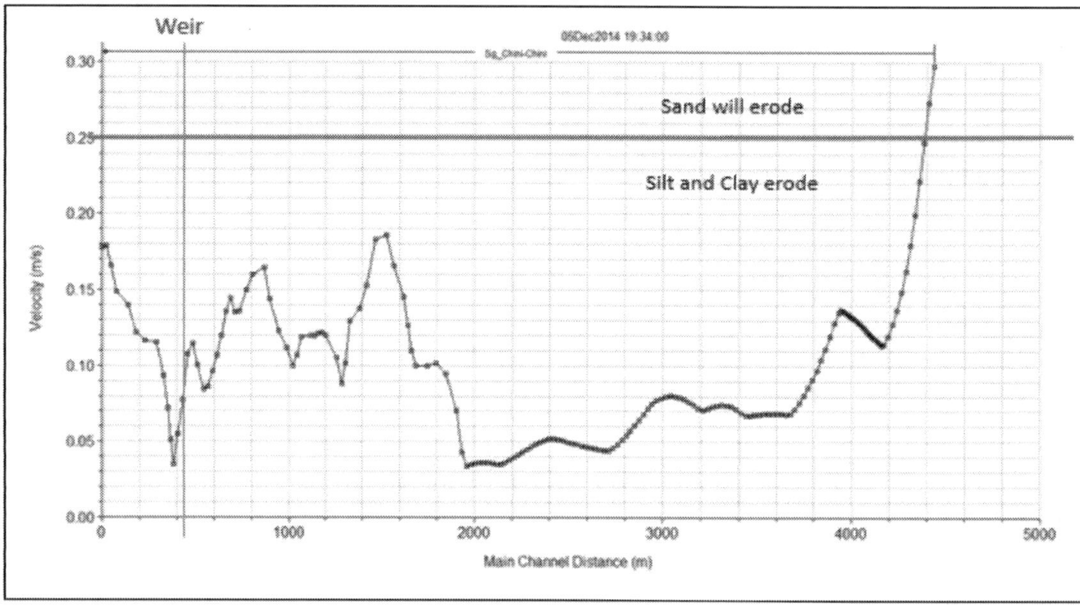

Figure 8. Susceptibility to Erode based on Hjulstrom curve.

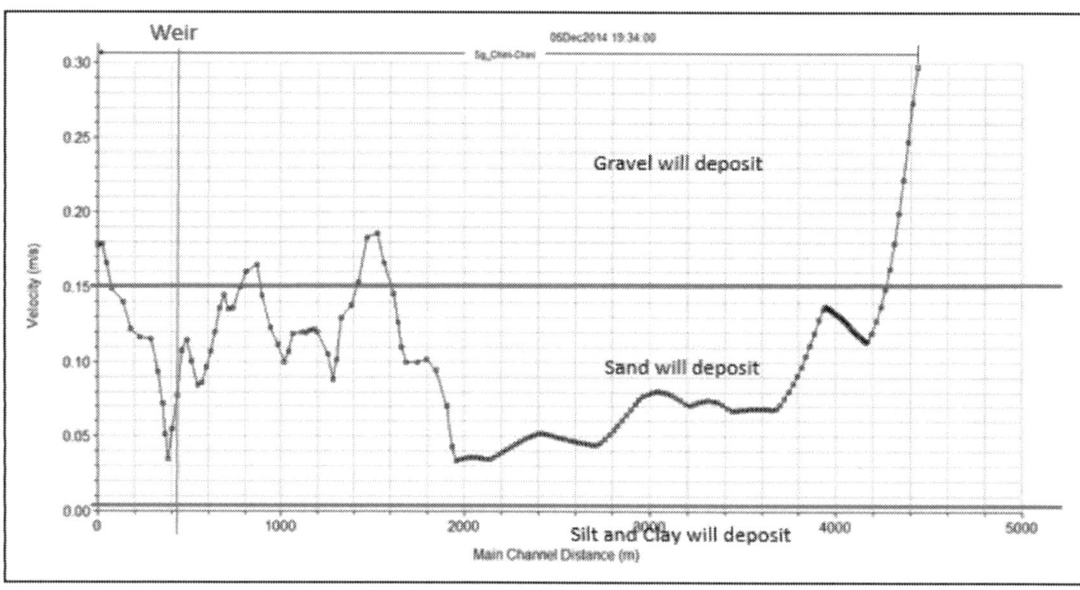

Figure 9. Susceptibility to Deposit based on Hjulstrom curve.

The erosional parts of the river mostly remain the same but the size of particles eroded is different for each event. As shown from the results, the silt and clay is potential to be eroded for the current scenario. However, sand and gravel has the potential to erode at the same part of the river especially during the high flow event.

3.5 Abatement measures for riverbed erosion and deposition of sediment

The sediment transport modeling indicated that the weir located at the downstream of Chini River is not subject to serious siltation problems (**Figure 6**); this is consistent with the site reconnaissance conducted during the project. Henceforth human intervention using structure or dredging is necessary to maintain navigation and passage of floodwaters that may otherwise causing excessive rise in water levels or velocities along the immediate up-stream reaches of Chini River.

To offset these estimated sediment loads discharging into Chini River, a system of sediment traps is proposed for the removal of portion of the tributary contributions. These sediment traps should have the flexibility to enable the sediment passing the trap to be adjusted if the removal is creating a sediment deficit or scour condition downstream. A de-silting program for the river channel is also required to pre-empt any excessive rise of floodwater levels that may otherwise cause overbank flooding in the areas.

The river sediment transport modeling using HEC-RAS indicated that Chini River will be subject to varying degrees of bed lowering in the short and long terms, which if not abated will undermine the stability of riverbanks. The areas of concern include the downstream reaches of the weir. However the threat of riverbank collapse to hinterlands is not imminent and it is proposed to introduce a monitoring program on riverbank stability or a detailed study to assess the need for mitigation.

3.6 Future studies on sediment transport dynamic in Sg Chini

It would be interesting to identify some possible scenarios for future studies. The following are prospective options to improve the results from the current study for future investigations:

1. Study of similar case with 2 or 3 other software and compare their results with observed data;
2. Study of similar case with 1D, 2D and 3D models and see if they offer the same results or if they differ and to what extent;
3. Study of two or more cases with the present model and by comparing results trying to find out under which circumstances the model perform well;
4. A combination of two or more of the above options;
5. It is suggested that a detailed and comprehensive study on cohesive sediment transport to be carried out using the more advanced numerical software to appraise further for the long term comportments of sedimentation and erosion assortment.

4. Conclusions

The HEC-RAS was engaged in the simulations and in general the model executed well, once all factors were set within their operative ranges. Conclusively, the current level of erosion is strongly related to the extent of land undergoing built up land use development, with the hilly areas being developed in the central portions of the Chini River catchment making major contribution to the sediment load in the Lake Chini system. More research in future is needed to imposed the details sediment transport dynamic in Chini River.

Acknowledgement

The author owes recognition to River Basin Research Centre, National Hydraulic Research Institute of Malaysia for their support and assistance in conducting the project.

References

[1] Ab. Ghani, A., Chang, C.K., Leow, C.S. & Zakaria, N.A. (2012). Sungai Pahang Digital Flood Mapping: 2007 Flood, International Journal of River Basin Management, Vol. 10, No. 2, pp. 139-148

[2] Sinnakaudan, S., Ab Ghani, A., S. Ahmad, M. S., & Zakaria, N, A. (2003). Flood Risk Mapping for Pari River Incorporating Sediment Transport. Environmental Modeling and Software, Vol. 18, No. 2, pp. 119- 130

[3] Chang, C. K., Ab. Ghani, A., Abdullah, R. & Zakaria, N.A. (2008). Sediment Transport Modeling for Kulim River: A Case Study. Journal of Hydro-Environment Research, Vol. 2, No.1, pp. 47-59

[4] Bettes, R. (2008). Sediment transport & alluvial resistance in rivers. R&D Technical Report W5i 609. Environment Agency, Rio House, Waterside Drive, Bristol.

[5] Nanson, G.C., Beach, H.F., 1977. Forest succession and sedimentation on a meandering-river floodplain, Northeast British Columbia, Canada. Journal of Biogeography 4 (3), 229–251.

[6] Howard, A.D., 1992. Modelling channel migration and floodplain sedimentation in meandering streams. In: Carling, P.A., Petts, G.E. (Eds.), Lowland Floodplain River, Geomorphic Perspectives. John Wiley, New York, pp. 1–41

[7] Sun, T., Meakin, P., Jossang, T., Schwarz, K., (1996_. A simulation model for meandering rivers. Water Resources Research 32 (9), 2937–2954

[8] Gilvear, D., Winterbottom, S., Sichingabula, H., (2000). Character of channel planform change and meander development: Luangwa River, Zambia. Earth Surface Processes and Landforms 25 (4), 421–436

[9] Güneralp, İ., Abad, J. D., Zolezzi, G., & Hooke., J. (2012). Geomorphology Advances and challenges in meandering channels research. Geomorphology, 164, 1–9. doi:10.1016/j.geomorph.2012.04.011

[10] Salo, J., Kalliola, R., Hakkinen, I., Makinen, Y., Niemela, P., Puhakka, M., Coley, P., 1986.River dynamics and the diversity of Amazon lowland forest. Nature 322, 254–2

[11] Ward, J.V., Tockner, K., Arscott, D.B., Claret, C., 2002. Riverine landscape diversity. Freshwater Biology 47 (4), 517–539

[12] Langbein, B.W., and Leopold. L.B. (1966). River Meanders – Theory of Minimum Variance. Physiographic and Hydraulic Studies of Rivers. Geological Survey Professional Paper 422-H

[13] Jabatan Pengairan dan Saliran Malaysia. (2003). Garis Panduan Pencerapan dan Analisa Pengangkutan Endapan Sungai, Kementerian Sumber Asli dan Alam Sekitar.

[14] HEC-RAS (2010) River Analysis System, Reference Manual, Version 4.1. U.S.Army Corps of Engineers, Hydrologic Engineering Centre, 411p

[15] Hjulstrom, F. (1935). Studies of the morphological activity of rivers as illustrated by the River Fyris: Upsala, Univ., Geol. Inst. Bull., v. 25, p. 221-527

[16] Rosgen DL (1996) Applied river morphology. Wildland hydrology. Pagosa Springs, Colorado

[17] Graf, W. H. (1984). Hydraulics of sediment transport. Water Resources Publication.

[18] Julien, P.Y., (1994) Erosion and Sedimentation, Cambridge University Press.

[19] Yang, C. T., Molinas, A., & Wu, B. (1996). Sediment transport in the Yellow River. Journal of Hydraulic Engineering, 122(5), 237-244.

[20] Liew, Y.S., Teo, F.Y. (2013) Assessment on the potential of erosion and sedimentation due to proposed flood mitigation works in Sarawak River, Malaysia. Proceedings of 2013 IAHR World Congress.

IOP Conf. Series: Materials Science and Engineering **136** (2016) 012073 doi:10.1088/1757-899X/136/1/012073

Hydrochemical Assessment of Surfacewater and Groundwater Quality at Bank Infiltration Site

M K N Shamsuddin[1], S Suratman[1], M F Ramli[1], W N A Sulaiman[1], and A Sefie[2]
[1]Hydrogeology Research Centre, National Hydraulic Research Institute of Malaysia, 43300 Seri Kembangan,Selangor, MALAYSIA
[2]Faculty of Environmental Studies, University Putra Malaysia, 43400 Serdang, Selangor, MALAYSIA

E-mail: nizar@nahrim.gov.my

Abstract: Groundwater and surface water quantity and quality are an important factor that contribute for drinking water demand and agriculture use. The water quality analysis was assessed using multivariate statistical analyses based on analytical quantitative data that include Discriminant Analysis (DA) and Principal Component Analysis (PCA), based on 36 water quality parameters from the rivers, lakes, and groundwater sites at Jenderam Hilir, which were collected from 2013 to 2014 (56 observations) . The DA identified six significant parameters (pH, NO2-, NO3-, F, Fe2+, and Mn2+) from 36 variables to distinguish between the river, lake, and groundwater groups (classification accuracy = 98%). The PCA had confirmed 10 possible causes of variation in the groundwater quality with an eigenvalue greater than 1, which explained 82.931% of the total variance in the water quality data set.

Keywords: Hydrochemical, water quality, multivariate analysis, principal component analysis, discriminant analysis.

1. Introduction

Bank Infiltration (BI) is a surface water (i.e. rivers and lakes) and groundwater process interact which surface water is forced to flow through aquifer into pumping well that are installed on the banks of river and lake [1]. BI which important systems for water supply in managed aquifer recharge (MAR), has been used for water supplies in Eroupe along the Rhine, Elbe, Danube, and Seine Rivers for over a centuryand it has provided the majority of the drinking water for large cities in Russia, USA, China and other countries over the last few decades [2].

The obvious advantage of BI is the conjunctive use of infiltrated surface water and groundwater from the alluvial catchments of intake structures, which ensure long-term productivity and stability of the water supply. Additionally, surface water contaminants can be significantly removed or degraded as the infiltrating water moves from the river/lake to the production wells due to a combination of physicochemical and microbiological processes [3, 4]. However, as noted by Schmidt et al. [5], poor surface water quality, heavy clogging of the riverbed, and accidental pollution have already greatly threatened BI systems. Importantly, the sustainability of BI is affected by particulate organic matter, which intensifies physical or chemical clogging of riverbed on various spatial and temporal scales [3]. The degree of these interactions significantly depends on meteorological, fluvial, anthropogenic, and geological processes [6]. Such interactions significantly affect the quantity and hydrochemical composition of water bodies on the local and regional scales. Therefore, the location of groundwater and surface water interaction should be identified and its mechanisms understood for effective environmental management. The hydrochemical composition of alluvial systems is largely influenced by precipitation, regional geology, aquifer–stream transfer, and anthropogenic contamination. Water quality offers insight into various flow pathways, geological settings, and physical and chemical processes in the water [7, 8, 9]. If water flux data are available, similarities in the water composition of neighboring groundwater and surface water bodies, such as in the iron ratio or the concentration of total dissolved solids (TDS), can be used to qualitatively or statistically infer groundwater–surface water interaction [e.g. 9,10,11], or to quantitatively estimate fractions of water from different sources or end-members with distinct hydrochemistry [5,12,13]. In Malaysia, the impact of pollutants on groundwater is highly influenced by surface water, the main source for treating domestic water supply. However, some aquifers are locally recharged, and a short period takes place between water entry into the aquifer and water distribution for consumption. In this case, the groundwater may retain surface water characteristics and contaminants. The pollution of rivers reduces their suitability and requires costly water treatment. It is very important to investigated the feasibility of BI as an alternative procedure for water demand purposely.BI methods improved the quality of surface water for drinking water use in Malaysia which is based on the percolation of river water through the ground into an aquifer.contaminants from the surface water are removed by filtration, adsorption, reduction, and biodegradation through percolation process [2,3,14].

Multivariate methods such as discriminant analysis (DA) and principal component analysis (PCA) have become increasingly popular in the qualitative and statistical analysis and interpretation of environmental data sets, especially where flux data are unavailable. These techniques simplify large data sets by grouping them into components or clusters based on the relationships between specified variables [9,15]. Both DA and PCA have been extensively used in hydrogeology to identify groundwater facies for the study of natural groundwater evolution [16,17,18,19,20] and anthropogenic contamination [18,19]. Several studies have also utilized multivariate statistics to classify surface water [21,22,23]. Guler and Tyhne [8] has classified both groundwater and surface water bodies on multi- and single-watershed scales respectively to characterize the hydrogeology of an area and identify localized areas of groundwater recharge and discharge. Kumar et al. [24] utilized a similar statistical methodology to identify the source of contaminated groundwater flowing to an urban reach of the Yamuna River in India, and to classify groundwater and surface water bodies according to hydrochemical similarities. This study aims to understand the hydrochemical influence on BI in the study area and to determine whether multivariate analysis can help to delineate the interaction between surface water and groundwater. It also aims to expand knowledge about this relation and convince water quality managers to use BI method to reduce certain parameters.

2. Methodology

2.1 Study Site

Located at the downstream confluence of the Langat and Semenyih rivers, the study area in Jenderam Hilir is a water reservoir surrounded by lakes. The flat Langat riverbank was chosen to test the effectiveness of BI. The study area is located in Selangor within the Langat Basin, extending between 2° 53′ 28.56″ N and 2° 53′ 39.75″ N and between 101° 42′ 03.78″ E and 101° 44′ 14.58″ E, covering an area of 10 km2 (Figure 1).

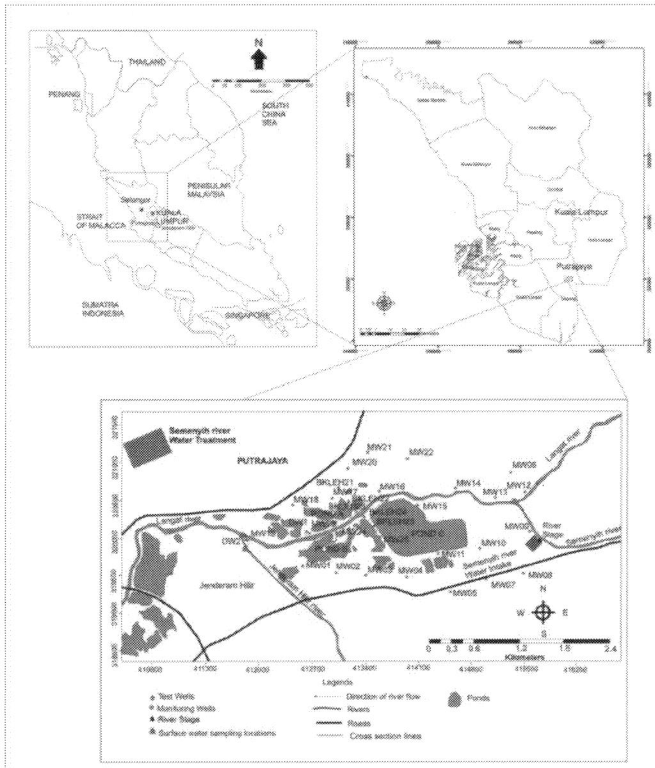

Figure 1. Location of study area, major river systems and location of 25 monitoring wells (MW) and test wells (DW).

2.2 Monitoring and Test Wells

Eight wells were drilled with a rotary drilling machine by wash boring the bedrock. Seventeen holes were drilled by using a motorized auger to a depth of 2 m to 3 m below the water table. Twenty-five monitoring wells (MW) were then constructed at the drilled locations, mostly on the downstream area of the confluence of the Semenyih and Langat rivers (Figure 1).

2.3 Data Collection and treatment

Fifty-six water samples were collected from rivers, lakes, and groundwater from December 2013 to October 2014 during rainy and dry season About 15 water samples collected from the upstream and downstream Langat and Jenderam Hilir Rivers were marked RW, and 13 water samples (LK) were obtained from three lakes (Ponds A, B, and C). Twenty-eight groundwater samples were also taken (from DW1 and DW2, the production well (PW1), and MW2, MW11, MW12, MW13, MW16, MW19, MW23, and MW25) and marked as GW. The sampling locations are shown in Figure 1. For groundwater sampling, the wells were pumped with submersible pumps for more than 30 min to reduce the volume of waste to dispose. Each sample bottle was filled with the water sample without entrapping air bubbles to prevent the biochemical and surface reactions during transportation and storage Each bottle was labeled with its corresponding sampling station and time and were kept at 4 °C to minimize microbial activity in the water [25].

The samples were prevent clogging during analysis with spectrometry instruments and to obtain dissolved ions for metal analysis [25]. The samples were then acidified with HNO_3 until pH<2 to prevent the precipitation of components, such as metal oxides and hydroxides, and retard biological activity. Calibrated Multi-parameter probes (SevenGo pro probe and SevenGo Duo pro probe, Mettler Toledo AG, Switzerland) were used to measure electrical conductivity (EC), TDS, and pH in situ.. Bicarbonate (titration using 0.02 N HCl) and chloride ions (argentometric method using 0.0141 N $AgNO_3$) were analyzed on-site by using unfiltered samples [25]. Meanwhile, the filtered samples were separated into two polyethylene bottles, one for the analysis of sulfate (SurfaVer 4 HACH method) and nitrate (NitraVer 5 HACH method) and the others to determine cations and metals which were analyzed using flame atomic absorption spectrometry (FAAS, Shimadzu AA6800) for cations, inductively coupled plasma mass spectrometry (ICP-MS, ELAN DRC-e, Perkin Elmer) as for trace metal. The sampling, preservation, and transportation to the laboratory of the water samples were based on the Standard Method for Water and Wastewater Analysis [25]. All of the reagents were from analytical or equivalent grade and contaminant-free. All the laboratory equipments were pre-cleaned with 5% (v/v) concentrated nitric acid (HNO_3) and then rinsed with distilled water to ensure that any contaminants or traces of cleaning reagents were removed before the analysis. Polyethylene bottles which are free from metal-containing material were used to collect the water samples to minimize interference of heavy metal analysis[25]. The accuracy of the result was determined by replicating samples (n^2) with relative standard deviation and assessed by diluting ICP Multi-Element Mixed Standard III (Perkin Elmer) in a series of concentrations with the same acid mixture used for sample dissolution. The water quality parameters that were considered for the analysis are turbidity, color, pH, EC, manganese (Mn), iron (Fe), nitrate nitrogen ($NO^{3-}N$), nitrite nitrogen (NO_2-N), TDS, total solids, chloride (Cl), ammonium (NH4), fluoride (F), carbonate (CO_3), bicarbonate (HCO_3), sodium (Na), calcium (Ca), potassium (K), aluminum (Al), magnesium (Mg), mercury (Hg), cadmium (Cd), selenium (Se), arsenic (As), chromium (Cr), silver (Ag), copper (Cu), zinc (Zn), sulfate (SO4), silica (SiO_2), cobalt (Co), barium (Ba), strontium (Sr), nickel (Ni), and phosphorus (P). All of the concentrations were expressed in milligrams per liter (mg/L).

2.4 Data analyses

The data set was screened according to the charge balance error (CBE) proposed by Freeze and Cherry [7] to identify electrically unbalanced samples:

$$CBE = \frac{\sum \upsilon mc - \sum \upsilon ma}{\sum \upsilon mc + \sum \upsilon ma} x100\% \qquad (1)$$

Where;

υ is the absolute value of the ionic valence,
mc is the molarity of the cationic species, and
ma is the molarity of the anionic species.

CBEs were calculated on each site by using the median concentration of the cations and anions. the inclusion of these samples in the calculations affecting the CBEs by less than 0.5%, the CBEs based on major ion concentrations were considered significant for this study. Although 36 water quality parameters were available, only 28 parameters consistently included in 56 sample data sets were chosen for multivariate analysis (data matrix: 28 × 56)The statistical analysis for water quality data sets were subjected to univariate analysis; range, mean, and standard deviation and multivariate analysis; and DA and PCA. DA was first applied to the groups or clusters among the stations, followed by PCA to extract and distinguish potential factors or sources of pollution contributing to variations in water quality measures.

2.5 Analytical Methods

DA is a multi-component statistical technique used to classify objects into mutually exclusive and exhaustive groups according to a set of independent variables.and and builds a discriminant function (DF) that operates on raw data for each group [26,27,28]:

$$f(\text{Gi}) = k_i + \sum_{j=1}^{n} (w_{ij} P_{ij}) \qquad (2)$$

Where;

i is the number of groups (G),
ki is the constant inherent to each group,
n is the number of parameters used to classify a set of data into groups, and
wj is the weight coefficient assigned by DF analysis (DFA) to a given parameter (pj).

DA was used to determine whether the groups differed in terms of the mean of a variable and use such variable to predict group membership [26]. All of the parameters (n = 28) were employed in the DA using the standard, forward-stepwise, and backward-stepwise modes. In the forward-stepwise mode, the variables were included step by step, beginning with the most significant variable until no significant changes were obtained.

PCA is an analytical technique for multivariate data.that reduces many variables to few variables, without sacrificing too much of the information. PCA provides information on the most significant parameters based on spatial and temporal variations that describe the whole data set by excluding the less significant parameters with minimum loss of original information [4,27,29,30]. The principal component (PC) can be expressed as

$$z_{ij} = a_{i1}x_{1j} + a_{i2}x_{2j} + \ldots + a_{im}x_{mj} \quad (3)$$

where

z is the component score,
a is the component loading,
x is the measured value of the variable,
i is the component number,
j is the sample number, and
m is the total number of variables

The PCs generated by PCA are sometimes not readily interpreted; therefore, rotating the PCs by varimax rotation is advised. The VF coefficients with correlations greater than 0.75 are considered as “strong” significant factor loadings, 0.75 to 0.50 as “moderate,” and 0.50 to 0.30 as “weak” [31]. The PCA explains relationships between measured variables, and thus be used to infer the hydrogeochemical processes that control water chemistry[32,33]. The PCA takes the data organized in a correlation matrix and rearranges them in a manner that better explains the structure of the underlying system that produced the data. Therefore, the matrix of correlation coefficients measures how well the variance of each constituent can be explained by its relationships with each of the variables.

3. Results and Discussion

3.1 Hydrochemical Characteristics

The descriptive statistics of the measured three-year data set are summarized in Table 1, including the univariate (mean, standard deviation, minimum, and maximum) overview of the chemistry of surface water and groundwater in the study area. The pH of the groundwater sample analyzed in this study within the permissible limit of drinking water standard of 5.5 to 9.0 [34]. Samples from surface waters show pH values that the mildly acidic surface water significantly increased the pH values of the groundwater. Meanwhile, EC for groundwater in the study area varies from about 62 μS/cm to 166 μS/cm with the mean value of 118 μS/cm. Low conductivity was observed in the Langat River upstream (SL1) and downstream (SL2) of the study area. High turbidity value for surface water in the study area ranges between 328 and 699 NTU. The groundwater samples in the study area show low turbidity, indicating that BI in Jenderam Hilir can significantly reduce the turbidity value. Sulfate concentration in the groundwater lies between 3.0 and 14.0 mg/l while in the surface water it ranges from 3.0 mg/l to 15.0 mg/l. Nitrate in natural water is due to organic sources or industrial and agricultural chemicals. Although nitrogen is an essential constituent of protein in all living organisms but the presence of , nitrate with a concentrations greater than 50 mg/l can cause cyanosis or blue baby syndrome among infants [35] and cancer in adults (WHO, 2008). The permissible limit for NO^{3-} in raw water is 10 mg/l (DSM, 2010).

The NO^{3-} concentration in groundwater and surface water in the study area varies from 1.80 mg/l to 39.0 mg/l. In Jenderam Hilir, high NO^{3-} concentration of 39.0 mg/l can be observed at MW16 for the groundwater and 16.0 mg/l at SL1 for the surface water. The high nitrate concentration of nitrate could have resulted from the seepage of the liquids due to the occurrence of sewage and septic tanks, industrial effluents, and agricultural chemicals nearby. As MW16 is near the oil palm estate. Meanwhile, all of the groundwater samples from eight wells showed Fe^{2+} concentration higher than the drinking water standard, i.e. 0.3 mg/l. The Fe^{2+} concentration of groundwater was within 14.0 mg/l to 39.0 mg/l. Fe is a common constituent of various primary minerals such as biotite, pyroxenes, and amphiboles. The geochemical conditions of the study area are assumed to result in a predominant distribution of ferrous and ferric irons, though the definite existence of these minerals in the area has not been confirmed. The Ca^{2+} concentration ranges from 1.50 mg/l to 32.0 mg/l. The permissible limit of

IOP Conf. Series: Materials Science and Engineering **136** (2016) 012073 doi:10.1088/1757-899X/136/1/012073

Ca^{2+} concentration is 75 mg/l in drinking water standard (WHO, 2008). The Na^+ concentration in the wells and surface water ranges is below the permissible limit of the Malaysia Drinking Water Standard (DSM, 2010). Ionic K^+ occurs at fairly low concentrations in groundwater, and the major sources of K concentration in the study area could be from rainfall, rivers, potash feldspar, and micaceous minerals. The K^+ concentration is in the range of show low K^+ concentrations.

3.2 Surface and Ground Water Classification

Figure 2 illustrates the Piper trilinear diagram for the data obtained from the chemical analysis of surface water and groundwater samples from the study area. The diagram consists of two lower triangles that show the percentage distribution, on the milliequivalent basis, of the major cations ($Ca^{2+,}$ $Mg^{2+,}$ Na^+, and K^+) and major anions ($SO_4^{2-,}$ Cl^-, CO^{3-}, HCO_3)..

Three chemical types of surface water and groundwater in the study area were recognized as (i) Calcium–Sodium–Sulfate–Chloride–Bicarbonate facies (Ca^{2+}–Na^+–SO_4—Cl^-–HCO^{3-}), (ii) Calcium–Chloride–Bicarbonate facies (Ca^{2+}–Cl^-–HCO^{3-}) and, (iii) Calcium–Magnesium–Chloride–Bicarbonate facies. The Ca^{2+}–HCO^{3-} water type prevails in all the groundwater and surface water samples. The major cations dominant in the groundwater are Ca^{2+} and $Mg^{2+.}$ In well MW16, the Na^+ and K^+ ions are dominant compared to the other major cations. The water classification shows that the surface water and groundwater in this study area are connected.

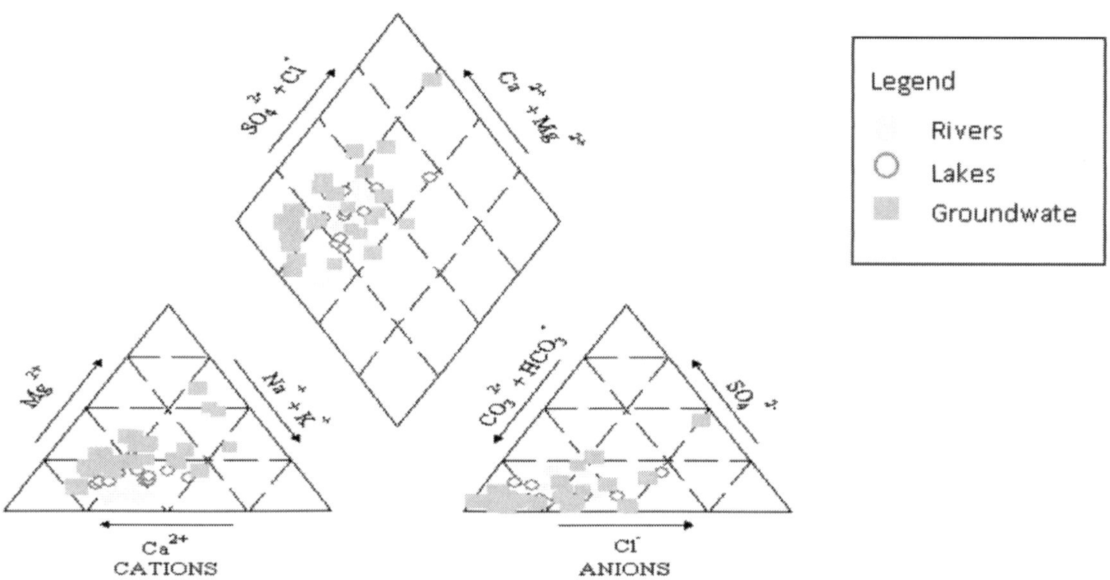

Figure 2. Piper trilinear diagram for the data obtained from chemical analysis of river, lake and groundwater sample from Jenderam Hilir.

Table 1: Descriptive statistics (mean, SD (standard deviation), minimum and maximum) of the surface water quality of Jenderam Hilir.

Parameter	pH	Turbidity	EC	TS	TDS	CO_3	Cl	SO_4	NO_2	NO_3	F	P	NH_4	Ag	Al
Unit		NTU	µS/cm	mg/l	mg/l	mg/l	mg/l	mg/l	mg/l	mg/l	mg/l	mg/l	mg/l	mg/l	mg/l
River (N=9)															
Mean	6.84	199.8	68.2	167	61	0.67	10.3	7.67	0.66	9.05	0.3	0.17	0.57	0.1	1.6
SD	0.39	229.92	60.22	98	23	0.24	6.23	3.69	1.35	4.25	0.1	0.26	0.75	0.1	2.7
Min	6.3	8	2	60	16	0.5	4	1.5	0	3.3	0.3	0.01	0.25	0	0.2
Max	7.7	699	161	370	98	1	24	15	3.9	16	0.5	0.81	2.4	0.3	11
Lakes (N=13)															
Mean	6.94	8.44	60.31	12.2	53	22.3	0.62	29.9	7.69	2.85	0	1.17	0.31	0.2	0.3
SD	0.18	9.3	39.94	12.6	27	11	0.22	12.9	4.23	1.69	0	0.91	0.11	0.3	0.2
Min	6.7	0.5	2	3.1	20	4	0.5	3	2	1.5	0	0.25	0.25	0	0.3
Max	7.3	25	147	38	108	40	1	53	18	7	0.1	2.6	0.5	0.8	00.7
Groudnwater (N=28)															
Mean	6.46	158.92	84.85	20/7	177	55.6	0.65	48.3	9.5	4.15	0.9	2.57	0.32	0.3	2
SD	0.48	223.14	67.14	16.2	195	27.2	0.23	29.8	7.57	3.94	4.6	7.51	0.12	0.3	4.7
Min	5.6	10	2	4.6	44	22	0.5	0.5	0.5	1.5	0	0.25	0.25	0	0.3
Max	8.1	1196	228	47	150	150	1	124	36	14	24	39	0.5	0.9	25

Parameter	As	Ba	Ca	Co	Cr	Cu	Fe	Hg	K	Mg	Mn	Na	Ni	Pb	Sr
Unit	mg/l	mg/l	mg/l	mg/l	mg/l	mg/l	mg/l	mg/l	mg/l	mg/l	mg/l	mg/l	mg/l	mg/l	mg/l
River (N=9)															
Mean	0	0.1	11	0	1.3	0.1	2.7	0.6	3.6	0.9	0.32	5.2	0.1	0	0
SD	0	0	5.1	0	4.9	0	1.6	0.3	1	0.3	0.48	2.8	0	0	0
Min	0	0.1	6.1	0	0	0.1	1.4	0.1	2.1	0.5	0.05	2	0.1	0	0
Max	0	0.1	24	0.1	19	0.1	7.9	1	4.8	1.5	1.5	10	0.1	0	0.1
Lakes (N=13)															
Mean	0.1	0.1	0	0.1	7.6	0	0.1	1.5	1.3	6.7	1.45	0.3	3.1	0.1	0
SD	0.1	0.1	0	0	3.4	0	0	1.7	2.5	12	0.46	0.2	0.9	0	0
Min	0	0.1	0	0.1	3.2	0	0.1	0.1	0.1	2.5	0.9	0.1	1.5	0.1	0
Max	0.3	0.3	0	0.1	15	0.1	0.1	4.9	9	46	2.6	0.6	4.9	0.1	0.1
Groudnwater (N=28)															
Mean	0.1	0.6	0.1	0.7	9.5	0	0.1	20	1.1	4.6	3.38	1.4	3.8	0.1	0.3
SD	0.1	1.6	0.1	1.7	7.2	0	0	9.7	1.6	2.3	1.6	1.5	3.6	0	1
Min	0	0.1	0	0.1	0.9	0	0.1	4.9	0.1	1.9	1.3	0.1	1	0.1	0
Max	0.3	7.7	0.2	7.9	32	0.1	0.1	39	8	12	7.8	6	20	0.1	5.2

3.3 Surface and Groundwater Spatial Pattern Discrimination

In this study, the DFs were performed based on the raw data from three types of water samples, and similar classification matrixes (CMs) treated as dependent variables while water quality parameters were treated as independent variables The accuracy of classification using the standard, forward stepwise, and backward stepwise mode DAs were 100% (28 discriminant variables), 92% (6 discriminant variables), and 98% (13 discriminant variables), respectively (Table 1). The Wilk's Lambda test for standard mode gave a Lambda value of 0.027. The null hypothesis states that the means of vectors of the three groups (RW, LK, and GW) are similar to one another. The alternative hypothesis states that at least one of the means of vectors is different from one another. Since the enumerated p-value is lower than the significant alpha level (i.e. 0.005), one should reject the null hypothesis and accept the alternative hypothesis. The risk of rejecting the null hypothesis while it is true is lower than 0.001%. Thus, the three groups of samples are indeed different from one another because of their different characteristics. DFs and CMs obtained from standard, forward stepwise, and backward stepwise modes of DA are shown in Table 2.

Table 2. Classification matrix for DA of spatial variations in surface water and groundwater interaction study at Bank.

Samples	Samples assigned by DA			Total	% correct
	GW	LK	RW		
Sampling DA mode (28 variables)					
GW	27	0	0	27	100.00%
LK	0	13	0	13	100.00%
RW	0	0	10	10	100.00%
Total	27	13	10	50	100.00%
Forward stepwise mode (6 variables)					
GW	26	1	0	27	96.30%
LK	0	13	0	13	100.00%
RW	0	3	7	10	70.00%
Total	26	17	7	50	92.00%
Backward stepwise mode (13 variables)					
GW	27	0	0	27	100.00%
LK	0	13	0	13	100.00%
RW	0	1	9	10	90.00%
Total	27	14	9	50	98.00%

Figure 3 presents the discriminant variables by DA into three groups, RW, LK, and GW, with 100% correct assignments. Using the forward stepwise mode, DA, pH, NO^{2-}, NO^{3-}, F^{-}, Fe^{2+}, and Mn^{2+} were found to be the significant variables. This result indicated that these parameters, which possess high variation, constituted the anaerobic redox parameters containing Fe^{2+} and $Mn^{2+.}$ In this study, redox process play an important role in changing the concentration of Mn^{2+} and Fe^{2+}, which represent the minor portion of reduction processes in the aquifer, whereas the presence of NO^{3-} is highly correlated with the existing major ion concentrations in the aquifer. In the backward stepwise mode of DA, 13 parameters (pH, TDS, EC, CO^{3-}, HCO^{3-}, NO^{2-}, NO^{3-}, P, NH_4^{-}, Al, Fe^{2+}, Mg^{2+}, and Na^{+}) were considered as the most significant parameters ($p<0.05$), as shown in their high spatial variation in the data set.

As identified by DA (backward step and forward step modes), box and whisker plots of selected discriminating parameters were constructed to evaluate different patterns associated with different water sample sources. Figure-3 shows the box and whisker plots of six variables of these water quality parameters during the period December 2013 to October 2014. The trend for NO_3^{-} and Fe^{2+} suggested that the average concentration in rivers was the highest, followed by the concentration in lakes. In groundwater, the average concentration was the lowest, while Mn^{2+} had the inverse trend. Within groundwater, most of the Mn^{2+} detected was naturally occurring. Six parameters that produced high variations based on the forward stepwise DA were used for further analysis.

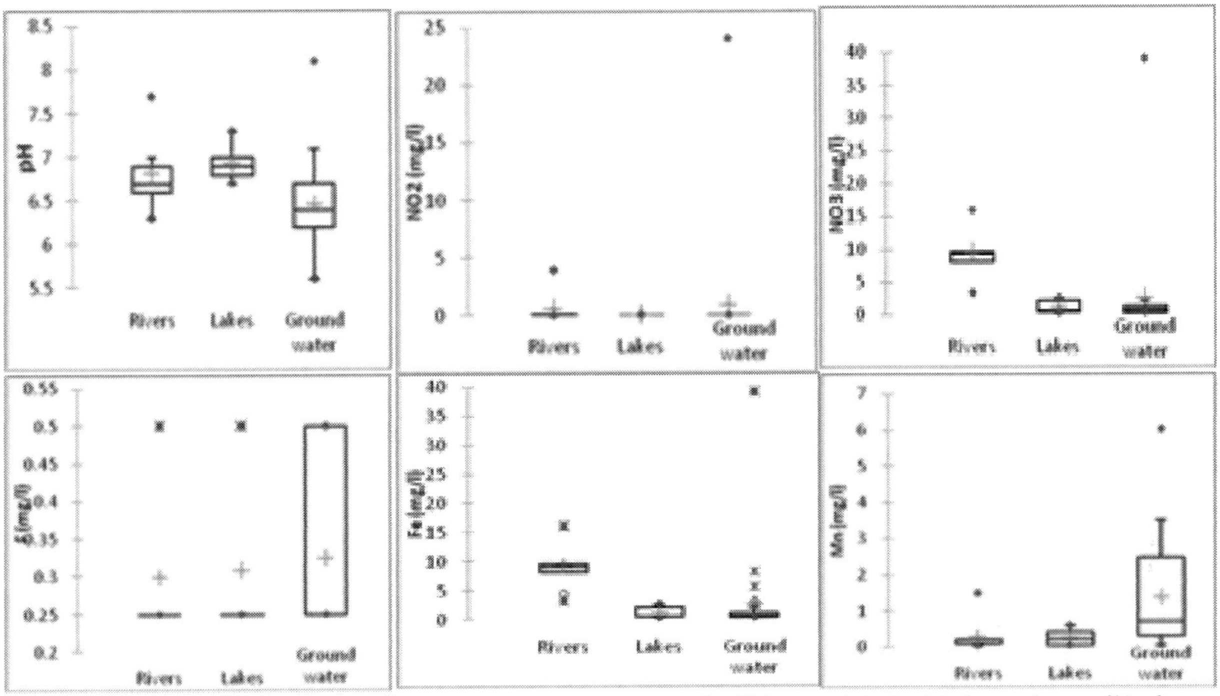

Figure 3. Box and whisker plots of some parameters separated by DA associated with the water quality data of Langat and Jenderam Hilir rivers, lakes and groundwater.

3.4 Correlation between Variables

The first step in factor analysis (FA) is to determine the parameter correlation matrix. The correlation matrix is used to account for the degree of mutually shared variability between individual pairs of water quality variables. PCA only considers the common hydrochemical variables of all samples taken from rivers, lakes, and groundwater. The correlation matrix of the 28 variables (Table-4) allows us to distinguish several relevant hydrochemical relationships (indicated by underlined values): (a) high and positive correlations between turbidity, total solid, NO^{2-}, NO^{3-}, NH_4, Al^{3+}, C^{2+}, and SO_4^{2-} (r=0.504–0.821) are attributed to river water hydrochemistry; (b) strong and positive correlations between HO^{3-}, F, P, As, Fe^{2+}, Mg^{2+}, Na^+, and Cl^- represent groundwater influence toward rivers (r = 0.516–0.905); and (c) high and positive K and Hg are characterized toward the water chemistry of lakes (r = 0.604). However, pH, SiO_2, total solid, HO_3, P, Ag, and Al have weak correlations with all the variables, which indicates that all variables have different origins. Only Ag shows good correlation with conductivity, due to the increase of conductivity along with the dissolution of metals through ion exchange or oxidation–reduction reaction in a groundwater aquifer system. Na shows moderate correlation with Cl (0.68), indicating that a major source of these elements was influenced by marine deposits (Kenny Hills Formation), a process during groundwater flow through rock in the study area. The Na^+ and Cl^- contents detected in certain samples may also suggest the dissolution of chloride salts. The dissolution of halite in water release is equal to the concentrations of Na^+ and Cl^- in the solution.

3.5 Source Identification Using PCA and Factor Analysis (FA)

PCA was employed in this study to investigate the compositional pattern of common hydrochemical variables of all samples taken from rivers, lakes, and groundwater. PCA then strengthens its role in identifying the dominant process that controls the chemical components in both surface water and groundwater. An eigenvalue gives a measure of the significance factor, where the highest eigenvalue resembles the most significant. Eigenvalues greater than 1 are considered significant. Table-3 summarizes the PCA results, including the loadings, eigenvalue, variance contribution rate of each Varimac Factors (VF), and cumulative variance contribution rate. The loadings with significant absolute value in each VF are highlighted.

To reduce the overlap of original variables over each principal component, a varimax rotation was conducted. The 10 factors are extracted to represent the contributions that influence the chemical composition of surface water and groundwater. The varimax rotated Factor Analysis (FA) results are presented in the Table-4 with probable source types. In the Table-4, the eigenvalue is set to 1.0 as a threshold in order to limit the number of extracted factors. Factor loadings equal to or greater than 0.5 are considered significant. Ten factors obtained explain 83% of the total variance.

In Table 3, VF 1 explains 17.171% of the variance, which is highly attributed to As, Fe, and Mg. This factor suggests the significant contribution of groundwater hydrochemistry and of redox conditions. VF2 elucidates 13.836% of the total variance, which has strong positive loading of F, P, and CO_3. This VF represents, during pre-monsoon, the high loading factor of F concentration. The important source of F in groundwater is fluoride-bearing minerals such as fluorite, apatite, amphiboles, and micas, which may be due to the ion exchange of F, leaching of F-containing minerals, higher evapotranspiration, and longer residence time of water in the aquifer. These factors are also significant in groundwater flow. VF3 describes 9.59% of the total variance provided by a strong positive loading of color, turbidity, and total solid, which represents the surface runoff of sources. VF4 (8.715% of total variance) has a strong positive loading of EC, Sr, and SiO_2. This VF can be explained by the groundwater contact with rock or weathered rock (Kenny Hills formations).

The hydrolysis and weathering of ranging silicate minerals to clay minerals such as kaolinite and illite are also explained by the fourth factor. Meanwhile, VF5, which contains 8.257% of the total variance, includes TDS, Ca, NH_4, and SO_4 as the highest mean concentrations found in river water and groundwater interaction, as well as in the discharge of untreated sewage and industrial waste into rivers. The association of TDS with higher Ca concentrations is caused by tanneries that use calcium carbonate to process raw hides and their effluent discharges into the river course. The substances in this VF are characteristic of input from surface water. They represent the substantial influence of surface water of the Langat River infiltration and control the groundwater quality. VF6 with 6.948% of the total variance has strong positive loadings of Na+ and Cl-, which indicate the groundwater interaction with rock (Kenny Hills formation) and are deposited into the marine environment. VF7 explains 5.638% of the strong positive loading of Al, NO_2, and NO_3, which represent agricultural wastes based on nitrate found in the upstream of Jenderam Hilir.

Table 3. Pearson correlation coefficients for 13 hydrogeochemical variables of water samples. Coefficients are significant at the 0.05 level and those higher than 0.70 were shown as bold fonts.

Variables	pH	Colour	Tur	EC	TS	TDS	Na	K	Mg	Ca	Mn	Fe	Al	Hg	Cl	F	P	Ag	As	Cr	Sr	SiO_2	CO_3	HCO_3	NH_4	SO_4	NO_2
pH																											
Colour	-0.01																										
Tur	-0.15	**0.65**																									
EC	0.07	0.21	0.25																								
TS	-0.16	**0.68**	**0.73**	0.3																							
TDS	-0.23	0.03	0.18	0.39	0.29																						
Na	-0.15	-0.14	-0.05	-0.04	0.04	0.15																					
K	0.06	-0.09	-0.11	0.12	-0.08	0.08	-0.04																				
Mg	-0.2	-0.02	0.09	0.39	0.17	0.43	0.23	0.09																			
Ca	0.08	-0.21	-0.09	0.49	0.06	**0.72**	0.08	0.14	0.28																		
Mn	-0.17	-0.12	0.06	-0.03	0.08	0.14	-0.14	-0.02	0.36	0.07																	
Fe	-0.23	0.15	0.25	0.43	0.3	0.33	0.09	-0.03	**0.8**	0.05	0.25																
Al	0.01	-0.06	0.18	0.14	0.23	0.03	0.1	-0.07	-0.05	0.02	-0.16	0.17															
Hg	-0.01	-0.06	-0.13	0.03	-0.09	-0.11	0.2	**0.64**	0.04	-0.18	-0.11	0.01	-0.03														
Cl	-0.19	-0.22	-0.09	-0.18	0.02	0.02	**0.68**	-0.05	0.27	-0.03	-0.11	0.12	-0.01	0.09													
F	0.2	-0.11	-0.16	0.33	-0.03	0.01	0.05	-0.08	0.06	0.2	-0.16	0.27	0.38	-0.04	-0.04												
P	0.19	-0.09	-0.1	0.41	0.01	0.05	0.04	-0.06	0.09	0.22	-0.14	0.31	0.4	-0.02	-0.1	**0.91**											
Ag	-0.02	-0.17	-0.1	**-0.78**	-0.14	-0.02	0.08	-0.14	-0.12	-0.11	0.02	-0.25	-0.13	-0.29	0.25	-0.37	-0.44										
As	-0.1	0.19	0.25	0.15	0.21	0.14	-0.04	-0.07	**0.51**	-0.08	0.2	**0.74**	0.16	-0.11	0.07	0.27	0.29	0.01									
Cr	0.11	-0.03	0.4	0.06	-0.04	-0.17	-0.06	-0.05	-0.16	-0.06	-0.08	-0.1	0.14	-0.08	-0.07	-0.09	-0.02	0	-0.09								
Sr	-0.17	-0.04	0.03	-0.15	-0.04	0.04	0.01	0.02	0.2	0.04	0.24	0.1	-0.05	-0.04	0.06	-0.02	-0.05	0.19	0.26	-0.03							
SiO_2	-0.34	0.12	0.1	0.21	0.04	0.21	-0.06	0.26	0.22	-0.07	0.22	0.16	-0.06	0.35	-0.06	-0.38	-0.28	-0.4	-0.12	-0.14	-0.13						
CO_3	0.19	-0.12	-0.14	0.27	0.01	0.1	0.14	-0.08	0.02	0.24	-0.18	0.23	0.38	-0.05	0.06	**0.95**	**0.86**	-0.3	0.25	-0.09	-0.03	-0.38					
HCO_3	0.23	0.2	0.25	0.33	0.37	0.49	0.13	0.05	0.29	0.34	0.03	0.28	-0.05	-0.06	0.14	-0.09	-0.06	-0.01	0.16	-0.03	-0.15	0.07	0				
NH_4	-0.11	-0.01	0.13	0.35	0.08	**0.57**	-0.02	0.17	**0.58**	0.41	0.08	0.32	-0.08	-0.02	-0.01	-0.14	-0.12	-0.1	-0.01	-0.04	-0.04	0.31	-0.15	0.36			
SO_4	0.01	-0.06	0.09	0.12	0.08	0.45	0.18	0.03	0.03	**0.52**	-0.06	-0.09	0.19	-0.17	0.08	0.16	0.17	0.15	-0.12	0.02	0.11	-0.12	0.21	0.16	0.23		
NO_2	0.02	-0.03	-0.01	0.14	-0.04	-0.01	0.01	-0.05	0.1	-0.17	-0.11	0.32	**0.5**	-0.01	0.01	0.22	0.23	-0.1	0.35	-0.02	-0.03	0	0.22	-0.04	-0.03	-0.02	
NO_3	0.08	-0.06	0.04	0.16	0.03	0.12	0.18	-0.09	-0.12	0.04	-0.2	0.1	**0.61**	-0.08	0.08	0.34	0.35	-0.11	0.19	0.1	-0.04	-0.13	0.42	0.06	-0.1	0.31	**0.82**

Table 4. Loading of 28 hydrochemical variables on nine significant varifactors (VF).Coefficients are significant at the 0.05 level and those higher than 0.50 were shown as bold fonts.

Variables	VF1	VF2	VF3	VF4	VF5	VF6	VF7	VF8	VF9	VF10
pH	-0.163	0.225	-0.117	-0.093	-0.046	-0.221	-0.004	**0.78**	0.087	0.136
Colour	0.015	-0.066	**0.877**	0.085	-0.129	-0.16	-0.049	0.09	-0.015	-0.122
Tur	0.145	-0.132	**0.836**	0.036	0.06	-0.039	0.058	-0.075	-0.066	0.397
EC	0.278	0.339	0.225	**0.633**	0.381	-0.137	0.032	0.101	0.088	0.149
TS	0.129	0.022	**0.894**	0.058	0.144	0.072	0.008	-0.03	-0.058	-0.053
TDS	0.249	-0.046	0.158	0.087	**0.842**	0.061	0.058	-0.04	-0.021	-0.168
Na	0.022	0.082	-0.015	-0.02	0.101	**0.889**	0.047	-0.065	0.076	-0.002
K	0.012	-0.075	-0.078	0.011	0.151	-0.092	-0.034	0.075	**0.893**	-0.01
Mg	**0.864**	-0.021	-0.024	0.123	0.293	0.199	-0.088	-0.046	0.057	-0.034
Ca	0.031	0.227	-0.116	0.064	**0.874**	-0.023	-0.132	0.081	-0.018	0.015
Mn	0.46	-0.184	-0.096	-0.1	0.103	-0.273	-0.202	-0.328	-0.08	0.001
Fe	**0.883**	0.178	0.185	0.176	0.047	0.078	0.143	-0.046	-0.019	-0.049
Al	-0.05	0.324	0.15	0.034	0.053	0.058	**0.694**	-0.153	-0.011	0.176
Hg	-0.021	-0.022	-0.048	0.199	-0.191	0.189	-0.022	-0.033	**0.866**	-0.063
Cl	0.155	-0.052	-0.09	-0.138	-0.018	**0.884**	0.01	-0.012	-0.006	-0.024
F	0.089	**0.952**	-0.075	0.063	0.027	-0.01	0.133	0.032	-0.035	-0.045
P	0.114	**0.916**	-0.036	0.141	0.053	-0.051	0.157	0.015	-0.011	0.021
Ag	-0.093	-0.39	-0.123	-0.782	0.018	0.177	0.003	0.11	-0.234	-0.065
As	**0.774**	0.236	0.226	-0.245	-0.143	-0.059	0.212	0.026	-0.044	-0.068
Cr	-0.091	-0.06	0.056	0.017	-0.062	-0.017	0.056	0.06	-0.051	**0.953**
Sr	0.287	0.057	-0.003	**-0.601**	0.083	-0.107	-0.097	-0.407	0.165	0.101
SiO_2	0.13	-0.458	0.044	**0.582**	0.079	-0.068	0.046	-0.328	0.3	-0.164
CO_3	0.041	**0.917**	-0.04	0.008	0.092	0.097	0.177	0.061	-0.043	-0.076
HCO_3	0.272	-0.14	0.311	0.09	0.433	0.163	0.001	**0.59**	-0.005	-0.065
NH_4	0.36	-0.279	-0.046	0.284	**0.605**	-0.021	-0.034	0.097	0.05	0.029
SO_4	-0.235	0.165	0.072	-0.247	**0.733**	0.105	0.176	-0.1	-0.008	0.066
NO_2	0.235	0.075	-0.072	0.052	-0.131	-0.053	**0.905**	0.043	-0.009	-0.056
NO_3	-0.054	0.251	0.015	-0.022	0.115	0.089	**0.902**	0.034	-0.04	0.062
Eigenvalue	4.802	3.877	2.687	2.44	2.312	1.945	1.579	1.272	1.195	1.112
Variability (%)	17.149	13.847	9.595	8.715	8.257	6.948	5.638	4.543	4.267	3.971
Cumulative %	17.149	30.996	40.591	49.306	57.563	64.511	70.149	74.692	78.959	82.931

Al possesses higher concentrations derived from mining waste [36]. This VF describes the lakes–groundwater–river water interaction. The presence of Al ions in the stream may result from industrial wastes, but is more likely to come from the wash water of drinking water treatments plants. This VF8 explains 4.543% of the strong positive loading of pH and HCO3. The relationship between HCO3 and pH reveals that the CO3–CO2 equilibrium is an important pH buffer system in the surface water and groundwater of the study area. HCO3 mainly originates from the dissolution of carbonate rocks present in the aquifer (Kumar et al. 2011), and from the reaction of CO2 with silicate minerals. These factors are strongly related to surface runoff. VF9 explains 4.267% of the total variances of high K and Hg loadings. These components reflect the signatures of natural water recharge and water–soil–rock. The existence of K is mainly due to the rock–water interactions on potassium bearing feldspars and clay minerals, such as illlite and biotite-rich minerals. Hg describes the lakes and groundwater interaction from a former mining area. Past mining pollution had strong positive loadings on Hg, representing heavy metal pollution that may have originated from metal smelting. The last principal component, VF10 (3.91% of the total variance), contributes high and positive Cr, a specific pollutant. This finding provides evidence of industrial pollution such as dyeing or paint-processing operations.

4. Conclusion

The result of the statistical analysis offers an overview of the main process responsible for the characteristic of the BI site in Jenderam Hilir; (i) The DA gave the best result of the redox parameters that contain Fe and Mn under anaerobic conditions, (ii) This study also found that the redox process had a significant role in influencing the variation of hydrogeochemistry concentrations in the studied aquifer systems, (iii) Mainly, the quality of groundwater, which was near the river bank of the Langat River and the lakes in Jenderam Hilir was affected by quality changes due to bank filtration, (iv) The study of rivers–lakes–groundwaters interactions at different sampling sites yielded an important data reduction, where six parameters (pH, NO2, NO3, F, Fe, and Mn) were close to 98% correct assignment, (v) On the groundwater flow paths, the effects of surface water fluctuations decreased with increasing distance from the river bank due to dilution, adsorption, reduction, and biodegradation, (vi) The variability of the concentrations in groundwater was also dependent on the changing water levels in the Langat River and the lakes, (vi) High water levels led to high infiltration potentials and less time for purification processes due to the shorter residence time in the aquifer. On contrary, the effects of back-flooding during high water levels were not significant. The temporal (residence time) and spatial dimension (distance of flow path) of the aquifer passage determined the balance in groundwater quality for its use as drinking water, (vii)The quality of Jenderam Hilir groundwater is mainly dependent on the aquifer`s material type, material quantity, and quality of infiltrating surface water from the Langat River.

References

[1] Ray C, Grischek J, Schubert JZ, Wang and TF, Speth (2002) A perspective of riverbank filtration. J Am Water Works Assoc 94(4); 149-160
[2] Tufenkij N, Joseph NR and Menachem E (2002) The Promise of Bank Filtration – A simple technology may inexpensively clean up poor-quality raw surface water. Environmental Science and Technology 1: 423-428
[3] Hiscock, KM and Grischek T (2002) Attenuation of groundwater pollution by bank filtration. J Hydrol 266:139–144
[4] Singh KP, Malik A, Singh VK et al. (2005) Chemometric analysis of groundwater quality data of alluvial aquifer of Gangetic plain, North India[J]. Anal Chim Acta, 550: 82– 91.
[5] Schmidt A, Gibson JJ, Santos R, Schubert M Tattrie K and Weiss H (2010) The contribution of grounfdwater discharge to the overallof to typical boreal lakes in alberta /canada estimated from a radon mass balance. Hydrol Earth Syst Sci. 14-7989, doi;10.5194/hess-14-79-2010
[6] Winter TC, Harvey JW, Franke OL and Alley WM (1998) Groundwater and surface water, in: A single resources, edited by; Survey, U.G. US government printing ofiice, Denver, Colorado.
[7] Freeze RA and Cherry JA (1979) Groundwater, Prenyice-Hall, New Jersey
[8] Guler C and Thyne GD (2004) Delineation of hydrochemical facies distribution in a regional groundwater system by means of fuzzy c-means clustering Water Resour Res 40: 1-11
[9] Kumar PJS, Jegathambal P and James EJ (2011) Multivariate and geostatistical analysis of groundwater quality in Palar River Basin International Journal of Geology 4 (5): 108-119p
[10] Burden RJ (1982) Hydrochemical variation in a water –table aquifer beneath grazed pastureland, J Hydrol 21: 61-75
[11] Taylor CB, Wilson DD, Brown LJ, Stewart MK, Buren RJ and Brailsford GW (1989) Sources and flow of north Canterbury Palins groundwater, New Zealand, J Hydrol 106: 311-340
[12] Vazquez-Sune E, Carrera J, Tubau I, Sanchez-Villa X and Soler A (2010) An approach to indetify urban groundwater recharge, Hydrol Earth Syst Sci. 12: 2085-2097, doi 10.5194/hess-14-2085-2010
[13] De Louw PGB, Van der Velde Y and Van der Zee SEATM (2011) Quantifying water and salt fluxes in a lowland polder catchment dominated by boil seepage; a probabilistic end-,member mixing approach. Hydrol Earth Syst Sci 15:2101-2117. Doi:10.5194/hess-15-2101-2011
[14] Irmscher R and Teermann I (2002) Riverbank filtration for drinking water supply- a proven method, perfect to face today's challenges. Water Sci Technol Water Supply 2(5–6):1–8
[15] Mencio A and Mas-pla J (2008) Assessment by multivarate analysis of groundwater-surfacewater interactions in urbanized Mediteranean streams, J Hydrol 352:355-366

[16] Banoeng-Yakubo B, Yidana SM and Nti E (2009) Hydrochemical analysis of groundwater using multivariate statistical methods-The Volta Region, Ghana. KSCE J Civ Eng 13(1): 55-63
[17] Cloutier V, Lefebvre R, Therrien R and Savard MM (2008) Multivarate statistical analysis of geochemical data as indicative of the hydrochemical evolutionof groundwater in a sedimentary rock aquifer system. J Hydrol 353:294-313.
[18] Demlie M, Wohnlich S, Wisotzky F and Gizaw B (2007) Groundwater recharge, flow and hydrogeochemical evolution in a complex volcanic aquifer system central Ethiopia, Hydrogeol J, 15, 1169-1181
[19] Koh DC, Chae GT, Yoon YY, Kang BR, Koh GW. and Park KH (2009) Baseline geochemical characteristics of groundwater in the mountainous area of Jeju Island, South Korea. Implications for degree of mineralization and nitratae contamination. J Hydrol. 376: 81-93
[20] Delgado-Outeirino I, Araujo-Nespereira P, Cid-Fernandez JA, Mejuto JC, Martinez-Carballo E and Simal-Gandra J (2011) Classification of thermal waters based on their inorganic fingerprint and hydrogeothermal modelling. Hydrol Earth Syst Sci Discuss 8:4559-4581. Doi 10.5194/hessd-8-4559-2011
[21] Alther GA (1989) A simplified statistical sequence applied to routine water quality analysis. A case history, Groundwater 17:556-561
[22] McNeil VH, Cox ME and Preda M (2008) Assesment of chemical water types and their spatial variation using multi-stage cluster analysis, Queensland, Australia, J Hydrol. 310: 181-200
[23] Panda UC, Sundaray SK, Rath P, Nayak BB and Bhatta D (2006) Application of factor and cluster analysis for characterization of river and estuarine water systems-A case study; Mahanadi River (india), J Hydrol 331: 434-445
[24] Kumar M, Ramanathan A and Keshari AK (2009) Understandng the extent of inteactions between groundwater and surface water trough major ion chemistry and multivariate statistical techniques, Hydrol Process 23: 297-310
[25] APHA (2005) Standard methods for the examination of water and wastewater, 21st ed. American Water Works Association, Water Environment Federation, Washington
[26] Johnson RA and DW Wichern (1992) Applied multivariate statistical analysis. 3rd ed. Prentice- Hall, Englewood Cliffs, New Jersey
[27] Singh KP, Malik A and Mohan D et al. (2004) Multivariate statistical techniques for the evaluation of spatial and temporal variation in water quality of Gomti River (India) a case study. J Water Research 38: 3980–3992.
[28] Wunderlin DA, Diaz MDP and Ame MV et al. (2001) Pattern recognition techniques for the evaluation of spatial and,temporal variation in water quality. A case study: Suquia river basin (Cordoba Argentina). J Water Research, 35: 2881–2894.
[29] Kannel PR, Lee S, Kanel SR and Khan SP (2007) Chemometric application in classification and assessment of monitoring locations of an urban river system. Analytica Chimica Acta 582: 390-399
[30] Juahir H, Zain MS and Yusoff MK (2010b) Spatial water quality assessment of Langat River Basin (Malaysia) using envirometric techniques. Environment Monitoring Assess 173: 625-641
[31] Liu CW, Lin KH and Kuo YM (2003) Application of factor analysis in the assessment of groundwater quality in a Blackfoot disease area in Taiwan. The Science of the Total Environment 313: 77–89
[32] Suk H and Lee K (1999). Characterization of a ground water hydrochemical system through multivariate analysis: Clustering into groundwater zones. Ground Water, 37: 358-366
[33] Helena B, Pardo R, Vega M, Barrado E, Fernandez JM and Fernandez L (2000) Temporal evaluation of groundwater composition in an alluvial aquifer (Pisuerga river, Spain) by principal component analysis. Water Research 34: 807–816.
[34] Department of Standards Malaysia (2010) Malaysian Standard: Drinking water-quality requirements (MS 2320:2010). Department of Standard Malaysia, 19p.
[35] Young CP, Oakes DB and Wilkinson WS (1976) Prediction of future nitrate concentration in Groundwater. Groundwater, 4(6). 426-438.
[36] Lim WY, Aris AZ and Praveena SM (2012) Application of the chemometric approach to evaluate the spatial variation of water chemistry and th identification of the sources of pollution in Langat River, Malaysia. Doi 10.1007/s1517-012-0756-6.

Simplified Method for Groundwater Treatment Using Dilution and Ceramic Filter

S Musa[1], N A Ariff[1], M N Abdul Kadir[1] and F Denan[1]
[1] Faculty of Civil and Environmental Engineering, Universiti Tun Hussein Onn Malaysia, 86400 Batu Pahat Johor, MALAYSIA

E-mail: sabariah@uthm.edu.my

Abstract..Groundwater is one of the natural resources that is not susceptible to pollutants. However, increasing activities of municipal, industrial, agricultural or extreme land use activities have resulted in groundwater contamination as occured at the Research Centre for Soft Soil Malaysia (RECESS), Universiti Tun Hussein Onn Malaysia (UTHM). Thus, aims of this study is to treat groundwater by using rainwater and simple ceramic filter as a treatment agent. The treatment uses rain water dilution, ceramic filters and combined method of dilute and filtering as an alternate treatment which are simple and more practical compared to modern or chemical methods. The water went through dilution treatment processes able to get rid of 57% reduction compared to initial condition. Meanwhile, the water that passes through the filtering process successfully get rid of as much as 86% groundwater parameters where only chloride does not pass the standard. Favorable results for the combination methods of dilution and filtration methods that can succesfully eliminate 100% parameters that donot pass the standards of the Ministry of Health and the Interim National Drinking Water Quality Standard such as those found in groundwater in RECESS, UTHM especially sulfate and chloride. As a result, it allows the raw water that will use clean drinking water and safe. It also proves that the method used in this study is very effective in improving the quality of groundwater.

Keywords: Groundwater treatment, dilution, filtration.

1. Introduction

The use of groundwater in remote areas and estates were not fully explored and utilized. the groundwater quality study are scares and the result is still not widespread. Arguably, there is still no party that conducted the study and obtain detailed data on the sources of groundwater [1]. Although the research and technology to treat groundwater is existed, but it still cannot improve the quality of groundwater and reduce the concentration of pollutants entirely. Raw water from various sources typically contain various contaminants and inappropriate for domestic and industrial use. Water treatment process is necessary to treat raw water to a level that can be used for certain purposes[2].

Study area was chosen as the location for this study caused by limited groundwater used in the area. Water wells are used for daily usage purpose in RECESS. However, the cleanliness and water quality of the well is not at a good level. Based on previous studies [3], several parameters were tested to determine the level of quality in groundwater in the study area. There are some parameters that should not be treated as already meet the standards set by the Ministry of Health [4] unless sulphate, chloride and turbidity need to be treated they have very high value in groundwater.

The presence of contaminants in the area is not caused by human activity which give a direct impact on groundwater or disposal of garbage and waste that can increase the concentration of contaminants in the groundwater. A result of the analysis finds that this area is a flat area and is located near to sea about 20 km. Thus, this area found a layer of soil sediments and fossils formed due to tidal effects for centuries. Acidity levels are also influenced by the existence of a patches saltwater identified exists and trapped in the clay layer which the movement to drain are limited [3]. Indirectly, it had some impact on the presence of undesirable concentrations.

When groundwater is contaminated, it is difficult to return to the level of clean water quality due to condition of its location far below the earth's surface. Therefore, in order to control and reduce the percentage of parameters that do not meet the standard requirements for water users in Research Centre for Soft Soil Malaysia (RECESS), UTHM. This study has selected certain parameters to be analyzed without disturbing the stability of other parameters using the method of dilution and ceramic filters. The selected parameters were pH, dissolved oxygen (DO), turbidity, manganese, chloride, sulphate and fluoride. Both of these methods has been tested for its effectiveness in reducing the percentage concentration of pollutants to groundwater.

This study aims to identify the effectiveness of the dilution of rainwater process to raw groundwater, to treat raw water using ceramic filter and to ensure the effectiveness of the combination of dilution and filtration ceramic process. This experimental study was set up by using the dilution and filtration methods to groundwater samples which were taken from well at RECESS.

2. Materials

2.1 Filtration materials

Raw water was treat flows down through ceramic media, filter model were consists of activated carbon, silica sand, LECA and stone mineral and then remain as a drinking water.

2.2 Ceramic media

Before water gone through the filter model, raw water were filtered first at the upper tank with ceramic media which is medium cell to provides biological filtration for raw water process. The ceramic filter media is a man-made material and will not alter the pH balance and increase the surface area available for bacteria colonisation enhancing biological filtration. Ceramic media filtration is the use of porous ceramic (fired clay) to filter microbes orother contaminants from drinking water. Pore size can be made small enough to remove virtually all bacteria and protozoa by size exclusive, down to 0.2μm, in therange referred to as microfiltration[5].

2.3 Filter model

A filter model was used where this model can be found at the market. This model consists of four layers which is, a first layer containing activated carbon, the second layer is silica sand, followed by a third layer which is LECA (Lightweight Expended Clay Aggregate) and the bottom layer is a mineral stone. This system model with a diameter of 8 cm and a height of 10.6 cm, 6.4 cm for the first stage, second stage, third and fourth stage is 1.4 cm. The ratio between the layers is 2: 1: 1: 1 as shown in Figure 1. It functions as a filter water flows down through these layers.

Figure 1. Water filter to reduce the percentage of unwanted parameters in the water.

2.4 Activated carbon

Activated carbon, also known as Granular Activate Carbon (GAC) is the most standard media for treating water containing chemicals and impurities (coal). Activated carbon can be created in a variety of different physical forms. Large surface area gives good absorption ability and is commonly used in the raw water intake, and gravity filters. GAC has long been applied to reduce the impurities found in water. It is widely introduced in the absorption and ability to suppress the materials and chemical solution were contained in the water because carbon has a strong attraction to absorb pollutants. The process water using GAC should be applied in the treatment process after using other media such as sand in the pre-treatment process to protect the carbon from contaminated before it reach time. Impurities and chemicals can cause color, taste and bad odors in the air [6]. Macro generated in this process is the active chemical that absorbs organic and other chemicals. This GAC media is commonly used from wood carcoal < 10mm diameter in size.

2.5 Silica sand

Sand is a material that has a variety of shapes, sizes and densities of approximately uniform, so it is widely used as filter media in most water treatment plants across the country. Silica sand used is composed of fine sand with a diameter between 0.60 mm and 1.18 mm. It was chosen because of its inert and durable. It is the unique nature of the removal of acidic substances and eliminates unpleasant taste in water [2].

During the filtering process is executed, the flow of running water will force the sand layer expands and yanks separate sand particles that move and interact with each other among the particles. This action has caused impurities trapped and cleaner water can be released.

2.6 LECA

LECA (Lightweight Expended Clay Aggregate) is a lightweight ceramic with honeycomb which produced by core natural clay for a temperature 1100-1200°C in a rotary kiln. Pellets are rounded and fall from kiln in a grade of approximately 0-32 mm with an average dry bulk density of about 350 kg/m^3. This material has been sieved at several different grades that can be adjusted in various conditions.

This clay is considered as very good aggregate for use in various applications with the advantages of lightweight, high permeability, high durability and excellent sound and thermal insulation properties. It is also an environmental friendly product where consists mainly of clay which occurs naturally, not susceptible to chemical attack, decay or frost and have long lifespan. Clay pellets are also widely used in water filtration systems because of its high surface area [7]. Clays have a relatively large surface and surface forces are important in determining the composition of the clay pieces during sedimentation [8]. Clay produces pure purification processin addition to adding calcium mineral content in the water. It also helps the production of calcium ion which enhances the taste of water.

2.7 Stone mineral

Mineral rock is type of granite rock were used in diameter between 10 to 15 mm where this material is often used in the refining process with sand. Irregular shape is very suitable for the preparation of natural porosity for water drainage. It is from the type of granite taken from the quarry. These stones will liberate minerals and oxygen as well as coordinating pH in the water to reduce the alkalinity of water that can help balance the pH in the body and makes the water more palatable. Before the filtration process conducted, these stones must be cleaned first to remove all elements of impurities like dust and silt so as not to affect the results obtained.

3. Methods

This study consists of two methods used as a treatment of groundwater quality. Figure 2 was shown the steps of methods used which are dilution, filtration and lastly combination of dilution and filtration process.

3.1 Dilution Method

For dilution process, raw rainwater was used as experimental medium of groundwater wells in RECESS. Water samples from well-mixed with rainwater by using certain ratio of 1:2 (1 litre = groundwater samples and 2 litre = rainwater), 1: 4 (1 litre = groundwater samples and 4 litre = rainwater) and 1:8 (1 litre = groundwater samples and 8 litre = rainwater). Before the rainwater mixed with raw groundwater samples, parameters that contained in rainwater were tested. The content of reduction in the parameters were measured on day 1, day 2 and day 3 to obtain the percentage of water parameters before and after experiments. Then, the percentage of parameters was compared to raw drinking water quality standards for groundwater outlined by the Ministry of Health [4].

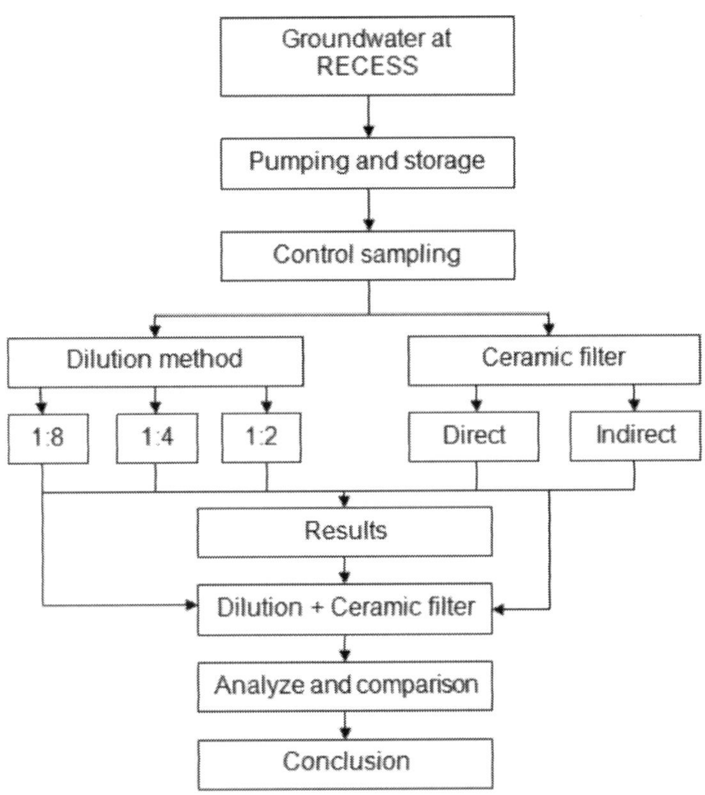

Figure 2. Flow of groundwater treatments.

3.2 Filtration Method

Beside filtration methods, this study was used a ceramic filter set were consists of ceramic media and filter model (Activated carbon, GAC, Silica sand and LECA). Experiments was conducted directly and indirectly sampling were taken from wells. Then, samples were directly flows into filter set and tested for each parameter. For indirect, samples were pumped into tank and store being let to deposited for three days and flows into filter set later. The reduction reading of parameters were measured on day 1, day 2 and day 3 to obtain the percentage of water quality improvement. Percentage of experimental parameters was compared with raw drinking water quality standards for groundwater outlined by Ministry of Health [4].

3.3 Dilution and Filtration

It is the combination of dilution using rainwater and then filtration methods. Samples were taken from the ratio of effective rainwater from dilution method before. Then, the process was conducted by dilution and then filtration processess. The observation were measured on day 1, day 2, and day 3 to obtain the percentage of changes of values. The percentage values were compared to raw drinking water quality standards for groundwater outlined by the Ministry of Health [4] and National Standard For Drinking Water Quality [9].

4. Results and discussion

4.1 Dilution and filtration methods

After dilution and filtration methods was analyzed to determine the water quality reduction of turbidity, manganese, fluoride, chloride and sulfate, the results of average comparison of those parameters are as shown in Table 1.

Table 1. Water quality measurement for dilution and filtration methods respectively.

pH				
Sample		Average	(KKM)	Remarks
Control	6.97		6.5 -. 9.0	√
Dilution	1:2	7.25		
	1:4	7.19		
	1:8	7.1		
Ceramic filter	Direct	7.56		√
	Indirect	7.73		
Turbidity				
Sample		Average	(KKM)	Remarks
Control	3.23		< 5 NTU (Class I)	√
Dilution	1:2	1.49		
	1:4	1.66		
	1:8	1.75		
Ceramic filter	Direct	1.28		√
	Indirect	1.15		

IOP Conf. Series: Materials Science and Engineering **136** (2016) 012074 doi:10.1088/1757-899X/136/1/012074

Table 1. Water quality measurement for dilution and filtration methods respectively..cont.

Sample		Average	(KKM)	Remarks
DO				
Control	6.8		8-10 mg/L	
Dilution	1:2	7.82		√
	1:4	7.87		
	1:8	7.89		
Ceramic filter	Direct	8.09		√
	Indirect	8.88		
Manganese				
Control	0.047		≤ 0.1 mg/L	
Dilution	1:2	0.042		√
	1:4	0.041		
	1:8	0.032		
Ceramic filter	Direct	0.017		√
	Indirect	0.008		
Fluoride				
Control	9.07		0.5-0.7 mg/L (Class II)	
Dilution	1:2	3.6		x
	1:4	2.25		
	1:8	1.16		
Ceramic filter	Direct	0.29		√
	Indirect	0.15		
Sulphate				
Control	1450.47		250 mg/L	
Dilution	1:2	315.61		x
	1:4	239.05		
	1:8	102.46		√
Ceramic filter	Direct	95.09		√
	Indirect	65.07		
Chloride				
Control	1518.49		250 mg/L	
Dilution	1:2	1853.8		x
	1:4	1585.3		
	1:8	1377		
Ceramic filter	Direct	1094.5		x
	Indirect	649.22		

Dilution and filtration treatments were purified the groundwater pH values. Based on these processes, the pH values of groundwater is to neutral after the treatment although, the raw groundwater is almost neutral and fresh. Then, this two processess can increase water to become more neutral until pH 7.65 as shown in Figure 3. This process flocculates the acidic water to be more neutral and natural by reaction of dilution and ceramic filtration actions.

The turbidity and manganese reduction have more effective reactions to treat groundwater quality by dilution and filtration methods until reaches the standard of groundwater and drinking water. Turbidity reduction of the groundwater can be seen from 3.23 NTU for control samples treatment to 1.63 NTU for dilution method and 1.22 NTU for filtration. Both treatments were able to reduce the turbidity although groundwater before treatment is still below the standards. After compared with the standard [4], it found that groundwater was used both treatments was better in class 1 with a minimum values. Groundwater samples before treatment showed a decrease of 1.60 and 2.01 for dilution and filtration methods respectively. These case cause by groundwater always clean from debris and any particular and also natural filtered by soil layer in the ground.

Both manganese concentrations before and after treatments was shows below the standards (<0.1 mg/l). Removal through dilution method is not so obvious that just as much as 0.038 mg/l from 0.047 mg/l. While the removal of a fairly significant effect for treatment of indirect filtration was found 0.013mg /l from 0.047 mg /l because the manganese from indirect samples was settled and helpful with sedimentation within 3 days process. The comparisons of resulted parameter was shown in Figure 3 to Figure 4.

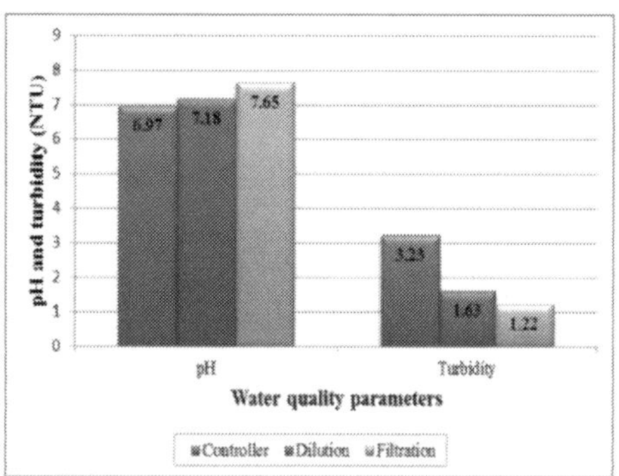

Figure 3. Average comparison of parameters pH and turbidity.

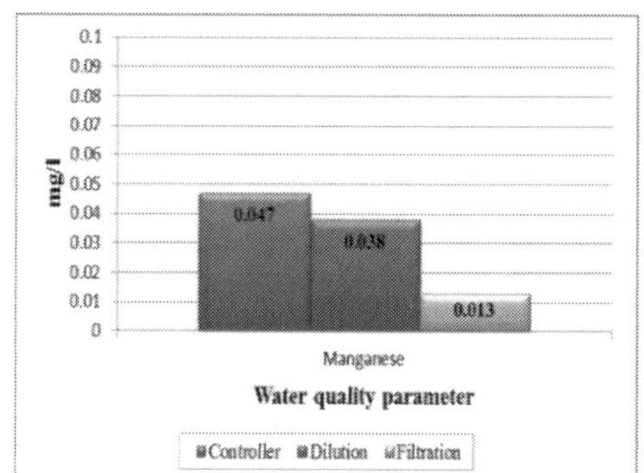

Figure 4. Average comparison of parameters manganese.

Figure 5 shows that the treatment of rainwater dilution and ceramic filtration toward changes in the level of removal of fluoride in groundwater quite stimulating results. However, the removal of fluoride in groundwater (rain water dilution) can be seen from 9.07 mg/l to 1.16 mg/l. It can be concluded that although the fluoride decreased, but the value is still exceeding the standards. This is because the content of rainwater in these experiments was not enough to reduce the parameter fluoride in groundwater. Beside, sample filtration method shows that rate of fluoride content in groundwater is better values. Ceramic filtration treatment was shown quite a big difference because the data obtained is lower than the standard groundwater (0.5-0.7mg/l). This result indicates that samples of water that has gone through a ceramic filter medium, no further treatment is required to reduce the fluoride content in the water of the earth.

Dissolved oxygen values for groundwater through rainwater dilution do not exceed the standards of the Ministry of Health, as shown in Figure 5. This result indicates that the dilution rainwater treatment was not effective in increasing the concentration of dissolved oxygen in the water. While the ceramic filtration treatment method was more effective because it can increase the concentration of dissolved oxygen in the water which is 8.49mg/l.

However, sulphate removal for dilution is not successfully comply with the standards under the standard [4] which is (250 mg/l) which is 219.04 mg/l as shown in Figure 6. Vice versa for ceramic filtration treatment in which the removal of successful compliance under the standard [4] (250 mg/l). Dilution process showed no positive impact on the reduction of the chloride. Despite the decline of the filtering process, but the chloride is still not under the standards.

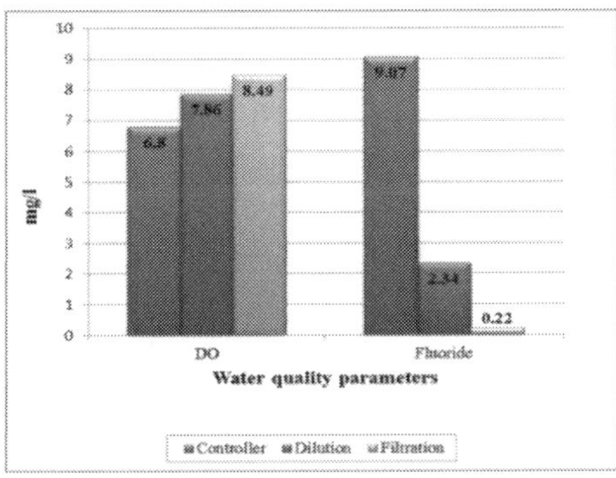

Figure 5. Average comparison of parameters DO and fluoride.

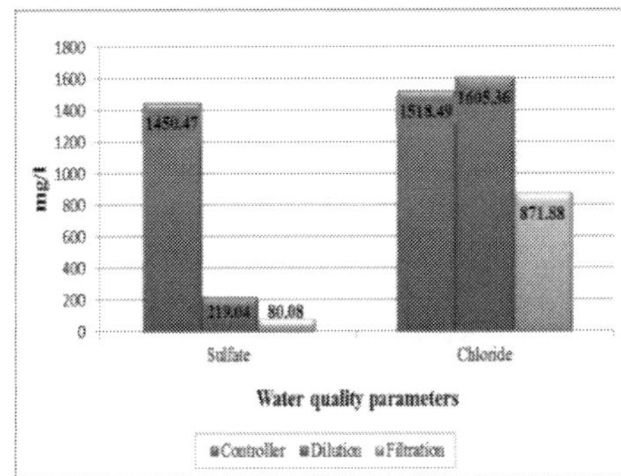

Figure 6. Average comparison of parameters sulfate and chloride.

4.2 Combination of dilution and filtration method
The best result for the combination of dilution and filtration process also was analyzed to groundwater samples in order to obtain the level of groundwater quality in higher and safer. Treatment using rain water dilution (1:8) and filtration using ceramic filter medium were combined in this study. Table 2 clearly shows the experimental results that the water quality has improved and can be eliminated for all parameters and successfully meet the standards set by the merger method of dilution and filtration. The analysis also proved that the groundwater is safe to drink directly after the it compared with the Ministry of Health [4] and the Interim National Water Quality Standard [9]. Combination of this method is more effective because the ceramic filter has been cleaned more water which was diluted with rainwater where rainwater dilution already helped to improve water quality.

Table 2. Results of the analysis combination of dilution and filtration teratments.

Parameters	Samples		Standard from MOH and INWQS	Remarks
	Controller	Dilution + Filtration		
pH	6.97	7.03	6.5-9.0 (class II)	Pass standard
Turbidity (NTU)	3.23	0.95	< 5 NTU (class I)	Pass standard
Dissolve oxygen (DO) (mg/l)	6.80	8.05	8-10 mg/l	Pass standard
Manganese (mg/l)	0.047	0.001	$\leq$0.1 mg/l	Pass standard
Fluoride (mg/l)	9.07	0.18	0.5-0.7 mg/l (class II)	Pass standard
Sulphate (mg/l)	1450.47	53.13	250 mg/l	Pass standard
Chloride (mg/l)	1518.49	245.09	250 mg/l	Pass standard

5. Conclusion

These treatments were demonstrated to identify the increasing of level of groundwater quality for entire parameters in this study. First method by using rainwater was known as dilution method shown that pH, manganese, turbidity and sulphate beyond the standard which was 57% achievement. Whereas, the second method for filtering using ceramic filter medium, the only chloride parameter that does not surpass the standard with 86%. However, third method for the combination of dilution and filtration treatments altogether was shown 100% effectively pass the standard during these treatments. Overall, this study was achieved through the three methods of treatments application. The effectiveness of treatment in this study has been shown to improve the quality of groundwater before the water supplied and distributed for use in RECESS, UTHM. This application method are also very economical, user friendly and easy to apply to all users.

Acknowledgements

The authors expresses profound gratitude to the Universiti Tun Hussein Onn Malaysia for funding this research (ERGS vot 039).

References

[1] M R Hamid 2005 Kajian Terhadap Air Bawah Tanah Kg. Sg. Ayam, Batu Pahat, UTHM, Tesis Sarjana Muda.
[2] Ariffin K S 2004 Mineral Perindustrian, EBS 425 – p. 1 of 7.
[3] Musa S, Hamdan R, Ahmad N A, Ibrahim M M, Denan F and Ariff N A 2015 Applied Mechanics and Materials, Vol. 773-774, pp. 1340-1344.
[4] National Standard For Drinking Water Quality 2000 Engineering Services Devision,Ministry of Health Malaysia.
[5] Dies R W 2003 Development of a Ceramic Water Filter for Nepal, Robert Dies, MIT. 2nd. ed.
[6] Viessmanc W J, Hammer M J, Perez E M and Chadik P A 2005 Water Supply and Pollution Control, Prentice Hall, 1st.ed.
[7] Tor Arne H, Klaas Van B, Steinar H, Ivar H, Magne M J, Mijnsbergen P G and Edda Lilja S 2000, Economic Design and Construction with Structural Lightweight Aggregate Concrete, Materials for Buildings and Structures.
[8] Mohd Yusof K N, Zainal Abidin M H and Abu Talib Z 2007, Ciri-ciri indeks dan mekanikal bagi tanah liat terstabil simen dan kapur di Semenanjung Malaysia, Persidangan Kebangsaan AWAM '07, Langkawi, Kedah.
[9] Department of Environment Malaysia, Development of Water Quality Criteria and Standards for Malaysia 1985.

Effectiveness Using Circular Fibre Steel Flap Gate As a Control Structure Towards the Hydraulic Characteristics in Open Channel

M R M Adib[1], A R M Amirza[1], T Wardah[2] and A Junaidah[2]
[1] Faculty of Civil and Environmental Engineering, Universiti Tun Hussein Onn Malaysia, 86400 Batu Pahat Johor, MALAYSIA
[2] Flood Marine Excellence Centre, Institute for Infrastructure Engineering & Sustainable Management, Faculty of Civil Engineering, Universiti Technology MARA, 40450 Shah Alam, MALAYSIA

E-Mail: adib@uthm.edu.my

Abstract. Hydraulic control gate structure plays an important role in regulating the flow of water in river, canal or water reservoir. One of the most appropriate structures in term of resolving the problem of flood occured is the construction of circular fibre steel flap gate. Therefore, an experiment has been conducted by using an open channel model at laboratory. In this case, hydraulic jump and backwater were the method to determined the hydraulic characteristics of circular fibre steel flap gate in an open channel model. From the experiment, the opening angle of flap gate can receive discharges with the highest flow rate of 0.035 m^3/s with opening angle was 47°. The type of jump that occurs at the slope of 1/200 for a distance of 5.0 m is a standing jump or undulating wave. The height of the backwater can be identified based on the differences of specific force which is specific force before jump, F_1 and specific force after jump, F_2 from the formation of backwater. Based on the research conducted, the tendency of incident backwater wave occurred was high in every distance of water control location from water inlet is flap slope and the slope of 1/300 which is 0.84 m/s and 0.75 m/s of celerity in open channel model.

Keywords: Circular fibre steel flap gate, hydraulic jump, backwater, opening angle, flooding.

1. Introduction

Hydraulic engineering is the application of the principles of fluid mechanics to manage the problem by gathering, storing, controlling, transportation, settings, measurements and water consumption [1]. Hydraulics is also a scientific study that related to the flow of a fluid including the flow which through the closed channels (pipes) and the area of open channel. Open channel can be defined as a channel that flow with a free surface of the boundary surface vulnerable into air [2].

Flap gates is a convenient system can be mounted on the end of sewer or above the wall of drain which is to allow the water stream only in one direction [3]. Top hinged flap gates is a water control structure that is used in tidal structure to prevent flooding in the shore areas due to high tide of sea water or surge flooding to allow water flow into the sea during high tides [4]. Therefore the effectiveness of a control gate structure can be determined against the characteristics of hydraulic, hydraulic jump and backwater that acting on the structure.

2. Hydraulic Characteristics

Tidal control gates can be interpreted as an opening which allows water flows through it in only one direction [5]. The circular fibre steel flap gate is one of the various type (Figure-1). Furthermore, the advantages of the circular fibre steel flap gate over the rest are their easily measured and controlled flow, their simple design, high flow discharge coefficient and their less expensive costs in large scales [6]. Nevertheless, the characteristic of

IOP Conf. Series: Materials Science and Engineering **136** (2016) 012075 doi:10.1088/1757-899X/136/1/012075

Fiber reinforced plastics (FRP) is similar with fibre steel since they have excellent durability and water or corrosion resistance [7]. Hydraulic assessment on the control flap gate was to study the pattern of the pressure distribution on the flap gate [8]. The different between static force and real energy was increasing as soon as the gate opened. This reduction is due to the power that based on the fact which water comes out under the gate has zero atmospheric pressure (Figure-2) [9].

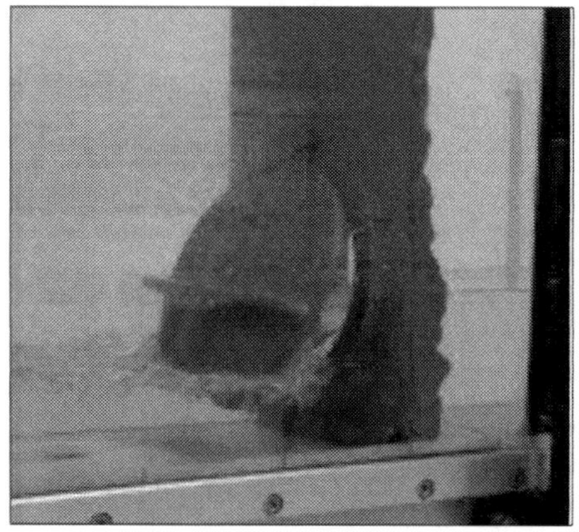

Figure 1. Fibre steel flap gate modeling.

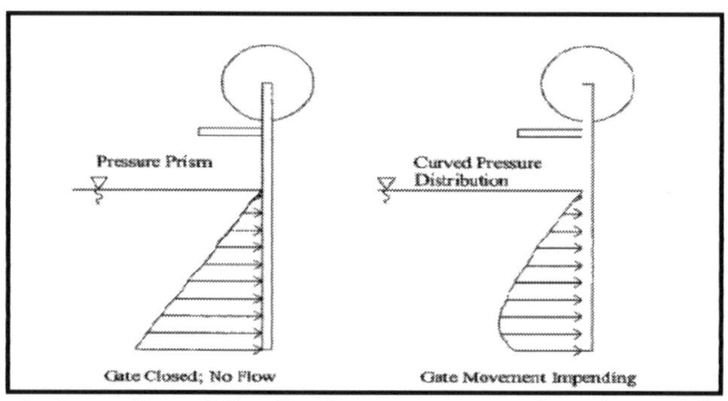

Figure 2. Operation of distribution pressure on faceplate of flap gate.

Hydraulic jump occurs when there is a friction between the upstream and downstream of a channels. They form transiently when deeper fluid gets pushed into a shallower and stagnant body of fluid [10]. Besides that, hydraulic jump also can be occurring when the supercritical flow met subcritical flow at a sufficient depth. Hydraulic jumps can also occur as steady phenomenon (in the lab frame) when a super-critical fluid ($Fr > 1$) transitions to sub-critical ($Fr < 1$) [11]. Hydraulic jumps have been extensively studied because of their frequent occurrence in nature and their use as energy dissipators in outlet works of hydraulic structures [12]. Two types of depth y_1 and y_2 is called as a depth classifiers ratio (Figure-3). The incidents of hydraulic jump are reviewed by using the principles of momentum.

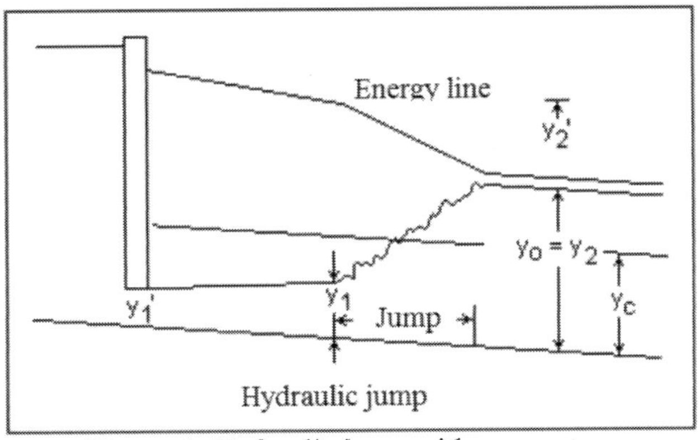

Figure 3. Hydraulic jump with parameters.

Backwater is an increase level of water greater than normal (Figure-4). However, differences in circumstances of hydraulic system able to change normal conditions of water movement that causes the water move in the opposite direction [13]. The amount of backwater caused by piers depends mainly on their geometric shape, their position in the stream, the flow rate and the amount of channel blockage [14]. Under the backwater effect of a hydraulic structure, downstream river course can have a flowless regime in which ponding can occur [15]. This is the case especially when there is a channel with very mild slope, an irrigation channel affected by gate operations or a diversion weir on the downstream of a dam [15].

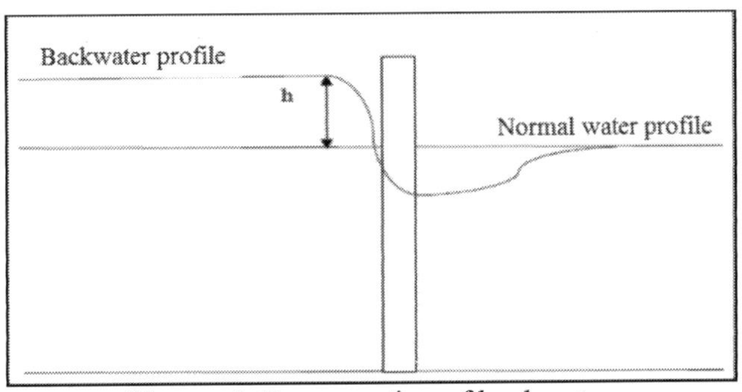

Figure 4. Cross-section of backwater.

3. Model development at open channel structure

This research had been prepared to ensure the experiment work smoothly. In this studies, the preparation of experiment is focusing on the laboratory test to analyzes the impact of backwater, hydraulic jump and hydraulic characteristic by using circular fiber steel flap gate as a control gate. During the experiment, the suitable slope surface can be determined.

The first stage that involved in this research is to find out the comprehensive understanding of the title given by literature review. The literature review that could be considered on this research is based on the principle of backwater flow, open channel concept, celerity toward flap gate in open channel, hydraulic jump and hydraulic characteristic.

Therefore, the preparation for the laboratory test must be find out and will be handle the measurements of the parameter of depth before and after jump occur, time of wave in one cycle and opening slope during experiment. The data collected from the laboratory test will be analyzed.

3.1 Structure model

Installation of the fibre steel flap gate can be described based on the prototype model and the function of the flap gate can be illustrate on (Figure-5).

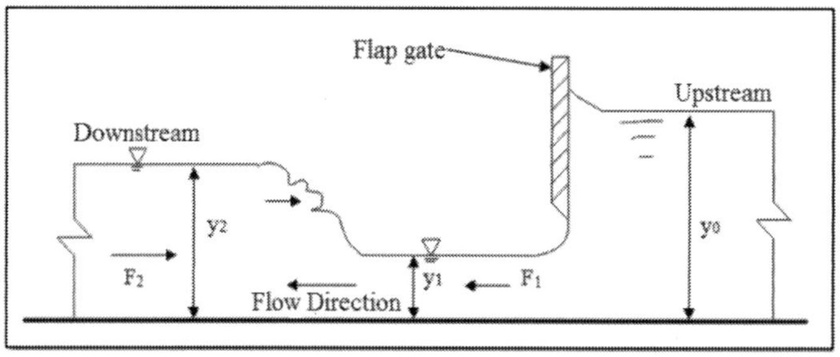

Figure 5. Prototype of model.

Water control gate that had been studied is a circular gate made of fibre steel reinforced material which has a hinge at the top. The dimensions of the control gate is shown (Figure-6). These circular flap gate provides greater water control due to the area in certain circumstances. As a result, opening force exerted on the flap gate was smaller [16].

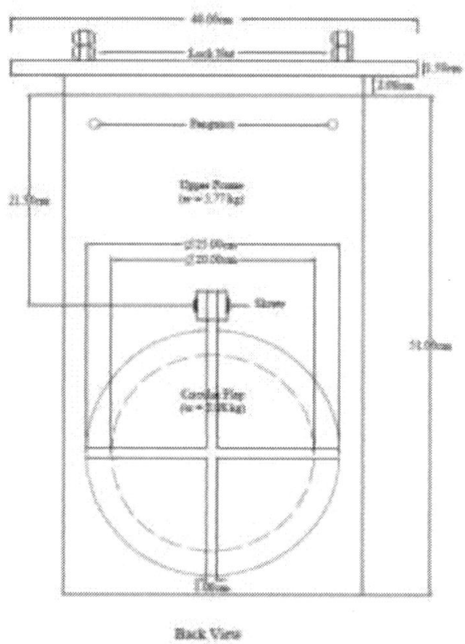

Figure 6. The circular fibre steel flap gate with dimension.

A wave that generate from backwater can define the velocity of celerity. There are some major formulas that are used compatible with the velocity of the celerity. Celerity is defined which involve of backwater wave toward the control gate such as:

$$C = \frac{L}{t} \qquad (1)$$

Where C and L is represent celerity of wave and length of the wave from crest to the crest, and t is the cycle time from crest wave to the another crest wave.

Specific force before and after jump that occur during the experiment based on the discharges value that is manipulated variation which is:

$$F_{1,2} = A_{1,2}h_{1,2} + \frac{Q^2}{gA_{1,2}} \qquad (2)$$

Where F and A is represent specific force and area of the circular gate, and h is the depth of the centroid point from surface to the base. However, Q and g represent discharge of the flow and gravity. (1 represent for before jump and 2 for after jump).

Specific energy is defined as energy flow per unit Joule Newton in any section of the channel calculated as the vertical distance between channel base and energy line. The equations involved are:

$$E = y + \frac{q^2}{2gy^2} \qquad (3)$$

Where E and y represent specific energy and depth of surface water to the base. Beside that, g and q represent gravity and discharge per width of the channel.

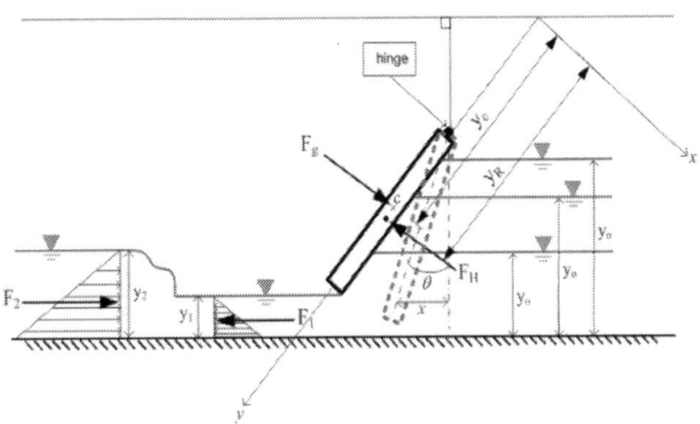

Figure 7. Schematic diagram of circular fibre steel flap gate.

3.2 Hydraulics performance

From experiment conducted, experiment data in raw represent basic parameter which been discussed in structure model. Raw data in schedule manner in Microsoft Excel for easier process of analysis shape as graphic graph. Any raw data from experiment that has been done will be compile using excel formula to analyse the graphic graph.

Therefore, from the data result. Distinctions in terms of the slope will be identified. The values of 16 different flow rates in between 0.005 m³/s and 0.035 m³/s were determined in this experiment with distance of control gate from water inlet in 5.0 m. Every flow rates will be test using different slope of channel which is flat, 1/300, 1/200 and 1/100.

3.3 Slope of flap gate

Figure-8 indicates that the relationship between opening angle of gate with discharge. Therefore, at the maximum discharge the highest opening angle of circular flap gate from the upstream channel was 47°. In brief, the opening angle of gates obtained within the range of 14.4° and 47°. The slope of zero indicates a greater value of opening angle compated to the others.

IOP Conf. Series: Materials Science and Engineering **136** (2016) 012075 doi:10.1088/1757-899X/136/1/012075

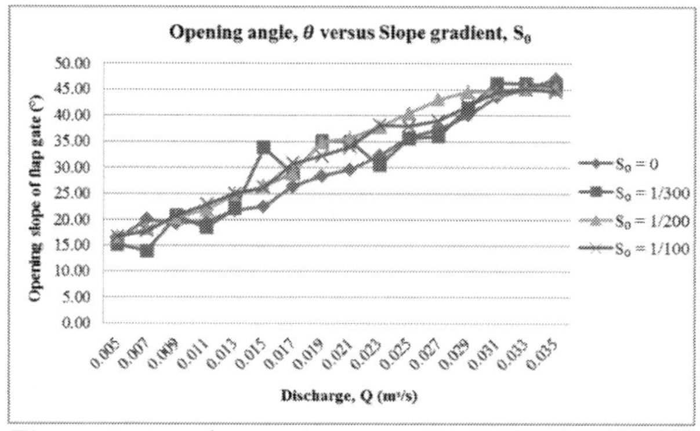

Figure 8. Opening angle versus slope gradient graph.

From the graph clearly shows that the greater opening of angle when the flow rate imposed higher. Hence, the slope gradient can influence the opening circular gate. This problem due to the mass and fibre steel flap gate balancing weight besides water pressure exerted on flap gate.

3.4 Hydraulic jump

Figure-9 indicates that depth of flow versus specific energy with different slope of flat, 1/300, 1/200 and 1/100 for distance 5.0 m. For the critical depth flow and critical specific energy are also included for facilitate the process of analyzing the graph of the depth of flow and specific energy. For the flat gradient, energy curve line does not applicable at all flow rates. This is because the depth of flow does not reach the critical depth flow point, y_c. The value of specific energy to flow in point y_1 is the lowest compared to y_0 and y_2 value. Hence, the hydraulic jump has occurred at the depth of y_1 due to the undulating wave which Froude number is between $1.0 < Fr < 1.7$. Thus, less energy is required through a stream.

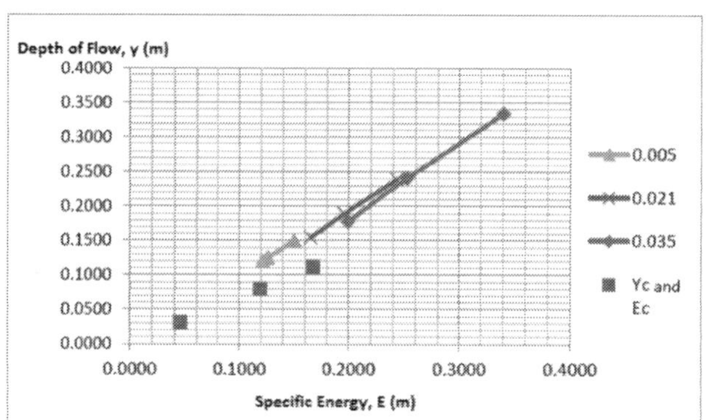

Figure 9(a). Graph Depth of Flow versus Specific Energy for flat slope.

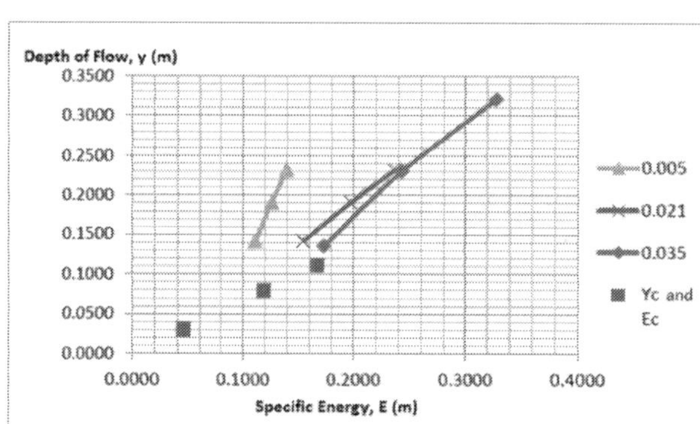

Figure 9(b). Graph Depth of Flow versus Specific Energy for 1/300 slope.

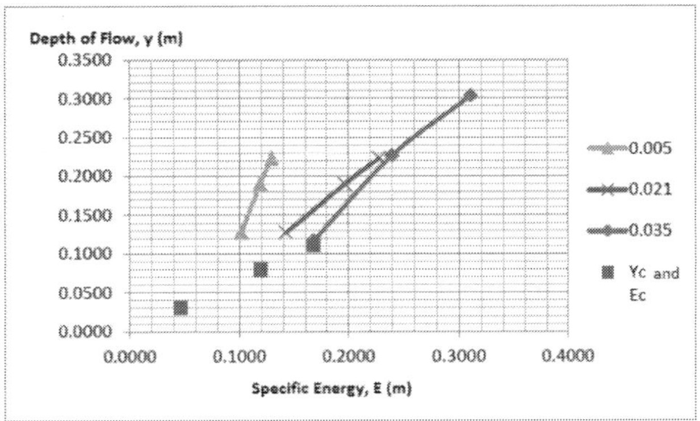

Figure 9(c). Graph Depth of Flow versus Specific Energy for 1/200 slope.

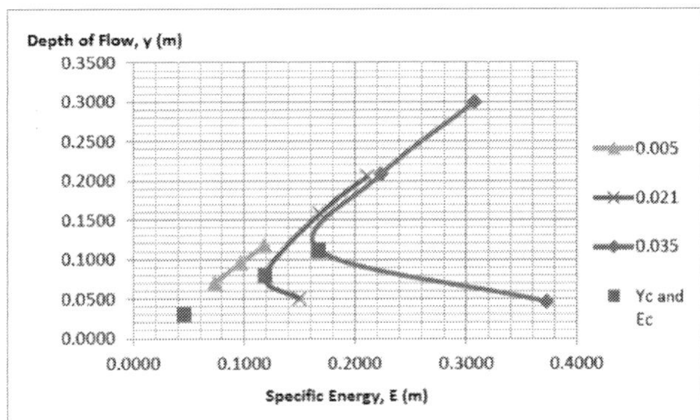

Figure 9d). Graph Depth of Flow versus Specific Energy for 1/100 slope.

Meanwhile, different situation for the slope of 1/100 because flow rate 0.021 m^3/s and 0.035 m^3/s have made an energy curve line occurred. The value of specific energy to flow in point y_1 is the lowest compared to y_0 and y_2 value. Hence, the hydraulic jump has occurred at the depth of y_1 due to the oscillating jump which Froude number is between $2.5 < Fr < 4.5$. Thus, less energy is required through a stream.

3.5 *Celerity waves*

Figure-10 indicate that difference slope occurs at a distance of 5.0 m, which resulted the celerity wave but at slope of 1/300 wave shown consistently from the flow rate of 0.005 m^3/s to 0.035 m^3/s. A celerity wave in a consistent state at the beginning discharge for flat slope but then decline began in the flow rate of 0.029 m^3/s. However, the slope of 1/200 and 1/100 decreased starting at 0.021 m^3/s for slope 1/200 and for slope 1/100 is at 0.017 m^3/s of flow rate. Thus, on the distance of 5.0 m from the upstream was discovered that the slope 1/300 a celerity wave can be generated.

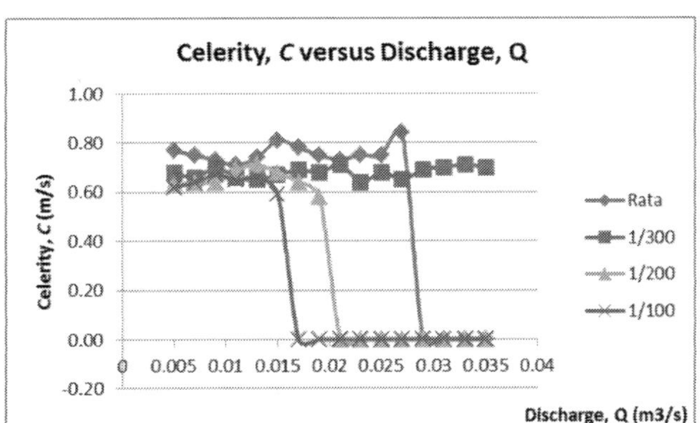

Figure 10. Graph Celerity versus Discharge on distance 5.0m.

3.6 Probability of backwater

The analysis of specific force was divided into two parts which are analyzed specific force before jump, F_1 and specific force after jump, F_2. The purpose of analyzed specific force was to review the probability of backwater wave occurring whether in weak or bumpy conditions

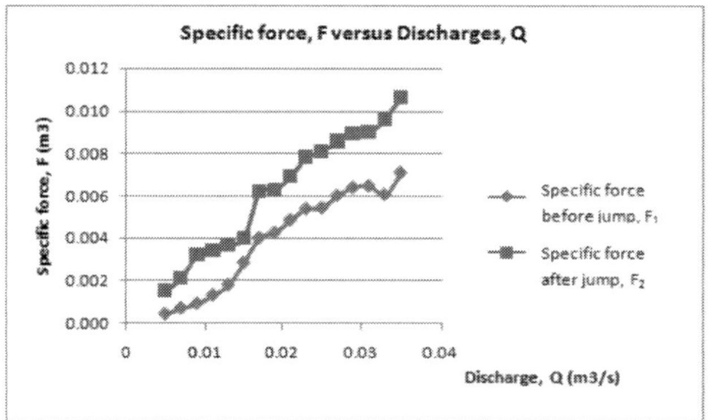

Figure 11. Specific force versus Discharge on different jump situation.

Since the analysis has been performed, it is found at the distance of 5.0 m was a suitable location for built control gate from the upstream channel. Thus, at the maximum discharge which is 0.035 m^3/s, the comparison of the differences specific force before and after jump during the experiment can determined the depth of the wave can be generate which is 0.004 m^3.

4. Conclusion

Therefore, the location to be constructed the control gate in order to reduce the risk of hydraulic jump and backwater occurring in the upstream channel is at the distance of 5.0 m with the slope is less than 1/300. While, clearly stated that the opening angle is greater if the flow rate imposed even greater. It shows that the flap gate can receive discharges with the highest flow rate of 0.035 m^3/s with opening angle was 47°. The type of jump that occurs at the slope of 1/200 for a distance of 5.0 m is a standing jump or undulating wave due to the Froude number is between $1.0 < Fr < 1.7$. In additional, celerity wave can be generated only occur in the slope of 1/300 for this distance and cannot form the celerity wave for slope less than 1/300. Moreover, the differences of specific force before and after jump were 0.004 m^3 for a maximum flow rate of 0.035 m^3/s.

Acknowledgement

The authors gratefully acknowledge with sincere thanks to Zulhelmi Yunos and Rahmawati Abdullah who worked diligently and for their selfless and consistent efforts to get through this task. We also express our sincere appreciations and gratitudes to the Department of Water And Environmental Engineering, Faculty of Civil and Environmental Engineering, UTHM, for or allowing the experiments conducted in the laboratory and also provide guidance on the use of open channel structure to prevent errors from occur.

References

[1] Prasuhn. Alan L (1987). *Fundamental of Hydraulic Engineering*, New York; Holt, Rirehart and Winston, 1987.
[2] Chow V.T. (1959). Open Channel Hydraulics, McGraw Hill International, 1959.
[3] Rampano, B (2009). Water Control Structures: Designs for Natural Resource Management on Coastal Floodplains. NSW Department of Industry and Investment, 2009.
[4] Lewin, J. (2001). *Hydraulic Gates And Valves In Free Surface Flow And Submerged Outlets.* Second Edition. London. Thomas Telford Publishing.
[5] Smith, D.J., & Kenneth GL Simpson (2011). Hydro-electric Dam and Tidal Gates, "Safety Critical Systems Handbook", Third Edition, 2011.
[6] Saeid Shabanlou, Eghbal Khorami, Ahmad Rajabi (2013). The Effect Of Various Upstream Arches Of Crest Of The Circular Crested Weir On Hydraulic Parameters., *Journal of flow measurement and Instrumentation.*, Vol 32., Aug 2013., Pages 103-106.

[7] Tomiyama, T., & Nishizaki, I. (2006). Applicability of fiber reinforced plastics to hydraulic gates. In *Third International Conference on FRP Composites in Civil Engineering (CICE 2006) Dec*, 2006, pp. 13-15.

[8] Raemy, F., and Hanger, W.(1997). Flap Gate for Hydraulic Head Control. *Proc., 27th Congr. Of the Int. Assn. for Hydr. Res., ASCE, Reston, Va.*, 1997, pp. 320-325.

[9] Burt, C. M., Angold, R., Lehmkuhl, M., & Styles, S. (2001). Flap gate design for automatic upstream canal water level control. *Journal of irrigation and drainage engineering*, *127*(2), 2001, pp. 84-91.

[13] Christina W.Tsai (2005). Flood routing in mild-sloped rivers-wave characteristics and downstream backwater effect. *Journal of Hydrology*. 2005, Volume 308 (*issues 1-4*), pages 151-167.

[10] Peregrine, D. H (1966). Calculation of the development of an undular bore, *J. Fluid Mech.*, vol. 25, 1966, part 2, 321330.

[11] Ratul Dasgupta, & Gaurav Tomar. (2015). Viscous Undular Hydraulic Jump of Moderate Reynolds number Flows, *Procedia IUTAM*, Volume 15, 2015, pages 300-304.

[12] Abbaspour, A., Davood Farsadizadeh, Mohammad Ali Ghorbani.(2013). Estimation of hydraulic jump on corrugated bed using artificial neural networks and genetic programming. *Journal Water Science and Engineering*., Vol 6, Issue 2, 2013, pages 189-198.

[15] Ceyhun Özçelik and Mustafa Doğan (2009). Investigation of wave attenuation mechanism under the downstream backwater effect.' *Journal of Flow Measurement and Instrumentation'*. Oct. 2009, Volume 20 (*issues 4-5*), Pages 180-188.

[14] Randall J. Charbeneau and Edward R. Holley(2001). Backwater effects of bridge piers in subcritical flow. *In Center for Transportation Research, The University Of Texas at Austin, Oct, 2001*, pp. 1-4.

[16] Schuster. J. C.(1951). Underdrain protection of 3 ½ inch reinforced concrete canal lining. Report HYD-309, United States Bureau of Reclamation, Pages 13, 1951.

Modeling the Influence of River Flow and Salt Water Intrusion in the Terengganu Estuary, Malaysia

H L Lee[1,2], F Tangang[1], M R Hamid[2], Y Benson[2] and M R Razali[2]
[1] The National University of Malaysia, Faculty of Science and Technology
[2] National Hydraulic Research Institute of Malaysia

E-Mail : hllee@nahrim.gov.my

Abstract. Salinity intrusion is a major concern when the freshwater extraction station is located in the estuary. This paper attempt to predict the salt intrusion length in the upper stretch of estuary, by applying different magnitudes of freshwater discharge at the river regime. The integrated two dimensional hydrodynamics model associated with advection dispersion model was performed to investigate the salinity intrusion. The model was well calibrated and verified by the measured data undertaken during dry season. The maximum salt intrusion length to the threshold of salinity density is 1.00 ppt on the existing condition was predicted at 9.97 km from the river mouth. Moreover, with the magnitude of 100.00 m^3s^{-1} and 30.00 m^3s^{-1} freshwater discharges at the upstream boundary (Kpg Tanggol), it was predicted the maximum salt intrusion length was 11.84 km and 21.41 km, respectively, from the river mouth. Therefore, it was determined the minimum freshwater discharge of approximately 100.00 m^3s^{-1} is required at the Kpg Tanggol river gauging station, in order to maintain the acceptable salinity levels at the Pulau Musang freshwater pump house. However, the actual water discharge at the Kpg Tanggol boundary station should be higher, since the minimum discharge does not take into consideration the amount of water extraction by the Pulau Musang and SATU pump stations. Further analysis is required to execute the consequences of water extraction toward the salinity intrusion in the Terengganu estuary that coupled with projected sea level rise.

Keywords: numerical model, salinity intrusion, diurnal tide, high water level, hydrodynamic.

1. Introduction

River discharge, tidal forces and the estuarine geometry are the main key parameters determining the salt water intrusion in an estuary [1]. Decades ago, to meet the domestic demands, freshwater reservoirs are constructed; however it treated the saltwater intrusion during the dry season and enhanced the tidal excursion length further upstream [2]. The contaminated freshwater by the salt water from the sea, is no longer useable and could interrupt the function of water intake stations that located in the estuary for drinking or agricultural purposes [3].

Therefore, it is crucial to identify the minimal freshwater discharge's quantity from the reservoir into the estuary, to ensure all the facilities particularly water intake stations located at the upper stretches of the estuary, able to function well without interrupted by saline water. In the water resources plan and management strategies, the numerical mathematical modeling tools are often being used to derive hydrodynamics behaviors and advection dispersion characteristics for salinity intrusion in the estuary [4,5].

Sea salinity intrusion is major concerns in the study area, as there are freshwater extraction pump houses located in the estuary. Surprisingly, the salt intrusion length into upstream of estuary still not yet been scientifically investigated by any researcher at the study area. Therefore, in order to examine the salt intrusion in the Terengganu estuary, the mathematical numerical model was used. The simulation of two dimensional depth averages hydrodynamic (HD) and advection-dispersion (AD) module was performance, which associated with boundaries condition that driving forces into the model. The objective of this study is to execute hydrodynamic regime and salinity intrusion in the Terengganu estuary for identifying the minimum freshwater

IOP Conf. Series: Materials Science and Engineering **136** (2016) 012076 doi:10.1088/1757-899X/136/1/012076

discharge requirement at the river gauging station at Kpg Tanggol, for maintaining an acceptable salinity levels during highest high water spring tide near the water intake at the Pulau Musang pump houses.

2. Material and methods

2.1 Study area

The Terengganu estuary is located at 103.13^0 E and 5.34^0 N, with the river basin is approximately 4,600 km^2 (Figure-1a). The main tributary is the Terengganu river with the river length of 65 km. Upper catchment of the Terengganu river encompasses of hydroelectric power freshwater reservoir namely Kenyir Dam, while the middle catchments are engaged with plantation activities and the lower catchment with an urbanized and industrialized setting close to the estuary draining into the South China Sea. Analysis on 11 years river discharge records from river gauge station located at the Kpg Tanggol (data obtained from the Department of Irrigation and Drainage Malaysia (DID)), which is located in the Terengganu river and approximately 30 km upstream from the river mouth, the annual freshwater was calculated approximately 10 X 10^9 m^3 year^{-1}, while the maximum, average and minimum freshwater discharge was analysed as 2,808 m^{3}s^{-1}, 266.2 m^{3}s^{-1} and 21 m^{3}s^{-1}, respectively. Higher freshwater runoff is discharged at the beginning and end of the year, with smaller volumes of river runoff is recorded from June to September. The study site is experiencing the wet season during end and early months of the year, where it is associated with the northeast monsoon [6,7,8], the average monthly rainfall during the wet season is ranging from 200 mm to 600 mm. Meanwhile, the dry season occurs from April to September with the average monthly rainfall ranging from 100 mm to 125mm. The tidal fluctuation at the study site is controlled by the diurnal tide, with the spring and neap tidal range approximately 3.07 and 0.7 m, respectively.

2.2 Data Collection

Bathymetry measurements on the shoreline and in the estuary were carried out by using Acoustic Doppler Current Profiler (ADCP) in both longitudinal and cross-sectional transects during the survey (Figure-1d). Water column current profiles were measured in the navigation channel located in the river mouth, by using a Nortek Aquadopp current profiler with a transducer head of 2 MHz during data campaign at July 2011 (Figure-1c). Meanwhile salinity levels at the Shah Bandar jetty and fisheries jetty was measured with XR-620 CTD sensor, the sensor were placed in a stainless steel pole and deployed at the bed layer at those stations from Dec 2011 to Sept 2012 (Figure-1c). All the equipment were calibrated and verified before been deployed at the site.

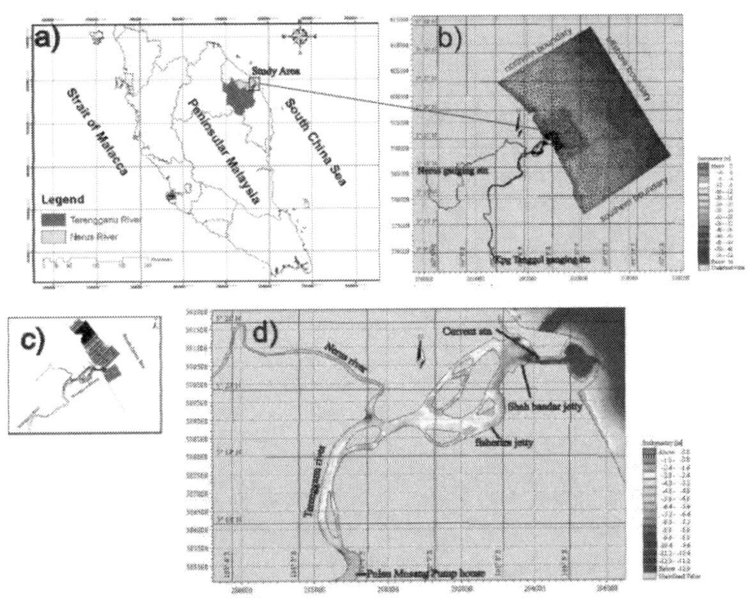

Figure 1. (a) Study area, (b) computational flexible mesh, (c) bathymetry survey data during data campaign and (d) bathymetry condiction in the numerical simulation model.

IOP Conf. Series: Materials Science and Engineering **136** (2016) 012076 doi:10.1088/1757-899X/136/1/012076

2.3. *Hydrodynamic Module*

The integrated two dimensional models DHI's Mike 21 HD (hydrodynamics) and AD (advection dispersion) flexible mesh finite volume method was used to investigate the salinity intrusion in the estuary. The offshore bathymetry for the model was generated by digitizing the Malaysian Naval chart No: RMAL 654, while the stretches in the nearshore and the Terengganu estuary were measured during the bathymetry survey in the year 2011 and 2012 (Figure-1d). All the topographical data of the study area has been converted to local datum mean sea level (MSL) for HD and AD models uses. The water depths in the estuary vary from 1.57 m to 12.77 m from the MSL (Figure-1c). The triangulation pattern of mesh was kept from dense to course meshes from estuary to deep sea, with the total number for surface meshes were 5906 (Figure-1b). There were five boundaries, namely offshore boundary associated with two lateral boundaries north and south, to transfer tidal fluxes from offshore to the study area. While another two boundaries are located at the freshwater river gauging station at the Kpg Tanggol and Sg Nerus station respectively, for inducing river discharge fluxes from highland into estuary. Those fresh water inputs are obtained from river gauged station has been incharged by local river authority DID (Figure-1b). The hydraulic simulation is based on the two dimensional incompressible Reynolds averaged Navier-Stokes equations, subject to the assumptions of Boussinesq and of hydrostatic pressure solution [9].

3. Result And Discussions

3.1 *Model calibration: current speed, direction and water levels*

Model calibration is essential part in mathematical modeling and generally calibrated with the dynamic parameters, such as current speed, directions, salinity levels and water levels [10]. This study has calibrated the water level, current speed and current direction that were measured in the navigation channel during the data campaign in July 2011. The comparison of measured (dotted black color) and simulated (solid red in color) for the tidal currents, current directions and water levels are given in Figure 2a, 2b and 2c respectively, evidence of the simulated data was good agreement with the measured data. As, the measured location of current speed and direction was in the navigation channel (Figure-1c), the highest tidal and river current was observed during spring tide, with tidal current magnitude was higher than river current. This phenomenon, perhaps allows a better penetration of the saline water further upstream during high water spring (HWS) compared to neap tide. This finding agreed well with the empirical analytical formula constructed by the [11] for maximum HWS salinity intrusion.

3.2 *Hydrodynamic model simulations*

The integrated HD and AD model simulation was simulated for the existing condition of bathymetry. It was observed the study site dominated by diurnal tide, where in daily, approximately 30% of the tidal period was in flood tide and the remainder 70% of the tidal period was in ebb tide (Figure-3a and 3b). During the flood tide, offshore water bodies are flushed towards southern stretches of the shoreline and saline water is intruded into the Terengganu estuary (Figure-3a). Meanwhile during ebb tide, the brackish water from the estuary is flushed out from estuary and the entire coastal water bodies are flowed northwards with a longer duration of tidal period, approximately 17 hours (Figure-3b).

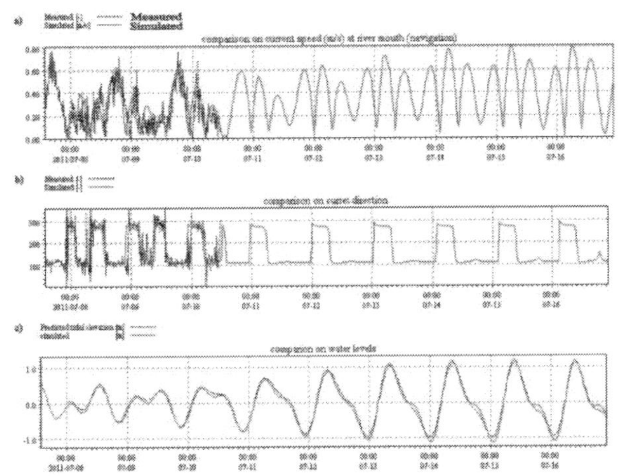

Figure 2. The solid red line represents the simulated results, while the solid black line illustrated the measured data. (a) is calibration for current speed, (b) is current direction and (c) is water leves at the study area.

Results on the maximum and mean current speed in a tidal cycle, which covered both neap and spring tide is shown in Figure-4. Maximum current speed was observed at the few stretches in the estuary, particularly at the navigation channel, the Manir village and the Kpg Cina areas with the magnitude approximately 0.8 ms^{-1} (Figure-4a). Meanwhile the highest mean current speed was defined in the navigation channel and the Manir village with magnitude of 0.3 to 0.4 ms^{-1}, in general the dominant mean current speed in the entire estuary was noted vary from 0.1 to 0.3 ms^{-1} as presents in Figure-4b.

3.3 AD model simulations

Analysis on the spatial salinity concentration at two fixed stations located at the Shah Bandar jetty and fisheries jetty as shown in Figure-5a and 5b respectively, illustrates the measured and simulated results for both stations were presented in good agreement with each other. The concentration of salinity was higher during spring tide compared to neap tide, this finding was consistence with current speed results (Figure-3a), where magnitude of tidal current was recorded higher during the spring tide compared to neap tide. Thus, this phenomenon enables saline water to penetrate further upstream and result in the high concentration of salinity inland (Figure-5a). This correlation has demonstrated that the model is a reasonably efficient tool for predicting salinity intrusion in the study area.

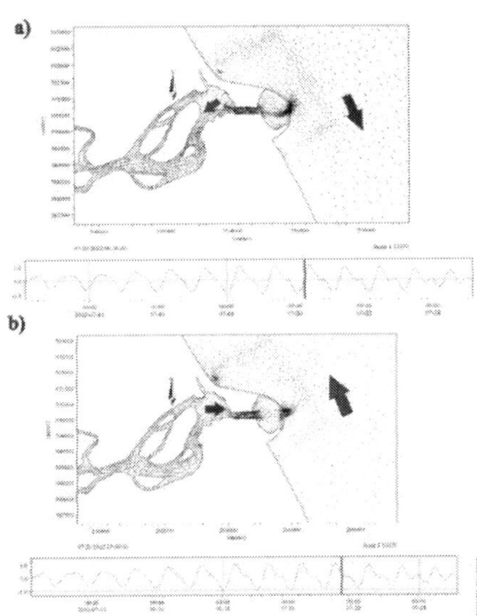

Figure 3. Current flow condition during flood tide (a) and ebb tide (b).

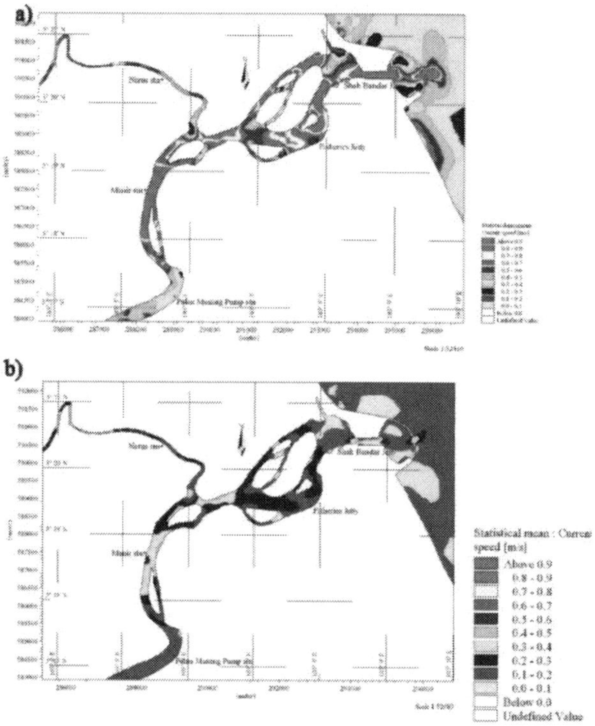

Figure 4. Maximum a) and mean b) current flow at the study area.

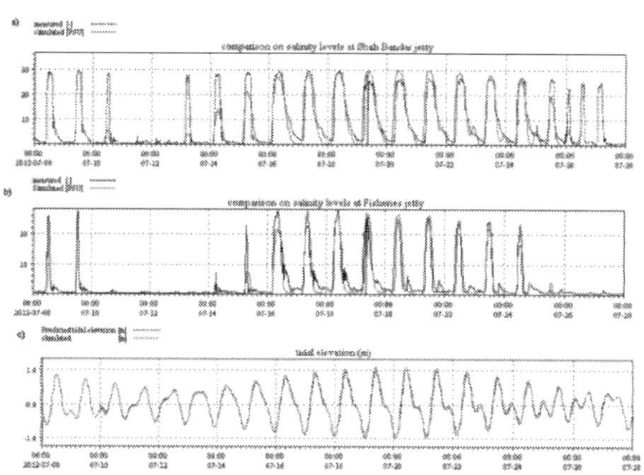

Figure 5. calibration for salinity at the Shah Bandar Jetty a) the fisheries jetty b) and water levels fluctuation c). The solid red line represents the simulated results, while the solid black line illustrated the measured data.

To predict the maximum salinity intrusion into upstream reaches of the study site, simulation has taken into consideration all the offshore boundaries that are subjected to the tidal during the highest high water spring (HHWS), while river discharge boundary condition for the Nerus river station was implied with existing condition during the HHWS. Nevertheless, for the Kpg Tanggol boundary condition, it was subjected to three scenarios, namely existing condition during HHWS, constant river discharge with 30 m^3s^{-1} and 100 m^3s^{-1} respectively. The predicted maximum salinity intrusion to the threshold of 1.00 ppt on the existing condition was predicted at 9.97 km from the river mouth (Figure-6a). Similarly, Savenije (2010) reported the typical length of the tidal excursion for diurnal tide is 10 km. Perhaps this is due to shorter duration of high water and longer period of low water, 30% and 70% (Figure-3) of daily tidal period respectively, which might contribute to shorter length of tidal excursion. Higher concentration of salinity was identified from the river mouth to the fisheries jetty with concentration approximately 30.00 ppt. Salinity levels from the fisheries jetty to junction of the Nerus river was vary from 15.00 to 28.00 ppt. Finally, from the the Nerus junction to the Manir village was predicted 1.00 to 15.00 ppt. Salinity concentration of 1.00 ppt halocline was approximately 3.03 km downstream from the water withdrawal station for drinking and agricultural purposes at the Pulau Musang pump house. Thus, under the current situation, the withdrawal station is safe to extract the water from the Terengganu river for daily activities.

As analyzed previously, the minimum freshwater discharge was detected as low as 21 m^3s^{-1}. However, by performing freshwater runoff at the Kpg Tanggol station boundary condition with constant 30 m^3s^{-1} magnitude of river runoff coupled with the highest high spring tide, the salinity intrusion able to penetrate 21.41 km from the river mouth. When this phenomenon takes place, it could interrupt the water intake station located at the Pulau Musang pump house due to high contamination of saline water, approximately 30.0 ppt (Figure-6b).

Result by applying constant freshwater discharge, 100 m^3s^{-1}, at the freshwater boundary in the Kpg Tanggol station, illustrated the halocline of 1.00 ppt was penetrated further inland with intrusion length was 11.84 km (Figure-6c). The Pulau Musang Pump house is predicted 1.16 km downstream of 1.00 ppt halocline; it is secured from salt contamination and is saved to extract river water for drinking and agricultural use. According to Shaha and Cho (2009) the recommended threshold values of salinity for drinking and agriculture purpose are less than 0.50 and 1.00 ppt respectively. It should be noted that freshwater discharge 100 m^3s^{-1} was the minimum requirement of the river runoff at the Kpg Tanggol river gaging station, the actual water discharge upstream of the water intake should be higher, which should takes into considering the water extraction by the water intake stations located at the downstream and associated with the sea level rise factors [12,13,14].

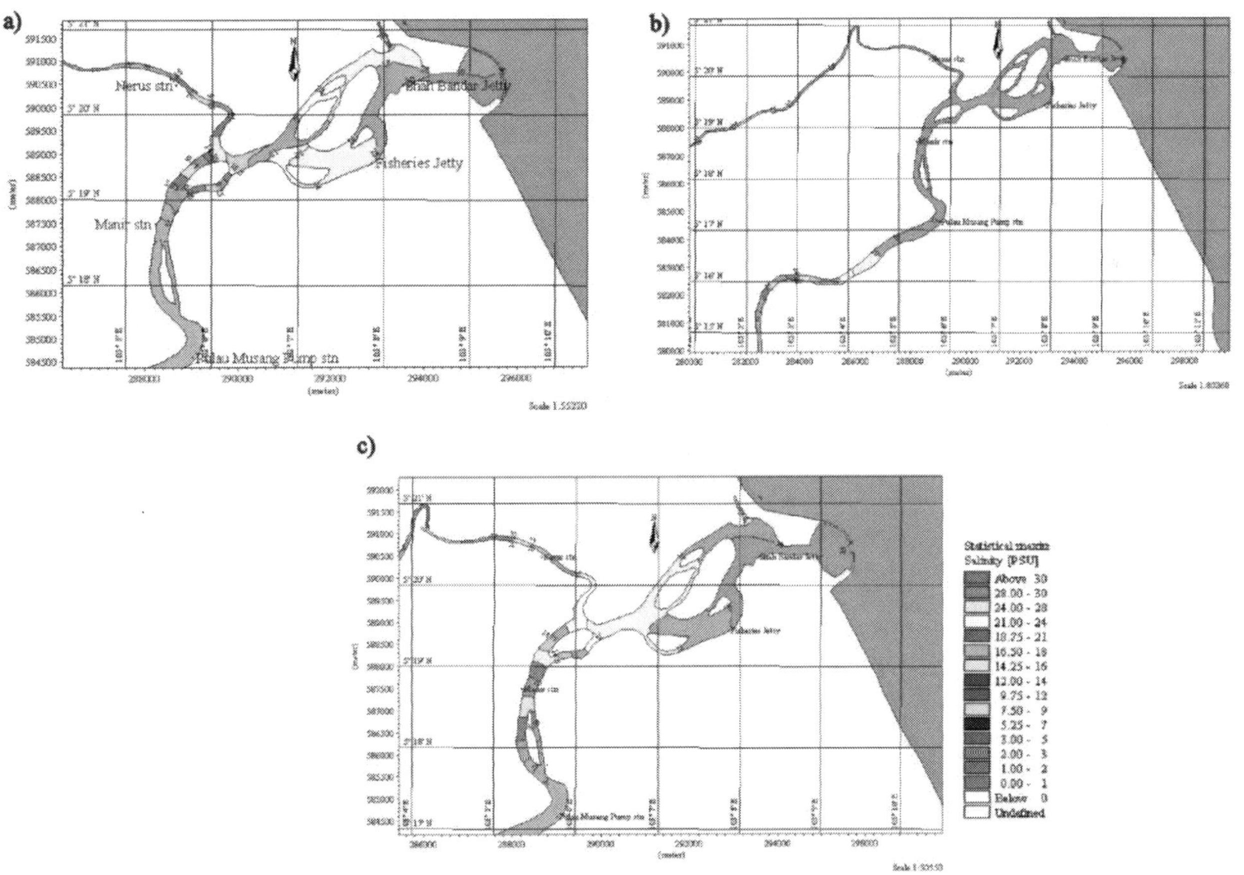

Figure 6. simulated salinity intrusion during HHWS at existing condition a), freshwater discharge at Kpg Tanggol boundary station with 30 m^3s^{-1} b) and Q=100 m^3s^{-1} c)

4. Conclusions

An integrated two dimensional models HD (hydrodynamics) and AD (advection dispersion) flexible mesh numerical model were developed to predict the possible of the salinity intrusion in the Terengganu estuary. It was noted during the spring tide, higher tidal velocity with magnitude approximately 0.8 ms^{-1} manages to penetrate saline water further inland. The maximum salinity intrusion was observed at 9.97 km inland from the river mouth. The severity of saltwater intrusion was observed with magnitude of 21.41 km from river mouth, when 30 m^3s^{-1} of freshwater discharge was applied constantly at the freshwater boundary at the Kpg Tanggol river gauging station. Nevertheless, when performed freshwater discharge with values of 100 m^3s^{-1} at the freshwater boundary, it has identified the water intake station is secure from the salinity contaminant. However, it should be noted that the actual water discharge upstream of the water intake should be higher than 100 m^3s^{-1}, which takes into considering the water extraction by the water intake stations located at the downstream and associated with the sea level rise factors at the nearshore area.

References

[1] Nguyen, A.D., Savenije, H.H.G., Pham, D.N.P. and Tang, D.T. (2007). Using salt intrusion measurements to determine the freshwater discharge distribution over the branches of a multi-channel estuary: The Mekong Delta case. *Estuarine, Coastal and Shelf Science* 77, pp. 433-445

[2] Qiu, C. and Zhu, J.R. (2013). Influence of seasonal runoff regulation by the Three Gorges Reservoir on saltwater intrusion in the Changjiang River Estuary. *Continental Shelf Research* 71, pp. 16-26

[3] Shaha, D.C. and Cho, Y.K. (2009). Comparison of Emprical models with intensively observed data for prediction salt intrusion in the Sumjin River Estuary, Korea. *Hydrology and Earth System Science Discuss.* 6, pp. 1879-1905.

[4] Trang, T.N., Long, N.H., Hai, N.X. (2014). Paddy soil salinization process in the Tien Hai District, Thai Binh Province under the sea level rise impacts. ARPN Journal of Agricultural and Biological Science (9), No 11

[5] Panda, U.S., Mahanty, M.M., Rao, V.R., Patra, S., Mishra, P. (2015). Hydrodynamics and water quality in Chilika Lagoon-A modeling approach. *Procedia Engineering* 116,pp. 639-646

[6] Suhaila, J. and Jemain, A.A. (2009). Investigating the impacts of adjoining wet days on the distribution of daily rainfall amounts in Peninsular Malaysia. *Journal of Hydrology* 368, pp. 17-25.

[7] Sultan, K. and Shazili, N.A. (2010). Geochemical baselines of major, minor and trace elements in the tropical sediments of the Terengganu River basin, Malaysia. *International Journal Of Sediment Research* 24, pp. 340-354.

[8] Tangang, F.T., Juneng, L., Salimun, E., Kwan, M.S., Loh, J.L., Muhamad, H. (2012). Climate change and variability over Malaysia: gaps in science and research information. *Sains Malaysiana* 41(11), pp. 1355-1366.

[9] DHI, 2007a. (2007). Mike 21 & Mike 3 Hydrodynamic and Transport Module: Sceientific Document.

[10] Shukla, V.K., Konkane, V.D., Nagendra, T., Agrawal, J.D. (2015). Dredged material dumping site selection using mathematical models. *Procedia Engineering* 116, pp. 809-817.

[11] Savenije, H.H.G. (2012). Salinity and Tides in Alluvial Estuaries, second ed. Elsevier BV.

[12] Graas, S. and Savenije, H.G. (2008). Salt intrusion in the Pungue estuary, Mozambique: effect of sand banks as a natural temporary salt intrusion barrier. *Hydrology and Earth System Sciences Discussions* 5, pp. 2523-2542.

[13] Ali, E., Kavvas, L.M., Fauzi, M. (2011). Sea level changes along the Peninsular Malaysia and Sabah and Sarawak coastlines. *World Environmental and Water Resources Congress 2011*: 1292.

[14] Bhuiyan,M.J.A.N. and Dutta, D. (2012). Assessing impacts of sea level rise on river salinity in the Gorai river network, Bangladesh. *Estuarine, Coastal and Shelf Science* 96, pp. 219-227.

Numerical Simulation of Evacuation Process in Malaysia By Using Distinct-Element-Method Based Multi-Agent Model

M S Abustan[1], N A Rahman[2], H Gotoh[3], E Harada[3] and S H A Talib[1]
[1] Faculty of Civil and Environmental Engineering, Universiti Tun Hussein Onn Malaysia, 86400 Batu Pahat Johor, MALAYSIA
[2]School of Civil Engineering, Engineering Campus, Universiti Sains Malaysia, 14300 Nibong Tebal, Penang, MALAYSIA
[3]Department of Civil and Earth Resources Engineering, Kyoto University, Katsura Campus, Kyoto, 615-8584, JAPAN

E-mail: sallehb@uthm.edu.my

Abstract. In Malaysia, not many researches on crowd evacuation simulation had been reported. Hence, the development of numerical crowd evacuation process by taking into account people behavioral patterns and psychological characteristics is crucial in Malaysia. On the other hand, tsunami disaster began to gain attention of Malaysian citizens after the 2004 Indian Ocean Tsunami that need quick evacuation process. In relation to the above circumstances, we have conducted simulations of tsunami evacuation process at the Miami Beach of Penang Island by using Distinct Element Method (DEM)-based crowd behavior simulator. The main objectives are to investigate and reproduce current conditions of evacuation process at the said locations under different hypothetical scenarios for the efficiency study of the evacuation. The sim-1 is initial condition of evacuation planning while sim-2 as improvement of evacuation planning by adding new evacuation area. From the simulation result, sim-2 have a shorter time of evacuation process compared to the sim-1. The evacuation time recuded 53 second. The effect of the additional evacuation place is confirmed from decreasing of the evacuation completion time. Simultaneously, the numerical simulation may be promoted as an effective tool in studying crowd evacuation process.

Keywords: Numerical simulation, evacuation process, distinct element method, microscopic model, crowd behavior simulator.

1. Introduction

Following from the 2004 Indian Ocean Tsunami tragedy, the government of Malaysia had enforced the relevant disaster management agencies to look in detail the matter. The initial measures taken was the needs to readdress the tsunami mitigation problems towards better mechanism in such a manner that tsunami rescue works could be performed in an efficient and effective manner [4,5]. In conjunction to that, the tsunami emergency plan had been established which covered: (1) evaluation awareness, preparedness of the community, local and state agencies in combating tsunamis; (2) enhancement of public awareness through education of community; and (3) the formulation of tsunami evacuation planning. Among the mentioned countermeasures, a greater attention is directed to a good planning of tsunami evacuation.

Evacuation planning is a procedure that involves the movement of people to a safer place. Currently, evacuation drills have been put in place and conducted annually by the Malaysian Meteorological Department (MMD) since 2006. The drill is one of the good indicators that an evacuation will go smoothly and successfully. Moreover the drill also is an important awareness tool to familiarize the public with evacuation procedure. However, drills alone is not enough in preparing an effective tsunami evacuation planning since it is quite time consuming, costly and furthermore, the drill is quite rigid which cannot deals with various conditions on site flexibly. Hence, in conjunction to have a better evacuation planning, perfomed the numerical simulation of current conditions of evacuations process on selected areas by considering different kind of scenarios [3]. By conducting evacuation process for different kind of scenarios, better planning of evacuation may be derived.

This paper focuses on the evacuation process optimization through numerical simulation by using Distinct-Element-Method (DEM) based multi-agent models namely Crowd Behavior Simulator for Disaster Evacuation model (CBS-DE) [1,2]. The author investigate current condition of evacuation process against tsunami and try to figure out the weaknesses in its possession. Hence, from the weaknesses highlighted, the suggestion of improvements will be made.

2. Crowd Behavior Simulator For Disaster Evacuation (CBS-DE)

Fundamental equations

The equation of motion for simulating each individual behavior which consists of translational and rotational equations are governed by the following equations:

$$m_{hi}\dot{\boldsymbol{v}}_i = \mathbf{F}_{mi} + \mathbf{F}_{ci}\ ;\ I_{hi}\dot{\boldsymbol{\omega}}_i = \mathbf{T}_i \tag{1}$$

where m_{hi} is the mass of the person i, $\boldsymbol{v}_i$ is the velocity of the person i, '$\cdot$' is time-derivative, Fmi is the autonomous driving force of the person i, $\mathbf{F}_{ci}$ is the interacting force acting on the person i, I_{hi} is the moment of inertia of the person i, ω_i is the angular velocity of the person i, T_i is the torque acting on the person i.

Autonomous driving force

The autonomous driving force is $\mathbf{F}_{mi}$ terms are:

$$\mathbf{F}_{mi} = m_{hi}\mathbf{a} \tag{2}$$

where a denote as the acceleration vector. The equilibrium walking velocity is affected by number density of human. Therefore, the magnitude of the maximum equilibrium walking velocity vmax is given as follows:

$$\boldsymbol{v}_{max} = \boldsymbol{v}_{limit} - \gamma \cdot \boldsymbol{c}_k \tag{3}$$

where $\boldsymbol{v}_{limit}$ is the magnitude of the specific equilibrium velocity, γ is the proportional coefficient depend on the congestion condition in designated area, $\boldsymbol{c}_k$ is the number density of persons inside the psychological perception area.

Interacting forces

The estimation of the interacting force is made by Voigt model and by referring to perception domain (Fig. 1). The total interacting force acting on the person i is described as follows:

$$\mathbf{F}_{ci} = \mathbf{F}_{d(p/p)i} + \mathbf{F}_{d(W/p)i} + \mathbf{F}_{ps\_i} \tag{4}$$

where $\mathbf{F}_{d(p/p)i}$ is the physical repulsive force acting on the person i, $\mathbf{F}_{d(W/p)i}$ is the physical repulsive force between the virtual wall element (W) and the person i, $\mathbf{F}_{ps\_i}$ is the psychological repulsive force acting on the person i.

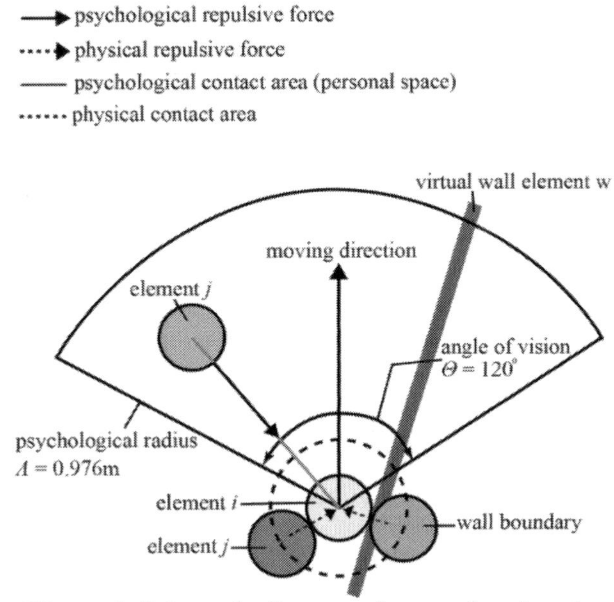

Figure 1. Schematic diagram of perception domain.

3. Results and discussion

EVACUATION PROCESS

The Miami Beach of Penang Island is selected as the site of the present study. The length of shoreline is approximately 270 meters and the width is about 30 meters. Fig. 2 shows the computational domain of the study which is illustrated by a bird's-eye view. The Miami Beach is located in sea-front delta with a hill on its back. The hill has an approximately 23meters in height above the sea level. And this height is considered as adequate level of setting an evacuation place against tsunami. Table 1 shows the distribution of population observed by field survey in the Miami Beach at a peak time.

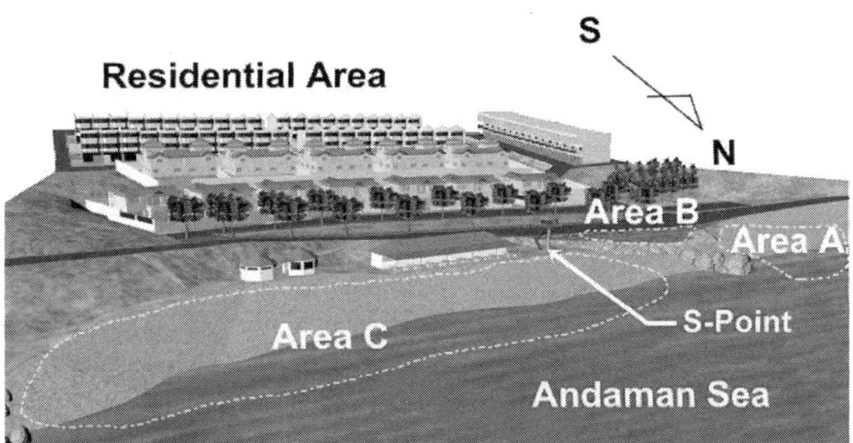

Figure 2. Bird's eye view of computational domain.

Table 1. Distribution of population observed during field survey in the Miami Beach (Peak time).

Category	Sex	Age	People	Ratio [%]	Velocity [m/s]
1	Male	10-39	93	27.93	**1.38**
2	Female	10-39	53	15.92	**1.14**
3	Male	40-69	4	1.20	**0.99**
4	Female	40-69	93	27.93	**1.20**
5	Male	>70	41	12.31	**1.04**
6	Female	>70	3	0.90	**0.89**
7	Child	5-9	46	13.81	**1.06**
		Total	**333**	**100 %**	

Simulation setup

The evacuation process is started after receiving the verbal evacuation signal propagated concentrically from the S-point with speed of 1.50 m/s and completed after all people reach at altitude of 7 meters above the sea level which is based on the observed tsunami record in 2004. The initial arrangements of the persons are assigned randomly around the beach area. In the present simulation, we do not consider a special attractive force between flocks, such as a family, friends and a couple. To simulate a flock behavior, a psychological attractive force between persons should be incorporated. Here, we assumed that each person walks independently without any special psychological attractive force.

Investigation of evacuation process

For the initial condition of evacuation planning, the evacuation place is designated in the end of the main road leading from coast as shown in Fig. 3. This simulation is named as the sim-1. The people are instructed to evacuate toward the designated evacuation place. The illustration snapshots of the sim-1 are shown in Fig. 4. The graph of the time taken versus accumulative number of people who have completed the evacuation process is plotted as shown in Fig. 5. The time required to complete the evacuation process of the sim-1 is 312 seconds.

From sim-1, the efficiency of evacuation process can be obtained. Therefore the deprivation area for quick evacuation has been disclosed. The interruption area for smooth evacuation was detected in the east end of the Area C and in the west end of the Area A. The significant factor of the delay in the evacuation process was the initial position of persons who are remote from the designated evacuation area. The proposals toward the quick evacuation are shown in the following section.

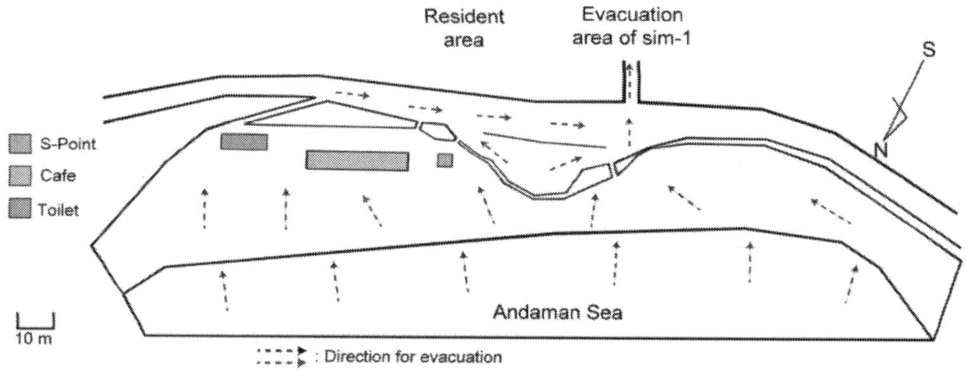

Figure 3. Evacuation route and movement direction of people for the sim-1.

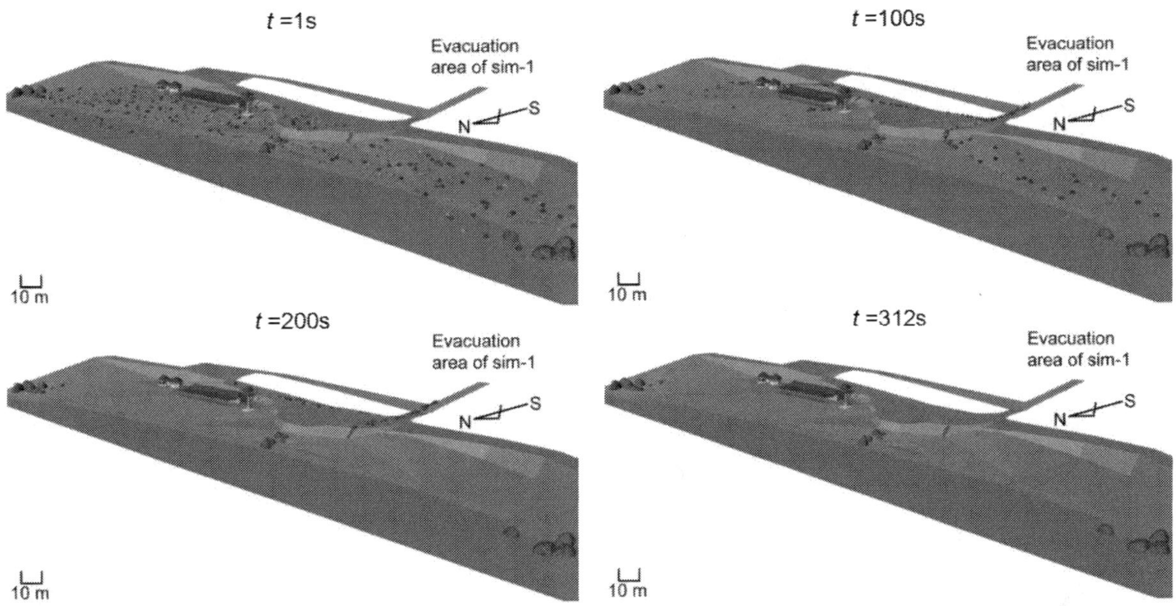

Figure 4. The snapshots of the sim-1.

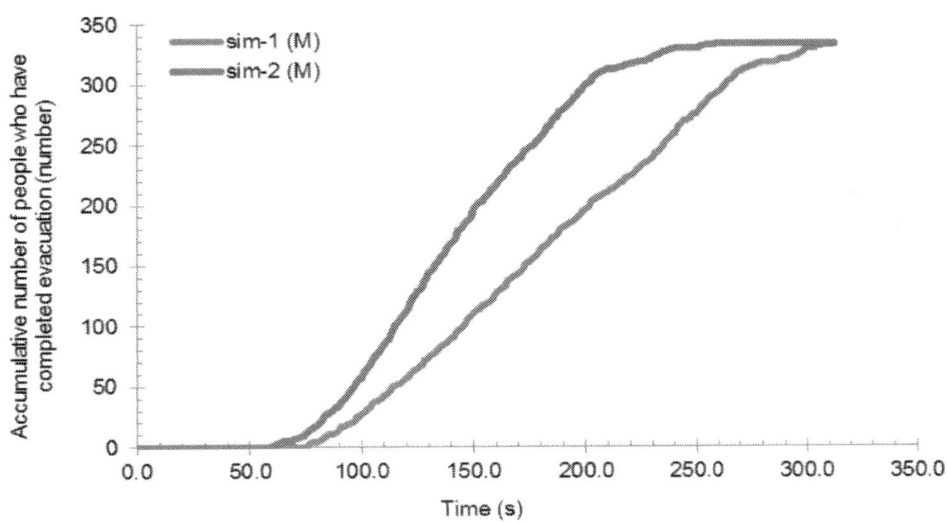

Figure 5. Time series of the accumulative number of persons who complete evacuation process for the sim-1 and the sim-2.

Improvement of evacuation planning

The late persons who are in the inconvenient area have been investigated in the alternative scenario. The alternative scenario is identified as the sim-2 for shortening the path between original location of people and the evacuation place, and in the same time, the completion time of evacuation is expected to be reduced. The evacuation route of the sim-2 is shown on Fig. 6. In the sim-2, persons arranged in the Area C are specified to evacuate to the new evacuation place, which is placed at the east end of the residential area. Whereas, the persons who in the Area A and B, evacuate to the same evacuation place as the sim-1. The illustration snapshots of the sim-2 are shown in Fig. 7. Fig. 5 shows the accumulated number of people who have completed the evacuation for the sim-2. From the

IOP Conf. Series: Materials Science and Engineering **136** (2016) 012077 doi:10.1088/1757-899X/136/1/012077

viewpoint of the evacuation completion time, a decrease of 53s is found in the sim-1 compared with the sim-2. From these results, the effect of the additional evacuation place is confirmed from decreasing of the evacuation completion time.

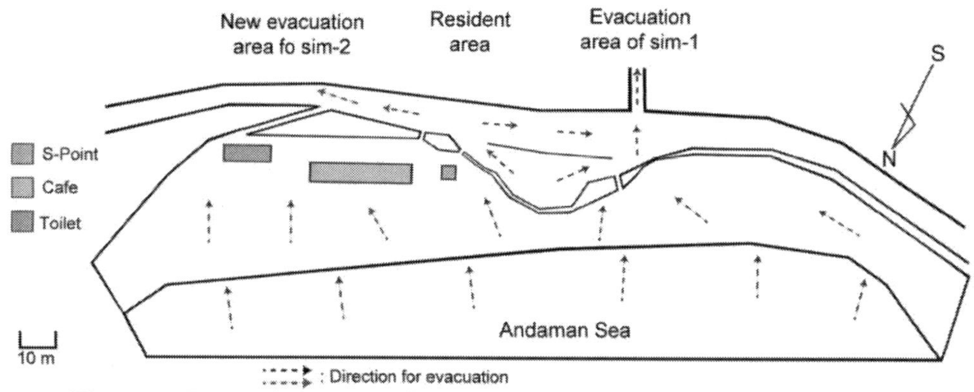

Figure 6. Evacuation route and movement direction of people for the sim-2.

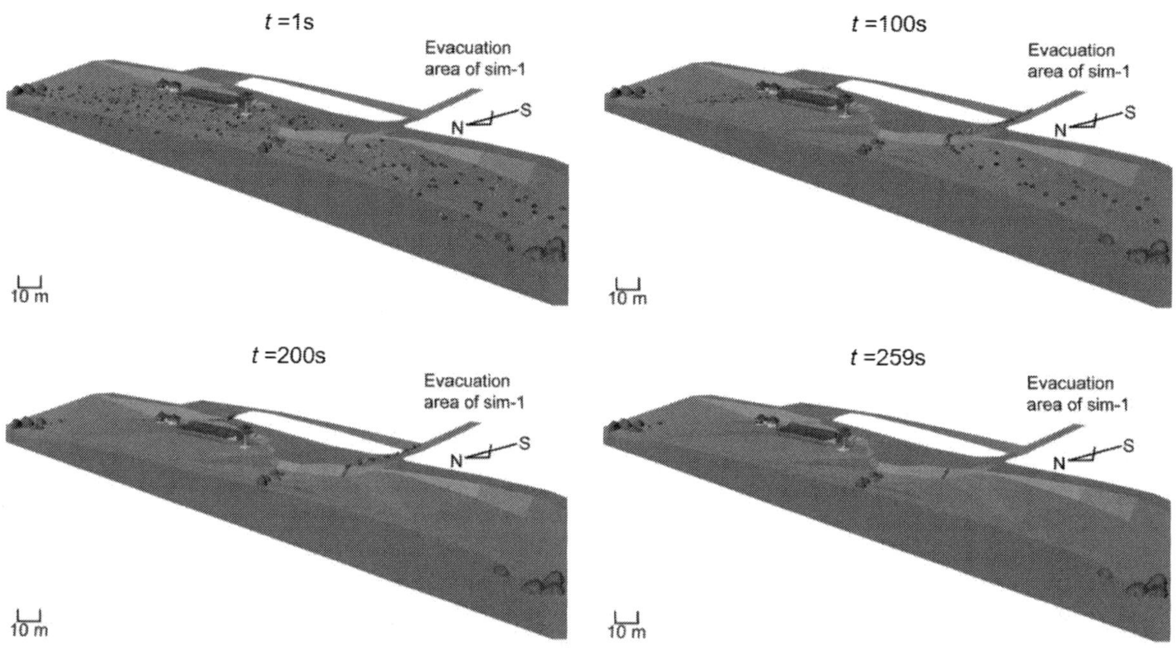

Figure 7. Snapshot of the sim-2.

<u>Assessment of initial position of visitor to the evacuation place</u>

Three main factors that influence the evacuation process are as follows: (1) distance from initial position of person to the evacuation area; (2) type of human; and (3) surrounding area. In practice, the information of human type composition can be useful to develop an effective evacuation planning. Children and elders have slower average walking velocities in comparison to adults. Therefore, new evacuation area can be proposed to be built closer to the area that has higher percentage of elders and children composition to ensure quick evacuation process. The moving distance of each person for the duration of n-th computational time marching steps can be given by :-

$$L_n = \sum_{k=1}^{n} \left(\boldsymbol{u}_{hi} \cdot \boldsymbol{t}\right)_k \Delta t \tag{5}$$

in which, n is the time step, t is unit tangent vector along to moving direction. The improvement by proposed the new additional evacuation area closest at Area C will be the reasonable choice.

4. Conclusion

In this study, the process of the evacuation planning has been shown through the numerical simulation for the evacuation against tsunami at the Miami Beach, Penang, Malaysia by using the CBS-DE simulator based on the Distinct Element Method (DEM). In the process of the evacuation planning, the inconvenient area has been revealed firstly, then the alternative planning for the improvement of the inconvenient area has been investigated, and the effect of the alternative planning has been shown quantitatively.The evacuation time had been decrease 53 second after introduced new evacuation area.

From the simulation results, the positions of each person have been tracked with detail information as discussed previously. Through the detail information obtained, a realistic virtual space by CG can be made. As shown in Fig. 8, the typical snapshots of evacuation process have been captured. This kind of the CG will be considered to be a useful tool for explaining the evacuation planning to the public, because CG movie has an advantage in the visual sense and people will be able to image evacuation process easily.

Figure 8. Typical snapshot of CG movie for evacuation process at Miami Beach.

References

[1] Gotoh, H., Harada, E., Kubo, Y., & Sakai, T. (2004). Particle-system model of the behavior of crowd in Tsunami flood refuge. Annual Journal of Coastal Engineering JSCE, 51, 1261-1265.

[2] Gotoh, H., Harada, E., & Andoh, E. (2012). Simulation of pedestrian contra-flow by multi-agent DEM model with self-evasive action model. Safety Science,50(2), 326-332.

[3] Harada, Eiji, Hitoshi Gotoh, and Noorhazlinda Binti Abd Rahman. "A switching action model for DEM-based multi-agent crowded behavior simulator." Safety science 79 (2015): 105-115.

[4] Kiyono, J., Miura, F., & Takimato, K. (1996). Simulation of emergency evacuation behavior in a disaster by using distinct element method. In Proceedings of civil engineering conference (Vol. 537, pp. 233-244).

[5] Kiyono, J., Miura, F., & Yagi, H. (1998). Simulation of evacuation behavior in a disaster by distinct element method. In Proceedings-Japan Society of Civil Engineers (pp. 365-378). Doboku Gakkai.

[6]. Liu, C., Ooeda, Y., & Sumi, T. (2008). A Model for Pedestrian Movement with Obstacle Evasion Using Personal Space Concept. Doboku Gakkai Ronbunshuu D, 64(4), 513-524.

[7] Tatebe, K., Tsujimoto, M., Shida, K., (1994). Method for judging the beginning point of avoiding behavior and avoidance distance between pedestrian and standing obstacle. Journal of Architecture, Planning and Environment Engineering AIJ 465, 95-104 (In Japanese).

A Study of Flood Evacuation Center Using GIS and Remote Sensing Technique

A A Mustaffa[1], M F Rosli[1], M S Abustan[1], R Adib[1], M I Rosli[1], K Masiri[2] and B Saifullizan[1]

[1] Faculty of Civil and Environmental Engineering, Universiti Tun Hussein Onn Malaysia, 86400 Batu Pahat Johor, MALAYSIA.

[2] Center for Diploma Studies, Universiti Tun Hussein Onn Malaysia, 86400 Batu Pahat Johor, MALAYSIA.

E-mail: Mustafa@uthm.edu.my

Abstract. This research demonstrated the use of Remote Sensing technique and GIS to determine the suitability of an evacuation center. This study was conducted in Batu Pahat areas that always hit by a series of flood. The data of Digital Elevation Model (DEM) was obtained by ASTER database that has been used to delineate extract contour line and elevation. Landsat 8 image was used for classification purposes such as land use map. Remote Sensing incorporate with GIS techniques was used to determined the suitability location of the evacuation center from contour map of flood affected areas in Batu Pahat. GIS will calculate the elevation of the area and information about the country of the area, the road access and percentage of the affected area. The flood affected area map may provide the suitability of the flood evacuation center during the several levels of flood. The suitability of evacuation centers can be determined based on several criteria and the existing data of the evacuation center will be analysed. From the analysis among 16 evacuation center listed, there are only 8 evacuation center suitable for the usage during emergency situation. The suitability analysis was based on the location and the road access of the evacuation center toward the flood affected area. There are 10 new locations with suitable criteria of evacuation center proposed on the study area to facilitate the process of rescue and evacuating flood victims to much safer and suitable locations. The results of this study will help in decision making processes and indirectly will help organization such as fire-fighter and the Department of Social Welfare in their work. Thus, this study can contribute more towards the society.

Keywords: Remote sensing, evacuation center, GIS, flood affected area.

1. Introduction

Flood is a natural disaster that could bring damage to the properties, infrastructures, animals, plants and even human lives. Flood is the most significant natural hazard in Malaysia in terms of the affects towards the population, frequency of the occurrences, areas damaged, the flood duration itself and its social and economic damage. Having 189 river basins throughout Malaysia, including Sabah and Sarawak, the rivers and their corridors of flood plains fulfill a variety of functions both for human use and for the natural ecosystem. These are fundamental parts of the natural, economic, and social system. In another perspectives, despite their numerous contribution for our daily lives, rivers itself might be the largest threat to its entire corridor areas [1].

The issue of flood especially involving the emergency response solution can be approached from the aspect of the evacuation center itself by the determinations of the effectiveness of every evacuation centers and this requires multiple solutions especially in using GIS and Remote Sensing technique. GIS was the technological developments which began to develop and start to be used in Malaysia in recent years. It was widely used in many sectors, especially those who required a lot of data management. There are various definitions and purposes of GIS applications, but based on the field this study, GIS is defined as a database system that is based on computer which were designed to support the acquisition, storage, analysis and display of data or information depending on the selected location [2]. Through GIS, the information needed can be obtained quickly and accurately as well having the benefits of the ability to connect spatial data and attribute data in order to make a more effective GIS data for analysis purposes [3]. Remote Sensing are defined as the process of sensing the Earth's surface from space by using the characteristics of electromagnetic waves emitted, reflected or disclosed by the detected objects, for the purpose of improving natural resource management, land use and environmental protection [4].

Flood is a natural phenomenon where the land is covered by water that overflow by the inland water bodies or the tidal waters. There are several definitions of floods that we can take into accounts so that it can help us to see the overall picture of how flood occurs. Flooding is a natural disaster caused by factors applicable climatology or climatic factors such as temperature state, a sprinkling of rain, evaporation, wind movement and the state of the natural earth [5]. Meanwhile, another theoretical explanation that commonly used to highlight the flood situation usually caused either by the continuous rain that causes greater quantity than usual or overflow of river water to the river bank or both happen at once [6].

2. Evacuation Center

Evacuation is a risk management strategy that may be used to mitigate the effects of an emergency on a community. It involves the movement of people to a safer location and their return. For an evacuation to be effective it must be appropriately planned and implemented [7]. Evacuation Centers will be defined by the relevant jurisdictional emergency management plan, and established (activated) by responsible authority on advice of the agency controlling or coordinating the emergency. It is acknowledged that terminology is different in each state and territory [8].

Planning for an evacuation center needs to address many public health concerns. These include the physical amenities and space required for well-being, minimizing of the risk of communicable disease outbreaks, and the need to promote the health of evacuees to prevent the acute exacerbation of chronic diseases [9]. An Evacuation Center is established to provide shelter to people who are directly affected by an emergency situation and for a people that do not have anywhere else to go to. For example they have been evacuated from their homes or cannot access their homes because of the incident. Evacuation Centers are usually established in halls or school gymnasiums to provide basic shelter. As the accommodation is fairly basic, it is recommended that people seek shelter with family or friends or private accommodation if they are able to [10]

3. Methodology

Data collection is one of the important phases in designing a GIS system. Primary data consist of the data from the remote sensing images. The data was obtained from the satellite image and composed into the form of value image data. The primary data of digital elevation model (DEM) is a data which is a type of elevation representation of the terrain surface. The data is in form of raster that contain value that represent each elevation point on the earth surface. Secondary data is the data that consist on the study area data such as placement location, the land used and the population data to support the primary data. The secondary data will provide support to any limitation in the primary data. Then, interview data will be provide consist of the data that from the conversation or interview with the department of welfare. The purpose of this interview was to determine the worst area levels of flood in Batu Pahat.

In the process of mapping, DEM data that obtained from the ASTER database is used in the process of contour extraction. Contour extraction process is a process that extracts the elevation value in DEM data to create elevation value or generate the contour of the study area and this process is conducted by using ArcGIS software. From ArcGIS the contour interval is set to 2 meter high according to the flood occurrences in the study area. The contour data is then analyse for each of it elevation value to contour map of the study area. For 2 meter, 4 meter and 6 meter high to determine the flood water level area. DEM data also had been used to determine the watershed of the study area. To produce a watershed of the study area, stream network of the area is needed for generate multiple basin in Batu Pahat area. From the basin and watershed delineated, the process of incorporating the value of curve number based on different types of land use can be done. After the watershed is produced the slope generation is processed from the DEM data to determining the peak discharge of the study area.

To determine the road access of each place in Batu Pahat the topographic map was used in the process of digitization to produce a road vector. Batu Pahat map act as a base map to digitize the road vector by using ArcGIS software. The road map is needed to be positioned in the process of geocoding that is the process that placed the map into the correct position and scale for the use of the next process. All the geocoding process will use the topographic map as the reference to get the exact position.

The satellite data used in this study is the Landsat 8 image. Landsat 8 image is used in making the land use image of classification. Before the process take place, the Landsat 8 image is needed to be placed into the correct position in the process of geocoding. The process of classification is to register the difference land use type into specific classes such as urban, forestry, vegetation, water body and open area. The land use map produce in the process of classification will help in differentiate the land use type for the study area.

All maps from different sources undergone different analysis. Flood map analysis is used to determine the level of flood of the study area. Flood map is produce by the process of contour selection of the contour and elevation value of the study area. Flood affected area analysis is the analysis of area that affected with the certain level of flood. Flood evacuation center analysis is the analysis of the suitability regarding the location of the evacuation. The effectiveness analysis is to determine the effectiveness of the evacuation center itself to support the flood victims according to the population of the area. The new location of evacuation center analysis is to determine the new location of evacuation center that can be used for the time of emergency. Figure-1. Show the flowchart of methodology.

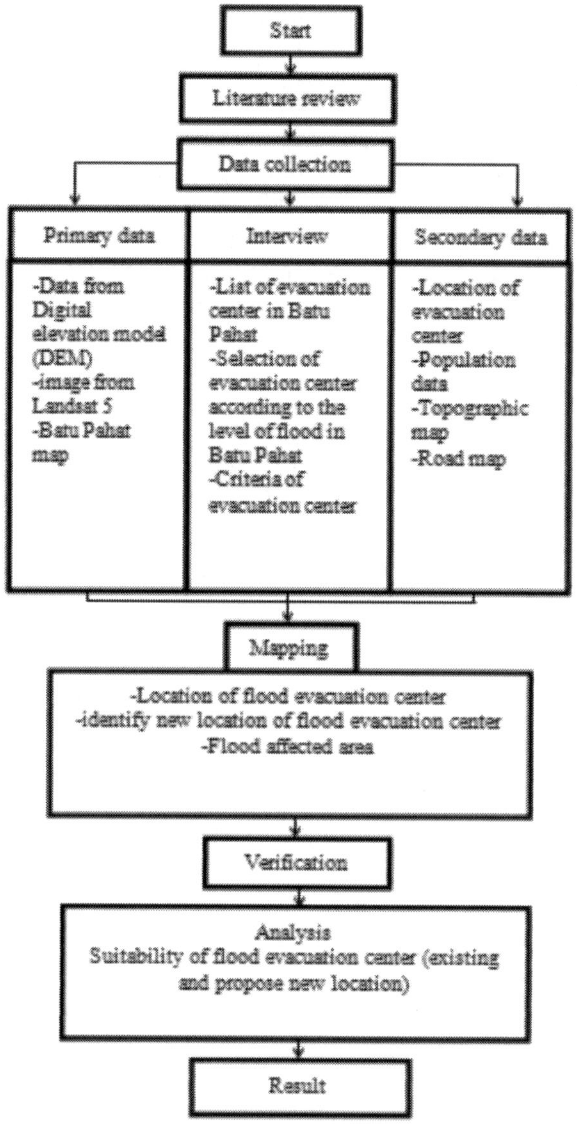

Figure 1. Flow chart of methodology.

4. Result and Discussion

Flood map is used to determine the level of flood of the study area. Flood map is produce by the process of contour selection of the contour and elevation value of the study area. The contour consists of the value of each elevation in the study area. Figure-2 show the flood map of Batu Pahat.

4.1 Flood Affected Area Analysis

Flood affected area analysis is the analysis of area that affected with the certain level of flood. This analysis consist the use of overlaid land use map with contour and elevation value. Both data need to be overlaid to show the relationship that can be produce by it. The land use map shows the characteristic of the land use classes and the contour will develop the flood level. Table-1 shows that the area coverage for the area affected by 2 metre flood from the mean sea level is 2.79% from the total area of Batu Pahat. This shows that 52.243km^2 area from 1872.56 km^2 area of Batu Pahat had been submerged by 2 metre flood.The area coverage for the area affected by 4 metre flood from the mean sea level is 5.56% from the total area of Batu Pahat. This shows that 104.114 km^2 areas from

1872.56 km^2 area of Batu Pahat had been submerge by 4 metre flood. The area coverage for the area affected by 6 metre flood from the mean sea level is 11.18% from the total area of Batu Pahat.

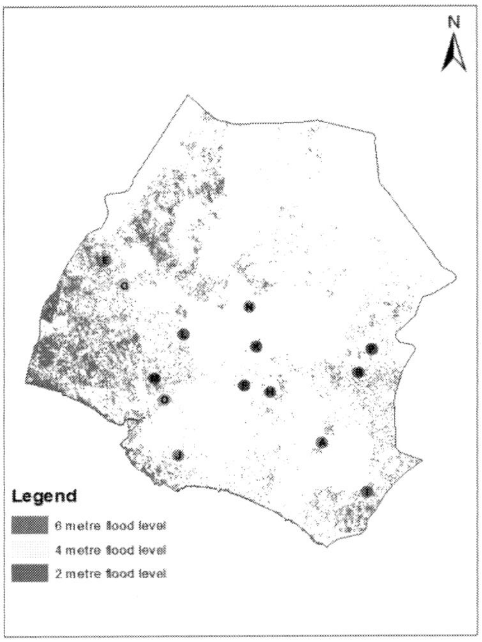

Figure 2. Flood affected area of Batu Pahat.

Table 1. Flood affected area analysis.

Flood affected area level (m)	Percentage(%)	Area affected (km^2)
2	2.79	52.243
4	5.56	104.114
6	11.18	209.49

This shows that 209.49km^2 area from 1872.56km^2 area of Batu Pahat had been submerge by 6 metre flood.

4.2 Flood Evacuation Center Analysis

Based on Table-2, it shows the evacuation center is place at the suitable distance from the flood affected area. Based from the interview data obtained from the Department of Public Welfare, one of the criteria of evacuation center is the building itself should be in the location that are reachable and safe from the flood affected area. Some of the evacuation center pass the criteria but there are certain of it did not because of the position of the building itself was placed in the affected area of flood. In this analysis we have determine that S. A. Tiga Serangkai, S. K.Tiga Serangkai and S. K. Bukit Kuari is not suitable for selected evacuation center for 4 meter and 6 meter flood. Meanwhile S. K. Sri Merlong and S. K. Kota Dalam are not suitable for selected evacuation center for 6 meter flood. Meanwhile Balairaya Parit Lapis Laman, S. K. Parit Sulong, S. A. Sengkuang, S. K. Sri Laksana, S. K. Sri Gading, S. K. Sungai Suloh, S. K. Sri Bengkal, S. A. Sri Binjai, S. K. Parit Lapis, S. K. Seri Telok, and S. M. Tinggi is suitable for selected for evacuation center in term suitability of the location in the study area. From this analysis it has been shown that the entire evacuation center listed is suitable for 2 metre flood.

Table 2. List of flood evacuation center affected.

No.	Evacuation center	Condition
1 (A)	Balairaya Parit Lapis Laman	Not affected
2 (B)	S.K. Bukit Kuari	Affected > 4 m
3 (C)	S. K. Tiga Serangkai	Affected > 4 m
4 (D)	S. K. Parit Sulong	Not affected
5 (E)	S. A. Tiga Serangkai	Affected > 4 m
6 (F)	S. A. Sengkuang	Not affected
7 (G)	S. K. Sri Laksana	Not affected
8 (H)	S. K. Sri Gading	Not affected
9 (I)	S. K. Sri Merlong	Affected > 6 m m)
10 (K)	S. K. Sungai Suloh	Not affected
11 (L)	S. K. Sri Bengkal	Not affected
12 (M)	S. A. Sri Binjai	Not affected
13 (N)	S. K. Parit Lapis	Not affected
14 (O)	S. K. Seri Telok	Not affected
15 (P)	S. M. Tinggi	Not affected
16 (Q)	S. K. Kota Dalam	Affected > 6 m

4.3 Road Access For The Evacuation Center

While, Figure-3 shows that all the evacuation selected have a road connected to it but the access of the road depend on the flood level from mean sea level. Based on the figure when flood is on 2 meter from the mean sea level, all the roads are able to access to the evacuation center and placed in a safe from flood affected area. When the flood level rises for 4 meter there are certain roads that have been affected and submerge by flood. From the analysis we found that S. A. Tiga Serangkai, S.K. Tiga Serangkai, S. K. Sri Merlong, S. K. Sungai Suloh, and S. K. Bukit Kuari does not have a proper road access for the process of delivering helps and needs of the victims. Most of the road within 1 km radius area is submerge by 4 meter flood and there is no other road that can be access to the evacuation center. For the 6 meter flood, the road that connected to the evacuation center such as S. K. Parit Sulong, S. K. Sri Laksana, and S. K. Parit Lapis are unable to be access. Most of the road within 1 km radius area is submerge by 6 meter flood and there is no other road that can be access to the evacuation center. So the evacuation centers that have the suitability to be selected as the evacuation center in term of road access for the three flood level is Balairaya Parit Lapis Laman, S. A. Sengkuang, S. K. Sri Gading, S. K. Sri Bengkal, S. A. Sri Binjai, S. K Seri Telok, S. M. Tinggi, and S. K. Kota Dalam. From this analysis the suitability of the evacuation center in term of road access is determined. For the conclusion, there are nine evacuation centers that are not suitable to be evacuation center and another seven of them suitable for being selected to be evacuation center.

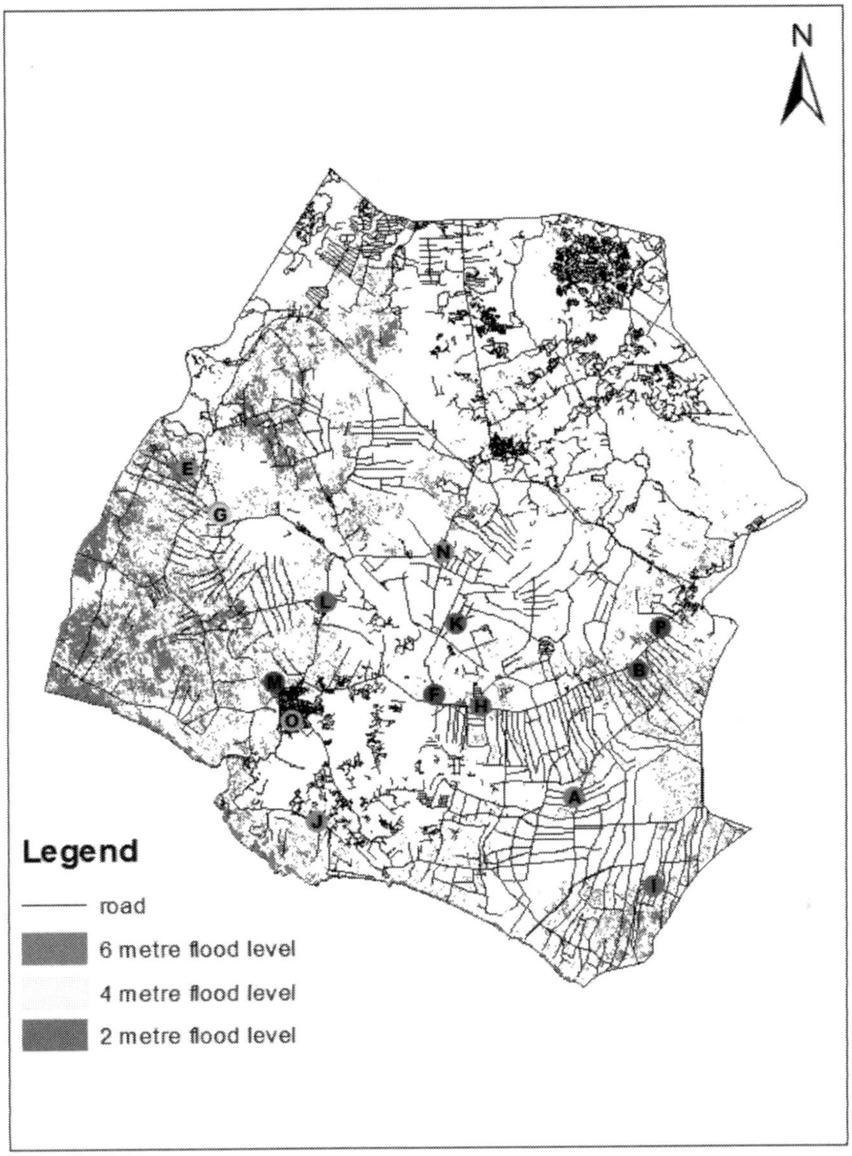

Figure 3. Road accesses for flood evacuation center on the flood map.

4.4 Evacuation Center Effectiveness Analysis

From the survey conducted by GEMA Terengganu the capacity that can be hold by the evacuation center such as Multipurpose Hall and other bigger building is 800 to 1000 capacities, School or High School is 300 to 500 capacities and for community center is 80 to 120 capacities. Multipurpose hall is the best building for being selected as an evacuation center. Based on the population data for Simpang Kiri, Lubok, Bagan, Perserai and Sungai Kluang, the total population of 64476 people are affected to the flood so it is 64 evacuation centers to hold the capacity of it. As for the Seri Medan that has average flood condition, 22501 populations of people should be hold by the evacuation center from the county of Seri Medan and the county that have low condition of flood. From the analysis at least 10 evacuation center should be open for emergency use in the county that have low flood condition such as Sri Gading, Linau, Minyak Beku, Kampung Bahru, Simpang Kanan, Chaah Baru, Tanjung Semberong and Sungai Punggor and Seri Medan should open 2 evacuation center of its own.

Table 2. List of purpose of evacuation center.

County	Population	Evacuation Cen.		
		A	B	C
Sri Gading	63903	3	1	2
Linau	8325	3	0	3
Minyak Beku	16,634	1	1	0
Kampung Bahru	11,627	1	0	1
Sg. Kluang	15,844	1	1	0
Seri Medan	22,501	4	4	0
Simpang Kanan	135,911	3	1	2
Bagan	7,440	0	0	0
Lubok	6,594	0	0	0
Perserai	9,143	0	0	0
Simpang Kiri	25,455	0	0	0
Chaah Baru	5,426	0	0	0
Tg. Semberong	34,456	0	0	0
Sg. Punggor	10,132	0	0	0

A – Existing, B – Non-suitable, C – Ready to used.

4.5 Propose of the New Location of Evacuation Center

This analysis was carried out to determine the new location of evacuation center that can be used during emergency time. The new location was determined based on several criterias as evacuation center that related with the effectiveness. It also had been determined by the suitability of the location and the road access that connected to it. Table-3 shows the new location of the proposed evacuation center on the map.

Table 3. Location of proposed evacuation center.

County	Type of Evacuation Center
Chaah Bahru	Multipurpose hall
Tanjung Semberong	SMK Dato' Seth
Seri Medan	Multipurpose hall
Linau	Multipurpose hall
Sri Gading	UTHM
SimpangKanan	SMK Dato' Bentara Luar
SimpangKanan	SMK Tun Aminah
Minyak beku	Multipurpose hall
Kampung Baharu	SMK Tunku Mahmood
Sungai Punggor	Multipurpose hall

5.0 Conclusion

This research had managed to improved several aspect on the evacuation center selection issues. In order to find the best solution for the evacuation selection location issues, a method by using GIS and Remote Sensing software have been created which is important contribution to the process of rescue and selection of the location for evacuation center. The creation of the method in software has brought new approach and technique for evacuation center site selection especially for fire fighter and Department of Public Welfare and also can make the decision makers to speed up the decision making process.

The evacuation site selection begins with the analysis of the flood map process which could determine the level of flood from the contour and elevation value of the study area. From the flood map the process continue with the analysis of the flood affected area analysis. This analysis determines the type of the land use affected by the flood with different level from the mean sea level. There are several classes of land use type such as urban, vegetation, open, area and forestry the flood affected the land use classes. All of this analysis is to determine the area of the flood effect and the level of flood from mean sea level.

The processes continue to the location analysis of the evacuation center. The analysis determined the suitability of the location and position of the evacuation center that exist in the study area. The flood map used to helps in determination of the suitable and non-suitable evacuation center from the map. The evacuation center that is expose to the flood will be classified a non-suitable evacuation center and the spared from flood will be classified as suitable evacuation center. The analysis continues to the road access analysis that is to determine the road access of the evacuation center. The analysis had showed that the not all suitable evacuation center could have the proper road access so the suitable evacuation center will be filter down and only the suitable evacuation center with the proper road access will be selected to be used in time of emergency. All the analyses is to determine the suitability of the existing evacuation center in the study area.

The process continues with the effectiveness of the evacuation center analysis. This analysis is to determine the effectiveness of the evacuation center in providing a enough capacity of the population in the area. The criteria are important to achieve the effective evacuation center for the comfort of the flood victims.

From all the process, the new selection of evacuation center can be propose to provide the suitable and effective type of evacuation center for the use of emergency time. It is needed that to propose 1 evacuation center on each suitable county in the study area which have a good location that a near to residential area, have high elevation, can hold enough capacity of people, and have a proper road access for the use of delivering help.

From this study all the objective of determine the flood affected area and analysing the existing and propose flood evacuation center are achieve. The result of the analysis shows the completion of each objective and provided a data that can be used in the future studies.

Acknowledgements
The author would like to express gratitude to the Ministry of Higher Education Malaysia, and Universiti Tun Hussein Onn Malaysia for their support with grant Vot 1526.

References
[1] Jabatan Pengairan & Saliran Malaysia (2007) Flood and Drought Management in Malaysia. Malaysia: Ministry of Natural Resources & Environment.
[2] Taher Buyong (1995), Prinsip-prinsip GIS. Universiti Teknologi Malaysia: Lecture notes
[3] Suhaimi Ahmad (1999). Aplikasi GIS Dalam Membantu Pengurusan Pelajar Asrama. UTM: Tesis Sarjana Muda.
[4] United Nations (1986), Principles Relating to Remote Sensing of the Earth from Space. United Nation: 95th plenary meeting.
[5] Balek J. (1983), Hydrology and water resources in tropical regions. *Developments in Water Science*.
[6] Balkema A. A. Rotterdam & Brookefield (1993), Hydrology and Water Management of Deltaic Areas. Netherland: *Center for Civil Engineering Research and Codes.*
[7] Bradley (2013), Evacuation Planning, 3rd, Ed. Australia, Commonwealth of Australia
[8] Australian Red Cross, (2012). Evacuation Center Field Guide. Australia: Australian Red Cross
[9] North Bay Parry Sound District Health Unit NBPSDHU (2012), Evacuation Center Plan, United States: North Bay Parry Sound District Health Unit.
[10] ADAS (2007). Impact of 2007 summer floods on agriculture, Battlegate Road Boxworth, Cambridge: *Australian Diver Accreditation Society.*

Natural Radium Detection and Inventory Flux of Isotopes in Particulate and Dissolved Phases of Seawater at Kapar Coastal Area Caused by Coal-Fired Power Plant

N Mohamed[1], N A N Ariffin[1] and C A R Mohamed[2]
[1] Department of Environmental Technology, Faculty of Applied Science, UiTM Shah Alam Selangor, MALAYSIA
[2]Marine Ecosystem Research Centre, Universiti Kebangsaan Malaysia, 43600 Bangi, Selangor, MALAYSIA

E-mail: noradila@salam.uitm.edu.my

Abstract. Distribution of ^{226}Ra and ^{228}Ra radioactive in marine have been studied at Kapar coastal area that closed to Sultan Salahudin Abdul Aziz Shah (SJSSAS) power station. The concentration level of ^{226}Ra and ^{228}Ra were measured in seawater include total suspended solids (TSS_{rw}) and dissolved phases from September 2006 to February 2008. The measurement technique used for ^{226}Ra and ^{228}Ra was using cation exchange column and counted using Liquid Scintillator Ciunter (LSC). The radioactivities of $^{226}Ra_{sw}$ and $^{228}Ra_{sw}$ in the dissolved phase of seawater ranged from 1.29 ± 0.52 mBq/L – 3.69 ± 1.29 mBq/L and 2.12 ± 0.71 mbq/L – 17.07 ± 6.03 mBq/L respectively. The measurement of radioactivities of radium isotopes in the particulate phase of seawater ranged from 15.62 ± 1.99 Bq/kg – 241.76 ± 100.23 Bq/kg ($^{226}Ra_{tsw}$) and 7.19 ± 3.21 Bq/kg – 879.66 ± 365.74 Bq/kg ($^{228}Ra_{tsw}$). Radium isotopes inventory in this study showed that suspended solid have higher inventory value than seawater and sediment. Study also found that suspended solid play an important role for flux contribution at seawater. Based on the finding, the radioactivity concentration of ^{226}Ra and ^{228}Ra is higher in particulate phase than in dissolved phase.

Keywords: Radium, coal-fired, seawater.

1. Introduction

Industrial revolution has introduced coal as crucial global energy sources. Coal is the world's most abundant and accessible source of fossil energy. It is also believed that coal is a source in circulation pattern of environmental marine of coal power station where natural radionuclide such as radium isotopes associated with coal [1]. Radium isotopes, ^{226}Ra ($T_{1/2}$ = 1600 years) and ^{228}Ra ($T_{1/2}$ = 5.75 years) are affected by vertical mixing, horizontal circulation and adsorption from particulate in seawater. ^{226}Ra and ^{228}Ra are among the most important isotopes in the environment from radio protection points of view. ^{226}Ra is powerful tool for studying geo-hydrological processes. It is used intensively as tracers for tracking groundwater sources that discharge into the coastal ocean [2,3,4] and mixing processes between the coastal zone and open ocean [5].

Inventory is a parameter that used to estimate and evaluate the incremental transfer/accumulation of element such as radionuclides, metals and others that mainly occur in

IOP Conf. Series: Materials Science and Engineering **136** (2016) 012079 doi:10.1088/1757-899X/136/1/012079

seawater sediments. Inventory of radionuclides in sediment is measured by measuring the radioactivity of radionuclide in the sediment by using following equation [6,7]:

$$I\,(\text{Bq/m}^2) = \text{Radioactivity}\,(\text{Bq/m}^3)\text{ x Average depth (m)} \qquad (1)$$

Flux is a parameter used for assessing the rate or velocity/collection of elements in a dissolved substances, suspended matter and sediments by dividing inventory with time of sampling. Equation used to estimate radionuclide flux is based on equation as follow:

$$F = I.\lambda \qquad (2)$$

Where
F=flux,
I=inventory,
λ=constant value

Since the study on ^{228}Ra and ^{226}Ra as geochemical tracer in marine are poorly known in Malaysia, this study carried out to observe the distribution of radium isotopes in marine phases at Kapar coal-fired power plant. The main aims of this study are to determine the concentration of radium isotopes and flux in three different phases of seawater at Kapar coastal area.

2. Materials and Methods

2.1 Study Area

Kapar Coal-fired Powerplant known as Sultan Salahudin Abdul Aziz Powerplant (SSAAPP) is one of the largest electric coal powerplants in Malaysia with capacitiy of 2420 MW and produced 23% energy at the maximum demand (TNB 2014). It was opened in March 1987 and the only power station in Malaysia with triple fuel firing capability which is gas, oil and coal. The study station is located by seaside near Kapar in Klang (Table 1).

Table 1. Sampling station at Kapar coastal water.

Site	Station	Latitude	Longitude
Kapar Coastal	1	03° 5' 55.4"	101° 17' 59.7"
	2	03° 7' 25.8"	101° 18' 1.8"
	3	03° 6' 37.2"	101° 08' 40.4"
	4	03° 5' 53.0"	101° 19' 18.6"
	5	03° 6' 27.1"	101° 19' 43.8"
	6	03° 5' 55.4"	101° 17' 59.7"

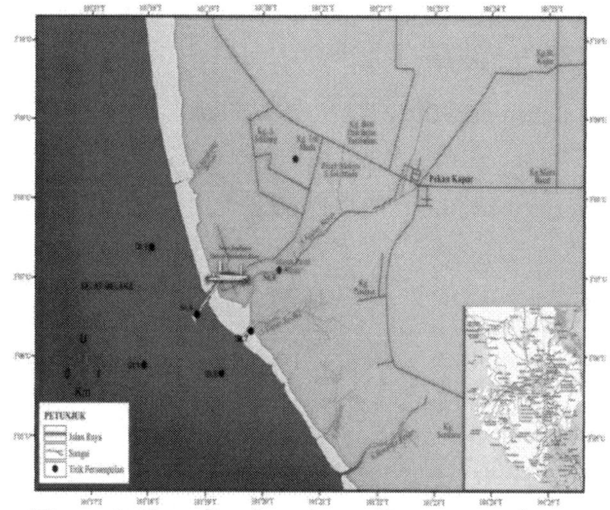

Figure 1. Sampling location at Kapar coastal area.

2.2 Methods

Six stations of seawater samples were collected from coastal area of Kapar power plant. Seawater samples were filtered through 0.45µm pore size of pre-weighed membrane filter paper. Seawater samples were acidified with concentrated HNO_3 to pH 2. Then, 1 ml of barium and carrier solutions (25mg/l) and Na_2CO_3 was added into the samples. Ammonia solutions were added continuously to the samples and were stirred vigorously until pH 10. After supernatant appear, siphon out the supernatant from the filtrate, and dissolve the carbonate with HNO_3 and $HClO_4$. Centrifuged was used to remove supernatant and then dissolved in 20 ml of 1% $HClO_4$ and stand in warm for 1-2 hours.

For particulate phases (total suspended solid and sediment) were weighted after dried into the oven at 70ºC. 1 ml of barium carrier (25mg/ml) was added into each sample into the Teflon beaker. Then the samples were evaporating to dryness with mixture consisting 10 ml of HNO_3, 5 ml of $HClO_4$ and 1 ml of HF on the hot plate. A few millimeters (0.5ml) of H_2O_2 were added at digestion stage. Then the residues were dissolved in 5 ml of concentrated HNO_3 and 1ml of $HClO_4$and stand in warm condition.

After dissolved and particulate phases done, the samples were purified using the cation exchange column. 200 ml of 2M HCl were used to elute Ba (Ra) from the cation resin and dry the Ba (Ra) aqueous on the hotplate until dryness. Then, 20 ml of 0.5M HCl and 1 ml of H_2SO_4 were added into the residue until white Ba (Ra) SO_4 precipitate appeared. Milipore filter (25mm diameter, 0.45µm pore size) was used to filter the precipitate.

Then, dried and weighed to calculate the chemical yield. The precipitates together with filter paper were transferred into a 20 ml glass vial followed by adding Instagel® XF and Ultimate™ Gold AB. The vials stored for over 21 days after well mixing using ultrasonic cleaner [8,9,10]. The vials were counted using liquid scintillation counter (LSC).

3. Results and Discussion

3.1 Distribution of ^{228}Ra and ^{226}Ra in dissolved, total suspended solids and sediment in seawater

The activity of ^{226}Ra and ^{228}Ra concentration in dissolved varied widely from 1.41 mBq/L to 3.69 mBq/L and 6.01 mBq/L to 17.07 mBq/L respectively. The activities of ^{226}Ra concentration in Kapar coastal area have been found relatively low level and almost similar than other surface area samples elsewhere in Malacca Straits [11,12] and world oceans. Mean concentration of ^{226}Ra and ^{228}Ra is differing for each sampling time.

ANOVA analysis showed significant difference 95% confidence for ^{228}Ra (p=0.000) and ^{226}Ra (p=0.001) with different sampling location for dissolved phase. Pearson correlation analysis showed a strong negative correlated between mean of radium activity and sampling occasions at Kapar (r =0.741) for ^{228}Ra. This is probably due to high dilution from different source (Figure 2).

Table 2. Mean radioactivity of $^{226}Ra_{sw}$ and $^{228}Ra_{sw}$ at Kapar coastal area.

Time	^{226}Ra(mBq/L±1 σ)	^{228}Ra(mBq/L±1 σ)
Mar 07	2.34 ± 0.91	11.02 ± 4.44
Aug 07	1.65 ± 0.59	8.08 ± 3.33
Nov 07	1.86 ± 0.64	7.67 ± 3.07
Feb 08	1.55 ± 0.59	7.44 ± 2.92

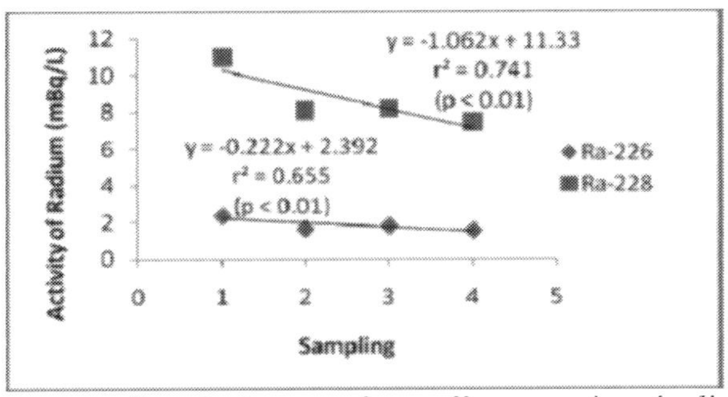

Figure 2. relationship between radium isotopes and sampling occasions in dissolved phase in seawater (mean values of four times sampling).

Radium isotopes activity in total suspended solids in seawater (TSS_{sw}) was higher than dissolved phase with mean activity of ^{228}Ra in TSS_{sw} range from 17.73±6.33Bq/kg-879.66±365.74Bq/kg. Meanwhile for ^{226}Ra in TSS_{sw} range from 5.62±1.99Bq/kg – 241.76±100.23Bq/kg.

In total suspended solids (TSS_{sw}) phase, radium isotopes activities was varied based on its sampling location. This was supported by ANOVA result that found significant 95% confidence level p=0.002 for ^{226}Ra and p=0.003 for ^{228}Ra (Figure 4). ^{228}Ra and ^{226}Ra activity in particulate phases were varied from 3.15 to 4.83. It found that ^{228}Ra are more abundance than ^{226}Ra in dissolved phase. It also found that most of the sampling stations showed high activity values of $^{228}Ra/^{226}Ra$. It may due to enrichment of ^{228}Ra in removal of ^{228}Ra from water column onto the particles which is caused from coal burning operation of Kapar power station.

3.2 *Inventory and flux of ^{226}Ra and ^{228}Ra at Sultan Salahudin Abdul Aziz Power Plant coastal area.*

Inventory of ^{226}Ra showed value range from 2.53 × 10³ Bq / m² - 3.63 × 10³ Bq / m² (March 2007) , 2.59 × 10³ Bq / m² - 3.62 × 10³ Bq / m² (August 2007) , 1.56 × 10³ Bq / m² - 1.74 × 10³ Bq / m² (November 2007) and 1.61 × 10³ Bq / m² - 2:52 × 10³ Bq / m² (February 2008) . It was found that at the time of sampling in November and February showed a low inventory as both the months was dry season (northeast monsoon). Flux also showed inconsistencies as inventories as the flux value changed based on the sampling location.

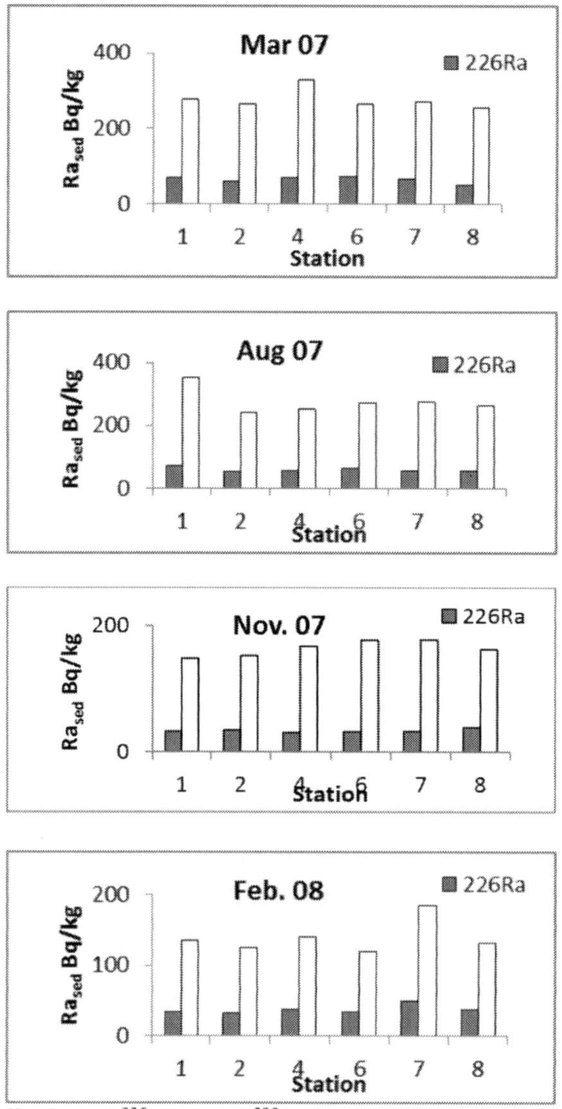

Figure 3. Distribution of $^{226}Ra_{sed}$ and $^{228}Ra_{sed}$ radioactivity at Kapar coastal area.

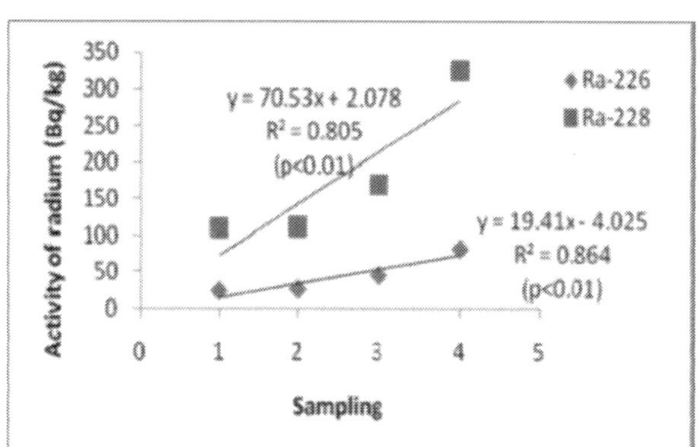

Figure 4. Relationship between radium isotopes and sampling occasions in total suspended solids (TTS_{SW}) phase in seawater (mean values of four sampling).

The average flux of sediment in the range of 6.78×10^3 Bqm $^{-2}$ yr $^{-1}$ to 7.13×10^3 Bqm^{-2} yr $^{-1}$ for $^{226}Ra_{sed}$ and 28.01×10^3 Bqm $^{-2}$yr $^{-1}$ to 55.49×10^3 Bqm $^{-2}$yr $^{-1}$ for $^{226}Ra_{sed}$ (Table 3). ANOVA analysis showed a significant difference between the inventory and the flux at the 95% significance level in all research stations with $p = 0.000$.

First sampling recorded highest inventory and fluxes in radium isotopes. It may explained that the time and location of sampling not only active with the biological activities but also active with physical activities such as flow and water flow , wave , tidal , and wave which accelerates the transport of biological accumulation elements to the area [7].

Resulted from three sampling phase showed that inventory is in order of suspended solid>Sediment>Seawater. Suspended solid phase capable of trapping and adsorb the chemicals in the water column [13] .The content of suspended particulate matter in the estuary has always been high and is constantly changing depending on the tide and are affected by the wind. This situation will result in an inventory of the sediment surface higher than the inventory in seawater.

Total radium isotopes flux in seawater and suspended solid recorded high value of during February for 226Ra is 17.21 x 104 Bq/m2/yr and 67.82 x 105 Bq/m2/yr for 228Ra. Study found that total flux for 228Ra is higher than 226Ra. This is caused gap in half-life of both isotopes which 226Ra is 1602 and 5.75 years for 228Ra. Studies conducted by Phuah et al.,[9] found that the amount of suspended material plays a crucial factor for contribution of flux in Straits of Malacca (West Coast) or South China Sea (East Coast). The values of radium isotopes flux from total suspended solid was higher than others samples. It believed that particles play an important role in a source of radium isotopes in estuaries or sea.

Table 3. Radium isotopes flux at Kapar coastal area.

Sample type	**March 2007**		**August 2007**		**November 2007**		**February 2008**	
			Flux ($Bq/m^2/yr$)					
	^{226}**Ra**	^{228}**Ra**	^{226}**Ra**	^{228}**Ra**	^{226}**Ra**	^{228}**Ra**	^{226}**Ra**	^{228}**Ra**
Seawater	35.71	194.39	27.87	87.89	39.37	150.19	25.67	141.76
Suspended solid	4.79×10^4	2.18×10^5	3.15×10^4	2.22×10^5	9.11×10^4	3.38×10^5	16.45×10^4	6.50×10^5
Sediment	13.07×10^3	55.49×10^3	10.92×10^3	50.57×10^3	6.78×10^3	32.87×10^3	7.55×10^3	28.01×10^3
Total	$\mathbf{6.10 \times 10^4}$	$\mathbf{27.37 \times 10^5}$	$\mathbf{4.24 \times 10^4}$	$\mathbf{27.27 \times 10^5}$	$\mathbf{9.79 \times 10^4}$	$\mathbf{37.10 \times 10^5}$	$\mathbf{17.21 \times 10^4}$	$\mathbf{67.82 \times 10^5}$

4. Conclusions

Highest values of radium were detected at total suspended solids compared with dissolve phase in seawater. Radium isotopes inventory and flux were varied based on sampling location and monsoon season. Isotopes inventory show in order of Suspended solid>Sediment>Seawater. The mean concentrations of flux for one year are 9.34×10^4 for ^{226}Ra and 3.99×10^6 for ^{228}Ra.

Acknowledgement

Authors would like to thank to all person and organization that involved in this study, particularly special thanks to Universiti Kebangsaan Malaysia and Universiti Teknologi MARA Shah Alam and also to Ministry of Education for this research grant and funded by Fundamental research Grant Scheme (FRGS) [File No: 600-RMI/FRGS 5/3 (43/2013)].

IOP Conf. Series: Materials Science and Engineering **136** (2016) 012079 doi:10.1088/1757-899X/136/1/012079

References

[1] Cevik U, Damla N. & Nezir S. 2007. Radiological characterization of Cayirhan coal-fired power plant in Turkey. *Journal of Fuel* 86: 2509-13.

[2] Charette M.A, Buesseler KO, Andrews JE. Utility of radium isotopes for evaluating the input and transport of groundwater–derived nitrogen to a Cape Cod estuary. Limnol. Oceanography 2001; 46: 465-70.

[3] Burnett W.C, Dulaiova H. Estimating the dynamics of groundwater input the coastal zone via continuous radon-222 measurements. Journal of Environmental Radioactivity 2003; 69: 21-35

[4] Moore W.S. Large groundwater inputs into coastal waters revealed by 226Ra enrichments. Nature 1996; 380: 612-14.

[5] Moore W.S. Determining coastal mixing rate using radium isotopes. Continental Shelf Research. 2000; 20: 1993-2000.

[6] Mohamed, C. A. R., Narita, H., Harada, K. & Tsunogai, S. 1996. Sedimentation of natural radionuclides on the seabed across the northern Japan Trench. *Geochemical Journal* 30: 1-9.

[7] Zal U'yun W. M. 2009. Geokronologi sedimen dengan menggunakan siri pereputan uranium-torium sebagai penunjuk pencemaran di persekitaran marin. Tesis Doktor Falsafah. Universiti Kebangsaan Malaysia.

[8] Chalupnik, S., & Lebecka, J., 1990. Determination of radium isotopes in liquids by means of quantulus liquid scintillation spectrometer. Proceedings of 14th europhysics conference on nuclear physics, Bratislava (p. 327). Singapore: *World Scientific*.

[9] Chalupnik, S., & Lebecka, J., 1993. Determination of ^{226}Ra, ^{228}Ra and ^{224}Ra in water and aqueous solutions by liquid scintillation counting. *Proceedings of liquid scintillation conference* (pp. 397±403). RADIOCARBON.

[10] Kim, Y. J., Kim, C. K., & Lee J. I., 2001. Simultaneous determination of ^{226}Ra and ^{210}Pb groundwater and soil samples by using liquid scintillation counter – suspension gel. *Journal of Radiation and Isotopes* 54: 275-281.

[11] Nioo, S. Y., Ahmad, Z., Mohamed C. A. R. 2005. Pencirian aktiviti ^{226}Ra dan ^{228}Ra dalam sistem sungai, muara dan marin di Malaysia. Tesis sarjana. Universiti Kebangsaan Malaysia.

[12] Phuah, C. S., Ahmad, Z. & Mohamed C. A. R. 2004. Taburan Keradioaktifan ^{226}Ra dan ^{228}Ra dalam sistem marin di Semenanjung Malaysia.

[13] Turner, A. & Millward, G. E. 2002. Suspended particles: Their role in estuarine biogeochemical cycles. *Estuarine, Coasta*l and *Shelf Science* 55: 857-883.

Analysis the Impact of Bridges Existance for the Segamat River Using Infowork RS

M S Adnan[1], N C Ali[2], Y Erfen[2], S N Rahmat[1], M A M Razi[1] and S Musa[1]
[1]Micro Pollutant Research Center, Faculty of Civil and Environmental Engineering, Universiti Tun Hussein Onn Malaysia, 86400 Batu Pahat Johor, MALAYSIA
[2] Faculty of Civil and Environmental Engineering, Universiti Tun Hussein Onn Malaysia, 86400 Batu Pahat Johor, MALAYSIA

E-mail: shalahudin@uthm.edu.my

Abstract: Malaysia is experiencing two monsoonal seasons, which have induced heavy rainfall. The increasing of rainfall intensity and longer duration of rainfall has caused a flood. In 2007, Segamat city was hit by the flood due to increasing of rainfall and this has caused a huge damage in term of infrastructure as well as human life. Thus, this study was carried out with the aims to analyze the flood and develop a flood map for the Segamat River. The hydrological data, such as rainfall and streamflow data were retrieved from the Department of Irrigation and Drainage Malaysia (DID). While, for map preparation such as base map, landuse map and river map were delineated based on the topography map. Additionally, contour map was obtained from MACGDI. InfoWork River Simulation (RS) software was used for analysis and develops the flood map. Moreover, two scenarios were carried out to determine the effect of hydraulic structure presence in the waterway. The first scenario is the flood map without a bridge, while the second scenario is the flood map with a bridge. Based on the results, it clearly shows that the flood map with a bridge has higher flow, velocity and water table compared to the flood map without a bridge. In addition, a flood-prone area based on the developed map was identified. Thus, the generated flood map could serve important information to related agency in proposing a suitable design the mitigation works for this area.

Keywords: Infowork RS, flood map, segamat river, flood analyses.

1. Introduction

Flood is a natural disaster that normally occurred due to several factors such as heavy rainfall, poor drainage system, mismanagement of land use and so forth. The increasing of temperature and a warmer climate globally may lead to unstable pressure in the atmosphere that has caused more frequent and intense floods for many areas around the world. Changes in world climatic could affecting the hydrological water balance and geomorphologic right to complexities in balancing the societal and environmental needs [1].

The flood cannot be avoided since the amount of precipitation is out of the human control. However, as a human, we could control the magnitude of the flood by mitigating it by using several approaches such as structural or non-structural measures. The mitigation works normally will be carried out with the aim to find a determined development and application of new tools and knowledge to limit the possibility and effects of floods. During a flood, a few hours can make all difference between disaster and massive disaster. The development of approaches to manage flood risks could lead to optimal utilization of the available mean for preventing flooding and optimizing the mitigation of the flood consequences. The selection of the flood mitigation either by structure or non-structural measures based on the purposes and aims of the project. The example of structural measures for flood mitigation could be a river bund, flap gate, pump house and many more. While, flood modeling, education and laws are non-structural measures that could be carried out.

Based on a report from the Department of Irrigation and Drainage, Malaysia [2] has stated that about 29,000 km2 or 9% of total land area and more than 4.82 million people, which is 22% is affected by flood annually with the damage caused by a flood is estimated about RM915 million. Residents who lived in low land, close to river bank and flood prone areas were identified as most vulnerable to flood hazard. The low land area suffers from severe flood damages every year, it has resulted in loss of life, damage to property and had negative impacts on human welfare.

The advanced technology in computer simulation has become a powerful tool in model the flood map. Computer model could give major advantages to the modeler since it likely to produce more accurate results, designs with cost saving by avoiding over and under sizing. The application of computer model in producing flood map was well recognized. Until today, many kinds of flood related maps were developed and produced such as flood map, flood risk map, inundation map, flood prone area and many more. Jones [3] from USGS has mentioned the important of flood map in minimizing the consequences of floods. He has stated that the maps generally available today are maps used for planning. They are maps of theoretical floods, not maps of flooding forecast for an approaching storm. The U.S. Geological Survey (USGS) and the National Weather Service (NWS) have developed a way to bring flood forecasting and flood mapping together, producing flood maps for tomorrow's flood today and getting them on the Internet in time for those in harm's way to react.

Effective modelling and mapping solutions are critical to our improved understanding of flood risks. Flood maps that represent information in graphs, the visualization can be supplemented with numerical data such as data of water level, flow. The accuracy of the results depends on the validity of the equations used to describe the physical processes and the quality and detail of the input data to which the equations are applied [4]. They also mentioned that DEM resolution has significant effect on simulation results. Flood simulation characteristics that are affected are inundation extent, flow velocity, flow depth and flow patterns across the model domain.

InfoWorks RS is one of the available commercial software that could model the flood map. This hydrodynamic modeling software could be used to analyze the flow in open channels, floodplains, hydraulic structure and embankment. This application is good enough in giving some thought to the related agencies and person involved in flood mitigation project to evaluate the potential impacts prior implementing prevention measures to avoid the anticipated adverse. Thus, in this paper, the application of InfoWork RS in modeling and analyzing the flood occurrence will be discussed. In addition, two scenarios were simulated to understand differences in water level as well as the flow if there is a hydraulic structure and no hydraulic structure in the study area.

2. Methodology

2.1 Description of Study Area

Segamat River is one of the Muar River tributary that flow passing through the Segamat Town. The total length of this river approximately about 23 km with typically 14m wide and 14 m above the sea level (Figure 1). This river originates from Gunung Besar and joining Muar River at Kampung Tebing Tinggi. Segamat River catchment areas comprise of 685 km2 and an average annual rainfall of 1813 mm, but can reach up to 600 mm per day during the monsoon season, between November through March [5]. During 2006-2007 floods, the water level has risen dramatically beyond the danger level in less than 72 hours and cause of 13 roads that connect the Bandar Segamat were closed due this catastrophic flood.

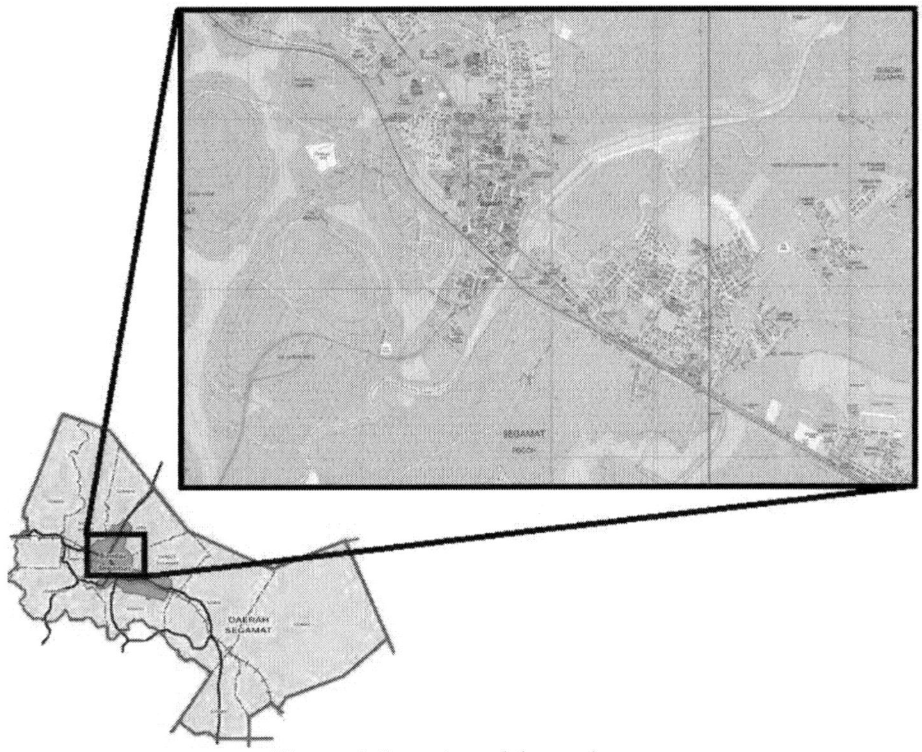

Figure 1. Location of the study area.

2.2 History of Flood in Johor

Johor state has been hit by major floods in 19-31 December 2006 the first flood of 12-17 January 2007 and the second flooded involving district of Johor Bahru, Kota Tinggi, Batu Pahat, Muar, Segamat, Kluang, Pontian and Mersing. The floods caused by series of storm events generated by Northeast Monsoon that caused heavy rains that fell in continuous and comprehensive for several days in Peninsular Malaysia especially in the state of Johor.

During this time, heavy rain has resulted in a massive increase of water level, which results in flood that has paralyzed almost all the daily activities which involved eight districts. It was reported that, most of the areas were observed to be under water due to the increasing of water, which resulted in more than 103,000 populations have to be evacuated to evacuation centers. Unfortunately, this flood event also killed at least 17 lives during the second flood occurred where when the people began to move back home after the first flood from was receded. Figure 2 shows the flood affected area in Segamat Town during the flood event.

Based on records and statistics from DID, most of the water level of major rivers in the state of Johor was recorded a significant increase of water and was categorized as dangerous level. Based on the record for Segamat River at Segamat Town, the maximum recorded water level is about 11.7 m which is above the alert level at 8.53. While for others gauge stations like Buloh Kasap and Bukit Kepong also had shown an increase of water level above a dangerous level at 11.2 m and 4.69 m respectively [6].

2.3 Data Management

Data management is one of the most important issues in producing good and acceptable results in modeling. Thus, in this study, combination of both approaches which are primary and secondary data collection were conducted to ascertain the data is enough and available to carry out the task. The data collection was divided into two parts which are (i) hydrology and hydraulic data and (ii) related maps. For hydrology and hydraulic data, the data was retrieved from DID which is the respective agency to monitor and collect the data such as rainfall, temperature, streamflow and sediment. While for map, the map was provided by Malaysian Centre for Geospatial Data Infrastructure (MaCGDI) and delineation based on paper map.

Hydrological model development involves with two main processes which are GIS application and hydrological data. In GIS data processes, the related map were developed and produced such as river network, contour, road and landuse maps. While, for hydrological data the maximum and minimum of annual rainfall as well as flood events are needed. On the other hand, river cross sections, river basin maps, sub-catchment maps, water level and sediment yield are being prepared. Moreover, latest data was collected by conducting a test and measurement at site which was used for calibration and validation purposes.

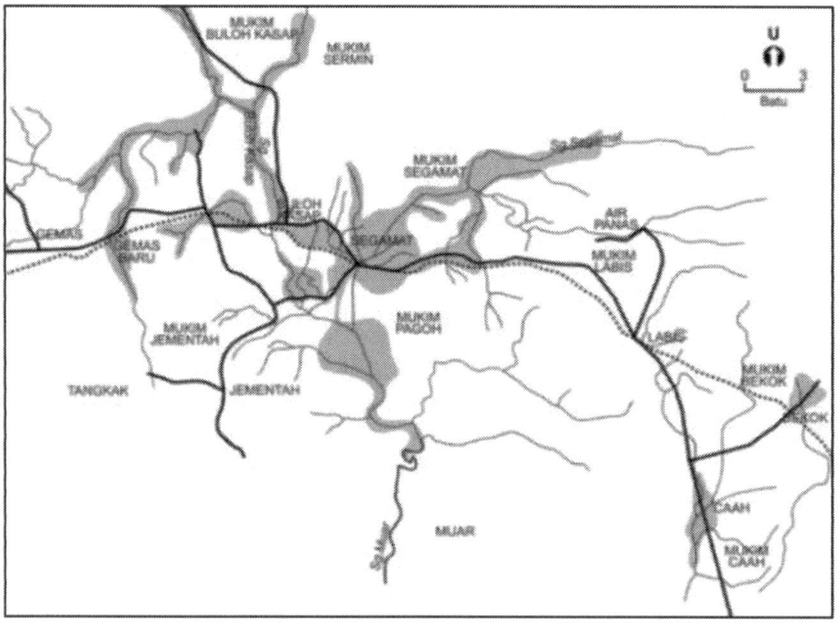

Figure 2. Flooded areas in Bandar Segamat (DID, 2007).

2.4 InfoWorks RS

InfoWorks River Simulation (RS) is one of the modeling programs that combine the advanced flow simulation engine, both hydrological and hydraulic models, GIS functionality and database storage within one single environment. The basic system architecture is an "Integrated Network Model" links data storage using a GIS to hydrologic or hydraulic modeling software suite embedded in InfoWorks RS. It will allow planners and engineers to carry out fast, precise modeling of the key elements of river and channel systems and to view the model data and results in new ways. InfoWorks RS also brings full model management in the water engineering field, allowing a full check trail to be maintained by the modeling process from source data to final outputs.

InfoWorks RS provides the ability to review present and historical model network versions and attribute data by storing information within a database environment. As well as providing full information of each alteration made to the network, it also provides version one dimension, date stamps, and modeler details. The data views show the assurance levels that have been attached to each asset data attribute within the system. A compare function allows the assessment of two model versions and the creation of a highly detailed report outlining the differences, including changes to the data flags recitation poise and source of all data items [7].

The Wallingford Software's InfoWorks RS is an example of a one-dimensional hydrodynamic model used for prediction of discharge and water level for a wide range of rivers, reservoirs, complex floodplains and narrow estuaries under both steady and unsteady conditions. In additional, data from a wide range of sources can be brought together within the software in a rapid and flexible way. InfoWorks computes flow depths and discharges using a method based on the equations for shallow water waves in open channels the Saint-Venant equations [8].

2.5 Development of Flood Map for the Segamat River

Geographical Information System (GIS) is a powerful tool in assisting a modeler to create a base map and others maps that can be put into a series of layer for comprehensive analysis. Soil and Water Assessment Tool

(SWAT) was coupled with GIS to delineate the catchment area. The Triangular Irregular Network (TIN) is a digital data structure for the representation of a surface as shown in Figure 3. TIN is a vector based representation of the physical land surface, and often derived from the elevation data such as a contour. TIN is made up of irregularly distributed nodes and lines with three dimensional coordinates (x-axis, y-axis, and z-axis). The digital data of contour and river are used for the formation of TIN for this study area. The TIN formation is through 3D analysis extension which is provided in ArcMap.

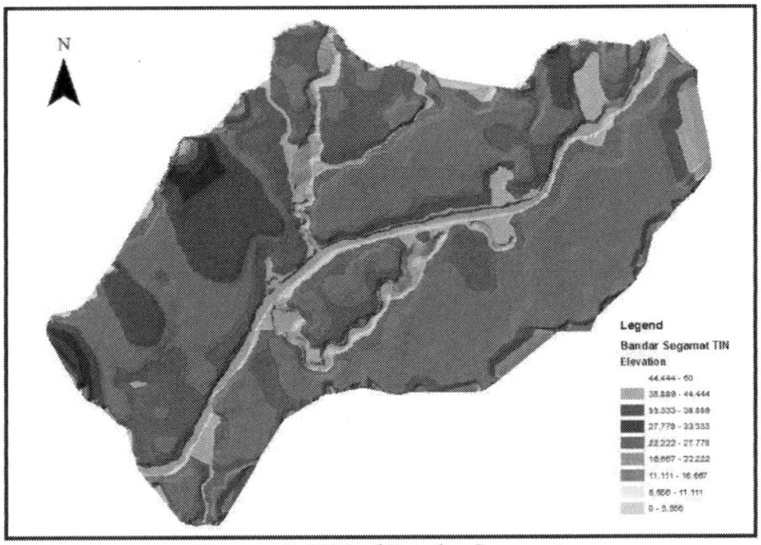

Figure 3. TIN of Bandar Segamat.

TIN model which was developed earlier was used to produce a DEM. Through the ArcToolbox that has been provided by ArcMap, TIN model is converted to DEM in raster format. The Digital Elevation Model is the simplest form of digital representation of topography and the most common. This data is used as input to quantify the characteristics of the land surface and formed the basis for dynamic flood mapping. There are several different methods have been used to create DEM series. A watershed is the area of land where all of the water that is under it or drains off of it goes into the same place. DEM is the main data required in the develop watershed. The formation of the watershed is important to identify the inlet and outlet of a catchment, as well as the amount of rainfall received by a river. In ArcMap there are special extensions for producing the watershed, which is using ArcSWAT. Through the ArcSWAT and raster data from the DEM, watershed can be determined.

The main input for the hydraulic model is the river cross section and floodplain, it also depicts the shape of the channel in which a river flows. The cross sectional area is the wet area of the river normal to the direction of flow. Even among different types of the rivers, common characteristics of cross sections are observable from headwaters to the river mouth [9]. Eventhough this river is 23km long, the analyzes that will discussed in this paper only covered 4.3 km which covered the area close to Segamat Township area (Figure 4). The selection of this area is primarily due to objective of this study which is to determine the most impact area at this town to avoid the losses and saving a life. Moreover, two scenarios will be discussed the impact of flood if there are a hydraulic structures and no structure in the waterway during a peak discharge.

3. Results and Discussions

Based on the rainfall analysis for the Segamat Basin, annual rainfall is about 1800 mm/yr. The hydrological data of 2007 were used to validate the results. As mentioned earlier, two scenarios will be discussed to determine the differences between having hydraulic structure and no hydraulic structure. Within 4.3 km reach, there are three (3) bridges were observed as shown in Figure 4.

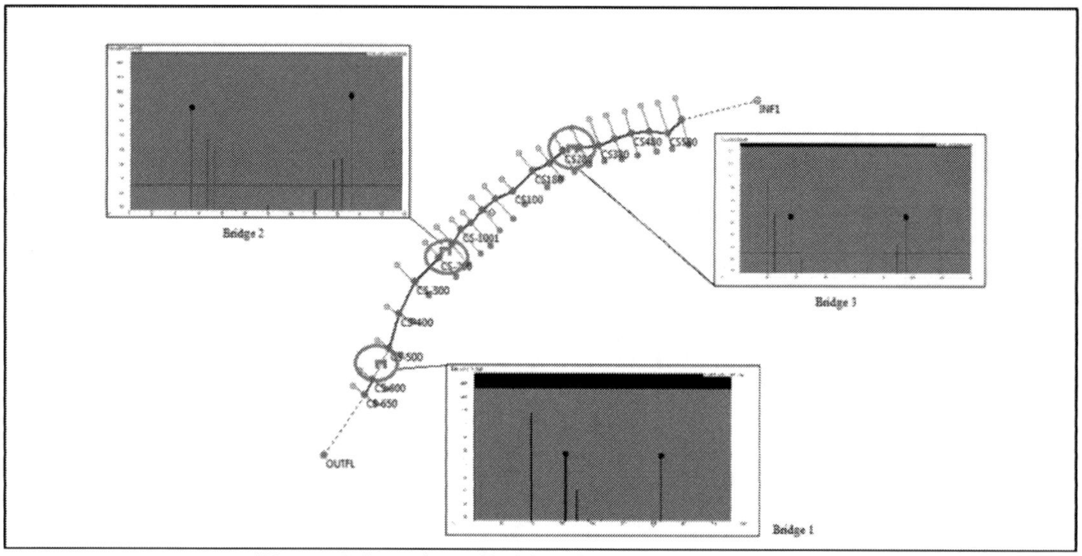

Figure 4. Location of bridges in study area (Segamat Town).

3.1 With a bridge

By having the bridge in the river has caused a flow disruption as shown in Figure 5. The hydrograph shows that the water level was significantly increased during this event and the increment within a range of 5 – 15 meter from normal depth. The water level start to increase since 6th Dec 2007 and reach the first peak on 7th Dec 2007 with a water level is 16.175 m. The second peak was recorded 4 days after which is on 11th Dec 2007 with a water level reading about 20.543 m. Both of water level reading is above the danger level based on DID historical data. While Figure 6 shows the longitudinal cross section from upstream to downstream during the peak flow. Based on this figure, it clearly shows that during the peak flow, the water level has surpassed the left and right river bank that caused the water flow to the lowland area on the surface.

Table 1 shows the value of maximum water level, flow and velocity for every bridge during peak hours. Based on this result, it shows that the existing of the bridge in waterway could trigger the increasing of water level and slow down the flow. Bridge number 3 shows the highest water level which is 20.467m, while bridge number 1 recorded the lowest water level. In terms of flow and discharge, the highest flow value is 2048.414 m3/s for bridge number 1 while the lowest value was calculated at bridge number 3. This could give some important information to us, the water level and flow was influenced by the existing of the bridge and the condition of river width. This condition correlates with the shape and width of the river. It was observed that the upstream area the terrestrial and the terrain of this river are uneven and the width is narrower compared to downstream areas. This finding corresponding with the study conducted by [10] in Jhalem River. They had mentioned that the increase in valley slope, discharge increases and the water level decreases. On the other hand, with the increase in valley width (in terms of entrenchment ratio) both discharge and water level decrease. It is also concluded that with an increase in valley slope, the overall flood severity in a river valley would increase due to fast flooding. The existing of bridges in the river has become an obstacle for flood flow to get through since the river width was narrow down and there are might be debris that was flowing together that block the flow when this debris stuck at the bridge piers.

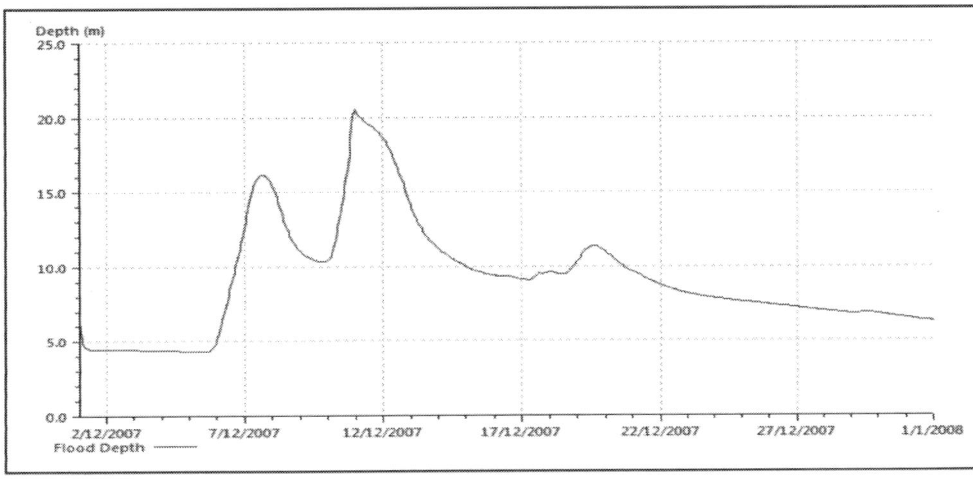

Figure 5. Water level of Segamat River at Segamat Town for event with bridge.

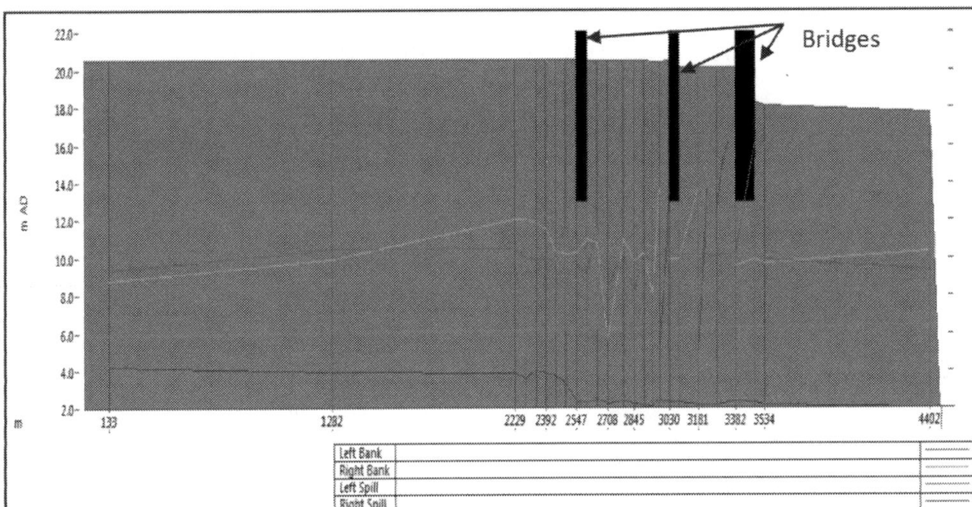

Figure 6. Long section view of the Segamat River for Unsteady event with bridge.

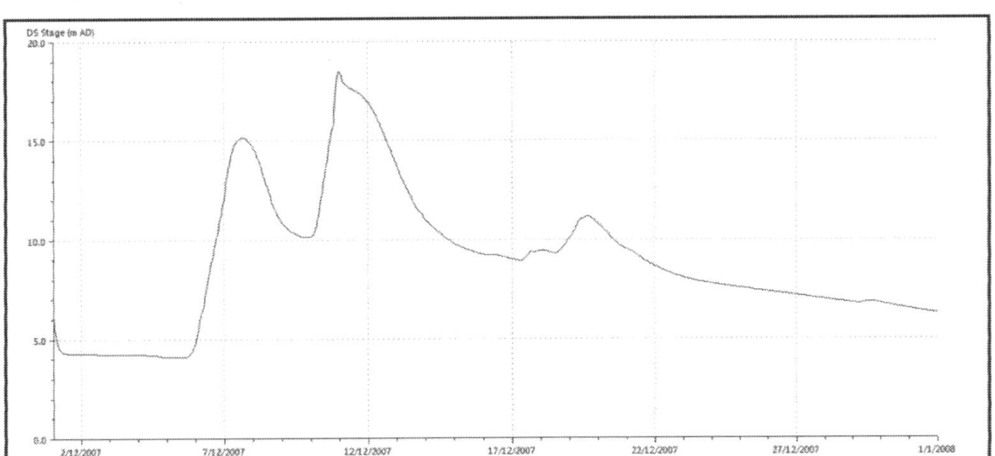

Figure 7. Water level of Segamat River at Segamat Town for event without a bridge.

Table 1. Analyses data for each bridge (with bridge).

Bridge No.	Stage (m)	Flow (m^3/s)	Velocity (m/s)
1	18.321	2048.414	0.711
2	20.154	1870.607	0.241
3	20.467	615.016	0.049

3.2 Without a bridge

For the second scenarios, the assumption was made by stating that there is no bridge was constructed to determine the difference of water level, discharge as well as velocity. The presence of the bridges act as a barrier has caused disruption to the water flow. With the absence of bridges, flood water flow passes the city faster compared with the presence of bridges. This has resulted in lesser effected areas hit by the flood. Based on hydrograph as shown in Figure 7, the maximum water level is about 18.54 m which is lower compared with the water level with the existing of bridges. This has strengthened our finding that the existing of hydraulic structure in the river could increase the water level.

The absence of bridges has also increased the water velocity; reaching 0.962 m/s. Water velocity tends to be increased if there is no obstacle in the river. For verification, the selection of one month simulation was carried out to determine the effect of the bridge on the velocity. Based on the Figure 8, it clearly shows that the velocity was higher for scenario having no bridge. The big difference was observed on 11th December with a difference of 0.227 m/s. A difference of ARI is also being conducted to understand the effect of a difference with the presence of bridge on flood mapping. 100 ARI was selected to model the flood map. Based on the results, it shows that velocity for scenarios without a bridge simulated higher value. Figure 8 clearly shows the affecting area by the two scenarios mentioned earlier. The area of the flood was increased when there is a structure being built in the river. The spreading of water is wider in the map having a bridge compared to no bridge.

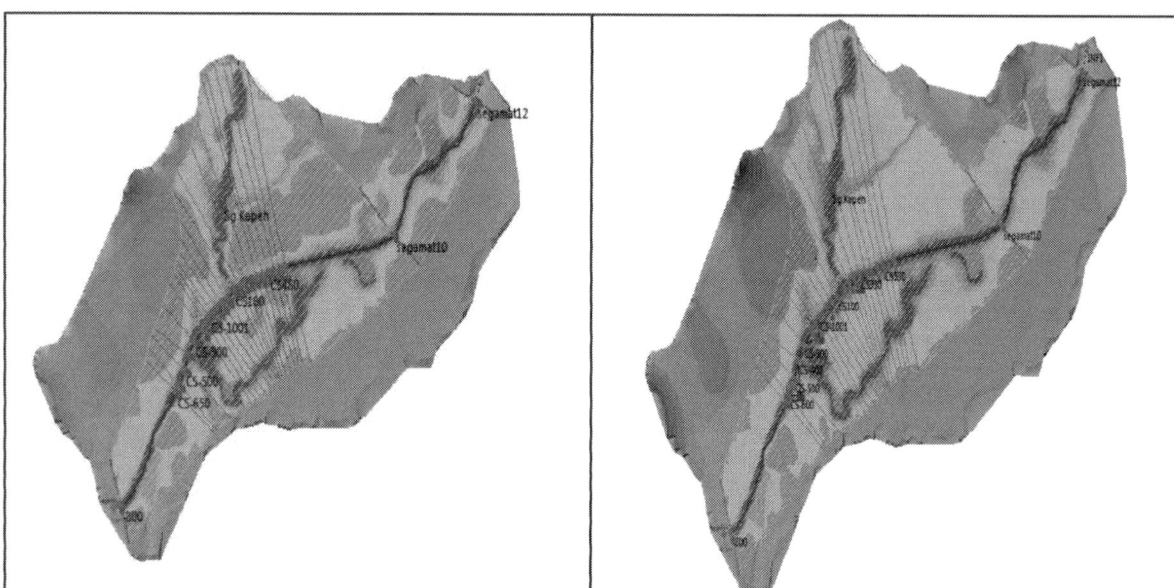

Figure 8. Flood mapping for both scenarios. (a) without bridge (b) with bridge.

4. Conclusion

This paper investigates the effect of the existence of a bridge for Segamat River at Segamat Town. It is concluded that the existing of bridge in the river influenced the water level, flow discharge, velocity as well as the affected area by the flood. The presence of the bridge has led to the increasing of water level, reduced the flow and velocity and wider affected area compared to no bridge. The flood water tends to spread and dispersed with the presence of the bridge. This can be seen clearly during the high flow especially during flood events. Thus, a finding from this study could be applied by the related agencies to plan flood mitigation project in this basin by identifying the possible locations of flood prone area.

Acknowledgements

The author gratefully acknowledges the financial support provided by the Ministry of Higher Education of Malaysia under a Grant RAGS, R017.The researcher expresses his sincere gratitude and appreciation to the University Tun Hussein Onn Malaysia in providing all the necessary facilities to achieve the current outcomes.

References

[1] Edna. M. R. (2007). Floodplain inundation simulation using 2d hydrodynamic modelling approach. International institute for geo-information science and earth observation enschede, the netherlands.

[2] Department of Irrigation and Drainage Malaysia, (2012). MSMA 2nd Edition: Urban Stormwater Management Manual for Malaysia, Malaysia,

[3] Jones, I.J. (2004). Mapping a Flood...Before it Happens. Fact Sheet USGS 2004-3060 june 2004.

[4] Hailea, A.T. & Rientjes, T.H.M. (2005). Effects of Lidar DEM Resolution in Flood Modelling: A model sentitivity Study for the City of Tegucigalpa, Honduras. Isprs wg iii/3, iii/4, v/3 workshop "laser scanning 2005", enschede, the netherlands, september 12-14, 2005.

[5] Liew, Y.S and Teo, F.Y. (2013). Impact of Climate Change on Flooding Scenario in Segamat River Basin, Malaysia. Proc. of the World Engineers Summit (WES2013). Pp 153-158.

[6] Bahagian Hidrologi dan Sumber Air. (2007). Ringkasan laporan banjir Januari 2007. Negeri Johor. Jabatan Pengairan dan Saliran Malaysia.

[7] Hassan. A. J. (2002). Penggunaan teknologi terkini ke arah pengurusan banjir yang bersepadu. Conference on pengurus kanan jps, (2002). Terengganu.

[8] Kulu, J.K.A., Darrien Y.S.Mah, Frederik Josep Putuhena &Salim Said (2007). Post-Flood Forensic Analysis of Sungai Maong Using Infoworks ™ River Simulation (rs). Journal - the institution of engineers, malaysia (vol. 68, no.4, december 2007).

[9] Ehsan, S. & Marx, W. (2011). Impact of river valley shape on flow characteristics in case of flooding. Pak. J. Engg. & appl. Sci. Vol. 8, jan., 2011 (p. 9-20).

[10] Jha. A. K., Bloch R., & Lamond J. (2011). Cities and Flooding: A Guide to Integrated Urban Flood Risk Management for the 21st Century.Washington: The World Bank.

IOP Conf. Series: Materials Science and Engineering **136** (2016) 012081 doi:10.1088/1757-899X/136/1/012081

Water Quality of Hills Water, Supply Water and RO Water Machine at Ulu Yam Selangor

N Ngadiman [1], N 'I Bahari[2], M Kaamin[1], N B Hamid[1], M Mokhtar[1] and S Sahat[1]

[1] Civil Engineering Department, Center of Diploma Studies, Universiti Tun Hussein Onn Malaysia, 86400 Batu Pahat Johor, MALAYSIA

[2] Faculty of Civil and Environmental Engineering, Universiti Tun Hussein Onn Malaysia, 86400 Batu Pahat Johor, MALAYSIA

E-mail: nryati@uthm.edu.my

Abstract. The rapid development resulted in the deterioration of the quality of drinking water in Malaysia. Recognizing the importance of water quality, new alternatives for drinking water such as mineral water processing from reverse osmosis (RO) machine become more popular. Hence, the demand for mineral water, natural spring water or water from the hills or mountains rose lately. More consumers believed the quality of these spring water better than other source of drinking water. However, the quality of all the drinking water sources is to meet the required quality standard. Therefore, this paper aims to measure the quality of the waters from hills, from RO machine and the water supply in Ulu Yam, Selangor Batang Kali, Malaysia. The water quality was determined based on following parameters: ammoniacal nitrogen (NH_3), iron (Fe), turbidity (NTU) and pH. The results show that the water from hills has better quality compared to water supply and water from RO machine. The value of NH_3 ranged from 0.03 mg/L- 0.67 mg/L; Fe was from 0.03mg/L – 0.12 mg/L, turbidity at 0.42 NTU – 0.88 NTU and pH is at 6.60 – 0.71. Based on the studied parameters, all three types of water are fit for drinking and have met the required national drinking water quality standard.

Keywords: Water quality standard, hills water, supply water, RO water machine.

1. Introduction

Water is a natural resource that is vital for life. Sufficient clean water is essential for our healthy living as well as the health of the environment. Malaysia is blessed with rich water resources and receives an average annual rainfall of 3000 mm that contributes to an estimated annual water resource of some 900 billion mm^3. All Malaysians have the right of access to safe, adequate and affordable water supply. One of the water related problems in Malaysia is the demand for drinking water that is rapidly rising due to equally increasing population [1]The current demand for drinking water is about 11 billion m^3, which is predicted to increase to about 18 billion m^3 in the year 2050 [2]. About 99% of Malaysia's raw water supply are derived from surface water sources primarily rivers and stream that originate and flow from the highlands, whereas, the 1% is is from groundwater source [3]. The majority of the potable water is tapped from the rivers, which are fed by the rain [4]. In view of the drinking water problem, the security and sustainability of water resources and to ensure adequate and safe water supply for all are crucial task for the authority [5].

However, water quality deterioration on an ongoing basis will have an impact on human health. Despite the enforcement of the Environmental Quality Act (EQA) in 1974, the water quality of Malaysian inland water (especially rivers) is following a deteriorating trend [4]. According to [4], reductions of good quality raw water is reflected in the increasing number of polluted and slightly polluted rivers. Land use changes is a leading cause of river pollution problems in Selangor State. The quality degradation of the Selangor River will still be expected in the years to come since pollutant loads are not envisaged to be handled effectively [6]. Therefore, monitoring the drinking water quality is significantly important because the possibility of the water source for drinking (from the river and ground water) becoming polluted is very high.

1.1 Water Quality Evaluation

Water quality is divided into three main categories: physical, chemical and biological [7]. Physical measurement involves turbidity, color, odor, and taste.. Normally physical parameters are recommended for use in monitoring water quality for small communities. In most cases the physical properties of drinking water are easily measured even by ordinary people [8]. Chemical parameters include the organic and inorganic forms of metals and nutrients. Measurement of ammonia nitrogen (NH_3-N) and Iron (Fe) are used for chemical parameters. Meanwhile, inorganic parameters were measured by pH, iron, sulfate and chloride. The biological parameters include total and fecal coliform, hetorotrophic plate count (HPC) and E. coli counts.

A Water Quality Index (WQI) developed to indicate overall water quality conditions based on water quality variables considered: dissolved oxygen, specific conductivity, turbidity, total phosphorus, and fecal coliform [9]. The WQI provides a useful way to predict changes and trends in the water quality by considering multiple parameters [10]. National Drinking Water Quality Standards as shown in **Table 1** was formulated in 2000 based on World Health Organisation (WHO) guideline. This standard was used as the water quality benchmarks of treated water and raw water which is either obtained from a water treatment plant or raw water in Malaysia [11].

Geographical locations may affect the quality of portable water, which its mineral contents are very dependent on the mineral compositions of the soil and pollutants such as heavy metal (Azlan et al. 2012). Further research should be done to identify the contents of water quality from difference water sourced. study of spring water are critical for everyday use for water is safe and does not cause any health problems for the long term or short term. Therefore, objectives of this study to identify the contents of water quality based on the experimental parameters of spring water, tap water and water RO machine for drinking water and measure the water quality based on Malaysian drinking water quality standards. Spring water security are critical for safety use purpose and contaminated spring water could cause health problems for the long term or short term.

Meanwhile, RO water machine is reverse osmosis membrane process cellophane, it's like separating the purified water from contaminated water. RO is when pressure is applied to the side, while the membrane to force water into the area, the dirt is pushed from the side and swept out. RO also can act as an ultra-filters remove particles such as certain microorganisms that may be too large to pass through the pores of the membrane. Even RO drinking water has undergone a filtration before purchasing by costumer but the water. Water characteristics profile developed based on the result of this study. The water characteristics consist; Ammoniacal Nitrogen (NH_3), Iron (Fe), Turbidity (NTU) and pH.

Turbidity occurs when pollutants such as clay, silt, organic matter, microorganisms and domestic or industrial wastewater mixed with water. Basically, turbidity can be seen with the naked eye and can be measured by the ability of light to penetrate into the water. Turbidity indicates the amount of particles suspended in water [14]. The turbidity meter is used with reference to the turbidity standard. Ammoniacal Nitrogen (NH3-N) is identified as one of the main pollutants to render many of the rivers polluted but it was not considered in the EQA as a monitoring parameter until the new regulations published in 2009 [4]. Ammoniacal nitrogen (NH3) is a parameter that determines the quality of water based on the growth of algae that contribute to the eutrophication process.

Table 1. National Drinking Water Standard (KKM, 2009).

PARAMETER	RECOMMENDED RAW WATER QUALITY	DRINKING WATER QUALITY STANDARDS
	Acceptable Value (mg/litre (unless otherwise stated))	Maximum Acceptable Value (mg/litre (unless otherwise stated))
Total Coliform	5000 MPN / 100 ml	0 in 100 ml
E.coli	5000 MPN / 100 m	0 in 100 m
Turbidity	1000 NTU	5 NTU
Color	300 TCU	15 TCU
pH	5.5 - 9.0	6.5 - 9.0
Free Residual Chlorine	-	0.2 - 5.0
Combined Chlorine	-	Not Less Than 1.0
Temperature	-	-
Clostridium perfringens (including spores)	-	Absent
Coliform bacteria	-	-
Colony count 22°	-	-
Conductivity	-	-
Enterococci	-	-
Odour	-	-
Taste	-	-
Oxidisability	-	-
Total Dissolved Solids	1500	1000
Chloride	250	250
Ammonia	1.5	1.5
Nitrat	10	10
Ferum/Iron	1.0	0.3
Fluoride	1.5	0.4 - 0.6
Hardness	500	500
Aluminium	-	0.2
Manganese	0.2	0.1
Chemical Oxygen Demand	10	-
Anionic Detergent MBAS	1.0	1.0
Biological Oxygen Demand	6	-

2. Methodology

2.1 Water Sampling

The water for spring water collected from water flow from the hill in Ulu Yam as shown in Figure-1. Water from the top of the hill flows out through the provided pipes. This area located on a hill bordering between Hulu Selangor and Gombak district, about 16 kilometers from Kuala Lumpur. There are four outlet pipes that drain water from the hill. Water samples were collected 6 different times. The second area is the RO machine water around Ulu Yam. There are two RO machines in the area. Water samples were also taken at 6 different times. The final field for the study was tap water supplied in Ulu Yam resident area.

2.2 Water Analysis.

There are four laboratory tests conducted at Laboratory of Environmental, Faculty of Civil and Environmental Engineering (FKAAS) which is turbidity, Ammoniacal Nitrogen (NH3), Iron (Fe) and pH. Ammoniacal Nitrogen test was done to check on water odor. Figure-2 shown the Ammoniacal Nitrogen test set. Spectrophotometer DR6000 machine used to analysis Ammoniacal Nitrogen, Iron and Turbidity. Four different apparatus were used, which are Hach DR 5000, Hach DR2100, DO meter and pH meter

Figure-1. Collecting water sample at hill area Ulu Yam, Selangor

Figure 2. Ammoniacal Nitrogen test set.

3. Result and discussion

The result obtained from turbidity analysis are shown in Figure-3. As can been seen from the graph, the water samples water hill (0.42 NTU) turbidity value is lower than RO water sample (0.88 NTU) and supply water (0.76 NTU). The turbidity of RO water samples RO is the highest since the machine is not well maintained in accordance with the schedule set. However, all the value are still under safe limits for consumption of less than 1000.00 NTU.

Figure-4 provides the value of Ammoniacal Nitrogen (NH_3) recorded for each of sample and the standard value NH_3 for the Ministry of Health (MOH) raw and treated water. Comparing the results, it can see that the value of Ammoniacal Nitrogen (NH_3) reading for RO water samples (0.67 mg/L) slightly higher compare to treated supply water (0.21 mg/L) and hill water (0.03 mg/L). These findings indicate the growth of algae in RO water samples are high. Meanwhile, the NH_3 reading for hill water samples are very low show of the least algae growth in hill water. This experimental evidence showed

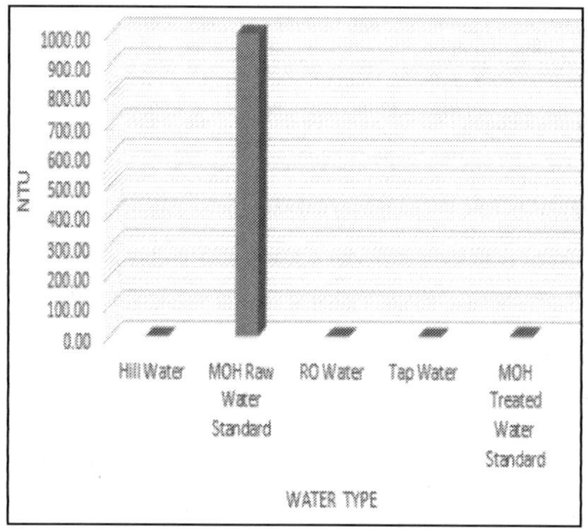

Figure 3. Turbidity (NTU) graph for water sample.

the occurrence of algae growth will be given the odor to the water. Therefore, experimental evident of less odor in hill water indicate there almost none of algae growth in hill water.

Figure-4 provides the value of Ammoniacal Nitrogen (NH_3) recorded for each of sample and the standard value NH_3 for the Ministry of Health (MOH) raw and treated water. Comparing the results, it can see that the value of Ammoniacal Nitrogen (NH_3) reading for RO water samples (0.67 mg/L) slightly higher compare to treated supply water (0.21 mg/L) and hill water (0.03 mg/L). These findings indicate the growth of algae in RO water samples are high. Meanwhile, the NH_3 reading for hill water samples are very low show of the least algae growth in hill water. This experimental evidence showed the occurrence of algae growth will be given the odor to the water. Therefore, experimental evident of less odor in hill water indicate there almost none of algae growth in hill water.

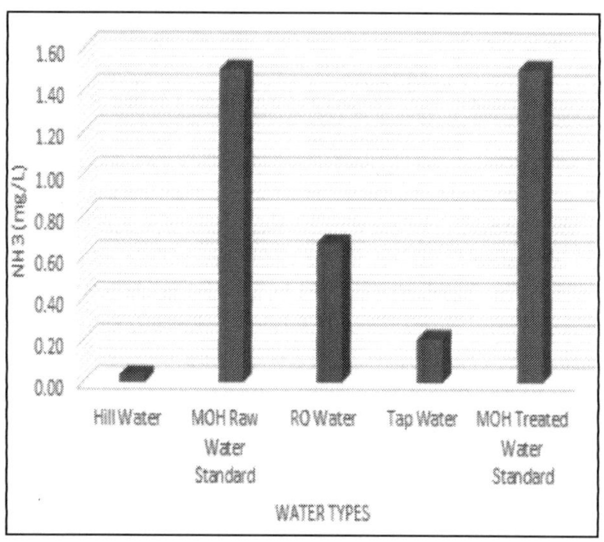

Figure 4. Results of Ammoniacal Nitrogen.

The result obtained from the Fe analysis shown in Figure-5. From the graph in figure 7, it is apparent that the hills water (0.03mg/L) has the lowest of Fe reading compared to supply water (0.08mg/L) and RO water (0.12mg/L). Overall, the value of Fe for all the water sample still under MOH standard. Meanwhile, Figure-6 illustrated the pH reading for the water samples. The pH value was recorded from the supply water is 7.01 or neutral. pH analysis indicate that tap water is free of alkaline and acid due to well treated before distribution to the consumer. pH value for hill water (6.76) and RO water (6.60) are almost neutral. The low pH reading of RO water is due to the process of osmosis occurs when the water is filtered repeatedly may cause lime content on the water is reduced from neutral.

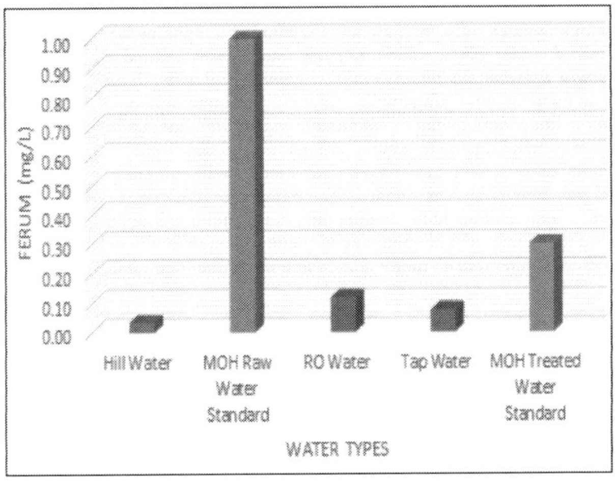

Figure 5. Result of Fe reading for each water sample.

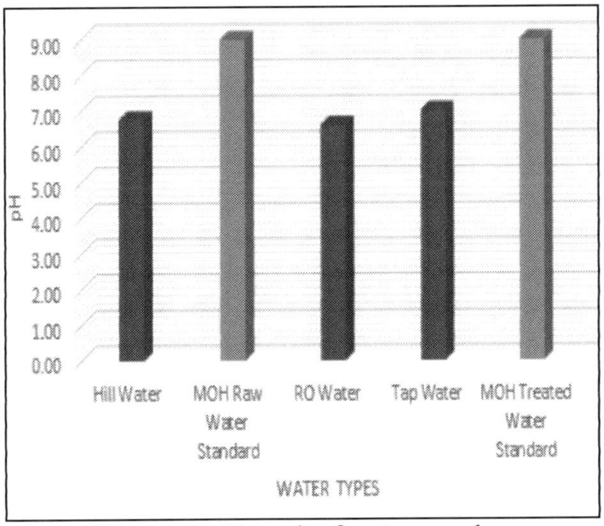

Figure 6. pH result of water samples.

Based on value for all the parameters, the hills water is best drinking water quality compared to supply water and RO water machine. The hill water has the lowest value of NTU, NH_3 and Fe and almost neutral for pH reading. Suprisingly NH_3 was found in tap water. This may occur because at that time the area experienced water rationing. Water flow through the pipe only once for every two days and give the odor to the water in the pipe. Meanwhile, the value of iron come from existing corrosion of underground pipes. RO water machine are still at a satisfactory level by the quality of treated drinking water to drink according to the standards for Drinking Water Quality of the Ministry of Health. However, the NH_3 values for RO water is the highest compared to hill water and tap water sample. Environmental conditions of the machines are less clean possibility cause of odor in the water. The iron content in the RO water high due to corrosion of the supply pipe that may have been long and not be replaced. This study support previous research that geographical location greatly influenced the in water sample content (Azlan et al. 2012) .In general, it seems that the results indicate all the water sources are safe for drinking because it still under MOH standard.

Water quality criteria should be more stringent to ensure water security. Although overall the samples of water in Ulu Yam is clean but still need control measures to be conducted as to avoid from water deterioration. Serious care should be a concern for tap water because the water is a major source of locals for everyday use. Running regular maintenance need for the existing pipe for making sure the piping system are clean and run properly. For turbidity control, the maintenance of the area around the pipe or piping system should be reviewed. This is because the turbidity can occur when there is a problem of leaking pipes. In addition, the screening machine for reverse osmosis systems also needs to be maintained so that the pH value is not affected. RO machine water is unfavorable in terms of mineral content in which all the minerals have been filtered by the machine.

Every day, more and more people come to get hill water either for drinking or daily use. Therefore, periodic reviews are proposed from time to time so that all consumers can know the quality of the water either is drinkable or not. Our government, especially the Department of Irrigation and Drainage is responsible for ensuring that the consumer drinks the water is safe and high quality. Legal action could be imposed on offenders who taking out the trash in the vicinity of hill water or anything to irresponsible in maintaining water RO machine. In addition, carry out awareness programs to educate the people about water quality and management at many levels, such as school and selected community. Expected from this program is for knowledge enhancement through engagement and educational outreach programme. So our young generation can start the initiative to conserve water such as hill water will not only in Ulu Yam but elsewhere as well.

4. Conclusions

As the conclusion, overall, the water quality of all the water samples are in good condition and suitable for use as drinking water. This can be proved by the average amount was recorded for all samples where water still according to the standards of the drinking water quality index by the Ministry of Health (MOH). Poor quality of drinking water will give the negative effect on local residents. Contaminated water can also be a source of various diseases to people in the area. Therefore, drinking water quality assessment is very important and needs to be conducted regularly.

References

[1] Al-Ajlouni, M.F., Rakmi Abd Rahman, Abdul Ghani Rafek, Mazlin Mokhtar, Noor Ezlin Ahmad Basri. (2010). Prediction for Natural Recharging In Langat Basin and Ukm Campus as Case Study. ARPN Journal of Systems and Software. Volume 1 No. 3; 93-100

[2] Economic Planning Unit – EPU (2000). "National Water Resources Study 2000 – 2050". Government of Malaysia, Department of Prime Minister's Office 2000.

[3] Azlan, A., Khoo, H. E., Idris, M. A., Ismail, A., & Razman, M. R. (2012). Evaluation of Minerals Content of Drinking Water in Malaysia. *The Scientific World Journal,* 2012, 403574. http://doi.org/10.1100/2012/403574

[4] Mamun, A.A & Zaki Zainudin (2013). Sustainable River Water Quality Management in Malaysia. *IIUM Engineering Journal*, Vol. 14, No. 1. pp 29-42

[5] Ministry of Mineral Resources and Environment Malaysia (MMREM). (2012). National Water Resources Policy. Perputakaan Negara Malaysia

[6] Fulazzaky, M.A., Teng Wee Seong & Mohd Idrus Mohd Masirin. (2010). Assessment of Water Quality Status for the Selangor River in Malaysia. *Water Air Soil Pollut* 205:63–77

[7] Cairncross, S., & Feachem, R. (1993). Environmental Health Engineering in the Tropics. 2nd ed. Chichester, U.K.: John Wiley & Sons.DOI 10.1007/s10661-010-1411-xDOI 10.1007/s11270-009-0056-2

[8] APHA (1998). "Standard Method for the Examination of Water and Wastewater", American Public Health Association, Washinton, D.C.

[9] Said, A., David K. Stevens & Gerald Sehlke. (2004). Environmental Assessment: An Innovative Index for Evaluating Water Quality in Streams. *Environmental Management* Vol. 34, No. 3: 406–414

[10] Hafizan, J., Sharifuddin M. Zain, Mohd Kamil Yusoff ·T. I. Tengku Hanidza, A. S. Mohd Armi, Mohd Ekhwan Toriman & Mazlin Mokhtar. (2011). Spatial water quality assessment of Langat River Basin (Malaysia) using environmetric techniques. *Environ Monit Assess* 173:625–641

[11] Akta Kualiti Alam Sekeliling 1974 (Akta 127) & Peraturan-peraturan dan perintah-perintah (2002). International Law Book Services.

IOP Conf. Series: Materials Science and Engineering **136** (2016) 012082 doi:10.1088/1757-899X/136/1/012082

Variations of Roughness Coefficients with Flow Depth of Grassed Swale

N Mustaffa[1], N A Ahmad[1] and M A M Razi[1]
[1] Faculty of Civil and Environmental Engineering, Universiti Tun Hussein Onn Malaysia, 86400 Batu Pahat Johor, MALAYSIA

E-mail: eyramusseth@gmail.com

Abstract: Grassed swales are the best management practice (BMP), which has been widely used to reduce the peak flow, reduce water pollution through vegetated filtration, and improve the groundwater recharge. Universiti Tun Hussein Onn Malaysia (UTHM) is using the approach of grassed swales recommended by the Department of Irrigation and Drainage Malaysia (DID) for reducing the risk of flooding and controlling the water pollution. This paper investigates the variations of roughness coefficients with the flow depth of grassed swales in the campus of UTHM. Fieldwork was carried out on the grassed swale to collect the hydraulic data, which including the levelling work, measuring the flow depth and flow velocity of the swale. The flow depth of swale was taken at three points divided along the width of swale and the flow velocity is captured three times at each of the point. The variations of roughness coefficients of grassed swales are presented in Manning's equation, and the results reveal that the n value increases with the increasing of flow depth. Manning's coefficient value found in this study is in the range of 0.110 to 0.756, which are higher than the value proposed by the Urban Stormwater Management Manual for Malaysia (MSMA). The relationships of flow depth and velocity at each section of the swale are portrayed in graphs, which show that the velocity increases with the decreasing of flow depth. The outcomes of this study can be concluded that the variation of Manning's coefficient value is influenced by the swale profile, flow depth, flow velocity, and as well as the vegetation used in the grassed swale concerned.

Keywords: Hydraulic parameters, swale profile, vegetation.

1. Introduction

Grassed swale is a vegetated, open channel management practices designed specifically to treat and attenuate stormwater runoff for a specified water quality and quantity volume [1]. Grassed swales are increasingly being employed in developed countries under temperate climate as a stormwater best management practice (BMP) for runoff quantity and quality control. As runoff travels through a swale, the vegetation reduces peak velocity while infiltration reduces flow volume. Attenuation of runoff flow promotes pollutant removal [2].

Establishment of grassed swale is a potential solution wherever stormwater needs to be transported from impervious surfaces, slowed down, and allowed to infiltrate into soils. When properly designed to accommodate a predetermined storm event volume, a grassed swale results in a significant improvement over the traditional drainage ditch in both slowing and cleaning of water [3].

The urbanization has caused an increment in the imperviousness which produced increased peak flow and more runoff volume [4][5][6]. Schueler also mentioned that the occurrence of flooding generally due to the change of catchment hydrology via an increase in the impervious area and reduction in catchment storage [7]. Roesner et al. has claimed the effect to a reduction of catchment response time due to the development was the increase of maximum flow discharge by a factor of 2 to 10 or even more into the conventional drainage system thus increases the frequency of significant floods [8]. Depressional storage in urban areas can be reduced by a factor of 5 to 10 depending on the original state of the watershed and the degree of imperviousness generated from the urbanization [9].

Swales are environmentally-friendly drainage system, which is derived from the Manual Saliran Mesra Alam Malaysia (MSMA). MSMA is required in order to provide a systematic construction so that the swales can be function properly to reduce the risk of flooding. This study was conducted in the campus of Universiti Tun Hussein Onn Malaysia (UTHM). The aim for this study is to investigate the variations of roughness coefficients with the flow depth of grassed swales in UTHM. The variations of roughness coefficients of grassed swales are presented in Manning's equation. Figure-1 shows the grassed swale in UTHM. The hydraulic parameters involved in this study are cross-sectional area (A) of swale, flow velocity (V), flow depth (Y), wetted perimeter (P), hydraulic radius (R), and longitudinal slope (S). The findings of this study give an extensive understanding to the hydraulic characteristics of grassed swale, where the suitable swale design can be proposed especially in the application of sustainable stormwater drainage system within UTHM campus.

Figure 1. Grassed swale in UTHM.

The roughness components in vegetated open channels are conceptually divided into three components, which are form roughness, soil grain roughness, and vegetative roughness. In most vegetated open channels, vegetative roughness characterized by vegetation density, grass height, and type of vegetation dominates the hydraulic resistance of channel [10]. Hydraulic resistance of the watercourse determines the water level and the flow distribution in the basin. Such resistance is commonly represented by parameters such as Manning's roughness coefficient (n), Chezy's resistance factor (C), or the Darcy-Weisbach friction factor (f), among which Manning's n is the most frequently used in the computation of open channel and overland flows [11][12]. Reliable results of flood routing and inundation simulation rely on an accurate of the resistance coefficient.

2. Methodology

2.1 Manning's Roughness Coefficient

The Manning's equation is an empirical formula estimating the average velocity of a liquid flowing in a conduit that does not completely enclose the liquid, such as open channel flow. All flow in so-called open channels is driven by gravity. Roughness coefficients represent the resistance to flood flows in channels and flood plains. The results of Manning's equation, an indirect computation of stream flow, have applications in flood-plain management, in flood insurance studies, and in the design of bridges and highways across flood plains [13]. Manning's equation is:

$$n = \frac{A}{Q} R^{2/3} S^{1/2} \quad (1)$$

Where;

n = Manning's roughness coefficient
A= cross-sectional area (m2)
Q=flow discharge (m3/s)
R= hydraulic radius (m)
S= longitudinal slope

Hydraulic roughness is the measure of the amount of frictional resistance water experiences when passing overland and channel features. An increase in this n value will cause a decrease in the velocity of water flowing across a surface [14]. Diaz claimed that the variations in the n values diminish when the slope increases [15]. According to Warmink et al., the uncertainty due to bed form roughness in the main channel and vegetation roughness in the floodplains were shown to be the major contributors [16]. Differ from the study by Ding et al., where the parameter identification method based on optimal control theories for identifying Manning's n in shallow water equations is presented to obtain realistic and accurate flows in a natural environment [17].

For intermediate flows in which the flow depth is greater than the height of vegetation (the grasses submerged), Ree and Palmer have showed the n values decrease as average velocity increases [18]. The decrease of n is regarded as a result of the increase of plant bending and submergence when velocity increases. For unsubmerged vegetation, Temple et al. hypothesized that an increase in flow depth less than that required to top the vegetation causes little change in the mean velocity [10]. Therefore, flow resistance tends to increase with the depth. According to DID , the roughness coefficients, n, varies with the type of vegetative cover, longitudinal slope, and average flow depth [19]. The n value must be adjusted for varying flow depths. Manning's roughness coefficients for swales are provided in Table-1.

Table 1. Values of Manning's roughness coefficient [19][20].

Drain/Pipe	Manning's (n)
Grassed Drain	
Short Grass Cover (< 150 mm)	0.035
Tall Grass Cover (≥ 150 mm)	0.050

Vegetation

Swales are broad and shallow channels designed to store and convey runoff at a non-erosive velocity, as well as enhance its water quality through infiltration, sedimentation, and filtration. Swales may be covered by dense vegetation, usually grass to slow down flows and trap particles and remove pollutant [21].

Swales can use a variety of vegetation including grass, sedge and tufted grass. Grassed swales are useful in residential areas but need to be mown and maintained regularly. The grass species chosen for lining of swales must be sturdy, drought resistant, easy to establish, and able to spread and develop a strong turf layer after establishment. The most common grass that used for swales in UTHM is cow grass or called as Axonopus Compressus, which depicted in Figure-2.

Proper planting equipment and methods must be used in establishing an effective grass lining. The grass origins and strains are known to be adaptable to the site so that the grass can be tolerant to flooding and be able to survive and continue to grow after the inundation period. Grass height of 50 mm to 150 mm should be maintained with a maximum grass height not exceed more of 150 mm [22]. Heights, types, and conditions of the vegetation are the factors that contribute to changes in the flow characteristics of the swale and its roughness coefficient. As the submergence starts to occur in the swale, the vegetative roughness coefficient tends to remain constant or rise [11].

Figure 2. Cow grass (Axonophus Compressus).

Fieldwork
The length of swale is 100 meters and was divided into three sections (Section A, Section B, and Section C). The distance from Section A to Section B is 50 meters and distance from Section B to Section C is 50 meters as well. Each section is divided into three points, labeled as Left, Center, and Right, where measured from the left bank to the right bank of the swale according to each top width of the section. The flow depth and top width of swale for three sections are shown in Table-2. The measurements of flow velocity have been taken three times at each point by using the current meter. Levelling has been done on the swale beforehand to obtain the swale profile.

Table 2. Flow depth and top width of the swale.

Section	Flow depth, Y (m)			Top width, T (m)
	Left	Center	Right	
A	0.360	0.390	0.385	1.8
B	0.375	0.380	0.320	2.2
C	0.210	0.225	0.205	2.1

3. Results and Discussions
Table-3 shows the characteristics of swale for Section A, Section B, and Section C. The values of Manning's roughness coefficients, n are calculated by using the Equation 1. The cross-sectional area (A) of swale is determined based on the average of flow depth and subsection width of the swale by applying the mid-section method. The flow discharge (Q) of swale is the multiplication of average velocity with cross-sectional area. The hydraulic radius is a cross-sectional area divided by wetted perimeter, which obtained by hydraulic equation. According to Table-3, the values of wetted perimeter considerably vary due to the flow depth at each section of swale, which shows the irregularities of swale. The relationship of flow depth and velocity in each section of the swale is shown in Figure-3.

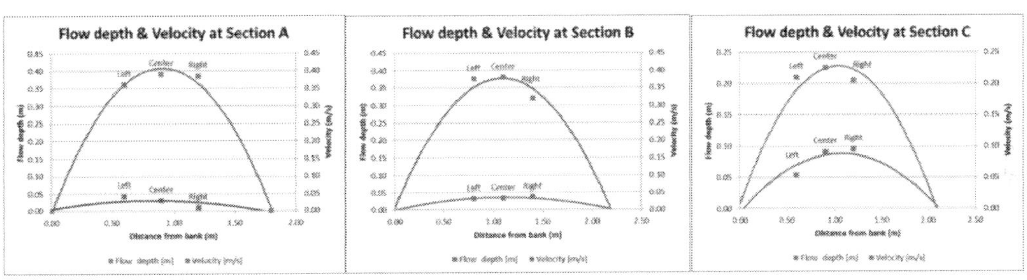

Figure 3. Flow depth and velocity of swale at Section A, Section B, and Section C.

Table 3. The characteristics of swale for each section.

Section	Wetted perimeter, P (m)	Area, A (m^2)	Average velocity, V (m/s)	Flow discharge, Q (m^3/s)	Hydraulic radius, R	Slope, S	Manning's, n
A	2.277	0.452	0.020	0.0091	0.193	0.002	0.756
B	2.220	0.254	0.026	0.0058	0.279	0.002	0.462
C	1.817	0.116	0.048	0.0075	0.204	0.002	0.110

Based on Figure-3, at Section A, the flow depth of 0.390 m has captured lower velocity of 0.030 m/s compared to the flow depth at 0.360 m has captured the velocity of 0.042 m/s. At Section B, the flow depth of 0.380 m has captured lower velocity of 0.033 m/s compared to the flow depth at 0.320 m has captured the velocity of 0.037 m/s. Meanwhile at Section C, the flow depth of 0.225 m has captured lower velocity of 0.091 m/s compared to the flow depth at 0.205 m has captured the velocity of 0.095 m/s. From these results it shows that the decrease of flow depth will increase the flow velocity of swale.

The variations of roughness coefficients are plotted to show the relationship of flow depth with Manning's roughness coefficients at each section of the swale as shown in Figure-4. Based on Figure-4,

Section A has the deepest flow depth (0.170 m) with the highest value of n (0.756), followed by Section B which has flow depth of 0.161 m with n of 0.462. Section C has the lowest value of n (0.110) with the shallow flow depth (0.098 m). These results show that the roughness coefficients will increase with the increase of depth. This is contrary to the finding by Ahmad et al., where the n value is slightly decrease with increment in flow depth [23]. However, they have observed that the n value is slightly increase with increment of flow depth higher than 0.10 m. Arcement and Schneider also found that the n value decreases with increasing depth [13]. They added that if the channel banks are much rougher than the bed or where dense brush overhangs the low-water channel, the n value is not constant with the flow depth.

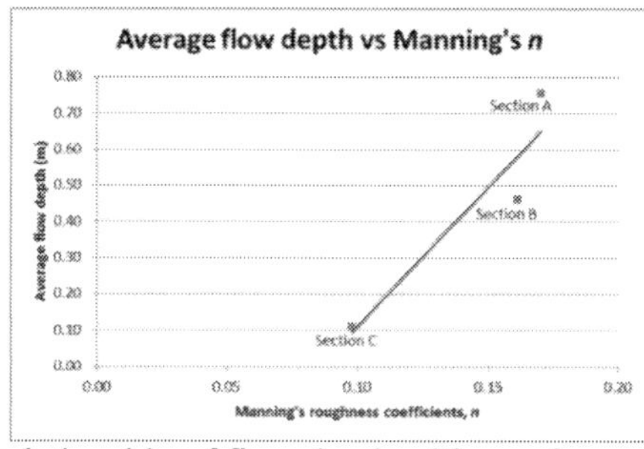

Figure 4. Relationship of flow depth with roughness coefficients.

Arcement and Schneider have produced an adjustment factors for channel n values [13]. The channel irregularities, alignment, obstructions, vegetation, and meandering have influenced the roughness of a channel. The value for n must be adjusted accordingly by adding increments of roughness to the base value for each condition that increases the roughness.

As for irregularity wise, Chow and Benson & Dalrymple showed that severely eroded and scalloped banks can increase n values by as much as 0.020 [20][24]. Larger adjustments may be required for very large, irregular banks that have projecting points.

Habitually, the greater roughness is associated with alternating large and small cross sections and sharp bends, constrictions, and side-to-side shifting of the low-water channel, and not affected significantly by relatively large changes in the shape or size of cross sections if the changes are gradual and uniform. A maximum increase in n of 0.003 will result from the usual amount of channel curvature found in designed channels and in the reaches of natural channels used to compute discharge [24].

Arcement and Schneider stated that the effect of obstructions on the roughness coefficient is a function of the flow velocity [13]. When the flow velocity is high, an obstruction exerts a sphere of influence that is much larger than the obstruction because the obstruction affects the flow pattern for considerable distances on each side. The sphere of influence for velocities that generally occur in channels that have gentle to moderately steep slopes is about three to five times the width of the obstruction.

Aldridge and Garrett have modified an adjustment values for factors that affect the roughness of a channel [25]. In wide channels having small depth-to-width ratios and no vegetation on the bed, the effect of bank vegetation is small, and the maximum adjustment is about 0.005. If the channel is relatively narrow and has steep banks covered by dense vegetation that hangs over the channel, the maximum adjustment is about 0.030. According to Chow, meanders can increase the n values by as much as 30 percent where flow is confined within a stream channel [20]. The meander adjustment should be considered only when the flow is confined to the channel. There may be very little flow in a meandering channel when there is flood-plain flow.

4. Conclusions

The values of Manning's roughness coefficient, n found in this study is in the range of 0.110 to 0.756, which are higher than the values proposed by the Urban Stormwater Management Manual for Malaysia (MSMA). According to MSMA (2012), the value n for the short grass cover is 0.035 and for the tall grass cover is 0.050. This may be due to several factors in terms of the swale cross-sections, the flow depth and velocity, the irregularities of the swale, and the height of vegetation at each section of the swale. The high value of n showed the less movement of flow within the swale. The swale should be maintained by mowing the vegetation to prevent any blockage for the water flows. Based on the modified adjustment values for factors that affect the roughness of a channel by Aldridge and Garrett, the channel conditions is very large with the n value adjustment in the range of 0.050 to 0.100, where turf grass growing and the average flow depth is less

than half the height of the vegetation [25].
Vegetation with suitable characteristics plays as a factor of roughness in a channel which act as an agent to slow down the flow velocity, thus reduced the flow discharge. An effective grassed swale would convey the stormwater runoff to the detention pond or river, which then prevent flooding from occurred. Swales should be able to carry their design flow without overtopping or eroding. If the flow velocity is too high for grass cover in the swale but the slope and cross section cannot be adjusted, the swale can be reinforced with rip-rap or turf reinforcement matting, which can withstand a higher flow velocity. Water velocities associated with a two-year design storm should not scour out vegetated materials nor should be matted down by water. In that situation, the grass must be in good and firm seed bed. It is recommended to water the grass as required until it is established and fertilizes according to the needs of the grass and soils.

Acknowledgements
The authors would like acknowledgment the Ministry of Higher Education Malaysia and Research, Innovation, Commercialization, and Consultancy Management Office (ORICC) of Universiti Tun Hussein Onn Malaysia (UTHM) for financially under FRGS Fund (Vot 1414) for supporting this study.

References
[1] United State Environmental Protection Agency (US EPA). (2012). Report of the Method 1664 Validation Studies. EPA-821-R-95-036, U.S. Environmental Protection Agency, Washington, D.C.
[2] Yu, S. L., Kuo, J. T., Fassman, E. A., and Pan, H. (2001). Field Test of Grassed Swale Performance in Removing Runoff Pollutant. Journal of Water Resources Planning and Management, May/June 2001, 186-171.
[3] Stevens, D. (2011). For Three Weeks I Owned the University of Illinois. Lulu.com, 135-136.
[4] Barber, M. E., King, S. G., Tonge, D. R. and Hathhon, W. E. (2003). Ecology Ditch: A Best Management Practice for Storm Water Runoff Mitigation. Journal of Hydrologic Engineering, Vol. 8, No. 3, May 1, 2003, 111-122.
[5] Seilheimer, T. S., Wei, A., Fraser, P. C. and Eyles, N. (2007). Impact of Urbanization on the Water Quality, Fish Habitat, and Fish Community of a Lake Ontario Marsh, Frenchman's Bay. Urban Ecosystems.(2007) 10, 299-319.
[6] Ouyang, T., Zhu, Z. and Kuang, Y. (2006). Assessing Impact of Urbanization on River Water Quality in the Pearl River Delta Economic Zone, China. Environmental Monitoring and Assessment (2006) 120, 313-325.
[7] Schueler, T. R. (1987). Controlling Urban Runoff: A Practical Manual for Planning and Designing Urban BMPs. Metropolitan Washington Council of Governments, Washington, D.C.
[8] Roesner, L. A., Bledsoe, B. P. and Brashear, R. W. (2001). Are Best-Management-Practice Criteria Really Environmentally Friendly? Journal of Water Resources Planning and Management, 127, 150-154.
[9] Viessman, W. and L. Lewis. (1996). Introduction to Hydrology, 4th ed. Harper Collins College Publishers, New York.
[10] Temple, D. M., Robinson K. M., Ahring, R. M., and Davis, A. G. (1987). Stability Design of Grass-lined Open Channels. Agricultural Handbook, 667. USDA: Washington, D.C.
[11] Wu, F., Shen, H. W., and Chou, Y. (1999). Variations of Roughness Coefficients for Unsubmerged and Submerged Vegetation. Journal of Hydraulic Engineering 125(9), 239-245.
[12] Kirby, J. T., Durrans, S. R., Pitt, R., and Johnson, D. (2005). Hydraulic Resistance in Grass Swales Designed for Small Flow Conveyance. Journal of Hydraulic Engineering 131(1): 65-68.
[13] Arcement, G. J. Jr. and Schneider, V. R. (1989). Guide for Selecting Manning's Roughness Coefficients for Natural Channels and Flood Plains. U.S. Geological Survey Water Supply Paper 2339. United States Government Printing Office, Washington.
[14] Barnes, H. H. Jr. (1967). Roughness Characteristics of Natural Channels. U.S. Geological Survey Water Supply Paper 1849. United States Government Printing Office, Washington.
[15] Diaz, R. G. (2005). Analysis of Manning Coefficient for Small-Depth Flows on Vegetated Beds. Hydrological Processes, 19, 3221-3233.
[16] Warmink, J. J., Straatsma, M. W. and Huthoff, F. (2013). The Effect of Hydraulic Roughness on Design Water Levels in River Models. Comprehensive Flood Risk Management – Klijn & Schweckendiek (eds). 2013 Taylor & Francis Group, London, ISBN 978-0-415-62144-1.
[17] Ding, Y., Jia, Y. and Wang S. S. Y. (2004). Identification of Manning's Roughness Coefficient in Shallow Water Flows. Journal of Hydraulic Engineering, ASCE, 501-510.

[18] Ree, W. O., and Palmer, V. J. (1949). Flow of Water in Channels Protected by Vegetative Linings. Tech. Bull. No. 967, Soil Conservation Service, U.S Department of Agriculture, Washington, D.C.
[19] Department of Irrigation and Drainage (DID). (2000). Urban Stormwater Management Manual for Malaysia. Government of Malaysia.
[20] Chow, V. T. (1959). Open Channel Hydraulics. McGraw-Hill Bok Company. New York.
[21] Department of Irrigation and Drainage (DID). (2012). Urban Stormwater Management Manual for Malaysia. 2nd Edition. Government of Malaysia.
[22] Water Sensitive Urban Design (WSUD). (2004). Technical Guidelines for Western Sydney. Section 5.
[23] Ahmad, N. A., Ghani, A. A., Zakaria, N. A., and Azamathullah, H. M. (2011). Hydraulic Characteristics for Flow in Swales. 3rd International Conference on Managing Rivers in the 21st Century: Sustainable Solution for Global Crisis of Flooding, Pollution, and Water Scarcity.
[24] Benson, M. A., and Dalrymple, T. (1967). General Field and Office Procedures for Indirect Discharge Measurement. U.S. Geological Survey Techniques of Water-Resources Investigations Book 3, chap. Al, 30 p.
[25] Aldridge, B. N., and Garrett, J. M. (1973). Roughness Coefficients for Stream Channels in Arizona. U.S.Geological Survey Open-File Report, 87 p.

Hydrodynamic and Salinity Intrusion Model in Selangor River Estuary

N F Haron[1] and W Tahir[1]
[1]Flood Control Research Centre, Faculty of Civil Engineering, Universiti Teknologi MARA, 40450 Shah Alam, Selangor, MALAYSIA

E-mail: neemzay@yahoo.com

Abstract: A multi-dimensional hydrodynamic and transport model has been used to develop the hydrodynamic and salinity intrusion model for Selangor River Estuary. Delft3D-FLOW was applied to the study area using a curvilinear, boundary fitted grid. External boundary forces included ocean water level, salinity, and stream flow. The hydrodynamic and salinity transport used for the simulation was calibrated and confirmed using data on November 2005 and from May to June 2014. A 13-day period for November 2005 data and a 6-day period of May to June 2014 data were chosen as the calibration and confirmation period because of the availability of data from the field-monitoring program conducted. From the calibration results, it shows that the model was well suited to predict the hydrodynamic and salinity intrusion characteristics of the study area.

Keywords: Delft3D-FLOW, hydrodynamic model, salinity intrusion, calibration, estuary.

1. Introduction

Estuaries are semi-enclosed coastal bodies of water where fresh water and salt water meet and mixed [1]. Based on mixing characteristics, estuaries can be classified as vertically-mixed, slightly-stratified, highly-stratified or saline-wedge. The fresh water from river lighter than salt water, thus it had a tendency to remain on top of the salt water. In order to study on salinity behaviour, there are several research focus on salinity intrusion in estuary [2][3][4][5][6][7] using various methods including hydrodynamic model.

The application of a three-dimensional hydrodynamic and salinity transport model is required to identify environmental impacts related to management alternatives. The motivation for the application of a 3D hydrodynamic model and associated data acquisition and analysis is to better understand how salinity distributions respond to changing freshwater and pollutant loading over varying time scales.

Numerical models are often resorted to the study of water movement, or circulation, in estuaries and tidal rivers. Because of that, computer-modelling works was chosen to assist in predicting the hydrodynamic and salinity transport in Selangor River Estuary. Delft3D-FLOW was applied to the study area to simulate the hydrodynamics and salinity intrusion. Delft3D-FLOW has been extensively used to carry out simulations of flows, sediment transports, waves, water quality, morphological developments and ecology in coastal, estuary and river.

The main objective of this research is to develop a proper numerical modelling that projected the salinity intrusion of Selangor River Estuary. The purposes of this investigation are to develop a comprehensive hydrodynamic and salinity intrusion model for Selangor River Estuary by using Delft3D-FLOW and to evaluate hydrodynamic and salinity transport model for different sets of data.

A hydrodynamic model is set up and developed in this study on the behaviour of the saline intrusion and movement in the estuarine system. The model is calibrated and validated by using a Delft3D-FLOW. The model covers from river mouth up to Kg. Asahan (34 km from river mouth) in order to observed the maximum salinity intrusion length in the estuary. It is because Ishak et al. determined that the maximum limit of salt water intrusion reached some 16 km from the river mouth which was recorded during a low river discharges (6.0 m3/s) on a medium tide (tidal range 3.2 m), but this was reduced to about half the distance

IOP Conf. Series: Materials Science and Engineering **136** (2016) 012083 doi:10.1088/1757-899X/136/1/012083

during high river discharges [8]. During this maximum intrusion, the salt water did not reach Kg. Kuantan, the attraction centre for the fireflies display which dwells on mangrove species Sonneratia Caseolaris (Berembang trees).

2. Methodology

2.1 Study Area

This study is focus on construct a hydrodynamic model and salinity intrusion in Selangor River Estuary (Figure 1). This research is emphasizing on the trend of the salinity transport in the estuary, determine the factors contributed to the trend observed and extrapolate the data collection based on the effects of those continuing factors imposed to the river. Selangor river has interacted many researchers to investigate about the river and the surrounding area along the river basin such as siltation modelling [9], flooding issues [10][11], case study of fireflies [12], land use changes [13], analytical salt intrusion model [14], and many more.

The Selangor River Estuary is located in Selangor river basin, situated in the upper part of the Selangor state. Selangor River starts from Fraser's Hill and stretches for 110 km to the coast of Kuala Selangor. The basin covers an area of 2,200 km2 contains nine sub-basins, which are Ulu Selangor, Kerling, Kuala Kubu, Sg. Tinggi, Rantau Panjang, Batang Kali, Tanjung Karang, Rawang and Kuala Selangor. It is the third largest river basin in Selangor after Sg. Langat and Sg. Bernam basins. The main uses of the river were listed in the Sg. Selangor Basin Management Plan 2007-2012 as water supply, ecosystem, tourism and recreation, sand mining, aquaculture, and inland navigation. The river is also famous for tourist attractions such as the firefly sanctuary at Kg. Kuantan.

To ensure that these uses can be sustained, the plan unveiled four main policies, which comprised ensuring water supply, ensuring clean water, protection against floods, and conservation of the fireflies as being observed in many research of Selangor river.

Salinity in the study area is influenced by the mixing of fresh water discharges from the Selangor river with salt water from the Malacca Strait. The quantity of fresh water discharged at any particular time is influenced by past and present precipitation, amount of surface water storage, evaporation and other hydrologic factors. The quantity of available salt water in the study area is primarily dependent on transport of ocean water through Selangor River Estuary inlet by tidal and low frequency coastal sea level forced currents.

Figure 1. Study Area. [15]

2.2 Field Data Collection Program

Field data are required to specify initial conditions, boundary conditions, and forcing function distributions in time and space for the calibration, validation and application of a three-dimensional hydrodynamic and salinity intrusion model of the study area. The objectives of the field program are to obtain calibration and verification data sets which include significant high freshwater inflow events as well as to provide information on representative mean seasonal conditions for the construction of hypothetical event scenarios. Sequences of scenarios will be used to predict and quantify the spatial and temporal extent of physical

IOP Conf. Series: Materials Science and Engineering **136** (2016) 012083 doi:10.1088/1757-899X/136/1/012083

processes that directly contribute to biological impacts within the study area. However, due to some problem regarding on measurement equipment constraints, and time constraints, the data collection for validation purposes only have been done. Meanwhile, data for calibration have been provided from a consultant, Mr. Abdul Jalil Hassan measured water level at Kg. Asahan from 15 November to 27 November 2005 since the station in the region of the study area.

A survey was conducted from 31 May to 5 June 2014 for Selangor River Estuary to collect data needed to validate the Selangor River Estuary hydrodynamic and salinity intrusion model as presented in Figure-2.

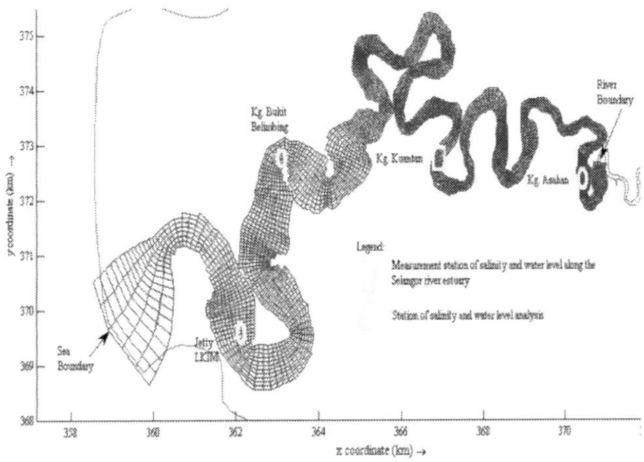

Figure 2. Location of salinity and water level for Selangor River Estuary model.

2.3 Selangor River Estuary Model Calibration

A common technique in flow modelling is to start the calibration process with a 2-dimensional model. Therefore, in this study, water level calibration was carried using data obtained from a consultant, Mr. Abdul Jalil Hassan who was measured water level at Kg. Asahan from 15 November to 27 November 2005 since the station in the region of the study area. The most suitable Manning's coefficient n value used in the calibration is 0.02. Besides, Figure-3 presents the comparison between observed and simulated water level not much difference and have the same pattern. Thus, the calibration of water level is acceptable.

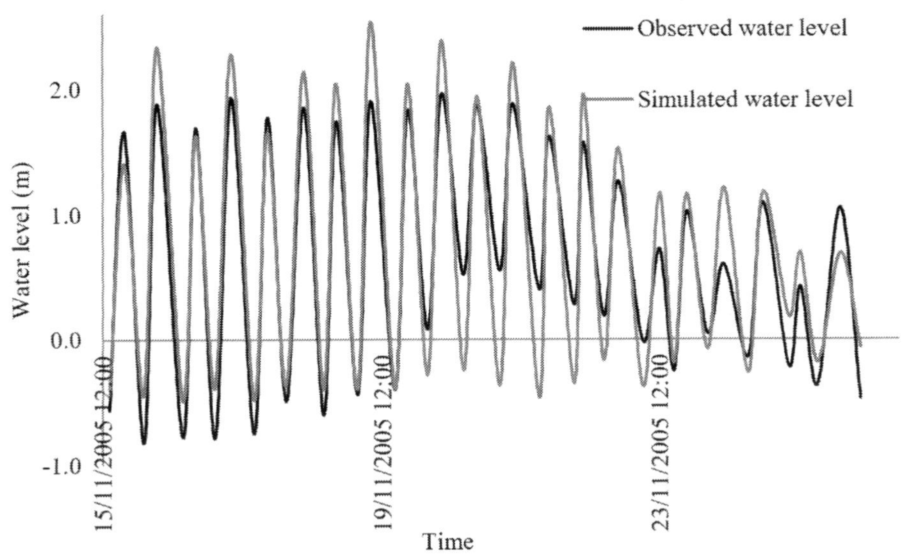

Figure 3. Water level comparison between observed and simulated for $n = 0.02$ at Kg. Asahan station on November 2005.

2.4 Selangor River Estuary Model Validation

Due to the limited availability of data the model is only validated for salinity. Salinity levels are controlled by flow velocities and water levels, so a validation of only salinity is assumed to be sufficient.

In Table-1 indicates salinity comparison between observed and simulated salinity in May until June 2014 at station Kg. Bukit Belimbing and Jetty LKIM, Kuala Selangor. Meanwhile, in Figure-4 and Figure-5 show the graph simulated salinity versus observed salinity from May 2014 until June 2014 at Kg. Bukit Belimbing station and Jetty LKIM Kuala Selangor station for validation purpose respectively. In both figures, the observed salinity lower than simulated salinity for several times due to the heavy local rain in Kuala Selangor during the measurement. As a result, the discharge from the river and rainfall have deviated the reading of salinity. However, from the conditions it is shown that the high freshwater discharge will dilute the concentration of salinity. Thus, salinity level will reduce.

3. Preliminary Simulation

Preliminary simulation was done for different flow (low streamflow and high streamflow) as shown in Figure-6 to Figure-9). All the figures show the salinity level and water level at Jetty LKIM, Kg. Bukit Belimbing, Kg. Kuantan and Kg.Asahan varies with different flow conditions where salinity increases while the water level decreases during low river discharge and vice versa during high river discharge. Besides, the salinity level becomes zero at river upstream (Kg. Asahan, 34 km from river mouth) as the maximum limit of salt water intrusion reached about 16 km from the river mouth which was recorded during a low river discharges as stated by Ishak et al. (2002). As a result, the model was able to predict the values and the extent of the salinity intrusion along the Selangor River Estuary.

Table 1. Salinity comparison between observed salinity and simulated salinity in May 2014 and June 2014 at station Kg. Bukit Belimbing and Jetty LKIM, Kuala Selangor.

Station	Date and Time	Observed Salinity (ppt)	Simulated Salinity (ppt)
Jetty LKIM	31/5/2014 10:11	2.7	2.6
Jetty LKIM	1/6/2014 14:34	0.1	0.3
Jetty LKIM	2/6/2014 14:31	0.1	0.7
Jetty LKIM	3/6/2014 14:32	0.3	1.1
Kg. Bukit Belimbing	4/6/2014 9:38	0.1	0.9
Jetty LKIM	4/6/2014 9:51	19.4	21.8
Jetty LKIM	4/6/2014 13:37	4.2	6.0
Jetty LKIM	5/6/2014 9:43	10.3	19.6
Kg. Bukit Belimbing	5/6/2014 9:57	0.1	0.1
Jetty LKIM	5/6/2014 14:16	2.6	4.8

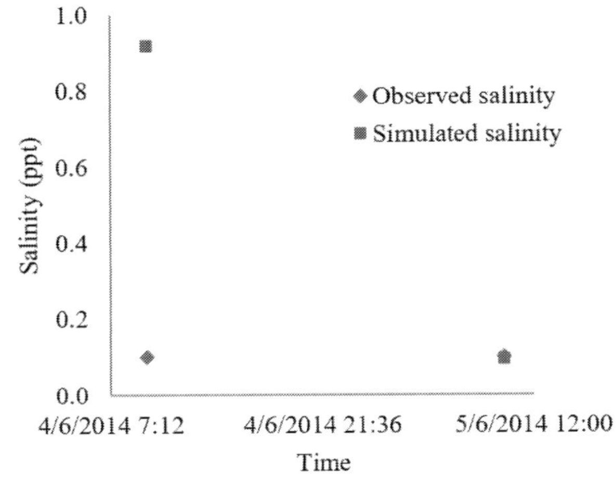

Figure 4. Simulated salinity versus observed salinity from May 2014 until June 2014 at Kg. Bukit Belimbing station for validation.

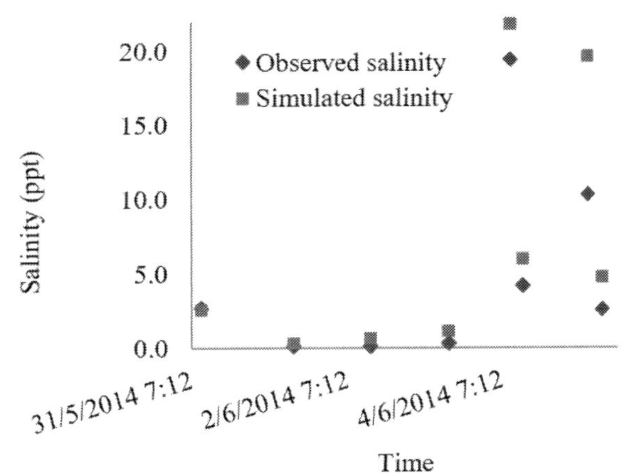

Figure 5. Simulated salinity versus observed salinity from May 2014 until June 2014 at Jetty LKIM Kuala Selangor station for validation.

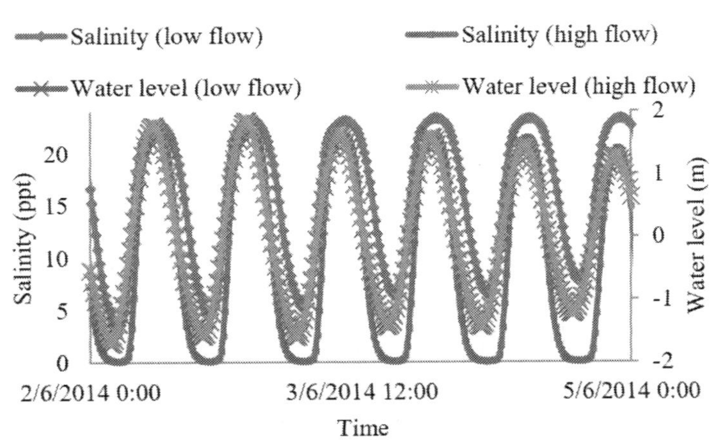

Figure 6. Comparison of salinity and water level during low streamflow and high streamflow at Jetty LKIM.

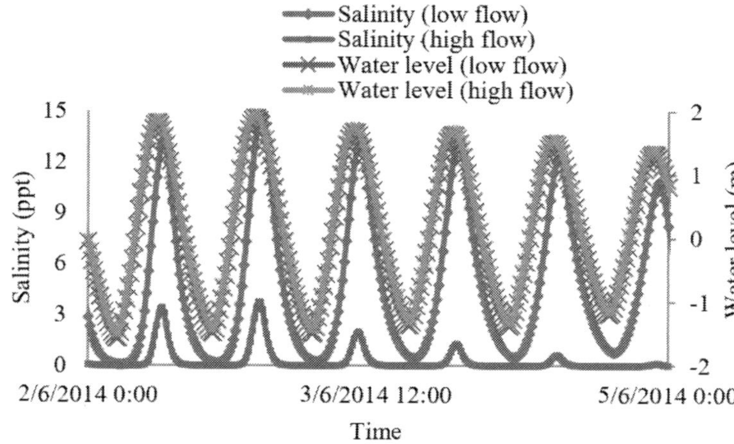

Figure 7. Comparison of salinity and water level during low streamflow and high streamflow at Kg. Bukit Belimbing.

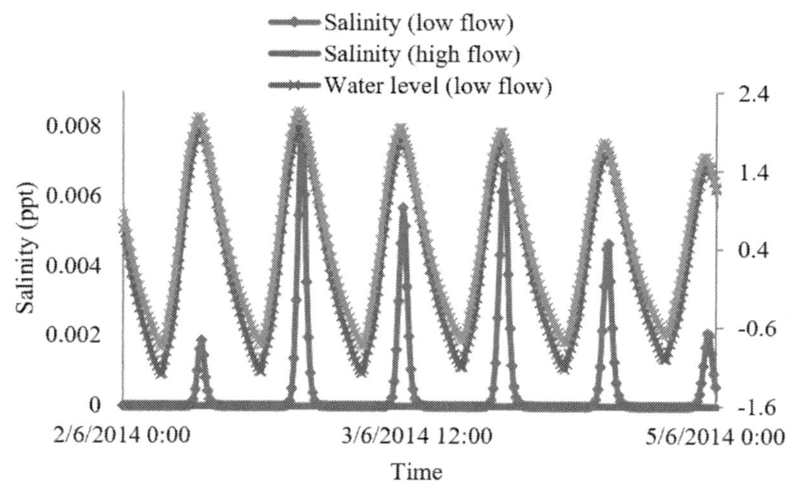

Figure 8. Comparison of salinity and water level during low streamflow and high streamflow at Kg. Kuantan.

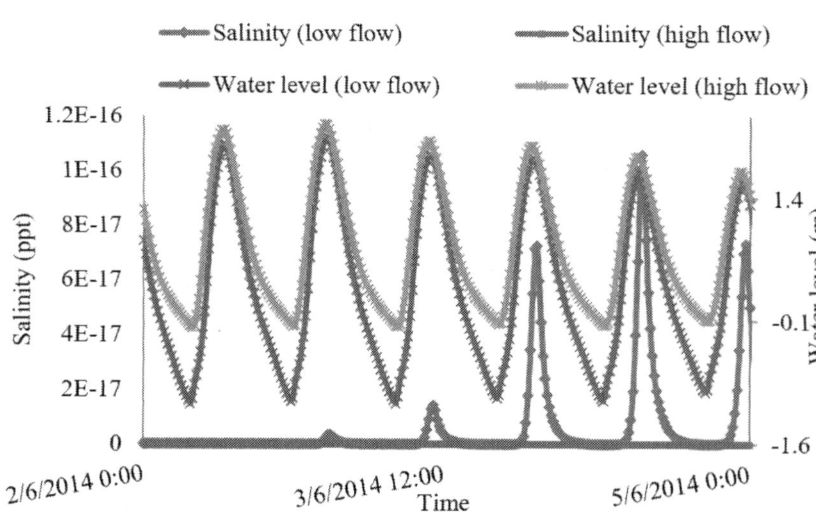

Figure 9. Comparison of salinity and water level during low streamflow and high streamflow at Kg. Asahan.

4. Conclusions

The Selangor River Estuary model was calibrated using data on November 2005 and validated using data from May to June 2014. The hydrodynamic and salinity intrusion model was applied for preliminary simulations for different flow scenarios at different locations along the Selangor River Estuary. From the simulations, it can be indicated that the water level will increase as the freshwater inflow increase, consequently the salinity level will decrease in the estuarine system. It is anticipated that the calibrated and validated model will be useful as a resource management tool for addressing alternate strategies to mitigate adverse impacts of high freshwater inflow events specifically in Selangor River Estuary.

References

[1] Cao, J., Li, R. & Zhu, Y., 2008. Study on Saltwater Intrusion in the Yangtze River Estuary by 3D Numerical Model. In Education Technology and Training, 2008, and 2008 International Workshop on Geoscience and Remote Sensing, ETT and GRS 2008. pp. 81–84.

[2] Nuryazmeen, F.H., Tahir, W. & Koon, L.W., 2013. Extreme estuarine flooding leading to estuary transverse flow salinity intrusion. International Journal of Civil & Environmental Engineering IJCEE-IJENS, 13(02), pp.54–58.

[3] Nuryazmeen, F.H., Tahir, W. & Mohamad, I.N., 2014. Potential of Estuary Transverse Flow Salinity Intrusion Due to Extreme Estuarine Flooding. In R. Yusoff et al., eds. InCIEC 2013. Springer Singapore, pp. 343–352. Available at: http://dx.doi.org/10.1007/978-981-4585-02-6_30.

[4] Nuryazmeen, F.H. & Tahir, W., 2014a. Laboratory investigations on estuary salinity mixing: preliminary analysis. International Journal of Sciences: Basic and Applied Research (IJSBAR), 13(1), pp.36–41.

[5] Nuryazmeen, F.H. & Tahir, W., 2014b. Physical model of estuarine salinity intrusion into rivers: A review. Advanced Materials Research, 905(2014), pp.348–352. Available at: http://www.scientific.net/AMR.905.348.

[6] Nuryazmeen, F.H. et al., 2015. Salinity Velocity Pattern in Estuary Using PIV. In S. H. A. Bakar et al., eds. Proceedings of the International Symposium on Flood Research and Management (ISFRAM 2014). Springer Singapore, pp. 221–243.

[7] Nuryazmeen, F.H. & Tahir, W., 2016. Simulation of Estuary Transverse Flow Salinity Intrusion During Flood Event : Case Study of Selangor River Estuary. In Wardah Tahir et al., eds. ISFRAM 2015 (Proceedings of the International Symposium on Flood Research and Management 2015). p. Chapter 12

[8] Ishak, A.K. et al., 2002. Salinity intrusions into the Selangor river estuary and its effect on the mangrove species sonneratia caseolaris. In Proceedings of the Seminar R and D MINT 2002: Strengthening R and D Culture for Technology Generation. pp. 123–130.

[9] Ariffin, J. & Abdul Talib, S., 2006. Siltation modeling for Selangor dam in Hulu Selangor (Report), Selangor, Malaysia, pp.1-24.

[10] Hassan, A.J., 2006. Pemodelan sungai dan dataran banjir untuk penjanaan peta risiko banjir : Kajian kes Sg. Selangor. Universiti Sains Malaysia, pp.1-249

[11] Hassan, A.J., Ab. Ghani, A. & Abdullah, R., 2006. Development of flood risk map using GIS for Sg. Selangor Basin. Proceeding 6th International Conf. on ASIA GIS, 9-10 Mac, UTM, pp.1–11.

[12] Kirton, L.G. et al., 2012. Monitoring populations of bioluminescent organisms using digital night photography and image analysis: A case study of the fireflies of the Selangor River, Malaysia. Insect Conservation and Diversity, 5(3), pp.244–250.

[13] Sulaiman, N. & Mohamad, J., 2014. Landuse conflicts in the Sungai Selangor watershed. In UM-KU, ed. Newsletter: JSPS Asian Core Program (IWM). Universiti Malaya, pp. 4–6.

[14] Gisen, J.I.A., 2015. Prediction in ungauged estuaries (Thesis). Delft University of Technology, pp.1-142.

Flood Water Level Mapping and Prediction Due to Dam Failures

S Musa[1], M S Adnan[1], N A Ahmad[1] and S Ayob[1]
[1] Faculty of Civil and Environmental Engineering, Universiti Tun Hussein Onn Malaysia, 86400 Batu Pahat Johor, MALAYSIA

E-mail: sabariah@uthm.edu.my

Abstract. Sembrong dam has undergone overflow failure. Flooding has been reported to hit the town, covering an area of up to Parit Raja, located in the district of Batu Pahat. This study aims to identify the areas that will be affected by flood in the event of a dam failure in Sembrong Dam, Kluang, Johor at a maximum level. To grasp the extent, the flood inundation maps have been generated by using the InfoWorks ICM and GIS software. By using these maps, information such as the depth and extent of floods can be identified the main ares flooded. The flood map was created starting with the collection of relevant data such as measuring the depth of the river and a maximum flow rate for Sembrong Dam. The data were obtained from the Drainage and Irrigation Department Malaysia and the Department of Survey and Mapping and HLA Associates Sdn. Bhd. Then, the data were analyzed according to the established Info Works ICM method. The results found that the flooded area were listed at Sri Lalang, Parit Sagil, Parit Sonto, Sri Paya, Parit Raja, Parit Sempadan, Talang Bunut, Asam Bubok, Tanjung Sembrong, Sungai Rambut and Parit Haji Talib. Flood depth obtained for the related area started from 0.5 m up to 1.2 m. As a conclusion, the flood emanating from this study include the area around the town of Ayer Hitam up to Parit Raja approximately of more than 20 km distance. This may give bad implication to residents around these areas. In future studies, other rivers such as Sungai Batu Pahat should be considered for this study to predict and reduce the yearly flood victims for this area.

Keywords: Semberong dam, flood map, dam failure.

1. Introduction

According to DBP [1], dam is a structure used to hold the current or water flow. Water retained in dams usually have specific functions according to the type of dams built. Among the functions available in Malaysia is to supply water to generate hydro power, mitigating floods, irrigation and also as recreation , for example Kenyir dam in Terrengganu was built to generate hydropower and Bekok dam in Johor as a flood insulator.

Sembrong dam is located about 10 km from the town of Ayer Hitam. It was a rainy mitigation dam, but was used by Syarikat Air Johor as a source of water supply. The Sembrong dam has a catchment area of about 130 km^2. Sources of water supplied to the dam are from Sembrong's river and Merpo's river which flows from the north-west and east. The outflow from the reservoir is released into river of Sembrong which is a tributary of Simpang Kanan's river which then flows into Batu

Pahat's river before flowing into the Strait of Malacca. The dam was built in 1981, began operations in 1984 and is still functioning until now. The construction cost RM 24 million [2] and [3].

Normally, dam failure will be analyzed based on two conditions, namely, the flow of the dam failure during good weather or sunny day with the situation at the level of an ordinary reservoir and receiving normal inflows or the flow of dam failures in the rain or flooding. Dam failure in good weather is generally considered to have the greatest potential for loss of life events, due to the element of surprise while failure during rainy conditions are deemed to have a low potential for loss of human life for residents of the area would potentially be more alert.

Sembrong River Basin located in Kluang district covers an area of almost 200 km^2 in the catchment area of Sungai Batu Pahat. The catchment area is the largest in the state of Johor. Sungai Batu Pahat formed from two rivers namely Sungai Simpang Kanan and Sungai Simpang Kiri. Sungai Simpang Kiri's length is 57 km from north near Sungai Simpang Kanan, Chaah while starting from the junction of the Sembrong and Bekok river. Bekok river flows as far as 55 km to meet Sembrong and Simpang Kanan river [4].

The outflow from the reservoir is released into Sungai Sembrong, which is a tributary of Sungai Simpang Kanan which later went into Sungai Batu Pahat before flowing into the Strait of Malacca. Sembrong dam has a catchment area of about 130 km^2. These dams provide flood protection in the area downstream of up to 100 years return period. The dam consists of three main dam embankment and four saddle dam embankment, fence for flood control and an un-gated spillway. The water flows out of the dam through the un-gated spillway. Figure 1 shows the un-gated spillway at Sembrong's dam.

Figure 1. Un-gated spillway at Sembrong dam.

1.1 Dam failures

According to [5], dam failure is the failure of a major disaster. Features of dam failure is a case of the release of water held back by the sudden, rapid and uncontrolled current. Dam failure is usually the secondary effects of large-scale rain and flooding, it occurs when too much water flows into the spillway system. These events may occur along with a warning or not. In addition, poor maintenance can cause failure of the dam.

Dam failure generally can be associated with one or more of the following reasons, overflow caused by floods that exceed the capacity of the dam; the act of intentionally; the failure of structural materials used in the construction of the dam; cracks in concrete ponds or dams; soil erosion in the dam embankment; and inadequate care or maintenance.

Flood maps is one of the most effective approaches to describe the pattern of flooding, especially in the approach to reduce the incidence of flooding. The last study was done by [5]. Sembrong River flood map was generated to get contour maps so the lowlands can be determined. The data collection work were carried out including the position, topography height, time and details of a study observation. SURFER 9 software was used to generate a map of the flooding catchment area at Sembrong's river. Figure 2 shows the mapping results.

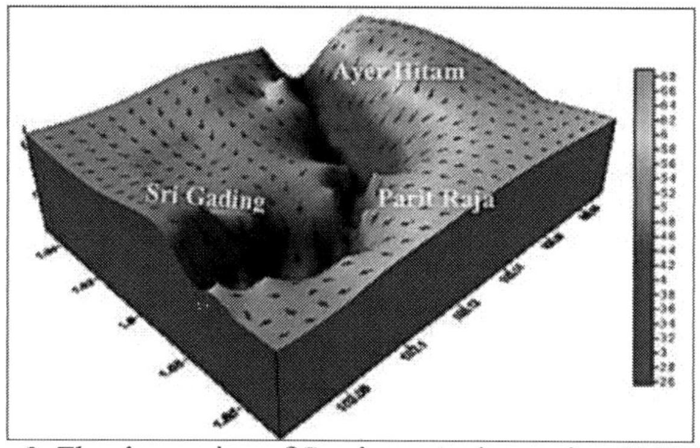

Figure 2. Flood mapping of Sembrong's river using SURFER 9.

A map that has been produced is a map of failure of the dam at the reservoir level PMF (Probable Maximum Flood). PMF can be defined as the expected flood of combination meteorlogy worst conditions and critical depending on hydrological conditions that are reasonable in the drainage basin under study [6].

2. Materials and methods

This study was conducted and analyzed based on the drainage system, hydrological data and mapping sources.

2.1 Data collection

This section describes the data collection and inspection of the work that was involved in the process of preparing this analysis. Effectiveness of analysis conducted is highly dependent on the quality of data for the creation of flood maps. Topographic map is one of the necessary data for this study. This map is available from the Department of Survey and Mapping (JUPEM) in digital form and hardcopy with the scale of 1:50,000 and 20 m contour interval. This map serves as the main base map outlining the boundaries of the study and catchment. While maps from Geographical Information System software was used as a reference and as a guide to its original position for Sungai Sembrong sheet and Sembrong dam. The map was taken from GIS software into the Infoworks ICM software as the basis to start development work on the river model. Figure 3 showed that the information from GIS software indicated Sembrong's river and Sembrong's dam in a scale of 1:125 000.

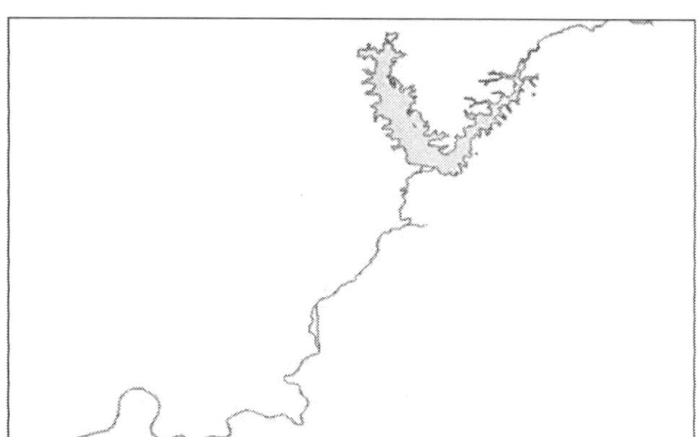

Figure 3. Information from GIS software indicate Sembrong's river and Sembrong's dam in scale 1:125 000.

For hydrological data, there are two important data required for this study which were cross-sectional river engineering survey and the maximum flood discharge from the Sembrong dam. River engineering study data are required to determine the depth of the cross section of the Sembrong River while the data of Probable Maximum Flood conditions was taken as maximum water discharge for this study.

2.2 Data analysis

The process of data analysis were the development model of the river, entering data into Infoworks ICM software and generate maps using ArcGIS. The river model development was done entirely through Infoworks ICM software. Using this software, the model was developed by taking data from GIS software. Map from GIS software layer was used as the basis for the model development work.

River model with establishment continued by inserting the river reach line data and data for the right and left bank. Once all the data is entered, the model was built using ICM Infoworks in full. At this stage all nodes will be linked in order to form the surface of the river that flows from the Sembrong dam as a starting point and end at the end point which is at the junction of the Bekok's and Simpang Kanan's river. Figure 4 shows the river cross sectional data that was added in infoWorks ICM software and Figure 5 shows the completed Sembrong's river streams.

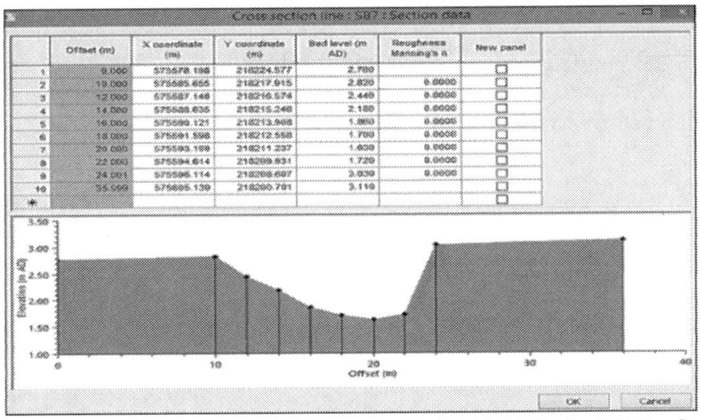

Figure 4. River cross-sectional data was added to InfoWorks ICM software.

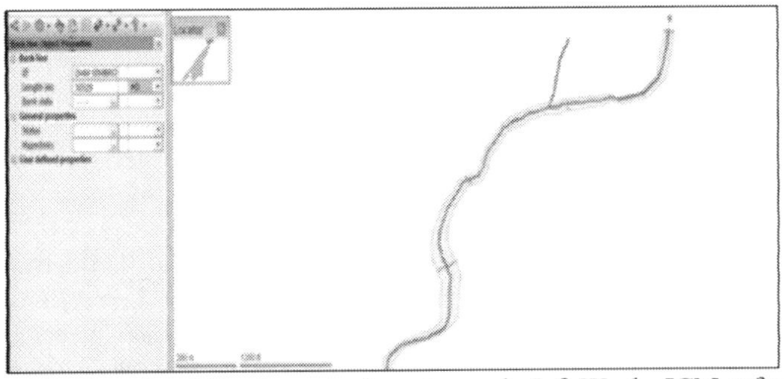

Figure 5. Completed Sembrong's river stream in InfoWorks ICM software.

The analyzed data continued by the data in the software development Infoworks ICM made based on information taken from HLA Associates Sdn. Bhd. The data entered was to identify areas affected by flooding in the event of a dam failure which is the Sembrong Dam. The data at issue is the flow rate data against time for the Probable Maximum Flood case scenario. Figure 6 shows the example of data that were inserted in the software.

Figure 6. Example of inserting flow rate data against time data in the software.

3. Results and discussion

Results for the extent of river overflow can be viewed in detail through the cross section for each node that was constructed in the early stages of development of the river. There are 11 nodes that represent a cross section that has been selected for Sembrong River, where, node structural bench, SB-1 (point downstream) and node structural bench, SB 107 (point upstream). Results for the cross river section was shown in Figure 7, where the graph shows the increase and decrease rate by time and water flow rates that have been entered into the software.

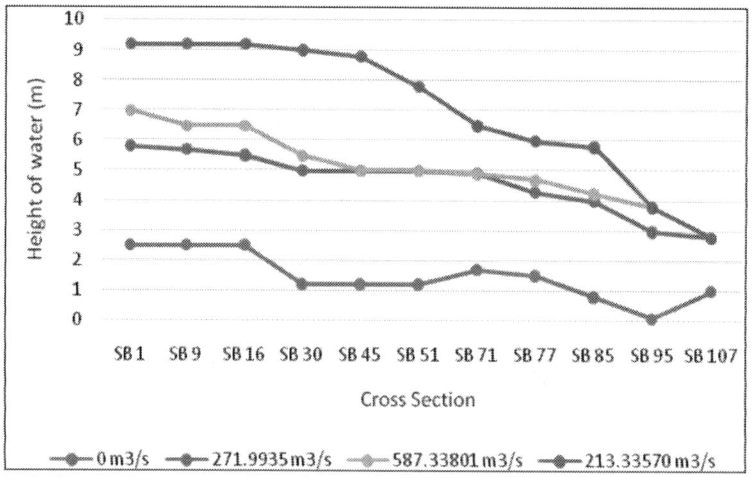

Figure 7. Result for each cross section shows the rate of increase and decrease of water flow.

Apart from cross-river view, the results of the analysis can also be seen from the map that has been built. The results of the maps that have been built showed that thereare 11 villages or regions that were expected to be flooded as Sri Lalang, Parit Sagil, Parit Sonto, Sri Paya, Parit Raja, Parit Sempadan, Talang Bunut, Asam Bubok, Tanjung Semberong, Sungai Rambut and Parit Haji Talib. Table 1 shows the details of the affected areas for flowrate of 587.34m^3/s while Figure 8 shows the flood map of the study.

Table 1. The flooded area information based on studies.

Areas	Maximum flood depth (m)	Type	Number of population
Sri Lalang	1 – 1.2	Main	4690
Parit Sagil	1 – 1.2	Main	1006
Parit Sonto	< 0.5	Main	2505
Sri Paya	1 – 1.2	Main	1397
Parit Raja	1 – 1.2	Main	2052
Parit Sempadan	< 0.5	Main	2851
Talang Bunut	1 – 1.2	Main	1025
Asam Bubok	< 0.5	Main	468
Tj. Semberong	< 0.5	Main	182
Sungai Rambut	1 – 1.2	Main	463
Parit Haji Talib	1 – 1.2	Main	1488

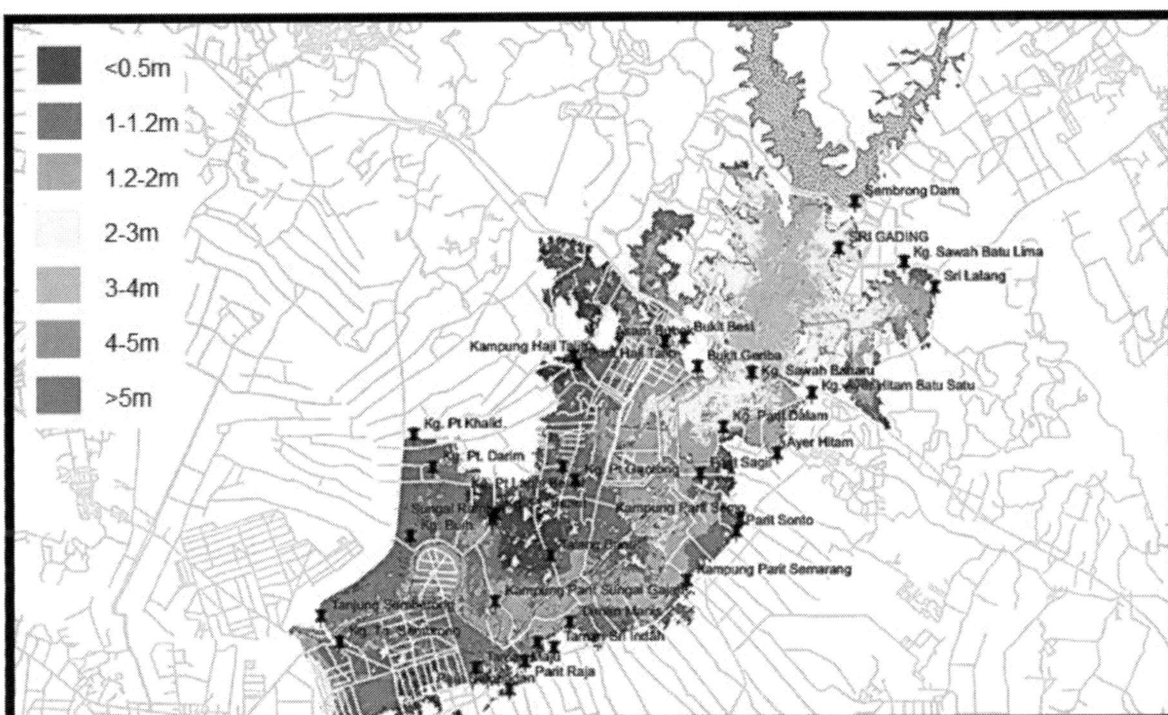

Figure 8. Flood map of the studies area.

IOP Conf. Series: Materials Science and Engineering **136** (2016) 012084 doi:10.1088/1757-899X/136/1/012084

4. Conclusion

In general, this study aims to investigate or determine the areas that will be affected by flood in the event of a dam failure. The areas chosen for this study are the area that includes Sembrong Dam catchment area. However, only Sembrong river that serve to map flooded area of Sungai Sembrong. Flood areal are maps that has been produced 587.34 m^3/s flow rates PMF (Probable Maximum Flood) conditions for 100 years return period as shown in Figure 9.

The study has shown where it can be seen betweenn the main areas to be inundated by flood water at the maximum flow rate was released as Sri Lalang 1 to 1.2 m, Parit Sagil 1 to 1.2 m, Parit Sonto less than 0.5 m, Sri Paya 1 to 1.2 m, Parit Raja of 1 to 1.2 m, Parit Sempadan less than 0.5 m, Talang Bunut 1 to 1.2 m, Asam Bubok less than 0.5 m, Tanjung Semberong less than 0.5 m, Sungai Rambut of 1 to 1.2 m and Parit Haji Talib also 1 to 1.2 m. However, the study also found that the probability of the region have begun to flood at a flow rate 216 143 m^3/s, especially near the banks of the river based on the results of the analysis obtained by cross-view Sembrong river. The flood that might occurred may caused bad effects to the residents of certain area that were affected.

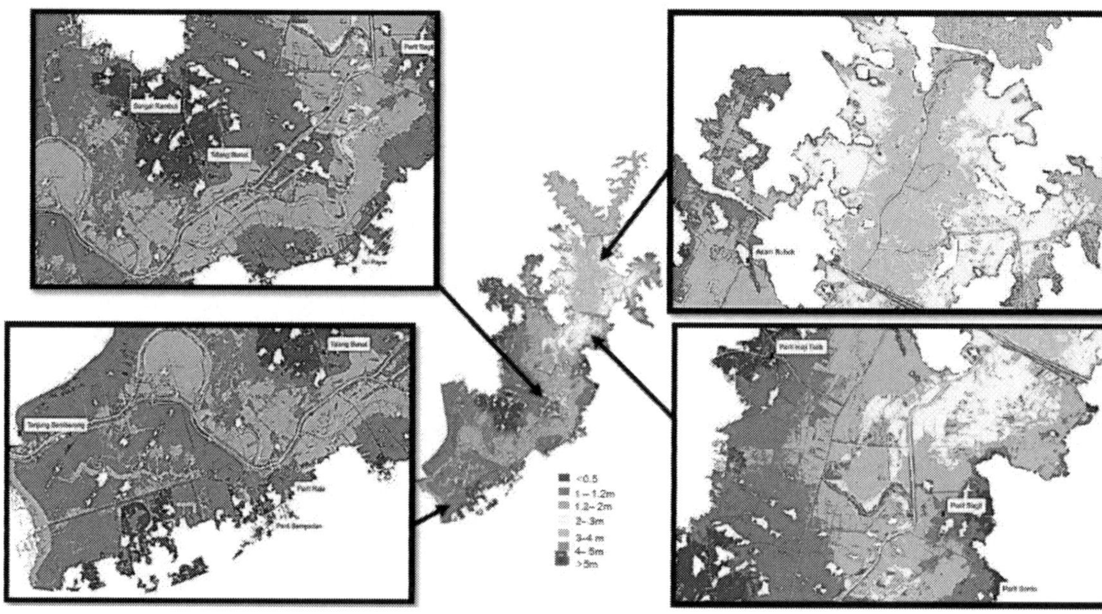

Figure 9. Flood arial mapping at Q = 587.34m^3/s.

References

[1] Dewan Bahasa dan Pustaka 2005, Kamus Dewan, Edition 5, Malaysia: Dewan Bahasa dan Pustaka.

[2] Shafie A 2009 Technical Report, A Case Study on Floods of 2006 and 2007 in Johor, Malaysia, Jabatan Pengairan dan Saliran Johor Bahru.

[3] Department of Irrigation and Drainage Malaysia 2015 Sembrong Dam, Johor, The official Portal for Department of Irrigation and Drainage Malaysia.

[4] Razi M A, Adnan M S, Ahmad M A, Tan L W, Mazlan M and Junaidah A 2014 Hydrodynamic Modeling and Flood Mapping of Sembrong Catchment Area, 13th International Conference on Urban Drainage, Sarawak, Malaysia, p. 9.

[5] County P 2013 Multi-Jurisdictional Hazard Mitigation Plan- Dam failure, Hazard Profile, pp.C6.

[6] Graham W J 2000 Should Dams Be Modified For The Probable Maximum Flood, Journal of the Amarican Water Resources Association, Vol. 36 (5), pp. 953-964.

Temporal Trends of Discrete Extreme Events – A Case Study

S N Rahmat[1], N Jayasuriya[2], M Bhuiyan[2] and M S Adnan[1]
[1] Faculty of Civil and Environmental Engineering, Universiti Tun Hussein Onn Malaysia, 86400 Batu Pahat Johor, MALAYSIA
[2] Senior Lecturer, School of Civil, Environmental and Chemical Engineering, RMIT University, GPO Box 2476, Melbourne, Victoria 3001, AUSTRALIA

E-mail: nazahiya@uthm.edu.my

Abstract: Investigating trends in discrete events is essential for the study of changing patterns of extreme events. Temporal trends in the inter-arrival times of occurrence of drought events were examined for 21 selected stations across Victoria, Australia. In the present study, the Standardize Precipitation Index (SPI) was applied for 12-month time scale to identify drought. A drought event here is defined as a period in which the SPI is continuously negative and reaching a value of -1.0 or less. Often, nonparametric tests are commonly used to test for trends including in discrete events. However, discrete events are not constant because of the presence of zero values or non-normality of data. The methodology applies to long-term records of event counts and is based on the stochastic concepts of Poisson process and standard linear regression. Overall, of the 21 stations, 15 showed statistically significant increasing frequency indicates those events are becoming more frequent. Only one station gave insignificant result. The remaining 5 stations showed the time between events was significantly increasing designates droughts are becoming less frequent.

Keywords: Parametric trend test poisson process standardize precipitation index (SPI).

1. Introduction

Measuring trends in extreme events has received much attention in recent years. The International Panel on Climate Change (IPCC), in its assessment of the vulnerability of the regional impacts of climate change, reports the potential risk of variations in extreme events [1]. Parametric and non-parametric methods are the most commonly used techniques to detect significant trends in hydro-climatologic time series [2]. While nonparametric techniques, such as a Mann-Kendall test for trend or a Sequential Mann-Kendall test for a change point have been widely used and applied in the case of extremes, they were not developed for dealing with discrete events as such [3].

When common statistical tests are used to test for trends in discrete events, fundamental assumptions are often violated because of the use of counted data, the presence of zero values, or non-normality of data, where little is known about the distribution of the underlying dataset [4]. An alternative approach common in hydrology has been to apply regression analysis with a trend variable to hydrologic time series transformed to an approximate normal distribution. For instance, Ross [4] applied the logarithmic transformation to stream flow, as consistent with a lognormal distribution.

Time series of climatological extreme events, however, become counting processes once a threshold level is identified. A counting process is simply a stochastic process by which a running count of the number of events occurring in a prescribed time increment is observed [5,6]. Keim and Cruise [6] applied Poisson process and linear trend analysis to detect changes in precipitation. They concluded that the methods provided in the paper to examine trends in rainfalls could also be applied to the temporal and spatial frequencies of other extreme events or any discrete random event. The Poisson distribution has also been fitted to the frequency of hurricanes by [1,7,8].

The trend of Standardised Precipitation Index (SPI) was investigated for more than 100 years of data using non-parametric trend technique Mann Kendall (MK) to detect wet (increasing) and dry (decreasing) periods across Victoria, Australia [1]. Out of 70 stations, 21 stations showed significant decreasing trend, indicating an increase in magnitude of dry periods. This finding highlights the need to look at the patterns as to whether the intervals between extreme (drought) events are becoming shorter and frequent. Morover, Australia's climate has always been variable and in particular, prone to drought.

Using a method proposed by Keim and Cruise [6], this present study will test for any gradual trend in the rate of occurrence of drought through an analysis of processed inter-arrival times using parametric method, Poisson process and standard linear regression.

2. Data and Methods

The locations of the 21 rainfall stations in Victoria, Australia and their characteristics for this study are shown in Figure 1 and Table 1. The number of years of available data varied from one station with a minimum of 60 years to two stations with a maximum of 160 years of monthly data.

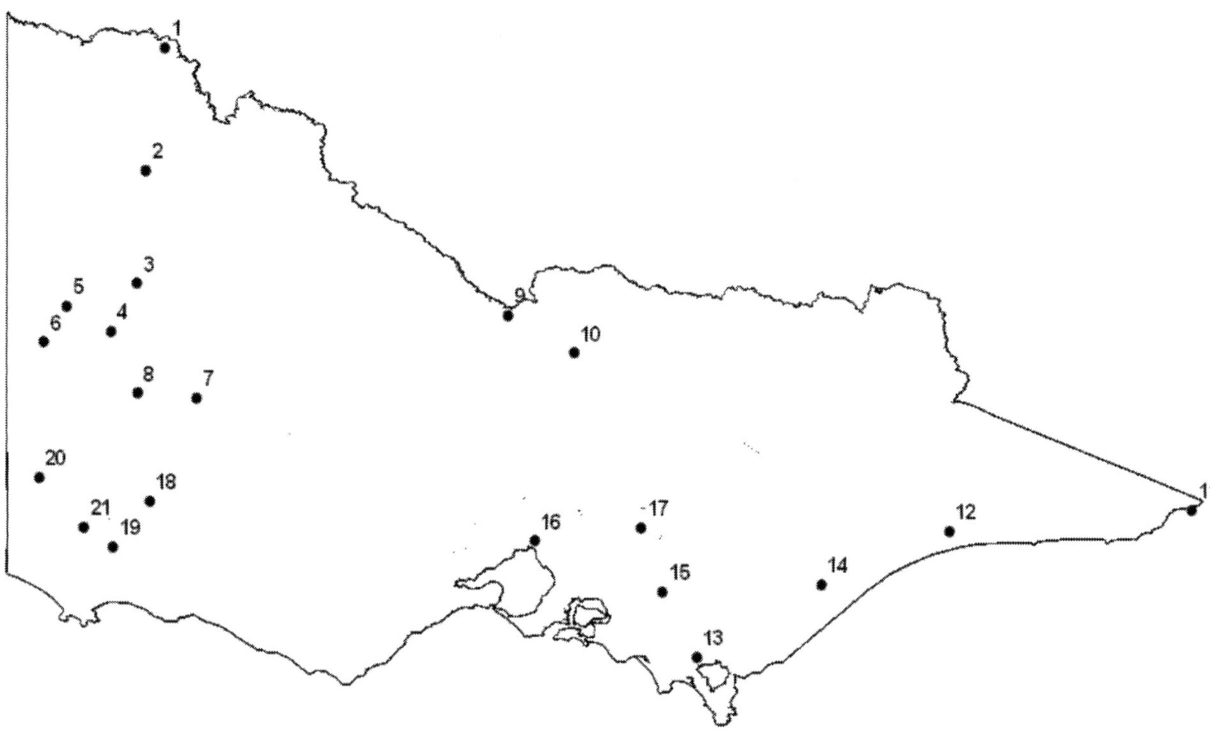

Figure1. Locations of the study sites.

2.1 The Poisson process

A Poisson process is intended to represent events that occur at random or in more general terms, a stochastic model for a series of events is called a counting process [3]. A counting process is generally a stochastic process by which a running count of the number of events occurring in a given time increment is observed [5]. In the Poisson distribution, there is equality of mean and variance represented by the parameter λ. If the arrival rate remains constant with time, then the process is homogeneous. Another very important property of the Poisson process is that the inter arrival times between realizations of successive events are known to be exponentially distributed [4].

Table 1. Description of selected rainfall stations for the analysis.

Stn. No.	Stn. Name	No of years of data
1	Irymple	103
2	Walpeup Research	74
3	Rainbow	115
4	Nhill (Woorak)	82
5	Yanac North	115
6	Kaniva	130
7	DrungDrung	83
8	Natimuk	106
9	Echuca Aerodrome	132
10	Tatura	71
11	Gabo Island Lighthouse	146
12	NowaNowa	64
13	Foster (Post Office)	128
14	East Sale Airport	69
15	Warragul	124
16	Melbourne Regional Office	157
17	Warburton	97
18	Cavendish (Post Office)	128
19	Branxholme (Bassett)	67
20	Dergholm (Hillgrove)	113
21	Merino	114

3. Results and Analysis

3.1 Temporal trends in drought events

In the present study, the arrival rate of drought events for all stations were tested to classify its distribution. As mentioned earlier, the most important counting process is the arrival rate of events follows a Poisson distribution which is, mean and variance must be equal or nearly equal. However, arrival rates are not always temporally constant and always showing non homogeneous Poisson process. The Poisson process relies upon two underlying assumptions; the number of events occurring in time is independent of its previous history, and no more than one event can occur in a short interval of time [9,10]. In this case, the Poisson model represents the number of randomly occurring infrequent storm events over a specified threshold.

Time series of the SPI at 21 stations selected across Victoria, are analyzed for temporal trends. The droughts events and the inter-arrival time were identified based on the SPI value, which is continuously negative and reaching a value of -1.0 or less. As for examples, the results for five stations are plotted in Figure 2. As proposed by Cox and Lewis [11], the series of inter-arrival times begins with the onset of an event because the length of time from some arbitrary point to the first event does not belong to the same distribution of intervals between events. The number of droughts events recorded varies due to the time series selected which ranging from 14 to 24. It can be seen that for each station inter-arrival times have high variability. For Irymple station, the inter-arrival times range from 10 to 99 months, while for Rainbow, Edenhope, Dookie and MRO, the inter-arrival times range from 6 to 144, from 5 to 135, from 12 to 140 and from 10 to 194 months, respectively.

3.2 Trend analysis of inter-arrival times of droughts using linear regression

In the present study, the natural log of the intervals between successive events (Yi), or successive groups of events (in months), on the independent variable (Xi), which is the cumulative frequency of months from the occurrence of the first event was identified. Advantages of using linear regression on transformed interarrival time data generated by a Poisson process include successive events are known to be independently distributed, and the probability distribution is moderately non-normal [5,6].

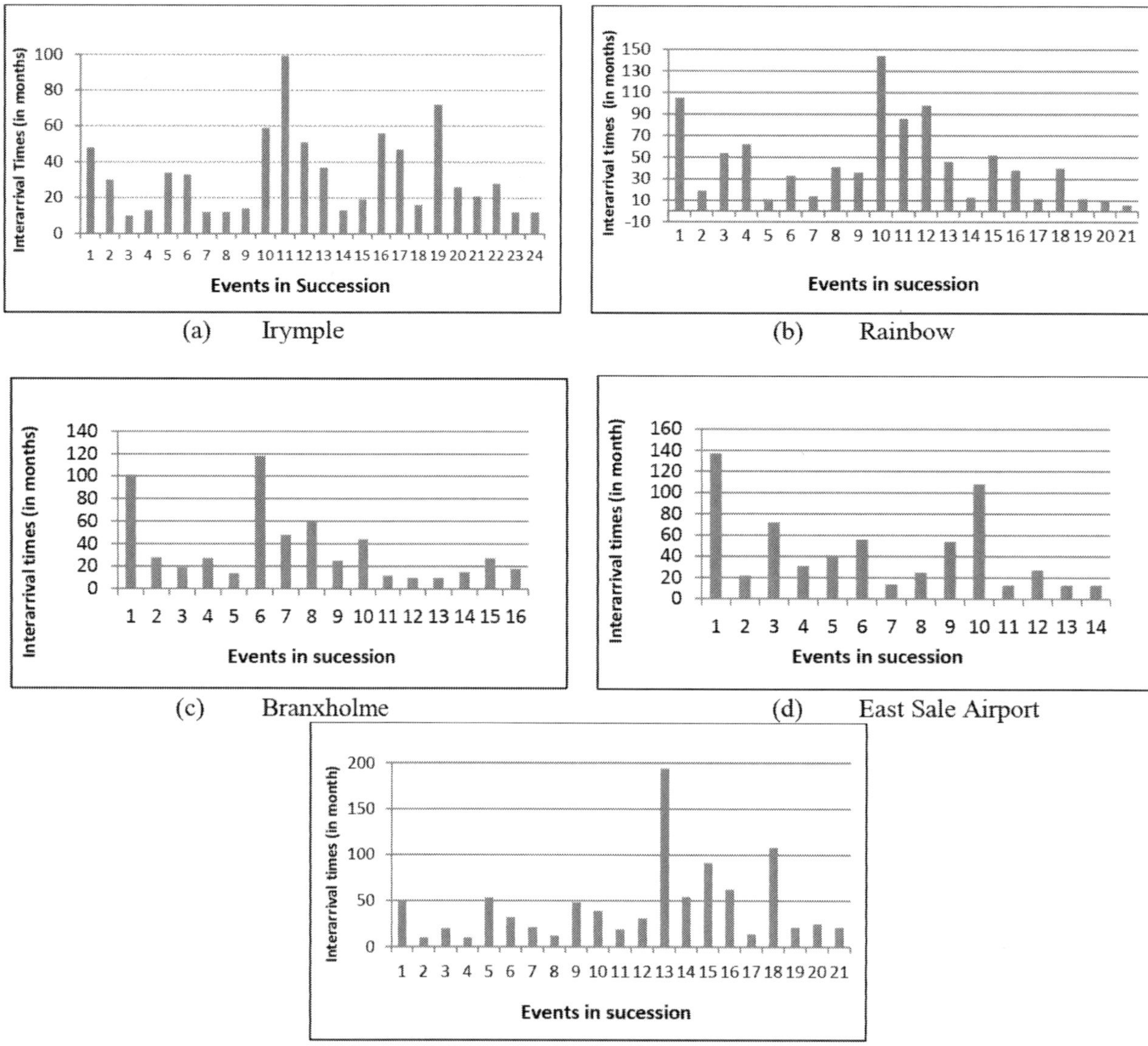

(a) Irymple (b) Rainbow

(c) Branxholme (d) East Sale Airport

(e) Melbourne Regional Office

Figure 2. Recurrence intervals of droughts for all stations.

As for example, the SPI values for 12-month time scale and linear regression analysis for Irymple station are given in Figure 3. The identification of the drought events was based on SPI drought class threshold (SPI = -1 and less) (see Figure 3a). A number of major droughts events were identified including the World War II drought, the 1982-1983 drought and the recent drought (Big Dry). Figure 3b displays the number of months between each drought event. Irymple station shows variability of inter-arrival times between events, with intervals 11 and 19 being the longest of the series. Those two events were in 1951 and 1985, respectively. Events in succession are plotted across the x axis and the number of months from the end of one event to the beginning of the next event (in sequence) is shown on the y axis. The x axis in Figure 3c shows the intervals from the occurrence of the first event in the series. In Figure 3d, the graph shows the number of months transformed into ln and then plotted on the y axis. The data are then plotted with the linear least squares fit.
A similar approach was applied to other four stations and the linear regressions of the log transformed intervals are shown in Figure 4. Table 2 shows the sample sizes and significance levels of the linear regression analysis. A positive sign for the Pearson, r signifies that the time between events is increasing and those events are becoming less frequent. In contrast, a negative sign indicates that the intervals between events are becoming shorter and that the frequency of events is temporally increasing. As shown in Table 2, the slope (r = 0.04)

depicts that the events are becoming less frequent but is insignificant. All the other three stations (Edenhope, Dookie and MRO) are similar, showing decreasing patterns of frequencies with r values of 0.01, 0.21 and 0.28, respectively. In contrast, the Rainbow station (r = -0.31) has an insignificantly increasing pattern of frequencies. This trend analysis was then applied for the rest of the 16 stations and the results are given in Table 2. Overall, of the 21 stations, 15 show statistically significant increasing frequency and those events are becomimg more frequent. The stations showing significantly increasing frequency are Walpeup Research, Nhill, Kaniva, Drung Drung, Natimuk, Echuca, Tatura, Gabo Island, Nowa Nowa, Foster, East Sale, Warragul, Avenel, Heathcote, Newstead, Cavendish, Branxholme and Dergholm. The remaining 5 stations (Irymple, Yanac North, Melbourne Regional Office (MRO), Warburton and Merino) show the time between events is increasing indicates droughts are becoming less frequent.

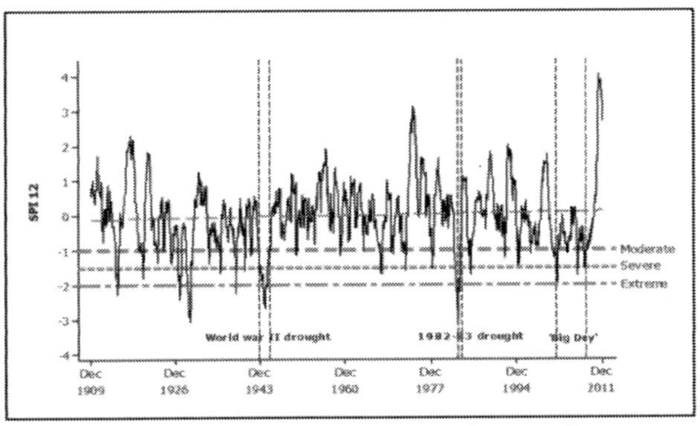

(a) The SPI 12-month time scale

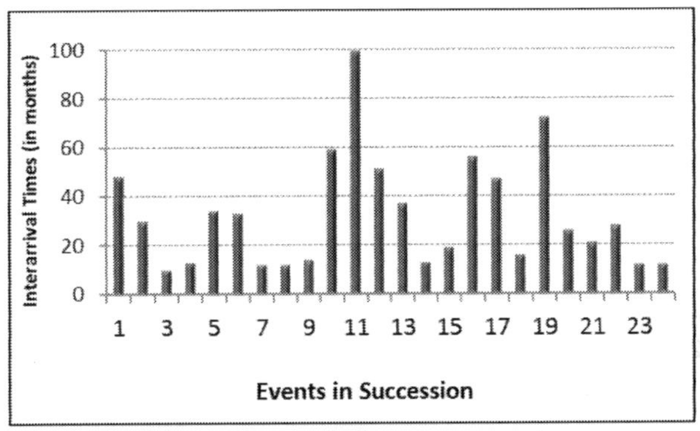

(b) Intervals between events

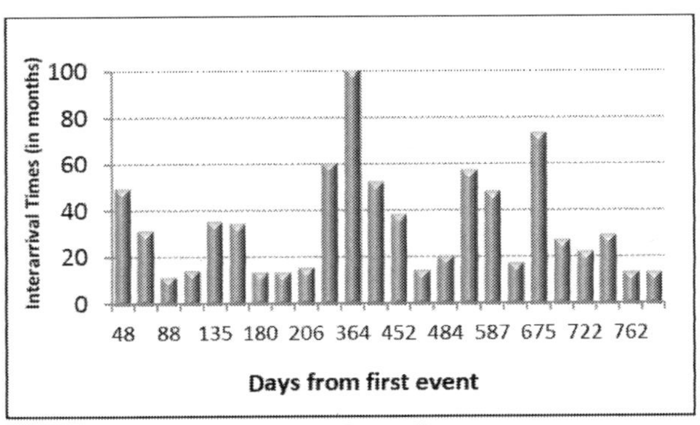

(c) Intervals from first events

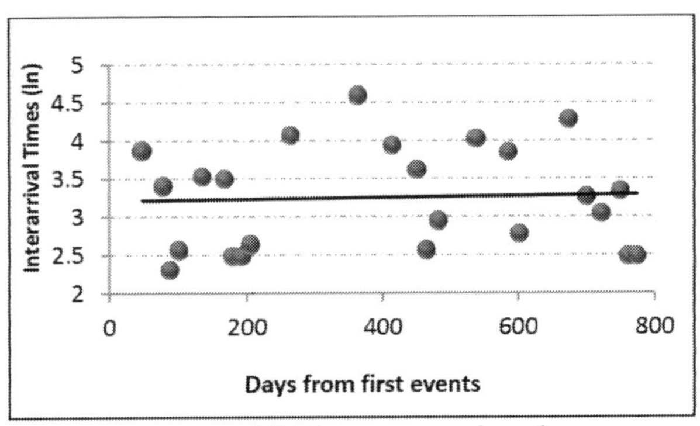

(d) Linear regression slope

Figure 3. Time series of droughts for Irymple station.

Overall, the following steps need to be carried out to identify the trend for any discrete random events:

1) The arrival rate of drought events for all stations are tested to classify its distribution.
(2) Once data are identified as Poisson, the inter-arrival times are known to follow the exponential distribution.
(3) These data are transformed by converting the raw values into natural logarithms (ln) so that they follow the normal distribution.
(4) The data are then plotted with the linear least squares fit.
(5) Pearson's correlation coefficient, r is identified to indicate whether the time between events is increasing, or decreasing.

IOP Conf. Series: Materials Science and Engineering **136** (2016) 012085 doi:10.1088/1757-899X/136/1/012085

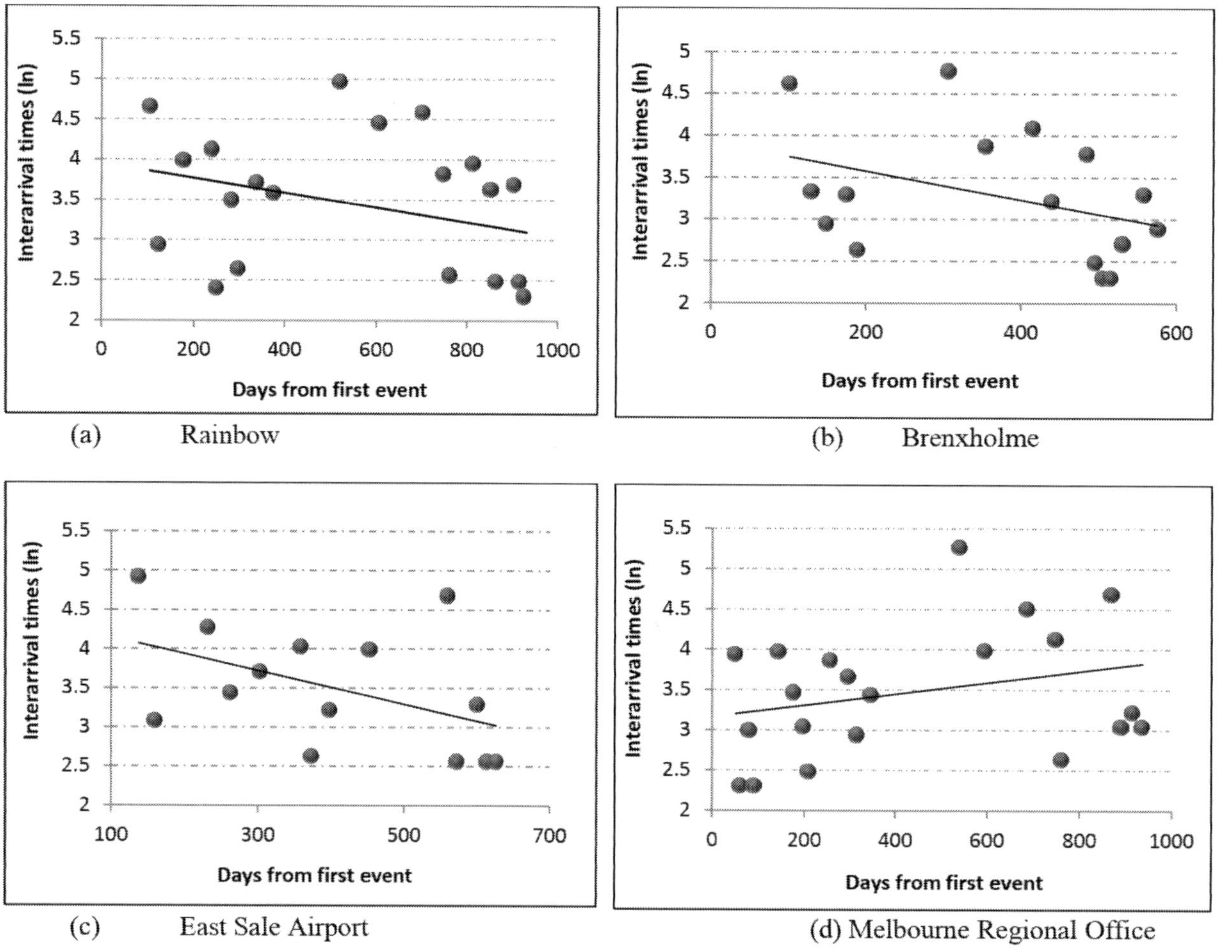

Figure 4. Linear regression of drought events at all stations.

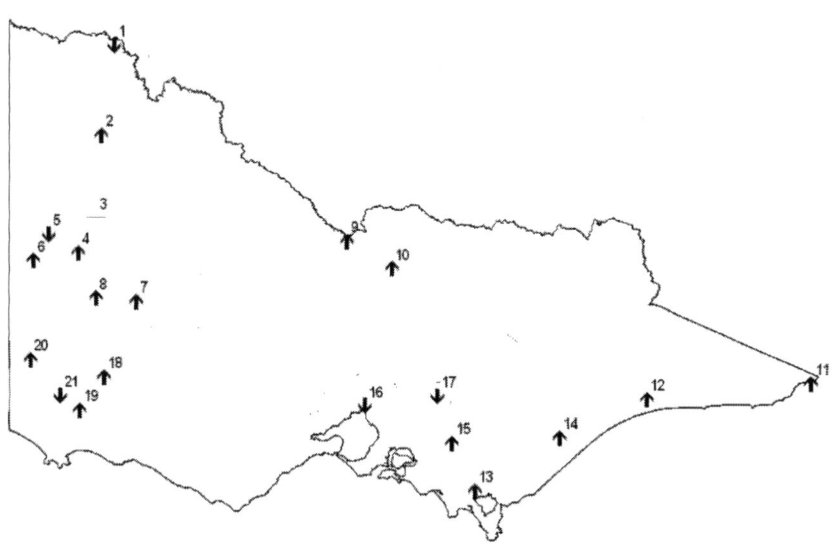

Figure 5. Trend analysis for all 21 stations (- : No trend; ↑: Increasing frequency; ↓: Decreasing frequency).

Table 2. Results of linear regression analysis, r = Pearson's correlation coefficient, p = significance of slope

No	Station	Events	r	p
1	Irymple	24	0.04	0.8
2	**Walpeup Research**	**15**	**-0.34**	**0.01**
3	Rainbow	21	-0.31	0.2
4	**Nhill (Woorak)**	**18**	**-0.12**	**0.04**
5	Yanac North	24	0.11	0.17
6	**Kaniva**	**33**	**-0.03**	**0.03**
7	**Drung Drung**	**20**	**-0.04**	**0.05**
8	**Natimuk**	**25**	**-0.06**	**0.03**
9	**Echuca Aerodrome**	**35**	**-0.01**	**0.02**
10	**Tatura**	**14**	**-0.65**	**0**
11	**Gabo Island**	**42**	**-0.2**	**0**
12	**NowaNowa**	**14**	**-0.45**	**0.01**
13	**Foster (Post Office)**	**32**	**-0.2**	**0**
14	**East Sale Airport**	**14**	**-0.46**	**0**
15	**Warragul**	**30**	**-0.05**	**0.01**
16	Melb. Reg. Office	21	0.28	0.2
17	Warburton	23	0.24	0.36
18	**Cavendish**	**36**	**-0.04**	**0.01**
19	**Branxholme**	**16**	**-0.38**	**0**
20	**Dergholm**	**27**	**-0.11**	**0.01**
21	Merino	30	0.08	0.05

*Results in boldface indicate significant increasing frequency.

4. Conclusions

Measuring trends of extreme events or disasters are real challenges in water resources management. However, it is essential as extreme events are becoming more common due to the impact of climate change. Also, the ability to detect long term variations and trends is critical to begin to detect any observed changes.

The trend of the SPI was investigated to detect trends for 70 selected stations across Victoria, Australia. 21 stations showed significant decreasing trend, showing an increase in magnitude of dry periods. Further, the trends were identified to measure the intervals between extreme (drought) events using parametric method. The parametric method, Poisson process is found to be appropriate for measuring trend in discrete random events. The advantage to using parametric method is that the test is generally more efficient.

Overall, of the 21 stations, 15 showed statistically significant increasing frequency and those events are becoming more frequent and one station gave insignificant value. The remaining 5 stations showed the time between events is increasing indicate that droughts are becoming less frequent. Based on these results, some of the study areas in Victoria may be facing frequent drought. Therefore, it is essential to use an appropriate methodology to develop suitable strategies to mitigate the impacts of future droughts and properly understand past droughts to be able to forecast the future wherever possible.

Acknowledgements
The authors are grateful for the funding provided by the Ministry of Higher Education under the Fundamental Research Grant Scheme (Vote No. 1232).

References

[1] Otto, M. (1998). Multivariate Methodsn: In: R. Kellner, J. M. Mermet, M. Otto and H. M. Widmer, Eds., Analytical Chemistry. Wiley-VCH, Weinheim, pp.916.

[2] Tabari, H., S. Marofi, A. Aeini, P. H. Talaee and K. Mohammadi (2011). Trend analysis of reference evapotranspiration in the western half of Iran. Agricultural and Forest Meteorology, 151(2), pp.128-136.

[3] Allen, L. J. S. (2010). An introduction to stochastic processes with applications to biology. Chapman & Hall/CRC Press, Boca Raton, Florida, pp.461.

[4] Ross, S. M. (2014). Introduction to probability models. 11th Edition. Elsevier, San Diego, California, pp.759.

[5] Katz, R. W. (2002). Stochastic Modeling of Hurricane Damage. Journal of Applied Meteorology, 41(7), pp.754-762.

[6] Keim, B. D. and J. F. Cruise (1998). A Technique to Measure Trends in the Frequency of Discrete Random Events. Journal of Climate, 11(5), pp.848-855.

[7] Bove, M. C., J. J. O'Brien, J. B. Eisner, C. W. Landsea and X. Niu (1998). Effect of El Niño on U.S. Landfalling Hurricanes, Revisited. Bulletin of the American Meteorological Society, 79(11), pp.2477-2482.

[8] Elsner, J. B. and B. H. Bossak (2001). Bayesian Analysis of U.S. Hurricane Climate. Journal of Climate, 14(23), pp.4341-4350.

[9] Ashkar, F. and J. Rousselle (1983). The effect of certain restrictions imposed on the interarrival times of flood events on the Poisson distribution used for modeling flood counts. Water Resources Research, 19(2), pp.481-485.

[10] Bezak, N., M. Brilly and M. Šraj (2014). Comparison between the peaks-over-threshold method and the annual maximum method for flood frequency analysis. Hydrological Sciences Journal, 59(5), pp.959-977.

[11] Cox, D. R. and P. A. W. Lewis (1996). The Statistical Analysis of Series of Events. Springer, London and Colchester, pp.279.

Estimation of Soil Erosion Rates in Oil Palm Plantation with Different Land Cover

S Sahat[1], Z Yusop[2], M Askari[2] and A D Ziegler[3]
[1] Center of Diploma Studies, Universiti Tun Hussein Onn Malaysia
[2] Water Research Alliance, Universiti Teknologi Malaysia, Malaysia
[3] National University of Singapore

E-mail: suhailasahat@gmail.com

Abstract: Soil losses from hill slopes in oil palm plantation in Sedenak Estate, Johor were measured using runoff plot and rainfall simulator. The plot was designed to be removable but the size was fixed at 0.8 x 3.75m. Four types of surface covers were investigated for the plots, i.e. half bare soil and half grass cover (HGC), half bare soil and half dry frond (HDF), fully grass cover (FG), and fully bare soil (BS). The influence of initial soil moisture, saturated hydraulics conductivity, Ks, bulk density and slope on rates of soil loss were also evaluated. The rainfall simulator produced rainfall intensities between 90 and 160 mm/hr with durations from 45 to 60 min per run. BS plot exhibited the highest Ks value among all plots but the percentage of initial soil moisture on this surface was low. BS plot recorded the highest runoff coefficient (C) and soil loss values of 73.6 ± 4 percent and 5.26 ± 3.2 t/ha respectively, while the lowest was from plot FG with 41.7 ± 5.7 percent and soil loss of 2.85 ± 2.1 t/ha. Meanwhile, the results suggested that the ground cover had the ability to reduce soil loss by 67% and 17%, respectively for plots BS-HGC and BS-HDF. Overall, soil erosion control such as surface is effective measures in reducing level of runoff and soil erosion.

Keywords: Rainfall simulator, soil loss, soil erosion plot, surface cover, saturated hydraulic conductivity.

1. Introduction

Erosion and sedimentation are the major issues when it comes to managing water quality and constitute the largest portion of non-point source pollution in the tropical region [1,2,3]. High soil erosion rate often leads to river constriction, increased flood frequency and magnitude, threatened aquatic habitats, increased rate of nutrients losses and interference with recreation opportunities. Sediment would absorb nutrients and thus increase rate of nutrient losses[4,5]. Studies on erosion and sedimentation in the past have mostly been confined to forestry related activities and urban catchments [6,7,8]. Activities associated with oil palm plantation, especially during land preparation or replanting, could have significant impact on soil health both in terms of quality and quantity.

Lord and Clay [9] stated that soil erosion is the major cause of soil degradation through increased leaching of reducing nutrients and organic matter, at the same time modifying the physical properties such as bulk density and infiltration rates. Soil losses from oil palm plantation during its development to maturity stage could be as high as 4 ton/ha/yr [10]. The rapid expansions of oil palm cultivation have put this crop as commercial cultivation after slow growth in 1917. Even so, after 50 years have passed, agricultural growth continues to involve massive investments from the government. Besides Malaysia known as a major producer of rubber crops and also cocoa, oil palm has also significantly contributed to the country's agricultural industry. This has given massive job opportunities for Malaysians when FELDA was first established on July 1, 1956 when the Land Development Act (1956) was enforced [11].

Soil erosion in oil palm plantations has caused water pollution in local rivers where turbidity and total solids are suspended high in the area. This may cause adverse impact on life which cannot live in such conditions. If precautions are not taken seriously, it will affect the supply of fresh water resources. The highest rate of erosion and sedimentation from oil palm plantation mostly occur during land preparation, involving forest clearance or replanting [12,13]. Malaysia's climate with frequent rainfall will worsen this situation. In matured oil palm plantations, soil erosion losses chiefly depending on the slope and soil management practices, including maintenance of unpaved roads and the harvesting paths. The degree of erosion rates is highly spatial in major landscape units such as harvesting paths and less disturbed surfaces, even after the trees have become established. Hartemink [14] reported that erosion rates in matured oil palm plantations in Malaysia range from 7.7 to 14 t/ha/yr. However, a quantitative estimate of erosion rates in oil palm plantation is still very scarce. Therefore, the objective of this paper is to quantify runoff and erosion from four types of surface covers in oil palm plantation on harvesting path and less disturbed area using rainfall simulator

Study Area

The study was carried out in an oil palm plantation at Sedenak Estate (01° 43' 35" N and 103° 32' 42"). The estate is managed by Mahamurni Plantation Sdn.Bhd, a wholly owned subsidiary of Kulim (M) Bhd. The oil palms at time of investigation were 10 years old. The palms are of PAMA/FELDA clone, planted with density of 143 stands per hectare arranged in an equilateral triangle with 9 m gap, alternating North-South rows. The catchment's topography is undulating and the soil belongs to Renggam series [15]. Renggam series are characterized by deep profile with sandy clay texture and in brownish yellow colour. The soil is well drained and suitable for oil palm, cocoa, rubber, coconut and fruit plantation [15].

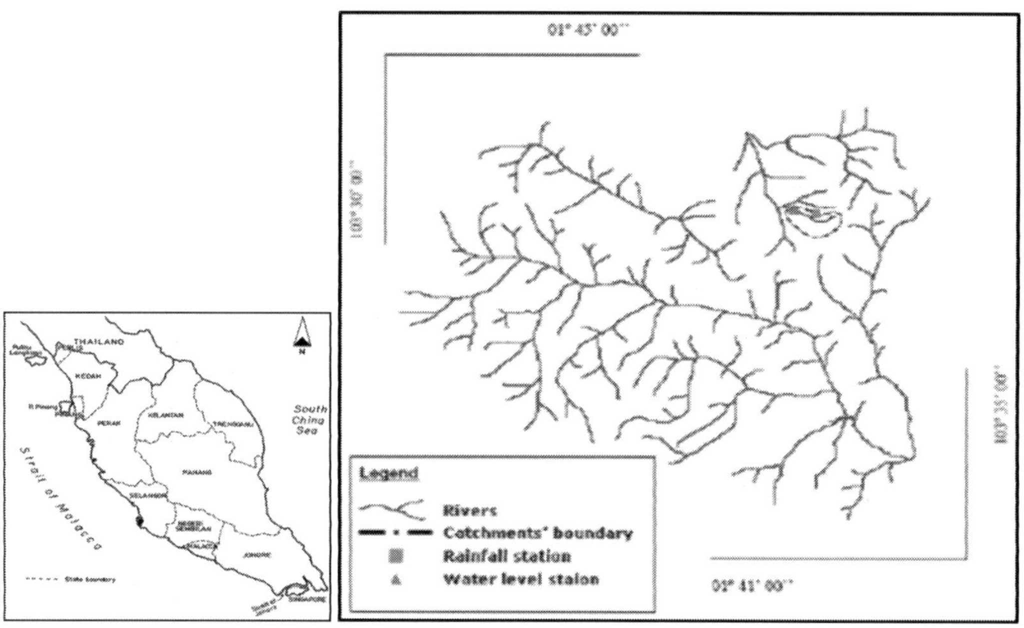

Figure 1. Location of the study area.

2. Methodology

2.1 Simulation Treatments

The rainfall simulator was run 15 times on harvesting path and less disturbed from February to May 2011. The harvesting paths consist of grass of 0.1 to 0.15 m height at the time of fieldwork. Prior to simulation, the harvesting path was less disturbed by farm workers. The oil palms were harvested every two weeks, during this time surface were disturbed by harvesting machines and workers footsteps. This will interfere with the structure of soil surface in terms of compaction and infiltration [16]. Harvesting machine may lead to the formation of tyre ruts on soil surface and could enhance the generation of Horton overland flow, HOF [17]. The four types of surface are function as runoff plots to measure erosion rates on that particular surface. The criteria for the surface selection were based on slope, surface cover and proximity to water resources and each plot is close to each other's. Table 1 shows the plot dimension, surface description and types of surface cover.

Table 1. Description of four erosion plots [18].

Surface Type	No. of Simulation	Surface Description	Ground Cover (%) and Type of Grass Cover
Bare Soil	4	100% harvesting path	Null
Fully Grass Cover	4	100% harvesting path	100% *Paspatum Conjugtum and Centotheca Lappacea*
Half Grass Cover	3	50 % undisturbed area	50% *Centotheca Lappacea*
Half Dry Frond Cover	3	100% harvesting path	50% Dry Frond

2.2 Measurement of physical properties

Prior to rainfall simulation, the soil physical properties for each plot were determined. Information of soil physical properties are important to determine the characteristics of the soil condition. Surface bulk density (ρb) and soil moisture was determined by sampling the upper 5 cm with 110 cm3 soil ring sample. Core sample was oven dried for 24 hours at 105°C. The plot slopes were determined by an Abney level. Saturated hydraulic conductivities (Ks) were estimated from infiltration measurements taken in situ using model 2800 K1 Guelph Permeameter.

2.3 Rainfall simulator and plot design

The rainfall simulator consists of 4 vertical steel rods with 4.3 m high to mount two 60° full cone nozzles with 70μm orifice diameter (Figure 2). Water was pumped from a storage container of 2000L to the simulator through 2.5 cm diameter PVC hose by two units of 750 W centrifugal pumps with pressure head of 172 kPa (25psi). The storage container is refilled by pumping water from a nearby river. This operating pressure produces rainfall energy flux densities (EFD) of 1000 to 1900 Jm-2h-1. Steel plates were erected to form rectangular. Runoff was sampled at the lower end of the plot. The simulated rainfall was measured for 45 and 60 minutes with manual gauges placed on the ground using 8 small cups with surface diameter of 6.3 cm. The volumes of water collected were measured every 5 min. EFD of the simulated rainfall was using formula (1) [19].

$$EFD = \frac{R_i m v^2}{2V_{D_{50}}} \quad (1)$$

Where Ri is event rainfall intensity (mh-1), VD50 is volume (m3) of the median-diameter (D50) raindrop and m for mass (kg). In equation 1, m is the mass (kg) of the D50 drop, which is estimated as

$$V_{D_{50}} = \frac{4\pi D_{50}}{6} \quad (2)$$

$$m = (\rho_w - \rho_{air})V_{D_{50}} \quad (3)$$

where ρw (1000kg m-3) and ρair (1.29 kg m-3). Factor v in Equation 1 is the fall velocity (m s-1) of the D50 drop, determined by the equation[19],

$$v = V_{max}\left(1 - e^{-\frac{D_{50}}{b}}\right) \quad (4)$$

where Vmax = 9.5 (m s-1), b = 1.77 and β = 1.147 [16]. D50 = 1.024 mm ((Median drop size was estimated from nozzle manufacturer engineering data). Natural rainfall and artificial rainfall obviously have different terminal velocity and drop size distribution. Therefore, artificial rainfall energy was calculated using those equations. Ziegler [19] emphasised that same value of rainfall intensity from different simulator architecture will cause different EFDs.

(a) (b)

Figure 2. (a) Half Dry Frond surface (b) Rainfall simulation design with water flow from the nozzle into trenches at the plot outlet

2.4 Simulation data collection

For each plot, the simulations were replicated for dry and wet conditions. Wet simulation was performed one day following the dry simulations. Instantaneous discharge and sediment output during each experiment were measured at time to runoff (TTRO). Discharge was determined by dividing the volume of a collecting bucket (30 litres) with the time to fill up the bucket. Suspended solids sediment concentration was determined by decanting 100 ml of runoff and followed by oven dried at 105°C for 24 hours. Instantaneous discharge and sediment output values were adjusted to rates per unit area. Cumulative discharge (Q_{cum}) was calculated as total runoff volume prior to any time t, divided by EFD since TTRO contribute differently to sub processes controlling runoff generation and sediment transport. At the beginning of rainfall, sediment is detached by raindrop impact and material is transported downslope via rainsplash. In essence, energy prior to TTRO contributes to the sediment supply that will be transported throughout the event after runoff commences.

3. Results and Discussions

As mentioned before, simulations were run on different day; therefore giving different values of soil moisture contents. Table 2 shows that FG recorded the highest value of soil moisture content followed by HGC. This could affect the infiltration of the water into the soil surface [20]. Bulk density was measured during each simulation exercise from the soil sample taken from the plots. HDF recorded the highest value of bulk density (see Table 2). BS and surface type like HGC had almost similar bulk density but slightly higher for HGC. Of all the plots, FG had the lowest bulk density. Meanwhile, Ks was the highest on the BS and lowest for FG. Besides BS having the highest Ks, it also had the highest value of bulk density. According to Penna [20], the increase in Ks can lead to increase in bulk density. Moisture content and soil properties can also indirectly affect on Ks.

Table 2. Soil and Rainfall Characteristics of the four erosion plots [18].

Surface Type	Slope (°)	Rainfall Intensities (mm/hr)	Soil Moisture (%)	Bulk Density (gcm^{-3})	K_s (mm/hr)	EFD ($Jm^{-2}h^{-1}$)
Bare Soil (BS)	23	103 ± 52	31.8 ± 7.2	1.34 ± 0.1	7.3	1010
Full Grass Cover (FG)	15	90 ± 45	48.7 ± 12.1	1.01 ± 0.1	1.4	1880
Half Grass Cover (HGC)	12	85± 50	36.2 ± 10.9	1.36 ± 0.1	3.0	1860
Half Dry Frond (HDF)	23	81 ± 48	40.1 ± 15.3	1.61 ± 0.2	7.3	1332

Note: values are mean ± standard deviation; EFD is energy flux density of rainfall intensities

3.1 Effect of Crops Cover

The ability of crops cover in reducing soil loss can be seen in Figure 3 which shows substantial reduction in soil loss with increasing undergrowth cover. Nevertheless, different levels of crop cover produced different situation. A fully covered plot (FG) recorded the lowest soil loss of 51.9 x 10-3 t/ha compared to BS plot with 96.9 x 10-3 t/ha. Meanwhile, runoff plot with half surface cover (HGC) recorded 0.002 t/ha to 0.006 t/ha soil loss, which was two times higher than at FG plot. Whilst the surface slope on FG plot was higher (Table 2) compared to HGC plot, the crops cover played a major role in controlling soil erosion. Besides intercepting soil particles, the root held the soil particles and stabilized the soil structure from the effect of raindrops and overland flow. Among the three runoff plots, HDF was the least effective in reducing erosion and recorded between 0.009 t/ha to 0.03 t/ha soil loss, or two to four times higher compared with FG and HGC.

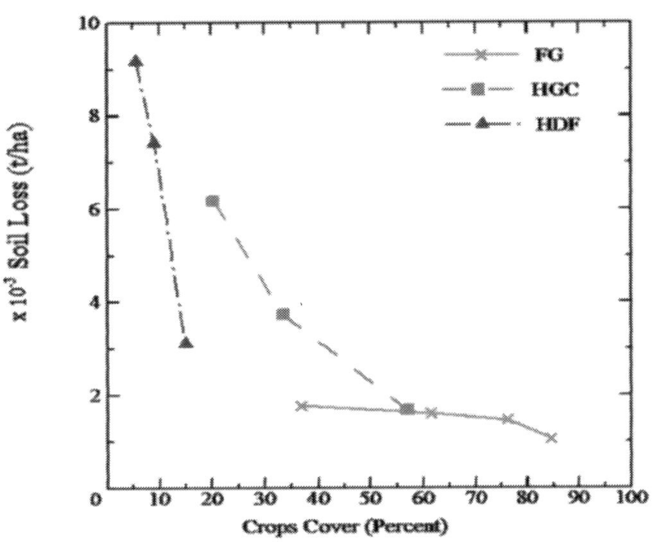

3.2 Runoff Coefficient (C)

Runoff coefficient (C) generally increases with the increase of soil moisture (Figure 4), except for plot HGC, which remained steady with 60% of rainfall appeared as runoff. Among the four surfaces, BS plot recorded the highest C due to higher rainfall and higher Ks value (see Table 2) and bare surface condition. Furthermore, it had steeper slope. Other factor that contributed to higher C readings for BS plot was EFD. According to Dalen et al. [21] and [22], rainfall kinetic energy was found to influence the soil surface structure during storm events. A lower kinetic energy has a lower capacity to reduce the detachment of soil particles by raindrops which reduce erosion rates. The C value for FG plot was higher compared to HGC and HDF plots even though this surface was fully covered by grasses.

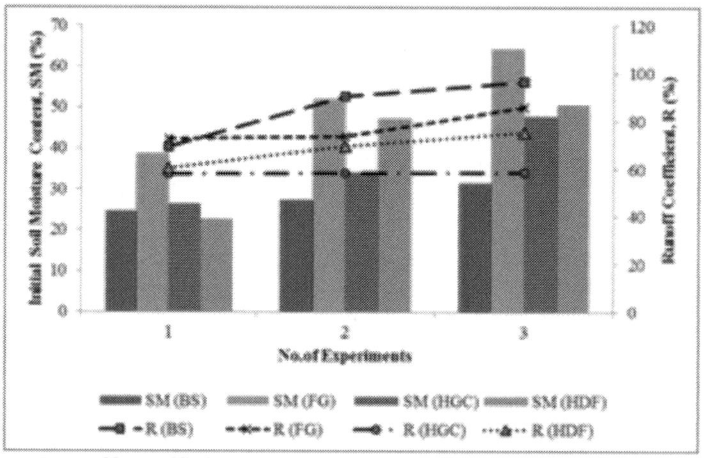

Figure 4. Runoff Coefficient (line) and initial soil moisture (bar) for all plots

3.3 Influence of Runoff Coefficient (C) on Soil Loss

Table 3 shows the means C and soil loss for all plots and the correlation between these variables. BS plot recorded the highest C value and soil loss of 73.6±4 percent and 5.26±3.2 t/ha respectively, while the lowest was from FG plot with average C of 41.7±5.7 percent and soil loss of 2.85±2.1 t/ha. The effectiveness of grass in FG plot was evident when the erosion rate reduced by 83% compared to value for plot BS. FG plot was located underneath the oil palm trees and could have higher content of organic matter. According to Ariza [23] organic matter can improve the structure of the soil. Meanwhile, between BS-HGC and BS-HDF plots, the results suggested that the ground cover had the ability to reduce soil loss by 67% and 17% respectively. The soil loss reduction was higher for HGC plot compared to HDF plot. Again this shows that grass and its root system are effective in reducing surface erosion [24]. Although runoff is important for transporting sediment, in this study, its correlation against soil loss was not significant ($p > 0.05$) with low Pearson coefficient values (0.01 to 0.45). This was rather unexpected as soil loss is the product of sediment concentration and runoff volume. However, it is still possible to argue in term of dilution of sediment concentration with time while the runoff rate is relatively constant. This may happen due to i) limited particle on the slope surface and ii) ponding effect due to much higher rainfall intensity compared to the infiltration capacity of the soil moisture. Moreover, the rainfall was the second highest (Table 2).

Table 3. Soil Loss and Runoff Properties for different plots.

Plot	BS	FG	HGC	HDF
Runoff Coefficient (%)				
Mean	73.6	41.7	45.6	53.4
Standard Deviation	4.7	5.7	18.7	18.8
Soil Loss ($x10^{-3}$ t/ha)				
Mean	5.26	2.85	3.12	4.48
Standard Deviation	3.2	2.1	1.4	2.6
Pearson coefficient (*r*)	0.01	0.45	0.15	0.23
***p*-value**	0.43	0.312	0.46	0.29

Figure 5 shows the soil loss response to runoff for the four plots. Error bars are also given at 95% Figure 5 shows the soil loss response to runoff for the four plots. Error bars are also given at 95% confident interval. It was evident that despite a relatively constant runoff in BS and FG plots, the soil loss showed a gradual reduction. Such condition could occur due to water ponding which covered the soil surface from direct rain drop impacts. Under such high rainfall intensity, a bare surface could generate less erosion compared to vegetated surface. As such, the erosion rate from bare surface could be lower compared to partially covered surface. The mean runoff from BS plot and HDF plot were 4.20 mm and 2.98 mm respectively (Table 4). It was observed that dry frond was able to reduce sediment loss by 41%. HDF and HDF plots are not recorded readings for runoff in the fifth minute due to time to runoff (TTRO) is more than five minutes.

Soil loss from FG plot decreased gradually with time except for a small peak at 65 minute due to sudden increase in runoff. Nevertheless, compared to other runoff plots, FG still had the lowest mean soil loss of 3.71 x 10-3 t/ha. Munoz et al. [25] found that a reduction in soil loss from vegetated surfaces associated with increasing hydraulic resistance by the vegetation which helps to slow down the runoff velocity. As such, larger soil particles tend to be redeposited along its journey from hill-slope to water course. The runoff from FG and HGC plots were quite uneven. According to Fox et al. [26], low infiltration can cause higher ponding depth that lead to the re-deposition of soil particles.

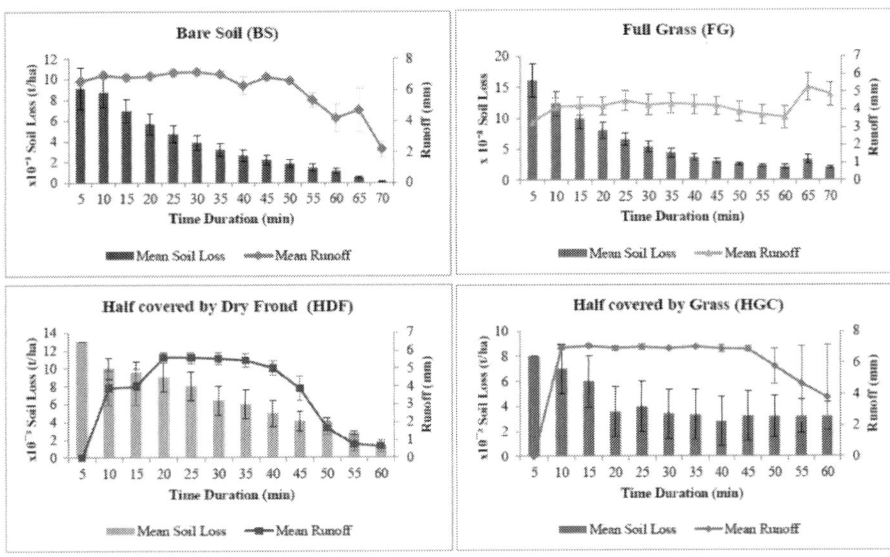

Figure 5. Relation between Soil Loss (t/ha) and Runoff (mm) against time.

Table 4. Statistics for analysis of variance on runoff and soil loss.

Variables	BS	FG	HGC	HDF
Runoff (mm)				
Mean	4.20	6.02	3.50	2.98
Standard Deviation	0.50	0.84	1.37	0.50
Confidence Level (95.0%)	0.29	0.84	1.37	0.50
p -Value between group	3.1962E-06			
Soil Loss (t/ha) x 10^{-3}				
Mean	5.88	3.71	3.64	6.93
Standard Deviation	4.32	2.95	2.22	3.57
Confidence Level (95%)	2.49	1.70	1.28	2.06
p -Value between group	0.027			

4. Conclusions

The study shows that areas with no protection from cover crops are exposed to high rates of soil erosion. This condition usually occurs during site clearing and dry season when soil becomes loose and less compact. When rain falls, there is no protection for the surface against raindrops and will cause the soil particles to be easily to wash away by runoff. Raindrops rupture the soil surface, causing the soil particles to detach and then splashed to short distance. Continuous high intensity rainfall with slope surface conditions makes this situation worse. Slopes are known to produce high runoff velocity.

Energy generated by high flow velocity can easily erode the soil surface. In addition, areas with low saturated hydraulic conductivity generate runoff on bare surface more quickly. Sediment transport on bare soil surface is initially high but declines over time when the loose soil particles on the soil become depletes. Cover crops such as grass could reduce the rate of soil erosion by 67%, whereas 17% dried fronds. Dried fronds can be used as surface flow breakers and to lower down the velocity of surface runoff. The arrangement of the dried frond also plays an important role in intercepting soil loss. Dried fronds layout position perpendicular to the groundwater surface flow is a better method to intercept soil loss compared to arrangement in parallel position. Therefore, it is important to ensure that the cover crops always thrives in agricultural areas, particularly at oil palm plantation and to stack dried frond with good arrangement for reducing the risk of soil erosion.

Acknowledgement

This research is funded by the Ministry of Agricultural of Malaysia (Science Fund Project, Grant no. 05-01-06-SF0418 (79328)). We are grateful to Associate Prof. Ziegler A.D from National University of Singapore (NUS) for his kind guidance in our fieldwork and technicians in Hydrology Laboratory from University Teknologi Malaysia (UTM) for their help in completing this project research

References

[1] Hediger, W. (2003). "Sustainable farm income in the presence of soil erosion: an agricultural Hartwick rule." Ecological Economics 45(2): 221-236.

[2] Hartemink, A.E. (2006). Soil Erosion: Perennial Crop Plantations. ISRIC–World Soil Information, Wageningen, Netherlands.: pp: 1613-1614 harvest areas." California Environment 1(4): 4-6.

[3] Morgan, R. P. C. (2005). Soil erosion and conservation. Soil Science (Vol. 1, p. 304). Longman. Retrieved from http://agris.fao.org/agris-search/search/display.doi=1994/US/US94076.xml;US9422179

[4] Nagano, T., N. Yanase, K. Tsuduki& S. Nagao (2003), Particulate and dissolved elemental loads in the Kuji Riverrelated to discharge rate. Environment International 28/7, pp. 649–658.

[5] Yusop, Z., Abdul Rahim N. and Douglas, I. (2006). Dissolved and undissolved nutrient loading from secondary forest catchments of Malaysia. Forest Ecology and Management. 224: 26-44

[6] Douglas, I. Spencer, T, Greer, T., Bidin, K. Sinun, W. & Wong, W. M. (1992) The impact of selective commercial loggingon stream hydrology, chemistry and sediment loads in the UluSegama rain forest, Sabah, Malaysia. Phil. Trans. Roy.Soc. Lond. B335,397-406

[7] Yusop, Z., Anhar S., (1994). Effect of selective logging methods on suspended solids concentrations and turbidity level in streamwater. Journal of Tropical Forest Science. 1(2) : 201-214

[8] Sammori, T., Yusop Y., Ksaran, B., Noguchi, S., Tani, M. (2004). Suspended solids discharge from a small forested basin in the humid tropics. Hydroogical.Process. 18, 721-738

[9] Lord.S and Clay.L (1999), Environmental Impacts of Oil Palm – Practical Considerations in Defining Sustainability for Impacts on the Air, Land and Water. New Britain Palm Oil LTD Centre, Dami Oil palm Research Station

[10] MPOB (Malaysia Palm Oil Board) (2008). www.mpob.gov.my

[11] Basiron, Y. (2007), Palm oil production through sustainable plantations. Eur. J. Lipid Sci. Technol., 109: 289–295.

[12] DID, Drainage and Irrgation Department (1986). Sungai Tekam Experimental Basin. Transitional Report July 1980 to June 1983. Water Resources Publication, no. 16. Drainage and Irrigation Department, Ministry of Agriculture, Kuala Lumpur, Malaysia

[13] Ismail, W. R. (1997). The impact of hill land clearance and urbanization on runoff and sediment yield of small catchments in Pulau Pinang, Malaysia. IAHS publication, (245), 91-100.

[14] Mualem, Y. and S. Assouline (1986). "Mathematical Model For Rain Drop Distribution And Rainfall Kinetic Energy." Transactions of the American Society of Agricultural Engineers 29(2): 494-500.

[15] DOA, Department of Argricultural (1993)."PanduanSiri-siri Tanah di Semenanjung Malaysia." Department of Agricultural, Kuala Lumpur, Malaysia.

[16] Reid, K. D., Wilcox, B. P., Breshears, D. D., & MacDonald, L. H. (1999). Runoff and erosion in a pinon-juniper woodland: influence of vegetation patches. Soil Science Society of America Journal, 63(6), 1869-1879. Retrieved from http://essm.tamu.edu/people/wilcox/Publications/019.pdf

[17] Ziegler, A. D., Sutherland, R. A., & Giambelluca, T. W. (2001). Acceleration of Horton overland flow and erosion by footpaths in an upland agricultural watershed in northern Thailand. Geomorphology, 41(4), 249-262.

[18] Sahat S., Yusop Z., A. M. (2012). Effectiveness of Different Surface Cover Types in Reducing Runoff and Soil Erosion in Oil Palm Catchments (p. 151). LANGKAWI.

[19] Ziegler, A. D., Sutherland, R. A., & Giambelluca, T. W. (2000). Runoff generation and sediment production on unpaved roads, footpaths and agricultural land surfaces in northern Thailand. Earth Surface Processes and Landforms, 25(5), 519-534.

[20] Penna, D., van Meerveld, H., Gobbi, A., Borga, M., & Fontana, G. (2011). The influence of soil moisture on threshold runoff generation processes in an alpine headwater catchment. Hydrology and Earth System Sciences, 15(3), 689-702.

[21] Dalen, E. N., Kirkby, M. J., Bracken, L., Chapman, P. J., & Irvine, B. (2008). Factors influencing runoff generation , and estimates of runoff in a semi-arid area , SE Spain, 1-6.BHS 10th National Hydrology Symposium, Exeter, 2008.

[22] Reany,S.M., (2003). Modelling ruoff generation and connectivity for semi-arid hillslopes and small cathments. Ph.D.tesis, University of Leeds.

[23] Ariza, C., 2004. Vegetation monitoring in a semi-desert afforestation project.M.Sc

[24] Ayed G. Mohammd, Mohammad A. Adam (2010).The impact of vegetative cover type on runoff and soil erosion under different land uses.Catena 81 (2010) 97–103.

[25] Munoz Carpena, R., Parsons, J. E., & Gilliam, J. W. (1999). Modeling hydrology and sediment transport in vegetative filter strips. Journal of Hydrology, 214(1), 111-129.

[26] Fox D.M, Bissonnais Y. Le , Bruand A. (1998). The effect of ponding depth on infiltration in acrusted surface depression.Catena 32 _1998. 87–100.

The Assesement of Rip Current at Kerachut Beach Using Hydrodynamic Modelling

W A H W Azhary[1], N A Awang[1] and M R A Hamid[1]
[1]Coastal Management and Oceanography Research Centre, National Hydraulic Research Institute of Malaysia

E-mail: wanhafiz@nahrim.gov.my

Abstract. KerachutBeach is a beautiful beach in Penang National Park (PNP). However this beach is categorisedas one of dangerous beach for swimming activities in Malaysia due to the drowning incidents reported almost every year. The steep beach slope and rip current were among the factors that lead to this incident. Using bathymetry profile, current, tidal and sediment data collected at site incorporated with UKMO wave data analysis,the hydrodynamic pattern was simulated using Mike 21 modelling software. Result from the model showed the evidence of rip current existence along the coastline. It showed that this rip current eventsoccurred during spring tide phase when the flow change from Flood to Ebb. During this period, the current tend to move parallel to the shoreline with maximum speed of 0.3m/s which is capable to swipe away a swimmer. The bathymetry profile at Kerachutis very steep and dangerous to swimmers since there is a 4 meter sudden plunge just meters away from the shoreline.

Keywords: Malaysia, drowning, rip current, bathymetry, hydrodynamic.

1. Introduction

Rip currents occur as a result of interaction between waves, current, water level and near shore bathymetry. According to [1], rip currents flow faster near the water surface. This current are ubiquitous along wave dominated sandy beaches that exhibit three-dimensional surf zone sandbars [2]. Rip currents most typically form at low spots or breaks in sandbars, and also near structures such as groins, jetties and piers. It can be very narrow or extend in widths to hundreds of yards. The seaward pull of rip currents varies: sometimes it ends just beyond the line of breaking waves, but sometimes it continue to push hundreds of meters offshore. The illustration for rip currents is shown in Figure-1.

Figure 1. Rip current formation along the shoreline.

IOP Conf. Series: Materials Science and Engineering **136** (2016) 012087 doi:10.1088/1757-899X/136/1/012087

According to [3], most of the fatal drowning each year in US were associated by the effect of rip currents. These rip currents are identified as being responsible for more than 100 deaths annually in US [4]. Breaking waves generate narrow currents sometimes directed seaward through the surfzone because of an excess of momentum flux [5]. Swimmers being caught in a rip current are easily transported off their feet todeeper waters seaward.

It is reported that drowning incident occur every year at this beach stretch. The Kerachut beach is among the most dangerous stretches in Malaysia. The latest mishap occurred in January 2014 when five of team building members were swept away by strong waves with three of them were found dead. The drowning incident is also reported almost every year at Kerachut Beach. The objectives of this study are to identify the occurrences of rip current along shoreline of Kerachut and investigate the other factors that lead to these drowning incidents. Collection of raw data at site and computer modelling are used for analysis purpose.

2. Materials and methods

2.1 Study Area

Kerachut is located on the north-western tip of Penang Island, Malaysia at 5°27'3.36" N and 100°10'57.52" E. It is among the beaches managed by PNP including Monkey Beach, TelukAling, TelukDuyung and TelukKampi. PNP is governed under the Department of Wildlife and National Parks of Malaysia (Perhilitan). This park is famous among the tourist for its rainforest and beautiful beaches. It is a great place to go trekking, camping, fishing, wildlife spotting and beach activities. The total length of Kerachut beach is 558 m with the slope gradient of 15° [6]. The main transport to Kerachut is either by boat which takes 15 minutes ride from PNP jetty or by one and half hour trekking from PNP entrance. The sand colour ranges from white to dark brown and formation of sandbar tend to change prior to the monsoon season.

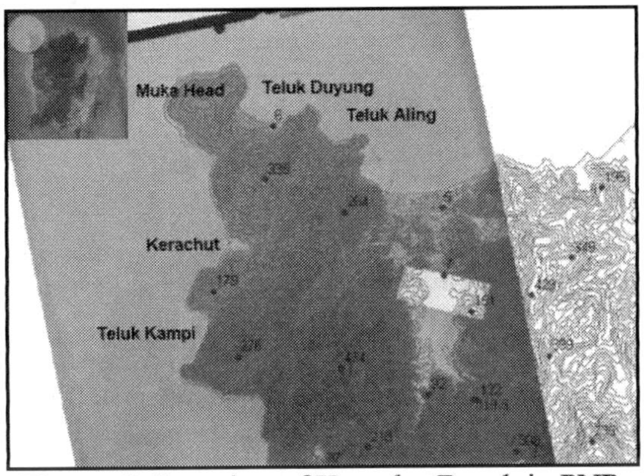

Figure 2. Location of KerachutBeach in PNP.

Kerachut is also known for its two layers of lake or lagoon of salty sea water and fresh water known as Meromictic Lake. This lake has two layers of water with different density that do not mix with each other [7]. According to [8], it is one of the few Meromictic Lakes in Asian Region and covers 10 hectares of PNP. During high tide, sea water could be seen on the top of the layer and fresh water would replace when the low tide occurs due to lower density of fresh water.

The Kerachut Turtle Conservation Centre represents the centrefor sea turtles research study in Penang Island. The beach was identified as the most preferred landing area for Green Turtle in Malaysia. According to [6], a total of 43 Green Turtle nests were recorded at Kerachut and TelokKampi from December 2009 to December 2010. The hatchery at Kerachut is the only hatchery in Penang which purposes to incubate turtle eggs and produce hatchlings to be released into the sea.

IOP Conf. Series: Materials Science and Engineering **136** (2016) 012087 doi:10.1088/1757-899X/136/1/012087

2.2 Field Data Collection

The field data collection works were carried out from 27th May 2014 until 20th June 2014. The information forms the basis for the required modelling and assessment purpose.

2.3 Bathymetry

Hydrographic survey is the science of marine surveying that determines the position of points on the earth's surface and also depths of the sea [9]. An Odom Hydrotrac II Single Frequency precision. echo sounder was utilized in the survey. Bathymetric survey at the site covers an area of 4000 meters by 1500 meters with 50 meters line spacing and 250 meters check lines (Figure-3). The objective of the survey is to produce an accurate bathymetric profile at Kerachut and nearby beach.

Figure 3. Transect line for bathymetry survey.

2.4 Current Measurement

The currents are measured with Acoustic Doppler Current Profiler (ADCP), a device that utilizes sonar to measure the velocity of moving water particles. The ADCP records water velocities with respect to the depth. Therefore, it can produce a profile of water currents over a range of depths. The current measurements were conducted at two locations for three weeks deployment period as DID Guidelines covering both spring and neap tide. The ADCP positions were indicated in Firure-3 and Table- 1. The data obtained from the measurement during both tidal periods are depicted in Figure-4 and Figure-5.

Table 1. Location of ADCP Deployment.

Location	Latitude (N)	Longitude (E)
ADCP 1	5.46219°	100.16867°
ADCP 2	5.44586°	100.15699°

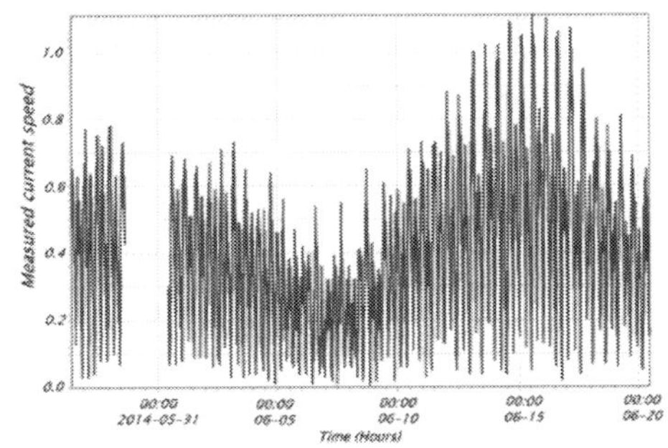

Figure 4(a). Measured current speed at ADCP 1.

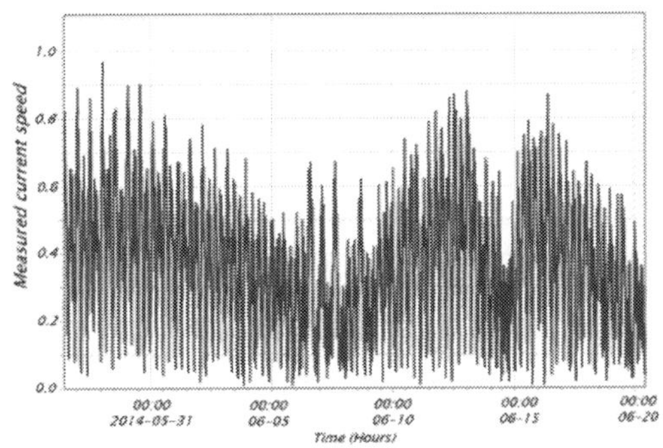

Figure 4(b). Measured current speed at ADCP 2.

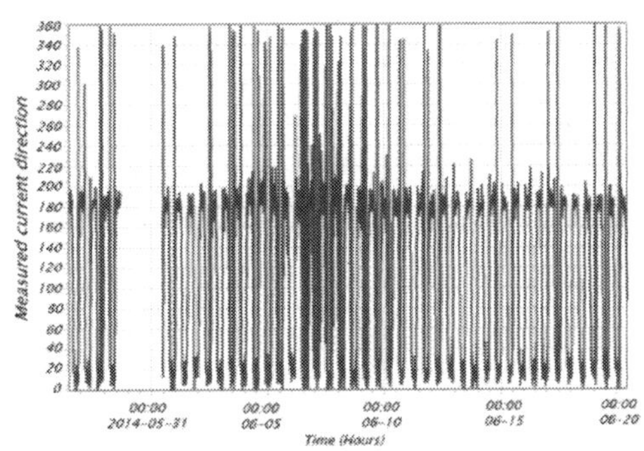

Figure 5(a). Measured current direction at ADCP 1.

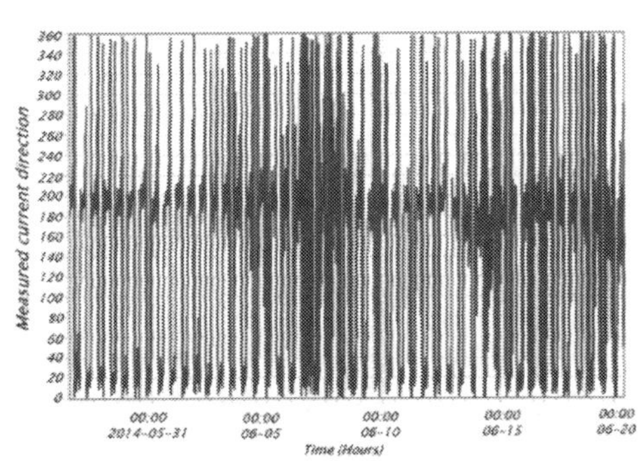

Figure 5(b). Measured current direction at ADCP 2.

Based on ADCP 1 data, the range of current speed is 0 - 1.10 m/s with the dominant current coming from 0° - 30° and 170° - 200° direction. There were missing data from 29th -31st May 2014 due to disturbance of the sensor during deployment period. For ADCP 2, the range of current speed is 0 - 0.95 m/s with the dominant direction coming from 0° - 30° and 170° - 220°.

2.5 Water Level

Water levels are produced by a combination of two major force components - gravitation forces and centrifugal forces. These forces induce a variation of the tidal levels due the shear effect of the winds and regional barometric pressure fields (in many cases producing an elevation due to a low or a reduction due to a high pressure field). The variation of the water levels due to climatic conditions is usually described as residual levels. One unit of tide gauge was installed at PNP jetty for two weeks duration for model calibration purpose.

2.6 Bed Samples

A total of 20 bed samples were collected within Kerachut and TelukKampi shore. These samples provide information on the type and distribution of bed materials present within the study area. A mechanical grabber was used to obtain seabed samples. The location of each sample was determined using a handheld GPS device.Since the position was marked, the grabber was released and retrieved after touching bed. Each sample was stored in a clear plastic bag which was then placed in another plastic bag before being tied and labelled. All samples were kept in a cool dry place before being sent to a laboratory for Particle Size Distribution (PSD) analysis. It can be inferred from the bed sampling exercise and analysis that the bed material within the study area primarily consists of clay and silt which particle size is <63 μm.

2.7 Wave Analysis

The meteorological conditions at Kerachut are governed by the Northeast (NE) and Southwest (SW) monsoon. The NE monsoon occurs between November to Marchwhile SW monsoon occurs betweenMay to September [10]. Thus, April and October are inter-monsoon periods.

To derive the wave conditions at Kerachut, offshore wind and wave data were purchased from the United Kingdom Meteorological Office (UKMO). Figure-6 shows the location of the UKMO station and data covered from 28th May 1999 to 30th June 2014. The wave climate at the offshore of Kerachut is composed of locally generated wind waves and swell waves approaching from Indian Ocean and Andaman Sea.

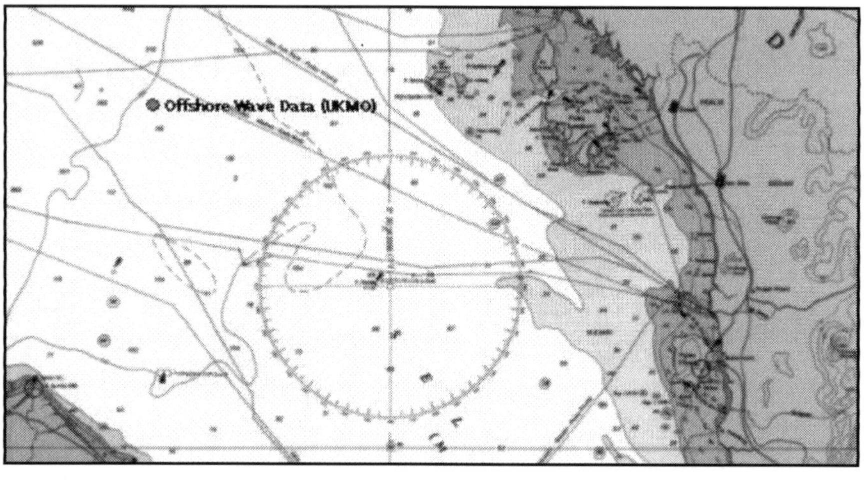

Figure 6. Location of wave data point (UKMO).

Figure-7 shows the time series of Significant Wave Height (*Hs*) of the purchased UKMO data while Figure-8 shows the wave rose of the same Hmo-parameters. From the wave rose, it is clearly seen that thelong period swell waves are approaching mainly from westerly directions during SW monsoon.

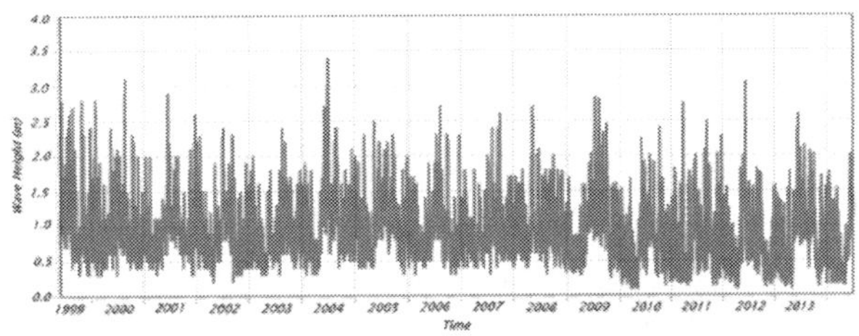

Figure 7. Time Series of Significant Wave Height (Hmo) Data (1999-2014).

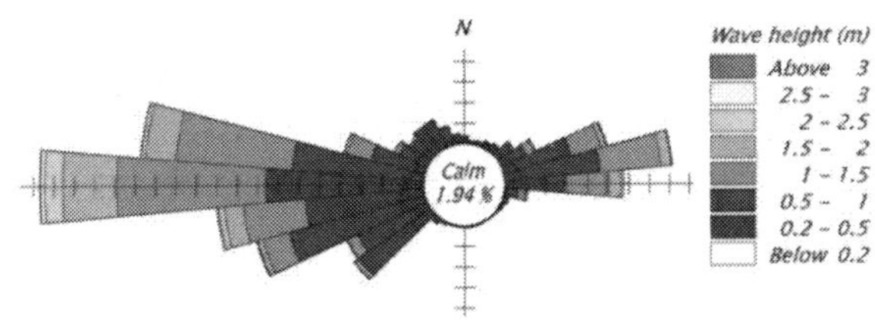

Figure 8. Offshore wave rose for the UKMO data.

The wave data used was from the offshore waves and far away from study area. Hence, numerical modelling of MIKE 21 NSW was used to transform this waves into the near shore environment. The model takes into account the effects of refraction and shoaling due to varying depth, local generation of wind waves, the influence of directional spreading and energy dissipation due to wave breaking and bottom friction.

The model parameters include bottom roughness (Nikuradse roughness parameter), $k_N = 2$ mm; and wave breaking parameters, $\gamma_1 = 1.0$, $\gamma_2 = 0.8$ and $\alpha = 1.0$. The bottom roughness and wave breaking parameters are very important in the surf zone. Parameters related to the discretisation of the wave spectra in the model are maintained at well-proven default values. Input for bathymetry, wave properties along the boundary, wind condition, bottom roughness, tidal elevation and wave breaking parameters were used to setup the hydrodynamic model.

2.8 Hydrodynamic Modelling

The modelling is carried out using the hydrodynamic module in MIKE 21 modelling system, MIKE 21 HD. This is a basic module of the entire MIKE 21 software. It provides the hydrodynamic basis for the computations performed in most of the other modules. The objectives of the modelling is to establish the existing hydraulic regime for the study area and to assess the rip current occurences along the Kerachut shoreline. The water levels and flows are resolved on a rectangular grid covering the area of interest when provided with the bathymetry, bed resistance coefficients, wind and wave field and hydrographic boundary conditions.

This study adhere the 'Guidelines for Preparation of Coastal Engineering Hydraulic Study Reports' prepared by Department of Irrigation and Drainage (DID), Malaysia. This guidelines were introduced to ensure that all the hydraulic and coastal modellings comply with the Malaysian's standard requirements.

The important criteria in modelling work is to define the model area that needs to be set up. The hydrodynamic model extends from Goh Don, Thailand to Bagan Datoh, Malaysia, which covers approximately 378 km x 351 km. The model contains four layers with an outer 270 m grid (coarse grid) down to 10 m grid (local grid) in order to enhance the resolution in the vicinity of the site. All nested grids are dynamically linked and solved simultaneously. The bathymetry data for the model setup is based on the Admiralty Charts from Royal Malaysian Navy, digital C-Map data from Mike 21 and NAHRIM bathymetry survey. The bathymetry profile for the model is shown as Figure-9.

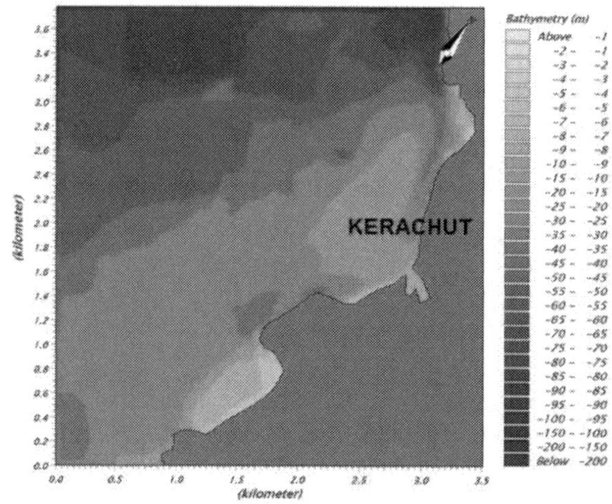

Figure 9. Bathymetry for Nested Hydrodynamic model: local grid.

The bed roughness is prepared based on the waves with ripples height of 4 m to 7 m and lengths of 250 m to 450 m exist in the Malacca Strait. The Manning number can be expected to vary between 17 and 61 over the entire study area. In the calibration, Manning numbers of 25 to 55 was used based on Different Range Of Water Depth.

3. Result and Discussion

The model was analysed for spring and neap tidal cycle of 27th May 2014 until 20th June 2014 covering a total of 25 days period. This period was sufficient enough to cover one completed tidal cycle and can be assumed to represent a hydrodynamic condition for the whole study area. The acceptance criteria for the calibration is based on. For the calibration parameters to be acceptable, the average absolute difference between the models calculated and observed should be less than the percentage stated in Table-2.

Table 2. Root Mean Square (RMS) acceptance for hydrodynamic model.

Parameter	Acceptance Criteria
Tidal range	RMS Error 10%
Current speeds	RMS Error 20%
Current direction	RMS Error 20°

The model was calibrated and verified using the NAHRIM measured water level. The extracted water levels from the model simulation suit very well with the observed data (an average difference of less than 10%). The current patterns in the Malacca Straits flow to the South during flood and went North during Ebb with the acceptance criteria was less than 20% in speed and 20° in direction.

Based on modelling result, it can be said that KerachutBeach is having semi-diurnal tides with two high waters and two low waters in a day with maximum tidal range of 2.90 meters during spring tides. It was within the Highest Astronomical Tide (HAT) predicted by [11] using Kedah Pier Station which predicted 3.09meters range of water height.

The Significant Wave Height (*Hs*) show relatively small during the 30 years return period simulations. Analysis of wave model showed the maximum of 1.0meter height along the Kerachut. Wave plays major role in sediment transport based on its frequencies. The sediment transport may actively occurs during May to September when the wave direction coming from 270°(West). This large reduction in wave height is caused by sheltering effect of the Langkawi Island and the dissipation of wave energy in the shallow coast of Penang Island.

The current pattern of Kerachut shoreline were plotted to determine the presence of rip current along the shoreline. Based on the analysis, it clearly showed the evidence that Kerachut is having rip current during the period of flood to ebb flow of spring tide. Typical plot of current pattern during this period is shown in Figure-10.

During this period, the rip current at Kerachut will flow to the NE and SW direction which parallel to the shoreline. The changes of bathymetric profile along the surf zone created the rip current to these directions as the sea water flow is influenced by tidal fluctuation. However, the mean current speed at Kerachut is generally moderate which in the range of 0.1 to 0.3 m/s and rip current will only occurred during spring tide when the flow change from Flood to Ebb. During other periods, the rip current effect can be considered as safe.

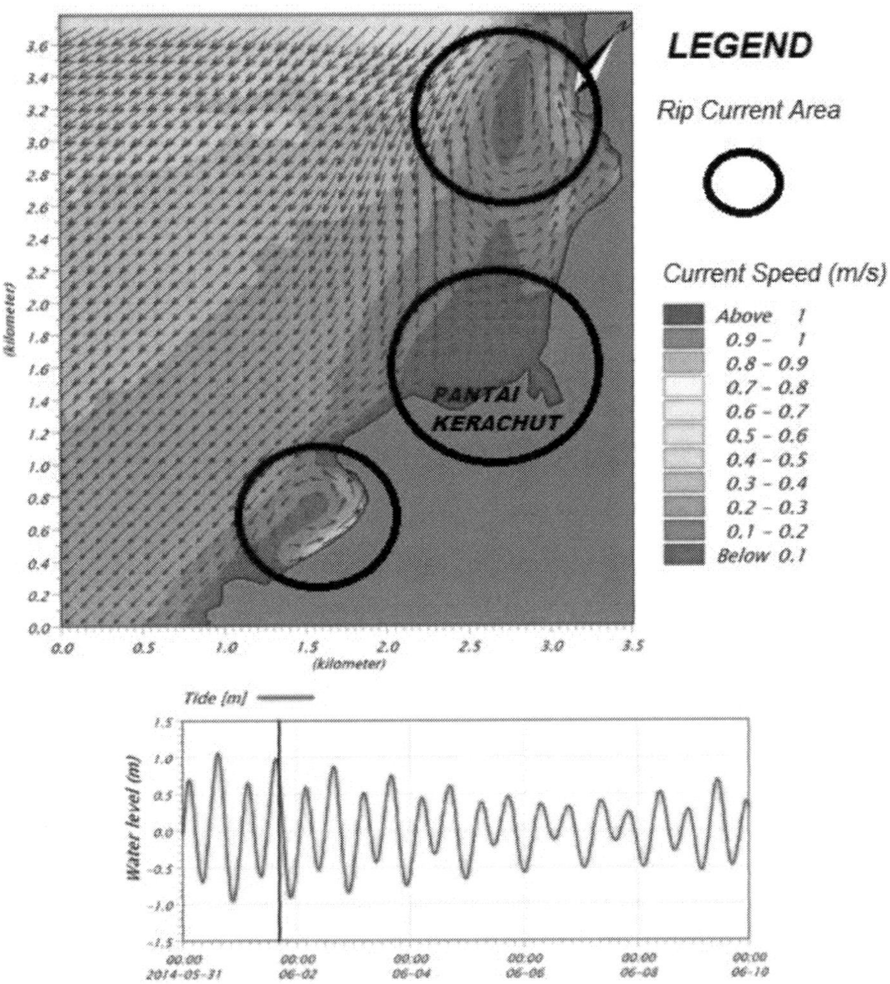

Figure 10. The rip current area during Flood and Ebb flowat Kerachut shore.

The bathymetry and beach profile transect for KerachutBeach and TelukKampiBeach is shown in Figure-11.

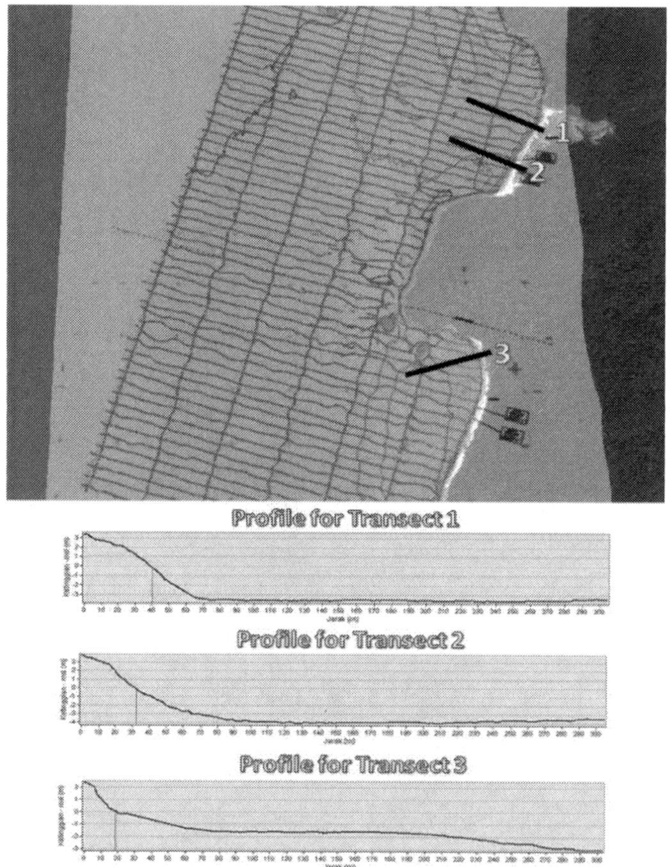

Figure 11. Shoreline profile at Kerachut and Teluk Kampi.

Based on the analysis of bathymetric data in Figure-11, it shows that this beach is also dangerous for swimming activities due to its steepness and some large scale of rocky features spotted on the seabed. As enter a few meters to the beach, there will be a plunge at a very sharp angle with 4 to 5 meters in depth. The strong wave and monsoon impact may create such a dangerous morphology profile along the coastline as the bathymetry contour was constant along the shoreline. In addition, the muddy properties which dominated the seabed make it more dangerous to the swimmer. Only a small stretch area of PantaiTelukKampi (Transect 3) is identified safe for swimming activities due to its gentle and mild profile of the beach.

Conclusion

Based on the extensive data collection and hydrodynamic modelling that have been carried out, it proved that KerachutBeach should be avoided for swimming activities.Modelling result shows that rip currents occurred during spring tide when the flow change from Flood to Ebb. This rip current tend to flow parallel to the shoreline in the direction of NE and SW with the magnitude of 0.3m/s. In addition, the coastal profile is very steep with sudden plunge of 4 to 5 meter by just meters away from the shoreline. The steepness of bathymetry profile, rip current occurrence and some rocky material scattered along the seabed makes this beach too dangerous for the swimmers. Swimming should be prohibited to all visitors except at the beach in front of PNP Guard Post at TelukKampi. According

[12], it is difficult to educate public on how to identify rip current. Hence, safety briefing related to current and wave pattern, tides and monsoon impact must be assigned to PNP officers and rangers on duty to enhance understanding about the hydrodynamic behaviour along the coastline and ensure the effective monitoring of the visitors.

References

[1] Haas, K.A. and Svendsen, I.A. (2002). Laboratory measurements of the vertical structure of rip currents. *Jurnal of Geophysical Research*, 107: 3047; doi: 10.1029/2001JC000911

[2] Castelle, B., Reniers, A., MacMahan, J. (2013). Numerical modelling of surfzone retention in rip current systems: on the impact of the surfzone sandbar morphology, *Jurnal of Coastal Dynamic*. p. 295-304

[3] Gensini V. A., Ashley W. A. (2009). An examination of rip current fatalities in the United States. *Nat Hazards*: DOI 10.1007/s11069-009-9458-0

[4] Alvarez-Ellacuria, A.; Orfila, A.; Olabarrieta, M.; Medina, R.; Vizoso, G., and Tintore., J. (2010). A Nearshore Wave and Current Operational Forecasting System. *Journal of Coastal Research*, 26(3), 503–509. West Palm Beach (Florida), ISSN 0749-0208

[5] Bowen, A.J., 1969. Rip Currents, 1, Theoretical Investigations. *Journal of Geophysical Research*, 74(23), 5467–5478

[6] Sarahaizad, M.S., Shahrul Anuar, M.S., Mansor, Y. (2012). Nest Site Selection and Digging Attempts of Green Turtles *(Chelonia Mydas, Fam.Cheloniidae)* at Pantai Kerachut and Telok Kampi, Penang Island, Peninsular Malaysia. Malays. *Appl. Biol.* (2012) 41(2): 31–39

[7] Wetzel, R.G. (2001). Limnology: Lake and river ecosystems.3rdEdition.New York: Academic Press.

[8] Hong, C.W. & Chan,N.W. (2009). The Potentials, Threats and Challenges in Sustainable Development of Penang National Park. *Malaysian Journal of Environmental Management* 11(2) (2010):95-109

[9] Aklima, C. A., Rusli, O. (2011). Hydrographic Survey using Real Time Kinematic Method for River Deepening. *Geoinformation Science Journal*, Vol. 11, No. 1, 2011, pp: 1-14

[10] Ahmad, J.S., Mohd Zaki, M.A., & Zubaidi, J. (2011). Potential Climate Change Impact on Water Resources in Malaysia, *Malaysia Water Research Journal, Vol 1(2011)* : p.5-14

[11] Hydrographic Directorate, Royal Malaysian Navy, Tide Tables for Malaysia, Volume 1 2015

[12] Fletemeyer, J. and Leatherman, S. (2010). Rip current and beach safety education. *Journal of Coastal Researches*, 26: 1-3

Isolation and Identification of Concrete Environment Bacteria

J M Irwan[1], L H Anneza[1], N Othman[2], T Husnul[3] and A F Alshalif[1]
[1]Research Centre for Susitainable Construction, Faculty of Civil and Environmental Engineering, Universiti Tun Hussein Onn Malaysia, 86400 Batu Pahat Johor, MALAYSIA
[2]Micropollutant Research Centre, Faculty of Civil and Environmental Engineering, Universiti Tun Hussein Onn Malaysia, 86400 Batu Pahat Johor, MALAYSIA
[3]School of Industrial Technology, University Sains Malaysia, Penang, MALAYSIA

E-mail: irwan@uthm.edu.my

Abstract. This paper presents the isolation and molecular method for bacteria identification through PCR and DNA sequencing. Identification of the bacteria species is required in order to fully utilize the bacterium capability for precipitation of calcium carbonate in concrete. This process is to enable the addition of suitable catalyst according to the bacterium enzymatic pathway that is known through the bacteria species used. The objective of this study is to isolate, enriched and identify the bacteria species. The bacteria in this study was isolated from fresh urine and acid mine drainage water, Kota Tinggi, Johor. Enrichment of the isolated bacteria was conducted to ensure the bacteria survivability in concrete. The identification of bacteria species was done through polymerase chain reaction (PCR) and rRDNA sequencing. The isolation and enrichment of the bacteria was done successfully. Whereas, the results for bacteria identification showed that the isolated bacteria strains are *Bacillus sp* and *Enterococus faecalis*.

Keywords: Bioconcrete, effective microorganism, enrichment, 16S rDNA gene sequencing.

1. Introducion

Bacteria is a living micro-organism that are everwhere. From the smallest grain of soil to the small drop of water. Bacteria can also be found in organic matters such as in live bodies of plants and animals. These living micro-orginisms are known to live in harsh environemts and thrive in these living conditions [1]. The bacteria species isolated from different parts of the world and different environmet have unique characteristics. The uniqueness of these bacteria are slowly being discovered and used to improve problems of human kind. Some groups of bacteria have capabality in improving crops, health, degraded organic contaminants and others. An attempt of bacteria in concrete able to improve concrete properties and healing process. Concrete is an important building material. However, the drawback of this material is the deterioration in which intervention in terms of chemicals are normally used. The chemicals are used to repair cracks, reduce water permeability of concrete, early hardening and many more. [2] had stated in his study that bacteria are able to promote preciptation of calcium carbonate in the form of calcite. The formation occurs as a by-product of a common microbial metabolic process which help in increasing alkalinity and produce microbial calcite precipitation. Many reseearchers added bacteria into concrete to allow precipitation of calcium carbonate within the concrete to improve its properties [3,4,5]. The bacteria used by previous researchers were isolated from local sources of temperate countries that different in environmental factors. Therefore, this study aims to isolate, enriched and identify a local source of bacteria which has potential to improve concrete properties.

Bacteria identification based on 16S ribosomal gene rDNA sequencing and polymerase chain reaction (PCR) is a popular technique of bacteria identification method [6,7,8,9,10]. This method for bacteria identification is the preferred method due to its speed in obtaining results and accuracy. The use of 16S rDNA gene sequencing are more commonly being employed in rountine laboratory diagnostics particularly after adopting PCR in biomedical science [10].

According to [9], PCR based method is a cheaper and more faster method to identify bacteria, the PCR typing assays is design from genome sequence data which uses the concept of observing variation in gene content of strains within a given species.

2. Experimental Procedure

2.1 Bacteria isolation and enrichment

2.1.1 Isolation

The bacteria used in this study are isolated from fresh urine and acid mine water, Kota Tinggi, Johor, Malaysia. Samples from acid mine water was taken over the dry season to avoid auto dilution on the sample via rainfall. Sample was collected from the discharge point of the mining retention pond. A PTFE bottle was used in order to keep the sample collected. To ensure that the sample collected is maintained at the right condition, the PTFE bottle containing the sample was kept in a box with ice cubes before being transported back to the lab. Upon arriving at the lab, the sample was kept at 4^0C until enrichment process. Bacteria isolated from fresh urine was temporarily name *Ureolytic bacteria* while bacteria isolated from acid mine water was named *Sulphate reduction bacteria*. The isolation process of both *ureolytic* and *sulphate reduction bacteria* (SRB) were conducted following a process of enrichment, serial dilution streaking plate, strain purification and gram staining. Isolation and purification method of bacteria was done according to [11]. All process and media used were autoclave at 121^0C for 15 minutes for sterilization.

2.1.2 Enrichment

The enichment of bacteria was conducted after the bacteria were isolated. The enrichment process is conducted to ensure the bacteria survivability in concrete. The process involves adding bacteria into media composition. A control flask without addition of urine and acid mine water were also prepared for ureolytic and sulphate reduction bacteria respectively. Composition of *sulphate reduction bacteria sample* (SRB) consists of 25 ml (nutrient broth) added with 10 ml ($MgSO_4.7H_2O$) and 1 ml (water from Sg pelepah Kota Tinggi). U*reolytic bacteria* composition including 25 ml (nutrient broth), 10ml (urea 40%) and 1 ml (urine). Two environmental condition namely pH and oxygen concentration were controlled in the isolation process. Realising that concrete environment is a high alkaline environment, the pH value of the isolation flask is maintained in alkaline condition for ensuring self survival of the isolated strains. The high pH value contributes to high urease activity which leads to high carbonate precipitation rate [12]. The pH of each sample were adjusted to an alkaline condition by adding NaOH until the pH value reaches the range of 9-11.

Oxygen concentration in the isolation flask was monitored to be in anaerobic condition. This is to ensure only anaerobic or facultative bacteria will be isolated. Adjusment of the oxygen is needed due to lack of oxygen in concrete material. Nitrogen gas was purged in the enrichment flask before enrichment started to remove the remaining oxygen in the flask. The concentration of oxygen was monitored with a dissolved oxygen meter to ensure very low oxgen or 0 ppm oxygen.The flask was shaked at room temperature. The adjustment of pH and anaerobic condition were done every day over the enrichment period of 20 days. The set-up of the enrichment flask as shown in Fig.1

Figure 1. Enrichment flask.

2.1.3 Dilution and Pure Streaking Plate Method

Along 20 days of enrichment, the flask with sample will turn turbid. Every 10 days 0.1 ml of the culture was taken and added to 9.9 ml of sterile distilled water. The serial dilution process was done for both samples using yeast extract and sulphate API agar which were prepared for *Ureolytic* and *Sulphate reduction bacteria* respectively. Each of the dilution was aseptically plated on yeast extract and sulphate API agar which were appropriately labeled. The plates were incubated at 37^{0}C in the incubator until the desired strain growth. Colonies appeared were picked up and cultured. Fig. 2 illustrates agar with colonies of bacteria.

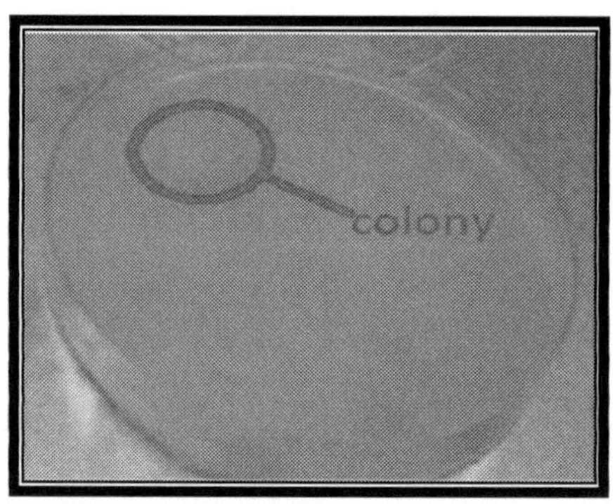

Figure 2. Plate with colony.

2.1.4 Strain Purification and gram staining

Purification step was performed until pure strain was obtained. Purification involves a streak plate method was done to ensure only single strain obtained in each plating. Gram staining is used to investigate the major categories of microorganism or examine the morphology. In this study simple stains including crystal violet, iodine and safranin were applied. Two groups of bacteria can be distinguished with gram staining. Using this method, gram positive bacteria cell wall will retain crystal violet while gram negative change.

2.2 Bacteria identification process

2.2.1 Preparation of Lysogeny Broth

The bacteria sample is first inoculated into Lysogeny Broth (LB) and incubated for 18 hours with temperature of 37° C and 400 RPM. The samples were then centrifuged to collect pellet for gDNA extraction.

2.2.2 Genomic DNA isolation of bacteria

The DNA extraction of the bacteria used in this study was done based on modified phenol-chloroform protocol. Based on this method, 1.5ml of bacterial cell suspensions was centrifuged at 13,000 RPM, 4°C for 2 min.The bacterial pellets were resuspended with TEN extraction buffer (0.1M NaCl, 10mM Tris-Cl; pH 8.0, 1mM EDTA; pH 8.0). They were then mixed by gently pipetting up and down till homogenized. RNase (10mg/ml) was added to the tube followed by the addition of lysozyme (50µg/ul). The tubes were gently inverted and incubated at 37°C for 30 min. After that, TENST buffer (0.1M NaCl, 10M Tris-Cl; pH 8.0, 1mM EDTA; pH8.0, 0.12% TritonX-100, 1.6% N-lauryl sarcosine) was added to the tubes followed by further 1 hour incubation at 37°C. Proteinase K (10mg/ml) was later added, followed by incubation at 55°C for 1 hour. The almost clear suspensions were treated with 15% SDS prior to further 30 min incubation at 55°C. An equal volume of phenol/chloroform/isoamyl alcohol (25:24:1) was added and mixed gently followed by centrifugation at 13,000 RPM, 4°C; for 5 min. The supernatants basically, the clear aqueous layer was transferred to sterile 1.5ml tubes. This processes were repeated until the supernatants were clear without any white protein precipitate. The recovered supernatants were added to an equal volume of chloroform/isoamyl alcohol (24:1) with careful mixing prior to centrifugation at 13,000 RPM, 4°C; for 5 min. The clear aqueous layers were transferred to new tubes. 0.1 volumes of sodium acetate were added to the supernatants followed by the addition of absolute ethanol and DNA precipitation is allowed to take place overnight at -20°C. DNA was pelleted by centrifugation at 13,000 RPM, 4°C; for 10 min and the supernatants were discarded. This is then followed by 70% ethanol wash, recentrifuged at 13,000 RPM, 4°C; for 5 min, discarded the supernatants and air dried. The dry DNA pellet was resuspended with Tris-HCl, pH 8.0 buffer and kept overnight at -20°C.

3. Results and Discussion

3.1 Bacteria

The bacteria strains were successfully isolated and enrichment process was conducted. The bacteria was purified by conducting serial dilution streaking plate. The pure strain of bacteria obtained was analyze by gram staining. The Gram staining results obtained showed that both bacteria was gram positive. Gram positive bacteria denoted that bacteria remain coloured with gram staining even after washing with alcohol or acetone. The characteristic for Gram positive bacteria is absent of outer membrane, 20-30mm thick cell wall, the wall contains 70-80 % murein and teichoic acids. The Gram-positive lineage of bacteria are species which have a common phylogeny and cell wall structure, among the many Gram positive bacteria are *Bacillus, Clostridium, Lactobacillus, Mycoplasma* species.

3.2 DNA Quantification of bacteria

DNA quantification was quantified spectrophotometrically using a Nanodrop 2000 spectrophotometer (Fisher Scientific). Absorbance measurements were taken at 260nm and 280nm and used to calculate the DNA purity and concentration. DNA samples were considered free from protein contamination if the A_{260} to A_{280} ratio was between 1.8 and 2.0. The result of the quantification is in Table 1.

Table 1. The concentration and the quality of DNA using Nanodrop.

No.	Sample ID	Nucleic Acid Conc.	Unit	260/280	260/230
1	S	165.1	ng/µl	1.85	2.15
2	U	159.7	ng/µl	1.84	2.12

3.3 Polymerase Chain Reaction (PCR)

PCR primers have been employed to confirm the presence or absence of microorgisms.Primers which are used in both directions;forward and reverse such as in this study are longer primers. This forward and reverse primers sequence are designed based on the short general primers. The PCR reaction condition was a 10X PCR Buffer, 25 mm $MgCl_2$, 2.5 mm dNTPs, 10 µmol each primer (27F; 1492R), 5U Taq DNA polymerase. Universal 16SrDNA PCR was performed with forward and reverse primers, which are 27 (forward primer)5'-AGAGTTTGATCMTGGCTCAG-3' and 1492 (reverse primer) 5'-GGGTTACCTTGTTACGACTT-3'. After 35 cycles consisting of denaturation at 95° C (30 Seconds), Annealing at 50 ° C (30 Seconds) and extension at 72 ° C (1 min and 30 seconds), the PCR product is analyzed on an agarose gel. A band indicating a fragment was shown by agarose gel electrophoresis. A PCR product of about 600 kb in size was successfully amplified from the two samples. The agarose gel product is as shown in Figure 3.

Figure 3. Agarose gel of PCR products.

3.4 Molecular Identification

After purification of the PCR product using the Qiagen PCR Purification Kit.Identification. The DNA isolated. Analysis of DNA sequencing was performed using the BLASTX software (BLAST), National center for biotechnology information. Both DNA sequences for sulphate reduction and ureolytic bacteria resulted to tree of phylogenetic as presented in Figure 7 and 8. The Figures will result to the most promising strains.

3.5 DNA sequences for bacteria

The DNA sequences for both *sulphate reduction bacteria* and *ureolytic bacteria* are provided in 3.5.1 and 3.5.2. The basic unit of DNA structure is nucleotide, composed of phosphate, deoxyribose sugar and a nitrogen base. Each deoxyribose sugar bonds convalently in a repeating pattern with two phosphates, One of the bond is to number 5' (five primer) another bond is to 3' carbon on deoxyribose. This specifies the order and direction of each strand. The nitrogen bases, purines and pyrimidines attach by covalent bonds at the 1' position of the sugar. Pairing of purines and pyrimidines is dictated by the formation of hydrogen bonds.Thus DNA sequence, the purine Adenine (A) pairs with pyrimidine Thymine (T) and the purine Guanine (G) pairs with pyrimidine Cytosine (C) [13]. These the bases that made up a strand of DNA. The DNA sequence is then compared to similar or any other DNA sequences by uploading the sequence to BLAST programs, which is a program that is supported by the National Center for Biotechnology Information (NCBI).

3.5.1 DNA Sequences for Sulphate Reduction bacteria

GNAGTTGCGGGCAGCTATACATGCAGTCGAGCGAATGGATTAAGAGCTTGCTCTTAT
GAAGTTAGCGGCGGACGGGTGAGTAACACGTGGGTAACCTGCCCATAAGACTGGGA
TAACTCCGGGAAACCGGGGCTAATACCGGATAACATTTTGAACCGCATGGTTCGAAA
TTGAAAGGCGGCTTCGGCTGTCACTTATGGATGGACCCGCGTCGCATTAGCTAGTTG
GTGAGGTAACGGCTCACCAAGGCAACGATGCGTAGCCGACCTGAGAGGGTGATCGG
CCACACTGGGACTGAGACACGGCCCAGACTCCTACGGGAGGCAGCAGTAGGGAATC
TTCCGCAATGGACGAAAGTCTGACGGAGCAACGCCGCGTGAGTGATGAAGGCTTTCG
GGTCGTAAAACTCTGTTGTTAGGGAAGAACAAGTGCTAGTTGAATAAGCTGGCACCT
TGACGGTACCTAACCAGAAAGCCACGGCTAACTACGTGCCAGCAGCCGCGGTAATAC
GTAGGTGGCAAGCGTTATCCGGAATTATTGGGCGTAAAGCGCGCGCAGGTGGTTTCT
TAAGTCTGATGTGAAAGCCCACGGCTCAACCGTGGAGGGTCATTGGAAACTGGGAG
ACTTGAGTGCAGAAGAGGAAAGTGGAATTCCATGTGTAGCGGTGAAATGCGTAGAG
ATATGGAGGAACACCAGTGGCGAAGGCGACTTTCTGGTCTGTAACTGACACTGAGGC
GCGAAAGCGTGGGGAGCAAACAGGATTAGATACCCTGGTAGTCCACGCCGTAAACG
ATGAGTGCTAAGTGTTAGAGGGTTTCCGCCCTTTAGTGCTGAAGTTAACGCATTAAG
CACTCCGCCTGGGGAGTACGGCCGCAAGGCTGAAACTCAAAGGAATTGACGGGGGC
CCGCACAAGCGGTGGAGCATGTGGTTTAATTCGAAGCAACGCGAAGAACCTTACCAG
GTCTTGACATCCTCTGAAAACCCTAGAGATAGGGCTTCTCCTTCGGGAGCAGAGTGA
CAGGTGGTGCATGGTTGTCGTCAACTCGTGCCNTGAAATGTTGGGTTAAGTCCCGCA
ACCGAGCGCAACCCTTTGACCTTAGTTGCCATCATTAAGTTTGGGCACTCTAAAGGTG
ACTGCCGGTTGACAAACCGGAAGGAAGGTGGGGATGACGTCAAATCATCCTTGCCC
CTTATGACCTGGGCTACCAACGTGGCTACAATGGACGGGTACAAAGAGCTTCCAA
GACCCGCGAGGTGGAAGCTTAATTCCTTAAAAACCGTTTTCCANTTCCGGAATGGAA
GGCCGGCAACCCGCCCACATNGAAACCTGGGAATCCCTTAGNAACNNNTTCTGTTCC
CCTTATGCGGCTGGCTCCAAAAGGTTACCCCACCGACTTCGGGTGTTACAAACTCTCG
TGGTGTGACGGGCGGTGTGTACAAGGCCCGGGAACGTATTCACCGCGGCATGCTGAT
CCGCGATTACTAGCGATTCCAGCTTCATGTAGGCAGTTGCAGCCTACAATCCGAACT
GAGAACGGTTTTATGAGATTAGCTCCACCTCGCGGTCTTGCAGCTCTTTGTACCGTCC
ATTGTAGCACGTGTGTAGCCCAGGTCATAAGGGGCATGATGATTTGACGTCATCCCC
ACCTTCCTCCGGTTTGTCACCGGCAGTCACCTTAGAGTGCCCAACTTAATGATGGCAA
CTAAGATCAAGGGTTGCGCTCGTTGCGGGACTTAACCCAACATCTCACGACACGAGC
TGACGACAACCATGCACCACCTGTCACTCTGCTCCCGAAGGAGAAGCCCTATCTCTA
GGGTTTTCAGAGGATGTCAAGACCTGGTAAGGTTCTTCGCGTTGCTTCGAATTAAACC
ACATGCTCCACCGCTTGTGCGGGCCCCCGTCAATTCCTTTGAGTTTCAGCCTTGCGGC
CGTACTCCCCAGGCGGAGTGCTTAATGCGTTAACTTCAGCACTAAAGGGCGGAAACC
CTCTAACACTTAGCACTCATCGTTTACGGCGTGGACTACCAGGGTATCTAATCCTGTT
TGCTCCCCACGCTTTCGCGCCTCAGTGTCAGTTACAGACCAGAAAGTCGCCTTCGCCA
CTGGTGTTCCTCCATATCTCTACGCATTTCACCGCTACACATGGAATTCCACTTTCCTC
TTCTGCACTCAAGTCTCCCAGTTTCCAATGACCCTCCACGGTTGAGCCGTGGGCTTTC
ACATCAGACTTAAGAAACCACCTGCGCGCGCTTTACGCCCAATAATTCCGGATAACG
CTTGCCACCTACGTATTACCGCGGCTGCTGGCACGTAGTTAGCCGTGGCTTTCTGGTT
AGGTACCGTCAAGGTGCCAGCTTATTCAACTAGCACTTGTTCTTCCCTAACAACAGA
ATTTACGACCCGAAAGCCTTCATCACTCAAGCGGCGTTGCTCCGTCAGACTTTCGTC
CATTGCGGAAGATTCCCTACTGCTGCCTCCCGTAAGAATCTGGGCCGTGTCCCAGTCC
CAGTGGGGCCGATCCCCTTCCCAGGCCGGCTACCCATCTTTGCCTTGGTGAGCCGTTA
CCTCACCAACTAGTTAAGGGGAAGCGGGGCCATCCTAAATGGAAACCCAAACCCGC
CTTTCAATTTCGAACCATGGGGTTCAAAAAGTTTCCGGGGTNTATCCCC

IOP Conf. Series: Materials Science and Engineering **136** (2016) 012088 doi:10.1088/1757-899X/136/1/012088

3.5.2 DNA Suquences for Ureolytic bacteria

NNNGANNTNGGGCTGCTATAATGCAGTCGACGCTTCTTTCCTCCCGAGTGCTTGCACTCA
ATTGGAAAGAGGAGTGGCGGACGGGTGAGTAACACGTGGGTAACCTACCCATCAGAGGG
GGATAACACTTGGAAACAGGTGCTAATACCGCATAACAGTTTATGCCGCATGGCATAAGA
GTGAAAGGCGCTTTCGGGTGTCGCTGATGGATGGACCCGCGGTGCATTAGCTAGTTGGTG
AGGTAACGGCTCACCAAGGCCACGATGCATAGCCGACCTGAGAGGGTGATCGGCCACAC
TGGGACTGAGACACGGCCCAGACTCCTACGGGAGGCAGCAGTAGGGAATCTTCGGCAAT
GGACGAAAGTCTGACCGAGCAACGCCGCGTGAGTGAAGAAGGTTTTCGGATCGTAAAAC
TCTGTTGTTAGAGAAGAACAAGGACGTTAGTAACTGAACGTCCCCTGACGGTATCTAACC
AGAAAGCCACGGCTAACTACGTGCCAGCAGCCGCGGTAATACGTAGGTGGCAAGCGTTG
TCCGGATTTATTGGGCGTAAAGCGAGCGCAGGCGGTTTCTTAAGTCTGATGTGAAAGCCC
CCGGCTCAACCGGGGAGGGTCATTGGAAACTGGGAGACTTGAGTGCAGAAGAGGAGAGT
GGAATTCCATGTGTAGCGGTGAAATGCGTAGATATATGGAGGAACACCAGTGGCGAAGG
CGGCTCTCTGGTCTGTAACTGACGCTGAGGCTCGAAAGCGTGGGGAGCAAACAGGATTA
GATACCCTGGTAGTCCACGCCGTAAACGATGAGTGCTAAGTGTTGGAGGGTTTCCGCCCT
TCAGTGCTGCAGCAAACGCATTAAGCACTCCGCCTGGGGAGTACGACCGCAAGGTTGAA
ACTCAAAGGAATTGACGGGGGCCCGCACAAGCGGTGGAGCATGTGGTTTAATTCGAAGC
AACGCGAAGAACCTTACCAGGTCTTGACATCCTTTGACCACTCTAGAGAATAGAGCTTTC
CCTTCGGGGACAAAGTGACAGGTGGTGCATGTTTGTCGTCAGCTCCGGGCCGTGAGATGT
TGGGTTAAGTCCCGCAACGAGCGCAACCCTTATTGGTAGTTGCCATCATTTAGTTGGGGC
ACTCTAACCGAGACTGCCCGGTGACAAACCCGAAGAGANNNNATCTATCCACCTTAGGC
GGCTGGCTCCAAAAGGTTACCTCACCGACTTCGGGTGTTACAAACTCTCGTGGTGTGACG
GGCGGTGTGTACAAGGCCCGGGAACGTATTCACCGCGGCGTGCTGATCCGCGATTACTAG
CGATTCCGGCTTCATGCAGGCGAGTTGCAGCCTGCAATCCGAACTGAGAGAAGCTTTAAG
AGATTTGCATGACCTCGCGGTCTAGCGACTCGTTGTACTTCCCATTGTAGCACGTGTGTAG
CCCAGGTCATAAGGGGCATGATGATTTGACGTCATCCCCACCTTCCTCCGGTTTGTCACCG
GCAGTCTCGCTAGAGTGCCCAACTAAATGATGGCAACTAACAATAAGGGTTGCGCTCGTT
GCGGGACTTAACCCAACATCTCACGACACGAGCTGACGACAACCATGCACCACCTGTCAC
TTTGTCCCCGAAGGGAAAGCTCTATCTCTAGAGTGGTCAAAGGATGTCAAGACCTGGTAA
GGTTCTTCGCGTTGCTTCGAATTAAACCACATGCTCCACCGCTTGTGCGGGCCCCCGTCAA
TTCCTTTGAGTTTCAACCTTGCGGTCGTACTCCCCAGGCGGAGTGCTTAATGCGTTTGCTG
CAGCACTGAAGGGCGGAAACCCTCCAACACTTAGCACTCATCGTTTACGGCGTGGACTAC
CAGGGTATCTAATCCTGTTTGCTCCCCACGCTTTCGAGCCTCAGCGTCAGTTACAGACCAG
AGAGCCGCCTTCGCCACTGGTGTTCCTCCATATATCTACGCATTTCACCGCTACACATGGA
ATTCCACTCTCCTCTTCTGCACTCAAGTCTCCCAGTTTCCAATGACCCTCCCCGGTTGAGC
CGGGGGCTTTCACATCAGACTTAAGAAACCGCCTGCGCTCGCTTTACGCCCAATAAATCC
GGACAACGCTTGCCACCTACGTATTACCGCGGCTGCTGGCACGTAGTTAGCCGTGGCTTT
CTGGTTAGATACCGTCAGGGGACGTTCAGTTACTAACGTCCTTGTTCTTCTCTAACAACAA
AGTTTTACGATCCGAAAACTTTCTTCACTCCAGCGGGGGTTGCTCGGGCAAAATTTTCGT
CCATTGGCGAAAAATTCCCTACTGGCTGCCTCCCCGTAAGGAATCTGGGGCCGGGGCTCC
AATCCCAAGTGTGGGCCGAATCACCCCTTCCAAGTGGGGNNTAGGCATCGGGGGCCTTGG
GGAAGCCCGTTACCTCCACCAACTTAGNTTAATGGACNCCGGGGGGTCCTCCATTTANGG
GGACCCCCAGA

Figure 4. Phylogenetic position of *Bacillus SP* and closely related taxa (*Sulphate reduction bacteria*).

4. Conclusion

The bacteria in this study was isolated from fresh urine and acid mine drainage water.The isolated bacteria was enriched to suit concrete environement. After enrichment, the bacteria was purify by serial dilution and gram staining was conducted. The Gram staining results revealed that both bacteria was gram positive. The bacteria was identified through PCR and 16S rDNA sequencing as *Bacillus sp* and *Enterococcus faecalis*.

Acknowledgement

This research was supported by Universiti Tun Hussein Onn Malaysia (UTHM) and The Ministry of Higher Education Malaysia through Fundamental Research Grant Scheme (FGRS), Vot 1211.

References

[1] Siddique, R. and Chahal, N.K. Effect Of Ureolytic Bacteria On Concrete Properties. *Construction and Building Materials*. 2011, 25, 3791-3801.

[2] Chahal, N., Siddique, R. Permeation properties of concrete made with fly ash and silica fume: Influence of Ureolytic bacteria. *Construction and Building Materials,* 2013, 49, 161-174.

[3] Ramachandran, S.K., Ramakrishnan, V. and Bang, S.S. Remediation Of Concrete Using Micro-Organisms. *American Concrete Institute: Materials Journal*. 2001.

[4] Ghosh, P and Mandal, S. Development Of Bioconcrete Material Using An Enrichment Culture Of Novel Thermophilic Anaerobic Bacteria. *Indian Journal Of Experimental Biology*, 2006, 44, 336-339.

[5] Muynck, W.D., Cox, K, Belie, N.D. and Verstraete, W. Bacterial Carbonate Precipitation As An Alternate Surface Treatment For Concrete. *Construction and Building Materials*, 2008, 22, 875-885.

[6] Miyoko Yanagi and Kazuhide Yamasato. Phylogenetic analysis of the family Rhizobiaceace and related bacteria by sequencing of 16S rDNA gene using PCR and DNA sequencer. *Federation of European Microbiological Societies* 0378-1097/93/s06.00, 1993

[7] Kit Boye, Estrid Hogdall and Martin Borre. Identification of bacteria using two degenerate 16S rDNA sequencing primers. *Microbial Res.* 154, 1999, 23-26

[8] Jessica M.E.van den Oever, Sahila Balkassmi, E.Joanne Verwij, Maarten van Iterson, Phebe N.Adama Van Scheltema, Dick Oepkes, Jan M.M. van Lith, Mariette J.V.Hoffer, Johan T.den Dunnen, Egbert Bakker and Elles.M.J.Boon. Single Molecule Sequencing of free DNA from maternal plasma for noninvasive trisomy 21 detection. *Clinical Chemistry*. 58:4, 2012, 699-706.

[9] Joy.Y.Yang, Shelise Brooks, Jennifer A. Meyer, Robert R.Blakesley, Adrian M. Zelazny, Julia A.Segre, Evan S.Snitkin. Pan-PCR, a Computational method for designing bacterium-typing assays based on whole-Genome sequence data. *Journal of clinical microbiology*. Vol.51(3), 2013, 752-758.

[10] John E.Moore, Junhua Huang, Pengbo Yu, Chaofeng Ma, Peter JA Moore, Beverly C.Millar, Colin E.Goldsmith and Jiru Xu. Identification employing 16S Rdna PCR,Gene sequencing and total antibiotic susceptibility techniques. *Ecotoxicology and Environmental Safety*. 108, 2014, 281-286.

[11] Irwan, J.M, Faisal, A Alshalif, Othman, N and Asyraf, R, M. Isolation of Ureolytic and Sulphate reduction bacteria: Acclimitize to concrete environment.*Interntional conference on civil, Biological and Environmental Engineering (CBEE-2014), Istanbul, Turkey*. May 27-28, 2014.

[12] Tugba, Onal.Okyay and Debora.Frigi, Rodriques. Optimized carbonate micro-particle production by *Sporosarcina pasteurii* using response surface methodology. *Ecological Engineering* 62, 2014, 168-174.

[13] Talaro.K.P. and Chess.B. Foundations in Microbiology.8th Edition. Mc.Graw Hill. 2012. Pg 254-258

Characteristics and Thermal Behaviour of Low Rank Malaysian Coals towards Liquefaction Performance via Thermogravimetric Analysis

M A M Ishak[1], K Ismail[2], W I Nawawi[1], A H Jawad[1], M F Abdullah[1], M N Kasim[1] and A Y Ani[1]
[1]Faculty of Applied Sciences, Universiti Teknologi MARA, 02600 Arau, Perlis, MALAYSIA
[2]Faculty of Applied Sciences, Universiti Teknologi MARA, 40450 Shah Alam, Selangor, MALAYSIA

E-mail: azlanishak@perlis.uitm.edu.my

Abstract. In this study, thermal behaviour of two low-rank Malaysian coals namely Mukah Balingian (MB) and Batu Arang (BA) were obtained under pyrolysis conditions via Thermogravimetric analysis (TGA) at a heating rate of 20°C min^{-1}. The thermal characteristics of the coals were investigated prior to direct liquefaction in order to determine the liquefaction performance, i.e. coal conversion and oil yield. The differential weight loss (DTG) results for both coals showed that there are three main stages evolved which consists of moisture, volatile matter and heavier hydrocarbons that correspond to temperature range of 150, 200-500 and 550-800°C, respectively. Apparently, the DTG curves of BA coal reveals a similar pattern of thermal evolution profile in comparison to that of the MB coal. However, the calculated mean reactivity of BA coal is higher than that of MB, which implied that BA would probably enhance coal conversion and oil yield in comparison to MB coal. Interestingly, results showed that under the same liquefaction conditions (i.e. at 4MPa pressure and 420°C), conversion and oil yield of both coals were well correlated with their reactivity and petrofactor value obtained.

Keywords: Coal liquefaction, low rank coal, thermal behavior, TGA

1. Introduction

The incentive to develop coal liquefaction arose mainly from the increasing demand for liquid fuels, particularly for transport applications and also due to the depleting of other fossil fuels such as petroleum and natural gas. Previously, most researchers correlated the liquefaction performance i.e. coal conversion and/or oil yield obtained with parameters based on the chemical structure, reactivity or petrological composition of coals [1, 2]. Several correlations have been proposed for predicting the thermal behaviour of coal during combustion and devolatization. Relationships exist between characteristic temperatures of the thermogram during coal combustion [3]. A number of correlations were established between volatile matter/fixed carbon (VM/FC), carbon/hydrogen (C/H) and (carbon+hydrogen)/oxygen ((C+H)/O) ratios and coal reactivity [4]. The fuel ratio (FC/VM) is used in many commercial specifications for power station contracts. The evaluation of the inverse of fuel ratio, i.e. VM/FC with the coal reactivity is an important parameter for the characterization of a particular fuel [5].

Artanto et al. [2] studied the effect of coal rank and H/C atomic ratios towards liquefaction conversion using 35 ml tubular batch-wise reactor at temperatures of 320, 350 and 405°C at 6 MPa under nitrogen and hydrogen gas atmospheres. The comparison was made between a sub-bituminous coal and brown coals originating from Indonesia. They found that the reactivity of the coals generally followed the correlation between total conversion and H/C atomic ratio, with conversion increasing with increasing H/C atomic ratio of the coal. It has been shown that, characterization of the evolved volatile matter can be done by leading the gaseous material into a GC, IR or MS [6]. These methods are usually cumbersome. Furthermore, they have limitations in as far as tar measurements are concerned. It is almost impossible to ensure the complete and quantitative transfer of the condensable tar components to the analytical instrument without loss or change. Thus, in this present study thermogravimetric analyzer (TGA) was used as a tool to obtain thermal evolution profiles of coals, which correspond to characteristics or thermal behaviour of the fuel. The thermal characteristics obtained under pyrolysis mode of two low-rank Malaysian coals namely Mukah Balingian (MB) and Batu Arang (BA) were investigated prior to direct liquefaction at established optimal conditions. The liquefaction results i.e. coal conversion and oil yield will determine the correlation between reactivity and thermal behaviour of coals with liquefaction performance.

2. Methodology

2.1 Coal preparation, coal assays and reactivity analysis

The samples used in this study were low-rank, Mukah Balingian (MB) and Batu Arang (BA) originating from Sarawak and Selangor, Malaysia, respectively. The procedure for coal preparation has been reported earlier [7]. Table 1 shows the ultimate, proximate and petrography analyses of the raw MB and BA coal samples. The procedure of ultimate and proximate analyses has been reported elsewhere [8]. For the reactivity study, pyrolysis mode was applied by heating the sample from ambient temperature to a final temperature of 1000°C under a nitrogen gas stream, with heating rate of 20°C/min. The gas stream flow rates of nitrogen were fixed at a flow rate of 100 ml/min. The mass loss data with temperature were recorded and were then converted into derivative thermal gravimetric (DTG) using the analyzer software provided.

2.2 Coal liquefaction procedure

The detail procedure for coal liquefaction has been reported elsewhere [8]. Briefly, the coal liquefaction experiments were carried out in duplicate (within 2% error) in a 1-litre high-pressure high-temperature batch-wise reactor system. In this study, optimal liquefaction conditions obtained from previous investigation was chosen i.e. 1:10 coal-to-solvent ratio, temperature at 420°C, 4 MPa pressure, 30 min reaction time and with stirring [7]. The results of coal liquefaction were reported as percent conversion and product yield i.e. oil.

3. Results and Discussions

3.1 Characteristics of Mukah Balingian and Batu Arang coals

Table 1 shows the characteristics of Mukah Balingian (MB) and Batu Arang (BA) coals, which include ultimate, proximate and petrographic analyses, respectively.

Table 1: Characteristics of MB and BA coal samples.

Ultimate analysis (wt% daf)	MB	BA	Proximate analysis (wt% db)	MB	BA	Petrographic analysis (vol. %)	MB	BA
Carbon	63.9	73.8	Vol. Matter	44.7	44.5	Vitrinite	60	48
Hydrogen	5.1	5.5	Fixed carbon	35.6	49.3	Liptinite	31	44
Nitrogen	1.9	1.0	Ash content	4.2	6.2	Inertnite	8	7
Sulphur	0.5	0.7	Fuel ratio[1]	1.1	1.1	Min. matter	1	1
Oxygen[2]	28.6	19.0						
H/C ratio	1.0	0.9	Cal. Value	24.6	27.1	Vit. Ref.	0.4	0.4

[1] = fixed carbon/volatile matter; [2] = by difference; daf = dry-ash-free; db = dry-basis
Cal. Value = Calorific value (MJ/kg); Vit. Ref. = Vitrinite reflectance

From Table 1, it can be seen that MB and BA coal samples have relatively high oxygen and volatile matter contents. The petrographic analysis of these coals show a vitrinite reflectance value of 0.40% and thus, can be categorised as low rank coal i.e., sub-bituminous C rank [7]. It has been suggested that low rank coals are normally composed of small aromatic clusters and contain many cross-links and functional groups and thus, are very reactive and undergo fast and extensive bond breaking during liquefaction [9]. Interestingly, it shows that most of the analysis parameters of the MB coal were comparable to that of the BA coal, although both coal samples are from different deposition, i.e MB is located in Sarawak and BA is in Selangor. The differences between the two coals can be seen with carbon and distributions in maceral contents (petrographic analysis). BA coal contains ca. 10% higher carbon and ca. 13% higher liptinite content with total reactive macerals of ca. 92%, in comparison to MB coal. It has been established by King et. al. [10] that under a wide range of liquefaction conditions the more reactive maceral is liptinite followed, in decreasing order, by vitrinite and inertinite. Thus, the reactivity of BA coal is expected to be higher than MB coal during the liquefaction process and contribute higher coal conversion and oil yield.

3.2 Thermal behaviour of Mukah Balingian and Batu Arang coals

Figures 1, 2 and 3 show the the thermal behaviour of decomposition of MB and BA coal samples under pyrolysis conditions via thermogravimetric analyser (TGA), respectively.

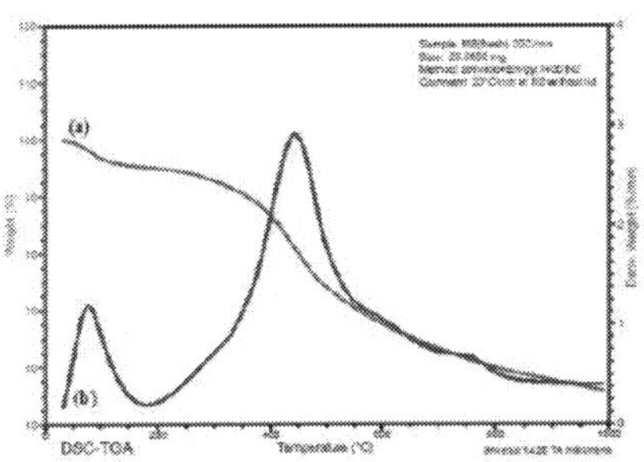

Fig. 1: TG (a) and DTG (b) profiles for the thermal decomposition of pyrolysed MB coal.

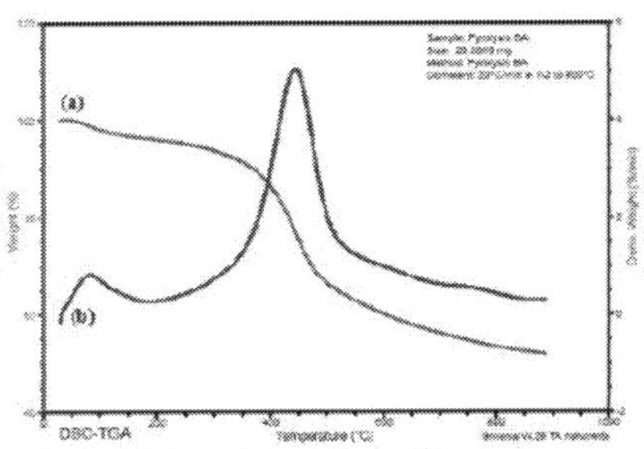

Fig. 2: TG (a) and DTG (b) profiles for the thermal decomposition of pyrolysed BA coal.

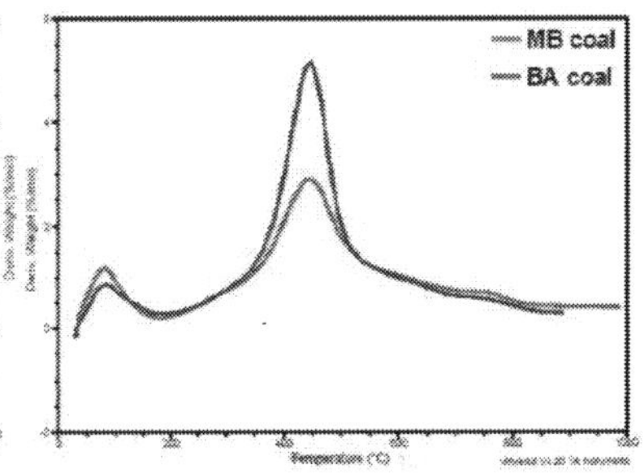

Fig.3: Superimposed DTG profiles for the pyrolysed MB and BA coal samples.

The pyrolysis of MB coal via TGA at a heating rate of 20°C min-1 is shown in Figure 1. The TG curve (Figure 1(a)) of the coal shows the weight loss profile with the weight decreasing as temperature is increased from ambient to 1000°C. The weight loss profile resembles that of a first order reaction. Figure 1(b) shows the differential weight loss (DTG) for MB coal that consists of three main stages. This conforms to previous findings as reported by Probstein and Hicks [11], Kastanaki et. al. [12], Serio et. al. [13] and Radovic et. al. [14]. The first stage pyrolysis, which occurs at temperatures ranging from ambient to 150°C, involves the dehydration of water and releasing of gas composed of oxides of carbon from the coal. The second stage pyrolysis, which occurs at temperatures of 200 – 550°C, however, is due to the release of volatile matter such as hydrocarbon gases, light oils and tars, and is of interest in this study. Finally, the third stage pyrolysis shows the appearance of some minor curves at temperatures ranging from 550 – 650°C and 700 – 800°C, that were attributed to the release of heavier hydrocarbons and non-condensable gases, mainly hydrogen, and from thermal decomposition of carbonates that are abundant in low rank coals. Also from the DTG profile of MB coal, it can be estimated that the softening temperature of this coal is around 350°C. Merrick [15] suggested that with extraction using liquid solvents, the preferred extraction temperatures lie in the range where the coal starts to decompose thermally, and typically the extraction is carried out at about 350 – 450°C. Moreover, Van Krevelen [16] suggested that temperatures around 350°C were found to be an indicative value for softening temperature in order to characterise coal for direct liquefaction. Thus, it can be suggested that MB coal would be a good feedstock and suitable for liquefaction and/or gasification processes and, hence, optimise the utilisation of low rank Malaysian coal. Apparently, the DTG curve of BA coal as shown in Figure 2(b) reveals a similar pattern of evolution profiles in comparison to that of MB coal (Figure 1(b)). However, when the DTG curves of both coal samples are superimposed, as shown in Figure 3, the peak height (R) for the second thermal peak of BA coal is higher than MB coal

Table 2: DTG reactivity of MB and BA coal samples during pyrolysis.

Coal Sample	Peak temperature T_{max} (°C)	Peak height R (mg min^{-1})	Mean reactivity Rm (mg min^{-1} $°C^{-1}$)
MB	446.6; 607.8; 754.2	0.584; 0.211; 0.138	0.184
BA	446.4; 610.4; 751.6	1.085; 0.194; 0.115	0.290

MB = Mukah Balingian coal; BA = Batu Arang coal.

Previously, it has been reported that the DTG peak height (R) and maximum peak temperature (Tmax) is directly and inversely proportional to the reactivity, respectively [17]. The overall reactivity of both MB and BA coal samples was calculated based on the mean reactivity, Rm by using a parameter (R Tmax-1)100 to calculate for each peak, adding the share of any secondary peaks or shoulders present in the thermal profile with the exception of the peak due to released moisture [18]. In this calculation the rate of weight loss (mg min-1) was used, giving the dimensions for the mean reactivity as mg min-1 °C-1. The mean reactivity (Rm) values for both coal samples are displayed in Table 2. Hence, based on the calculated mean reactivity (Rm) value in Table 2, it shows that the reactivity of BA coal is higher than that of MB coal. This indicates that, under the same liquefaction conditions, BA coal would probably enhance coal conversion and oil yield in comparison to the MB coal. Thus, it can be suggested that BA coal would also be a good feedstock and suitable for liquefaction and/or gasification processes.

3.3 Correlation between coal reactivity and thermal behaviour with liquefaction results

Table 3 and Figure 4 show the liquefaction results of MB and BA coal samples, respectively. From Table 3 and Figure 4, it can be seen that the coal conversion of 91.2%, which was obtained for BA coal is comparable with the total amount of its reactive macerals (vitrinite + liptinite), i.e. 92%. The percent of oil yield obtained on the BA coal was ca. 75%. In order to evaluate the effect of intrinsic coal reactivity and contribution of macerals toward coal conversion and oil yield in BA coal, liquefaction results of MB coal at the same conditions were compared. It can be seen that the coal conversion and oil yield obtained for BA coal were higher than those of MB coal, i.e., ca. 7 and 8%, respectively. This observation seems to agree with the higher reactivity of BA coal as revealed from its DTG profile shown in Figure 3. Indeed, the calculated mean reactivity value of BA coal is also higher than that of MB coal (see Table 2). In terms of contribution of macerals; since the inertinite content of both coals is comparable (i.e., 7 and 8%) and less likely to be involved in the conversion processes of coal [19,20], the higher coal conversion and oil yield obtained for BA coal was attributed to liptinite. From Table 3, it shows that liptinite content of BA coal was ca. 13% higher than that of MB coal. Thus, this showed that the liptinite was more significant in determining the formation of oil yield during liquefaction. Furthermore, it has been established that liptinite is more reactive than vitrinite [10]. This showed that the intrinsic coal reactivity and reactive coal macerals especially liptinite were the contributing factors in determining higher coal conversion and oil yield during liquefaction. Hence, these results thus far showed that the use of thermogravimetric analysis (TGA) in determining the sample reactivity under pyrolysis or combustion mode was found to be useful in correlating the coal reactivity with liquefaction performance.

Table 3: Liquefaction[1] results and macerals content of MB and BA coals.

Results	MB coal	BA coal
Coal conversion	84.0%	91.2%
Oil	67.7%	75.3%
Asphaltene	9.7%	2.5%
Preasphaltene	6.7%	13.4%
Vitrinite	60%	48%
Liptinite	31%	44%
Inertinite	8%	7%

[1]conditions: at 420 °C, 4 MPa, 30 min, 1:10 coal-to-solvent ratio and with stirring at 500 rpm.

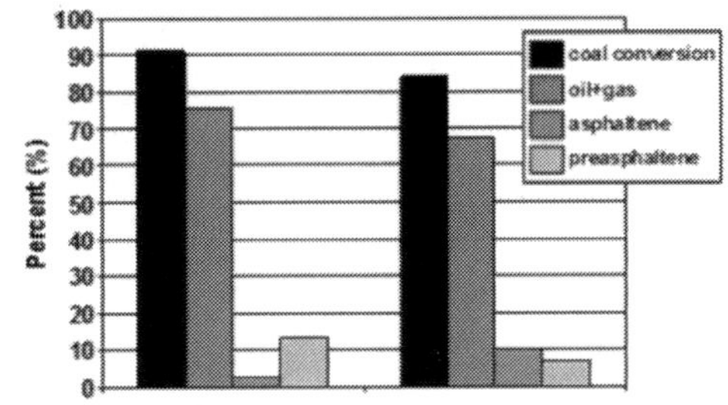

Figure 4: Liquefaction results of MB and BA coal samples.

Conclusion

The thermal characteristics of two low-rank Malaysian coals were investigated prior to direct liquefaction. The differential weight loss (DTG) results for both coals showed that there are three main stages evolved which consists of moisture, volatile matter and heavier hydrocarbons that correspond to temperature range of 150, 200-500 and 550-800°C, respectively. Apparently, the DTG curve of both coals reveals a similar pattern of evolution profiles, however when the curves are superimposed the peak height for the second evolution profile of BA coal is higher than MB coal. The higher coal reactivity of BA is in agreement with liquefaction results obtained on this coal with comparison to MB coal. Thus, the usage of TGA to predict sample reactivity under pyrolysis or combustion mode was found to be useful and the usage of the instrument prior to liquefaction is highly recommended.

Acknowledgment

This research work was funded by the Ministry of Education, Malaysia under the FRGS grant scheme. The authors also thank the management of Universiti Teknologi MARA for supporting the research work.

References

[1] Redlich P J, Carr R M, Jackson W R, Larkins F P and Chaffee A L 1990, *Fuel.* 69 764-770
[2] Artanto Y, Jackson W R, Redlich P J, and Marshall M 2000, *Fuel.* 79, 1333-1340
[3] Morgan P A, Robertson S D and Unsworth J F 1986, *Fuel.* 65, 1546-1551
[4] Ghetti P 1986, *Fuel.* 65, 836-839
[5] Zaidi S A H 1995, *Fuel Processing Technology.* 41, 253-259
[6] Slaghuis J H and Morgan P M 1991, *Thermochimica Acta.* 175, 135-140
[7] Ishak M A M, Ismail K, Abdullah M F, Kadir M O A, Mohamed A R and Abdullah W H 2005, *Coal Preparation Journal.* 25, 221-237
[8] Ali S N, Yusop M F, Ismail K, Ghani Z A, Abdullah M F, Ishak M A M and Mohamed A R 2014, *Energy Procedia.* 52, 618-625
[9] De Marco Rodriguez I, Chomon M J, Caballero B, Arias P L , and Legarreta J A 1998, *Fuel Processing Technology.* 58, 17-24
[10] King H H, Dyrkacz G R and Winans R E 1984, *Fuel.* 63, 341-345
[11] Probstein R F and Hicks R E 1982 *Synthetic Fuels* (Mc Graw-Hill)
[12] Kastanaki E, Vamvuka D, Grammelis P and Kakaras E 2002, *Fuel Processing Technology.* 77-78, 159-166
[13] Serio M A, Hamblen D G, Markham J R and Solomon P R 1987, *Energy and Fuels.* 1, 138-152
[14] Radovic L R, Walker Jr P L and Jenkins R G 1983, *Fuel.* 62, 849-856.
[15] Merrick D 1984 *Energy Alternatives Series* (Macmillan Publishers), 168-230
[16] Van Krevelen D W 1993 *Coal – Typology, Physics, Chemistry, Constitution* (Amsterdam: Elsevier Science).

[17] Bonfanti L, Davini P and Ghetti P 1994, *Salsomaggiore Terme*. 2, 10
[18] Ghetti P 1996, *Thermal Analysis of Biomass*. 75, 5
[19] Teichmüller M 1989, *International Journal of Coal Geology*. 12, 1-87
[20] Merrick D 1984, *Coal Combustion and Conversion Technology, Energy Alternative Series* (London: MacMillan Publishers)